Ecology of Australian Temperate Reefs

THE UNIQUE SOUTH

EDITORS: SCORESBY SHEPHERD AND GRAHAM EDGAR

CSIRO PUBLISHING

National Library of Australia Cataloguing-in-Publication entry
Ecology of Australian temperate reefs : the unique South /
edited by Scoresby Shepherd and Graham Edgar.
9781486300099 (hardback)
9781486300105 (epdf)
9781486300112 (epub)
Includes bibliographical references and index.
Reefs – Australia, Southeastern.
Reef ecology – Australia, Southeastern.
Marine biology – Australia, Southeastern.
Shepherd, Scoresby, editor.
Edgar, Graham, editor.
577.7890994

Published by

CSIRO PUBLISHING
36 Gardiner Road, Clayton VIC 3168
Private Bag 10, Clayton South VIC 3169
Australia

Telephone: [+613] 9545 8555
Local call: 1300 788 000 (Australia only)
Fax: +61 3 9662 7555
Email: csiropublishing@csiro.au
Website: www.publishing.csiro.au

Cover images by GJ Edgar

Set in 10/13 Minion and Stone Sans
Edited by Peter Storer Editorial Services
Cover design by Andrew Weatherill
Text design by James Kelly
Typeset by diacriTech, Chennai, India
Printed by Ingram Lightning Source

Mar26_RP_ILS

Contents

Acknowledgements

We are deeply indebted to the many biologists and others for commenting on sections or chapters of the manuscript. Without their help and advice this book would have been much the poorer. Unless otherwise noted, all material has been written by the main author SA Shepherd, with comments and additional material by GJ Edgar. However, some fields are so specialised that it has been more prudent to involve specialists in these fields to contribute some sections and these are noted at the start of the relevant section. We are especially grateful to these contributors, who are also listed for each chapter on the next page.

We are most grateful to those photographers who generously supplied photographs of species or habitats, or figures. They are: Bob Baldock, Yvonne Bone, the late Kevin Branden, James Brook, Simon Bryars, CSIRO (courtesy David Griffin), the late Peter Clarkson, Helen Crawford, Sabine Daume, Paul Fitzgerald, Lisa Gershwin, Adrienne Grant, Andrew Green, Adele Harvey, Nigel Holmes, Juan-Carlos Livore, Andy Lowther, George Matsumoto, Leo Mantilla, Dave Muirhead, Andrew Newton, Craig Proctor, Steve Reynolds, Craig Sanderson, Brian Saunders, Fiona Scott, David Staples, Ben Stobart, Craig Styan, Jan Watson and Carolina Zagal.

Lastly we thank our supporting institutions: the South Australian Research and Development Institute (Aquatic Sciences), and the Institute for Marine and Antarctic Sciences, University of Tasmania, for their continuing support and patronage.

Contributors

Chapter 3: Dr Joyce Richardson (brachiopods), Email: podsl@bigpond.com; Dr Jan Watson (hydrozoans), 7 Bampi St, Clifton Springs, Victoria 3222, Email: jwa6259@bigpond.net.au; the late Dr Patricia Kott (ascidians)

Chapter 4: Dr Craig Sanderson (giant kelp), Institute of Marine and Antarctic Studies, University of Tasmania, Hobart, Tasmania, Email: craig.sanderson@utas.edu.au

Chapter 6: Dr Sabine Daume (encrusting coralline algae), Marine Stewardship Council, 2200 Powell St, Suite 725, Emeryville, CA 94608, USA, Email: sdaume@scscertified.com

Chapter 10: Craig Proctor (blue-ringed octopus), CSIRO, Marine and Atmospheric Research, Hobart, Tasmania, Email: craig.proctor@csiro.au

Chapter 13: Darren James (succession in shallow reef habitats), Email: darren.james@dse.vic.gov.au

Chapter 14: Dr Adrienne Grant (zoanthids and stony corals), University of Sydney, Email: Adrienne.grant@sydney.edu.au; Dr Carolina Zagal (Stauromedusae), Tasmanian Aquaculture and Fisheries Institute, University of Tasmania, Private Bag 49, Hobart, Tasmania 7001, Email: Carolina.zagal@utas.edu.au

Chapter 15: Dr DA Staples (pycnogonids), Museum Victoria, GPO Box 666, Melbourne, Victoria 3001

Chapter 16: Dr Brian Saunders (zebra fish), Email: coffinbay@harboursat.com.au

Chapter 17: Dr Simon Bryars (ambush predators), Email: sbryars@bigpond.com

Reviewers and others who assisted in various ways

The following most helpfully reviewed or advised on the chapters below:

Introduction: Dr Ib Svane prepared the food energy pyramid

Chapter 1: Dr Vic Gostyn

Chapter 2: Dr JAT Bye and Dr Paul Van Ruth (upwellings)

Chapter 3: Dr Shirley Sorokin (sponges) and Dr Fred Gurgel (algae)

Chapter 4: Dr Craig Sanderson, Janine Baker, who also prepared Fig. 3.1 and Dr Hugh Kirkman (*Ecklonia*)

Chapter 5: Janine Baker, Dr Craig Sanderson and Dr Greg Collings

Chapter 6: Dr Andrew Irving and Janine Baker

Chapter 7: Dr Michael Hawkes (*Delisea*), Dr Andrew Irving and Janine Baker

Chapter 8: Dr Paris Goodsell, Dr Marty Deveney (European fanworm) and Janine Baker; Dr Pat Hutchings (polychaetes etc.)

Chapter 9: the late Peter Clarkson (cowries) to whom the chapter is dedicated

Chapter 10: Dr Natalie Moltschaniwsky, Sylvia Rapo (pygmy squid), Dr Karina Hall (cuttlefish). Craig Proctor (blue-ringed octopus) who deeply thanks Dr Alan Butler, Zoology Dept, University of Adelaide, for supervision of his research.

Chapter 11: Jodie Haig; Dr Jane Williamson, Dr Jan Watson, Dr Andrew Morgan, Dr Mark O'Loughlin and Dr Maria Byrne

Chapter 12: Dr Juan Livore

Chapter 13: Dr Matt Edmunds for his many reports, Dr AJ Butler and the late Dr PM Mather (nee Kott) (ascidians)

Chapter 14: Dr Adrienne Grant for many helpful comments; Dr Jan Watson (hydrozoans), Dr Lisa Gershwin (jellyfish), Dr Michela Mitchell and Dr Konstantin Tkachenko (anemones)

Chapter 15: Dr Matt Hoare, Dr Rick McGarvey and Dr Adrian Linnane (lobsters); Dr Andrew Levings (giant crab)

Chapter 16: Dr Peter Coulson and Janine Baker

Chapter 17: Dr Paul Rogers and Steve Reynolds (advice)

Chapter 18: Dr PD Shaughnessy

Chapter 19: Dr Marty Deveney and Dr Ib Svane, who also prepared Fig. 19.4

Chapter 20: Janine Baker, James Brook, and Dr Hugh Kirkman, Dr Rick Stuart-Smith, Dr Ben Stobart and Dr Jon Nevill

Preface

In this book we have sought to summarise what is known about the life history, ecology and behaviour of organisms – from algae to seals, sharks and even whales – that live in or visit the rocky coastal regions of southern Australia. Most is known about the shallow reef life to ~20 m deep, but remarkably little about life below that. Nevertheless, we have included snippets of information gleaned from studies extending over shelf waters. We have brought together field, laboratory and fisheries information in order to present as complete an account as possible, not of every species, but of a selected few that have been well studied, and provide a flavour of the wealth of knowledge available on the rich southern marine flora and fauna. Subtidal reefs are incredibly diverse and dynamic, and our fascination grows with our understanding of the ecology and life history of just a few of the many players in the system.

Where possible, we have illustrated our accounts with current theory, but with caution because so many ideas are fashionable for a time and then become dated. It is often better to let the accounts speak for themselves, and avoid interpretation. Conceptually, we have emphasised ecology and behaviour, but have included aspects of morphology, evolution and development where they contribute to a better understanding. Much of the literature on particular groups has developed its own specialist jargon, which impedes understanding and renders the work inaccessible to all except the few elite specialists in the field. We have consciously tried to avoid jargon, and use non-specialist terms and simple language wherever possible. We have used common names of species (where they exist) and, to avoid ambiguity where different names are used, we give the scientific name when the species is first mentioned.

The book is aimed at several different kinds of readers. Firstly, our fellow marine ecologists are becoming increasingly narrow in their field of expertise, and they may wish to expand their knowledge into related or overlapping areas. This book will provide easy access into some of those fields. Secondly, students, whether under- or postgraduate, may wish to look over the fence into neighbouring gardens and appreciate some of the beauty hidden there. Thirdly, there is a growing band of interested and highly skilled amateur divers and underwater photographers who have long had access to books describing the flora and fauna of temperate waters, and now wish to flesh out their basic knowledge with a richer understanding. Lastly, fishery biologists, managers and aquaculturists may find that the natural history of the species they work with, and knowledge of their habitats, may help them in their work or in understanding the ecosystem that they help to manage or wish to have well managed.

We apologise to authors or readers whose work is not cited, or is dealt with too briefly to do it justice. In this kind of book many references must be omitted, especially in fields where the literature is profuse. Inevitably we will have overlooked studies of seminal value, omitted a reference to a local journal or thesis, or favoured work with which we are familiar. One of us (SAS) has training in algal biology from the eminent Professor HBS Womersley, as well as in zoology. The other (GJE) has very wide experience of the temperate Australian marine life, and both of us have logged many thousands of hours underwater around southern Australia, as well as broadening our knowledge in other temperate regions, such as the Mediterranean, Pacific, Atlantic and Indian Oceans.

Introduction

The southern coast of Australia has been rightly dubbed 'The unique South'. It has a long coastline of more than 5500 km, with a vast array of habitats, from exposed coasts to sheltered bays, and, in contrast to the tropics, the great majority of its species are endemic – occurring in no other country. This alone renders an understanding of their natural history and ecology all the more significant.

Four major systems can be recognised along southern Australian coasts – estuaries, two benthic systems (reefs and soft bottoms) and the pelagic system. Estuaries, characterised by fresh or brackish waters, are scattered along the southern coast, except in the arid central regions. Soft sediments dominate the benthos on Australia's continental shelf. Reef systems occur intermittently from the intertidal zone to the shelf edge, and the pelagic ecosystem, containing all the life forms that inhabit the water column, extends from coastal to oceanic waters.

The four systems are not independent, and interactions between them can be very important. Yet, they are very different – almost with different ecologies – and ecologists have tended to focus on habitats or organisms living in one or the other of the systems. In this book, our focus is on reef systems, mainly in the more easily accessible photic zone and adjoining deeper water, and the animals and plants associated with them. Reefs are a prominent habitat type along much of the coast, and provide substrate for the most highly productive ecosystems in the region. Because the links with soft sediments and the pelagic systems can be quite strong, such as via import and export of nutrients, pelagic larval stages of benthic species and so on, we consider in some detail these connections, where relevant.

Marine life has been observed and described since ancient times. In the 5th century BCE, Herodotus, Aristotle and others began to describe various forms of marine life. For Australia, much of the early work in marine science during the 19th and 20th centuries was devoted to taxonomic descriptions of the rich flora and fauna, and we can now safely assert that most of the prominent benthic reef organisms have been described, thanks to the monographic works of Womersley, Kott, Poore, Gomon and others – although serious gaps still remain in some phyla. On the other hand, the meiofauna (animals < 0.5 mm) and microbial life are virtually unknown.

Ecology is the study of the natural history of organisms, and the relations between them and their environment. It has two approaches – a reductionist one, in which individual species or habitats are examined one at a time – or a holistic one, in which the whole system is examined, and its properties revealed, at that higher level. Reductionism is essentially a watchmaker's view of the system, which, like a watch, can be taken apart and each part looked at in isolation from other parts, whereas a holistic approach tries to reveal the working of the ecosystem as a whole. The often-stated aphorism that 'the whole is greater than the sum of its parts' is a simple way of putting the principle that when the parts are put together the whole may behave quite differently than its parts. We clearly need both approaches: the former to understand the role and importance of all the components in the system, and the latter to understand feedback mechanisms and to manage the many and increasing pressures that are imposed on our seas – including overfishing, pollution, and climate change.

Reductionist ecology uses observation, measurement and experiments to help understand the parts, and depends on a good understanding of the natural history of the species. As Paul Dayton (2003) stated: 'creative ecology is based on a deep sensitivity to natural patterns and processes'. A holistic approach, dependent on information from the parts, tries to reveal the system's properties underlying the complexities. This can only be done by using models, and has given rise to a new discipline, called systems ecology. In this book we emphasise the parts because they are best known, and touch on holism in the few cases where it has been unravelled.

The ecology of reef systems and their flora and fauna have been much studied over the past 50 years, from the intertidal to deep water (see Fairweather and Quinn 2007 for a historical summary), but the amount of study declines almost exponentially with depth, being highest in the intertidal zone, moderate in the upper sublittoral zone, dominated by brown algae, and least in the deeper lower sublittoral zone, dominated by red algae and animal life. The advent of SCUBA in the 1960s made the sublittoral zone accessible to direct observation for the first time. But, as yet, no attempt has been made to synthesise the ecology of our temperate Australian reef ecosystems, although some tantalising glimpses have been provided in books published in the last three decades or so, for example, Clayton and King (1981), Andrew (1999), Edgar (2001), Connell and Gillanders (2007). We have not been able to include colour pictures of all the species that are considered and, for many of these, the recent books of Edgar (2008), Gomon *et al.* (2008), Gowlett-Holmes (2008) among many earlier ones will be invaluable.

In this book, after considering the evolution of the southern seas of Australia, and the origins of its life, we follow roughly the organisation of the food energy pyramid (see diagram), starting with the primary producers, mainly benthic algae, then the various levels of consumers and lastly the top predators. In general, we cite review papers where possible on individual topics to reduce the number of papers cited, but at the same time providing the reader with an entry to the literature. Primarily we describe the ecology of southern species, including their behaviour and life history, occasionally straying, when relevant, into related topics such as chemistry, anatomy and evolution.

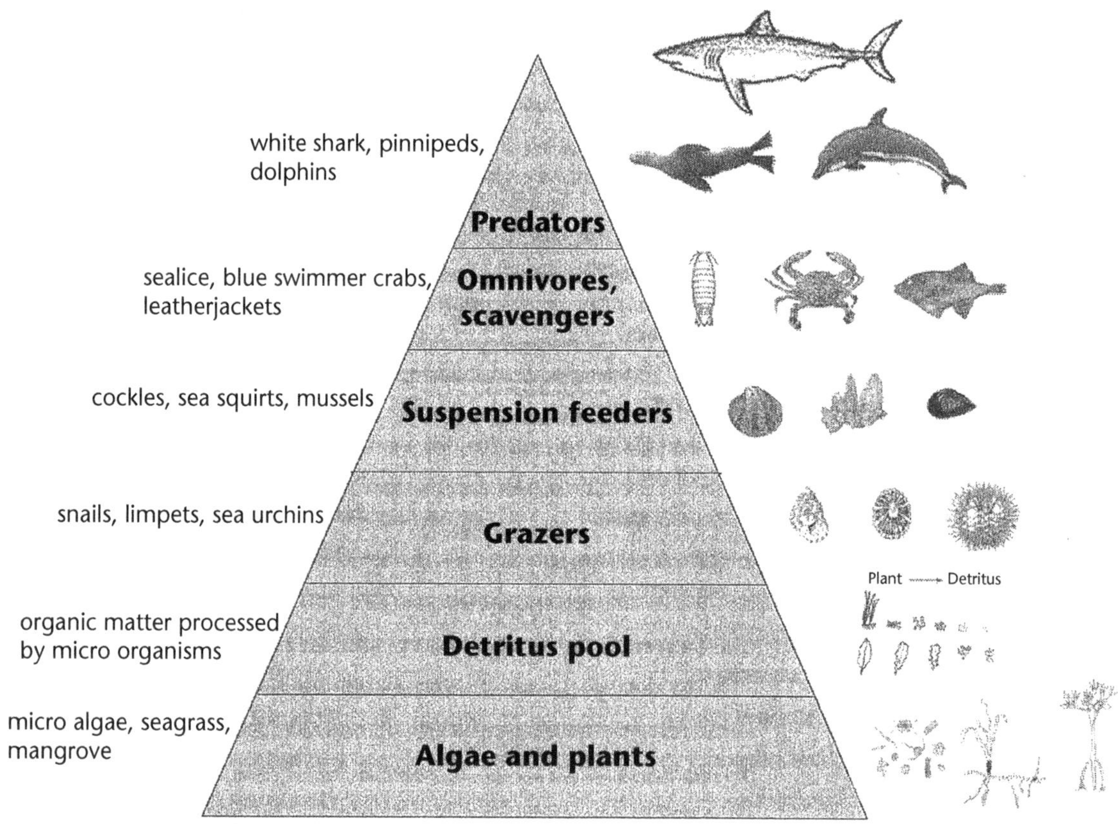

A general food energy pyramid, illustrating the various trophic levels from marine plants and detritus at the base to top predators. Of course, ecosystems vary, and the relative biomass of feeding types at different levels will vary accordingly. The pyramid shape symbolises the decline in biomass with each higher trophic level (after Shepherd *et al.* 2008).

REFERENCES

Andrew N (Ed.) (1999) *Under Southern Seas: the Ecology of Australia's Rocky Shores.* University of NSW Press, Sydney.

Clayton MN, King RJ (Eds) (1981) *Marine Botany: an Australian Perspective.* Longman Cheshire, Melbourne.

Connell SD, Gillanders BM (2007) *Marine Ecology.* Oxford University Press, Melbourne.

Dayton PK (2003) The importance of the natural sciences to conservation. *American Naturalist* **162,** 1–13.

Edgar GJ (2001) *Australian Marine Habitats in Temperate Waters.* Reed New Holland, Sydney.

Edgar GJ (2008) *Australian Marine Life: the Plants and Animals of temperate Waters.* Reed New Holland, Sydney.

Fairweather PG, Quinn GP (2007) Introduction: Australasian marine ecology in a global context. In *Marine Ecology* (Eds SD Connell and BM Gillanders) pp. 1–15. Oxford University Press, Melbourne.

Gomon M, Bray D, Kuiter R (2008) *Fishes of Australia's Southern Coast.* Reed New Holland, Sydney.

Gowlett-Holmes K (2008) *A Field Guide to the Marine Invertebrates of South Australia.* Notomares, Hobart.

Shepherd SA, Bryars S, Kirkegaard I, Harrison P *et al.* (Eds) (2008) *Natural History of Gulf St Vincent.* Royal Society of South Australia, Adelaide.

Colour plates

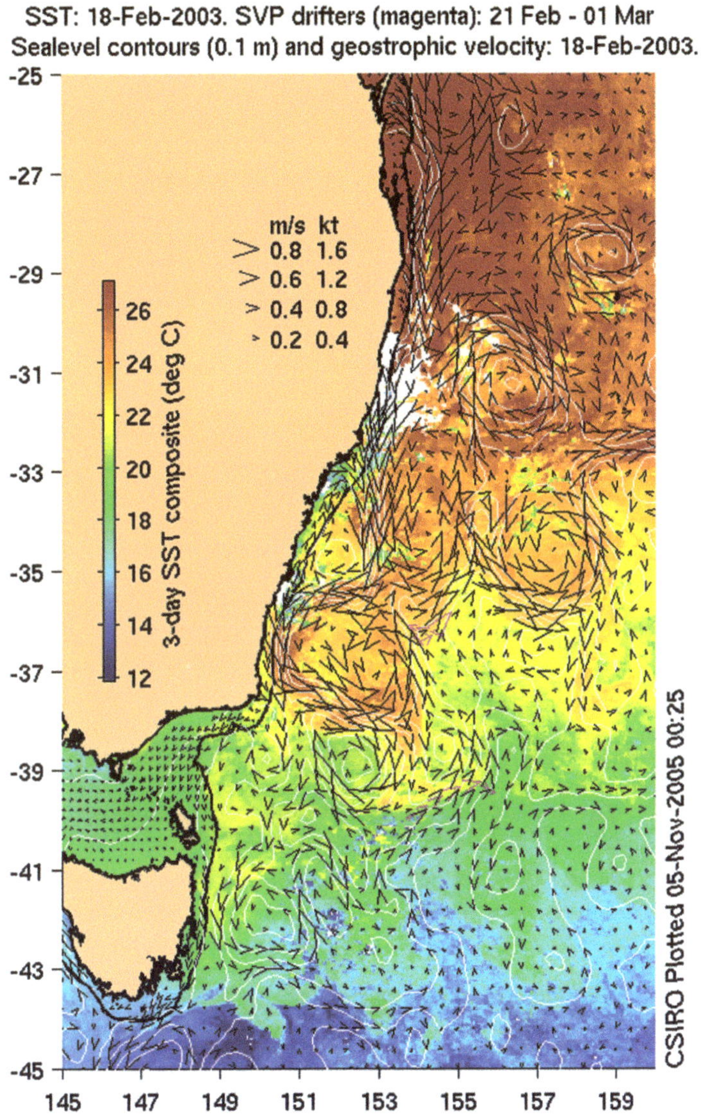

Plate 1. Satellite sea surface temperatures of SE Australia (courtesy David Griffin, CSIRO).

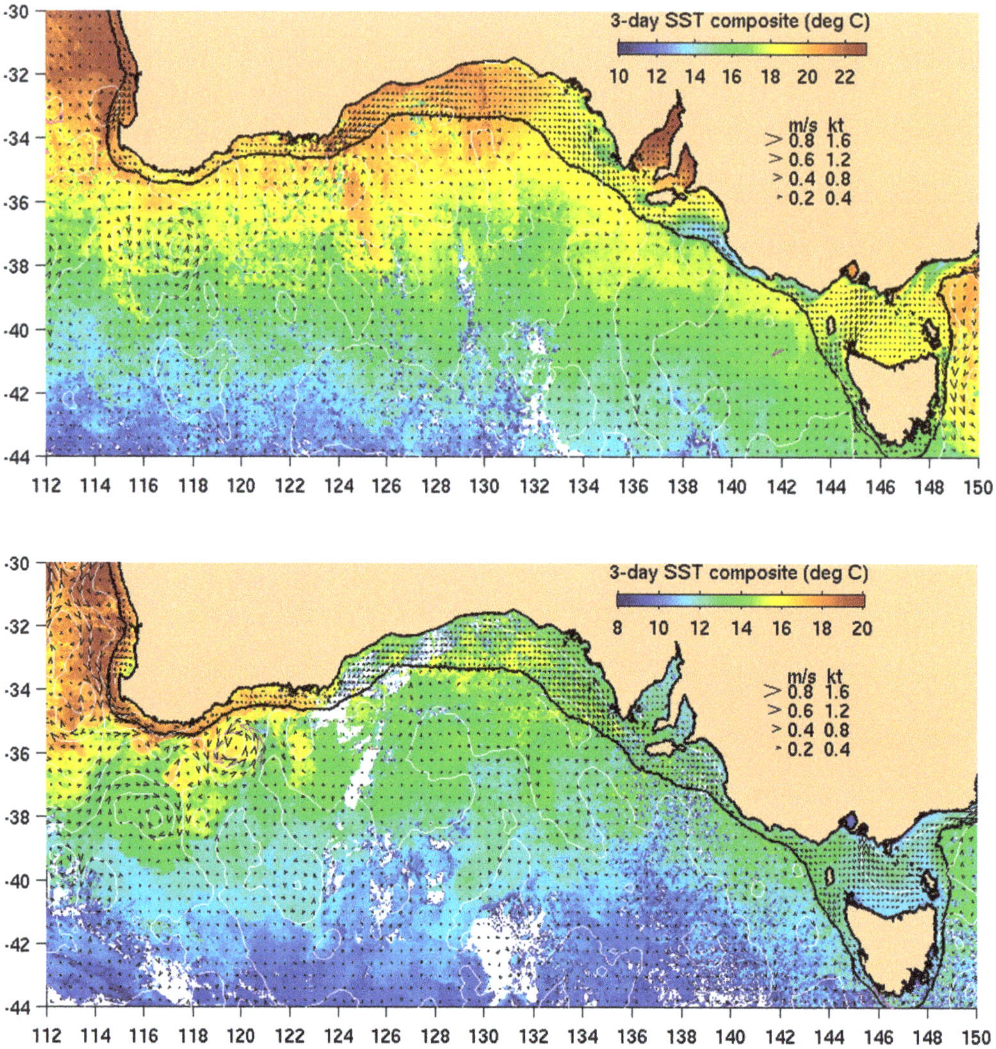

Plate 2. Satellite sea surface temperatures of southern Australia (a) (upper) in summer, and (b) (lower) in winter (courtesy David Griffin, CSIRO).

Plate 3. Macroalgal community with dominant kelp, *Ecklonia radiata*, the green alga, *Caulerpa brownii*, some red algae, and low fucoid algae at lower right (photo James Brook).

Plate 4. Giant kelp, *Macrocystis pyrifera*, canopy in SE Tasmania (photo Craig Sanderson).

Plate 5. The introduced kelp, *Undaria pinnatifida*, in SE Tasmania (photo Graham J Edgar).

Plate 6. A dense mixed forest of the fucoid canopy species, *Cystophora moniliformis*, (centre), and *Seirococcus axillaris* (upper left), and an understorey of the red alga, *Asparagopsis armata* and other species (photo Bob Baldock).

Plate 7. The fucoid, *Sargassum lacerifolium* (photo Bob Baldock).

Plate 8. The fucoid, *Xiphophora gladiata*, with the green alga *Ulva* sp. in foreground (photo Fiona Scott).

Plate 9. A dense forest of the fucoid, *Phyllospora comosa* (photo Bob Baldock).

Plate 10. Forest of corkweed, the fucoid, *Scaberia agardhii*, with a single plant of the fucoid, *Caulocystis cephalornithos* in centre (photo Bob Baldock).

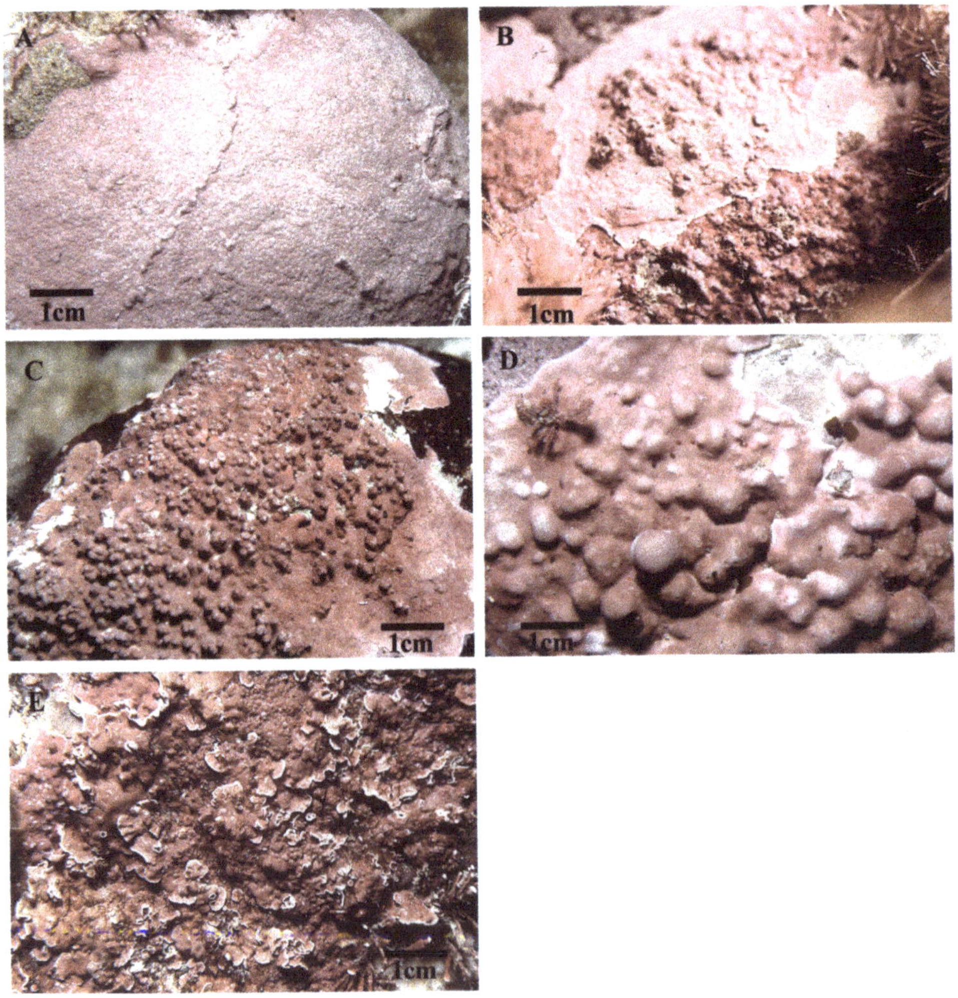

Plate 11. Growth forms of non-geniculate coralline algae and intergradations at West I., South Australia: (A) encrusting; (B) encrusting–warty; (C) warty–fruticose; (D) lumpy; (E) foliose (photos Sabine Daume).

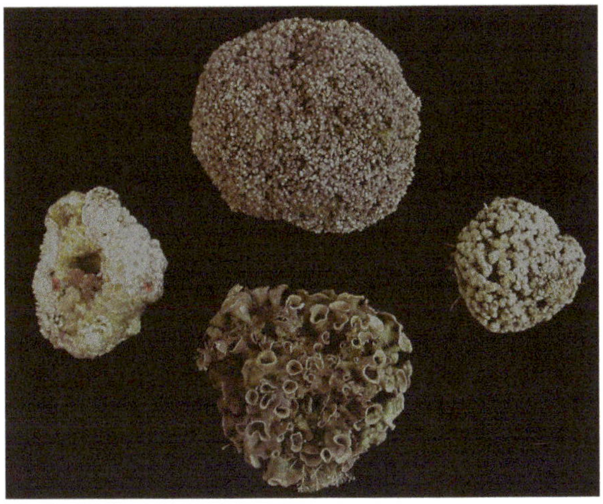

Plate 12. Four growth-forms of rhodoliths from Western Port, Victoria. (Left) encrusting; (top) fruticose; (right) 'open' fruticose; (bottom) foliose (photo Adele Harvey).

Plate 13. The prostrate red alga, *Peyssonnelia inamoena* (photo Bob Baldock).

Plate 14. The understorey erect coralline alga, *Cheilosporum sagittatum* (photo Bob Baldock).

Plate 15. Common crustaceans and a polychaete. (Top left) tanaid; (bottom left) isopod; (top right) polychaete – Opheliidae; (mid-right) caprellid; (bottom right) euphausiid or krill (photo Leo Mantilla).

Plate 16. European fanworm, *Sabella spallanzanii* (photo Jan Watson).

Plate 17. Two warreners, *Turbo undulatus*, grazing on a blade of the kelp, *Ecklonia radiata* (photo Helen Crawford).

Plate 18. A turban, *Turbo torquatus*, with large female epifaunal bonnet limpets, *Sabia australis*, and tiny males near the female spires (photo Dave Muirhead).

Plate 19. A juvenile greenlip abalone, *Haliotis laevigata*, fleeing from an 11-armed sea-star, *Coscinasterias muricata* (photo Kevin Branden).

Plate 20. Sea-hare, *Aplysia parvula* (photo Dave Muirhead).

Plate 21. The cowry, *Zoila friendii*, grazing on the sponge, *Spirastrella* cf. *papillosa* (photo Peter Clarkson).

Plate 22. Two ovulids, *Primovula heleneae* on their host, the gorgonian coral *Euplexaura* sp. in upper Spencer Gulf (photo Nigel Holmes).

Plate 23. Aggregation of doughboy scallops, *Mimachlamys asperrima*, with their epizoic crellid or myxillid sponges (photo Craig Styan).

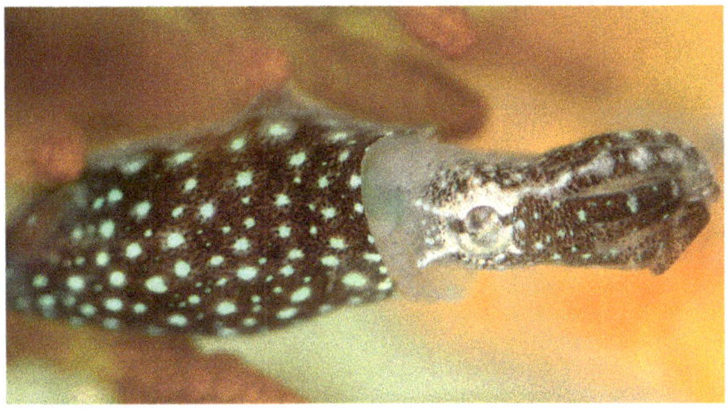

Plate 24. Pygmy squid, *Idiosepius notoides* (photo Brian Saunders).

Plate 25. Several giant cuttlefish, *Sepia apama*, at spawning site in upper Spencer Gulf. Large male in centre with female below (photo Graham Edgar).

Plate 26. Blue-ring octopus, *Hapalochlaena maculosa* (photo Craig Proctor).

Plate 27. Orange feather-star, *Comanthus trichoptera* (photo James Brook).

Plate 28. Cluster of the northern Pacific sea-star, *Asterias amurensis* (photo Graham Edgar).

Plate 29. The viviparous brittle-star, *Amphipholis squamata* (photo Brian Saunders).

Plate 30. The eastern egg urchin, *Holopneustes purpurascens* feeding on *Ecklonia* blade (photo Graham Edgar).

Plate 31. Four purple urchins, *Heliocidaris erythrogramma*, with different colours (photo Juan Livore).

Plate 32. The southern sea cucumber *Australostichopus* cf. *mollis* (photo Ben Stobart).

Plate 33. Barrens formed by the black urchin *Centrostephanus rodgersii*. The tape shown was used to monitor urchin density (photo Andrew Green).

Plate 34. Mini-barrens formed by the purple urchin *Heliocidaris erythrogramma* in SE Spencer Gulf (photo Steve Reynolds).

Plate 35. Colonies of the ascidian bluebells, *Clavelina moluccensis* (photo Dave Muirhead).

Plate 36. Colonies of the bryozoan, *Bugula dentata*, growing on vertical face with sponges and ascidians. The species is closely related to *B. neritina* (photo Andrew Newton).

Plate 37. A cluster of the stalked bryozoan, *Parmularia smeatoni* (photo Jackie Bone).

Plate 38. Late successional stage of growth on a concrete artificial reef after eight years in upper Port Phillip Bay, showing sponges, native oysters, purple sea-urchins, a blacklip abalone and sea-star, *Tosia australis* at top right (photo Andrew Newton).

Plate 39. The magnificent hydroid, *Ralpharia magnifica* (photo Andrew Green).

Plate 40. The red thimble jelly, *Turritopsis* cf. *nutricula* (photo Lisa Gershwin).

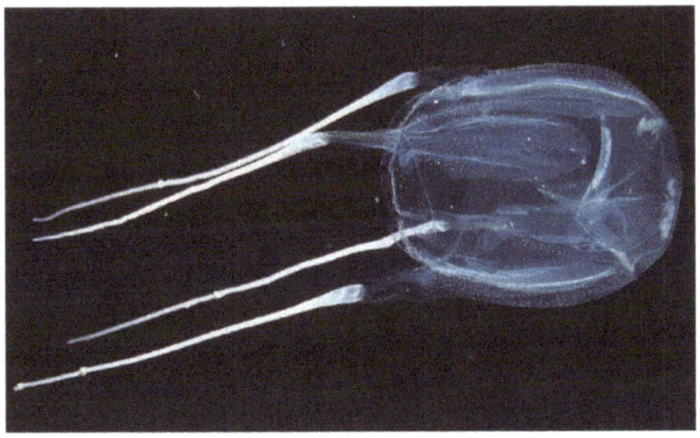

Plate 41. The box jelly, *Carybdea rastoni* (photo Graham Edgar).

Plate 42. The moon jelly, *Aurelia* sp.7. in Coffin Bay (photo Brian Saunders).

Plate 43. The lion's mane jelly, *Cyanea rosella*, with juvenile trevalla hovering above (photo Andrew Green).

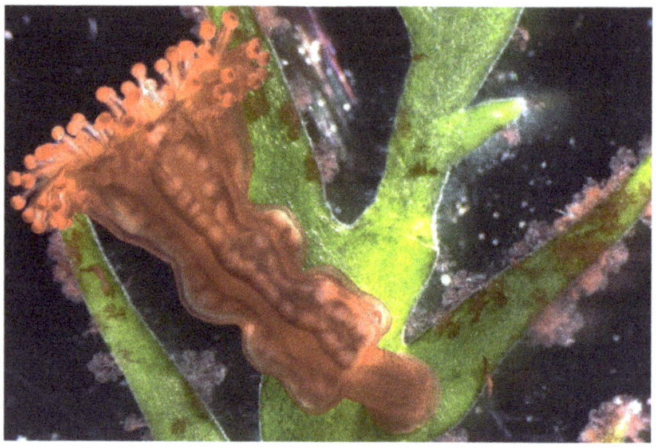

Plate 44. The Stauromedusa, *Depastromorpha africana* (photo Carolina Zagal).

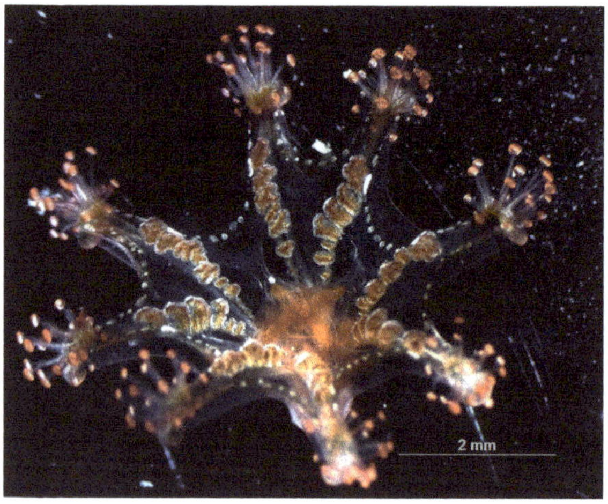

Plate 45. The Stauromedusa, *Lucernariopsis tasmaniensis* (photo Carolina Zagal).

Plate 46. The white-striped anemone, *Anthothoe albocincta*. Note the thread-like acontia extruded from the mouth of the anemone at left (photo Dave Muirhead).

Plate 47. Polyps of the robust zoanthid, *Zoanthus robustus* (photo Dave Muirhead).

Plate 48. The green and blue coral, *Plesiastrea versipora*, showing the different colours of colonies on a cave floor in southern New South Wales. Green and blue colours are due to coral pigments. The coral tentacles contain brown pigmented symbiotic algae, sometimes giving a brownish colour, but which is not always apparent in a photo (photo Adrienne Grant).

Plate 49. The Gabo Island soft coral, *Drifa gaboensis* (photo Dave Muirhead).

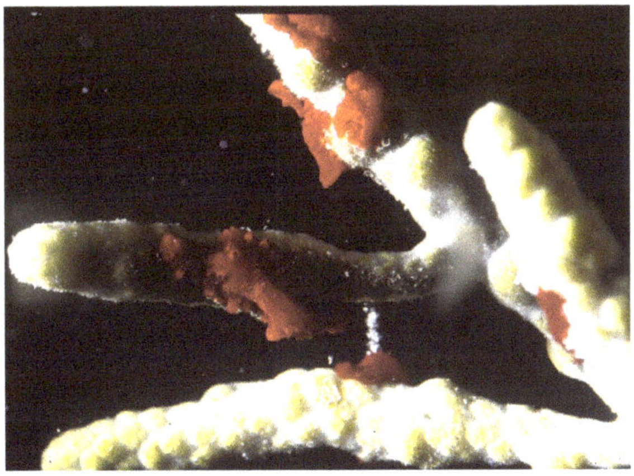

Plate 50. The ctenophore, *Coeloplana scaberiae*, on its brown algal host, *Scaberia agardhii* (photo George Matsumoto).

Plate 51. The cleaner shrimp, *Periclimenes aesopius* (photo Dave Muirhead).

Plate 52. The pea crab, *Pinnotheres hickmani*, symbiotic with the blue mussel, *Mytilus galloprovincialis* (photo Graham Edgar).

Plate 53. The European green crab, *Carcinus maenas* (photo Graham Edgar).

Plate 54. The giant crab, *Pseudocarcinus gigas* (photo Leo Mantilla).

Plate 55. Evan's sea spider, *Anoplodactylus evansi* (photo David Staples).

Plate 56. The weeping toado, *Torquigener pleurogramma* (photo Dave Muirhead).

Plate 57. A blue morwong, *Nemadactylus valenciennesi*, at a cleaning station, being cleaned of ectoparasites by three western cleaner clingfishes, *Cochleoceps bicolor* (photo Paul Fitzgerald).

Plate 58. A rainbow cale, *Odax acroptilus*, posing among *Ecklonia*, and waiting to be cleaned by an approaching cleaner, a female pencil weed whiting, *Siphonognathus beddomei* (photo Dave Muirhead).

Plate 59. The harlequin fish, *Othos dentex* (photo Simon Bryars).

Plate 60. A female southern blue devil, *Paraplesiops meleagris*, laying a cluster of eggs on the roof of a cave (photo Simon Bryars).

Plate 61. A New Zealand fur seal feeding her pup (photo Andy Lowther).

Plate 62. A southern right whale and her calf in the Great Australian Bight (photo Steve Burnell).

PART 1

THE SOUTHERN OCEAN FROM ITS BEGINNINGS TILL NOW

'The surface of the earth binds geography and geology together in an indissoluble union rather like that of man and wife. Geography, like a prudent woman, has taken to herself an 'elder than herself', (though)... she does not flaunt the assertion that she is a woman with a past.'

Geologist Charles Lapworth

The first three chapters provide the backdrop – an account of the origins and evolution of the Southern Ocean, as Australia drifted north, and the large-scale patterns that give the 'unique south' its present character. The unique features of the southern Australian flora and fauna – its high endemism, and species richness – can only be understood when we are made aware of the geology of the Australian continental plate, as described in Chapter 1. The plate has enjoyed a long period of stability with only minor vertical earth movements, as Australia broke free from Antarctica and drifted northwards. However, sea level has gone up and down with the alternate freezing and melting of polar ice caps, creating gulfs, bays and peninsulas.

The long isolation and stability of the southern Australian region, with the longest stretch of an east–west coast in the Southern Hemisphere, the temperate climate and the topographic complexity of the coastline, have created a vast range of habitats. The ocean currents that have washed its shores for 70 million years, the tidal currents and the variable wave climate have all played a role in further diversifying a bewildering complexity of habitats (Chapter 2). Together these factors have contributed in a variety of ways to the evolution and diversification of a rich marine flora and fauna, as described in Chapter 3.

1 Geological history and climate change in southern Australia

'The history of any one part of the Earth, like the life of a soldier, consists of long periods of boredom and short periods of terror.'

Geologist DV Agar

OVERVIEW

This chapter examines the past history of the Southern Ocean from its beginnings some 80–100 million years ago (Ma) to the present. As Australia broke free from Antarctica and drifted north, the Southern Ocean widened and then, as the Tasmanian gateway and Drake Passage opened, the Circumpolar Antarctic Current developed, insulating Antarctica, and making way for its glaciation. During the past 65 million years (My), Earth's climate has gone from super-greenhouse to icehouse conditions as atmospheric carbon dioxide (pCO_2) concentration has declined. Southern Australian waters at first cooled with the global temperature decline, but then gradually warmed to the present temperate temperature regime as Australia drifted towards the tropics. Superimposed on these changes were four global chills and some warming episodes. Milankovitch cycles related to Earth's orbit, and small changes in insolation, underlay climatic oscillations, greatly magnified by pCO_2. Southern Ocean waters have also been influenced by the episodic Leeuwin Current and ENSO events affecting upwellings. Sea levels have fluctuated with global temperatures and the extent of polar ice sheets. The past history of the Southern Ocean can now be used to predict the consequences of anthropogenic increases in pCO_2. These include: higher sea temperatures, rising sea levels, and changes in southern wind patterns and ocean currents.

INTRODUCTION

The marine flora and fauna of southern Australia are undeniably rich, but understanding why they are rich is not easy. This is a question of history, so we need to examine the geological past, the palaeogeography and climate of the Southern Ocean since its beginnings some 80–100 million years ago (Ma). Another reason why geological history is important is the present concern about climate change. As climatic records are short (~100 years), geological records are invaluable in testing theories and developing models about long-term changes. This chapter gives a simplified thumbnail sketch of geological events over the past 80 million years (My), and examines some of the key factors that have shaped the environment in which the present flora and fauna of southern temperate reefs have evolved. The Southern Ocean has played a major role, and still does, in the global climate system, so we will need to consider in some detail its evolution and that of its 'parental' neighbouring land mass, Antarctica. The chapter concludes with a summary of predicted oceanic and climatic changes in southern coastal regions with global warming.

The modern fauna and flora had its origins in the near-total extinction event at the end of the Permian, 250 Ma, when 95% of species disappeared (Benton and Twitchett 2003), but events in the last 80 My, as Gondwanaland fell apart, have been the most crucial for the modern fauna.

The Earth's history has been marked by continuous climatic fluctuations, which find their origin in four kinds of events: periodicities of the Earth's orbital cycles (Milankovitch cycles), tectonic processes, occasional aberrant climatic extremes, and the rare global catastrophe. Of all these, only the Milankovitch cycles have an astronomic regularity, and establish an underlying cyclic pattern. Their periodicities, ranging from a few to many thousands of years (ky), are of three types, founded on: (a) the eccentricity of the Earth's orbit round the sun, going from nearly circular to elliptical, with periods of 400 and 100 ky, and with normally slight effects on insolation; (b) the obliquity of the Earth's axis (period 41 ky), which mainly affects seasonal climatic contrasts from the tropics to high latitudes; and (c) precession; that is, the wobble of the Earth's axis (periods of 19 and 23 ky), which, modulated by orbital eccentricity, increases seasonal contrast in one hemisphere and decreases it in the other (Zachos *et al.* 2001).

The major tectonic processes were the break-up of Gondwanaland, and the developing rift between the Antarctic plate and the Australian and South American plates, as the last two plates gradually drifted northwards. The major extreme climatic aberrations were:

- the late Paleocene warm phase (super-greenhouse), in which sea temperatures increased by 5–8°C according to latitude, with globally higher precipitation and humidity and other consequences
- the extreme glaciation of Antarctica ~34 Ma
- the later brief (200 ky) glacial maximum.

The major global catastrophic event was the extensive volcanic eruptions combined with a huge meteor impact in Mexico ~66.4 Ma. These caused both a mass extinction of life on land, and also the collapse of oceanic marine ecosystems. The near complete cessation of oceanic productivity in the sea for ~10 ky gave rise to a 'Strangelove' ocean, so called because of its supersaturated state and domination by inorganic $CaCO_3$. Modelling showed that the event would have caused a two- to three-fold increase in atmospheric CO_2 (see below), and led to ~3°C warming of surface waters (Zachos *et al.* 1989).

The above events were all accompanied by increased turnover of marine species and the evolution of new species.

THE CAENOZOIC

Birth of the Southern Ocean

Our story starts with the separation of Australia from Antarctica, which began ~100 Ma, and saw the formation of the Southern Ocean. The major tectonic, climatic, oceanographic and biotic events from ~65 Ma to the present, with sources, are summarised chronologically in Table 1.1. About 100 Ma, Australia, New Zealand, Antarctica, South America, Africa and Maria Byrd Land together formed Gondwanaland, with break-up first occurring as the New Zealand plateau drifted north. By the end of the Cretaceous and beginning of the Cainozoic (65 Ma), Antarctica was still linked tenuously to Australia, whose southern coastline then lay at 60–65°S (Fig. 1.1). Numerous shallow marine transgressions penetrated successively from south-west (SW) Australia eastwards along southern Australia in the succeeding 5–10 My, but Tasmania and the South Tasman Rise remained barriers to deep water circulation until the mid-Eocene (~42 Ma). Two events – the opening of Drake Passage between South America and Antarctica and the accelerated widening of the oceanic gap between the south Tasman Rise and Antarctica at ~41 Ma – gave birth to the proto-Circum-Antarctic Current (CAC), also known as the West Wind Drift, and led to a marked fall in southern shallow sea-water temperatures. The opening of these two ocean gateways and

Table 1.1: Cainozoic palaeo-events, as reviewed by McGowran *et al.* 2000.

Ma	Epoch	Climatic and tectonic events	Oceanographic, biotic events	References
100	Mid-Cretaceous	Global temperatures at a max. Antarctic–Australia rifting has begun		Crame 1999
70–65	Palaeocene	Meteor impact. Supergreenhouse. Shallow Australia–Antarctic opening	Mass extinction event. Marine productivity collapse. Five shallow marine transgressions.	Keller *et al.* 2004 Zachos *et al.* 1989 McGowran 1991 Brinkhuis *et al.* 2006
55	Eocene	Massive methane/CO_2 release		Sluijs et al. 2006 Pearson and Palmer 2000
50–49		Global cooling (Chill 1) Proto-Circum-Antarctic Current	Shallow marine gateway Archaic whales appear	Crame 1999 Lear *et al.* 2000
44–2		Sea-floor spreading rates increase	Intermittent proto-Leeuwin Current	
42–1		Flooding of Australian continental margin in three transgressions	Drake Passage opens, and CAC begins	Scher and Martin 2006 McGowran and Li 1998
39			Leeuwin Current strong	
36		Ephemeral ice-sheets appear	Spheniscid penguins appear	Zachos *et al.* 2001
34	Oligocene	Cooling trend (Chill 2). pCO_2 decline. Glaciation of Antarctica increases	Tasmanian gateway widens Leeuwin Current weak Baleen whales appear	Pälike *et al.* 2006 Coxall *et al.* 2005 McGowran *et al.* 2004
24	Miocene	Warming cycles begin	Leeuwin Current weak or closed down	
22–20		Australia–Indonesia deep-water gap closes	Seals, sea lions appear. East Australian Current develops	
16–14		Warm phase followed by cooling phase. (Chill 3) East Antarctic icesheet grows		DeConto and Pollard 2003
10–7		Brief episodic warm phases		
6		West Antarctic icesheet grows		
5	Pliocene	Continuous El Niño. Then, episodic, brief warm phases	Hominids appear.	Pearson and Palmer 2000
5–3		Northward drift of Australia. El Niño now episodic Major global cooling. Chill 4 at 2.4 Ma	Restricted Indonesian through-flow. Panama seaway closes. Leeuwin Current strong	Fedorov *et al.* 2006
1.6	Pleistocene	Bipolar glaciation onset 27 glacial–interglacial cycles	Leeuwin Current episodic, but strengthens in last	Drexel and Preiss 1995

(*Continued*)

Table 1.1: (Continued)

Ma	Epoch	Climatic and tectonic events	Oceanographic, biotic events	References
400 ky		4 glacial–interglacial cycles Strong greenhouse gas cycles	interglacial Hominids arrive (50 ky BP)	Petit *et al.* 1999 McGowran *et al.* 1997
10ky	Holocene	Last interglacial. Sea-level fluctuates (–130 m to +3.5 m) El Niño frequency increases		Belperio *et al.* 1983 Haq *et al.* 1987 Moy *et al.* 2002
2100	Future	Continuous El Niño	Leeuwin Current weakens Upwelling intensity increases. Sea level rise 1 m per century.	Collins 2005 Overpeck *et al.* 2006. Rahmstorf 2007

the CAC together played a crucial climatic role, because they cut off Antarctica from a warm, south-flowing tropical current along South America. This led to the thermal insulation of Antarctica and saw the growth of the Antarctic ice-sheets 36–34 Ma (see below), which were key elements affecting the scale of climate change then and later, up to the present time (Fig. 1.2).

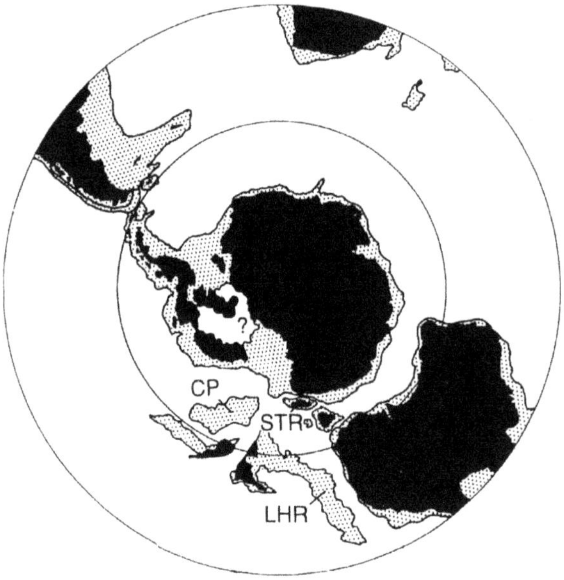

Figure 1.1: Reconstruction of positions of Australia and Antarctica 70 Ma. Present day continents are in black, and continental shelf stippled. CP – Campbell Rise; LHR – Lord Howe Rise; STR – South Tasman Rise. Inner circle is 60°S, and outer circle 30°S, palaeolatitude (after Crame 1999).

Climate change: super-greenhouse to icehouse

During the Palaeocene (~55 Ma) there was a dramatic burst of intense warming – termed the late Palaeocene thermal maximum – when polar temperatures were 18–20°C, and alligators roamed the Arctic. This period was a paradise for warm-water animals and a pivotal time in global climate evolution, because a long-term cooling trend then began that ultimately ended with a cooler Earth, with contrasting tropical regions and polar glaciations. Analysis of deep-sea cores and the stable isotope record preserved in them from the Pacific, Atlantic and Southern Oceans (Lear *et al.* 2000) provided the data for estimating past sea temperatures and atmospheric CO_2 (pCO_2). The results (Fig. 1.3) show that after an initial warming trend in the mid-Palaeocene, global deep-sea temperatures declined over the next 60 My, while pCO_2 declined in an oscillating manner. Deep-sea temperatures are plotted because they are better indicators of global change than sea-surface or terrestrial temperature data. Four major chills, each related to increased polar glaciation, occurred during the period, with a series of intervening warmer phases particularly during the Miocene and early Pliocene. More detailed analysis of temperature oscillations during the four chills showed that these oscillations were determined strongly by Milankovitch cycles, mainly the orbital eccentricity cycles with periods of 100 ky and 400 ky. However, the role of declining pCO_2 (see below), as a

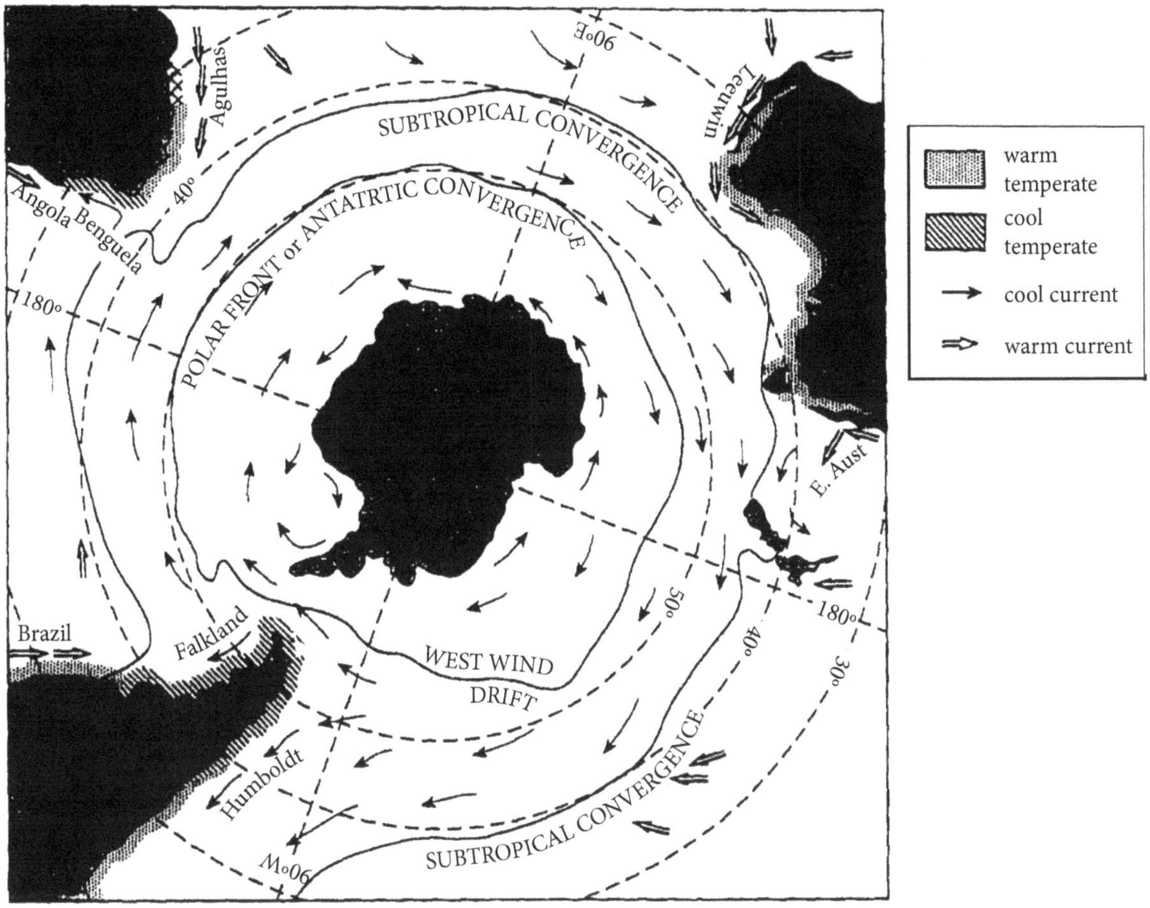

Figure 1.2: The modern Southern Ocean showing directions of flow of cool and warm currents in relation to the continents, and their present day biogeographic provinces, tropical to cool (based on Morgan and Wells 1991, and McGowran *et al.* 1997).

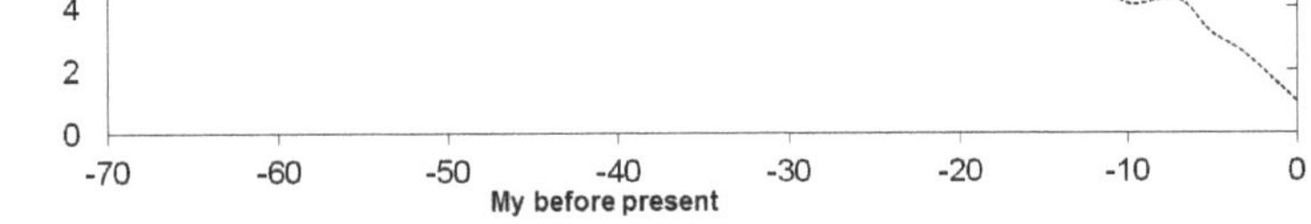

Figure 1.3: Deep-sea temperatures for the past 68 My (redrawn from Lear *et al.* 2000) (dotted line), and sea-surface temperatures for southern Australia (continuous line) calculated for a present day latitude of 35°S, and adjusted for the latitudinal northward movement of Australia from 64°S 68 Ma. The four warming events occur at 50–60, 45, 38–40 and 15 Ma, respectively, and the three Chills, 1,2 and 3 at 50, 34 and 12 Ma, respectively (redrawn from Frakes *et al.* 1994 and McGowran *et al.* 2000).

primary cause amplifying Earth's orbital forcing and increasing polar ice-sheet formation, was emphasised by DeConto and Pollard (2003).

Changes in the Earth's obliquity (period 40 ky) acquired new significance from the Oligocene, and precessional forcing (period 19 and 23 ky) from the Pliocene, when the climatic response to their forcing became remarkably amplified as long-term global cooling brought into play new feedback mechanisms. The precession of the equinox and changes in Earth's obliquity both induced high-latitude variation in summer solar radiation and, coupled with ice-albedo effects from northern glaciation, led to drastic oscillations between prolonged glaciations and warm, but brief, interglacial periods, which persist to this day.

A continent adrift

As global cooling continued, Australia kept gradually migrating northwards towards warmer tropical waters (Fig. 1.4), leading some to suggest a neat, climatic balancing act in which the warming and cooling tendencies cancelled each other out. However, this is at best a crude approximation, because it is evident from Fig. 1.3 that, as Australia drifted, the southern region was continuously subjected to strong temperature oscillations of several degrees magnitude at time scales of a few to tens of My.

The substantial growth of the Antarctic icesheets from ~14 Ma greatly influenced oceanic and atmospheric circulation in the Southern Ocean, leading to conditions likely similar to those of today. By now Australia was roughly at its present latitude, and the only later significant tectonic event was the closure of the Panama gateway at ~4–4.5 Ma. This event changed oceanic circulation and steepened sharply the equatorial–polar temperature gradient, likely contributing further to increased bipolar glaciation.

Further to the above events, several oceanic or climatic features have been pivotal influences on the southern marine climate and on the shape of southern coastlines. These are: the Leeuwin Current, changing sea levels and El Niño events, and we shall consider them in turn.

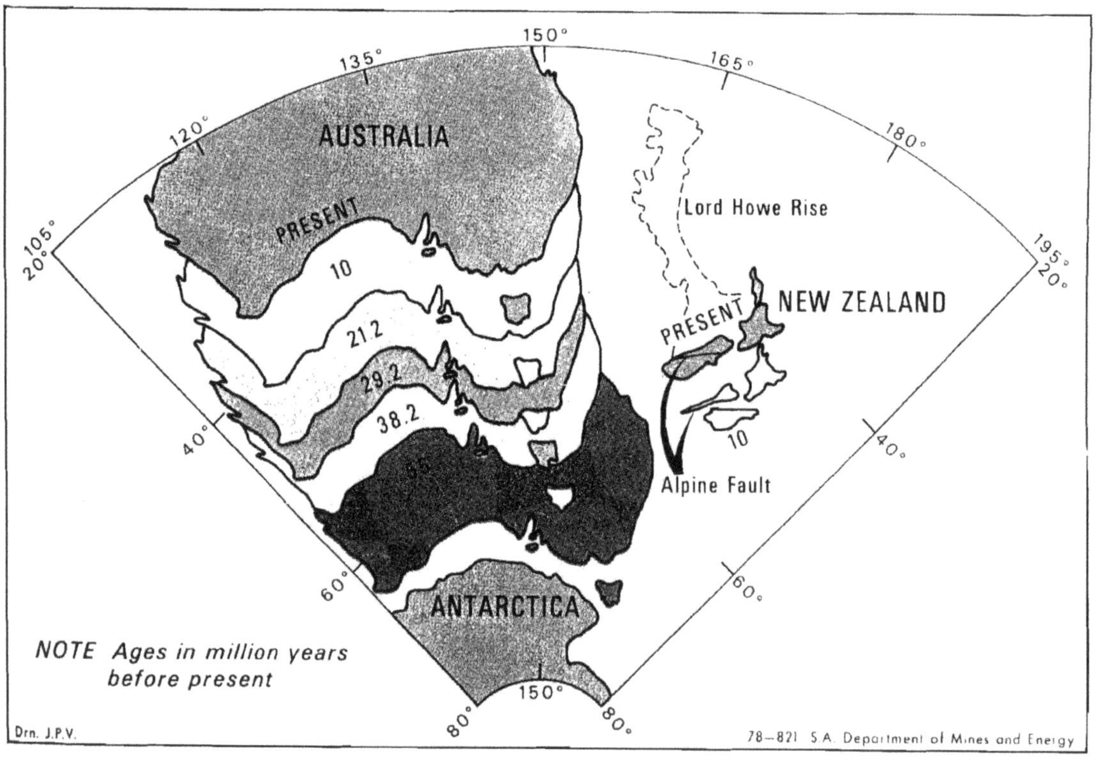

Figure 1.4: Australia's drift northwards in the last 65 My (from Ludbrook 1980).

Leeuwin Current

A major event biogeographically in the middle–late Eocene for southern Australia was the development of the warm proto-Leeuwin Current. This current hived off from the Pacific south equatorial current via the Indonesian through-flow, and then flowed down the west Australian coast and eastwards along southern Australia as the Southern Ocean gradually widened (Fig. 1.2). In the succeeding epochs, the current was episodic, sometimes strong and at other times weak, or even closed down. During cool periods, the subtropical convergence and the West Wind Drift shifted northwards, and the Leeuwin Current shut down (Fig. 1.5). Hence, from the late Eocene onwards it appears to have been cyclic in strength at a timescale of ~1 My, as inferred from the intermittent presence of fossil tropical foram assemblages found in southern Australia (McGowran *et al.* 1997, 2000; McGowran 2009). It was at about this time also (41–39 Ma) that major marine transgressions took place along southern Australia, and established marine environments in the Eucla Basin (the Wilson Bluff Transgression), and later in the St Vincent, Murray and Otway Basins (the Tortachilla Transgression) (Fig. 1.5).

During the Pleistocene, the strength of the Leeuwin Current continued to be driven by its interaction

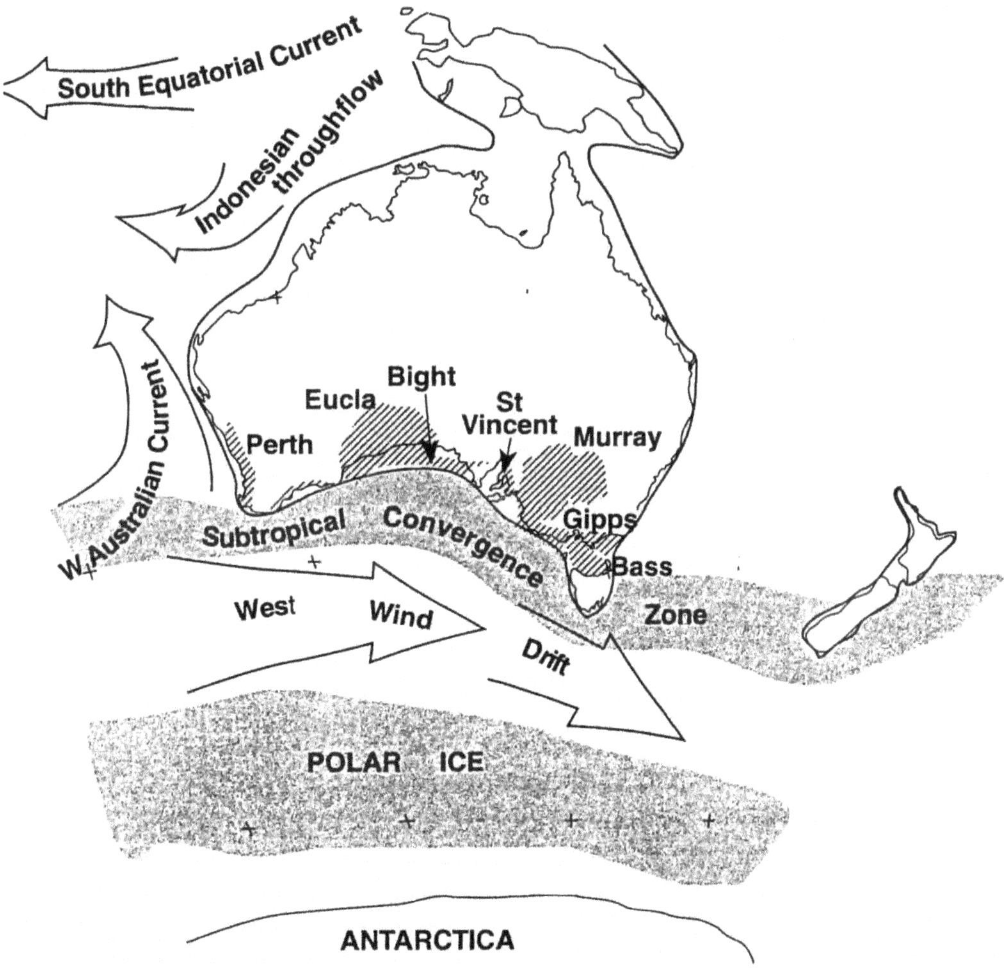

Figure 1.5: Australia with the major basins facing the Southern Ocean during the Cainozoic, showing oceanic conditions in the Southern Ocean during cooling periods. In these conditions, the West Wind Drift and Subtropical Convergence shift northwards, and the Leeuwin Current, normally flowing south along the west Australian coast, shuts down. Marine transgressions over the southern basins occurred from the late Eocene and reached their height in the early Miocene (from McGowran *et al.* 2000).

with the West Wind Drift. Over the past 500 ky, during glacial periods, the Subtropical Convergence Zone moved north and the West Wind Drift with it, cooling southern waters, while the Leeuwin Current weakened. During the briefer interglacial periods, as the Subtropical Convergence and West Wind Drift moved south, the Leeuwin Current again strengthened. Thus, during this later epoch, the Current turned on and off with a period of ~100 ky (Fig. 1.5).

During the early Miocene (~22–20 Ma) the Australian plate, now sliding north at a rapid half-rate of ~20 mm year^{-1}, collided with South-East Asia, restricting the Indonesian through-flow to several deep but narrow channels, and deflecting warm water down the east Australian coast, as well as causing the development of the Pacific equatorial counter-current (Kennett 1985). McGowran *et al.* (2000) speculated that the Indonesian constriction may also have indirectly had a sluicing effect on the Leeuwin Current, causing it to flow more strongly down gradient along the west Australian coast.

Sea level changes

Sea level changes are a key factor controlling directly coastal hydrographic patterns, and, indirectly, the distribution of benthic animals and plants. The extent of exposure of continental shelves has at times exerted a strong influence on marine animal migration patterns.

Over the past 60 My, global (or eustatic) sea levels fluctuated up and down over a vertical distance of ~300 m, at time-scales of a few thousands to millions of years, and strongly affected the existence and extent of shallow seas on the Australian continental shelf, and the diversity of marine habitats. Figure 1.6 shows the major trends in sea level during the Cainozoic, on which could be superimposed many more minor fluctuations that have occurred over time-scales of ~100 ky. In addition, tectonic upwarping and downwarping of the land, subsidence and coastal erosion have also occurred in southern Australia, complicating the isolation of the different effects. Generally, sea levels have shown a downward trend in concert with major global sea temperature declines, as is expected from the thermal contraction of sea water, but this does not account for the magnitude of the changes that have occurred, so we must also examine the contribution of polar glaciations.

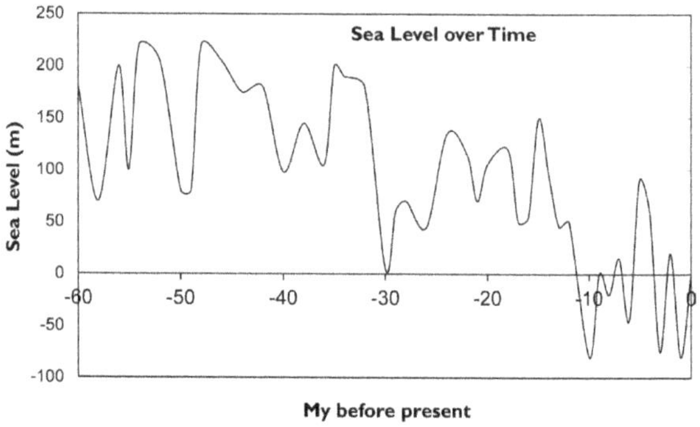

Figure 1.6: Plot of large-scale changes in global sea-level in the last 60 My (redrawn from Haq *et al.* 1987).

Polar glaciation and ice cap formation depress sea levels substantially, and melting ice caps raise them. Some correlations in sea level changes and these glaciation events are conspicuous. For example, during Chill 2, 35–30 Ma, and during Chill 3, 15–10 Ma, sea levels plummeted ~200 m, in concert with the development of the Antarctic ice caps. Conversely, ~20 Ma, during the Miocene warming, extensive marine transgressions in southern Australia reached their maximum extent in the Perth, Bremer and Eucla Basins in south-west (SW) Australia, and in the Murray, Otway and Gippsland Basins in south-east (SE) Australia (Ludbrook 1980). Since the Pleistocene (1.6 Ma), sea levels have see-sawed dramatically at a timescale of ~100 ky due to polar glaciations, and in so doing have extensively sculpted much of the southern coastline, and left stranded beach lines, and remnant submerged reefs as records of those past events.

The most recent rise in sea level began during the Holocene ~12 ky ago, and by ~7 ky ago reached the present sea level, although continuing to a maximal high stand of 2–3 m above present sea levels ~2–4 ky ago in upper Spencer Gulf, likely due to later coastal upwarping (Belperio *et al.* 1983). Since 1870, sea levels have risen by 20 cm (= 1.7 mm year^{-1}), but since 1993, the rate has doubled to ~3 mm year^{-1}, a rate attributed equally to the thermal expansion of water and melting of polar icecaps (Poloczanska *et al.* 2007).

El Niño Southern Oscillation (ENSO) events

El Niño (warm water) events originate in the equatorial Pacific, and with its sister event, La Niña (cool water), form a low-frequency, irregular, climatic oscillation between the two states with a 2–7 year frequency. It is the strongest natural climatic fluctuation globally, and influences the world's climate patterns. During an El Niño event, west Pacific trade winds relax, equatorial upwellings in the eastern Pacific cease, and tropical warm waters expand in northern Australia and spread far to the east Pacific. Hot and dry conditions are pronounced over much of Australia and Indonesia, while floods occur in Ecuador and Peru. In the early Pliocene (~5 Ma), El Niño conditions were continuous in the equatorial Pacific, enhancing the then warm conditions by causing the absence of stratus clouds from the eastern equatorial Pacific, and lowering the planetary albedo (Wara *et al.* 2005; Fedorov *et al.* 2006).

The continuous El Niño also caused the shutdown of major global coastal upwellings off South Africa and South America, and substantial changes in global rainfall patterns. Paradoxically, in southern Australia upwelling intensity off SE South Australia tended to increase, while the Leeuwin Current weakened during El Niño events (Middleton *et al.* 2007; Chapter 2), so increasing the summer temperature contrast between warm shallow inshore and cool offshore waters.

During the Holocene epoch, evidence from lake core samples in Ecuador has shown that the frequency of moderate to strong El Niño events has increased more than 10-fold in the last 10 ky, with a peak occurring ~1200 years ago (Moy *et al.* 2002). Although El Niño events may have existed during the early Holocene, they were apparently not strong enough to trigger alluvial deposition in the lake drainage basin. A plot of El Niño event frequency over time (Fig. 1.7) shows oscillations at ~2 ky superimposed on the gradually increasing frequency. It is noteworthy that the oscillations correspond with cyclic changes in insolation and the carbon cycle. The tendency for stronger and more frequent El Niños than La Niñas since the mid-1970s (McPhaden *et al.* 2006) raises the question whether El Ninõ events will become a dominant feature as greenhouse gas concentrations increase.

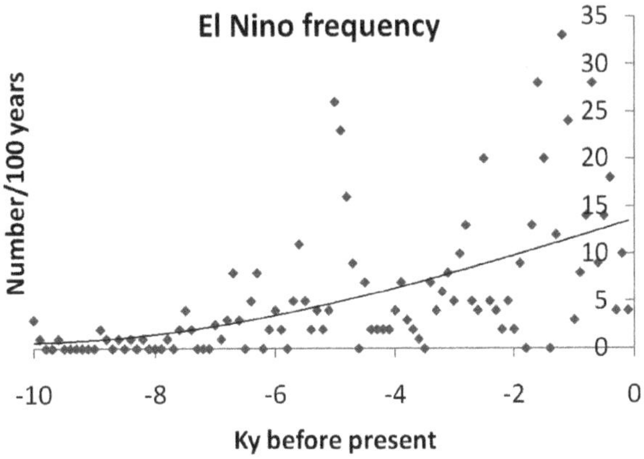

Figure 1.7: Frequency of moderate to strong El Niño events over the past 9000 years, as recorded in lake sediments from Laguna Pallacocha, Ecuador (redrawn from Moy *et al.* 2002).

This much debated question has been addressed by numerous global climate modellers (Collins 2005), and one of the best such models at predicting El Niño events has predicted that El Niño-like conditions (with more droughts) will become more frequent, and La Niña events will become stronger. In other words, climatic extremes will occur more often (Timmermann *et al.* 1999; Abram *et al.* 2007). If El Niño events become continuous, as some climatologists believe, a return to the warmer conditions of the early Pliocene seems inevitable.

The dark side of carbon dioxide: climatic effects

Carbon dioxide's heat-trapping properties are a chance product of the molecule's affinity for infrared light. Infrared rays from the sun are absorbed by CO_2 and emitted as heat. Trace concentrations of CO_2 in the atmosphere (pCO_2), together with mainly water vapour and methane create the greenhouse effect and keep Earth from ice-age frigidity. Hence changes in pCO_2 have been a major forcing mechanism on global climate over at least the past 60 My. In the early Palaeocene, pCO_2 levels were extraordinarily high at ~3500 ppm, apparently due to very high outgassing of CO_2 from north Atlantic rifting, volcanic activity and magmatism in the Himalayas, and release of methane from wetlands and deep-sea sediments. After the north Atlantic volcanism ended 54–53 Ma in the early

Eocene, pCO_2 levels fell, and cooling began. However, global warmth increased and reached a maximum (the super-greenhouse) towards 50 Ma when pCO_2 levels had fallen to 700–900 ppm (Fig. 1.8). For this period, McGowran (1989) proposed the operation of a reverse greenhouse effect, caused by the high productivity of siliceous plankton, and the resultant high rates of burial of organic carbon in the deep sea.

The next 45 My saw an erratic decline both in pCO_2 and deep-sea temperatures (see Fig. 1.3). At the Eocene–Oligocene boundary (~34–33 Ma) the sharp fall from 1500 to 400 ppm occurred at about the same time as a steep fall in global temperature (Fig. 1.3), followed quickly by the appearance of permanent ice sheets in Antarctica and a sharp drop in sea level (Fig. 1.6). This precipitous fall in pCO_2, preconditioned by a Milankovitch cycle in Earth's obliquity, played a primary role in ice sheet development, which then caused the drop in sea level (DeConto and Pollard 2003; Coxall *et al.* 2005). By ~25 Ma, the pCO_2 level had fallen to ~200–300 ppm, and then remained fairly constant until the late Miocene (4–2 Ma), when it fell from 280 to 210 ppm concurrently with the onset of major Arctic glaciation (Fig. 1.8).

Throughout the Oligocene (33–23 Ma) pronounced oscillations, with prominent 405 ky and 1.2 My Milankovitch modulations, occurred in temperature and CO_2 levels, with many minor glaciations in the cooler phases. The evidence suggested that the cycles were driven by oscillations in organic productivity in response to changes in solar insolation. These in turn

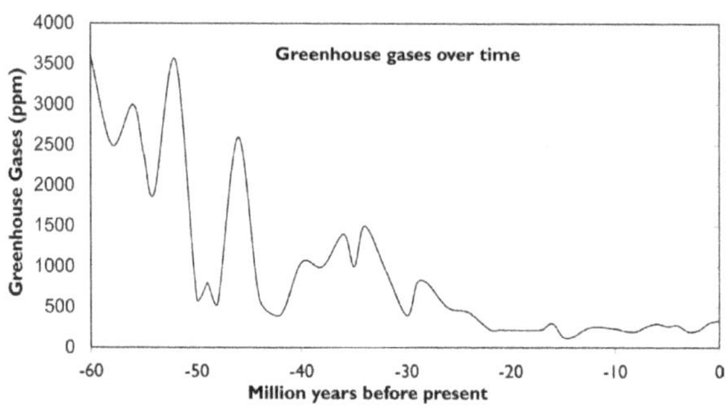

Figure 1.8: Record of greenhouse gases, comprising mainly atmospheric carbon dioxide (pCO_2), for the past 60 My as measured in boron and alkenone proxies in deep-sea sediments from the Pacific (redrawn from Pearson and Palmer 2000).

drove changes in the carbon cycle, such that increased pCO_2 paced increasing temperatures, consistent with its role as a climate-forcing mechanism (Pälike *et al.* 2006). But what caused the cyclic declines in pCO_2?

Here, the Southern Ocean has played a central role in reducing pCO_2 globally, by its high uptake of CO_2 via a biogeochemical cycle (Marinov *et al.* 2006); that is:

High CO_2 + Nutrients and Iron → Phytoplankton blooms → Increased ocean productivity + high CO_2 drawdown

By ~2.7 Ma, the warm conditions of the early Pliocene had ended with a fall in high latitude sea temperatures, and large ice sheets were well established on northern continents. Analysis of Antarctic ice cores has given a very detailed picture of the dramatic glacial–interglacial cycles in the last 420 ky, and illustrates how Milankovitch cycles, greatly amplified by greenhouse gases (CO_2 and methane – CH_4), have driven temperature changes (Petit *et al.* 1999; Shackleton 2000; Federov *et al.* 2006; Jouzel *et al.* 2007). Figure 1.9 shows the sawtooth rise and fall of Antarctic air temperature in unison with pCO_2 over the last 800 ky, and also the recent 100 ppm upsurge in pCO_2 during the last two centuries – an effect only now beginning to be felt (see below). Prior to this last anthropogenic increase, the correlation between temperature and greenhouse gas (CO_2 + CH_4) concentration was remarkably high ($r^2 = 0.71$). During this period there were eight complete glaciation cycles, with interglacial peaks ~100–120 ky apart. The periodicities of the Milankovitch cycles driving the oscillations were: orbital eccentricity (period 100 ky) contributing 37%; and obliquity (period 40 ky) 23%. The pCO_2 peaked at ~280 ppm during each interglacial peak and fell to low levels of ~180 ppm during the glaciations. Both the increases and falls in pCO_2 lagged the temperature oscillations by a few ky. The lags in the increase in pCO_2 implied that the Milankovitch cycles initiated the temperature rise, whereupon the greenhouse gases acted as an amplifier. The lag in the decline, due to polar ice-albedo feedbacks, then acted to prolong the cycle. Polar ice volumes also lagged the falls in pCO_2, air- and deep-sea temperatures (Shackleton 2000).

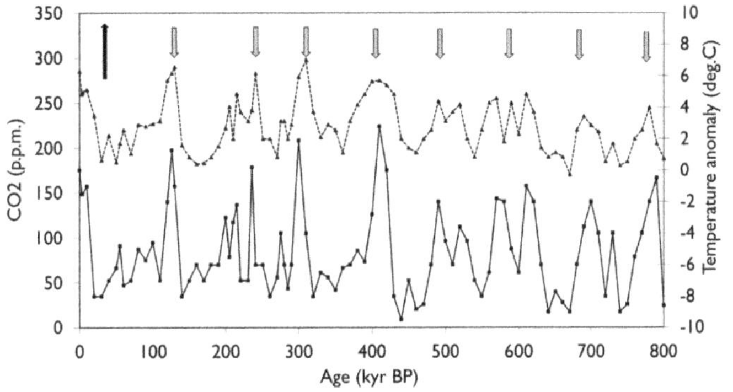

Figure 1.9: Record of atmospheric concentration of carbon dioxide in parts per million (p.p.m.) from air trapped in Antarctic ice cores over the past 800 000 years (upper), and fluctuations in temperature anomaly, based on deuterium records for the same period (redrawn from Lüthi *et al.* 2008). The solid arrow at extreme left marks the post-industrial rise in pCO_2 levels towards current levels of 380 ppm, and arrows in grey mark the Milankovitch cycles amplified by pCO_2 with high correlation between pCO_2 and Antarctic temperatures. The lag in pCO_2 is due to ice-albedo effects prolonging the cycle.

Overall, greenhouse gases contributed about half (2–3°C) of the temperature change, and the ice-albedo feedback much of the remainder. The ice-albedo effect is the reflective blanket of ice and snow over the polar regions, reducing the amount of sunlight that the Earth absorbs (just as a white T-shirt keeps us cooler on a hot, sunny day). As temperatures cool, ice and snow cover grows and amplifies climate change. The Southern Ocean also seemed to play an important role, possibly through deep ocean circulation and sea-ice extent, in regulating the long-term changes in pCO_2, although the mechanisms remain unclear. In all, the ice-core record reinforced conclusions from studies of deep-sea sediments, and suggested that the Antarctic climatic sequence of events applied widely to the temperate Southern Hemisphere. The succession of climatic changes through each cycle of the 400 ky period was strikingly similar, showing oscillations of similar magnitude, but always within stable limits. However, the Holocene, the last interglacial period, has already lasted 11 ky, which is by far the longest stable warm period yet experienced in the last half My.

In the last 200 years, industrial activity, culminating in a present day annual addition of ~17 billion tonnes of CO_2 to the atmosphere, mainly from coal and oil burning caused the unprecedented global addition of ~100 ppm (36% increase) of pCO_2, of which ~50 ppm was added since 2000. The pCO_2 level is predicted to increase further to ~500 ppm in the next 50 years and to > 560 ppm (Hobday *et al.* 2006), or even to 800 ppm (Feely *et al.* 2004) by 2100 before it can be halted. The predicted changes resulting from such increases are summarised below.

THE APPROACHING GREENHOUSE

'A smooth forehead betokens
A hard heart. He who laughs
Has not yet heard
The terrible tidings.'

Berthold Brecht

The global climate is currently changing rapidly, and is predicted to continue heating up for the next century. Average global temperatures rose by ~0.6°C in the 20th century, more than at any period in the last millennium, and are predicted to increase by 2–5°C by 2050 (Stern 2006) or by 6°C by 2070 (IPCC 2007). Globally, recent sea surface temperatures have risen by ~0.2°C per decade, but the increase has been variable – up to 3°C in the last decade in some tropical regions and close to zero in waters washing southern Australia (Behrenfeld *et al.* 2006). Although the role of greenhouse gases in causing warming trends has been known for many decades (e.g. Sawyer 1972), it is only recently that the increasing greenhouse gas levels since industrialisation have become accepted as the cause of present warming and other climatic shifts (Poloczanska *et al.* 2007). But it should be noted that recent predictions are very conservative, because numerous feedbacks occur, which will enhance the rates of warming, but cannot be included in models because of uncertainty in the magnitude of their effect. The main positive feedbacks are:

- Higher temperatures increase the levels of atmospheric water vapour, and this is expected to amplify warming estimates by 40–50%.
- The capacity of the ocean to take up pCO_2 declines as more CO_2 is taken up into solution, due to chemical feedbacks (the Revelle factor – see Sabine *et al.* 2004).
- Ocean warming inhibits mixing in the water column, reduces nutrient levels, phytoplankton productivity and the rate of pump down of CO_2 (Doney 2006; IPCC 2007, and see below).
- The ice-albedo feedback – as ice and snow melt the dark ground exposed absorbs sunlight rather than reflecting it, and becomes warmer.
- There are numerous terrestrial feedbacks as forests disappear, exposing barren land and releasing CO_2 and methane (reviewed by Lovelock 2006).

Another reason why the predictions are conservative is that they assume stabilisation of CO_2 emissions by 2050 – shown by Raupach *et al.* (2007) to be illusory. CO_2 emissions are accelerating globally, with a growth rate increasing from 1% per annum. during the 1990s to 3% since 2000. They will keep increasing into the foreseeable future because the developing countries (China, India, and those in Africa and South America) with 80% of the world's population contribute 73% of the increase in global emissions, which already far exceed the IPCC's stabilisation trajectory of 650 ppm.

Climatic trends

The major climatic future trends in southern Australian waters, as reviewed by Howard *et al.* (2009), are described in the sections below.

Sea temperatures

Waters around Australia will warm by 1–2°C by 2030, and by 2–3°C by 2070, while temperatures on land are expected to be double the latter figures. However, heating will not be uniform. The East Australia Current will flow much further south, and warming will be greater in SE Australia. Already mean sea temperatures to 50 m depth off SE Tasmania have increased by 0.9°C since 1945, with a mean winter–spring increase of 1.8°C between May and November (Harris *et al.* 1988; Chapter 12). On the west Australian coast, sea temperatures have increased 1.4°C since 1795, inferred from a 200–year coral core (Kuhnert *et al.* (1999), while direct measurements in SW Australia have shown a rise of 0.3–0.4°C a decade since 1970 (Caputi *et al.* 2009; Thompson *et al.* 2009). In contrast, sea temperatures along the remaining southern Australian coast may be seasonally hotter inshore in bays and gulfs, but summer upwellings (see above) will increase in intensity and maintain cooler waters offshore. Hotter conditions on land strengthen longshore winds and increase upwelling intensity, an effect which has already increased upwelling intensity on African, and North and South American coasts (McGregor *et al.* 2007), and will do so in southern Australia. Overall, average ocean sea temperatures along the southern coast are expected to increase by only 1–1.2°C by 2070.

The general consequences of increasing temperature, and the accompanying oceanic stratification, are a fall in marine productivity as phytoplankton biomass declines, mainly in tropical and mid-latitudes (Behrenfeld *et al.* 2006). Such decline is normally expected to change the taxonomic composition of pelagic and benthic food webs, and reduce fishery yields, but its extent in a region with strong or increased upwelling intensity may be minimal. In addition, the life history of benthic species will be affected. For example, a 2–4°C temperature increase in SE Australia will seriously compromise larval development of the purple urchin (Byrne *et al.* 2009).

Wind strength

The westerly wind belt has a dominant influence on the Tasmanian climate, and drives coastal upwellings along its east coast (Harris *et al.* 1988). During La Niña events, westerly winds fail, and so weaken the Tasmanian east coast upwellings. Climate change models predict that these westerly winds will weaken as far south as 50°S, as the westerly wind belt migrates towards the Antarctic, and will strengthen there: an effect substantially due to the hole in the ozone layer (Thompson and Solomon 2002). The weaker westerly winds, together with the greater strength of the East Australia Current, will also further increase sea temperatures. In the region from the eastern Great Australian Bight to western Victoria, however, SE winds will be stronger, especially with an increasing frequency of El Niños, and this will strengthen upwellings, and keep oceanic waters cooler.

Ocean currents

As indicated above, the East Australia Current will strengthen, and extend south of SE Tasmania. Stronger east to SE winds, increased summer upwellings in the eastern Great Australian Bight and increased frequency of El Niño events will likely strengthen the west-flowing Flinders Current. On-shelf surface flows have typically been ~11 cm s^{-1} to the north-west (NW) in summer, from Bass Strait towards the eastern Bight, and reversing in winter with speeds of 15–19 cm s^{-1} to the SE (Middleton and Platov 2003; Cirano and Middleton 2004; Olsen and Shepherd 2006). Summer flow rates will likely increase and winter flow rates decrease. But what of the Leeuwin Current?

The CSIRO model (Hobday *et al.* 2006) predicts no substantial change in the Leeuwin Current. However, as stated above, the Leeuwin Current typically weakens during an El Niño; hence it might be predicted that, with an increasing frequency of El Niños and fewer but stronger La Niñas, the Leeuwin Current will oscillate between a prevailing weak flow interspersed with occasional strong flows, as shown by the recent unprecedented 'marine heat wave' on the western coast (Pearce *et al.* 2011).

CO_2, pH and calcium carbonate saturation

In the last 200 years, the oceans have absorbed 40–50% of the CO_2 released into the air, so helping to moderate future climate change. But, as pCO_2 increases still more, the ocean uptake will increase, hydrolysis of CO_2 will increase, and hydrogen ion concentration (H^+) will rise, leading to a further drop in pH (acidification). Already, at the ocean surface pH has fallen by 0.1 since pre-industrial times (equating to a 30% increase in H^+), and will fall another 0.3–0.4 by 2100, with a 40% reduction in aragonite (a form of calcium carbonate) saturation (Orr *et al.* 2005; reviewed by Doney *et al.* 2009). This process is irreversible over the next century, and it will take thousands of years for ocean chemistry to return to pre-industrial conditions. What biological impacts will this have in temperate waters?

One impact relates to the reduction in biological calcification under more acidic conditions. Experimental evidence from calcareous benthic organisms has shown a reduction in calcifying rates of up to 45% (Feely *et al.* 2004), and, in one study at 2 × pCO_2, juvenile echinoderms stopped growing and produced more fragile and brittle skeletons (Shirayama and Thornton 2005). Many other benthic groups with calcareous skeletons (e.g. coralline algae, gorgonians, bryozoans, echinoderms and molluscs) would suffer similar effects. How acidification affects shallow-water reef communities is well illustrated by a naturally occurring volcanic carbon dioxide vent in the Mediterranean, where calcareous algae, sea urchins, and grazing molluscs are all absent (Hall-Spencer *et al.* 2008).

A second impact is that the saturation horizon for calcium carbonate in oceanic waters rises with increasing CO_2. Put simply, this means that marine species above the horizon can calcify, while below it they cannot. Already the saturation horizon is at ~700 m depth in the Pacific, and is rising at a rate of 1–5 m $year^{-1}$ (Coontz 2007). If CO_2 emissions continue on current trends, the aragonite saturation horizon will rise to the surface by 2100, making aragonite skeletons of marine species unstable in the water column over the entire Southern Ocean (Royal Society 2004). A third impact is the reduced solubility of oxygen in sea water. A decrease of 0.25 in pH will cause a reduction of 50% in the oxygen-carrying capacity of sea water, which will affect the growth and survival of species with a high oxygen demand such as fish (Australian Academy of Science 2013).

Sea level

Predicting future sea levels is complicated by our inability to predict future rates of melting of polar ice caps and, for this reason, past modelling predictions have all under-estimated sea level rises of recent decades. The complete melting of global ice sheets would increase sea levels by ~70 m, but this could take a millennium (Overpeck *et al.* 2006). Using an empirical approach and recent historical data from 1880–2000, Rahmstorf (2007) calculated a present day sea level increase of 3.4 mm year^{-1} per °C increase in mean global air temperature, and predicted a mean sea level rise of 0.9 m ± 0.4 m by 2100. However, he noted that this linear rate would only apply for the next few centuries, due to the millennial-scale lag time in ice sheet melt rates to reach an equilibrium state. Melting ice-sheets presently contribute ~0.35 mm year^{-1} to sea level rise, but this could increase if melting increases bed flows within the ice-sheets and accelerates ice flows to the ocean, as is happening in Greenland (Shepherd and Wingham 2007). Palaeoclimatic data have shown sea level increases of 10–30 m per °C increase in global temperature. For example, 3 Ma during the Pliocene, sea levels were up to 35 m higher than today when Earth was 2–3°C warmer (Fig. 1.6), and during the last ice age (20 ky ago) when temperatures were 4–7°C colder, sea level was 120 m lower. Rahmstorf (2007) also warned that sea level rises could be even greater than the rates given above, due to still poorly understood feedbacks, a warning supported by the more recent analyses of Kopp *et al.* (2009).

CONCLUSIONS

Some 60–90 Ma saw the birth of the Southern Ocean, as Australia broke off from Antarctica, and started drifting north towards Indonesia. Over that period, Earth experienced erratic cooling, as shown by deep-sea temperatures, in concert with a decline in pCO_2. Southern Australian ocean water temperatures oscillated downwards during global Chills and increased during warm periods, but declined overall in concert with the global pattern until ~30 Ma, and then increased in an oscillating manner, as Australia drifted towards tropical Indonesia.

Reviewing the whole 65 My time period, Pearson and Palmer (2000) and McGowran *et al.* (2000) recognised, among nine abrupt changes, four sudden Chills in sea temperatures (Table 1.1), with intervening warm periods. Each Chill was accompanied by sharp reductions in pCO_2, and, for Chills 2–4, by ice volume increases in the Arctic or Antarctic. Accumulating evidence on palaeo-levels of pCO_2, air and sea temperatures, sea level and the extent of polar ice sheets have shown that their oscillations are all tightly linked, and preconditioned by Milankovitch cycles affecting insolation.

Although other complex feedbacks, brought about by the drifting continents, have also played a role in these long-term changes, the pCO_2, with its known effect on temperature, has been the fundamental forcing mechanism greatly amplifying orbital changes in insolation, and causing Caenozoic climate change. The last 3 My, and especially the four glacial cycles in the last 400 ky with finer temporal resolution, illustrate the climate forcing by pCO_2, with deep ocean circulation and ice-albedo effects also playing a part.

During the evolution of the Southern Ocean: (a) the Leeuwin Current has turned on and off according to the latitude of the subtropical convergence and the West Wind Drift; (b) sea level has oscillated according to Earth's temperature and accumulated polar ice sheets; and (c) ENSO events have varied in intensity and frequency. The past behaviour of these climatic features provides a basis on which to predict the consequences of greatly enhanced forcing by pCO_2 in the next century. In the Southern Ocean these include: increased sea temperatures, rising sea levels, changing ocean currents and upwellings, and increasing acidification of sea water.

REFERENCES

Abram NJ, Gagan MK, Liu Z, Hantoro WS *et al.* (2007) Seasonal characteristics of the Indian Ocean dipole during the Holocene epoch. *Nature* **445**, 299–302.

Australian Academy of Science (2013) Acid test for the seas. <www.science.org.au/nova/106/106print.html> (Accessed 30 January 2013).

Behrenfeld MJ, O'Malley RT, Siegel DA, McClain CR *et al.* (2006) Climate-driven trends in contemporary ocean productivity. *Nature* **444**, 752–755.

Belperio AP, Hails JR, Gostin VA (1983) A review of Holocene sea levels in South Australia. In *Australian Sea Levels in the Last 15000 Years: a Review. Monograph Series. Occasional Paper 3.* (Ed. D Hopley) pp. 37–47. Department of Geography, James Cook University, Townsville.

Benton MJ, Twitchett RJ (2003) How to kill (almost) all life: the end Permian extinction event. *Trends in Evolution and Ecology* **18**, 358–365.

Brinkhuis H, Schouten S, Collinson ME, Sluijs A *et al.* (2006) Episodic fresh surface waters in the Eocene Arctic Ocean. *Nature* **441**, 606–609.

Byrne M, Ho M, Selvakumaraswamy P, Nguyen HD *et al.* (2009) Temperature, but not pH, compromises sea urchin fertilization and early development under near-future climate change scenarios. *Proceedings Biological Sciences* **276**, 1883–1888.

Caputi N, de Lestang S, Feng M, Pearce A (2009) Seasonal variation in the long-term warming trend in water temperature off the Western Australian coast. *Marine and Freshwater Research* **60**, 129–139.

Cirano M, Middleton JF (2004) Aspects of the mean wintertime circulation along Australia's southern shelves: numerical studies. *Journal of Physical Oceanography* **34**, 668–684.

Collins M (2005) El Niño- or La Niña-like climate change? *Climate Dynamics* **24**, 89–104.

Coontz R (2007) Wedging sustainability into public consciousness, *Science* **315**, 1068–1069.

Coxall HK, Wilson PA, Pälike H, Lear CH *et al.* (2005) Rapid stepwise onset of Antarctic glaciation and deeper calcite compensation depth in the Pacific Ocean. *Nature* **433**, 53–57.

Crame JA (1999) An evolutionary perspective on marine faunal connections between southernmost South America and Antarctica. *Scientia Marina* **63** (Suppl. 1), 1–14.

DeConto RM, Pollard D (2003) Rapid Cenozoic glaciation induced by declining atmospheric CO_2. *Nature* **421**, 245–249.

Doney S (2006) Plankton in a warmer world. *Nature* **444**, 695–696.

Doney SC, Fabry VJ, Feely RA, Kleypas JA (2009) Ocean acidification: the other CO_2 problem. *Annual Review of Marine Science* **1**, 169–192.

Drexel JF, Preiss WF (Eds) (1995) *The Geology of South Australia. Volume 2. The Phanerozoic.* Bulletin 54. Geological Survey of South Australia. Department of Mines and Energy, Adelaide.

Fedorov AV, Dekens PS, McCarthy M, Ravelo AC *et al.* (2006) The Pliocene paradox (mechanisms for a permanent El Niño). *Science* **312**, 1485–1489.

Feely RA, Sabine CL, Lee K, Berelson W *et al.* (2004) Impact of anthropogenic CO_2 on the $CACO_3$ system in the oceans. *Science* **305**, 362–366.

Frakes LA, Probst J-L, Ludwig J (1994) Latitudinal distribution of palaeotemperature on land and sea from early Cretaceous to Middle Miocene. *Comptes Rendus Academie Science Paris* **318**, série II, 1209–1218.

Hall-Spencer J, Rodolfo-Metalpa R, Martin S, Ransom E *et al.* (2008) Volcanic carbon dioxide vents show ecosystem effects of ocean acidification. *Nature* **454**, 96–99.

Haq BU, Hardenbole J, Vail PR (1987) Chronology of fluctuating sea levels since the Triassic. *Science* **235**, 1156–1167.

Harris GP, Davies P, Nunez M, Meyers G (1988) Inter annual variability in climate and fisheries in Tasmania. *Nature* **333**, 754–757.

Hobday AJ, Okey TA, Poloczanska ES, Kunz TJ, *et al.* (2006) 'Impacts of climate change on Australian marine life. Part A: executive summary, part B: technical report, part C: literature review'. CSIRO Report to Australian Greenhouse Office, Canberra. <www.eprints3.esrc.unimelb.edu.au/277/> (Accessed 30 January 2013).

Howard W, de Salao M, Cheal A, Lough J *et al.* (2009) Marine climate change in Australia: impacts and adaptation responses. 2009 Report Card. In *Marine Climate Change Impacts and Adaptation Report Card for Australia.* (Eds ES Poloczanska, AJ Hobday and AJ Richardson) p. 196. National Climate Change Adaptation Research Facility Publication 05/09, Southport. <www.nccarf.edu.au/userfiles/Report_card_websmall.pdf> (Accessed 30 January 2013)

IPCC (2007) *Climate Change 2007.* Intergovernmental Panel on Climate Change Reports. IPCC, Geneva. <www.ipcc.ch> (Accessed 30 January 2013)

Jouzel J, Masson-Delmotte V, Cattani D, Dreyfus G *et al.* (2007) Orbital and millennial Antarctic climate variability over the past 800,000 years. *Science* **317**, 793–796.

Keller G, Adatte T, Stinnesbeck W, Rebolloedo-Vieyra M *et al.* (2004) Chicxulub impact predates K-T boundary mass extinction. *Proceedings of the National Academy of Sciences of the United States of America* **101**, 3753–3758.

Kennett JP (Ed.) (1985) The Miocene Ocean: Palaeogeography and Biogeography. *Geological Society of America* Memoir **163**, 337 p.

Kopp RE, Simons FJ, Mitrovics JX, Maloof AC *et al.* (2009) Probabilistic assessment of sea level during the last interglacial stage. *Nature* **462**, 863–867.

Kuhnert H, Pätzold J, Hatcher B, Wyrwoll KH *et al.* (1999) A 200-year coral stable oxygen isotope record from a high-latitude reef off Western Australia. *Coral Reefs* **18**, 1–12.

Lear CH, Elderfield H, Wilson PA (2000) Cenozoic deep-sea temperatures and global ice volumes from Mg/Ca in benthic foraminiferal calcite. *Science* **287**, 269–272.

Lovelock J (2006) *The Revenge of Gaia*. Penguin Books, London.

Ludbrook NH (1980) *A Guide to the Geology and Mineral Resources of South Australia*. South Australian Government Printer, Adelaide.

Lüthi D, Le Floch M, Bereiter B, Blunier T *et al.* (2008) High-resolution carbon dioxide concentration record 650,000–800,000 years before present. *Nature* **453**, 379–382.

Marinov I, Gnanadesikan A, Toggweiler JR, Sarmiento JL (2006) The Southern Ocean biogeochemical divide. *Nature* **441**, 964–967.

McGowran B (1989) Silica burp in the Eocene ocean. *Geology* **17**, 857–860.

McGowran B (1991) Maastrichtian and early Cainozoic, southern Australia: foraminiferal biostratigraphy. In *The Cainozoic of the Australian Region* (Eds MA Williams, P De Dekker, AP Kershaw) *Geological Society of Australia* Special Publication **18**, 79–98.

McGowran B (2009) The Australo-Antarctic Gulf and the Auversian facies shift. In *The late Eocene Earth-Hothouse, Icehouse, and Impacts* (Eds C Kouberl and A Montanari). Geological Society of Australia Special Paper **452**, 215–240.

McGowran B, Li Q (1998) Cainozoic climatic change and its implications for understanding the Australian regolith. In *The State of the Regolith. Proceedings of the Second Australian Conference on Landscape Evolution and Mineral Exploration* (Ed. RA Eggleton) pp. 86–83. Geological Society of Australia Special Publication 20, Sydney.

McGowran B, Li Q, Cann J, Padley D *et al.* (1997) Biogeographic impact of the Leeuwin Current in southern Australia since the late Eocene. *Palaeogeography, Palaeoclimatology, Palaeoecology* **136**, 19–40.

McGowran B, Archer M, Bock P, Darragh TA *et al.* (2000) Australasian palaeobiogeography: the Palaeogene and Neogene record. *Memoirs of the Association of Australasian Palaeontologists* **23**, 405–470.

McGowran B, Holdgate GR, Li Q, Gallagher SJ (2004) Cenozoic stratigraphic succession in southeastern Australia. *Australian Journal of Earth Sciences* **51**, 459–496.

McGregor HV, Dima M, Fischer HW, Mulitza S (2007) Rapid 20th-Century increase in coastal upwelling off northwest Africa. *Science* **315**, 637–639.

McPhaden MJ, Zebiak SE, Glantz MH (2006) ENSO as an integrating concept in earth science. *Science* **314**, 1740–1745.

Middleton JF, Platov G (2003) The mean summertime circulation along Australia's southern shelves: a numerical study. *Journal of Physical Oceanography* **33**, 2270–2287.

Middleton JF, Arthur C, Van Ruth P, Ward TM *et al.* (2007) El Niño effects and upwelling off South Australia. *Journal of Physical Oceanography* **37**, 2458–2477.

Morgan GJ, Wells FE (1991) Zoogeographic provinces of the Humboldt, Benguela and Leeuwin Current systems. *Journal of the Royal Society of Western Australia* **74**, 59–69.

Moy CM, Seltzer GO, Rodbell DT, Anderson DM (2002) Variability of El Niño/Southern Oscillation activity at millenial timescales during the Holocene epoch. *Nature* **420**, 162–165.

Olsen AM, Shepherd SA (2006) Historic drift bottle experiments show reversing surface water masses in western Bass Strait waters: implications for lobster larval dispersal. *Transactions of the Royal Society of South Australia* **130**, 113–122.

Orr JC, Fabry VJ, Aumont O, Bopp L *et al.* (2005) Anthropogenic ocean acidification over the twenty-first century and its impact on calcifying organisms. *Nature* **437**, 681–686.

Overpeck JT, Otto-Bliesner BL, Miller GH, Muhs DR *et al.* (2006) Palaeoclimatic evidence for future ice-sheet instability and rapid sea-level rise. *Science* **311**, 1747–1750.

Pagani M, Caldeira K, Archer D, Zachos JC (2006) An ancient carbon mystery. *Science* **314**, 1556–1557.

Pälike H, Norris R, Herrie JO, Wilson PA *et al.* (2006) The heartbeat of the Pliocene climate system. *Science* **314**, 1894–1898.

Pearce A, Lenanton R, Jackson G, Moore J, *et al.* (2011) 'The 'marine heat wave' off Western Australia during the summer of 2010/11'. Fisheries Research Report No. 222. Department of Fisheries, Western Australia.

Pearson PH, Palmer MR (2000) Atmospheric carbon dioxide concentrations over the past 60 million years. *Nature* **406**, 695–699.

Petit JR, Jouzel J, Raynaud D, Barkov NI *et al.* (1999) Climate and atmospheric history of the past 420,000 years from the Vostok ice core, Antarctica. *Nature* **399**, 429–436.

Poloczanska ES, Babcock RC, Butler A, Hobday AJ *et al.* (2007) Climate change and Australian marine life. *Oceanography and Marine Biology - an Annual Review* **45**, 409–480.

Rahmstorf S (2007) A semi-empirical approach to projecting future sea-level rises. *Science* **315**, 368–370.

Raupach MR, Marland G, Ciais P, Le Quére C, *et al.* (2007) Global and regional drivers of accelerating CO_2 emissions. *Proceedings of the National Academy of Science U.S.* **104**, 10288–10293.

Royal Society (2004) *Ocean Acidification due to Increasing Atmospheric Carbon Dioxide*. Royal Society Policy Document 12/05, London.

Sabine CL, Feely RA, Gruber N, Key RM *et al.* (2004) The oceanic sink for anthropogenic CO_2. *Science* **305**, 367–371.

Sawyer JS (1972) Man-made carbon dioxide and the greenhouse effect. *Nature* **239**, 23–26.

Scher HD, Martin EE (2006) Timing and climatic consequences of the opening of Drake Passage. *Science* **312**, 428–430.

Shackleton NJ (2000) The 100,000-year Ice-Age Cycle identified and found to lag temperature, carbon dioxide, and orbital eccentricity. *Science* **289**, 1897–1902.

Shepherd A, Wingham D (2007) Recent sea-level contributions of the Antarctic and Greenland ice sheets. *Science* **315**, 1529–1532.

Shirayama Y, Thornton H (2005) Effect of increased atmospheric CO_2 on shallow water marine benthos. *Journal of Geophysical Research* **110**, C09S08.

Sluijs A, Schouten S, Pagani M, Woltering M *et al.* (2006) Subtropical Arctic Ocean temperatures during the Palaeocene/Eocene thermal maximum. *Nature* **441**, 610–613.

Stern N (2006) *The Stern Review on the Economics of Climate Change*. Cambridge University Press, Cambridge. <www.hm-treasury.gov.uk> (Accessed 19 March 2007).

Thompson DWJ, Solomon S (2002) Interpretation of recent Southern Hemisphere climate change. *Science* **296**, 895–899.

Thompson PA, Baird ME, Ingleton T, Doblin MA (2009) Long-term changes in temperate Australian coastal waters: implications for phytoplankton. *Marine Ecology Progress Series* **394**, 1–19.

Timmermann A, Oberhuber J, Bacher A Esch M *et al.* (1999) Increased El Niño frequency in a climate model forced by future greenhouse warming. *Nature* **398**, 694–697.

Wara MW, Ravelo AC, Delaney ML (2005) Permanent El Niño-like conditions during the Pliocene warm period. *Science* **309**, 758–761.

Zachos JC, Arthur MA, Dean WE (1989) Geochemical evidence for suppression of pelagic marine productivity at the Cretaceous/Tertiary boundary. *Nature* **337**, 61–64.

Zachos JC, Pagani M, Sloan L, Thomas E *et al.* (2001) Trends, rhythms, and aberrations in global climate 65 Ma to present. *Science* **292**, 686–693.

2 Oceanography and marine climate of southern Australia

OVERVIEW

The west and east coasts of Australia are washed by clear, southward-flowing, warm-water currents that are typically low in nutrients (oligotrophic), so that Australian waters are deserts in terms of productivity. Rainfall is also low, and nutrients entering the sea from rivers are very low. Hence, the productivity of coastal waters depends on tight recycling of nutrients and on coastal upwellings, which occur, mainly seasonally and spasmodically, on many parts of the coast. The major ocean currents are: the winter eastward-flowing Leeuwin Current (LC), which becomes the South Australian Current and then the Zeehan Current off Tasmania; the westward-flowing Flinders Current on the shelf-edge and slope; and the southward-flowing East Australia Current. Inshore of the ocean currents are coastal currents: the Capes Current and Cresswell Current on the south-west (SW) coast; reversing inshore currents on the south coast; Bonaparte's Tongue and gravity currents in Spencer Gulf; and the Bass Strait currents. The water circulation patterns within the major straits, gulfs and bays are described – that is, Cockburn Sound, Spencer Gulf, Gulf St Vincent, Port Phillip Bay, Western Port, and Bass Strait. Finally the wave, tidal and light climates for the region are outlined.

INTRODUCTION

Ocean currents, upwellings, tidal flows and wind-driven waves are important physical processes that together create the environment of reef organisms. Ocean currents are driven ultimately by the global climate and by gravitational forces acting on a rotating Earth. The oceans with their nutrients and trace metals are the medium for growth of phytoplankton and algae, which are at the base of the food web, while oceanic and local currents are responsible for the dispersal of the planktonic larval stages of many organisms at scales ranging from continental to local. Water currents produce directional, and wave action reversing, water movement underwater of differing intensity, and these have a multitude of effects on many aspects of the life of benthic plants and animals.

In this chapter, we consider first the large-scale picture – the ocean currents impinging on the southern Australian coastline, sea temperatures and salinity – and then the smaller scale physical influences – the tides, waves, local currents and light intensity. Our focus is on their significance for biological phenomena, such as biogeographic relations, biodiversity, distribution of species, composition of ecological communities, larval dispersal and population structure of species and their life histories in shelf waters < 50 m deep. Shelf waters go to 200 m depth, and near-shore coastal currents are intimately linked with the major shelf and shelf-slope currents. Hence, we shall consider briefly the oceanography of the shelf and slope in the region, with emphasis on coastal systems and effects. Australia's southern coasts are washed by three oceans – the Indian Ocean, Southern Ocean, and the Pacific. By oceanographers' convention, waters north of 40°S Lat. (about the latitude of the Subtropical Convergence) and west of Tasmania are in the South-East (SE) Indian Ocean, and those north and east of Tasmania are in the Tasman Sea in the Pacific Ocean.

OCEAN CURRENTS

The two major forces driving ocean currents around southern Australia are:

- global wind patterns resulting from different barometric pressure patterns, as they impinge on the Australian land-mass, producing the wind-driven currents
- differences in density of sea water leading to thermohaline circulation patterns.

In the Southern Hemisphere, the prevailing westerly winds drive the world's dominant current – the Antarctic Circumpolar Current or West Wind Drift – below ~40°S. In the mid southern latitudes, the combined effect of the southern westerly winds and the tropical easterly winds is to drive ocean currents in large circular gyres rotating anticlockwise. Due to Coriolis forces, these are more intense on the western coast of the Pacific Ocean than on the eastern coast of the Indian Ocean, and are called boundary currents. The East Australia Current (EAC) is a western boundary current and the West Australian Current (WAC) is an eastern boundary current. Inshore of the WAC is the thermally driven LC. Both the EAC and LC flow to the south, and both are highly variable. On the south coast of Australia, the LC turns east, and meanders easterly, being variously called the South Australian Current (SAC) off South Australia (Black 1853), and the Zeehan Current off western Tasmania (Baines *et al.* 1983). Although some have argued for a single name – the Leeuwin Current – for this long, but unstable, current (see Ridgway and Condie 2004), we have followed recent reviews, and used the above names for the geographically distinct sections of the current. Counter to these currents, the Flinders Current (FC) flows westwards along the shelf break and slope, while inshore, in shallower shelf waters, wind-driven coastal currents (CC), and sometimes eddy systems prevail. Figure 2.1 shows the main oceanic and coastal currents around Australia, except for the FC and SAC.

The hydrodynamics of coastal waters are complex because they are influenced both by large-scale factors – the ocean current systems – and finer-scale factors – local winds, tides, and coastal and seabed topography. First, we describe briefly the major currents, then the coastal currents, and lastly some major bays, such as Cockburn Sound and Port Phillip Bay, the South Australian gulfs, and Bass Strait. Together, these illustrate the principles of water circulation in shallow, coastal waters. For this chapter we have drawn on the reviews of Bunt (1987), Jeffrey *et al.* (1990), Cresswell (1991), Pearce (1991), Middleton and Bye (2007), Pattiaratchi and Middleton (2007), Suthers and Waite (2007), Binnie and Cann (2008) and James and Bone (2011), as well as other more specific papers referred to in the text.

Leeuwin Current, South Australian Current and Zeehan Current

Well offshore from Western Australia, the WAC flows northwards towards the equator, and inshore is the opposing, southward-flowing LC (Fig. 2.1; Plate 1). The latter current is notable and exceptional, because, unlike other southern hemisphere currents on western coasts, it transports warm tropical waters south from NW Australian waters, then around Cape Leeuwin, and suppresses coastal upwellings, which would otherwise occur.

The driving force of the LC is thought to result from the Indonesian Throughflow; that is, the warm, low-salinity water from the western Pacific that flows through the Indonesian Archipelago into the Indian Ocean NW of Australia, resulting in a rise of ~0.5 m in sea level. The water then flows 'downhill' along the west Australian coast where it is augmented by tropical water from the Indian Ocean, and so becomes stronger. This N–S pressure gradient is constant

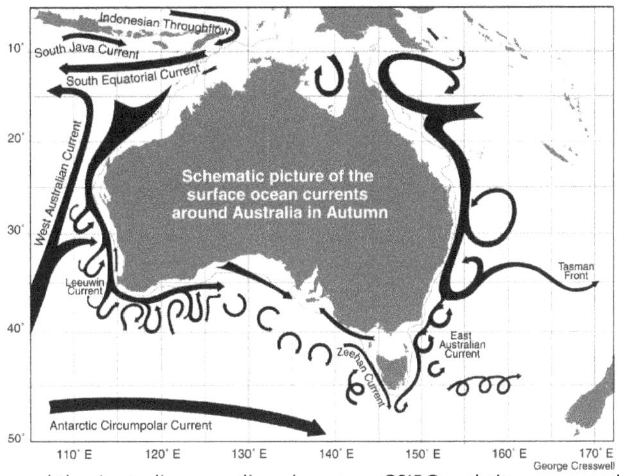

Figure 2.1: Major currents around the Australian coastline. (courtesy CSIRO website www.marine.csiro.au).

throughout the year, but does decline during El Niños, when relaxing trade winds cause the Indonesian Throughflow to weaken, and also during summer, due to adverse southerly winds (see below). Also, there is now evidence (Cresswell and Peterson 1993) that as the LC approaches ~30–32°S it is further strengthened by inflow from an offshoot of the WAC, which meanders eastwards and deflects southwards as it meets the LC. Off west to SW Australia, the LC is a strong current ~200 km wide in the north, narrowing to 50–100 km in the south as the shelf narrows, with many meanders and eddies, and extends down to 200 m depth. It is warm (> 24°C initially), of low salinity (~34.7%) and nutrient-poor (nitrate = 0.2–0.4 μm L^{-1}), and travels ~100 km d^{-1} (≈1.2 m s^{-1}). As it travels south to Cape Leeuwin it becomes cooler (20–22°C) and saltier (35.6–35.8%), while along the way eddies flowing clockwise, typically 50–100 km in diameter, form at the shelf break near sharp bathymetric changes – such as west of the Abrolhos Is, SW of Jurien Bay, at the Perth Canyon west of Rottnest I., and off Cape Leeuwin – and drift off to the west.

Recent studies (Caputi *et al.* 2009) have shown that since the 1950s a warming trend of 0.3–0.4°C every decade has occurred in the waters off SW Australia during the autumn to winter season. This trend appears to be due partly to the increasing frequency of El Niño events during that period (see Chapter 1), and the consequent weakening of the LC, and partly to the decreasing frequency of SW storms and westerly winds in the same period. Another factor is the Indian Ocean Dipole (IOD): an inter-annual oscillation inducing anomalous sea temperatures in the Indian Ocean. During the IOD's positive phase, easterly winds off southern Java are unusually strong, driving a stronger Indonesian Throughflow, and during the negative phase the opposite occurs, modulating the El Niño influence (Oppo and Rosenthal 2010). These shifts have had serious consequences for lobster fisheries (Chapter 15).

Underneath the LC is the little studied Leeuwin Undercurrent, which is a weak counter-current flowing northwards along the west Australian shelf slope, and is strongest in summer. The current is narrow and lies at 250–450 m depth, and can be considered as an extension of the Flinders Current that flows westwards below the LC from SE Australia, and is one of its sources (Woo *et al.* 2006) (see below).

From autumn to early spring, when the LC is flowing south strongly, it moves onto the continental shelf as it approaches the projecting coastline from Cape Naturaliste to Cape Leeuwin, before swinging eastwards into the Great Australian Bight (GAB). During the warmer half of the year, and most strongly during summer, the LC weakens because of seasonal adverse winds. Strong summer southerly sea breezes both retard the LC, deflecting it further offshore, and drive an inshore counter-current (the Capes Current – see below). The annual oscillation in strength of the LC is reflected in the mean sea level at Fremantle, and this has been found to be a convenient measure for determining its role in the life histories of important fisheries, such as bluefin tuna and western rock lobster (Pearce 1991). For example, during normal years with a strong LC, sea levels (and lobster puerulus settlement) are high, whereas during El Niño years the LC weakens, and sea levels (and puerulus settlement) fall.

As the LC turns east at Cape Leeuwin and heads into the GAB, it meets and mingles with subtropical water from a branch of the (salty) WAC that veers southwards towards Cape Leeuwin and continues eastwards. The WAC flows most strongly in summer. Together, the resulting current is called the LC, irrespective of season and origin.

The southern Australian coast from Cape Leeuwin to Portland (~2500 km) is the world's longest mid-latitude shelf, and has a varied topography and complex currents. The LC is strongest in winter, substantially due to forcing by strong westerly winds, with speeds averaging 0.2–0.3 m s^{-1}, and flows mainly inshore of and on the shelf break (Figs 2.1, 2.2). The LC then meanders across the Bight with anticyclonic eddies occurring off Albany and the South Australian gulfs. Drifter-buoy data suggest that waters from North West Cape in Western Australia take ~2 months to reach Cape Leeuwin, and then another 3 months to cross the GAB.

In the eastern GAB, wind forcing contributes ~42% to the total flow of the current, now called the South Australian Current (SAC), which may spread over the Flinders Current on the shelf slope, so weakening the strength of that current (Middleton and Bye 2007). The SAC continues to flow south-easterly down the Lacepede Coast and down the west coast of Tasmania, where it is called the Zeehan Current (ZC). Here it traces the shelf-edge, and continues around southern Tasmania towards Schouten I., bringing water up to 4°C colder than the East Australia Current (EAC) to SE Tasmania (Cresswell 2000), where the water is mixed with the remnants of the EAC. The southward flow of the ZC off western Tasmania is year round, although much stronger in winter (June to November) than in summer.

Flinders Current

Further offshore on the shelf slope, the Flinders Current (FC) flows from western Tasmania to Cape Leeuwin (Fig. 2.2). The current is strongest at ~600 m depth, but extends to 1200 m depth, with current speeds ranging from ~0.05 m s^{-1} in the east to 0.2 m s^{-1} in the west, where it joins the Leeuwin Undercurrent. The strength of the FC is affected by seasonally varying winds and is weakest in winter with maximum speeds of 0.1–0.15 m s^{-1}, and strongest in summer in terms of volume, although speeds are lower (0.05–0.1 m s^{-1}). The FC is driven by the anticyclonic wind field in the South Australian basin and also by the westward Tasman Outflow south of Tasmania (Middleton and Bye 2007).

East Australia Current

The East Australia Current (EAC) originates in the Coral Sea, and is fed by the waters of the South Equatorial Current, which flows westwards across the Pacific Ocean (Fig. 2.1). It is a western boundary current, 100 km wide and up to 1 km deep, pressed against the western side of the Pacific by the Earth's rotation, and flows down the east Australian coast, past Bass Strait to eastern Tasmania (Plate 2). The EAC is the largest current on Australia's coast, and brings warm, salty water southwards. As it flows, oceanic eddies, up to 200 km across, peel off, rotating in an anti-clockwise direction, and drift into the Tasman Sea. On the shoreward side, small clockwise eddies also break off, causing downwelling, and giving rise to northward inshore coastal currents, which may be strengthened by southerly winds (Middleton *et al.* 1996). Hence, drift bottles released in winter in western Bass Strait were sometimes carried eastwards through Bass Strait and northwards up the east Australian coast as far as Queensland (Olsen and Shepherd 2006).

The EAC is strongest in summer, with maximum longshore currents of ~90 cm s^{-1} at 30°S, and slowing to ~10 cm s^{-1} at 40°S off Tasmania. In winter the current is only about half as strong as in summer (Fig. 2.2; Ridgway and Dunn 2003). The most significant nearshore effects result from episodic incursions of the EAC and its eddies (Oke and Middleton 2001). Where

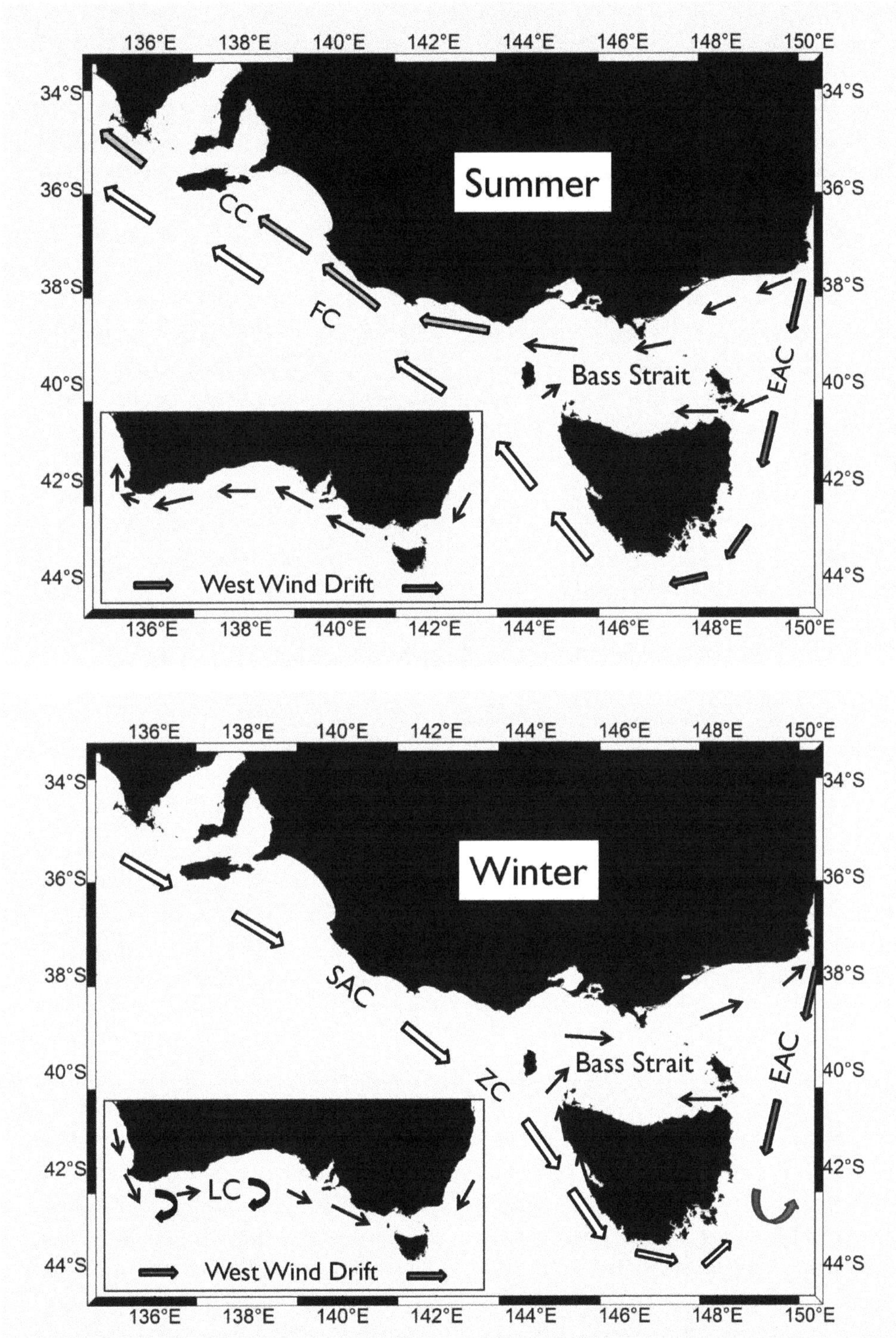

Figure 2.2: Major oceanographic currents in SE Australia in summer (upper) and winter (lower). CC = Coastal Current (Cresswell Current in SW Western Australia); eddies are shown by curved arrows. EAC = East Australia Current; FC = Flinders Current; LC = Leeuwin Current; SAC = South Australia Current; ZC = Zeehan Current. (After Olsen and Shepherd 2006).

the EAC moves away from the shore off the New South Wales coast or is compressed by jutting headlands, local upwellings occur (see below), and enrich coastal waters. The strength and pole-ward extent of the EAC varies from year to year, but over the last 300 years the magnitude, frequency and duration of the current have increased (Thresher *et al.* 2004). This is shown by trends of increasing mean summer salinity (from 35.1 to 35.35%) and mean summer temperature (from 16.2 to 17.0°C) measured off Maria I., eastern Tasmania, over the last 60 years (Harris *et al.* 1987; Johnson *et al.* 2005; Chapter 12).

The EAC is typically low in nutrients (nitrates < 1 μM), except when it draws up deep water from offshore and induces coastal upwellings, as described below.

Coastal currents

Currents on the broad, shallower parts of the continental shelf differ from those in deeper water. They are often nearly two-dimensional in that the waters are usually well mixed, and current speeds vary little throughout the water column. Hence, flows can be calculated with barotropic numerical models, using the more easily measured wind speeds and tidal forcing. Inshore currents, driven by winds and/or thermohaline forcing, occur on the west and east coasts of Australia, and along the southern coast from Cape Leeuwin to Bass Strait and Tasmania, and are therefore of great biological significance for coastal ecosystems and for transport of larvae of coastal species. Here we briefly describe the main coastal currents in shelf waters.

Capes Current

The Capes Current is a cool, inner shelf current, which flows northwards from Cape Leeuwin, and originates from deeper water offshore (Fig. 2.1). The current occupies the width of the inner shelf (20 km) to ~80 m depth, and broadens as it goes north past Perth. The current is colder (21.0–21.4°C), saltier (35.4–35.5%), but with only slightly more nitrates than the LC (~0.2–0.5, spiking to ~1–8 μM in late summer), and extends as far north as the Abrolhos Is (29°S). It is strongest from November to about March, averaging 5–10 cm s^{-1} when the southerly sea breezes are strong, and weakens as they subside. But what is the origin of the water?

Recent studies (Gersbach *et al.* 1999; Pearce and Pattiaratchi 1999) have shown that the cool water upwells from the bottom of the mixed layer of the LC, which is here centred over the shelf break at 200 m depth, and impedes the upwelling of nutrient-rich deeper waters. There are two main sources of the water: the continental slope south of Cape Naturaliste and the slope on the south coast, both from depths of ~60–100 m. Upwelling events are intermittent, according to wind strengths, and flush over the inner shelf waters five to nine times in a summer. These upwellings increase near-shore primary production in summer from Cape Leeuwin northwards, because they entrain some nutrient-rich water at the base of the LC. In this inner shelf region, even the LC may have unexpectedly high nitrate levels of 3.1 μM L^{-1} in winter, probably because it entrains some deeper nutrient-rich water from the shelf slope (Hanson *et al.* 2005).

Reversing southern coastal currents

Winter

In winter, inshore of the LC and SAC, an eastward coastal current flows on the shelf at speeds of 0.1–0.3 m s^{-1}, reaching up to 0.5 m s^{-1} further east off Kangaroo I. and the Lacepede Shelf. Modelling studies (Cirano and Middleton 2004) showed that in winter the westerly wind stress drives an onshore mass transport (termed Ekman transport), leading to coastal downwelling of bottom water over the shelf edge.

In the GAB, a pulse of warm, highly saline water, formed by evaporation at the head of the Bight during summer in shallow water, sinks and flows eastwards, ultimately mingling with the SAC (Herzfeld and Tomczak 1997). In Spencer Gulf, warm, salty water arising from summer evaporation forms a gravity current pulse that flows down the eastern side of the Gulf, and then south round Kangaroo I., cascading over the shelf and joining the eastward flow of the SAC at ~200 m depth (Cirano and Middleton 2004). This has sometimes been called Bonaparte's Tongue. Both pulses occur from April and continue through winter, and the outflow from Spencer Gulf enhances the quasi-stationary anticyclonic eddy SW of Kangaroo I. Much smaller pulses of dense water similarly flow out of Gulf St Vincent along the north-east (NE) side of Backstairs Passage into Encounter Bay from April to July (Ng *et al.* 1993).

Except during the seasonal, spasmodic upwellings (see below) nitrate levels throughout the whole southern Australian region are extremely low (< 1 µM) (Rochford 1980), leading to very low planktonic productivity, and among the clearest waters in the world.

Summer

During summer (from ~November to March) as the LC and the eastward CC weaken, the coastal winds reverse and blow strongly, but not continuously, from the SE.

Inshore of the LC off SW Australia, from Recherche Archipelago to Cape Leeuwin a coastal current (called the Cresswell Current) flows westwards in summer at speeds of up to 0.1 m s^{-1}. Although little is known about this current, it is similar to the Capes Current, which it may feed into, and likely causes the upwelling observed off Esperance, as well as anticyclonic circulation gyres (Pattiaratchi and Middleton 2007).

Within the GAB from Esperance to the Bonney coast, the coastal current flows at speeds of ~0.1 m s^{-1} westwards, although strong SE winds can generate currents up to 0.25 m s^{-1} off the Bonney coast, where the shelf is narrowest. In the upper central GAB, thermohaline forcing is important, because salinity (36.4%) and sea temperatures (~20°C) are high. Here, and towards the eastern GAB, an anticyclonic (anticlockwise) gyre is found, and its seaward arm forms part of the SAC flowing eastwards, overlying and opposing the deeper FC (Middleton and Platov (2003).

The summer–winter seasonally reversing coastal current off SE Australia has been known since the 1960s. Drift-bottle experiments during the 1950s showed summer surface currents of ~0.1 cm s^{-1} flowing to the west or NW, and winter ones of 0.15–0.2 m s^{-1} flowing to the SE from the eastern GAB to western Tasmania, similar to speeds of the now well-described coastal currents (Olsen and Shepherd 2006).

Inshore currents around Tasmania

Tasmanian coastal waters are directly influenced by three oceans – the eastern Indian Ocean to the west, the Pacific to the east, and the Southern Ocean to the south (see Cresswell 2000). In winter, the Zeehan Current veers around southern Tasmania and up the east coast as far north as Schouten I. before turning SE under the influence of the EAC. In summer, when the Zeehan Current is weak and the EAC is strong, the latter sweeps around southern Tasmania. In contrast, inshore waters off the north coast are less affected by Bass Strait currents, and have flushing times of weeks to months (see below). Tasmanian coastal waters, like those elsewhere in Australia, are poor in nutrients (< 1 µM), except in the SE, where subantarctic water is seasonally brought to the surface, as described below.

Upwellings and uplift

Western GAB to Western Tasmania

Here the term upwelling is used to refer to the raising of cooler nutrient-rich waters to the surface, whereas 'uplift' refers to the raising of cold water towards, but not reaching, the surface (Rochford 1991). Sometimes, in an upwelling region, when the driving mechanism weakens, uplift but not upwelling will occur.

In the western GAB shelf in the Baxter Section (east of Israelite Bay), the LC blocks incursions of nutrient-rich slope waters in winter, but when it weakens in summer these incursions are likely, and enough to promote abundant macrophyte and coralline algal growth (James *et al.* 2001).

Off SE Australia, the strong SE winds, which typically blow for 3–10 days and two to four times each summer, lead to strong upwellings of cold water of 11–12°C onto the shelf from Portland along the Bonney coast and Lacepede Shelf, into the eastern GAB; cold water uplift extends onto the shelf further into the eastern GAB as far west as Nuyts Archipelago (Herzfeld and Tomczak 1999) (Plate 1).

Upwelled water can be transported 20–40 km a day at speeds of 0.2–0.4 m s^{-1}, and supply pools of nutrient-rich water to coastal areas. These upwellings are due to simple Ekman dynamics; that is, a steady wind stress causes a net transport of surface water 90° to the left of the wind direction in the Southern Hemisphere. If the wind stress is alongshore with the coast on the right, then surface waters are driven offshore, and deeper water upwells on the coast. However, some upwellings occur off western Eyre Peninsula without upwelling-favourable winds, suggesting that intense oceanic anti-cyclonic gyres on the shelf may also cause them (Griffin *et al.* 1997). On the Bonney coast, water upwells from 250 m deep, and off southern Kangaroo I. from 150 m deep. Due to the wide continental shelf, upwelling west of Kangaroo I. is a two-step process in which water is first uplifted to form a sub-surface pool, and is then driven by the longshore shelf currents to the NW (Herzfeld and Tomczak 1999). The area around Flinders I. in the eastern Bight also shows evidence of a localised upwelling driven by the divergence of the prevailing flow past a shallow ridge projecting normal to the island coast (McClatchie *et al.* 2006) (Plate 1).

Eastern Australia

On the east coast of Australia, from northern NSW to the south of Eden, upwellings can occur seasonally at numerous points along the coast, but are not always predictable in their occurrence. This is because the mechanisms driving them are various and sporadic. Roughan and Middleton (2002) identified four different mechanisms that operated off this coast. Firstly, where the EAC flows near to the coast, as in the Smoky Cape/Laurieton region (~31°S), northern NSW, it accelerates due to a funnelling effect of the narrow shelf, and deeper water from the slope upwells into coastal waters (known as Ekman transport). Secondly, where the EAC separates from the coast (e.g. off Crowdy Head), current strength decreases slightly downstream of the point of separation, and again deeper water upwells, often up to 5°C cooler. A third mechanism is local wind forcing by seasonal north and NE breezes, but upwellings from these events are localised and short lived (1–2 weeks).

The magnitude of an upwelling event on these coasts can be estimated by comparing the concentration of nutrients brought to the surface (Hallegraeff and Jeffrey 1993). Off the NSW coast, nitrate concentrations range variously from 2–4 μM in winter to spring, compared with < 1 μM elsewhere. Hence, of the three mechanisms described above, the encroachment or funnelling effect appears to produce the strongest upwelling, whereas current separation induces a semi-permanent upwelling, as shown by the regular occurrence of algal blooms at Port Stephens.

Sporadic upwellings or uplift also occur off Eden and the NE Victorian coast for most of the year, except May and June, as the EAC separates from the coast (e.g. Plate 2) (Newell 1961; Rochford 1975). Off NE Victoria, the upwelled water from these sporadic events is derived from winter Bass Strait waters that have sunk into the eastern approaches. These waters are very low in nutrients, so provide little or no increase in productivity of coastal waters (Jeffrey *et al.* 1990). However, El Niño events off SE Australia have a greater effect. In the year before an El Niño, low air pressure over Australia generates strong NE winds along the southern New South Wales to NE Victorian coast, which induce offshore transport of surface water and the upwelling of deep offshore water with elevated nutrient levels (Hsieh and Hamon 1991).

Off eastern Tasmania in winter, cyclonic (clockwise) eddies develop and create a vortex uplift of deeper, nutrient-rich (2–4 μM), subantarctic waters to the surface, enriching small areas off the coast (Rochford 1984; Jeffrey *et al.* 1990) (Plate 2).

Productivity of upwellings

Upwelling or uplift of deep water, rich in nitrogen and silica, into the euphotic zone promotes primary productivity through phytoplankton growth, which in turn sequentially nourishes a flourishing zooplankton community and an enriched trophic system from sardines or pilchards through to tuna, sharks and whales. However, deep uplifted water that reaches the euphotic zone but not the surface, though highly productive, is missed by conventional satellite imagery that records surface chlorophyll-a levels. Hence, as shown by van Ruth *et al.* (2010a, b), the productivity of the eastern GAB is far greater than previously thought, and rivals that of other great upwelling systems, like the Humboldt and Benguela Currents. Nevertheless, the productivity of the region is very patchy. The highest productivity is in hotspots in the Bonney region of SE South Australia, off SW Kangaroo I., SW Eyre Peninsula and off Cape Adieu in the central GAB at ~132°E longitude. Mid-shelf and near-shore waters have intermediate productivity, while offshore waters have low productivity.

El niño effects

Long-term switches in high- and low-pressure regions in the central Pacific, called the El Niño-Southern Oscillation (ENSO), have far-reaching effects on climate globally, and also affect Australian waters. During an El Niño year, as discussed above, the LC on the west Australian coast weakens, and along the southern coast the signal decays even further (Li and Clarke 2004), and winter-time shelf-edge currents are weaker. Hence, off the Bonney coast in SE Australia, El Niño events lead to enhanced upwelling. More specifically, winter time downwelling during the onset of an El Niño is reduced, and in the following summer, upwelling increases, leading to reduced sea temperatures of up to 2°C compared with normal years (Middleton *et al.* (2007).

In the central Tasman Sea, summer sea temperatures are higher during El Niño years, and the warmer EAC waters penetrate well south past Maria I., Tasmania, as shown by higher summer temperatures and reduced biological productivity there (Jeffrey *et al.* 1990).

CIRCULATION WITHIN GULFS, BAYS AND BASS STRAIT

The hydrology of sheltered, inshore waters is very different from that of the adjoining open ocean. Whereas the latter is mainly influenced by large-scale currents driven by global and continental-scale forces, sheltered waters are much shallower and are most affected by the regional weather, such as local winds, land temperatures and tidal currents, which reverse every 6 h (semi-diurnal) on the east and southern coast, and every 12.5 h (diurnal) on the west coast. Topographic features, such as the extent to which coastal waters are enclosed, are also important, because they determine rates of exchange with the open sea, and hence the degree of influence of the weather on coastal waters. These principles are illustrated by examples of five well-studied inshore regions in different parts of southern Australia.

The hydrology of partly closed coastal waters is important both for biological reasons (e.g. extent of larval dispersal), and for anthropogenic reasons, because such sheltered coastal waters often have ports on their margin, and are population centres with consequent discharge of industrial wastes, nutrients, stormwater and sewage effluent into adjacent waters.

Cockburn Sound

Cockburn Sound is a coastal embayment of ~80 km^2 running from off Fremantle south to Cape Peron, bounded on the west by Garden I., and on the north by shallow banks (Fig. 2.3). The Sound is ~16 km long by 7 km wide, with a maximum depth of 22 m. The mainland and Garden I. are connected by a causeway, with two narrow openings, 300 and 600 m wide.

The Sound was much studied during the late 1970s, as industries developed onshore, and before construction of the causeway (see Steedman and Craig (1983). These studies showed that the water circulation is mainly wind-driven. The three major types of wind events are:

- the SW sea breeze that is dominant from late spring to autumn (annual occurrence ~35%)
- easterly winds in winter (annual occurrence ~18%)
- year-round low pressure systems, with winds from all directions (annual occurrence ~28%).

In general, water circulation within Cockburn Sound is in the direction of the wind over the eastern and western sides, with a return flow in the deeper middle part of the basin. Hence, during the summer sea breeze, the flow is northward along the coast and southward down the middle, whereas the winter storms blowing from the north to NW, cause flows in the opposite direction. Yet water flow speeds are very low (< 0.05 m s^{-1}) and, given the shallow depths and the shallow northern entrance, the Sound is almost a closed system.

The South Australian Sea

The South Australian Sea consists of Spencer Gulf, Gulf St Vincent and Investigator Strait (Bye 1976), as discussed below.

Spencer Gulf

This long, triangular water body, tapers from a mouth ~80 km wide to an apex 320 km to the NE, with an area of 21 700 km^2, and with depths ranging from 50 m at the entrance to 5–10 m at the top. The Gulf is an inverse estuary with high water temperatures and salinity (up to 28°C and 48‰ at the top in summer,

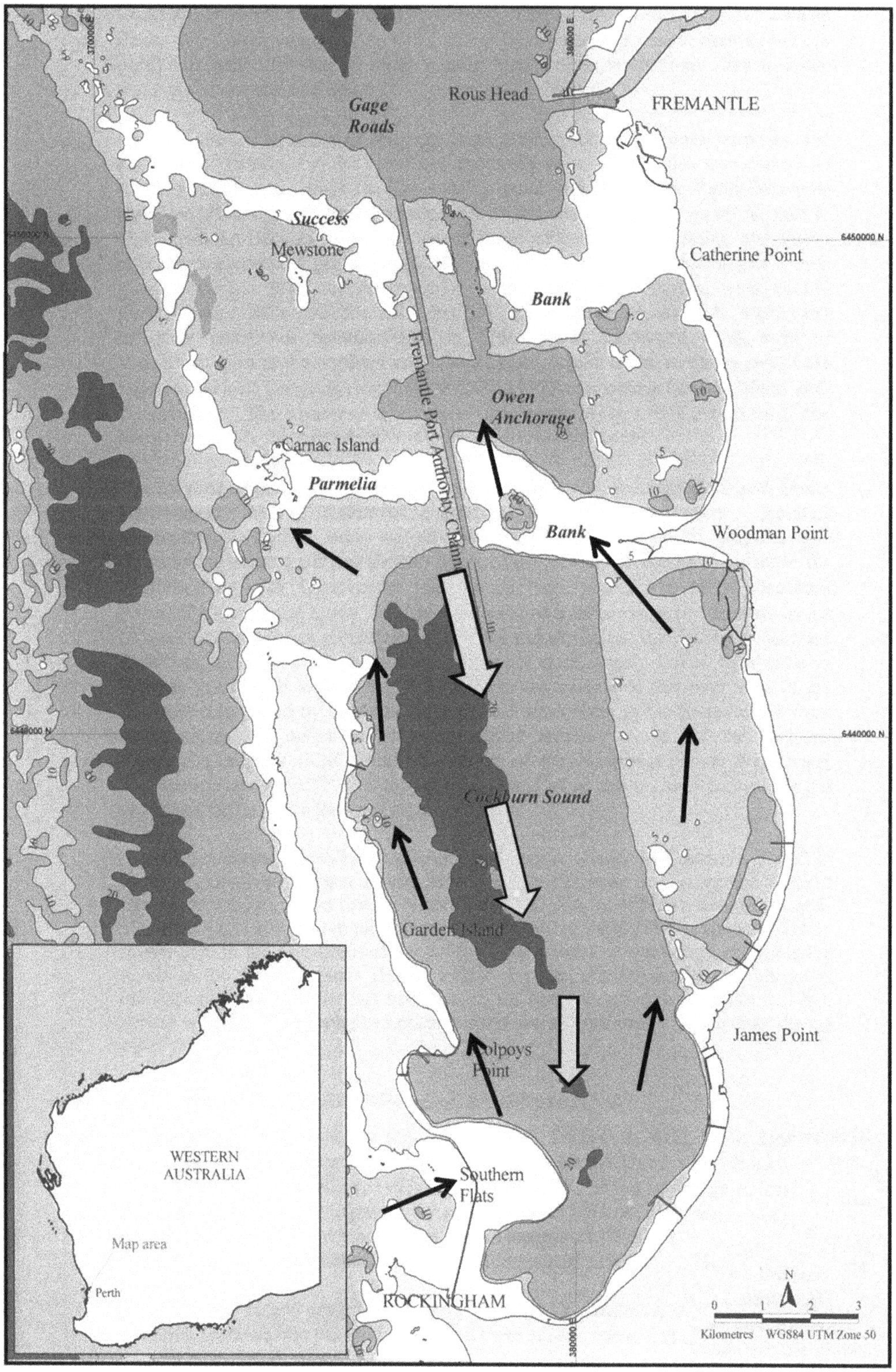

Figure 2.3: Circulation in Cockburn Sound in summer, with a northerly sea breeze. The fine arrows are the inshore surface currents, and the large block arrows are the bottom central southerly current. Depths > 20 m are in dark grey. Map courtesy DA Lord & Associates Pty Ltd (now Oceanica Consulting Pty Ltd).

compared with 21°C and ~35.8‰ at the entrance), due to high summer evaporation in the north. There is no significant fresh water input.

The Gulf has some remarkable hydrological features (Bullock 1975; Provis and Lennon 1983; Green 1984; Nunes Vaz *et al.* 1990; Petrusevics 1993), with the water circulation determined by three factors – thermohaline forcing, wind forcing and tidal currents.

Thermohaline currents are induced by horizontal pressure gradients, due to density variations, caused by the strong north–south temperature and salinity gradients within the Gulf. The general circulation (Fig. 2.4) shows that water enters the Gulf on the western side (the Port Lincoln boundary current), and leaves via the eastern side. Off Wallaroo, halfway up as the Gulf narrows, there is a cyclonic gyre, and the water turns back towards the south, so partly blocking off the water of the upper Gulf.

From Wallaroo north to Point Lowly, water exchange with the south is further reduced, although residual flows northwards on the western side are apparent. In the upper Gulf north of Point Lowly, exchange is reduced even further, although residual flows still tend to flow northwards up the western side and southwards down the eastern side. Large-scale eddies also occur SE of Point Lowly, and further north off Redcliff Point. When wind stress is imposed on the system, the above pattern of circulation tends to be accentuated, with, for example, residual current speeds (i.e. over and above tidal current speeds) typically doubling to ~0.16 m s^{-1} during SW winds of 14 m s^{-1}. Eddy diffusion coefficients, which indicate the rate of diffusion of salty water are very low north of Point Lowly (< 10 m^2 s^{-1}), but much higher further south, being constant at ~80 m^2 s^{-1}. Thus it is evident that the Gulf comprises three regions, with limited exchange between them – a southern region from the entrance to about Wallaroo (~34°S), a central region from 34°S to Point Lowly (33°S), and a northern region from 33°S northwards to the apex at Port Augusta.

The tidal system in Spencer Gulf strongly influences its circulation and flushing, due to the large tidal range, which increases up the Gulf to a maximum of ~4 m at the top (Fig. 2.4) (Schluter *et al.* 1995). Hence, tidal currents are strong, reaching > 1 m s^{-1} in the central and upper parts of the Gulf. The most famous feature of the tides is that the amplitude of the diurnal (solar) and semi-diurnal (lunar) constituents are almost equal, so that during neap tides twice a month the circulation system virtually closes down for a day as the pull of the sun and moon cancel each other out, and tidal flow ceases (Easton 1978). Another curiosity of the tidal system is that tidal currents are mostly diurnal at the mouth, but semi-diurnal in the central region, where there is a nodal or minimum tide region.

The exchange between Spencer Gulf and shelf waters is of great importance, because this is the only mechanism by which highly saline water generated in the upper Gulf after summer evaporation can escape to the open sea. Petrusevics (1993) showed that during summer at the entrance a sea temperature front is set up, in which Gulf waters are ~3°C warmer than the adjoining ocean waters. This front effectively isolates the Gulf from shelf waters in summer, and raises the question how the high salinity water reaches the open sea.

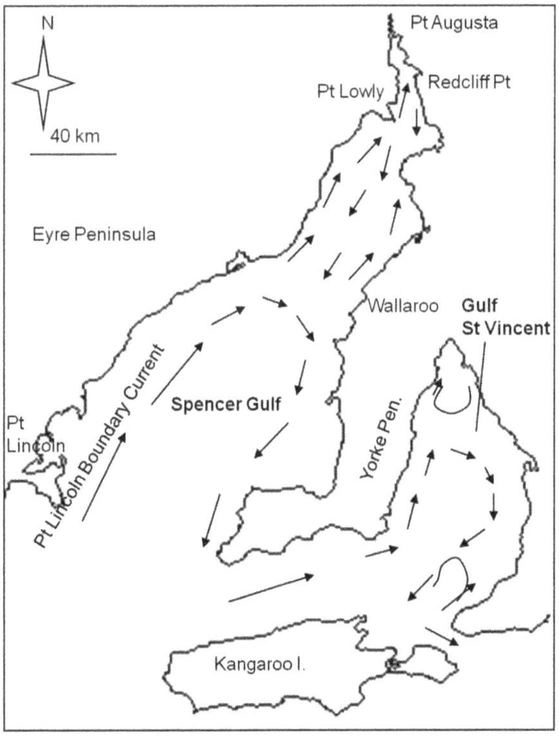

Figure 2.4: Circulation in Spencer Gulf, Investigator Strait and Gulf St Vincent (redrawn from Bye and Kämpf 2008).

Studies by Lennon *et al.* (1987) resolved the question. In autumn to winter, as Gulf waters cool, they become denser than waters on the shelf, and can then slip out of the Gulf along the sea floor, to be replaced by a surface inflow of less salty water. This outflow is the gravity current of salty water, alluded to above, which flows at ~0.1 m s^{-1} and joins the eastward flowing shelf coastal current. The gravity current flows for about 3 months – a sufficient time to discharge the salt that accumulated in the Gulf in the preceding summer. However, the flow is not constant, and it seems likely that it is modulated by the Gulf's strong spring–neap tidal cycle, which would produce salt water pulses 150 km apart that cascade over the shelf intermittently.

Gulf St Vincent and Investigator Strait

Gulf St Vincent (GSV) and Investigator Strait (IS) are together smaller than Spencer Gulf, with areas of ~6800 km^2, and 6100 km^2, respectively. Depths in GSV reach 40 m in the south and shallower to the north, while those in IS range from 50 m in the west to 30 m in the east. The oceanography is reviewed by Bye and Kämpf (2008) and Petrusevics (2008).

GSV is in many respects similar to Spencer Gulf. It is an inverse estuary, with similar flushing features, similar tides and a similar gravity current of salty water, but all these features are on a smaller scale than Spencer Gulf, as shown below. Temperature and salinity range from ~19°C and ~36‰ in IS to ~23.5°C and ~41‰, respectively, at the top of GSV in summer, and from ~14°C and ~36‰ in IS to ~12°C and 39.5‰ in upper GSV in winter. Like Spencer Gulf, there is no significant inflow of fresh water.

The general pattern of circulation, driven to a large extent by the wind, is that lower salinity water enters GSV on its western side, and flows down the eastern side, and out via Backstairs Passage (Fig. 2.3). There is a cyclonic eddy in the upper GSV, and another, called the Myponga eddy, off the lower Fleurieu Peninsula. Average residual current speeds are low, typically 0.02–0.06 m s^{-1}, depending on location (Bye and Kämpf 2008). These speeds are insignificant compared with the semi-diurnal tidal currents, which are dominant, and can reach speeds of > 1 m s^{-1} during spring tides, especially near the entrances (Noye and Grzechnik 1996). During neap tides, when the tidal currents shut down for 1–2 days, the water column stratifies, and a wedge of salt water forms on the bottom and escapes from the Gulf via the NE side of Backstairs Passage. This salt wedge was revealed for the first time during a neap tide in June 1995, when oceanographers were monitoring a legendary cross-Passage swim (Carvalho and Bye 1996).

Bass Strait

Bass Strait is mainly ~60 m deep, with shallower sills < 50 m deep to the west (King I.) and east (Flinders I.). During the last glacial maximum 20 000 years ago, when sea level was 120 m lower than at present (Binnie and Cann 2008), the middle of the Strait was a large inland lake. Then, as sea level rose ~14 000 years ago, Tasmania was left attached to the mainland only by a narrow eastern isthmus (Waters 2008), until it became fully separated some 8000 years later. The hydrology of Bass Strait is described by Newell (1961), Rochford (1975) and Sandery and Kaempf (2005). The currents are driven mainly by the West Wind Drift, local winds and the winter coastal current. In winter, currents are easterly at average speeds of 0.04 m s^{-1}, and the saltier north Bass Strait water flows eastwards, while subantarctic water, driven by the West Wind Drift, enters the Strait between King I. and Cape Grim (NW Tasmania) at a speed of ~0.11 m s^{-1}, and continues east- to north-eastwards (Fig. 2.2). The dense water then cascades over the shelf slope at several places into the Tasman Sea (the Bass Strait Cascade), and turns northwards as an inshore counter-current up the NSW coast (Godfrey *et al.* 1980; Gibbs *et al.* 1986a; Luick *et al.* 1994). This counter-current often entrains deeper water brought to the surface by upwelling (Newell 1961). More recently, autonomous gliders have discovered that mid-depth lenses of Bass Strait water, 40 km across to 300 m high, are captured within anti-cyclones and carried out of Bass Strait down the SE coast of Tasmania (Baird and Ridgway 2012).

Wind-driven surface waters in Bass Strait may be transported at much higher speeds than those given above, as shown by drift bottle experiments, in which drift bottles were carried at up to 0.2 m s^{-1} through Bass Strait. In summer, there appear to be weak flows of variable direction, but tending westwards, driven by SE to E winds and by tides, with possibly some inflow from the EAC moving westwards through Bass Strait (Fig. 2.2). However, in modelling studies, Fandry (1983) found that, in the absence of wind, tidal motion was weak in central Bass Strait due to opposing flows of tidally induced currents on the east and west sides of the Strait.

Winter flushing of Bass Strait is variable, being 20–40 days in the west, and up to 180 days from the waters east of the line from Cape Grim, NW Tasmania to Cape Howe (Sandery and Kaempf 2005). Nutrient levels are generally low (< 1 µM), but are higher (2–7 µM) in winter at the western and eastern edges, due to an influx from deeper water (Gibbs *et al.* 1986b).

Port Phillip Bay

Port Phillip Bay is a large, shallow marine embayment of ~1930 km^2, with a maximum depth of 24 m and a very narrow entrance to Bass Strait, called 'The Rip' (Fig. 2.5). Water circulation of the Bay is driven by tide and wind (Rosenberg *et al.* 1992), and its hydrology was reviewed by Harris *et al.* (1996). Tidal currents dominate the circulation in the lower part of the Bay, with speeds up to 2.5 m s^{-1} at the entrance, and to 0.8 m s^{-1} in the main channels, but weaker elsewhere (0.02–0.1 m s^{-1}), and mainly driven by the wind; current and wind strengths have been found to be highly correlated (Hinwood *et al.* 1979). Wind-driven currents form gyres in which currents flow in the direction of the wind at shallow depths, and against the wind in deeper areas, giving rise to overturning water. The modelled depth-averaged water currents

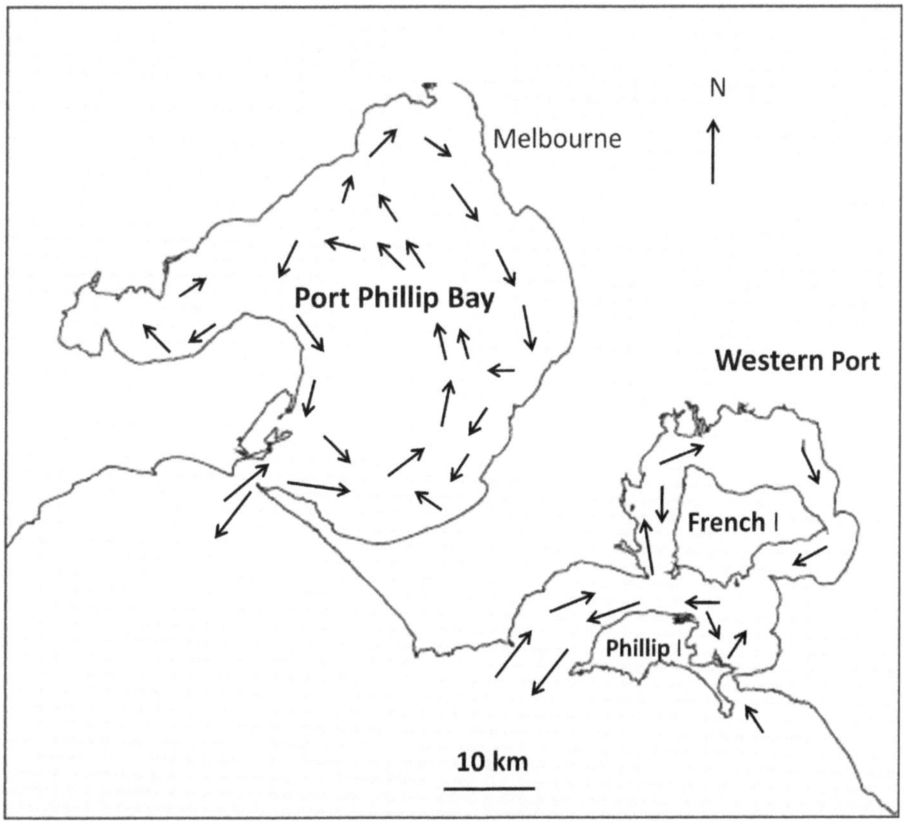

Figure 2.5: Modelled depth-averaged circulation in Port Phillip Bay, and circulation of Western Port derived from water chemistry (redrawn from Harris and Robinson 1979 and Harris *et al.* 1996).

throughout the Bay over 2 days with NW winds of 11.6 m s^{-1} are illustrated in Fig. 2.5. The flushing time for the Bay was estimated to be ~270 days, varying with position within the Bay (Walker 1999).

Western Port

Western Port is much smaller (270 km^2) than the neighbouring Port Phillip Bay, and has a wide entrance on the west side of Phillip I. and a narrow one on its east. Hence, the semi-diurnal tides are much larger than in Port Phillip Bay, and are the main driving mechanism of the circulation. Winds have a short-term (hours to days) influence on the tides. The hydrology of the Bay was reviewed by Shapiro (1975) and Harris and Robinson (1979). Water currents in Western Port channels can reach 1.3 m s^{-1}, but are weaker in the numerous bays and inlets, and residence times of the water range from months in the north and east of the Bay to < 1 day near the entrance. The net water circulation as deduced from the water chemistry (Fig. 2.5) shows that water circulates clockwise around Phillip I and French I., with a weak counter-clockwise gyre just north of the eastern entrance (Harris and Robinson 1979).

WAVE CLIMATE

> *'Air in motion, which is wind, in passing over the smooth surface of the water, may rub, as it were, upon that surface, and raise it into wrinkles, which, if the wind continues, are the elements of future waves.'*
>
> Benjamin Franklin, 1774

Waves are among the most important environmental factors influencing marine plants and animals, and it is the energy from wave action that transports food to sessile organisms, and uproots them during storms. A wave has three component dimensions:

- wavelength – the distance between crests
- height – the vertical distance from trough to crest
- period – the time in seconds for successive crests to pass a given point.

For deep water waves these measures are related by the two formulae:

$$L = VS \text{ and } L = 1.56\,S^2$$

where L is the wavelength in metres, V is the velocity in m s^{-1}, and S is the period in seconds.

These formulae are a good approximation where the depth is more than one half the wavelength. On a coast with sloping bottom, when the depth is less than the wavelength, the wave 'feels the bottom', slows down, and water particle movement becomes increasingly elliptical (Fig. 2.6), as wave energy is gradually dissipated by bottom friction; here the relationship between

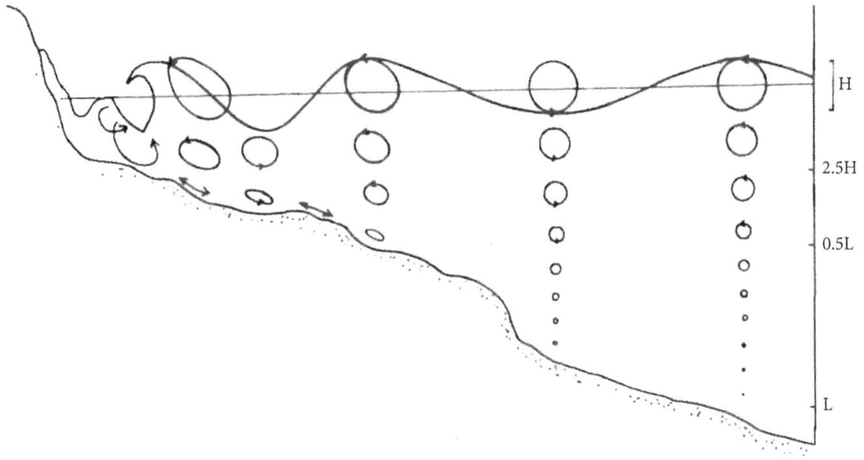

Figure 2.6: Movement of a water particle as a wave approaches a steeply sloping shore (redrawn from Bearman 1999). H = wave height; L = wavelength.

L, V and S is more complex. If the wave strikes a vertical cliff emerging from deep water, water particles simply move vertically next to the cliff face, where the sea is relatively calm. That is why penguins can safely leap out of the water onto a cliff ledge in conditions of strong swell.

Climatologists distinguish 'sea-state', which refers to locally wind-generated waves, from 'swell', which refers to long-period waves generated from distant regions (in temperate Australia from the Southern Ocean, Indian Ocean or the west Pacific). Wind-driven waves have periods of 2–8 s, and are important in bays, gulfs and coastal waters. Swell, consisting of high energy, deep-water waves, with periods of 8–15 s, is undoubtedly the major force along all the exposed temperate coasts of Australia, and is almost continuous throughout the year. Although swell height is < 1.2 m for 50–70% of the time in the deep sea off Australia (Morgan 1974), it increases in height and energy, as it 'feels' the bottom on the inner shelf. Hence, swell is perceived as being much higher near to the coast, especially where the continental shelf is narrow. Hence, for example, the coasts off SW Tasmania and SE South Australia with narrow shelves are notoriously the roughest coasts of Australia, with average swell heights of ~2 m near the coast for most of the year (e.g. Shepherd 1979).

Within bays and gulfs, sea surface waves are wind-generated (although residual swell may penetrate their entrances), and wave height and length depend on wind speed and fetch (the unobstructed distance over which the wind blows). If the fetch is great enough and the wind speed constant, an equilibrium is soon reached, in which energy is dissipated at the same rate as the wave receives energy from the wind. This is then a 'fully developed sea'. In practice, wind speed varies, provoking variation in wave size, and causing a more or less confused sea.

For comparisons between sites of the degree of exposure to wave energy, ecologists have sometimes used indices variously based on fetch and extent of openness and aspect of the site to swell (e.g. Barrett and Buxton 2002). Wave climate is now routinely measured by remote sensing, which provides a synoptic view of wide expanses of sea, and such data, and predictions based on them, are readily available on numerous internet sites (Bearman 1999).

TIDES

Tidal waves behave as very long, shallow-water waves, in which the back-and-forth orbital movements are flattened (Fig. 2.6), and are the tidal currents that accompany the rise and fall of the tide as the wave progresses. Each of the ocean basins develops a resonating frequency due to the basin geometry and the Coriolis force, which causes a unique pattern of high and low water for each ocean. Because of the many factors involved, tidal theory is complex, and it is sufficient for our purposes here to note that nodal points (called amphidromic systems) exist in each basin, round which the crest of a tidal wave circulates. One such node is off Tasmania, in SE Australia, and another off Perth in the Indian Ocean (Bearman 1999).

Normally, there are two semi-diurnal tides every 24 h 51 min., but in some circumstances one of the tides is negated, so that there is only one high and one low diurnal tide, as on the west coast of Western Australia. Another variant in the tidal pattern is during neap tides, when, in some places, the tide level stays at about mid-tide for a day or more, and tidal movement virtually shuts down. This is called a dodge tide and is a peculiar feature of the South Australian Sea.

Off southern Australia, the tidal range is mostly < 2 m, but ~1 m in SW Western Australia, increasing up the east and west coasts of Australia. Tidal features significant for circulation patterns and for benthic assemblages are:

- The tidal range and tidal currents increase as the tidal wave propagates into shallow waters, as in Spencer Gulf described above.
- In the open ocean, tidal currents are typically < 0.5 m s^{-1}, but speeds can vary according to coastal and bottom topography. Associated with large amplitude tidal currents are residual currents; that is, long-term net water movement in particular directions. Residual currents are important for the movement of sediments, and for flushing contaminants and salt wedges.

- Tidal currents promote vertical mixing of water in the water column, reducing stratification of different density layers, while, conversely, during neap or dodge tides, stratification may become marked, unless reduced by turbulent mixing by the wind.
- Fronts (i.e. boundaries) between water masses of different density can sometimes occur, where tidal currents are weaker or the water is deeper. Such fronts may have higher nutrient levels and so may be zones of higher productivity.
- Tidal current speeds are increased where the flow is restricted by the presence of islands or coastal topography, as in the SW Bass Strait between King I and Tasmania, Port Phillip Heads, Western Port, Backstairs Passage in SE Gulf St Vincent and in the upper Spencer Gulf.

Tidal currents are important biologically both for larval transport and dispersion, and for some fish when migrating to or from spawning or feeding grounds.

LIGHT CLIMATE

The amount of sunlight penetrating the sea is of critical importance for the growth of phytoplankton and benthic algae. Photosynthetically available radiation ranges from 350 to 750 nm (UV light to red light), and longer wavelengths are rapidly absorbed near the sea surface; blue-green light is the last to be absorbed, and can penetrate to ~200 m depth in the clearest water.

Marine waters were classified by Jerlov (1976) into five oceanic types (I, Ia, Ib, II and III), and nine coastal types to describe optically the range of waters from the clearest to the most turbid coastal waters. The depth of penetration of 1% of the surface light level, termed the euphotic zone, ranges from ~110 m for the clearest water to ~6 m for coastal type 9, and approximates to the maximum depth to which algae can grow. Low-nutrient oceanic waters, such as occur on Australian temperate coasts, have low levels of plankton and are among the clearest in the world, being usually Oceanic Type 1a (as in the eastern GAB) or 1b or II (as in western and northern GAB). In Oceanic Type 1a, benthic algae can grow to depths of ~70 m. As plankton abundance, suspended particles, or dissolved substances from land run-off increase, light penetration declines, so that in the most turbid inshore waters, such as those in northern Western Port (Coastal Type ~9), benthic algae can grow to only 4–5 m depth (see Chapter 7). As the amount of light reaching bottom habitats can vary seasonally and according to wave action and run-off, adaptability of algae to changes in light reaching the habitat is essential for their survival.

REFERENCES

Baines PG, Edwards RJ, Fandry CB (1983) Observations of a new baroclinic current along the western continental slope of Bass Strait. *Australian Journal of Marine and Freshwater Research* **34**, 155–157.

Baird ME, Ridgway KR (2012) The southward transport of sub-mesoscale lenses of Bass Strait water in the centre of anti-cyclonic mesoscale eddies. *Geophysical Research Letters* **39**, L02603.

Barrett NS, Buxton C (2002) 'Examining underwater visual census techniques for the assessment of population structure and biodiversity in temperate marine protected areas'. Tasmanian Aquaculture and Fisheries Institute Technical Report No. 11, Hobart.

Bearman G (Ed.) (1999) *Waves, Tides and Shallow-water Processes.* 2nd edn. Butterworth Heinemann and Open University, Oxford.

Binnie MN, Cann JH (2008) The distribution of foraminifera in Gulf St Vincent – indicators of sea-level change. In *Natural History of Gulf St Vincent* (Eds SA Shepherd, S Bryars, I Kirkegaard, P Harbison *et al.*) pp. 38–55. Royal Society of South Australia, Adelaide.

Black A (1853) *Black's General Atlas of the World.* Adam and Charles Black, Edinburgh.

Bullock DA (1975) The general water circulation of Spencer Gulf, South Australia, in the period February to May. *Transactions of the Royal Society of South Australia* **99**, 43–53.

Bunt JS (1987) The Australian marine environment. In *Fauna of Australia. Vol. 1A General Articles* (Eds GR Dyne and DW Walton) pp. 17–42. AGPS, Canberra.

Bye JAT (1976) Physical oceanography of Gulf St Vincent and Investigator Strait. *Transactions of the Royal Society of South Australia* **100**, 143–160.

Bye JAT, Kämpf J (2008) Physical oceanography. In *Natural History of Gulf St Vincent* (Eds SA Shepherd, S Bryars, I Kirkegaard, P Harbison, *et al.*) pp. 56–70. Royal Society of South Australia, Adelaide.

Caputi N, de Lestang S, Feng M, Pearce A (2009) Seasonal variation in the long-term warming trend in water temperature off the Western Australian coast. *Marine and Freshwater Research* **60**, 129–139.

Carvalho O, Bye JAT (1996) The first recorded successful cross-Backstairs Passage swim. *South Australian Geographic Journal* **95**, 70–74.

Cirano M, Middleton JF (2004) Aspects of the mean wintertime circulation along Australia's southern shelves: numerical studies. *Journal of Physical Oceanography* **34**, 668–684.

Cresswell GR (1991) The Leeuwin Current – observations and recent models. *Journal of the Royal Society of Western Australia* **74**, 1–14.

Cresswell GR (2000) Currents of the continental shelf and the upper slope of Tasmania. *Papers and Proceedings of the Royal Society of Tasmania* **133**, 21–30.

Cresswell GR, Peterson JL (1993) The Leeuwin Current south of Western Australia. *Australian Journal of Marine and Freshwater Research* **44**, 285–304.

Easton AK (1978) A reappraisal of the tides in Spencer Gulf, South Australia. *Australian Journal of Marine and Freshwater Research* **29**, 467–477.

Fandry CB (1983) Model for the three-dimensional structure of wind-driven and tidal circulation in Bass Strait. *Australian Journal of Marine and Freshwater Research* **34**, 121–141.

Gersbach GH, Pattiaratchi CB, Ivey GN, Cresswell GR (1999) Upwelling on the south-west coast of Australia – source of the Capes Current? *Continental Shelf Research* **19**, 363–400.

Gibbs CF, Cowdell RA, Longmore AR (1986a) Seasonal variations of density patterns in relation to the Bass Strait cascade. *Australian Journal of Marine and Freshwater Research* **37**, 21–25.

Gibbs CF, Tomczak M, Longmore AR (1986b) The nutrient regime in Bass Strait. *Australian Journal of Marine and Freshwater Research* **37**, 451–466.

Godfrey JS, Jones ISF, Maxwell JGH, Scott BD (1980) On the winter cascade from Bass Strait into the Tasman Sea. *Australian Journal of Marine and Freshwater Research* **31**, 275–286.

Green HS (1984) Fluid transport processes in upper Spencer Gulf. *Marine Geology* **61**, 181–195.

Griffin DA, Thompson PA, Bax NJ, Bradford RW, *et al.* (1997) Outbreaks and spread of the 1995 Australian pilchard mass mortality—no link found with phytoplankton nor ocean conditions. *Marine and Freshwater Research* **48**, 27–42.

Hallegraeff GM, Jeffrey SW (1993) Annually recurrent diatom blooms in spring along the New South Wales coast of Australia. *Australian Journal of Marine and Freshwater Research* **44**, 325–334.

Hanson CE, Pattiaratchi CB, Waite AM (2005) Seasonal production regimes off south-western Australia: influence of the Capes and Leeuwin Currents on phytoplankton dynamics. *Marine and Freshwater Research* **56**, 1011–1026.

Harris JE, Robinson B (1979) Circulation in Western Port, Victoria, as deduced from salinity and reactive silica distributions. *Marine Geology* **30**, 101–116.

Harris GP, Nilsson C, Clementson L, Thomas D (1987) The water masses of the east coast of Tasmania: seasonal and interannual variability and the influence on phytoplankton biomass and productivity. *Australian Journal of Marine and Freshwater Research* **38**, 569–590.

Harris GP, Batley G, Fox D, Hall D, *et al.* (1996) *Port Phillip Bay Environmental Study Final Report.* CSIRO, Canberra.

Herzfeld M, Tomczak M (1997) Numerical modelling of the sea surface temperature and circulation in the Great Australian Bight. *Progress in Oceanography* **39**, 29–78.

Herzfeld M, Tomczak M (1999) Bottom-driven upwelling generated by eastern intensification in closed and semi-closed basins with a sloping bottom. *Marine and Freshwater Research* **50**, 613–627.

Hinwood JB, Watson JE, Mills DE, Dandy GC, *et al.* (1979) *The Port of Melbourne Environmental Study.* Centre of Environmental Studies, University of Melbourne, Melbourne.

Hsieh WW, Hamon BV (1991) The El Niño–Southern oscillation in south-eastern Australian waters. *Australian Journal of Marine and Freshwater Research* **42**, 263–275.

James NP, Bone Y (2011) *Neritic Carbonate Sediments in a Temperate Realm, Southern Australia.* Springer, London.

James NP, Bone Y, Collins LB, Kyser TK (2001) Surficial sediments of the Great Australian Bight: facies dynamics and oceanography on a vast cool-water carbonate shelf. *Journal of Sedimentary Research* **71**, 549–567.

Jeffrey SW, Rochford DJ, Cresswell GR (1990) Oceanography of the Australasian region. In *Biology of Marine Plants* (Eds MN Clayton and RJ King) pp. 244–265. Longman Cheshire, Melbourne.

Jerlov NG (1976) *Marine Optics.* 2nd edn. Elsevier Oceanography Series. Academic Press, London.

Johnson C, Ling S, Ross J, Shepherd SA, *et al.* (2005) 'Establishment of the long-spined sea urchin (*Centrostephanus rodgersii*) in Tasmania: first assessment of potential threats to fisheries'. TAFI. FRDC Report. Project No. 2001/044. University of Tasmania, Hobart.

Lennon GW, Bowers DG, Nunes RA, Scott BD, *et al.* (1987) Gravity currents and the release of salt from an inverse estuary. *Nature* **327**, 695–697.

Li J, Clarke AJ (2004) Coastline direction, interannual flow, and the strong El Niño currents along Australia's nearly zonal southern coast. *Journal of Physical Oceanography* **34**, 2373–2381.

Luick JL, Kase R, Tomczak M (1994) On the formation and spreading of the Bass Strait cascade. *Continental Shelf Research* **14**, 385–399.

McClatchie S, Middleton JF, Ward T (2006) Water mass analysis and alongshore variation in upwelling intensity in the eastern Great Australian Bight. *Journal of Geophysical Research* **111**, C08007.

Middleton JF, Bye JAT (2007) A review of shelf-slope circulation along Australia's southern shelves: Cape Leeuwin to Portland. *Progress in Oceanography* **75**, 1–41.

Middleton JF, Platov G (2003) Hydrodynamic model of the Great Australian Bight. In *Trophodynamics of the GAB: Assessing the Need for an Ecological Allocation in the SA Pilchard Fishery.* (Eds T Ward, SD Goldsworthy SD and B Page) pp. 51–99. SARDI Publication No. RD03/0166-2. SARDI Aquatic Sciences, Adelaide.

Middleton JF, Cox D, Tate P (1996) The oceanography of the Sydney region. *Marine Pollution Bulletin* **33**, 124–131.

Middleton JF, Arthur C, Van Ruth P, Ward TM, *et al.* (2007) El Niño effects and upwelling off South Australia. *Journal of Physical Oceanography* **37**, 2458–2477.

Morgan GA (1974) *Sea-State Statistics for the Australian Region.* Australian Defence Scientific Service, Department of Supply, Adelaide.

Newell BS (1961) 'Hydrology of south-east Australian waters: Bass Strait and New South Wales tuna fishing areas'. CSIRO Division of Fisheries and Oceanography Technical Paper No. 10. CSIRO, Melbourne.

Ng S, Bye JAT, Petrusevics P (1993) The St Vincent Gulf Stream. *Bulletin of the Australian Meteorological and Oceanographic Society* **6**(3), 50–55.

Noye BJ, Grzechnik M (1996) Tides and currents of Gulf St Vincent. In *Proceedings of Ocean and Atmosphere Pacific International Conference* (Ed. TH Aung) pp. 253–258. Flinders University of South Australia, Adelaide.

Nunes Vaz RA, Lennon GW, Bowers DG (1990) Physical behaviour of a large, negative or inverse estuary. *Continental Shelf Research* **10**, 277–304.

Oke PR, Middleton JH (2001) Nutrient enrichment off Port Stephens: the role of the east Australia Current. *Continental Shelf Research* **21**, 587–606.

Olsen AM, Shepherd SA (2006) Historic drift bottle experiments show reversing surface water masses in western Bass Strait waters: implications for lobster larval dispersal. *Transactions of the Royal Society of South Australia.* **130**, 113–122.

Oppo DW, Rosenthal V (2010) The great Indo-Pacific com municator. *Science* **328**, 1492–1494.

Pattiaratchi CB, Middleton JF (2007) Physical oceanography of the SW region. In *The South-West Marine Region: Ecosystems and Key Species Groups.* pp. 10–147. Department of Environment and Water Resources, Canberra. <www.laptop.deh.gov.au/coasts/mbp/publications/south-west/pubs/sw-ecosystems-part1.pdf> (Accessed December 2012)

Pearce A (1991) Eastern boundary currents of the southern hemisphere. *Journal of the Royal Society of Western Australia* **74**, 35–45.

Pearce A, Pattiaratchi CB (1999) The Capes Current: a summer countercurrent flowing past Cape Leeuwin and Cape Naturaliste, Western Australia. *Continental Shelf Research* **19**, 401–420.

Petrusevics P (1993) SST fronts in inverse estuaries, South Australia—indicators of reduced gulf-shelf exchange. *Australian Journal of Marine and Freshwater Research* **44**, 305–323.

Petrusevics P (2008) Satellite remote sensing of Gulf St Vincent. In *Natural History of Gulf St Vincent* (Eds SA Shepherd, S Bryars, I Kirkegaard, P Harbison and JT Jennings) pp. 71–80. Royal Society of South Australia, Adelaide.

Provis NG, Lennon GW (1983) Eddy viscosity and tidal cycles in a shallow sea. *Estuarine, Coastal and Shelf Science* **16**, 351–361.

Ridgway KR, Condie SA (2004) The 5500-km-long boundary flow off western and southern Australia. *Journal of Geophysical Research, C, Oceans* **109**(4), C04017

Ridgway KR, Dunn JR (2003) Mesoscale structure of the mean East Australia Current system and its relationship with topography. *Progress in Oceanography* **56**, 189–222.

Rochford DJ (1975) Oceanography and its role in the management of aquatic ecosystems. *Proceedings of the Ecological Society of Australia* **8**, 67–83.

Rochford DJ (1980) 'Nutrient status of waters around Australia'. CSIRO Fisheries and Oceanographic Report 1977–1979. Cronulla.

Rochford DJ (1984) Nitrates in eastern Australian coastal waters. *Australian Journal of Marine and Freshwater Research* **35**, 385–397.

Rochford DJ (1991) 'Upwelling': does it need a stricter definition? *Australian Journal of Marine and Freshwater Research* **42**, 45–46.

Rosenberg M, Hodgkinson R, Black KP, Colman R (1992) *Hydrodynamics of Port Phillip Bay: Field Data Collection and Data Reduction.* Victorian Institute of Marine Science, Queenscliff.

Roughan M, Middleton J (2002) A comparison of observed upwelling mechanisms off the east coast of Australia. *Continental Shelf Research* **22**, 2551–2572.

Sandery PA, Kaempf J (2005) Winter-spring flushing of Bass Strait, south-eastern Australia: a numerical modelling study. *Estuarine, Coastal and Shelf Science* **63**, 23–31.

Schluter C, Bye JAT, Harbison P (1995) The most vigorous South Australian tide. *Transactions of the Royal Society of South Australia* **119**, 123–132.

Shapiro MA (1975) *Westernport Bay Environmental Study 1973–1974.* Ministry for Conservation, Melbourne.

Shepherd SA (1979) The ecology of a deep water red algal community off Cape Northumberland, South Australia. MEnvSt thesis. Centre for Environmental Studies, University of Adelaide, Adelaide.

Steedman RK, Craig PD (1983) Wind-driven circulation of Cockburn Sound. *Australian Journal of Marine and Freshwater Research* **34**, 187–212.

Suthers IM, Waite AM (2007) Coastal oceanography and ecology. In *Marine Ecology* (Eds DS Connell and BM Gillanders) pp. 199–226. Oxford University Press, Melbourne.

Thresher R, Rintoul SR, Koslow JA, Weidman C, *et al.* (2004) Oceanic evidence of climate change in southern Australia over the last three centuries. *Geophysical Research Letters* **31**, L07212.

van Ruth PD, Ganf GC, Ward TM (2010a) Hotspots of primary productivity: an alternative interpretation to conventional upwelling models. *Estuarine, Coastal and Shelf Science* **90**, 142–158.

Van Ruth PD, Ganf GC, Ward TM (2010b) The influence of mixing on primary productivity: a unique application of classical critical depth theory. *Progress in Oceanography* **85**, 224–235.

Walker SJ (1999) Coupled hydrodynamic and transport models of Port Phillip Bay, a semi-enclosed bay in south-eastern Australia. *Marine and Freshwater Research* **50**, 469–481.

Waters JM (2008) Marine biogeographical disjunction in temperate Australia: historical landbridge, contemporary currents, or both? *Diversity & Distributions* **14**, 692–700.

Woo M, Pattiaratchi CB, Schroeder W (2006) Summer surface circulation along the Gascoyne continental shelf, Western Australia. *Continental Shelf Research* **26**, 132–152.

3 Biogeography and biodiversity

'There is one radical distinction between different minds ... some minds are stronger and apter to mark the differences of things, others to mark their resemblances.'

Francis Bacon

OVERVIEW

This chapter reviews knowledge about the origins and biogeography of the southern Australian marine plants and animals. An ancient flora and fauna existed around the shores of Gondwana 100 Ma, and as Australia broke off and drifted northwards the southern coast was colonised by many of these species. South-west and south-east elements are still seen today, resulting from this ancient biota from Tethys Sea on the west, and palaeo-austral ancestry on the east of the Australian continent. Over the succeeding epochs the biota was enriched from the north, from extensive radiation along the 5500 km east–west coastline, from the West Wind Drift, and from cosmopolitan species arriving from diverse sources. Glaciations from the Pleistocene set up barriers, giving rise to division of populations of species, and later allopatric speciation. As a result of these and other factors, the region has an extraordinarily rich flora and fauna. Endemicity varies between phyla, depending *inter alia* on habitat extent and diversity, depth range and dispersal potential. It is high (> 70%) for red algae (Rhodophyta), brachiopods, echinoderms, molluscs, ascidians and fishes, moderate (50–60%) for brown algae (Heterokontophyta) and sponges, and low (< 40%) for bryozoans, green algae (Chlorophyta), and some decapods and hydrozoans.

INTRODUCTION

Biogeography is the study of the distributions of species in space and time, and has two distinct sub-disciplines – historical and ecological biogeography. Historical biogeography in the marine context examines the biota of a region in the light of its past history in terms of:

- location of the region in relation to other continents, and the way this has changed as continents have rifted apart
- water currents, wave exposure, sea temperatures and seawater chemistry and their changes over time
- the emergence of environmental barriers, such as widening oceanic waters, appearance of a peninsula or isthmus, or steepening temperature or other gradients. Populations may become split by the slow appearance of a barrier arising from the geological movement of a continent. This is an example of what

is termed vicariance. If a species happens to disperse across a barrier to form a new population, perhaps by hitching a ride on a drifting alga, or by the temporary collapse of a barrier, such as by a brief oscillation in sea level, it is termed jump-dispersal. The relative importance of the two models has been much debated, often on sparse evidence (Waters 2008); however, molecular analyses can now allow a better informed and more reliable inference on the appropriate model in a given case, as is illustrated below.

Up until ~100–120 million years ago (Ma), Australia, Antarctica and other continents formed a supercontinent called Gondwana (Chapter 1), and then the present-day continents broke free and began to drift apart, with Australia slowly splitting off from Antarctica from ~70 Ma, and drifting northwards from ~40 Ma. As it drifted, sea temperatures and ocean currents around its shores changed with shifts in global temperatures and the opening of Drake Passage and the Tasmanian gateway. All of these events are the stuff of historical biogeography, and the basis for the analysis of evolutionary processes in the biota.

On the other hand, ecological biogeography is concerned with the modern distribution of a flora or fauna and its component species. It seeks to explain distributions in terms of the ecology of species – their temperature tolerances, food and other habitat requirements, and larval dispersal. Jump-dispersal, or rafting, of a species across a geographic barrier would be an ecological phenomenon if it happened in the lifetime of scientific observers, but a historical one if it occurred in the distant past.

In this chapter we review the biogeography of the southern Australian marine biota. For benthic algae, we consider species associated with rocky substrata in the sublittoral zone down to ~50 m depth. This is nearly at the lower limit of the photic zone in clear oceanic waters, and as deep as direct studies using scuba have been conducted. For animals, we discuss several phyla for which sufficient historical and ecological information are available over their full depth range. For each major group, we review the role of historical processes in their evolution, if known, and then the ecological and geographical processes that have contributed to their present richness and distribution patterns.

Because many phyla show similar distribution patterns for groups of species within them, we shall first distinguish the four most common distribution types as follows (Fig. 3.1):

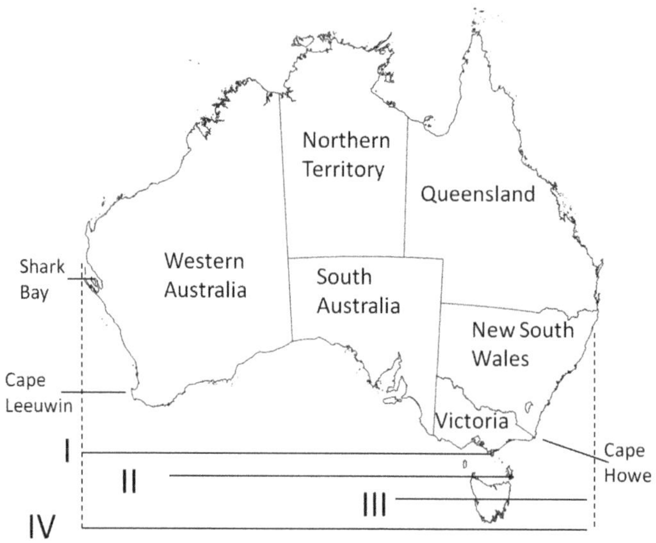

Figure 3.1: Map of Australia with State boundaries, showing the southern temperate (Flindersian) region, and four common distribution types, I–IV, for marine organisms (modified from Phillips 2001).

- **Group I.** This is a western group of species ranging from the West Australian coast somewhere between the Abrolhos Is, and Perth to the central South Australian coast to Kangaroo I. (or even to western Victoria).
- **Group II.** This is a southern group ranging from Cape Leeuwin to eastern Victoria (Cape Howe) and around Tasmania. This has historically been called the Flindersian province (see below).
- **Group III.** This is a SE group of species, ranging from mid- to eastern South Australia across Victoria and up the east coast of Australia to Sydney, or even the Queensland border.
- **Group IV.** This is the most widely distributed group of southern species, ranging south of ~30°S (i.e. south of Shark Bay or the Abrolhos Is and up the New South Wales coast, as far as southern or mid-Queensland coast).

Additional temperate groups occur off the NSW coast and off SW Western Australia from about Cape Leeuwin to Shark Bay, but species in these groups are vagrants on the south coast, so are not discussed further. Note that the boundaries given above are always approximate, because no two species have exactly the same distributional limit.

MARINE PLANTS

Macroalgae

Historical processes

Hommersand (2007) considered that the marine floras of the world originated not in warm to tropical waters like the marine faunas, but in cold- to cool-temperate boreal and anti-boreal waters. On this view, Australasia is seen as a major centre of origin and diversity for marine algae. Southern Australia is indisputably a biodiversity hotspot, possessing the richest algal flora (~1500 species) for its length of coastline in the world (Womersley 1990; Bolton 1994) (see below), so there is a rich field for studies on the origins and causes of such diversity.

Macroalgae are poorly preserved in the fossil record, except for some calcareous species, so molecular studies (in addition to morphological ones) have become crucial to understanding the history and genetic relations of families, genera and species. Such molecular studies, beginning in the 1990s, have already provided much evidence on algal phylogeny, and have become the backbone of theories and speculations on algal biogeography. Cladistic analysis of phylogenetic data typically gives series of nodes joining related groups, which are then interpreted in terms of separation of populations through emergence of barriers (vicariance) or dispersal events. The nodes can be dated if the genetic sequence evolved at a constant rate. This provides a molecular clock for estimating when, for example, sister species diverged. However, the accuracy of the molecular clock is still uncertain, and additional evidence is often needed. Hence, the occurrence of similar nodes for different species, specifying the same vicariance event, may provide corroborative evidence of the operation of that event, and strengthen the conclusion.

The algal floras of cool temperate regions in the world are typically rich, and algal diversity appears to be highest in those regions where sea temperature ranges from ~15–20°C (extremes 10–22°C) (Womersley 1990). In the Southern Hemisphere during the late Cretaceous (70 Ma), the east and west Australian coasts abutting Antarctica experienced mild mean temperatures of 12–14°C and higher sea levels. At that time, surface water currents originating in the tropics flowed far to the south, in the absence of a Circum-Antarctic Current and a closed Drake Passage (Chapter 1). Hence, benthic algae could evolve and disperse widely on antiboreal coasts. During the succeeding 40 million years (My), southern waters passed through an initial warming episode, followed by gradual cooling episodes and four major Chills (Chapter 1). The warming episode provided 'new adaptive zones, with many unoccupied niches available for colonisation by tropical macroalgal species', while the cooling events would have seen 'the extinction of much of the older tropical flora, and the ecological opportunity for the diversification of the temperate flora' (Phillips 2001).

The current state of knowledge on the historical origins of the present algal flora of southern Australia

(~1500 species), illustrated with examples drawn from the literature, can be summarised as follows:

- An ancient flora appears to have evolved along the northern edge of Gondwana (180–120 Ma). From this, a relict flora evolved and radiated around the southern Australian shores (e.g. Areschougiaceae, Table 3.1), while another relict flora evolved in the northern Tethys Sea, and eventually became the Atlantic and Indian Ocean floras as the Tethys Sea collapsed. An example of this is the Solieriaceae group (excluding Areschougiaceae), with representatives in northern seas (Fredericq *et al.* 1999).
- In the early Tertiary (from ~80 Ma), the original algal flora in the southern Tethys Sea was able to evolve over a long time period, in isolation from that of the west Pacific, because Australia, whose northern shores were at ~30°S, formed a barrier between the Tethys Sea and the west Pacific. As the Tasmanian gateway opened ~34 Ma, the two floras intermingled, with the western element expanding eastwards, and the eastern element westwards. These two elements can be recognised today, because both still have relationships with their respective floras of origin. For example, today some 140 species are common to Australia and New Zealand (Table 40 in Hommersand 2007). In general terms, the species of the South Island of New Zealand are more similar to the flora of Tasmania

Table 3.1: Some examples of families and genera of algae of southern Australia, their likely origins, present distributions, and the probable mechanisms of spread.

The term 'dispersal' means via propagules, water currents or rafting.

Algal families/genera	Origin	Present distributions	Mechanism
Areschougiaceae (e.g. *Areschougia, Rhabdonia,Erythroclonium*)	SE Tethys Sea[1] 180–120 Ma	SW and southern Australia	Dispersal
Callophycus	SE Tethys Sea[2]	Southern Australia, Phillipines, Indian Ocean, South Africa	Vicariance (collapse of Tethys Sea), dispersal
Some Fucales (e.g. *Hormosira, Xiphophora, Phyllospora, Scytothalia, Seirococcus*)	SE Tethys Sea[3] 120–80 Ma	Southern Australia and terminal groups in Europe, NE and NW Pacific	Vicariance, dispersal
Gigartina spp.	SE Tethys Sea 80–30 Ma[4]	Australia, New Zealand, South Africa, Pacific	Vicariance, dispersal
Melanthalia, Curdiea, Pterocladia	Southern Australia 80 Ma[5,6]	Southern Australia, New Zealand	Vicariance (Tasman Sea)/ dispersal
Delisea, Polyopes, Sarcodia, Plocamium, Phacelocarpus, Euptilota, Dasyclonium. etc.	Southern Australia[6]	Southern Australia, East Asia	Dispersal northwards during Chills in Miocene or Pliocene
Macrocystis pyrifera	North Pacific 1–2 Ma[7,8]	Tasmania, New Zealand, South America, NE Pacific	Recent dispersal by Circum-Antarctic Current
Durvillaea	Southern Australia[6,9]	Circumglobally in Southern Ocean	Dispersal by Circum-Antarctic Current
Lessonia	Problematic[6,10]	Tasmania, New Zealand, South America, NE Pacific	Dispersal by Circum-Antarctic Current; vicariance?

[1] Fredericq *et al.* (1999); [2] Kraft (1984); [3] Clayton (1994), Serrão *et al.* (1999); [4] Hommersand and Fredericq (2003); [5] Bailey and Freshwater (1997); [6] Hommersand (2007); [7] Nelson (1994); [8] Coyer *et al.* (2001); [9] Waters (2008); [10] Chin *et al.* (1991).

and Victoria, due to dispersal via the West Wind Drift, whereas those of the North Island are more similar to the flora of New South Wales, likely due to dispersal via the East Australia Current.

- Hommersand (2007) lists 19 genera of brown algae and 116 genera of red algae, whose species are spread throughout the Southern Hemisphere, with representatives in southern Australia, South America and South Africa, with a few extensions to the north Pacific. These genera originated mainly in the SE Tethys Sea, with ancestral species in Australia or New Zealand and sometimes Antarctica. Examples of common species are given in Table 3.1.
- Some southern Australian or New Zealand red algal genera have amphi-tropical distributions – that is, they are absent in the tropics but have NW Pacific representatives (Table 3.1). These genera migrated from Australasia to the north Pacific. Conversely, very few species have made the reverse north-to-south migration. A notable example of the latter is the kelp genus *Ecklonia*, with species in east Asia and South Africa, and a widely dispersed species (*E. radiata*) in Australia, South Africa, Oman (NW Indian Ocean), and New Zealand. The dispersal of *Ecklonia* to the Southern Hemisphere likely occurred in the last few million years.
- The Circum-Antarctic Current, which developed with the opening of Drake Passage some 40 Ma, dispersed, often by rafting, around the Southern Hemisphere many species that originated in eastern and southern Gondwana. A few other brown algal species, notable because they dominate some subtidal assemblages, either spread from the North to the South Pacific (e.g. *Macrocystis*) or evolved in the south (e.g. *Durvillaea* – see Table 3.1).
- Recent historical vicariant events, such as sea-level movements, causing the emergence of the Bassian Isthmus during the last glaciation, and earlier Chills (Chapter 1), would have caused disjunct distributions of warm-water species, as they were forced out of colder waters (Edgar 1986). These in turn can give rise to pairs of allopatric species, and contribute to further speciation (see below). Shifts or reversals in water currents may also have occurred during these events, and further imposed limits or disconnects to the ranges of species.

In summary, southern Australia may be seen both as a primary source of algal species for Pacific, Atlantic and Southern Ocean coasts, and as a sink for a few species from other places. Significant vicariant events were: the break up of Gondwana (separation of Australia from the Antarctic); the collapse of the Tethys Sea; the opening of the Tasman Sea; and the emergence of the Bassian Isthmus. The Chills of the Pliocene and Miocene allowed trans-tropical spread, so were important dispersal events. Lastly, the Leeuwin Current (LC) and East Australia Current (EAC), on Australia's west and east coasts, respectively, enabled dispersal southwards, while the Circum-Antarctic Current dispersed species eastwards (Kerswell 2006).

Ecological processes

The southern Australian coastline extends for ~5500 km from Cape Leeuwin to Cape Howe, and much of it comprises rocky shores falling steeply to reef substrata. This vast area of inshore reefs, together with often extensive subtidal reefs offshore (Bunt 1987), provide abundant habitat for macroalgae. Except in sheltered bays or gulfs and on the lee of promontories or islands, the coast is exposed to strong southern ocean swell (Chapter 2).

In a study of the present-day distribution of the southern Australian algal flora, Phillips (2001) found that four distinct floristic elements exist: a strictly southern endemic group; a more widely distributed temperate group; and small tropical and polar groups. Her analysis omitted the two red algal families, the Delesseriaceae and the Rhodomelaceae, which were not monographed until 2003 (Womersley 2003), but are included below. The southern endemic group is the largest, comprising 70% of the flora, the widely distributed temperate group is next with 22%, and the tropical and polar groups comprise 6% and 2%, respectively. The proportions of green, brown and red algae vary between these groups. Most green algae are in the widely distributed temperate element (45%), followed closely by the endemic element (40%); most red algae are in the endemic group (77%), while the polar group (41% brown algae and 37% red algae) is limited to SE and southern Tasmania.

The migration of tropical species into southern Australia has very likely been aided in the recent past by the warm LC (e.g. Walker 1991), and less recently during the last interglacial warm period ~120 000 years ago, when tropical corals, and warm-water forams and molluscs expanded their ranges into southern Australia (Kendrick *et al.* 1991; McGowran *et al.* 1997; Binnie and Cann 2008). Some tropical algal species are now relict in the South Australian Gulfs (Womersley 1990), notably *Dictyopteris australis* (see Phillips 2000), *Sargassum decurrens*, and *Asparagopsis taxiformis*. But what is the distribution pattern of the largest group – the endemics?

For this group of 771 endemic species, Phillips (2001) recognised the four distribution patterns in southern temperate waters as outlined above: that is, Group I with 245 species; Group II, the largest, with 398 species; Group III with 69 species; and Group IV with 59 species. These 771 species are thought to have evolved somewhere in the region, although, as noted above, the species in Group III and some in Group IV may have evolved from an ancestral Pacific palaeo-austral stock, and extended their ranges westwards to varying extents after the opening of the Tasmanian gateway. Goldberg and Kendrick (2005) also found a gradual floristic transition in nine adjoining sub-regions of southern Australia from the Abrolhos Is, Western Australia to SE Queensland, with adjacent sub-regions clustering separately from the east coast regions in a similarity analysis.

For all groups of species, it is widely accepted that the ranges of algal species are determined by sea temperature. Thus, Womersley (1990) considered that a shift of 5°C in summer or winter isotherms marked a change in the flora, although in any one region of southern Australia the winter–summer temperature range may be 12–20°C.

Recent understanding of coastal currents (Chapter 2) gives a complex picture of coastal sea temperatures. Seasonal northward-flowing inshore currents on the west and east coasts of Australia, the warm winter LC, seasonal upwelling events along the southern and east coasts, and inshore wind-driven coastal currents may all, by their moderating effect on sea temperature, play a role in extending the ranges of species. On the other hand, many barriers to dispersal exist along the coast, as noted variously by Edgar (1984, 1986), Womersley (1990) and Phillips (2001). They include: adverse currents during the reproductive season; long stretches of sandy bottom (e.g. the Coorong and the Ninety Mile Beach, eastern Victoria); the 850 km Nullarbor Cliffs; and abrupt coastal features, such as Wilsons Promontory, and the NE and NW corners of Tasmania. Some of these barriers are also associated with sharp changes in exposure to swell, as well as sea temperature. For example, Edgar (1984) noted a sharp transition in the algal flora from Bass Strait to the Tasmanian east and west coasts, where many 'Flindersian' canopy species (e.g. *Sargassum heteromorphum, Cystophora monilifera, C. polycystidea, C. botryocystis* and *Scaberia agardhii*) disappear. Even in suitable conditions, current hypotheses on biotic exchange (i.e. the arrival of new species) suggest that range extensions will be limited because invading species are not often successful, unless the recipient flora has suffered an extinction event (Vermeij 1991) – something which has not happened in southern Australia.

Seagrasses

No discussion of marine plants is complete without reference to the seagrasses, although they are sand-dwelling angiosperms and outside the scope of this book. Nevertheless, seagrasses occasionally grow on reefs, where their rhizomes or root systems attach themselves in holes or crevices mainly in limestone, as is the case for species of *Amphibolis*, and occasionally *Zostera*, *Posidonia* and *Halophila*.

Despite the rarity of seagrass fossils, it seems that seagrasses arose in the late Cretaceous in the Tethys Sea, and from there spread north and south. By the Eocene, seagrass species very close to those living today had already evolved, with temperate species spreading from west to east along southern Australia (Larkum and den Hartog 1989). However, den Hartog and Kuo (2006) caution that, given the paucity of fossil material, it is still premature to work out the history of the seagrass flora. Nevertheless, the distribution pattern of the temperate seagrass flora is substantially similar to the four groups given above, apart from the addition of a fifth group of eastern Australian species of tropical origin (Larkum and den Hartog 1989). At present 19 seagrass species have been described for temperate Australian waters, all of them endemic, although the taxonomy is still uncertain for a few species. The western group (I), with seven species of *Posidonia* and two of *Amphibolis*, is the largest, although we note that *Amphibolis antarctica* does extend to Bass Strait, with small patches in eastern Tasmania (Edgar 1986).

SPONGES (PORIFERA)

Sponges dominate benthic communities below the photic zone at 30– > 50 m depth on open coasts. In shallower water, they can be dominant on shaded surfaces, such as caves and crevices, where they compete with bryozoans and ascidians. Sponges have short-lived larvae that remain in the water column for minutes to a few days, rarely to a few weeks (Maldonado 2006), so their potential for dispersal is consequently low. However, sponge fragments can regrow vegetatively – a dispersal mechanism whose importance is virtually unknown.

The southern Australian sponge fauna is poorly known, with only ~360 species described out of an estimated 800 species. This account is largely taken from the reviews of Hooper and Lévi (1994), van Soest (1994) and Hooper and Wiedenmayer (1994). The NW and NE coasts of Australia have an estimated 1500 species each, and the Indo-West Pacific region reputedly has the highest sponge diversity in the world, suggesting a latitudinal gradient of declining diversity from the tropics to high latitudes. But what are the ancestral relations of the southern sponge fauna?

The fossil record of sponge spicules is impoverished, but existing information suggests that Demospongiae (the largest class of sponges) was widespread in the Tethys Sea in the early Tertiary (60–70 Ma). Analysis of a subset of the southern sponge fauna by Hooper and Lévi (1994) indicated two historical elements – a tropical Tethyan group and a cold-water Gondwanan group.

The only detailed study of a temperate Australian sponge fauna is that of Wiedenmayer (1989) who examined the affinities of the Bass Strait fauna, with some 600 species. The strongest affinity was with the Indo-West Pacific, with 33% similarity (20% of species being shared with NE Australia). Links with the west Pacific fauna were weak, with 3%, and somewhat more with New Zealand (10%). Overall, 58% of the southern species were endemic, and thought to be a relict palaeo-austral fauna. The cosmopolitan element was 11% – a figure questioned by Hooper and Lévi (1994) due to the uncertain taxonomy – and the Indo-Malay Group was 26%, implying colonisation from the tropical north. From a broader perspective Hooper and Lévi (1994) suggested that a distinctive eastern Australian group of species occurred, quite different from the tropical fauna, and with a northern boundary in northern New South Wales. This group had ~25% overlap with the Bass Strait fauna, and had 10–15% of species in common with New Zealand. On the western side of Australia, there was an extensive overlap region between the tropical NW fauna and the SW Australian fauna, complicated by pockets of endemism in the Abrolhos Is (23°S Lat.) and the Bunbury-Cape Naturaliste region (33–34°S). Thus the data suggested SW and SE elements, corresponding roughly to Groups I and III. In the eastern Great Australian Bight(GAB), shallow waters (< 85 m depth) have the highest diversity, which declines sharply with increasing depth (Sorokin *et al.* 2007).

BRACHIOPODS

Joyce R Richardson and SA Shepherd

Brachiopods range from shallow water to kilometres deep, and variously live on rocky, gravelly or sandy bottoms. The phylum is divided into two classes, the Articulata and Inarticulata, distinguished by differences in shell structure and larval type. The articulate brachiopods have very short-lived larvae, with a larval life of hours to 1–2 days, while the inarticulate ones have pelagic larvae that drift in the plankton for up to 6 weeks (Hyman 1959). Hence, the articulates have much more restricted distributions than the inarticulates. As so-called 'living fossils' that once dominated ancient seas > 600 Ma, brachiopods are notable for their survival with little or no change – a record described in the Inarticulata as static and in the Articulata as repetitive (Rudwick 1970).

The terebratellids in Australia and New Zealand have revealed more about the articulates than is known in any other part of the world (Richardson 1994, 1997). The terebratellids are more varied in lifestyle than the living sedentary members of older families, and the bryozoan sands of the Australian shelf contain a variety of free-living species, in some of which the pedicle has evolved as a free appendage. The same adaptations are known throughout articulate history in species that occupy carbonate sands. Substratum type is the prime factor determining distribution, and no evidence exists that depth, temperature, latitude or wave energy limit distribution.

Species indistinguishable from articulates in shallow seas are found in hydrothermal vent communities at similar depths as in the Tasman Sea (3550 m), where species have been collected. Dispersal is limited by the brief life of larvae, and the distribution of Tertiary and Recent faunas in the Southern Hemisphere is consistent with an ancestral fauna surrounding Gondwanaland. The ancient brachiopod fauna once showed a strong latitudinal gradient from the tropics towards the poles, but extensive glaciation in the late Palaeozoic (~327 Ma) preferentially eliminated much of the then tropical fauna (Powell (2007).

Thirty articulate genera have been recorded from Australian waters (23 species from southern Australia), of which 21 are endemic. Members of one genus in each of three inarticulate families are found in Australia – Lingulidae, Discinidae and Craniidae (Emig 1997). The genera from all inarticulate families are cosmopolitan – a consequence of a planktonic larva with the capacity to search and settle on substrata that the adults are specialised to occupy.

BRYOZOANS

At least 850 species of bryozoans have been described from southern Australia, from both hard substrata and sediments (Bock and Hageman in McGowran *et al.* 2000), but this represents only ~60% of the likely total (Gordon 1999). Bryozoans range from the intertidal to abyssal depths, although most occur on the continental shelf (to 200 m deep), and range vertically over that depth. Bryozoans are poor dispersers, producing short-lived larvae, which settle in a few days, so it may seem surprising that their species and genera are so widespread in the Southern Hemisphere. For example, Barnes and Griffiths (2008), in an analysis of the better-known subset of the southern Australian species, found that only ~38% were endemic. They attributed this low level of endemism to their slow evolutionary turnover rate and their transport round the Southern Hemisphere via the West Wind Drift, since the opening of the Drake Passage and the Tasmanian gateway 30–40 Ma. Bryozoans fossilise well, and many Tertiary (to 80 Ma) genera of bryozoans are well represented in the modern fauna of Australasia. It is probable that the modern fauna evolved after Gondwanan fragmentation, as the Southern Ocean opened.

Bryozoans do not show a strong, or indeed any, latitudinal gradient in the diversity of species from the tropics to cold waters. Diversity hotspots are equally present in the tropical Indo-West Pacific, and in temperate waters below ~30°S in southern Australian, New Zealand and Antarctic waters (Gordon 1999; Barnes and Griffiths 2008). Of the three orders in the phylum – the cheilostomes, ctenostomes and cyclostomes (Bock 1997) – ~90% are cheilostomes, and 70% of the species occur both in the SE and SW regions of Australia, indicating strong links between the two southern regions.

HYDROIDS

Jeanette E Watson

History

Hydrozoans are poor subjects for fossilisation because many are composed entirely of soft tissue, while the chitinous skeletons of others fragment and decay after death of the organism. Hydroids are predominantly sessile colonial marine organisms in the Superclass Hydrozoa. The oldest unequivocal records of hydroids are those from rocks of Lower Cambrian age at Heathcote and Lancefield 60 km north of Melbourne (Chapman and Thomas 1936). Other early records are from the Ordovician, early Silurian and Cretaceous, notably in families with more easily fossilised, calcareous skeletons – the Milleporidae and Stylasteridae. Despite their scant fossil record, hydrozoans almost certainly date back to the late Precambrian, as the fauna of the Flinders Ranges in South Australia includes other medusoid and pennatulid cnidarians (Ludbrook 1980). This, together with the Heathcote evidence, suggests that hydroids were almost certainly well established as an important and diverse component of the Ediacaran marine fauna. Their long geological history indicates an evolutionarily conservative group, as some of the Heathcote fossils are indistinguishable from present day hydroid families. Yet, while hydroid genera have radiated throughout the world oceans over geological time, only a few genera, (e.g. *Corystolona* and *Halicornopsis*) are endemic to southern Australia.

Ecology and distribution

Hydroids typically have an attached, benthic colonial form (the hydroid), and a medusoid, which is the solitary, sexual, free-swimming and dispersive stage of the life cycle (see Watson 1982 for a detailed account of hydroid reproduction). Dispersal is likely to range from a few metres to hundreds of kilometres, according to the particular life cycle of a species.

Hydroids range from the intertidal to abyssal depths, and occur worldwide. Out of the > 4000 species globally, ~150 species are known from tropical Australia, and ~250 species from southern Australia, with more species likely to be found. The tropical Australian element has strong Indo-Pacific affinities and is quite distinct from the southern temperate fauna, although there is some overlap of species. The faunal break between the tropical and temperate faunas is between latitudes 25°S and 29°S (i.e. in southern Queensland in the east, and the Abrolhos Islands in the west).

Across the southern continent, the shallow water (< 50 m) hydroid fauna is relatively uniform in species composition, and probably > 90% of known temperate species range throughout the southern Australian region. The hydroid fauna of coastal Tasmania to depths of ~50 m is almost identical with that of Victoria. The broad southern distribution of hydroids indicates their adaptation to southern temperate temperature ranges. Nevertheless, more species are known from SE Australia than from SW Australia, suggesting either a centre of origin of hydroids in the SE of the continent, or alternatively, a better knowledge of the Bass Strait fauna.

Some apparently endemic species are known only from the continental slope in GAB waters, but this may simply reflect collecting effort in deep water, rather than any evolutionary pattern. Interestingly, the hydroid fauna of seamounts 100 km SE of Tasmania at depths of 1000–2000 m consists of species known from deep water off New Zealand and Japan, indicating the capacity of deep water species to disperse very widely, likely via deep ocean currents.

About 70% of southern Australian hydroid species are common to New Zealand, mainly the South Island. Most of the trans-Tasman species have long-lived medusan or larval phases, suggesting dispersal from Australian waters over geological time, or rafting in recent times, via the West Wind Drift. Undoubtedly, however, some species have been exchanged in historical times on ships' hulls or in ballast water on ships plying the Tasman Sea.

ECHINODERMS

History

The echinoderms comprise five classes: Asteroidea (the sea-stars); Crinoidea (the feather-stars); Ophiuroidea (the brittle-stars); Holothuroidea (the sea-cucumbers); and Echinoidea (the sea-urchins). Of these, only the Echinoidea has left a substantial fossil record in the numerous basins of the Australian west and south coasts, which were flooded from the Paleocene onwards (after ~65 Ma) (Chapter 1; McGowran *et al.* 2000). The echinoid fauna has been strongly influenced by dispersal into the region from the Indian Ocean, as Australia drifted north over the last 40 My, and then by the periodic southerly movement of species down the western coast, and eastwards across the eastern Indian Ocean. For example, eight genera (e.g. *Echinometra*), originating in the Mediterranean, spread across the Indian Ocean to Australia, and thence down the west Australian coast. Invasions of tropical species were marked during the middle Eocene and Pliocene, and some genera that arrived then are still present today (e.g. *Phyllacanthus*, *Microcyphus* and *Amblypneustes*). The fossils of all these genera illustrate the role of the proto-Leeuwin Current, which periodically brought them into southern waters. In all, the several invasion events and the cooling of southern waters have given rise to two distinct faunal elements among the echinoids – a subtropical element on the east and west coasts, and a cool temperate one on the southern coast, with species in the living genera *Holopneustes*, *Ammotrophus* and *Echinocardium*.

Ecology

Echinoderms exhibit a range of life histories. Some have long-lived planktotrophic larvae, such as several sea-cucumbers whose larvae drift for weeks in the water column. Others have much shorter-lived lecithotrophic (non-feeding) larvae that may drift at sea for a few days. An example is the sea-urchin *Heliocidaris erythrogramma*, whose larvae settle after ~6 days. Hence, larval dispersal is variable both between and within the five echinoderm classes.

The diversity of echinoderms is highest in the Indo-West-Pacific, and O'Hara and Poore (2000) found a latitudinal decline in species richness from the subtropical–warm temperate regions on the west and east Australian coasts to cool temperate Tasmania. Further, as expected from this distribution pattern, the number of species in local areas throughout southern Australia was correlated with both summer maximum and winter minimum temperatures. The coasts near Perth and Sydney were overlap regions between the northern tropical fauna and the southern fauna, and many species of both groups had their southern and northern limits, respectively, at these points. Along the southern coast O'Hara and Poore (2000) noted that the fauna clustered into two groups – a SW group, more or less equivalent to Group I above, and a SE group equivalent to Group III. An overlap region occurred in the South Australian Gulfs, giving rise to a diversity peak there. Other locations of increased rates of change (i.e. species turnover) were in the region of the Coorong, Wilsons Promontory, eastern Victoria and northern Tasmania. Overall, about half of the species in the southern half of Australia are endemic to the region (see Table 3.2), and a very few are either cosmopolitan (e.g. *Echinocardium cordatum* and *Coscinasterias muricata*) or have a broad circumpolar distribution (e.g. *Parvulastra exigua*), suggesting a West Wind Drift dispersal as first proposed by Fell (1962). Ten Tasmanian species (besides two introductions from New Zealand, *Patiriella regularis* and *Astrostole scabra*) are common to New Zealand, implying that a trans-Tasman eastward migration was very likely (Edgar 1986).

There is little evidence of speciation in southern temperate waters, and most genera have only one or two species in them, but notable exceptions include the genera *Cenolia, Amphiura, Nectria, Meridiastra* and *Amblypneustes*, with five to nine species each (O'Hara and Poore 2000). Moreover, some recent molecular studies on sea-stars have provided some fascinating insights into cryptic speciation in *Meridiastra*. Species in this genus have short-lived lecithotrophic larvae, and the multi-coloured species, once known as *M. gunnii*, has proved to be three cryptic species: an east coast *M. oriens*; a western and central

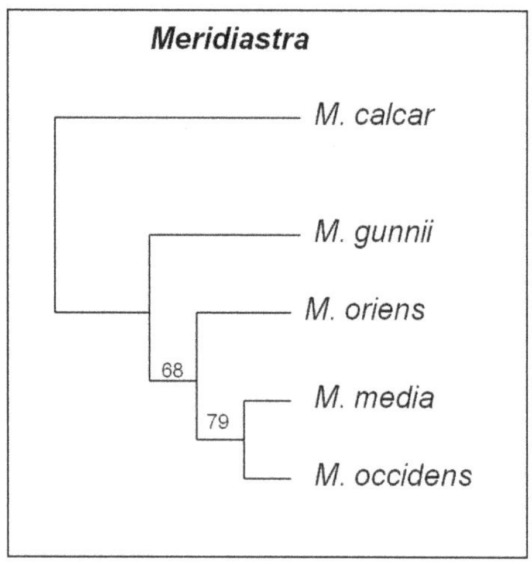

Figure 3.2: Maximum likelihood tree from DNA analysis of *Meridiastra* (formerly *Patiriella*). Numbers are bootstrap percentages from maximum parsimony percentages (redrawn from Waters *et al.* 2004).

Table 3.2: Estimated species diversity of marine plant and selected animal phyla in southern Australia and globally, with dispersal potential (DP).

Percentage endemism of southern Australian species and percentage of southern Australian species to global total are given. Low dispersal means from a few to tens of metres, and high dispersal a few to many kilometres. Global estimates from Bolton (1994) for algae, and from Gibbons *et al.* (1999) for animals. Na = not available.

Phylum	No. of species Southern Australiaí	Global	DP	Percentage endemic	Percentage global	References
Algae (total)	~1500	12 000	Low		~12.5	Womersley (1990)
Chlorophyta	124			30		
Heterokonto-phyta	231		Low	57		Womersley (1990)
Rhodophyta	~1140		Low	75		Womersley (1990)
Sponges	> 800	7400	Low	58	~11	Hooper and Lévi (1994)
Bryozoans	~850	10 000	Low	38	8.5	Barnes and Griffiths (2008)
Molluscs[1]	19 000	96 850	Varies	95	~8	Poore (1994)
Brachiopods	23	360	Low	87	6.4	Richardson (1997)
Echinoderms[2]	347	6200	High	56	5.6	O'Hara and Poore (2000)
Decapods[3]	392	N.a.	High	40	Na	O'Hara and Poore (2000)
Crustacea	~800	11 000			7.3	Poore (2004)
Cnidaria[4]– Scyphozoans	~34	200	High	> 50	~17	Southcott (1982), L Gershwin *pers. comm.*
– Hydrozoans	~250	> 4000	Varies	30	~6	JE Watson *pers. comm.*
Pycnogonids[5]	71	1250	Varies	30?	~11	D Staples *pers. comm.*
Ascidians	368	2650	Low	93	13.9	Kott (2005, 2006)
Fishes[6]	612	14 000	High	78	4.4	Gomon *et al.* (2008)

[1] Australia-wide, with < 50% in southern Australia; [2] analysed jointly with some decapod crustaceans; but if the southern region is defined as from Albany, WA to Eden, NSW, endemism is 74%; [3]only two out of 35 Orders of Crustacea; [4] the Class Scyphozoa; the Class Hydrozoa does not include Hydromedusae; [5] described southern species, and ~50% of the total (D Staples *pers. comm.*); [6] species occurring in < 50 m depth.

southern coast *M. occidens*; and a central southern to northern Tasmania *M. media* (Waters *et al.* 2004). The uniformly purple species, long known as *Patiriella brevispina*, is now synonymised with *M. gunnii*, which ranges across southern Australia. The molecular evidence suggested that the three new species (*oriens, occidens* and *media*) diverged in the late Pliocene, ~2.1–2.4 Ma at the onset of glaciation (Fig. 3.2). At this time, the Bassian Isthmus emerged and the LC may have stopped flowing along the south coast, further cooling southern waters, events which would have geographically isolated populations. Subsequently, after glaciation ceased, *M. oriens* is thought to have expanded its range southwards to southern Tasmania, and *M. gunnii* westwards into South Australia. But another species in the genus tells a different story. The tiny, intertidal *Parvulastra exigua* is cosmopolitan and occurs in three oceans, although it has non-dispersive larvae (it lays egg masses that develop after ~18 days directly into miniature adults). This species apparently evolved in South Africa, and during the Pleistocene dispersed by long-distance rafting, aided by the West Wind Drift, eastwards across the Indian Ocean to Australia and other oceanic islands. Within southern Australia, dispersal must be very sporadic, because minor genetic divergence among near populations has already evolved (see Waters 2004).

It is likely that cryptic speciation has occurred in other genera, as shown by the recently discovered species of the biscuit star genus *Tosia* (Naughton and O'Hara 2009). The new species *Tosia neossia* differs only subtly from *T. australis* in morphology, but unlike that species, is a brooder, producing benthic, non-dispersing larvae, and is limited to the cooler waters of SE Australia. It possibly evolved during past cooler climates.

The cosmopolitan genus *Coscinasterias* is widespread throughout the Mediterranean, Atlantic, Indian and Pacific Oceans, with two Australian (one tropical and one temperate) species. A recent molecular study of all four known species (Waters and Roy 2003a) enabled a reconstruction of lineages and divergence times, and indicated that the southern Australian species *Coscinasterias muricata* was basal, and that New Zealand was colonised during the Pleistocene from Australia. The other species did not arise by vicariance associated with opening of the Atlantic 100 Ma, but probably spread more recently by long-distance rafting (Waters and Roy 2003a). However, within southern Australia, cryptic speciation occurred with distinct lineages on the west Australian, Tasmanian and NSW coasts, suggesting divergence resulting from Plio-Pleistocene temperature changes (0.2–0.3 Ma) and the cyclic emergence of the Bassian Isthmus (Waters and Roy 2003b).

In summary, the richness of the southern echinoderm fauna can be viewed as a consequence of the presence of a very rich tropical fauna to the north, which periodically over the last 40 My exported 'recruits' to the southern coast via the EAC and the LC. Meanwhile, Australia slowly drifted northwards and maintained a favourable, warm to cool temperate environment with subtropical transitions on its east and west flanks.

MOLLUSCS

The biogeography of the modern molluscan fauna is fairly well known because a rich fossil fauna – mainly bivalves from soft sediments and a few gastropods – has been well preserved from the Cambrian (550 Ma) to the present. Here we summarise the historical development and affinities of the modern fauna during the Cenozoic era (from 70 Ma), drawn from the reviews of Darragh (1985), Wilson and Allen (1987), Jell and Darragh (1998), Ponder and Wells (1998), and McGowran *et al.* (2000). Examples of the radiation and origins of particular genera are also given.

History

Two provinces (regions with distinctive faunas) have been present since the mid- to late Eocene (~40 Ma):

- an Austral Indo-Pacific Province. This is an extension of the tropical Indo-West Pacific province along the west and south coasts of Australia. Many warm water Indo-Pacific taxa were able to penetrate south (see below) until the early Pleistocene (1.6 Ma).
- a SE Australian Province, containing four distinct elements:
  a. an Australia–New Zealand element comprising species common to the two countries, but unknown outside them. They probably originated in the southern circum-Pacific palaeoaustral fauna. Genera in this group are *Montfortula, Eucrassatella, Cymatiella, Penion* and *Maoricolpus* (after extinction in Australia, *Maoricolpus* is now a recent invader from New Zealand). The proportion of this element in the total fauna has ranged from 17 to 30% during the period.
  b. a Tethyan Indo-Pacific element originating in the Indo-Pacific basin, together with taxa from the Mediterranean to Indonesia to the Caribbean. Well known genera in the group are *Katelysia, Scutus, Concholepas* (now extinct in

Australia), *Cellana, Austroharpa, Thalotia* and *Zoila*.

c. a southern Australian endemic element. Notable genera are *Equichlamys, Neotrigonia, Plebidonax, Haliotis, Notocypraea, Conuber, Austroharpa* and *Amoria*. The proportion of this endemic element has ranged from 24 to 34% of the fauna from the Eocene to the Pleistocene.
d. a cosmopolitan element. These genera are widespread globally, and well known examples are *Glycymeris, Brachidontes, Modiolus, Chlamys, Spondylus* and *Pleuroploca*. During the Pleistocene (1.6 Ma), a few genera, *Pecten* and *Mytilus*, arrived from the Northern Hemisphere, and one, *Anadara*, came from South America. The element has comprised 15–25% of the molluscan fauna over the period.

In summary, we have an essentially southern fauna, in which the proportion of the tropical Indo-Pacific element has increased over the last 20 My from episodic invasions from the north via the LC and the EAC, and the proportion of the endemic element has also increased – both at the expense of the other elements. Thus, the modern southern Australian molluscan fauna can be seen as a merger of eastern and western faunal elements, as the Southern Ocean developed after the opening of the Tasmanian gateway. The eastern fauna pushed westwards, and the western fauna eastwards, but the former became dominant in southern Australia after a Pleistocene expansion of species (Darragh 1985). But, over the last 40 My, many warm-water genera appeared and later disappeared during the numerous warming and cooling events, resulting in substantial generic uniformity across southern Australia by the end of the Pliocene.

The present day Australian molluscan fauna is the richest in the world, with ~19 000 species, of which somewhat less than half are in temperate Australian waters. Species endemicity is high at ~95%, indicative of long isolation, and many species now range across southern Australia. Some species end their ranges at Cape Leeuwin in the west, and Cape Howe in the east, but others extend up the western and/or the eastern coasts. Thus, the present-day distribution of molluscs could also be fitted within the same groups noted for the marine flora, with most, however, being in Groups II and IV.

Molluscan examples

The evolution and radiation of molluscs in southern Australia can be illustrated by studies on five well-known groups – the abalone genus *Haliotis*, and the four genera, *Austrolittorina, Scutus, Zoila* and *Umbilia*, each with different life history characteristics.

The abalone *Haliotis*

Haliotis is first known from fossils in the Cretaceous ~66 Ma, and there are now 56 recognised species around the Pacific and Indian Oceans, with outlier species in the Mediterranean and the NE Atlantic (Geiger 2000; Geiger and Owen 2012). Of these, 10 species are endemic in southern Australia. Degnan *et al.* (2006) concluded from their work and that of others that the genus evolved in the tropical Indo-West Pacific,and from there spread to the Pacific and Indian Oceans, with a major phylogeographic break between the two oceans, which was attributed to low sea levels during Pleistocene glacial maxima. The New Zealand group of three species has an uncertain origin, and they either migrated south via the EAC from the tropical west Pacific, or evolved from an ancient lineage that became isolated when Gondwana broke apart. As an aside, Estes *et al.* (2005) have proposed the interesting thesis that the large size of commercially taken abalone evolved after the evolution of kelps in the Northern hemisphere ~25 Ma. The large size of greenlip *H. laevigata* and blacklip *H. rubra* first appeared in the Pleistocene (1–2 Ma), at about the same time as the giant kelp *Macrocystis* (a preferred food of *H. rubra*), and after the red algae (preferred by *H. laevigata*) had diversified, so a co-evolutionary scenario is plausible, although environmental factors, such as glacial-interglacial cycles, and coastal upwellings may also have played a role. With respect to the latter, Hearn (1986) suggested that the fast growth rate and size of exploited abalone populations in the regions of coastal upwellings, compared with other populations, are due to the much greater algal productivity of those nutrient-rich coasts.

The southern Australian group of haliotids appears to have evolved from an ancestor of the blacklip *H. rubra*, which migrated from the Indian Ocean. In the molecular tree (Fig. 3.3), *H. roei*, *H. coccoradiata* from SE Australia and *H. scalaris* are closely related, with *H. laevigata* and *H. cyclobates* a little more distant. It seems likely that the southern species evolved late in the Pliocene glaciation, through isolation of populations caused by glacial barriers or the Bassian Isthmus. The present distribution of six southern Australian species falls into one or other of the four groups given for marine flora above (i.e. I – *H. roei*; II – *H. cyclobates*, *H. scalaris* and *H. laevigata*; III – *H. coccoradiata*; IV – *H. rubra*), while two species, *H. elegans* and *H. semiplicata*, are restricted to the Australian lower west coast, and two, *H. brazieri* and *H. hargravesi* to the lower east coast (Lindberg 1992). The factors determining these diverse distributions are likely to be physiological ones, related to sea temperature, and life history ones, such as a very limited larval dispersal and highly specific larval settlement cues favouring speciation, or restricted ranges (Morgan and Shepherd 2006).

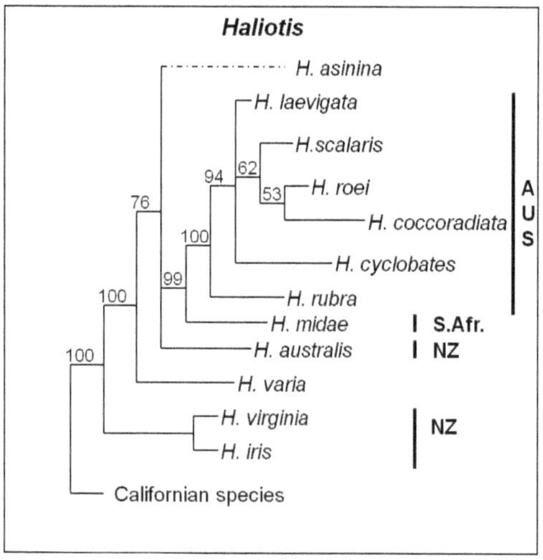

Figure 3.3: Bayesian majority-rule consensus tree for some Southern Hemisphere abalone, genus *Haliotis*. *H. varia* and *H. asinina* are tropical species. Six southern Australian species, a South African species and three New Zealand species are shown. Numbers represent posterior probabilities (redrawn from Degnan *et al.* 2006).

The genera *Austrolittorina, Scutus* and *Nerita*

The first two genera each have one species, and *Nerita* two, in southern Australia, and the same or different species occur in New Zealand. Species in the first two genera have a dispersive larval phase of a month or a little more, and *Nerita* a much longer one of ~6 months, so interesting questions arise whether their distributions and biogeographic structure are determined by temperature gradients, habitat barriers or vicariance (Waters *et al.* 2005, 2007). Wilsons Promontory was a major point of disjunction for all three genera between western and eastern populations, suggesting a vicariant mechanism – divergence during the late Pliocene when the Bassian Isthmus emerged. The Tasman Sea, however, was no barrier to the dispersal of *Nerita* from Australia to New Zealand, but was for *Austrolittorina* and *Scutus*. Further, the Coorong was a point of disjunction for *Scutus* populations, but not for the other genera. It seems that the genus *Nerita* spread from the tropics to southern Australia in the early Pliocene (5–6 Ma), and that eastern and western populations then diverged in the late Pliocene. Since then, the genus has been able to readily disperse to New Zealand due to its long-lived larval stage. On the other hand, *Austrolittorina* and *Scutus*, with much shorter larval lives, must have crossed the Tasman Sea ~25 Ma after the Oligocene marine inundation of New Zealand.

Cowries, *Zoila, Umbilia* and *Notocypraea*

These examples are southern cowry genera, and they all have direct development of eggs into miniature adults – a trait that likely evolved from planktotrophic ancestors during colder periods (Meyer 2003; Wilson and Clarkson 2004). Hence, their dispersal is very restricted. Species in *Zoila* and *Umbilia* are sponge feeders and have a wide depth range, often going to the edge of the shelf at 200 m deep, while species of *Notocypraea* live in shallow water to ~30 m deep and feed on sponge and bryozoans.

Fossil cowries date back at least 100 Ma. The genus *Zoila* has Tethyan–Indo-Pacific ancestry, which appeared 65–70 Ma, and *Umbilia* and *Notocypraea* a temperate palaeo-austral ancestry, appearing during the Oligocene ~30 Ma, and evolving in SE Australia. *Zoila* invaded the southern Australian coast at least twice during warm periods (in the Pliocene and Miocene), and the present species have evolved either from this early stock or from a third Pleistocene invasion. *Notocypraea* evolved in southern Australia, and has radiated into five species (Meyer 2003).

The present day distribution of species in all three genera reflects their past. Four of the six species of *Zoila* (*friendii, marginata, venusta* and *rosselli*) are endemic in temperate Australia, with a SW Australian distribution extending to the central south coast (Group I), and two species (*decipiens* and *perlae eludens*) are endemic to the NW shelf. The two species of *Umbilia* (*U. hesitata* and *U. armeniaca*) have Group III and Group I distributions, respectively, with a third species *(U. capricornica*) found only in Queensland. Many subspecies of the several species of *Zoila* are known, suggesting that over its history subpopulations of several species have become isolated, and evolved into separate subspecies. The five species of *Notocypraea* have a Group II or Group III distribution (see Gowlett-Holmes 2008). A molecular tree for *Zoila* and *Umbilia* is shown in Fig. 3.4.

The above examples show that the speciation and distribution of molluscs vary greatly between genera. However, some generalisations can be inferred from the evidence (Meyer 2003). Species with dispersive (planktonic) larvae have larger ranges and increased resistance to extinction, while species with poor dispersal (e.g. direct developing eggs) have smaller ranges, higher speciation rates and higher turnover rates (i.e. extinction rates). In addition to larval dispersal ability, topographic, coastal or oceanographic features may also present barriers to dispersal.

DECAPOD CRUSTACEANS

Crustaceans are the most diverse group of animals in marine habitats, comprising about a third of all marine species, but only a few groups are even moderately well known. Further, the fossil record is scant, so their evolution must be inferred from the few groups, such as the tiny ostracods (Order Ostracoda) that have left a long fossil record. Although crustaceans have a long history dating back to the Cambrian (Briggs 1983), here we shall fast forward to the more recent fossil ostracod evidence for southern Australia starting from the Paleocene (Warne in McGowran *et al.* 2000).

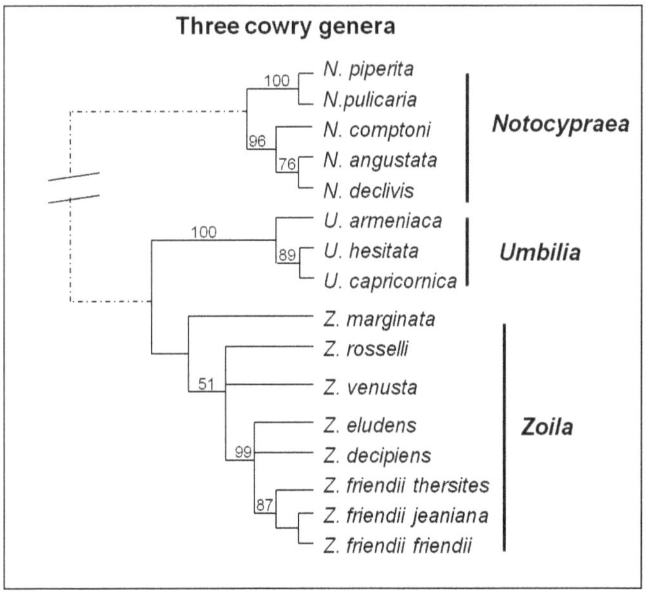

Figure 3.4: Phylogenies of three related genera of cowries, *Umbilia, Zoila* and *Notocypraea*. Maximum likelihood phylograms are presented for each group, with bootstrap values inserted above branches (redrawn from Meyer 2003).

Several patterns have emerged. First, numerous endemic genera originated in shallow seas of the SW Pacific and in newly opening seaways, such as that between Australia and Antarctica. Some of these genera underwent evolutionary radiation, notably around the Miocene–Pliocene boundary in SE Australia. At about the same time, other genera were evolving in the tropical Indo-West-Pacific, and some of them migrated to, and became dominant in, southern waters, most likely via the LC and EAC. The reverse also happened, in which endemic southern species dispersed towards the tropics (e.g. Feldmann 1986). Thus, southern crustaceans, like some other phyla, comprise two elements – an endemic group and a group of tropical origin.

The modern decapods – made up of prawns, shrimps, lobsters and crabs – have pelagic larvae, which drift or swim for anywhere from a few weeks, as in some crabs (e.g. Bryars and Svane (2008), up to ~1 year, as in the southern lobster (Sastry 1983; Olsen and Shepherd 2006).

Poore (2004) monographed ~800 species in just two of 35 orders of Crustacea and, of these two, some 392 species occur in southern Australian waters at depths of < 100 m. About 40% of these 392 species are endemic (Poore 1994). The biogeography of these species is similar to that of echinoderms, and were analysed together by O'Hara and Poore (2000) (see above). Their findings for decapods can be summarised as:

- more tropical than temperate decapods exist, producing a latitudinal decline in the richness of species down the east and west coasts of Australia. Little change in species occurs across the GAB, probably due to the warm LC extending the distributions of species eastwards.
- a SW and a SE Australian group of decapods are present, with ~20% of the species shared by the end-points at Sydney and Perth. A high turnover of species occurs in the SA Gulfs' region, as at the above end-points, with lesser changes in the Coorong and Wilsons Promontory. There is some evidence that the southern limit of a few species is imposed by minimum sea temperatures.
- few species generally occur per genus, indicating a low overall level of evolutionary speciation in southern waters. However, notable exceptions exist, as in the genera *Alpheus* and *Synalpheus*, with 36 species between them, and *Paguristes, Pilumnus, Upogebia, Portunus, Charybdis, Leptomithrax, Naxia* and *Halicarcinus*, with four to nine species each (Poore 1994; O'Hara and Poore 2000). These exceptions show that a few shallow-water groups have high endemism and have radiated extensively.

ASCIDIANS OR SEA SQUIRTS

SA Shepherd and the late P Kott

The Australian ascidian fauna is unusually rich, with 726 species currently known, of which about half (368 species) are found in southern temperate waters. The largest family is the Didemnidae, with 120 endemic southern species (Kott 2005). Ascidians are among the most important components of the benthic fauna, living on reefs and sandy substrata, and extending into abyssal depths. They are sessile animals, and either solitary or form colonies by the budding of zooids. Solitary forms are hermaphrodite, although male and female gonads mature at different times to avoid self-fertilisation. Gametes of each sex are usually liberated into the water and externally fertilised, whereas colonial species are usually internally fertilised and brood larvae within the colony. The larval stage is short – lasting a few minutes (in the case of the didemnids) to hours or days – and so dispersal is limited to a few metres, rarely to kilometres. For these reasons, the group is valuable for biogeographic studies. Although fossil ascidians date back to the Cambrian, the paucity of the fossil record makes discussion of their evolution speculative, so here we summarise current information on the biogeographic relations of the fauna.

Despite their short larval life, many ascidians have a broad southern distribution, and a few are even cosmopolitan, suggesting dispersal by rafting or human factors (ballast water, ships' hulls, etc.) (Primo and Vázquez 2008). Of the temperate Australian ascidian fauna, 93% are endemic (Kott 2005, 2006) and, of those discussed by Kott (1997), the distribution of 17% was in Group II type, 38% was in Group III type, 33% was in Group IV type, and the remainder were pan-tropical or cosmopolitan, ranging widely through the Indo-West Pacific region. However, these percentages may become modified with the discovery since 1997 of new species and range extensions of known species (Kott 2005, 2006).

Colonial species comprise the majority of endemic species from temperate Australia. Most of these are well-integrated colonies in the sub-order Aplousobranchia, especially the family Didemnidae, which is notable for the tiny and numerous zooids that replicate prolifically, resulting in fast growth and occupation of space to the exclusion of other organisms. This family has calcareous spicules embedded in the test, often forming skeletal support for sheet-like or massive three-dimensional colonies in which larvae are incubated and released for very short periods (< 10 min.) to seek settlement sites.

The tropical Indo-West Pacific has a greater diversity of ascidian species than southern Australia, giving rise to a latitudinal gradient in species richness (Kott 2008). Primo and Vázquez (2008) suggested that the distribution pattern of southern ascidians is due to three factors:

- a Gondwanan origin of many taxa, and their subsequent isolation and radiation in southern waters
- the extension southward of many Indo-West Pacific species by the warm LC and EAC – a view that accepts the tropical hotspot model (see below); that is, with a centre-of-origin in the Indo-West Pacific (Briggs 2003)
- rafting by means of the West Wind Drift, or by human-mediated dispersal of the 25 or so cosmopolitan or widely distributed Southern Hemisphere species.

The Tasman Sea has been a barrier to trans-Tasman dispersal, and the commonality of eight species to Australia and New Zealand may possibly have arisen by vicariance, as New Zealand split off from Australia (Kott 2007; Primo and Vázquez 2008).

FISH

Fishes evolved ~400 Ma and have since diversified globally into all aquatic habitats. The southern Australian species probably originated from Tethyan ancestral stock, which spread south as the Southern Ocean opened. The modern Indo-West Pacific region is the richest in the world in terms of fish species, and there are more than three times the number of species in northern Australia, than in southern Australia, where ~628 species live in inshore waters to ~100 m deep (Wilson and Allen 1987; Nelson 2006; Gomon *et al.* 2008). Of these, ~78% are endemic and the rest range more widely in New Zealand and other temperate or tropical waters. In a few cases, species occur both in SE Australia and Japan, but not in the intervening tropics (Randall 1981). These species probably crossed the tropics during the Pleistocene when tropical waters were much cooler than now.

Most species have pelagic larvae, which drift or swim in the water for weeks to several months, facilitating dispersal over tens to hundreds of kilometres. Some families with dispersive larvae have radiated explosively in southern Australia, and notably include the wrasses (Labridae) and leatherjackets (Monacanthidae), each with ~20 species.

A few families, such as the angler fishes (Antennariidae), weedfishes (Clinidae), gobies (Gobiidae), clingfishes (Gobiesocidae) and pipefishes (Syngnathidae), reproduce in ways that severely limit dispersal – they are variously viviparous, pouch brooders or tend demersal nests – and these families, with 10–40 species in each, have radiated more than nearly all other families in southern temperate waters.

The biogeography of fishes shows some similarities to, and some differences from, other groups (Wilson and Allen 1987). Centres of endemism occur in SW and SE Australia, and many species have expanded from these regions to different extents until they reach some barrier to dispersal. The various barriers noted are: lower sea temperatures east of Albany as the LC weakens (Kendrick *et al.* 2005); lack of shallow or suitable reef habitat along the Nullarbor Cliffs; the Coorong region; and the Ninety Mile Beach east of Wilsons Promontory. Of all the species found in < 50 m depth listed in Gomon *et al.* (2008): ~5% are subtropical species extending down the west coast and east on the south coast as far as the Recherche Archipelago; ~12% are in Group I as defined above; ~22% are in Group II (with a few species limited to the western or eastern part); ~17% are in Group III; ~5%

range from western Victoria to Queensland; ~18% are in Group IV; and another 18%, many of them pelagic species, are widespread in tropical and/or temperate waters or circum-global. Lastly, very few (~3%) are known only from the South Australian Gulfs or Tasmanian waters.

The rise in average sea temperatures (~0.7°C on the Tasmanian east coast since the 1950s; Chapter 2), as the East Australia Current has penetrated further south, has expanded the range of some northern species southwards or increased their abundance in Bass Strait and the east Tasmanian coast (Edgar 1984; Johnson *et al.* 2011; Last *et al.* 2011). Examples are the long-snout shovelnose ray *Aptychotrema rostrata*, one-spot puller *Chromis hypsilepis*, maori wrasse *Ophthalmolepis lineolata*, Australian mado *Atypichthys strigatus* and white ear *Parma microlepis*.

A remarkable feature of the southern fish fauna is that many species pairs occur, in which one of the pair has an eastern, and the other a western, distribution. The boundary between species in each pair varies, but is often in central to eastern Victoria (Edgar 1986). Wilson and Allen (1987) listed 18 such pairs, of which the following are notable: the blue groper (*Achoerodus* spp.); Australian salmon (*Arripis* spp.); blue devil (*Paraplesiops* spp.); talma (*Chelmonops* spp.); blue morwong (*Nemadactylus* spp.); and the rock blackfish (*Girella* spp.). These species pairs may have evolved during the Pleistocene as sea levels rose and fell during the glacial–interglacial cycles, or when the Bassian Isthmus periodically emerged. In both cases, populations became isolated and speciation occurred.

Odacine radiation

The radiation of an endemic Australian group of Odacines is discussed by Poore (1994) and Clements *et al.* (2004). This group of 12 species is restricted to temperate Australian and New Zealand waters, and comprises *Odax* (3 spp.), *Siphonognathus* (6 spp.), *Haletta semifasciata, Olisthops cyanomelas* (formerly in *Odax*) and *Neoodax balteatus*. Molecular studies on the group illustrate how the evolutionary origins of the group have been determined (Fig. 3.5). The odacines are derived from a labrid tribe, the Hypsigenyini, which appeared in the Eocene ~50 Ma in the Tethyan Sea in a shallow-water, cool-to-warm temperate environment before the development of the east Antarctic icesheet. The ancestral fish probably lived in the waters of west Antarctica, SE Australia and New Zealand. The two New Zealand ancestral members of the genus *Odax* diverged from the other Australian hypsigenyines ~38 Ma, when sea temperatures in southern Australia and New Zealand were temperate to subtropical. Eventually, the tribe became extinct everywhere except in southern Australia and New Zealand. The genus *Siphonognathus* (the weed whitings) diversified in the vast seagrass and algal habitats of southern Australia.

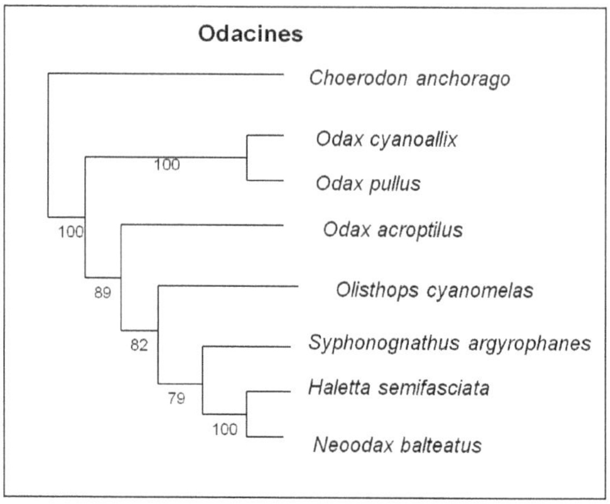

Figure 3.5: Maximum parsimony phylogeny of the odacines. Numbers are maximum parsimony posterior bootstrap proportions (redrawn from Clements *et al.* 2004).

Some of the odacines are herbivorous and some carnivorous – a feature that poses the question whether herbivory evolved once very early, or twice later – once in New Zealand (*O. cyanoallix*) and once in southern Australia (*O. cyanomelas*). Clements *et al.* (2004) discussed the problem and tentatively concluded that herbivory most likely evolved only once in ancestral odacines and that some species later reverted to carnivory. A corollary of the study was that the southern Australian species, formerly called *Odax cyanomelas*, proved to be in a separate clade from the New Zealand species, so its generic name has reverted to *Olisthops*.

Southern Australian biogeographic provinces

The phyletic biogeographic patterns outlined above show sharply divergent patterns between the marine plants and the animal phyla, which also differ from each other to a less degree. The algae show high endemism and more distinct distributional elements than any faunal group – an expected feature because shallow-water algal distributions are strongly influenced by temperature. In fact, many other organisms of the intertidal and shallow sublittoral also have very localised distributions (Edgar 1984).

Many animal phyla show distinct SW and SE elements, although the ascidians have more. These phyla reflect a Tethyan and/or Austro-Pacific ancestry, or alternatively invasions from the Indo-West Pacific via the west or east coasts of Australia – events that have occurred ever since the Eocene (McGowran *et al.* 1997). However, brachiopods have no localised elements at all. What are the reasons for these disparate patterns?

In general terms, they can be explained in terms of history (vicariance and geography) in combination with recent ecological factors (dispersal capability, temperature, depth range, habitat extent, geographic barriers, etc.), which vary between phyla. Thus, algal distributions are strongly influenced by temperature, and show more localised distribution patterns over southern Australia – perhaps not surprising, because the shallow sublittoral temperatures are more variable, and wave energies stronger than those in deeper water. Similarly, intertidal and sublittoral fringe organisms have more localised distributional patterns than those with greater depth ranges, and geographic boundaries are more evident (Edgar 1984, 1986).

Some phyla, or groups of species within a phylum, have distinctive life histories. These are of particular interest because they illustrate significant distributional mechanisms. Thus, ascidians, with short larval lives, have a mixture of elements, defying categorisation, and this may reflect the various ways in which some of them have dispersed – by vicariance, rafting, in ballast water, and so on. Many species, scattered among several phyla such as molluscs, echinoderms and fish, with very limited dispersal capacity, have tended to radiate more. Conversely, other species in these same phyla with pelagic larvae and a much enhanced dispersal ability have broader ranges and have radiated less. Lastly, the brachiopods are a unique group. They have vast depth ranges, albeit with variable larval dispersal and very specific substratum preferences, and have the broadest distributional range of any group of organism. In their case, the enormous expanse of available habitat in deep water determines their remarkable distribution.

As understanding of the distributions of species within many phyla has grown, so have concepts of provincial boundaries changed. Intertidal ecologists in the 1950s–1960s recognised a southern Flindersian province, a Maugean Province for the cooler waters of Tasmania, and a Peronian province for the east Australian coast (reviewed by Poore and O'Hara 2007). With more information on the distribution of benthic algae (Womersley 1990), and deeper water biota, it was evident that the old provincial boundaries became obscured with depth – that is, the similarities between provinces far outweighed the differences. For example, in the shallow sublittoral zone, the prominent canopy-forming algae, *Macrocystis, Durvillaea* and *Lessonia*, clearly demarcate Tasmanian waters, but when the deeper water red algae and animals, with much wider distributions, are given weight, the demarcation disappears. Thus, when the total biota is considered, a distinctive southern region is evident, although with fuzzy boundaries. Defined provinces and their sub-categories, bioregions, are very useful for management, but their artificiality must be kept in mind. Hence, a broad southern temperate province (historically called the Flindersian), with extensive transition zones on the west and east coasts, is now widely accepted (see Poore 1994; Poore and O'Hara 2007). But this depiction on such a broad canvas is of little help for management at a fine scale. For this reason, finer scale bioregions – identified by physical factors, such as exposure, substrata and topography, and by subtler differences in the biota, at a scale of a few to hundreds of kilometres – have been found far more relevant for biodiversity management of States' coastal waters, and for the designation of marine parks.

SOUTHERN BIODIVERSITY

The estimated number of species in the various phyla described above, their dispersal potential and percentage endemism are summarised in Table 3.2. Southern

Australia has a relatively high percentage of the known global species, ranging from 5 to 14%, and often twice the number in, for example, southern Africa, with its much smaller coastline length (Gibbons *et al.* (1999).

Many have speculated on the causes of the high endemism of species and the high diversity in the southern flora and fauna. The two aspects are intimately related, and the following historical and ecological factors have been widely quoted by biogeographers (e.g. Edgar 1986; Bolton 1994; Poore 1994; Phillips 2001; Poore and O'Hara 2007). These factors include:

- the long isolation (~80 My) of temperate Australia from other continents
- the relative stability of the environment in terms of sea temperature over the past 60 My (Chapter 1), and the lack of volcanic activity
- an absence of extinction events
- periodic invasions of Indo-West Pacific forms, especially during warmer periods via the LC and the EAC. This 'double jeopardy' of invasion from the north depends on the 'tropical hotspot' model, which views the tropics, notably the Indo-West Pacific, as both a 'cradle', where new species evolve, and a 'museum', where existing diversity is preserved (see below).
- the West Wind Drift, which has facilitated the circum-polar dispersal of some species
- cyclic changes in sea level, with the periodic emergence of the Bassian Isthmus and possibly other barriers, so giving rise to allopatric speciation
- the vast extent of rocky substrata, and the habitat diversity provided by different geological types (granites, shales, limestones, etc.), plus a diverse coastal topography, extending over 5500 km of coast.

Biogeographers have also developed models, which consider two of the above factors in more detail. These are the tropical hotspot model and the species-pool model, and we shall see how they apply to southern coasts.

Tropical hotspot model

Fossil evidence has shown that biodiversity hotspots have existed in different parts of the world during the last 50 My: first in the Mediterranean during the Eocene (~40 Ma); then in the Arabian Sea in the early Miocene ~20 Ma; and since then in the Indo-West Pacific (Renema *et al.* 2008). These hotspots have all coincided with major tectonic events – the last one being the collision of Australia with the SE Asia tectonic plate. This event resulted in a vast area of islands and shallow seas to the north of Australia, and provided a great diversity of ecological habitats.

The latitudinal diversity gradient, in which species diversity decreases away from the tropics, is perhaps the most striking of natural diversity patterns globally, and has fostered much debate. Many (> 30) hypotheses have been proposed to explain the immediate causes of diversity (see Jablonski et al. 2006; Mittelbach *et al.* 2007; Arita and Vázquez-Dominguez 2008). The model has been examined recently, using historical data on bivalve molluscs, which have left a rich fossil record and have proved to be an ideal system to explore the various hypotheses. Jablonski *et al.* (2006) showed, for bivalve genera and subgenera over the last 11 My, that several processes operated to cause the number of species to decline from the tropics towards the poles:

- greater speciation occurred in the tropics
- species periodically expanded their range from the tropics towards the poles, without losing their tropical presence (see Crame 2009)
- extinction rates were more than three times higher at high latitudes than in the tropics.

Together, these processes interacted and caused the observed latitudinal gradient in diversity. The modern bivalve fauna still reflects this, with 750–1200 species in the tropical provinces such as Southern China, Philippines and Queensland–Northern Territory, and only 191 in the Antarctic (Crame 1999). A similar pattern holds for the decapod crustaceans. Thus, the tropics have been well described as both a cradle and a museum, and in effect are an engine driving biodiversity. Most explanations as to why the tropics are a diversity engine relate to area and time – a larger tropical area fosters higher speciation rates, and the tropics have had a longer time in which to evolve new species (Arita and Vázquez-Dominguez 2008).

Yet the above model does not preclude the evolution of new species in temperate regions, and Clarke and Crame (1992) suggested that a range of brachiopods, molluscs, echinoderms and crustaceans all originated in the Eocene at high southern latitudes, and were exported to low latitudes during phases of global cooling. Now when we examine the two principal bases of high speciation – area and time – it is evident that both the temperate Australian region and the Southern Ocean have had vast areas and a long time to become engines driving speciation in their own right. Hence, in a few groups, we can now see different, or even opposite, patterns, with just as many or more species at intermediate or high latitudes compared with the tropics. Examples are the algae, bryozoans, some echinoderms (ophiuroids and holothurians), sponges and possibly brachiopods (Bolton 1994; Crame 1999). Some of these opposing patterns are due to modes of larval development. Cold waters at high latitudes favour direct development, and warmer waters at low latitudes favour planktotrophic larvae that disperse widely (Fernández *et al.* (2009). Hence, in the above examples, bryozoans, sponges and some molluscs with short-lived larvae have diversified greatly in temperate waters.

Species-pool model

This theory examines the last of the suggested causes of high diversity – habitat extent and its diversity. It predicts that, at a local scale, the larger the area of suitable habitat, the more species will be found there; it also predicts that the number of species locally will depend on the regional pool of species –that is, the richness of species at a larger scale. Hence, it relates local ecological factors to processes at a regional scale (Zobel 1997).

Goldberg and Kendrick (2005) applied the theory to the richness of southern Australian algal species. The 2nd Council of Nature Conservation Ministers Technical workshop in 1985 (CONCOM) recognised nine sub-regions from Shark Bay, Western Australia to SE Queensland. Goldberg and Kendrick (2005) found that the number of algal species recorded per sub-region was related linearly to the length of that sub-region's coastline (Fig. 3.6). This supported the view that the rich southern Australian algal flora was

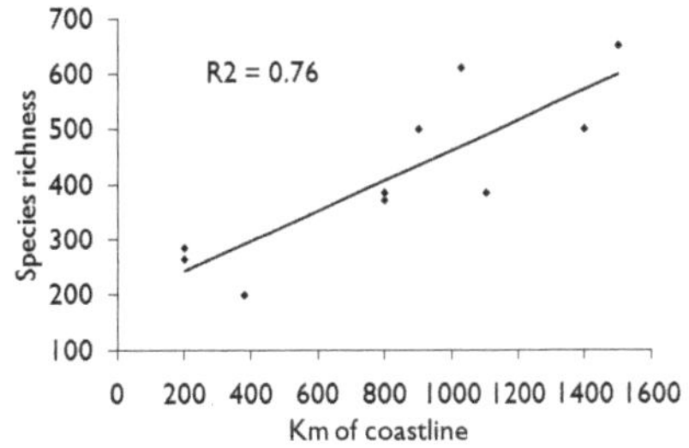

Figure 3.6. The richness of algal species for eight CONCOM regions of southern Australia plotted against length of coast (redrawn from Goldberg and Kendrick 2005).

due in part to the long length of coastline and, by implication, the greater range of habitats present. At the local scale of islands in the Recherche Archipelago, Goldberg and Kendrick (2005) also found that both the richness of regional species and local ecological factors (e.g. exposure to swell) determined the number of species present.

In summary, the high level of endemism and the richness of species of flora and fauna appear to be the outcome of many factors: Australia's long geological past; proximity to the species-rich Indo-West Pacific region; stable palaeo-environments with relatively slight temperature oscillations; benign coastal currents; absence of cataclysmic extinction events; and a long east–west coastline providing diverse habitats for the diversification of an ancient flora.

REFERENCES

Arita HT, Vázquez-Domínguez E (2008) The tropics: cradle, museum, or casino? A dynamic null model for latitudinal gradients in species diversity. *Ecology Letters* **11**, 653–663.

Bailey JC, Freshwater DW (1997) Molecular systematics of the Gelidiales: inferences from separate and combined analyses of plastid *rbc*L and nuclear SSU gene sequences. *European Journal of Phycology* **32**, 343–352.

Barnes DKA, Griffiths HJ (2008) Biodiversity and biogeography of southern temperate and polar bryozoans. *Global Ecology and Biogeography* **17**, 84–99.

Binnie MN, Cann JH (2008) The distribution of foraminifera in Gulf St Vincent – indicators of sea-level change. In: *Natural History of Gulf St Vincent* (Eds SA Shepherd, S Bryars, I Kirkegaard, P Harbison and JT Jennings) pp. 38–55. Royal Society of South Australia, Adelaide.

Bock PE (1997) Bryozoans (Phylum Bryozoa) In *Marine Invertebrates of Southern Australia Part III* (Eds SA

Shepherd and M Davies) pp. 319–394. SARDI Aquatic Sciences, Adelaide.

Bolton JJ (1994) Global seaweed diversity: patterns and anomalies. *Botanica Marina* **37**, 241–245.

Briggs DEG (1983) Affinities and early evolution of the crustacean: the evidence of the Cambrian fossils. In *Crustacean Issues I. Crustacean Phylogeny.* (Ed. FR Schramm) pp. 1–22. Balkema, Rotterdam.

Briggs DC (2003) Marine centres of origin as evolutionary engines. *Journal of Biogeography* **30**, 1–18.

Bryars S, Svane I (2008) Blue swimmer and sand crabs in Gulf St Vincent and Investigator Strait. In *Natural History of Gulf St Vincent* (Eds SA Shepherd, S Bryars, I Kirkegaard, P Harbison and JT Jennings) pp. 464–479. Royal Society of South Australia, Adelaide.

Bunt JS (1987) 2. The Australian marine environment. In *Fauna of Australia. Vol. 1A* (Eds DW Dalton and GR Dyne) pp. 17–42. Australian Government Publishing Service, Canberra.

Chapman F, Thomas DE (1936) The Cambrian Hydroida of the Heathcote and Monageeta Districts. *Proceedings of the Royal Society of Victoria* **48,** (2), 193–212.

Chin NKM, Brown MT, Heads MJ (1991) The biogeography of Lessoniaceae, with special reference to *Macrocystis* (Phaeophyta: Laminariales) *Hydrobiologia* **215**, 1–11.

Clarke A, Crame JA (1992) The Southern Ocean benthic fauna and climate change. *Philosophical Transactions of the Royal Society of London. Series B, Biological Sciences* **338**, 299–309.

Clayton MN (1994) Circumscription and phylogenetic relationships of the Southern Hemisphere family Seirococcaceae (Phaeophyceae) *Botanica Marina* **37**, 213–220.

Clements KD, Alfaro ME, Fessler JL, Westneat MW (2004) Relationships of the temperate Australasian labrid fish tribe Odacini (Perciformes; Teleostei) *Molecular Phylogenetics and Evolution* **32**, 575–587.

Coyer JA, Smith GJ, Anderson RA (2001) Evolution of *Macrocystis* spp. (Phaeophyceae) as determined by ITS1 and ITS2 sequences. *Journal of Phycology* **37**, 574–585.

Crame JA (1999) An evolutionary perspective on marine faunal connections between southernmost South America and Antarctica. *Scientia Marina* **63** (Suppl. 1) 1–14.

Crame JA (2009) Time's stamp on modern biogeography. *Science* **323**, 720–721.

Darragh TA (1985) Molluscan biogeography and biostratigraphy of the Tertiary of southeastern Australia. *Alcheringa* **9**, 83–116.

Degnan SM, Imron, Geiger DL, Degnan BM (2006) Evolution in temperate and tropical seas: disparate patterns in southern hemisphere abalone (Mollusca: Vetigastropoda: Haliotidae) *Molecular Phylogenetics and Evolution* **41**, 249–256.

den Hartog C, Kuo J (2006) Taxonomy and biogeography of seagrasses. In *Seagrasses: Biology, Ecology and Conservation.* (Eds AWD Larkum, RJ Orth and CM Duarte) pp. 1–23. Springer, Dordrecht.

Edgar GJ (1984) General features of the ecology and biogeography of Tasmanian rocky reef communities. *Papers and Proceedings of the Royal Society of Tasmania.* **118**, 173–186.

Edgar GJ (1986) Biogeographical processes in the Southern Hemisphere marine environment. In *Actas II Congreso sobre Algas Marinas Chilenas* (Ed. R Westermeier) pp. 29–46. Universidad Austral de Chile, Valdivia.

Emig CC (1997) Ecology of inarticulated brachiopods. In *Treatise on Invertebrate Paleontology Vol. 1. Part H. Brachiopoda.* (Ed. RL Kaesler) pp. 473–495. Geological Society of America and University of Kansas, Boulder, Colorado and Lawrence, Kansas.

Estes JA, Lindberg DR, Wray C (2005) Evolution of large body size in abalones (*Haliotis*): patterns and implications. *Paleobiology* **31**, 591–606.

Feldmann RM (1986) Paleobiogeography of two decapod crustacean taxa in the Southern Hemisphere: global conclusions with sparse data. In *Crustacean Issues 4: Crustacean Biogeography* (Eds RH Gore and KL Heck) pp. 5–19. Balkema, Rotterdam.

Fell HB (1962) West-wind drift dispersal of echinoderms in the Southern Hemisphere. *Nature* **193**, 759–761.

Fernández M, Astorga A, Navarette SA, Valdovinus C *et al.* (2009) Decons tructing latitudinal species richness patterns in the ocean: does larval development hold the clue? *Ecology Letters* **12**, 601–611.

Fredericq S, Freshwater DW, Hommersand MH (1999) Observations on the phylogenetic systematics and biogeography of the Solieriaceae (Rhodophyta, Gigartinales) inferred from *rbc*L sequences and morphological evidence. In *Proceedings of the 16th International Seaweed Symposium* (Eds J Jones and MT Brown) *Hydrobiologia* **389/90**, 25–28.

Geiger DL (2000) Distribution and biogeography of the recent Haliotidae (Gastropoda: Vetigastropoda) worldwide. *Bollettino Malacologico* **35**, 57–120.

Geiger DL, Owen B (2012) *Abalone: World-wide Haliotidae.* ConchBooks, Hackenheim, Germany.

Gibbons MJ, Abiahy BB, Angel M, Assuncao CML *et al.* (1999) The taxonomic richness of South Africa's marine fauna: a crisis at hand. *South African Journal of Science* **95**, 8–12.

Goldberg N, Kendrick G (2005) Linking contemporary patterns to historical events: a comparison of macroalgal diversity within the Recherche Archipelago, Western Australia, and with the south coast of Australia. In *Characterising the Fish Habitats of the Recherche Archipelago* (Eds G Kendrick and 12 co-editors) pp. 391–410.

FRDC Report No. 2001/060. University of Western Australia, Perth.

Gomon MF, Bray DJ, Kuiter RH (2008) *Fishes of Australia's Southern Coast.* Reed New Holland, Sydney.

Gordon DP (1999) Bryozoan diversity in New Zealand and Australia. In *The Conservation and Biodiversity of Invertebrates.* (Eds W Ponder and D Lunney) pp. 199–204. Royal Society of New South Wales, Sydney.

Gowlett-Holmes K (2008) *A Field Guide to Marine Invertebrates.* Notomares, Hobart.

Hearn WS (1986) Mathematical methods for evaluating marine fisheries. PhD thesis. University of New South Wales, Sydney.

Hommersand MH (2007) Global biogeography and relationships of the Australian marine macroalgae. In *Algae of Australia: Introduction.* (Eds PM McCarthy and AE Orchard) pp. 511–559. ABRS, Canberra and CSIRO PUBLISHING, Melbourne.

Hommersand MH, Fredericq S (2003) Biogeography of the marine red algae of the South African west coast: a molecular approach. In *Proceedings of the 17th International Seaweed Symposium* (Eds ARO Chapman, VJ Vreeland and IR Davison) pp. 325–336. Oxford University Press, Oxford.

Hooper JNA, Lévi C (1994) Biogeography of Indo-West Pacific sponges: Microcionidae, Raspailiidae, Axinellidae. In *Sponges in Time and Space. Proceedings of the 4th International Porifera Congress.* (Eds RWM van Soest, TMG van Kempen and JC Braekman) pp. 191–212. Balkema, Rotterdam.

Hooper JNA, Wiedenmayer F (1994) Porifera. In *Zoological Catalogue of Australia. Vol. 12.* (Ed. A Wells) pp. 1–33. CSIRO PUBLISHING, Melbourne.

Hyman LH (1959) *The Invertebrates: Smaller Coelomate Groups. Vol. V.* pp. 516–609. McGraw-Hill, New York.

Jablonski D, Roy K, Valentine JW (2006) Out of the tropics: evolutionary dynamics of the latitudinal diversity gradient. *Science* **314**, 102–105.

Jell PA, Darragh TA (1998) The fossil record. In *Mollusca: the Southern Synthesis. Fauna of Australia. Vol. 5.* (Eds PL Beesley, GJB Ross and A Wells) pp. 88–101. CSIRO PUBLISHING, Canberra).

Johnson CR, Banks SC, Barrett NS, Cazzasus F *et al.* (2011) Climate change cascades: shifts in oceanography, species' ranges and marine community dynamics in eastern Tasmania. *Journal of Experimental Marine Biology and Ecology* **400**, 17–32.

Kendrick GW, Wyrwoll K-H, Szabo BJ (1991) Pliocene-Pleistocene coastal events and history along the western margin of Australia. *Quaternary Science Reviews* **10**, 419–439.

Kendrick G, Harvey E, McDonald J, Pattiaratchi C *et al.* (2005) 'Characterising the fish habitats of the Recherche Archipelago'. FRDC Report 2001/060. University of Western Australia, Perth.

Kerswell AP (2006) Global biodiversity patterns of benthic marine algae. *Ecology* **87**, 2479–2488.

Kott P (1997) Tunicates (Sub-Phylum Tunicata) In *Marine Invertebrates of Southern Australia Part III.* (Eds SA Shepherd and M Davies) pp. 1092–1255. SARDI Aquatic Sciences, Adelaide.

Kott P (2005) New and little-known species of Didemnidae (Ascidiacea, Tunicata) from Australia (Part 3). *Journal of Natural History* **39**, 2409–2479.

Kott P (2006) Observations on non-didemnid ascidians from Australian waters (1). *Journal of Natural History* **40**, 169–234.

Kott P (2007) Taxonomic affinities of three stalked colonial species of the Ascidiacea (Tunicata) from the central coast of New South Wales and indications of a trans-Tasman connection. *Journal of Natural History* **41**, 633–645.

Kott P (2008) The Tunicata. In *Biology, Environment and Management of the Great Barrier Reef.* (Eds P Hutchings, O von Guldberg and M Kingsford). CSIRO PUBLISHING, Melbourne.

Kraft GT (1984) Taxonomic and morphological studies of tropical and subtropical species of *Callophycus* (Solieriaceae, Rhodophyta). *Phycologia* **23**, 53–71.

Larkum AWD, den Hartog C (1989) Evolution and biogeography of seagrasses. In *Biology of Seagrasses: a Treatise on the Biology of Seagrasses with Special Reference to the Australian Region.* (Eds AWD Larkum, AJ McComb and SA Shepherd) pp. 112–156. Elsevier, Amsterdam.

Last PR, White WT, Gledhill DC, and Hobday A *et al.* (2011) Long-term shifts in abundance and distribution of a temperate fish fauna: a response to climate change and fishing practices. *Global Ecology and Biogeography* **20**, 58–72.

Lindberg DR (1992) Evolution, distribution and systematics in Haliotidae. In *Abalone of the World: Biology, Fisheries and Culture.* (Eds SA Shepherd, MJ Tegner and SA Guzmán del Próo) pp. 3–18 Blackwell, Oxford.

Ludbrook NH (1980) *A Guide to the Geology and Mineral Resources of South Australia.* Government Printer, Adelaide.

Maldonado M (2006) The ecology of the sponge larva. *Canadian Journal of Zoology* **84**, 175–194.

McGowran B, Li Q, Cann J, Padley D *et al.* (1997) Biogeographic impact of the Leeuwin Current in southern Australia since the late middle Eocene. *Palaeogeography, Palaeoclimatology, Palaeoecology* **136**, 19–40.

McGowran B, Archer M, Bock B, Darragh TA *et al.* (2000) Australasian palaeobiogeography: the Palaeogene and Neogene record. *Memoirs of the Association of Australasian Paleontologists* **23**, 405–470.

Meyer C (2003) Molecular systematics of cowries (Gastropoda: Cypraeidae) and diversification patterns in the tropics. *Biological Journal of the Linnean Society. Linnean Society of London* **79**, 401–459.

Mittelbach GG, Schemske DW, Cornell HV, Allen AP *et al.* (2007) Evolution and the latitudinal diversity gradient: speciation, extinction and biogeography. *Ecology Letters* **10**, 315–331.

Morgan L, Shepherd SA (2006) Population and spatial structure of two common temperate reef herbivores: abalone and sea-urchins. In *Marine Metapopulations.* (Eds JP Kritzer and PF Sale) pp. 205–246. Elsevier, Amsterdam.

Naughton KM, O'Hara TD (2009) A new brooding species of the biscuit star *Tosia* (Echinodermata: Asteroidea: Goniasteridae), distinguished by molecular, morphological and larval characters. *Invertebrate Systematics* **23**, 348–366.

Nelson WA (1994) Distribution of macroalgae in New Zealand: an archipelago in space and time. *Botanica Marina* **37**, 221–233.

Nelson JS (2006) *Fishes of the World.* 4th edn. John Wiley & Sons, Hoboken.

O'Hara TD, Poore GCB (2000) Patterns of distribution for southern Australian marine echinoderms and decapods. *Journal of Biogeography* **27**, 1321–1335.

Olsen AM, Shepherd SA (2006) Historic drift bottle experiments show reversing surface water masses in western Bass Strait waters: implications for lobster larval dispersal. *Transactions of the Royal Society of South Australia* **130**, 113–122.

Phillips JA (2000) Systematics of the Australian species of *Dictyopteris* (Dictyotales, Phaeophyceae). *Australian Systematic Botany* **13**, 283–323.

Phillips JA (2001) Marine macroalgal biodiversity hotspots: why is there high species richness and endemism in southern Australian marine benthic flora? *Biodiversity and Conservation* **10**, 1555–1577.

Ponder WF, Wells FE (1998) Distribution and relationships of marine and estuarine fauna. In *Mollusca: the Southern Synthesis. Fauna of Australia. Vol. 5.* (Eds PL Beesley, GJB Ross and A Wells) pp. 77–82. CSIRO PUBLISHING, Canberra.

Poore GCB (1994) Marine biogeography of Australia. In *Marine Biology* (Eds LS Hammond and RN Synnot) pp. 189–212. Longman Cheshire, Melbourne.

Poore GCB (2004) *Marine Decapod Crustacea of Southern Australia.* CSIRO PUBLISHING, Melbourne.

Poore GCB, O'Hara T (2007) Marine biogeography and biodiversity of Australia. In: *Marine Ecology* (Eds SD Connell and BM Gillanders) pp. 177–198. Oxford University Press, Melbourne.

Powell MG (2007) Latitudinal diversity gradients for brachiopod genera during the late Palaeozoic time: links between climate, biogeography and evolutionary rates. *Global Ecology and Biogeography* **16**, 519–528.

Primo C, Vázquez E (2008) Zoogeography of the southern New Zealand, Tasmanian and southern African ascidian fauna. *New Zealand Journal of Marine and Freshwater Research* **42**, 233–256.

Randall JE (1981) Examples of anti-tropical and anti-equatorial distribution of Indo-West Pacific fishes. *Pacific Science* **35**, 197–209.

Renema W, Bellwood DR, Braga JC, Bromfield K *et al.* (2008) Hopping hotspots: global shifts in marine biodiversity. *Science* **321**, 654–657.

Richardson JR (1994) Origins and dispersal of a brachiopod family—the systematics, biogeography and evolution of the family Terebratellidae. *Proceedings of the Royal Society of Victoria* **106**, 17–29.

Richardson JR (1997) Brachiopods (Phylum Brachiopoda) In *Marine Invertebrates of Southern Australia Part III.* (Eds SA Shepherd and M Davies) pp. 999–1027. SARDI Aquatic Sciences, Adelaide.

Rudwick MJS (1970) *Living and Fossil Brachiopods.* Hutchinson University Library, London.

Serrão EA, Alice LA, Brawley SH (1999) Evolution of the Fucaceae (Phaeophyceae) inferred from nrDNA-ITS. *Journal of Phycology* **35**, 382–394.

Sastry AN (1983) Pelagic larval ecology and development. In *The Biology of Crustacea. Vol. 7. Behavior and Ecology.* (Eds FJ Vernberg and WB Vernberg) pp. 214–282. Academic Press, New York.

Sorokin S, Fromont J, Currie D (2007) Demosponge biodiversity in the benthic protection zone of the Great Australian Bight. *Transactions of the Royal Society of South Australia* **132**, 192–204.

Southcott RV (1982) Jellyfish (Classes Scyphozoa and Hydrozoa) In *Marine Invertebrates of Southern Australia. Part I.* (Eds SA Shepherd and IM Thomas) pp. 115–159. South Australian Government Printer, Adelaide.

van Soest RWM (1994) Demosponge distribution patterns. In *Sponges in Time and Space. Proceedings of the 4th International Porifera Congress.* (Eds RWM van Soest, TMG van Kempen and JC Braekman) pp. 213–223. Balkema, Rotterdam.

Vermeij GT (1991) When biotas meet: understanding biotic interchange. *Science* **253**, 1099–1104.

Walker DI (1991) The effect of sea temperature on seagrass and algae on the western Australian coastline. *Journal of the Royal Society of Western Australia* **74**, 71–77.

Waters JM (2004) Out of Africa: the slow train to Australasia. *Systematic Biology* **53**, 18–24.

Waters JM (2008) Driven by the West Wind Drift? A synthesis of southern temperate marine biogeography, with new directions for dispersalism. *Journal of Biogeography* **35**, 417–427.

Waters JM, Roy MS (2003a) Global phylogeography of the fissiparous sea-star genus *Coscinasterias. Marine Biology* **142**, 185–191.

Waters JM, Roy MS (2003b) Marine biogeography of southern Australia: phylogeographical structure in a temperate sea-star. *Journal of Biogeography* **30**, 1787–1796.

Waters JM, O'Loughlin PM, Roy MS (2004) Cladogenesis in a starfish species complex from southern Australia: evidence for vicariant speciation? *Molecular Phylogenetics and Evolution* **32**, 236–245.

Waters JM, King TM, O'Loughlin PM, Spencer HG (2005) Phylogeographical disjunction in abundant high-dispersal littoral gastropods. *Molecular Ecology* **14**, 2789–2802.

Waters JM, McCulloch GA, Eason JA (2007) Marine biogeographical structure in two highly dispersive gastropods: implications for trans-Tasman dispersal. *Journal of Biogeography* **34**, 678–687.

Watson JE (1982) Hydroids (Class Hydrozoa) In *Marine Invertebrates of Southern Australia. Part I.* (Eds SA Shepherd and IM Thomas) pp. 77–114. South Australian Government Printer, Adelaide.

Wiedenmayer F (1989) Demospongiae (Porifera) from northern Bass Strait, southern Australia. *Memoirs of the Museum of Victoria* **50**, 1–242.

Wilson BR, Allen GR (1987) Major components and distribution of marine fauna. In *Fauna of Australia. Vol.1A. General Articles.* (Ed. GW Dyne) pp. 43–68. Australian Government Publishing Service, Canberra.

Wilson BR, Clarkson P (2004) *Australia's Spectacular Cowries: a Review and Field Study of Two Endemic genera:* Zoila *and* Umbilia.. Odyssey Publishing, El Cajon.

Womersley HBS (1990) Biogeography of Australasian marine macroalgae. In *Biology of Marine Plants* (Eds MN Clayton and RJ King) pp. 368–381. Longman Cheshire, Melbourne.

Womersley HBS (2003) *The Marine Benthic Flora of Southern Australia: Rhodophyta - Part IIID.* ABRS, Canberra, and State Herbarium of South Australia, Adelaide.

Zobel M (1997) The relative role of species pools in determining plant species richness: an alternative explanation of species coexistence? *Trends in Ecology & Evolution* **12**, 266–269.

PART 2

THE ALGAE

Benthic macroalgae dominate upward-facing and sloping rocky bottoms in the photic zone throughout southern Australia. In the next four chapters, we consider the ecology of the different algal groups:

- **Canopy species.** These are of two types – the kelps, which are usually dominant on exposed coasts, and fucoid algae, which are usually dominant in less exposed conditions. Historically, the term 'kelp' has been used to describe both laminarian kelps and fucoids (Fraser 2012), but here we shall follow the common modern practice, and restrict the term kelp mainly to the laminarians, and the term fucoids to the Fucales.
- **The algal middle stratum/understorey.** This comprises green, brown and red algae living under the canopy or in gaps between patches of canopy species. The term 'algal turf' generally refers to filamentous algae a few centimetres high, which are common in gaps in the canopy.
- **Low-light algal assemblages**, mainly of red algae living on shaded surfaces or in deeper water, with which are included some encrusting and prostrate species.

Overall, the ecology of very few species of the rich algal flora is at all known, and most of these live in shallow water.

Other common terms used by marine ecologists describe subtidal zonation patterns, following the terminology developed by early European workers (see Shepherd and Womersley 1970). They distinguished upper, mid- and lower sublittoral zones to describe vertical algal zonation patterns. The upper zone is marked by algal turf, or the bull-kelp *Durvillaea* in exposed conditions, and the mid- and lower zones are dominated by brown and red algae, respectively. This zonation pattern does not apply to the whole southern region –it may vary from only two zones to as many as four – and, in any event, the terms are less used now, because ecologists have focused increasingly on habitats.

Habitats are defined as easily recognisable structural, biological or environmental conditions that affect a suite of species. Substratum type and dominant life forms are usually used to describe them (Ward *et al.* 1999). Species co-existing in a habitat may be called an assemblage or a community. The latter term is usually preferred where species interact, such as where structure is provided by a dominant species (e.g. an '*Ecklonia* community', as in Plate 3) and the former term where no such structure or interaction is implied, as in a 'red algal assemblage'.

At the algal assemblage level, algae can be conveniently categorised by their strata for descriptive purposes (see Figure). A floating canopy layer, with fronds buoyed up by gas floats, is provided by *Macrocystis* in the cooler waters of Victoria, Tasmania and SE South Australia, but is absent elsewhere (Plate 4). Below this floating canopy, there may be a second canopy layer of kelp or fucoid algae, usually 40 cm to 2 m in height and, in the absence of a floating canopy, this is termed the upper storey or stratum. Below this layer is a middle stratum, often called an understorey, of green, brown and/or red algae to 20–40 cm high, whose density depends on the light reaching the habitat. Hence, persistent *Macrocystis* beds will have a much sparser understorey than an ephemeral bed. The lowest strata are provided by an algal turf, usually filamentous algae 1–2 cm high, and lastly a prostrate level of encrusting or flattened species.

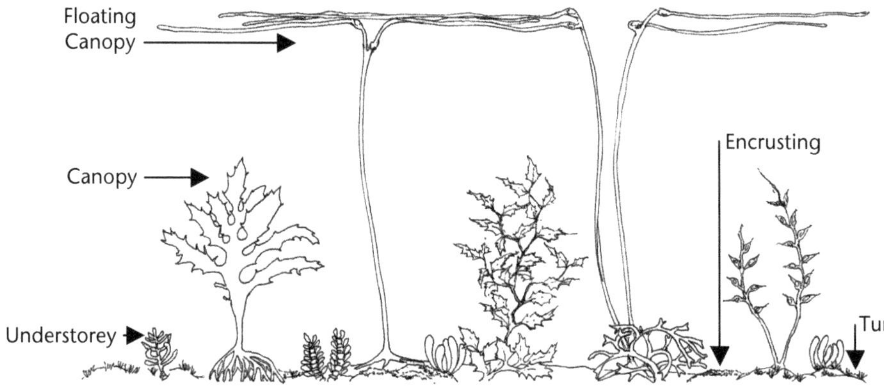

Schematic drawing of the five possible layers or strata of a macroalgal assemblage. (Artwork by Margaret Collings; after Turner and Collings 2008.)

REFERENCES

Fraser CI (2012) Is bull-kelp kelp? The role of common names in science. *New Zealand Journal of Marine and Freshwater Research* **46,** 279–284.

Shepherd SA, Womersley HBS (1970) The sublittoral ecology of West Island, South Australia. 1. Environmental features and the algal ecology. *Transactions of the Royal Society of South Australia* **94,** 105–138.

Turner DJ, Collings GJ (2008) Subtidal macroalgal communities of Gulf St Vincent. In: *Natural History of Gulf St Vincent* (Eds SA Shepherd, S Bryars, IR Kirkegaard, P Harbison and JT Jennings) pp. 264–278. Royal Society of South Australia, Adelaide.

Ward TJ, Vanderklift MA, Nicholls AO, Kenchington RA (1999) Selecting marine reserves using habitats and species assemblages as surrogates for biological diversity. *Ecological Applications* **9,** 691–698.

4 The kelps

'...The number of living creatures of all orders, whose existence intimately depends on the kelp, is wonderful.....I can only compare these great aquatic forests of the southern hemisphere, with the terrestrial in the inter-tropical regions. Yet if in any country a forest was destroyed, I do not believe nearly so many species of animals would perish as would here, from the destruction of the kelp. Amidst the leaves of this plant numerous species of fish live, which nowhere else could find food or shelter.'

Charles Darwin, *The Voyage of the Beagle*, 1860

OVERVIEW

Four species of kelp, including one introduced species, are large, canopy-forming brown algae in the Order Laminariales. They are often dominant in shallow waters in southern Australia. The smallest of these, *Ecklonia radiata*, has a broad southern range throughout the region from the Abrolhos Is, Western Australia, around Tasmania, and north to southern Queensland. The sporophyte's growth (elongation) rate ranges from 1 to 2.6 mm day^{-1} and its biomass production from 0.3–1.5 g dw day^{-1} (dry weight per day). Erosion of the blade is continuous but variable, and is increased by storm events. Growth rates vary with depth, temperature, light, blade length and nutrient supply. Individuals live for 1–5 years with half-lives ranging from 90–700 days. Spores settle on reefs throughout the year, sometimes with a seasonal peak and, after settlement, juveniles can reach full size in 6–9 months. Kelps can acclimate to different light regimes by increasing their photosynthetic efficiency, and achieve a more or less constant productivity throughout the year. The rate of carbon uptake is high, but only ~10% is converted to biomass – the rest being respired, exuded as mucus, released as spores or lost through erosion. The biomass of *Ecklonia* ranges from high levels of ~13 kg fresh weight m^{-2} down to average values of 2–4 kg m^{-2}, and the kelp grows variously from the shallow subtidal to 25–48 m depth on exposed coasts of much of southern Australia.

Two forms of *Macrocystis pyrifera* occur in southern Australia: the *angustifolia*-form in Bass Strait and SE South Australia, and the *pyrifera*-form in SE Tasmania. Growth of the latter form is rapid, and has a high productivity of 2.4 kg dw m^{-2} day^{-1} (= 24 kg fresh weight m^{-2} day^{-1}) at a density of four individuals per square metre. *Macrocystis* only grows in the region where temperatures are < 19°C, and has minimum nutrient (nitrogen) requirements of 0.1–0.4 µM; hence, it is dependent on specific local oceanographic conditions, such as upwelling zones for nutrients. *Macrocystis* is also vulnerable to storms, which can destroy whole forests, which may suddenly disappear after a storm or strong swell. Recruitment depends on transport of spores over long distances or on spores released by drifting algae. The *pyrifera*-form in SE Tasmania has seriously declined in area since the 1940s, and re-afforestation trials have had little success. *Macrocystis* is the most productive alga globally, and is critically important as a foundation species in coastal reef ecosystems, supporting a high diversity of species via the detrital food web.

Undaria pinnatifida was introduced to SE Tasmania in the 1980s, and has spread over > 300 km of coast, as well as to Port Phillip Bay, Victoria. It is a seasonal, rapidly growing, opportunistic species that appears in early spring, and dies after 6–9 months. It rapidly colonises disturbed areas, but does not outcompete native algal forests. *Undaria* has a broad temperature range of 10–20°C, and could spread throughout southern Australia over time. *Lessonia corrugata* is a long, strap-like, cool-temperate kelp, largely limited to southern Tasmania, that grows below the *Durvillaea* zone on open coasts. Its ecology is unknown, but that of related species in New Zealand and Chile suggest that it has a high growth rate of up to 10 mm day^{-1} and a net productivity of ~40 g dw m^{-2} day^{-1}.

INTRODUCTION

Large, canopy-forming brown algae are highly productive subtidally, and dominate many habitats throughout southern Australia. Like forests on land, they create a complex structure on reefs, providing habitat for understorey species, which may form a middle algal stratum, and for a lower stratum of prostrate or crustose species. One group of brown algae, the kelps in the Order Laminariales, is represented by three native, and one introduced, species in southern Australia. These are: giant kelp *Macrocystis pyrifera*, with two forms *pyrifera* and *angustifolia*; strapweed *Lessonia corrugata*; common kelp *Ecklonia radiata*, and the introduced *Undaria pinnatifida*. Only *Ecklonia* is widespread throughout the southern region, with the others being restricted to the cooler waters of SE Australia.

In this chapter, we consider the ecology of each of the above kelps in turn, starting with *Ecklonia*, which is a common canopy species in terms of cover and biomass over its geographic range, and has been the most extensively studied of the above species. The common kelp is the dominant canopy-forming species on the lower Western Australian coast, but a geographic transition occurs from these kelp-dominated to fucalean-dominated assemblages (see Chapter 5) in southern Australia (Wernberg *et al.* 2003). A further transition back to kelp-dominated communities then occurs from far eastern Victoria up the east Australian coast. For general references on the common kelp see Stewart *et al.* (1961), Mann and Kirkman (1981), Larkum (1986), Kirkman (1981, 1984, 1989), Hatcher *et al.* (1987), Santelices (1990), and Steinberg and Kendrick (1999).

THE COMMON KELP *ECKLONIA RADIATA*

The common kelp (Plate 3) has a range extending southwards from the Houtman Abrolhos Is on the west coast, around the southern Australian coast, including Tasmania, and up the east coast to southern Queensland (a maximum temperature range from 8–24°C). *Ecklonia* also occurs in South Africa, Oman and New Zealand. The depth range varies according to average water clarity, but in the clear ocean waters of the western coast of Australia and the Great Australian Bight the maximum depths recorded are ~50 m (Shepherd 1979; Shepherd and Womersley 1976; Staehr and Wernberg 2009) and 70 m off Tasman I., SE Tasmania. However, the species likely extends into even deeper water, as off Lord Howe I. where it is confined to deep water habitat.

The kelp grows on rocky bottoms, attached by a holdfast, with a stipe of variable length, and one primary and many secondary laminae (Fig. 4.1). The alga's growth form varies around southern Australia, and also with depth and water movement; stipes are longer and laminae smaller in eastern Australia compared with southern Australia (see Chapter 6). Some forms, notably those at sheltered sites, have spines growing over their laminae, while other forms are smooth. Presumably the spines aid gas and nutrient exchange in still conditions. With increasing depth, the primary lamina (or blade) tends to be longer, and the secondary laminae become fewer and shorter. This is shown by the data given in Table 4.1, summarising growth studies of the kelp, where, overall, the mean blade length was correlated with depth over a broad depth range of 1.5–21 m (r = 0.800; P = 0.005). However, this relation is not invariable, and in upper Western Port, where turbidity is moderate to high, the blades of shallow water *Ecklonia* at 3 m depth were on average 45% longer (max. length 1.5 m) than plants in deeper water – the longest recorded for mainland Australia, although in SE Tasmania they can exceptionally reach 2 m (Edgar 2008; Shepherd *et al.* 2009).

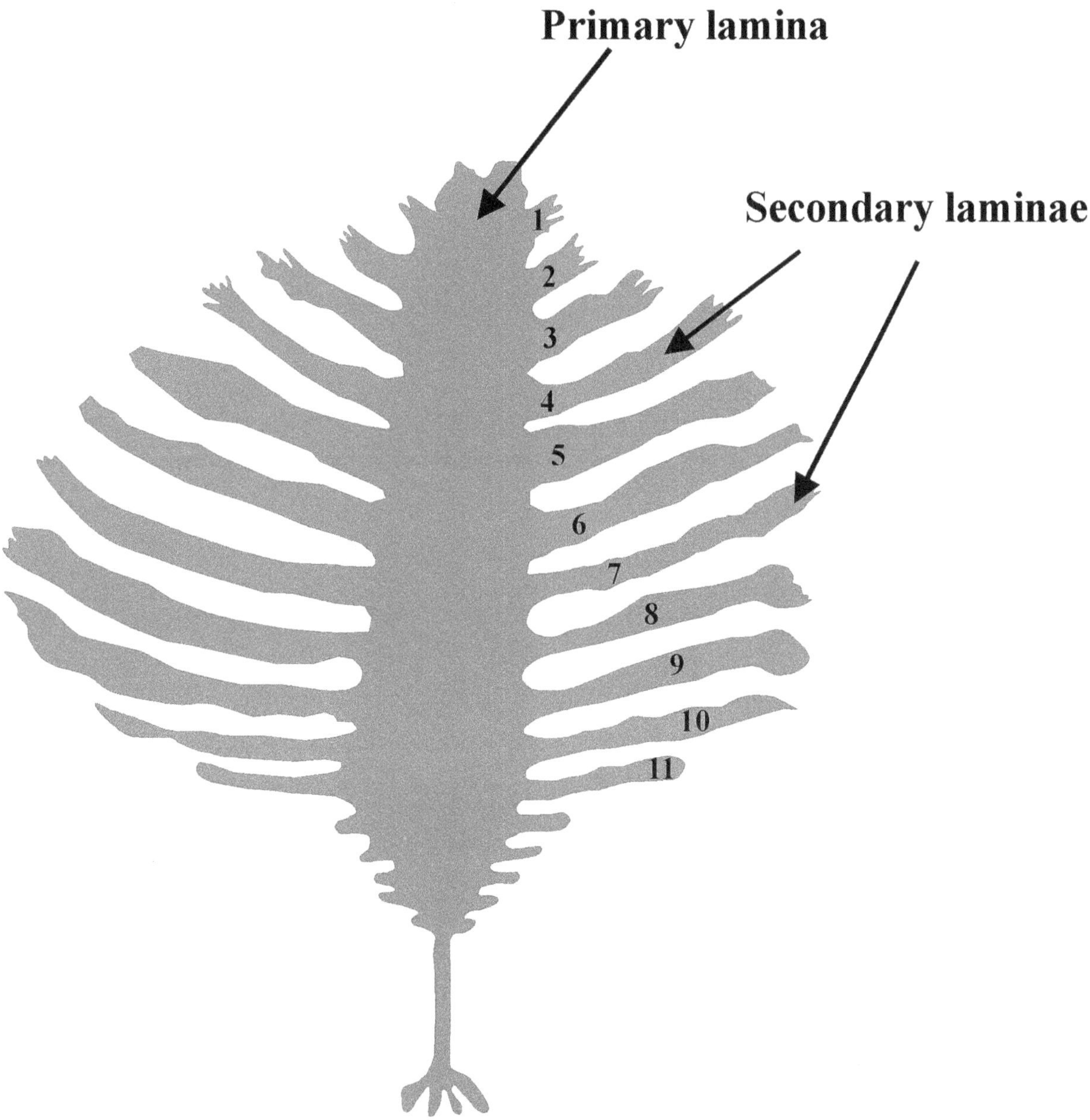

Figure 4.1: Sporophyte of *Ecklonia radiata*, showing the primary lamina or blade, and secondary laminae (courtesy Janine Baker).

Table 4.1: Mean growth rates, biomass increase, and survival of *Ecklonia radiata* sporophytes at sites around southern Australia, from west to east, with data on mean length of primary lamina, size and stipe length, with environmental data (annual temperature range, exposure and depth).

n.d. = no data. Numbers in brackets in Column 1 are cited authorities, and numbers in brackets in Columns 2, 3, 5, and 6 are standard errors.

Site (Authority)	Lamina length (cm)	Growth rate (mm day^{-1})	Biomass inc. rate (g dw day^{-1})	Stipe length (cm)	Algal density (no. m^{-2})	Survival (half life in days)	Environment: depth – m; exposure to swell; temp range
Abrolhos I. (site W), WA[1]	38(5)	1.3(0.2)		5.7(0.8)	0.6	272	2–5 m; Mod.; 20–25°C
Abrolhos I.(siteL)[1]	26(5)	0.75(0.1)		4.6(0.9)	0.06	90	2–5 m; Mod.; 17–23°C
Marmion Reef,WA[2]	50(3)	1.8(0.2)	0.5	7.4(1.4)	24.5(3.3)	~270[b]	5 m; Mod.; 16–24°C
Marmion Reef, WA[2]	26(5)	1.7(0.4)	0.4	7.3(0.7)	11.8(3.2)	n.d.	10 m; Mod.; 16–24°C
Five FathomBank, Garden I., WA[2]	74(13)	2.6(0.5)		10.1(3.3)	0.6(0.1)	n.d.	27 m; Mod.; 16–24°C
West I., SA[3]	36[a]	1.7(0.5)	0.6(0.2)	7[a]	14.3(1.3)	n.d.	3 m; sheltered; 13–20°C
West I., SA[3]	42[a]	1.4(0.4)	0.46(0.2)	17[a]	3.9(0.6)	n.d.	10 m; sheltered tomod.; 13–20°C
Cape Northum-Berland, SA[4]	72(6)	2.1(0.2)	0.2(0.03)	11.7(0.6)	0.04	326	15 m; Mod.; 12–17°C
George III Reef, Tas.[5]	78(2)	2.2(0.1)	1.0(0.1)	8.3(0.1)	0.7(0.1)	690	14 m; Mod. 11–17°C
Bicheno,Tas.[5]	55(6)	2.0(0.2)	1.5(0.2)	11.0(1.0)	3.9(0.6)	300	10 m; Mod. 12–20°C
Fairlight Bay, NSW[6]	34(7)	1.3(0.4)	0.3(0.1)	8.9(0.8)	22.5(0.3)	309[c]	1.5 m; sheltered; 14–24°C

[a]estimated by SAS from other data; [b]Hatcher *et al.* (1987); [c]from Andrew (1991) for a nearby site. [1]Hatcher *et al.* (1987); [2]Kirkman (1981, 1984, 1989); [3]Fairhead (2001); [4]Shepherd (1979), unpublished data; [5]Sanderson (1990); [6]Larkum (1986).

Ecklonia has a complex life cycle with alternating generations of a sporophyte – the conspicuous subtidal form of the kelp – and a microscopic form, the gametophyte. The sporophyte produces microscopic zoospores from sori scattered over the blade. After release, the zoospores settle onto reefs, and develop into microscopic male and female organisms, called gametophytes. These are far too small to be seen with the naked eye, but have been cultivated in the laboratory, where they produce gametes (sperm and ova) after 10–30 days (Novaczek 1984a). After fusion, the gametes sink to the bottom and grow into sporophytes. Gametophytes in the laboratory survived over a temperature range of ~8–24°C, with a slightly narrower optimum temperature tolerance, a range which matches the Australian geographic distribution of the kelp.

The dispersal capabilities of *Ecklonia* spores or gametes are not known. However, genetic studies indicate strong genetic differences at small scales, with genetic differences increasing with distance in accordance with an isolation-by-distance model (Coleman *et al.* 2009). Hence, spore dispersal is likely to be localised, and drifting sporophytes an important mechanism for dispersal.

Growth

Fronds

The growth of the common kelp has been examined at several sites over its geographic range, using the hole-punching method of Mann and Kirkman (1981). The kelp is essentially a moving conveyor belt, with blade growth occurring at the base of the primary lamina, and erosion happening at the tips of the primary and secondary laminae (Fig. 4.1). So, if small holes are punched through the primary lamina some 10 cm from its base in tagged individuals, and the movement of the hole from the base measured over time, growth (in terms of length and biomass increase) can be determined. However, this method cannot be used for very small recruits, and for these Hugh Kirkman (1981) fixed tags to their holdfasts. He found that those that had settled in deep shade under the kelp canopy grew very slowly, if at all, but when a gap appeared in the canopy, growth was rapid. In such gaps, at a depth of 5 m, small individuals to 15 cm long and 1–2 weeks old could grow to ~70 cm in 55 days, and reach the full size of 90 cm in 8–10 months. Similarly, Toohey and Kendrick (2007) found that juveniles reached full size in 6 months. Yet, in deeper water the situation may be very different and, at one site at 15 m depth in SE Australia, a 5-month-old plant 15 cm long took another 17 months to reach a length of 90 cm (Shepherd 1979).

Stipes and holdfasts

Throughout southern Australia, *Ecklonia* stipes range from ~6 to 11 cm long, with stipes tending to be ~10 cm long in New South Wales and Tasmania, and ~7 cm long elsewhere (Hatcher *et al.* 1987; Sanderson 1992; Fowler-Walker *et al.* 2005). In New South Wales, Tony Larkum (1986) noted that stipe growth was continuous throughout the year, and that on average stipes were ~9 cm long on algae ~3 years of age. Haptera (i.e. the roots of the holdfast) were added in mid-summer. At Cape Northumberland, South Australia, Shepherd (1979) found that stipe growth was irregular, but that they grew 2–3 cm in the first 3 months after settlement, and up to ~8 cm in one year, and little after that. In SE Tasmania, stipe lengths in deep water may be > 1 m. Craig Sanderson (1992) found that after the first year or two, stipes of tagged algae virtually ceased growing, suggesting that stipe length is genetically determined.

Factors affecting blade growth

The results of various studies on *Ecklonia* sporophytes, summarised in Table 4.1, show much variation in algal size, density, survival and growth rate over the geographic range of the species. The mean elongation rate of fronds was highly correlated with the mean normalised productivity (i.e. mg g^{-1} fresh weight day^{-1}) ($r = 0.91$; $P < 0.001$), confirming the value of measuring growth in terms of elongation rate (Hatcher *et al.* 1987).

The growth rates given in Table 4.1 varied with several environmental factors, including light, temperature and nutrients, which change seasonally and with depth. In addition, algae can seasonally store carbon as a reserve and later remobilise it for growth. The seasonal changes in growth rate recorded for eight studies are shown in Fig. 4.2, and, in some of them, the likely factors influencing growth rate were inferred, as discussed below.

Temperature

The Abrolhos Is are at the northern limit of the kelp's geographic range in Western Australia. Hatcher *et al.* (1987) found that the growth rate there was seasonally low from January to June, when sea temperatures were > 23°C, and that for the whole year the growth rate was inversely correlated with temperature. Further south at Marmion Reef, Hugh Kirkman (1984) found that growth was strongly seasonal, peaking in November and declining to low levels in May–Sept. Only during summer, as sea temperatures rose above 18.5°C, was growth inversely correlated with temperature. At temperatures < 18.5°C growth rates were high. A subsequent comparison of growth rates between two depths, 5 and 10 m, showed a similar seasonal spring peak in growth at 5 m, but a summer peak 2 months later at the 10 m site (Kirkman 1989). At Cape Northumberland, where temperatures were always < 18°C, growth was optimal in summer. At West I., South Australia, where sea temperatures are 20–21°C from December to March, growth rates also declined in summer both at 5 and 10 m depth, although without a phase shift as at Marmion Reef (Fairhead 2001; Fairhead and Cheshire 2004a, b).

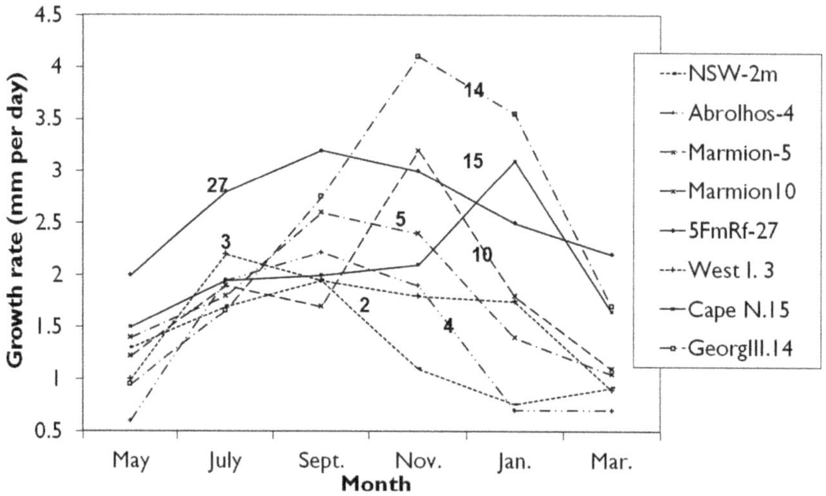

Figure 4.2: Seasonal growth rates of *Ecklonia radiata* at southern Australian sites. Depths in metres for which data obtained are given in bold. Data were extracted from Larkum (1986) for 'NSW-2m'; Hatcher *et al.* (1987) for 'Abrolhos-4' (Site W), 'Marmion-5', 'Marmion10', and '5FmRf-27' (Five fathom Reef); Fairhead (2001) for 'West I. 3'; and Shepherd (1979) for 'Cape N. 15' (i.e. Cape Northumberland).

The latter authors concluded, however, that one or more of several factors could be implicated in lower summer growth rates – either higher temperatures, low nitrate or the accumulation of reserve carbohydrate in the tissue, as noted by Stewart *et al.* (1961). Similar considerations apply to the trend of seasonally increasing growth rates with declining sea temperature, as shown in Fig. 4.2. The physiological mechanisms underlying these changes were due to changes in the kelp's metabolic balance (Staehr and Wernberg 2009). With increasing temperature, photosynthetic rates declined 50%, respiration rates 90% and, to compensate, the kelp increased its critical light demand. A conclusion from these studies is that temperatures above ~18°C become suboptimal for *Ecklonia*, and that 23°C is a critical threshold, above which growth is slow, tissue necrosis increases and algal survival is compromised.

Blade length

In the deep-water study off Cape Northumberland over > 2 years, Shepherd (1979) found that growth (elongation) rates were highest in summer and lowest in winter, with blade length accounting for 32% of the variation in growth, and temperature 8%. The following regression equation was fitted to growth rate data:

$$G = 0.725 + 0.0215\,L - 0.0465\,t$$
$$(r^2 = 0.37;\ P < 0.001)$$

where G = growth (elongation) rate in mm day^{-1}, L = mean length in cm, and t = temperature (°C). As light reaching the habitat and nutrients were highest in summer, the latter due to the seasonal upwelling, the role of temperature (which rarely exceeds 18°C) seemed minimal. A correlation between growth (elongation) rate and blade length is common to all kelps, due to the increased photosynthetic area available in longer blades. *Ecklonia* conformed to this pattern across southern Australia (Table 4.1), with the data showing a strong correlation between mean blade length and growth rates ($r = 0.827$; $P = 0.003$).

Light and swell

Craig Sanderson's (1992) study at two sites in Tasmania (George III Reef and Bicheno – see Fig. 4.13) showed strong seasonality in blade growth, with maxima in summer and minima in winter. At George III Reef, three factors affected growth rates – number of daylight hours (L_D), sea temperature (T) in °C at 20 m depth, and mean swell height (S_W). At Bicheno L_D was the only significant factor. The equations for the two sites were:

George III Reef:

$$G = 0.69 + 0.065\,L_D - 0.037\,T - 0.118\,S_W$$
$$(r^2 = 0.78;\ P = 0.0001)$$

Bicheno: $$G = -0.016 + 0.035\,L_D$$
$$(r^2 = 0.529;\ P = 0.017)$$

where G is the growth (elongation) rate in mm day^{-1}.

At these sites Craig Sanderson concluded that light (L_D) was the most important factor affecting the kelp's productivity, and that the negative effect of swell height (S_W), often associated with storms, was probably due to the fact that turbidity increased and reduced the amount of light reaching the habitat. The negative effect of temperature was minimal at George III Reef and non-existent at Bicheno.

Depth

A plot of average daily growth rates versus depth for all sites (Fig. 4.3) shows a trend of increasing growth rates, in terms of elongation rate, with depth. This is supported by the seasonal growth data (Fig. 4.2), which shows that seasonal growth curves for the three deepest sites peak at higher levels than the other sites. The increased growth rate with depth is in part a reflection of the changing morphology of the alga with depth. As discussed above, algae are longer and secondary laminae shorter in deeper individuals than in shallower ones. This is shown by the plot of biomass per cm length of the alga, which decreases with depth (i.e. with increasing depth, more growth is put into the primary lamina than into secondary laminae) (Fig. 4.3). A longer primary lamina, together with lower algal density with increasing depth means that self-shading becomes less likely. Hence, the changing growth form of the kelp with depth maximises the amount of light absorbed. Annual algal productivity, defined as daily biomass increase, is known for only a few sites (Table 4.1) and, as Sanderson (1992) concluded, there seems to be no significant trend with depth. The influence of depth on photosynthesis is further discussed below.

Nutrient uptake and storage of reserves

Neither the uptake nor storage of nutrients by *Ecklonia*, nor their relation with growth, has been specifically investigated in any study. The input of nutrients into coastal waters is seasonal or sporadic at probably all the southern sites, and related to local rainfall, river flow or upwelling events. At Marmion Reef, high growth rates in winter and spring coincided with higher nutrient levels, released during the seasonal high wave energy period, suggesting that nutrients were probably limiting for the rest of the year (Kirkman 1984; Lourey and Kirkman 2009). Similarly, Fairhead and Cheshire (2004a) suggested that the poor autumn growth rate at West I. was due to the depletion of nutrient levels in the water and/or in the kelp's tissues in that season, because both algal growth and nutrient levels increased markedly in winter with guano run-off from the adjoining land (Fig. 4.2).

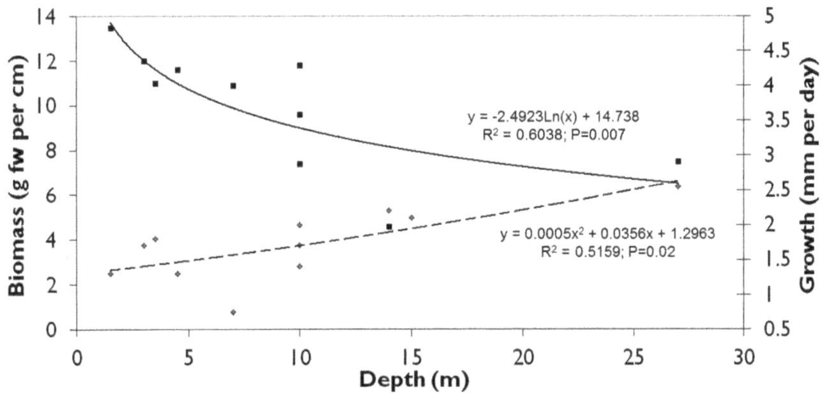

Figure 4.3: Plot of changes in mean growth rate (mm day^{-1}) versus depth (dashed line), and changes in biomass of primary and secondary laminae (g fresh wt cm^{-1}) versus depth (continuous line) for the 10 sites listed in Table 4.1.

In summer to autumn of each year, *Ecklonia* transfers carbon taken up during photosynthesis for storage as the two carbohydrates – laminaran and mannitol – in the fronds and stipe, and these can be remobilised later. Stewart *et al.* (1961) found that these storage reserves fell from ~50 to 27% of the kelp's dry weight in August and September, but were restored by the following June. Winter–spring is a period of rapid or increasing growth at many sites (Fig. 4.2), and Larkum (1986) concluded that some of the frond growth in Fairlight Bay, New South Wales, was due to the remobilisation of these reserves. Fairhead and Cheshire (2004a) also concluded that the higher winter–spring growth rate of the kelp at West I. was due to the use of these reserves.

Erosion of the blade

The primary and secondary laminae erode mainly at the tips – the oldest parts of the respective laminae – and this erosion goes on throughout the year, although not at a constant rate. Here, erosion includes breakage of the frond, grazing and bacterial action causing necrosis of the tissue. Loss of tissue by erosion was greatest in summer (October–March) in the Abrolhos Is, but in winter (April–July) on Marmion Reef, probably due to storms (Hatcher *et al.* 1987; Kirkman 1984, 1989). Similarly, at West I., Fairhead (2001) attributed the seasonal decline in weight of the kelp to storm events. In Tasmania, Sanderson (1992) found that loss by erosion was greatest in summer, and slightly higher at Bicheno than at George III Reef – a difference attributed by him to the shallower depth of the kelp, and greater abrasion at the former site.

The causes of erosion were also examined by Shepherd (1979) at Cape Northumberland. There, blade length and a storm index, based on swell height and sea state, had significant effects, but temperature had no effect. The following multiple regression equation was fitted to the data:

$$E = -0.21 + 0.016\, L_B + 0.040\, S$$
$$(r^2 = 0.136;\ P < 0.001)$$

where E = erosion rate in mm day^{-1}, L_B = mean blade length in cm and S = storm index. In this regression, blade length accounted for 10% of the variance and the storm index 4%. The tendency for increased blade erosion in longer kelps is likely the combined result of their greater age and a whiplash effect in severe surge conditions.

Photosynthesis and growth

An understanding of the growth of *Ecklonia* was greatly expanded by the work of Fairhead and Cheshire (2004b) who used a photo-respirometer (Cheshire Systems), consisting of sealed Perspex chambers and accompanying apparatus, to measure photosynthesis and carbon uptake of whole kelps. As the water in the chamber is changed hourly, differences in ambient temperature, light and nutrients are taken into account in the measurements. The growth of marine kelps is complex. The light environment changes daily and seasonally, and kelps have evolved the ability to change their photosynthetic apparatus to acclimate to different light levels. Anne Fairhead found that algae increased their photosynthetic capacity in winter, but not in spring, and suggested that there was an interactive effect of irradiance and temperature on the kelp's photosynthetic capacity – that is, the alga required both low light and low temperature to maximise photosynthesis. In this way, the kelp was able to perform optimally throughout the year in terms of productivity. So, although light levels were up to five times higher and day-length was 3 hours longer in summer than winter, productivity changed very little. In deeper water, the kelp had the same response – their photosynthetic efficiency was higher, and saturating irradiance levels were lower, so they could acclimate to lower light levels. From all this work, Fairhead (2001) showed that productivity (Fig. 4.4) varied little with season, despite a dramatic seasonal decline in light at 3 m depth, but less so at 10 m depth. However, some of the assimilated carbon from photosynthesis is used for maintenance, some is released in spores during reproduction, some is exuded as mucilage (to become dissolved organic matter), and some is eroded at the tips and taken up by secondary consumers. Lastly, some is stored in the alga for later transfer into growth, while the rest goes directly into growth. Hence, the classical hole-punching method estimates only a small fraction (~11%) of the total carbon that is taken up by the kelp, as shown in Fig. 4.5.

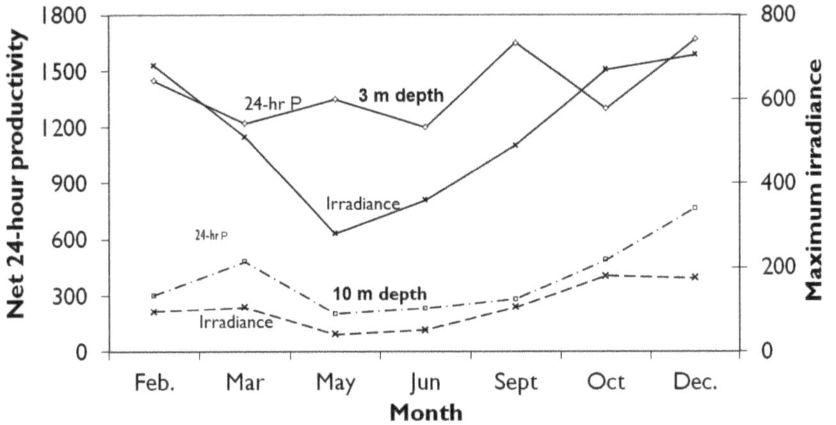

Figure 4.4: Seasonal change in mean daily productivity (24-h P) of *Ecklonia radiata*, and change in average maximum irradiance, at 3 m depth (continuous lines), and 10 m depth (dashes, dots) at West I. Productivity values (heavy lines) are in μmol O_2gwt day^{-1}, and irradiance values (lighter lines) are μmol photons m^{-2} s^{-1} (redrawn from Fairhead and Cheshire 2004b).

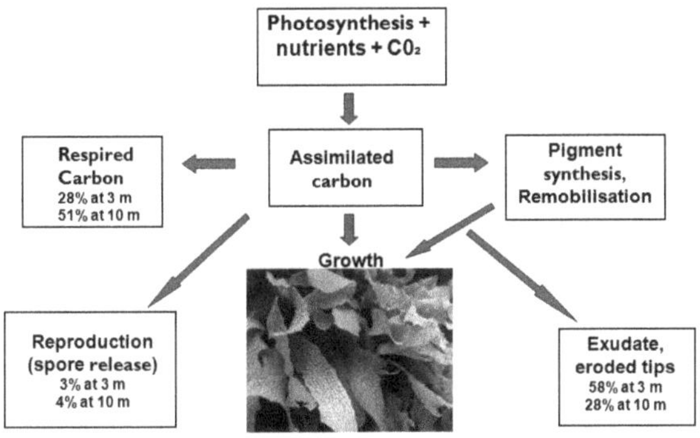

Figure 4.5: Annual carbon flow in an *Ecklonia radiata* population. Estimated percentages of assimilated carbon used in processes other than growth are shown for 3 m and 10 m depths (adapted from Fairhead (2001).

Reproduction

The sori from which algal spores are produced are visible on *Ecklonia* fronds as raised lighter areas at the centre of the blade and at the base of the secondary laterals. Novaczek (1984b) found that the kelp became fertile and released spores from an age of 8–12 months (but at a greater age in deeper water). In adults, sori were present and released zoospores year-round, although at lower levels from September to February. Similarly, Sanderson (1992) and Fairhead (2001) both noted a strong seasonality in spore release, with minima in late summer and maxima from late autumn through winter. In Tasmania, kelps became fertile and released spores after ~1 year.

Recruitment, dispersal and survival

Kirkman (1981) compared an *Ecklonia* forest to a stand of trees, with an understorey of a few seedlings ready to develop once an opening appeared in the canopy, usually after a storm. In the same way, newly settled *Ecklonia* sporophytes can remain dormant in low light conditions under a canopy for months, and then grow rapidly if a gap appears. In experiments mimicking storm events, Steve Kennelly (1987a, b) cleared away the canopy at different times of the year, and counted the number of young *Ecklonia* recruits. Although spores settled all year, the greatest number settled in winter (July–August) at a Sydney Harbour site. At other times of the year, an algal turf rapidly became dominant on the substratum, and inhibited the recruitment of *Ecklonia* sporelings. From year-long surveys at other sites over > 300 km of the New South Wales coast, the same pattern was always observed – a strong recruitment consistently occurred once a year, but in different months between June and December, according to site (Kennelly and Underwood 1992, 1993).

Dispersal of *Ecklonia* spores is not well known, but Dave Schiel (1988) concluded that it was very localised, with 75% of recruitment occurring within 8 m of the adult parent kelp. This is a greater dispersal distance than many fucoids – which Schiel attributed to the smaller size of laminarian propagules, which take longer to settle.

The survival of juvenile kelp has only been examined twice. Wood (1987) in experiments at 5 m depth at Marmion Reef found that after 6 months the survival of juveniles was 14% under natural, under-canopy conditions, and 44% when the canopy was removed and juveniles were exposed to full light. In the latter situation, fronds became heavily fouled by epiphytes and suffered extensive necrosis of the tissue. In a third treatment, the canopy was removed and shade cloth placed over the area. In this treatment, which reduced light reaching the bottom by 80%, juvenile survival was 76% over the same period. The shade cloth results suggest that in shallow water UV light is a potent factor reducing survival of juveniles exposed to full sunlight.

Further south in Hamelin Bay, SW Australia, Toohey and Kendrick (2007) mimicked the effect of a severe storm destroying an *Ecklonia* forest, by clearing the canopy from patches of forest on flat (i.e. without relief) and moderate relief (1 m) reefs over 30 days. The flat reefs were at ~5 m depth and the reefs of moderate relief were at 1–2 m depth. They found that on flat reef the survival of juveniles (defined as kelps < 40 cm) ranged from 85 to 95% with most of the mortality occurring in the first week after clearing. On reef of moderate relief, survival was much lower, ranging from ~50 to 83%, according to time of clearing. The surviving juveniles changed their photosynthetic state by adapting to the higher light conditions and grew to full size in ~6 months. The kelp canopy recovered rapidly from late summer–autumn clearings, but slowly from early-summer clearings. The lower survival on the shallower reefs was due to bleaching and death of juveniles in the stronger light environment, rather than to dislodgement in the higher water movement. The above two studies demonstrated not only the damaging effect of high light on juvenile fronds, particularly at shallow depths, but also the ability of a forest to recover quickly from storms.

The survival of adult *Ecklonia* is very variable (Table 4.1). Towards the NW limit of its range, it is virtually an annual kelp, with 92% of plants dying within a year (Hatcher *et al.* 1987), as it is in the NW Indian Ocean off Oman (unpublished observations). Further south on Marmion Reef, it lives for 2+ years, with a half-life of ~270 days. (The half-life is the average time taken for 50% of the plants to die). In Fairlight Bay, New South Wales, adult kelps live for 2–4 years, at other southern Australian sites for 2–3 years (Shepherd 1979; Fowler-Walker and Connell 2002), and in SE Tasmania for 5 years. Some of these estimates of longevity were based on ring counts in the stipe, shown by Novaczek (1981) to be annual.

Grazing

Few species are adapted to grazing on *Ecklonia* because the kelp tissues are well defended chemically (Chapter 5). The turbinid snail *Turbo undulatus* is sometimes found on the fronds of the kelp, which it grazes (Chapter 9). Sea urchins in the genus *Holopneustes* hide among *Ecklonia* fronds, and graze on them, excavating small holes in the fronds. The black urchin *Centrostephanus* and purple urchin *Heliocidaris* can also graze on the kelp, or feed on kelp drift (Chapter 11). The amphipod *Allorchestes compressa* feeds on decaying tips of the kelp among nearshore beds of drift kelp, but whether it feeds on attached plants is not known (Poore *et al.* 2007; Chapter 8). The most common grazer on *Ecklonia* is the herbivorous fish, herring cale (Chapter 16).

Abundance and productivity of *Ecklonia*

Ecklonia is a dominant species over most of its geographic range, with biomass (fresh weight) usually ranging from 0.5–3.5 kg m^{-2}. The highest biomass values occur in SW Western Australia and lowest values in SE Tasmania, where other large brown algae (e.g. *Durvillaea*, *Phyllospora*, *Lessonia* and *Macrocystis*) may be dominant to ~15 m. Plots of *Ecklonia* biomass versus depth for various sites in southern Australia are shown in Fig. 4.6. From these values, the productivity of the kelp at different sites can be readily derived using biomass increase rates in Table 4.1. For example, Kirkman (1989) calculated that at Marmion Reef the annual productivity was 4.4 kg dw m^{-2} $year^{-1}$ at 5 m depth and 1.8 kg m^{-2} $year^{-1}$ at 10 m depth (using a fresh weight:dry weight ratio of 7.1). Bearing in mind that these biomass values represent only ~11% of the total carbon uptake of the kelp, its ecological importance as a primary producer in reef ecosystems is profound.

Conclusions

The common kelp *Ecklonia radiata* has evolved a sophisticated strategy for maximising its productivity in different light, temperature and nutrient

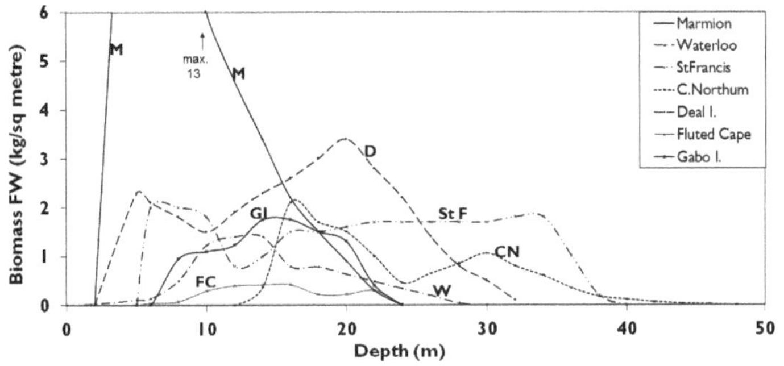

Figure 4.6: Plots of biomass fresh weight of the kelp *Ecklonia radiata* versus depth for various southern Australian sites. Data extracted from Kirkman (1989) for Marmion reef, Western Australia (M), Shepherd and Womersley (1981) for Waterloo Bay (W), Shepherd and Womersley (1976) for St Francis I., Transect A, (StF), Shepherd (1979) for Cape Northumberland (CN), and SA Shepherd (unpublished data) for Little Squally Cove, Deal I., Bass Strait (D), Fluted Cape, SE Tasmania (FC), and Gabo I. and SE NSW (GI).

environments. It acclimates to low light by increasing efficiency of light absorption, or changing its morphology, as in changes that occur with depth. Carbohydrates are stored in summer–autumn when light levels are high, and remobilised during winter–spring when they are low. Spores can be released year-round, and can survive for months on the bottom until conditions are optimal for rapid growth. Longevity varies from a year or so at the upper temperature edge of its range to 5 years or more in colder water. Overall, *Ecklonia* forests are resilient to disturbance, and juveniles soon replace adults after removal by storms. Carbon uptake is high, but ~90% of it is either respired, used in spore production, exuded as mucilage or eroded.

THE GIANT KELP *MACROCYSTIS PYRIFERA*

SA Shepherd and Craig Sanderson

The giant (or string) kelp *Macrocystis* (Plate 4) is aptly named because it is the largest marine alga known on Earth, and has a recorded maximum length in Tasmania of ~30 m, although fronds rarely exceed ~20 m (Cribb 1954). It is the only kelp that forms a forest with a dense floating surface canopy, growing in shallow rocky areas close to shore. There, as vividly described by McPeake *et al.* (1988) 'the spreading fronds form a golden-hued canopy that stands out against the vivid blue of the sea'. The giant kelp is anchored to the rocky bottom by a large, conical holdfast consisting of a mass of intertwining root-like branches (the haptera). From the holdfast emerges a cluster of stipes that branch three to six times a little above the holdfast to produce fronds 4–20 m long, which grow towards the surface. Each frond bears a series of blades, joined to it by bladders that keep the frond afloat. These stipes, fronds and blades form the conspicuous and extensive surface canopies that provide the main photosynthetic biomass of the plant. For a metre or two above the holdfast, other short stipes grow and produce small, branched blades, called sporophylls, at the base of fronds, and these bear the gamete-producing sori.

Many species of *Macrocystis* have been described globally on the basis of differences in the shape of holdfast, stems and blades, and two species have until recently been recognised in SE Australia – *Macrocystis angustifolia* and *Macrocystis pyrifera*. However, the classification of the species has been debated for > 100 years because of the kelp's great plasticity in form, and the most recent evidence, based on the genetic relatedness of different forms and their inter-fertility, suggests that only one variable species *M. pyrifera* exists after all (Coyer *et al.* 2001). Here we shall call the two local 'species' forms, and include both of them in the term *Macrocystis*. Most studies on giant kelp have been done in California, and this account draws on the reviews of Steneck *et al.* (2002), Reed *et al.* (2006), Schiel and Foster (2006), Graham *et al.* (2007a, b), and the Tasmanian studies of Sanderson (1992), Edyvane (2003), and Sanderson *et al.* (2004).

Distribution

Macrocystis is widely distributed on islands and land masses fringing the Southern Ocean, and along the temperate east coast of the Pacific from Tierra del Fuego to Alaska. Whether the kelp originated in the Northern Hemisphere and migrated to the South as recently as 10 000 years ago, as some claim, or the reverse, is still debated. Different growth forms tend to dominate different habitats. In southern Australia, the *pyrifera*-form is largely limited to the east and south coasts of Tasmania at depths to 10 m, occasionally to 25 m, although small forests occur on the west coast where there is some protection from swell. Intergrades between the *pyrifera* and *angustifolia* forms are found on the north coast of Tasmania and in D'Entrecasteaux Channel. However, forests of the *angustifolia*-form are notoriously transient, developing rapidly in good conditions, to be later ripped out by storms.

Forests of the *angustifolia*-form are found from Cape Jaffa in the SE of South Australia to Walkerville, in western Victoria, and on the NW coast of Tasmania. In western Victoria, the *angustifolia*-form occurs as patches of scattered individuals, and forests have been mapped in Discovery Bay, Point Addis and Barwon Bluff (Bull and Blake 2007). The dynamics of the *angustifolia*-form are better known in the SE of South Australia than elsewhere. Along much of this coast, fringing reefs occur up to 1 km offshore, partially protecting inshore lagoons and bays from extreme swell conditions. Inside the fringing reefs, *angustifolia* individuals may persist for years on reefs at depths of 2–10 m as scattered individuals at densities of up to 10 per 100 m^2. Outside the fringing reefs, forests can occur (Fig. 4.7) at depths of 10–25 m, rarely to 35 m. However, these forests are patchy in space and time. They appear rapidly after strong upwellings during El Niño events, and persist until winter storms remove them, but may re-appear the following spring. In exposed areas, forests may be present in only one out of every 4–5 years, or may come and go at longer time scales. For example, abalone divers reported that they were scarce from ~1995 to 2005, but abundant from 2006 to 2009 (J. Godden *pers. comm.*, Edyvane and Baker 1999).

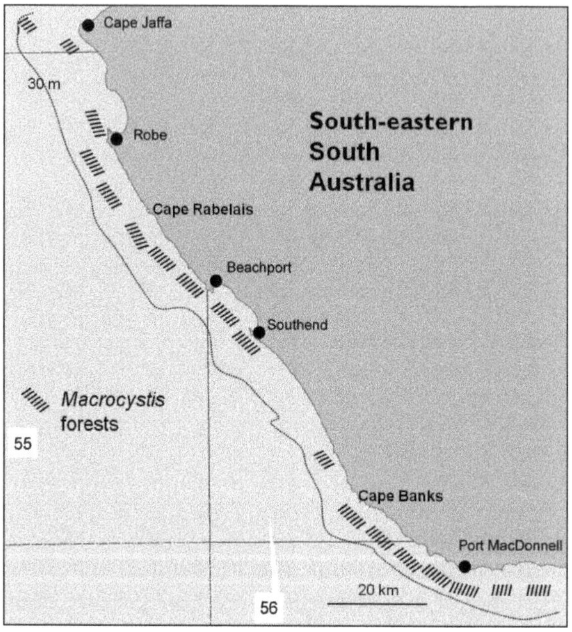

Figure 4.7: Map of SE of South Australia showing approximate recorded extent of *Macrocystis angustifolia*-form at depths of 10–25 m outside fringing reefs, and within the 30 m depth isobath. Inside the reefs, scattered plants of *angustifolia*-form are perennial right along the coast.

Life history

Like *Ecklonia*, the giant kelp has a two-phase life cycle, with alternating generations of the giant sporophyte and the microscopic gametophyte. However, much more is known about the stages of *Macrocystis* than about those of *Ecklonia*. The sporophyte becomes reproductively mature at 9–12 months (Santelices 1990), and releases into the water male and female zoospores, located in sporophylls – the skirt of small blades just above the kelp's holdfast (Fig. 4.8). A single giant kelp has many sporophylls at its base, and each sporophyll can release billions of zoospores throughout the year. Sanderson (1992) noted that the number of sporophylls reached a maximum of up to 12 per frond in summer–autumn, and suggested that this was the time of maximum spore release. Giant kelps first start to produce sporophylls before they are a year old.

The zoospores are the main dispersal stage of the giant kelp. After a short dispersal period of hours to

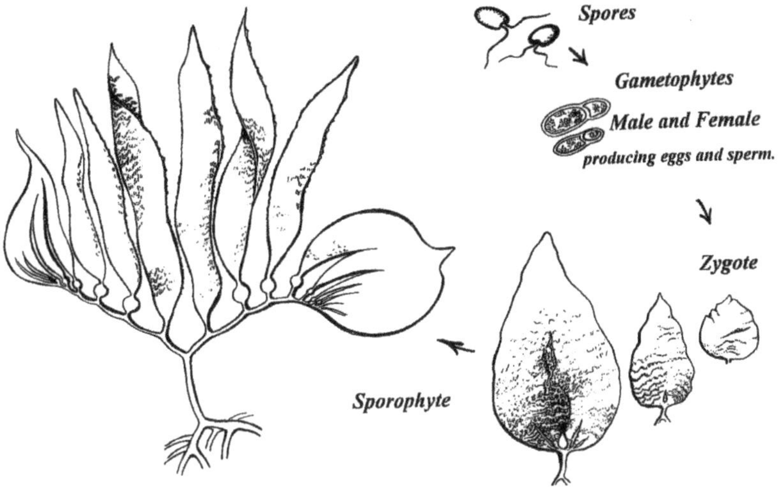

Figure 4.8: Life history of the giant kelp *Macrocystis* showing young sporophyte (left). Spores released from sporophytes settle and develop into gametophytes, producing eggs and sperm. Thence, zygotes develop to produce juvenile sporophytes (after Sanderson *et al.* 2004).

days, the zoospores settle on the bottom, and germinate into microscopic male and female gametophytes < 20 μM long – a process that can take up to 40 days. The male:female gametophyte ratio is ~1:1. The female gametophyte gives off a pheromone, which triggers the release of sperm from a neighbouring male gametophyte and guides it to the egg. The pheromone's ability to attract males acts over a very short distance of < 1 mm, so the embryonic sporophyte, called a zygote (which ultimately develops into the giant kelp) can only form at a site of high spore density (> 1 mm^{-2}). Young sporophytes (called recruits) can appear year-round, although in SE Australia the season of their appearance may be limited, as in the SE of South Australia, where they appear during the summer upwelling.

A curiosity of gametophyte fertilisation is the laboratory finding that *Macrocystis* sporophytes can develop from unfertilised female gametes, raising the possibility of a virgin birth that is, in the absence of high spore densities, a parthenogenetic alga may grow into an adult (Druehl *et al.* 2005). The likelihood of this event in nature is quite unknown.

Recruitment and dispersal

The recruitment of sporophytes has been shown in California to be greatest during times of high nutrients and low temperatures. The critical environmental conditions controlling sporophyte recruitment are light, nutrients and temperature at the stage of release and fertilisation of gametes. The minimum requirements that have been measured are > 4 μM photons m^{-2} s^{-1} for light, > 1 μM concentration of nitrate, and a temperature range of 11–19°C. Hence, successful recruitment can only occur when these environmental requirements are satisfied. This is termed the 'recruitment window', and it is only open during the annual summer upwellings on SE Australian and Tasmanian coasts, when giant kelp forests appear.

Knowledge of the dispersal stage is basic for understanding the dynamics of giant kelp beds, and how they can re-establish after destruction by storms. Spores are tiny, being only ~6 μm in diameter, but are flagellated and can swim, although at a slow rate of 10–12 mm min^{-1}. Field studies on spore dispersal from an isolated kelp bed (Fig. 4.9) (Reed *et al.* 2006) showed that spores settled in high densities at short distances of only a few metres from the edge of a kelp bed. This study was done in a coastal area where currents and swell were relatively low. The authors then modelled spore dispersal, simulating natural flow regimes of 3–4 cm s^{-1} near kelp

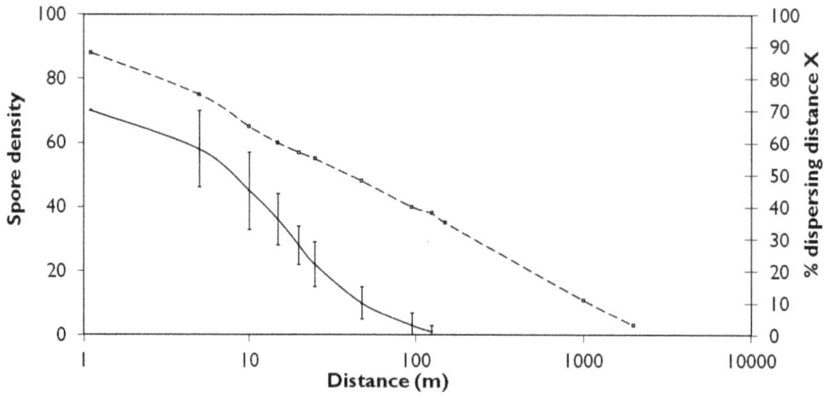

Figure 4.9: Settlement density of spores in numbers 2.5 mm^{-2} (± 1 SE) as a function of distance from the edge of a kelp bed (continuous line). This graph is the average of several field experiments. The dashed line represents the probability of spores dispersing at least as far as the distance on the *x*-axis, based on measured currents at 10 m depth near reefs with *Macrocystis* beds (redrawn from Reed *et al.* (2006).

beds – much faster flow conditions than those during the field experiment. In this case, spores could be carried up to ~10 km from a kelp bed (Fig. 4.9). This variability in possible dispersal distances shows that connectivity between kelp stands is greatly influenced by local currents, swell or wave conditions. Of course, if spores are released at the centre of a kelp forest they are much more likely to be retained within the forest than if released towards the edge, because drag from the kelp fronds slows down water movement within it. But, how can isolated giant kelp beds recover after they have been completely destroyed by winter storms, as happens annually for this species in the most exposed parts of its range?

From the extensive research on spore dispersal, four guiding principles have emerged:

- The density of settling spores on any reef is proportional to the size of the parental kelp bed upstream.
- The fecundity of kelps, the extent of synchrony of spore release and oceanographic conditions (as mentioned above) are all important for adequate densities of settlement.
- Where giant kelp forests are distributed over a narrow depth range along coasts with long-shore currents, stepping-stone exchange of spores among neighbouring kelp patches is a likely mechanism of dispersal.
- In addition to the above, drifting adult fronds or fertile fragments, torn off by storms, may disperse spores, often along a defined line of recruitment (e.g. Dayton *et al.* 1984).

If gametophytes or embryonic sporophytes could survive for extended periods at a resting stage, like seed-banks in soils, this would provide a mechanism for rapid recovery of a kelp forest when suitable conditions for growth were re-established. Some researchers have argued that gametophytes, which have broader temperature tolerances than spores, and the ability to survive long periods in the dark, are the most likely form to comprise a seed-bank, whereas others argue that embryonic sporophytes are better suited for persistence during unfavourable periods (tom Dieck 1993; Kinlan *et al.* 2003). Whichever the case, the persistence of one of the two dormant microscopic forms is needed to explain the reappearance many months later of forests separated by large distances – in one example ~100 km away from the nearest possible source forest (Ladah and Zerluche-González 1999). A similar situation exists in western Tasmania.

Growth and productivity

The growth of *Macrocystis* is more complex than that of *Ecklonia*. The young kelp starts out as a simple blade resembling a small *Ecklonia*. It then splits in two, and these split again repeatedly, but unequally, and give rise to stipes that elongate and produce blades. A bladder develops at the base of the blades to give buoyancy to the entire frond. In this way, the stipes grow apically towards the surface, producing terminal blades, which split into more blades. The blades also grow in length and width over time. At the base, the holdfast becomes larger as more roots (haptera) emerge from the stipe to fix it to the substratum, and the reproductive sporophylls also emerge above the holdfast as outgrowths from the base of the stipes. Growth has been estimated by measuring the rate of production of blades on fronds, and the elongation rate of the distal part of fronds where growth occurs (Fig. 4.10) (Sanderson 1992). Both measures of growth of the *pyrifera*-form showed strong seasonality at Craig Sanderson's two sites: Bicheno and George III Reef. The fronds grew fastest in length in winter, but the production of blades on the fronds was highest in spring. This asynchrony in timing of growth was thought to be due to seasonal differences in wave action and swell height, which also affect the plant's blade width and distance between blades.

Over the time of the study, the average growth of fronds at George III Reef was 5.25 cm day^{-1}, and the average production of blades was 0.53 blades day^{-1}. Growth rates at Bicheno (not shown) were similar. From this, Sanderson (1992) estimated an annual frond productivity of 6 kg fresh weight (fw) per frond, and, for a typical density of four fronds per square metre, an annual productivity of 24 kg FW m^{-2}. Of the 6 kg fw production per frond, ~20% was lost through erosion of the blades. But what caused the marked seasonality in the growth rate of the kelp?

Sanderson (1992) also examined the relation between growth rate and sea temperature and nitrates (plus nitrites) in the water during his study. Nitrate concentrations varied seasonally, with minima in summer and maxima in winter that is, inversely with temperature (Fig. 4.11). The rate of increase in blades of giant kelp declined with increasing sea temperature, and also when nitrates fell below a threshold of 0.3–0.5 μM, as illustrated in Fig. 4.12. The plot of frond growth versus nitrate was similar, but that of frond growth versus temperature showed little change with temperature, except for a decline over ~15°C; neither plot is shown here. Sanderson (1992) fitted the following multiple regression to the data on blade growth rates:

$$BG = 2.101 - 0.106\,t + 0.132\,N$$
$$(r^2 = 0.706; P = 0.01)$$

where BG = blade growth in number of blades added per day, t = sea temperature (°C) at 20 m in Storm Bay, and N = nitrate concentrations in μM at 10 m depth. Frond growth was not significantly related to any environmental factor, although wave action and day

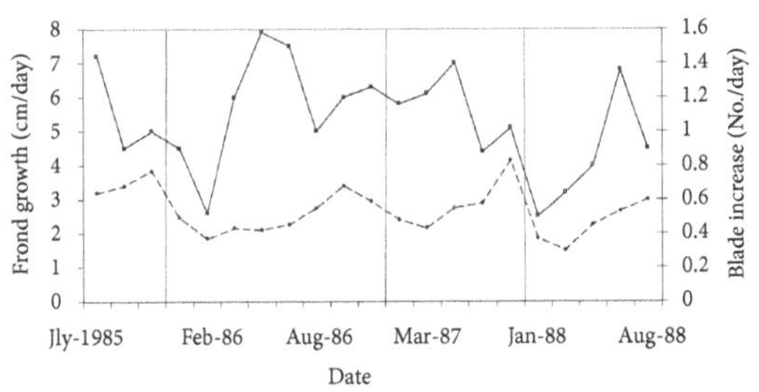

Figure 4.10: Seasonal growth of the giant kelp *Macrocystis pyrifera*-form at George III Reef, Tasmania. Frond growth rate is given in cm per day (continuous line), and blade increase rate in number of blades per day (dashed line) (redrawn from Sanderson 1992, omitting standard errors).

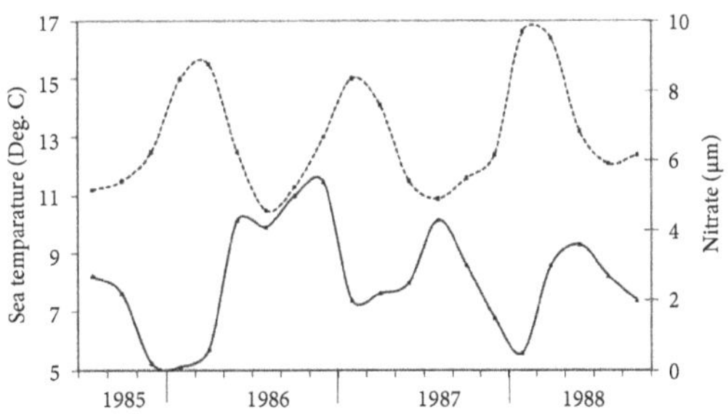

Figure 4.11: Seasonal changes in smoothed sea temperature (°C) (dotted line), and nitrate (μM) (continuous line) at 10 m depth in Storm Bay, southern Tasmania from 1985 to 1988 (redrawn from Harris *et al.* 1991).

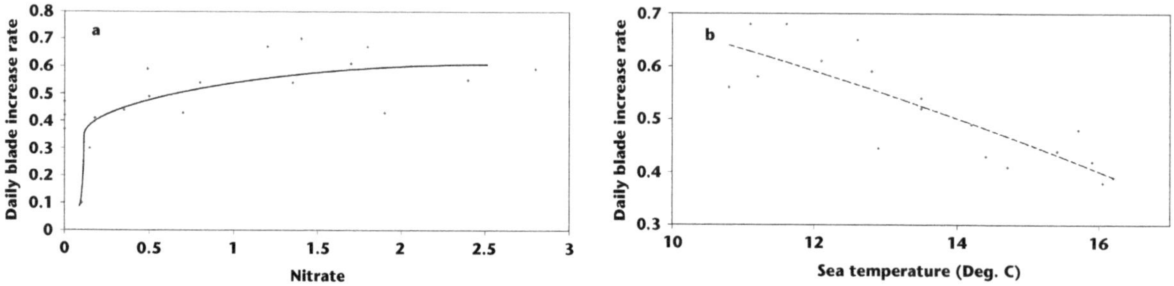

Figure 4.12: Plots of blade increase rate (number of blades per day) of *Macrocystis pyrifera*-form versus (a) nitrate levels (µM), and (b) sea temperature at 10 m depth, with curves of best fit (redrawn from Sanderson 1992).

length were almost significant. From these results, it became clear that the major factors affecting growth rates of the kelp were the availability of nutrients and sea temperature; kelp growth became limited below ~1 µM nitrate. These two factors are inversely related, as is evident from a perusal of Fig. 4.11 and, together with the fact that kelps have a low storage capacity for nutrients (exhausting them within 2–4 weeks), emphasise the significance of local oceanography on the growth of this kelp. Overall, biomass increase is tightly coupled to the nutrient supply.

Survival

Frond life was estimated to be 7–10 months by Cribb (1954), and 9–12 months by Sanderson (1992). The longevity of whole kelp plants is not known, but Californian and Argentinean studies have shown that in places of strong wave action they die within a year, whereas elsewhere, as in southern California, they may have a life span of 4–7 years (Graham *et al.* 2007a). The vulnerability of the kelp canopy to storms, and the kelp's dependence on nutrients and cool temperatures, result in great variability in its presence, survival, productivity and health. These aspects become significant when we now consider the causes of the giant kelp's decline in southern Tasmania.

Fate of the giant kelp in Tasmania

In the last 70 years, as the East Australia Current has extended down the east Tasmanian coast, and water temperatures have increased (see Chapters 2 and 12), an ongoing decline of the giant kelp has occurred, with some kelp beds disappearing and others shrinking in size. Edyvane (2003) and Johnson *et al.* (2011) reviewed this decline, and Fig. 4.13 shows the extent

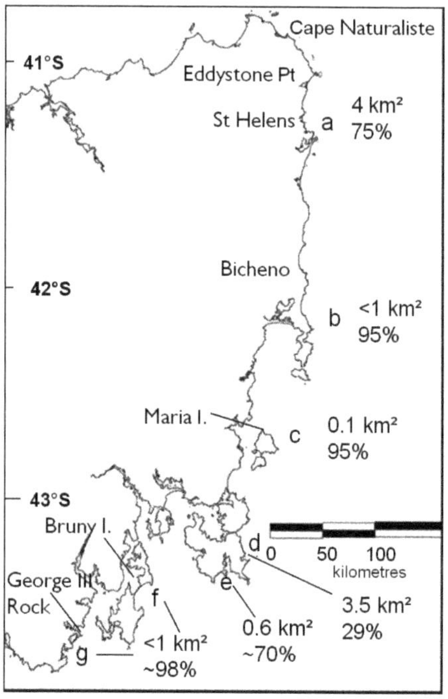

Figure 4.13: Map of east coast of Tasmania, showing the areal extent in 2007 (km^2) and estimated percentage decline in extent of visible surface canopy of *Macrocystis pyrifera*-form from the 1940s to 2007 for seven regions of the coast, marked a to g (data extracted from Johnson *et al.* (2011).

of decline by 2007, as estimated from historic and recent aerial photography for seven regions of the east coast. The average overall decline from 1946 to 2007 was 91%.

The northern limit of giant kelp beds is currently at Cape Naturaliste, NE Tasmania, where substantial beds persist. Historical surveys there have been patchy, and the overall decline cannot be determined, although some beds have declined by 50%. Along the remainder of the coast the greatest percentage decline has occurred in the Bicheno to Maria I. region. The Triabunna kelp harvesting company, Alginates Australia, annually harvested 3 000–17 000 t of kelp from 1965 to 1973, but then closed down as the kelp beds declined. However, repeated surveys of the kelp beds showed that the decline was patchy and sporadic, with intermittent periods of temporary recovery. So what were the cause(s) of decline?

The ultimate cause was undoubtedly the increased southerly penetration of the nutrient-poor East Australia Current, which caused a shift of subantarctic waters to the south (Johnson *et al.* 2011) and increased sea temperatures three to four times above the global average. However, the immediate causes were far more complex. Edyvane (2003) noted that major kelp losses coincided with the 1972, 1982–83 and 1987 El Niño events, when sea temperatures increased by up to 1°C, rainfall declined, and the severity and frequency of storms increased. El Niños became more frequent over the 1980–2000 period and the warming trend on the east coast (see Chapters 2 and 12), marked during the 1970s, continued and sometimes approached the kelp's upper temperature tolerance limit of ~20°C (Cribb 1954). Edyvane concluded that episodic storms and strong swell caused a high mortality of individual kelps, and even entire kelp forests, and that the reproductive success of the kelp had been dampened.

Sanderson *et al.* (2004) suggested that several other factors were implicated, for the Mercury Passage area in particular. They included: an increasing sediment load in coastal waters as a result of land-clearing and wood-chipping; increased urchin grazing, resulting from lobster over-fishing (Chapter 12); and increased competition from the exotic *Undaria* discussed below. Johnson *et al.* (2011) added yet another contributing factor – the southward invasion of the sea urchin *Centrostephanus*.

After successfully re-establishing giant kelp at Hope I., southern Tasmania, Sanderson *et al.* (2004) undertook re-afforestation trials in the Derwent Estuary and in Mercury Passage, where giant kelp forests once existed. Young sporophytes transplanted to sites in the upper parts of the Estuary failed, but those transplanted towards the Estuary mouth were modestly successful, and led to the development of small forests. In Mercury Passage, juvenile kelps and mature sporophylls (the reproductive blades) were transplanted onto some substrates where urchins occurred and onto others from which urchins had been cleared. In the former case, surviving transplants were eaten by urchins, whereas at the site without urchins some survived, although they were mainly stunted, possibly due to limiting nutrients.

Macrocystis – a foundation species

Like trees in a forest, the giant kelp modifies the environment such that it favours a remarkably rich diversity of species; hence, it has been aptly called a foundation species. The physical effects of a kelp forest are that it:

- dampens water movement, providing a unique sheltered habitat for animals on the forest floor
- shades the sea bed, but still permits a diverse underflora or turf of low-light species under the canopy
- due to its architecture, provides habitats in the upper canopy, among the mid-water fronds, and within the holdfast, for myriads of mobile pelagic and bottom-living animals.

The biological effects of the kelp are that it:

- extracts nutrients from the water for recycling
- provides a rich source of food as drifting kelp blades and dissolved organic matter for animals within and downstream from the forest
- functions as a refuge and nursery ground for juveniles to adults of many fish and other species.

An interesting ecosystem result of the dampening effect of *Macrocystis* on swell and current is that larvae drifting in coastal currents through a forest tend to be retained within the reef system, facilitating larval settlement on the bottom. This has been shown to enhance, for example, rock lobster puerulus recruitment to Tasmanian coastal reefs (Gardner *et al.* 1998; Mills and Crear 2004). Although a descriptive list of common species in *Macrocystis* forests in SE Tasmania is known (Sanderson *et al.* 2004), much is still to be discovered of the ecological services of giant kelp.

The few direct grazers on giant kelp include the urchins *Holopneustes inflatus* and *H. purpurascens*, which live in nests made by wrapping themselves in kelp blades for protection, and feed on the kelp blades (Chapter 11). Herbivorous amphipod grazers in the Family Ampithoidae also occur in the canopy and graze on the blades (Poore *et al.* 2007). Many other bottom-dwelling molluscs and echinoderms graze benthic algae and feed on detrital kelp. Notable among them are blacklip abalone, turbinid and other snails, the sea-star *Meridiastra calcar*, the long-spined urchin *Centrostephanus rodgersii*, and the purple urchin *Heliocidaris erythrogramma*. A notable carnivore in the kelp forest is the southern rock lobster, and there is now abundant evidence that the lobster can control density of the purple urchin (Chapter 12). However, the lobster has suffered extensive overfishing in SE Tasmania, leading to a likely cascade of increased urchin abundance and expanding barrens that inhibit the natural recruitment of the giant kelp (Chapters 12 and 19).

Conclusions

The giant kelp is a rapidly growing and highly productive alga that provides food resources and habitat for a diverse fauna. The morphology of the sporophyte is plastic, and two main forms, with some intermediates, occur over its southern range in SE Australia. The microscopic stages are long lived, and can be transported distances of 1 km or more. The kelp forest system is driven by oceanographic processes, with an optimal temperature range of 12–18°C, and nitrate levels above a threshold of ~1 μM. The kelp has declined seriously from the 1940s to 2007 in SE Tasmania due to multiple causes, including increasing temperature, which may have dampened the reproductive stage, increasing storm activity during El Niño events, and the expansion of sea urchin barrens. Experimental transplanting of kelp sporophytes enjoyed only a temporary success. In consequence of the above cumulative threats, in 2012, *Macrocystis* forests were declared an endangered ecological community under the federal *Environmental Protection and Biodiversity Conservation Act 1999.*

THE ASIAN KELP *UNDARIA PINNATIFIDA*

The Asian or Japanese kelp *Undaria pinnatifida* (Plate 5) was likely introduced to SE Tasmania in ~1982 on ships coming to Triabunna to transport wood-chips to Japan. The kelp's distribution and ecology was examined by Sanderson (1990, 1992), after he first discovered it in Mercury Passage in 1988, with later studies by Valentine and Johnson (2003, 2004) and Edgar *et al.* (2004). Little is known of the ecology of this species in Port Phillip Bay, where it spread in 1996, and later to Western Port and Eden, New South Wales (Campbell and Burridge 1998).

Undaria is unusual among kelps because it is seasonal and starts growing in late autumn to winter. Like other kelps, it has a two-phase life history with alternating generations of microscopic gametophyte and sporophyte stages. Sporophytes grow over a wide temperature range of 5–20°C, but optimally at 12–15°C. Gametophytes have an even broader tolerance range of 8–30°C, with an optimum range of 17–20°C, and are substantially more resistant to desiccation and reduced salinity than *Macrocystis* or *Ecklonia* (JC Sanderson and SI Blackburn *pers. comm.*). Hence, transport of gametophytes on boat hulls for long distances is quite probable.

In Tasmania, young sporophytes appear in winter, reaching maturity in < 6 months at an average maximum length of ~1.5 m (rarely to 1.8 m). Thus, the average growth rate is 7–9 mm day^{-1}, which is similar to its growth rate in New Zealand (Dean and Hurd 2007). The high seasonal growth rate is apparently dependent on high nitrate levels (up to ~5 μM) and low sea temperatures that co-occur in winter–spring (see Fig. 4.11). Individuals mature in summer and die off by early autumn. Apparently, gametophytes and/or zoospores settle on the bottom in the summer–autumn period, and are able to respond opportunistically to increasing nutrient levels, and the increased light after decline or loss of native canopy species. *Undaria* is found in highest biomass at ~4 m depth and extends from intertidal rockpools to ~12 m depth.

Undaria is a highly opportunistic species that can rapidly establish itself in disturbed habitats, where native species have been lost or grazed. Consequently, the species grows well in barrens grazed by the purple urchin *Heliocidaris*, in areas scoured by sand at reef edges and on unstable boulders, but rarely invades established native communities other than through the appearance of scattered individuals. Valentine and Johnson (2003) examined the invasive process of the kelp by experimentally clearing patches of native canopy algae (*Ecklonia* and fucoids) at different times of the year. They found that *Undaria* recruited in highest densities after winter clearings compared with summer clearings, but over the next 2 years the native species always recovered, although with some shift in species composition. They concluded that continual disturbance (probably at least every year or two) to native forests is needed for *Undaria* to become established – a conclusion supported by a later study following a natural dieback of the large fucoid *Phyllospora comosa*. *Undaria* rapidly invaded the dieback area, and became established at high densities, whereas it was absent where the native species remained intact.

By 1988, *Undaria* had spread from the region of Triabunna (~42° 32′ S.) for some 10 km southwards. By 1996, it had spread over > 80 km of coast, and by 2002 some 320 km along almost the entire east Tasmanian coast south to the D'Entrecasteaux Channel (Edyvane 2003; MES 2003), giving an average rate of spread of ~10 km $year^{-1}$ since 1982. The rapid spread of the kelp has been attributed to ocean currents, transport on ships' hulls or with fishing gear, and Sanderson (1990, 1992) predicted that, given the kelp's temperature tolerance, it could potentially spread across southern Australia from SW Western Australia to the mid New South Wales coast – a prediction partly fulfilled. The kelp is now harvested commercially in SE Tasmania.

The invasive success of this kelp appears to be due to two reasons (Dean and Hurd 2007). First, the species has a high rate of nutrient uptake – in this trait resembling small, fast-growing species such as sea-lettuce (*Ulva*) and *Enteromorpha*. This rate is much faster than in most other kelps. Second, the alga increases its pigment content in winter, enabling it to reach maximum growth rates in low light. These features, and its high spore production, enable *Undaria* to out-compete most other macroalgae in disturbed habitats.

The interactions of *Undaria* with native species, and its effect on local ecosystems, have also been examined. Although Valentine and Johnson (2003, 2004) concluded that *Undaria* would not displace native species in the absence of disturbance, Sanderson (1992) suggested that it might inhibit the recovery of *Macrocystis*, as, for example, in D'Entrecasteaux Channel, where *Undaria* has replaced *Macrocystis* at sites where the latter was once dominant. Furthermore, environmental disturbances appear to be increasing in frequency at regional scales due to the impact of climate change (rising temperatures) affecting native species and the overfishing of lobsters, which has favoured a sequence of increasing urchin densities, barrens formation and strong recruitments of *Undaria* (Johnson *et al.* 2004).

In the face of these combined threats in the region, marine reserves provide a potential tool to inhibit spread of *Undaria* through lobster–urchin interactions (Edgar *et al.* 2004). On this point, Sanderson (1992) suggested that increased urchin populations, dependent on *Undaria* as a food source, might overgraze other algae when *Undaria* disappeared in late summer. An alternative scenario is that the development of *Undaria* beds on urchin barrens may facilitate recovery of native algae, because urchins then cannot completely overgraze available algal food resources, thus providing a foothold for re-establishment of native species. This multi-species interaction appears to have happened recently in the Tinderbox Marine Reserve (GJ Edgar and NS Barrett *pers. comm.*), but is yet to be fully explored.

THE STRAPWEED *LESSONIA CORRUGATA*

Strapweed grows from a much-branched holdfast as a mass of long blades that stream out over the bottom. It typically lives in exposed to semi-exposed locations below the bull-kelp *Durvillaea* zone from just below low tide to ~8 m depth, but, in places of strong swell in southern Tasmania, to depths of ~20 m.

Lessonia is a cold-water genus of six species found variously in cold temperate waters of Tasmania, New Zealand, Chile and some subantarctic islands. Virtually no research has been undertaken on the ecology or productivity of *Lessonia corrugata* in southern Australia. By contrast, the ecology of other species of *Lessonia*, notably on the Chilean coast, where they are cultured on long-lines and harvested for export, has been well studied (see Edding and Tala 2003; Tala and Edding 2005, 2007).

Species of *Lessonia* typically grow from a basal holdfast disc, from which rises several stipes. Each stipe divides into two branches, with at least two strap-like blades per branch. The blades of a subtidal Chilean species *L. trabeculata* grow at average annual rates of 4 mm day^{-1}, mainly in spring and summer, reaching lengths of 60–90 cm, while erosion of the tips occurs at slightly lower rates. Fronds of *L. trabeculata* reach a total length of 2 m, at densities of > 500 blades m^{-2}, and exhibit very high net productivity (production minus erosion) of > 40 g dw m^{-2} day^{-1} in spring. In addition, about the same amount is produced and lost through erosion. *Lessonia* kelp is clearly an important contributor of algal detritus for molluscs, polychaetes, amphipods and fishes in inshore systems.

REFERENCES

Andrew NL (1991) Survival of kelp adjacent to areas grazed by sea urchins in New South Wales, Australia. *Australian Journal of Ecology* **19**, 466–472.

Bull D, Blake S (2007) *Shallow Water Habitat Mapping at Victorian Marine National Parks and Marine Sanctuaries. Vol. 1. Western Victoria.* Parks Victoria Technical Series No. 36. Melbourne.

Campbell SJ, Burridge TR (1998) Occurrence of *Undaria pinnatifida* (Phaeophyta: Laminariales) in Port Phillip Bay, Victoria, Australia. *Marine and Freshwater Research* **49**, 379–381.

Coleman MA, Gillanders BM, Connell SD (2009) Dispersal and gene flow in the habitat-forming kelp, *Ecklonia radiata*: relative degrees of isolation across an east-west coastline. *Marine and Freshwater Research* **60**, 802–809.

Coyer JA, Smith GJ, Andersen RA (2001) Evolution of *Macrocystis* species as determined by ITS1 and ITS2 sequences. *Journal of Phycology* **37**, 574–585.

Cribb AB (1954) *Macrocystis pyrifera* (L.) Ag. in Tasmanian waters. *Australian Journal of Marine and Freshwater Research* **5**, 1–34.

Dayton P, Currie V, Gerodette T, Keller B *et al.* (1984) Patch dynamics and stability of kelp forests of some California kelp communities. *Ecological Monographs* **54**, 253–289.

Dean PR, Hurd CL (2007) Seasonal growth, erosion rates, and nitrogen and photosynthetic ecophysiology of *Undaria pinnatifida* (Heterokontophyta) in southern New Zealand. *Journal of Phycology* **43**, 1138–1148.

Druehl ID, Collins JD, Lane CD Saunders GW (2005) An evaluation of methods used to assess intergeneric hybridization in kelp using Pacific Laminariales (Phaeophyceae). *Journal of Phycology* **41**, 250–262.

Edding ME, Tala FB (2003) Development of techniques for the cultivation of *Lessonia trabeculata* Villouta et Santelices (Phaeophyceae: Laminariales) in Chile. *Aquaculture and Research* **34**, 507–515.

Edgar GJ (2008) *Australian Marine Life: the Plants and Animals of Temperate Waters.* 2nd edn. New Holland, Sydney.

Edgar GJ, Barrett NS, Morton AJ, Samson AR (2004) Effects of algal canopy clearance on plant, fish and macroinvertebrate communities on eastern Tasmanian reefs. *Journal of Experimental Marine Biology and Ecology* **312**, 67–87.

Edyvane KS (2003) 'Conservation, monitoring and recovery of threatened giant kelp (*Macrocystis pyrifera*) beds in Tasmania'. Final Report to Environment Australia (Marine Species Protection Program). Department of Primary Industries, Water and Environment, Hobart.

Edyvane KS, Baker JL (1999) 'Marine benthic survey of the south-east coast (Murray Mouth – Portland)'. Report to Environment Australia (Marine Protected Areas Program): Project D801 (Stage 4). SARDI Aquatic Sciences, Adelaide.

Fairhead VA (2001) The ecophysiology and production of the kelp *Ecklonia radiata* (C. Ag.) J. Agardh, at West Island, South Australia. PhD thesis. University of Adelaide, Adelaide.

Fairhead VA, Cheshire AC (2004a) Rates of primary productivity and growth in *Ecklonia radiata* measured at different depths, over an annual cycle, at West Island, South Australia. *Marine Biology* **143**, 41–50.

Fairhead VA, Cheshire AC (2004b) Seasonal and depth related variation in the photosynthesis-irradiance response of *Ecklonia radiata* (Phaeophyta, Laminariales) at West Island, South Australia. *Marine Biology* **145**, 415–426.

Fowler-Walker MJ, Connell SD (2002) Opposing states of subtidal habitat across temperate Australia: consistency and predictability in canopy-benthic associations. *Marine Ecology Progress Series* **240**, 4956

Fowler-Walker MJ, Connell SD, Gillanders B (2005) Variation at local scales need not impede tests for broader scale patterns. *Marine Biology* **147**, 823–831.

Gardner C, Cawthorn A, Gibson I, Frusher S, *et al.* (1998) 'Review of the southern rock lobster *Jasus edwardsii* puerulus monitoring program: 1991-1997'. Tasmanian Department of Primary Industry and Fisheries Technical Report No. 52, Hobart.

Graham MH, Vasquez JA, Buschmann AH (2007a) Global ecology of the giant kelp *Macrocystis*: from ecotypes to ecosystems. *Oceanography and Marine Biology - an Annual Review* **45**, 39–88.

Graham MH, Halpern B, Carr M (2007b) Diversity and dynamics of Californian subtidal kelp forests. In *Food Webs and the Dynamics of Marine Benthic Ecosystems* (Eds TR McClanahan and GR Branch) pp. 103–134. Oxford University Press, Oxford.

Harris GP, Griffiths FB, Clementson LA, Lyne V *et al.* (1991) Seasonal and interannual variability in physical processes, nutrient cycling and the structure of the food chain in Tasmanian shelf waters. *Journal of Plankton Research* **13**, (Suppl.), 109–131.

Hatcher BG, Kirkman H, Wood WF (1987) Growth of the kelp *Ecklonia radiata* near the northern limit of its range in Western Australia. *Marine Biology* **95**, 63–73.

Johnson CR, Valentine JP, Pederson HG (2004) A most unusual barrens: complex interactions between lobsters, sea urchins and algae facilitate spread of an exotic kelp in eastern Tasmania. In *Echinoderms München. Proceedings of the 11th International Echinoderm Conference* (Eds T Hunzeller and JH Nebelsick) pp. 213–220. Taylor and Francis, Munich.

Johnson CR, Banks S, Barrett NS, Cazassus F *et al.* (2011) Climate change cascades: shifts in oceanography, species' ranges and subtidal marine community dynamics in eastern Tasmania. *Journal of Experimental Marine Biology and Ecology* **400**, 17–32.

Kennelly SJ (1987a) Inhibition of kelp recruitment by turfing algae and consequences for an Australian kelp community. *Journal of Experimental Marine Biology and Ecology* **112**, 49–60.

Kennelly SJ (1987b) Physical disturbances in an Australian kelp community. I. Temporal effects. *Marine Ecology Progress Series* **40**, 145–153.

Kennelly SJ, Underwood AJ (1992) Fluctuations in the distributions and abundances of species in sublittoral kelp forests in New South Wales. *Australian Journal of Ecology* **17**, 367–382.

Kennelly SJ, Underwood AJ (1993) Geographic consistencies of effects of experimental physical disturbance on understorey species in sublittoral kelp forests in central New South Wales. *Journal of Experimental Marine Biology and Ecology* **168**, 35–58.

Kinlan BP, Graham MH, Sala E, Dayton PK (2003) Arrested development of giant kelp (*Macrocystis pyrifera*, Phaeophyceae) embryonic sporophytes: a mechanism for delayed recruitment in perennial kelps? *Journal of Phycology* **39**, 47–57.

Kirkman H (1981) The first year in the life history and the survival of the juvenile marine macrophyte, *Ecklonia radiata* (Turn.) J. Agardh *Journal of Experimental Marine Biology and Ecology* **55**, 243–254.

Kirkman H (1984) Standing stock and production of *Ecklonia radiata* (C. Ag.) J. Agardh. *Journal of Experimental Marine Biology and Ecology* **76**, 119–130.

Kirkman H (1989) Growth, density and biomass of *Ecklonia radiata* at different depths and growth under artificial shading off Perth, Western Australia. *Australian Journal of Marine and Freshwater Research* **40**, 169–177.

Ladah LB, Zertouche-González JA (1999) Giant kelp (*Macrocystis pyrifera*, Phaeophyceae) recruitment near its southern limit in Baja California after mass disappearance during ENSO 1997–1998. *Journal of Phycology* **35**, 1106–1112.

Larkum AWD (1986) A study of growth and primary production in *Ecklonia radiata* (C. Ag.) J. Agardh (Laminariales) at a sheltered site in Port Jackson, New South Wales. *Journal of Experimental Marine Biology and Ecology* **96**, 177–190.

Lourey MJ, Kirkman H (2009) Short-lived dissolved nitrate pulses in a shallow Western Australian coastal lagoon. *Marine and Freshwater Research* **60**, 1068–1080.

Mann KH, Kirkman H (1981) Biomass method for measuring productivity of *Ecklonia radiata* with the potential for adaptation to other large brown algae. *Australian Journal of Marine and Freshwater Research* **32**, 297–304.

Marine Environmental Systems (MES) (2003) '*Undaria pinnatifida* or 'Wakame' community based survey 2001/2002'. Report to Natural Heritage Trust, Canberra.

McPeake RH, Glantz DA, Shaw CA (1988) *The Amber Forest.* Aqua Quest Publications, San Diego.

Mills D, Crear B (2004) Developing a cost-effective puerulus collector for the southern rock lobster (*Jasus edwardsii*) aquaculture industry. *Aquacultural Engineering* **31**, 1–15.

Novaczek I (1981) Stipe growth rings in *Ecklonia radiata* (C. Ag.) J. Agardh (Laminariales). *British Phycological Journal* **16**, 363–371.

Novaczek I (1984a) Response of gametophytes of *Ecklonia radiata* (Laminariales) to temperature in saturating light. *Marine Biology* **82**, 241–245.

Novaczek I (1984b) Development and phenology of *Ecklonia radiata* at two depths in Goat Island Bay, New Zealand. *Marine Biology* **81**, 189–197.

Poore AGB, Hill N, Sotka E (2007) Phylogenetic and geographic variation in host breadth and composition by herbivorous amphipods in the Family Ampithoidae. *Evolution* **62**, (1), 21–38.

Reed DC, Kinlan BP, Raimondi PT, Washburn L, *et al.* (2006) A metapopulation perspective on the patch dynamics of giant kelp in southern California. In *Marine metapopulations* (Eds JP Kritzer and PF Sale) pp. 353–386. Elsevier Academic Press, Burlington, MA.

Sanderson JC (1990) A preliminary survey of the distribution of the introduced macroalga *Undaria pinnatifida* (Harvey) Suringer on the east coast of Tasmania. *Botanica Marina* **33**, 153–157.

Sanderson JC (1992) Subtidal macroalgal studies in east and south eastern Tasmanian coastal waters. MSc thesis. University of Tasmania, Hobart.

Sanderson JC, Ibbott S, Foster J (2004) 'Restoration of string kelp (*Macrocystis pyrifera*) habitat on Tasmania's east and south coast'. Report to National Heritage Trust. Seacare & Marine Environmental Systems, Hobart.

Santelices B (1990) Patterns of reproduction, dispersal and recruitment in seaweeds. *Oceanography and Marine Biology - an Annual Review* **28**, 177–276.

Schiel DR (1988) Algal interactions on shallow subtidal reefs in northern New Zealand: a review. *New Zealand Journal of Marine and Freshwater Research* **22**, 481–489.

Schiel DR, Foster MS (2006) The population biology of large brown seaweeds: ecological consequences of multiphase life histories in dynamic coastal environments. *Annual Review of Ecology Evolution and Systematics* **37**, 343–372.

Shepherd SA (1979) The ecology of a deep water red algal community off Cape Northumberland, South Australia. MEnvSt thesis. University of Adelaide, Adelaide.

Shepherd SA, Womersley HBS (1976) The subtidal algal and seagrass ecology of St Francis Island, South Australia. *Transactions of the Royal Society of South Australia* **100**, 177–191.

Shepherd SA, Womersley HBS (1981) The algal and seagrass ecology of Waterloo Bay, South Australia. *Aquatic Botany* **11**, 305–371.

Shepherd SA, Watson JE, Womersley HBS, Carey JM. (2009) Long-term changes in macroalgal assemblages after increased sedimentation and turbidity in Western Port, Victoria, Australia. *Botanica Marina* **52**, 195–206.

Staehr PA, Wernberg T (2009) Physiological responses of *Ecklonia radiata* (Laminariales) to a latitudinal gradient in ocean temperature. *Journal of Phycology* **45**, 91–99.

Steinberg P, Kendrick G (1999) Kelp forests. In *Under Southern Seas: the Ecology of Australia's Rocky Reefs.* (Ed. N Andrew) pp. 60–71. University of New South Wales Press, Sydney.

Steneck RS, Graham MH, Bourque BJ, Corbett C *et al.* (2002) Kelp forest ecosystems: biodiversity, stability, resilience and future. *Environmental Conservation* **29**, 436–459.

Stewart CM, Higgins HG, Austin S (1961) Seasonal variation in alginic acid, mannitol, laminarin, and fucoidin in the brown alga, *Ecklonia radiata. Nature* **192**, 1208

Tala FB, Edding ME (2005) Growth and loss of distal tissue in blades of *Lessonia nigrescens* and *Lessonia trabeculata* (Laminariales). *Aquatic Botany* **82**, 39–54.

Tala FB, Edding ME (2007) First estimates of productivity in *Lessonia trabeculata* and *Lessonia nigrescens* (Phaeophyceae: Laminariales) from the southeast Pacific. *Phycological Research* **55**, 66–79.

tom Dieck I (1993) Temperature tolerance, survival in darkness of kelp gametophytes (Laminariales, Phaeophyceae): ecological and biogeographical implications. *Marine Ecology Progress Series* **100**, 253–264.

Toohey BD Kendrick GA (2007) Survival of juvenile *Ecklonia radiata* sporophytes after canopy loss. *Journal of Experimental Marine Biology and Ecology* **349**, 170–182.

Valentine JP, Johnson CR (2003) Establishment of the introduced kelp *Undaria pinnatifida* in Tasmania depends on disturbance to native algal assemblages. *Journal of Experimental Marine Biology and Ecology* **295**, 63–90.

Valentine JP, Johnson CR (2004) Establishment of the introduced kelp *Undaria pinnatifida* following dieback of the native macroalga *Phyllospora comosa* in Tasmania, Australia. *Marine and Freshwater Research* **55**, 223–230.

Wernberg T, Kendrick GA, Phillips JC (2003) Regional differences in kelp-associated algal assemblages on temperate limestone reefs in south-western Australia. *Diversity & Distributions* **9**, 427–441.

Wood WF (1987) Effect of solar UV radiation on the kelp *Ecklonia radiata. Marine Biology* **96**, 143–150.

5 Large brown algae: the fucoids

OVERVIEW

Fucoid algae are the dominant group of canopy species of brown algae on exposed to sheltered southern coasts. The life histories of several species of *Cystophora* and *Sargassum*, and *Xiphophora gladiata*, *Phyllospora comosa* and *Durvillaea potatorum*, are described. These species live for several years and have only one free-living generation. Fertilised eggs from sporophytes sink to the bottom and grow into the next generation of sporophytes. Fucoids mainly occur as mixed assemblages of many species, and these are described for specific sites from SW to SE Australia. The composition of fucoid species changes along an exposure gradient. In areas of strong water movement, *Acrocarpia paniculata* and *Scytothalia* are common and, in moderate to slight water movements, species of *Cystophora*, *Sargassum*, *Seirococcus*, *Caulocystis* and *Scaberia* tend to form a replacement series along a water movement gradient. In the cooler waters of SE Australia and Tasmania, *Durvillaea*, *Phyllospora* and *Xiphophora* are dominant in shallow water. The patch dynamics of fucoid assemblages are complex. In exposed conditions, disturbance from storms tears out plants and creates gaps, so making space for recruitment of many species. Disturbance declines with increasing shelter, and the colonisation of patches becomes increasingly dominated by local factors, such as the extent of spore dispersal, topography and competition.

INTRODUCTION

The most abundant canopy-forming brown algae in southern Australia are the fucoids (Order Fucales), comprising > 60 species, mainly in the genera *Cystophora* (23 species), *Sargassum* (19 species), *Myriodesma* (seven species), *Acrocarpia* (two species), *Caulocystis* (two species), *Xiphophora* (two species) and one species in each of the genera *Carpoglossum*, *Seirococcus*, *Scytothalia*, *Phyllospora*, and *Scaberia*. *Xiphophora gladiata* is limited to colder Tasmania waters and *Phyllospora comosa* to southern New South Wales, SE of South Australia, Victoria and Tasmania. A related cold-water species, the bull-kelp *Durvillaea potatorum*, once placed in the Fucales, but now in a separate order Durvillaeales, is also included. On the southern coasts of Australia, these species variously dominate the majority of reef habitats from the low intertidal to 10–30 m depth (to > 50 m in the clearest ocean waters of the Great Australian Bight), and from sheltered to moderately exposed waters. Not surprisingly, the composition of fucoid-dominated species assemblages varies greatly with depth and water movement gradients, with longitude from western to eastern Australia, and with latitude from warm to cool temperate environments. In this chapter, we discuss firstly the distinctive life history features of fucoids, then the ecology of some well-studied species, and lastly, the changing pattern of assemblages throughout the southern region. Information is derived from the reviews of Clayton (1990), Emmerson and Collings (1998), Goldberg and Kendrick (2007), and Turner and Collings (2008), the studies of Greg Collings (1996), Sarah Hotchkiss (1999) and David Turner (2004), and others referred to below.

LIFE HISTORIES

In contrast to kelps, fucoid algae do not have a free-living gametophyte stage. Some species are dioecious (unisexual male and female plants) and some are monoecious (bisexual plants). Fertile areas on fronds can be recognised by the swellings, which are called conceptacles, and are located on the ends of reproductive ramuli, called receptacles. In bisexual algae, such as some species of *Cystophora*, the conceptacles produce both sperm and eggs (Fig. 5.1). Eggs adhere to the parent alga during fertilisation, and then the fertilised eggs drift off and sink to the bottom. Fertilised eggs should be strictly called zygotes, but the more general

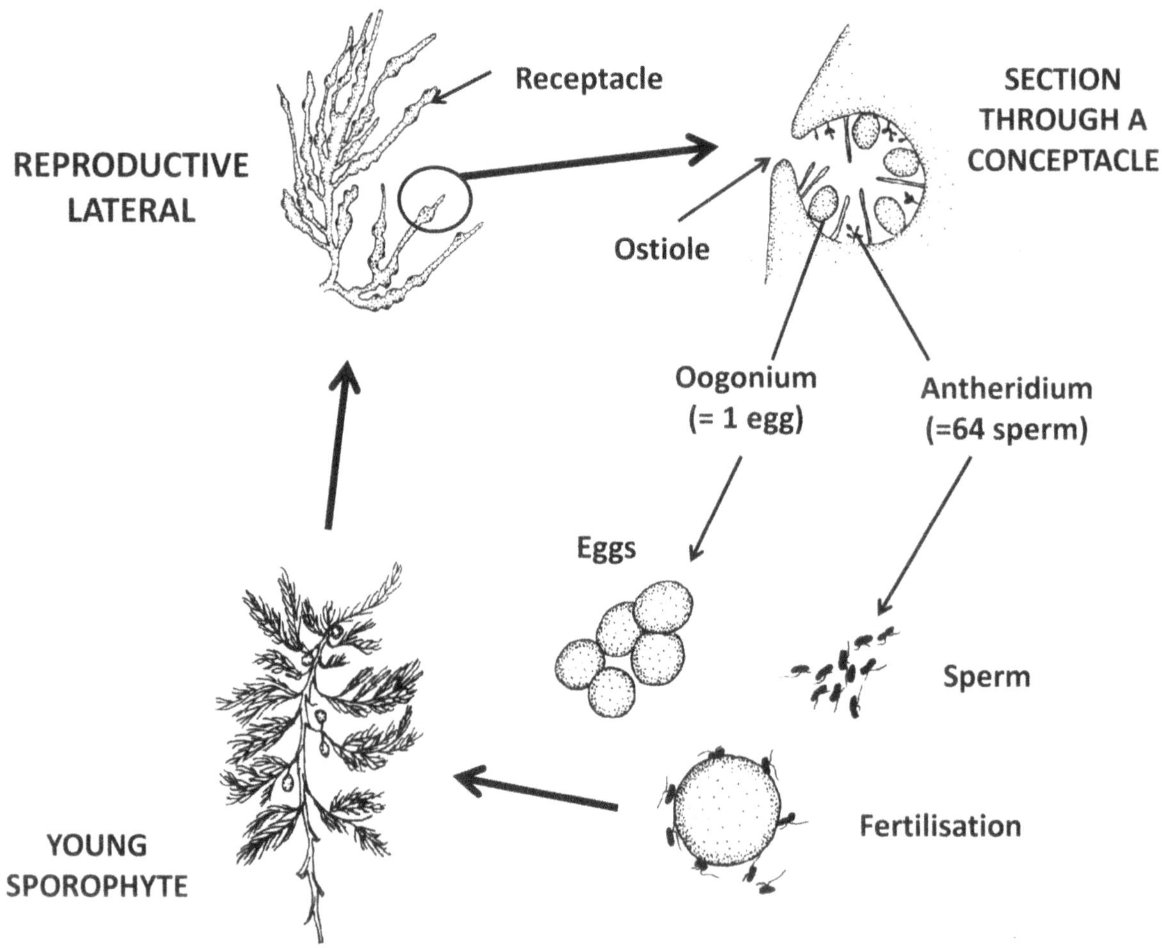

Figure 5.1: Reproduction in species of *Cystophora*. Swollen ramuli contain the receptacles, and gametes (eggs and sperm) are produced in the conceptacles embedded in the receptacles. The fertilised egg sinks to the bottom and grows into the next generation of sporophyte (Courtesy Sarah Hotchkiss 1999).

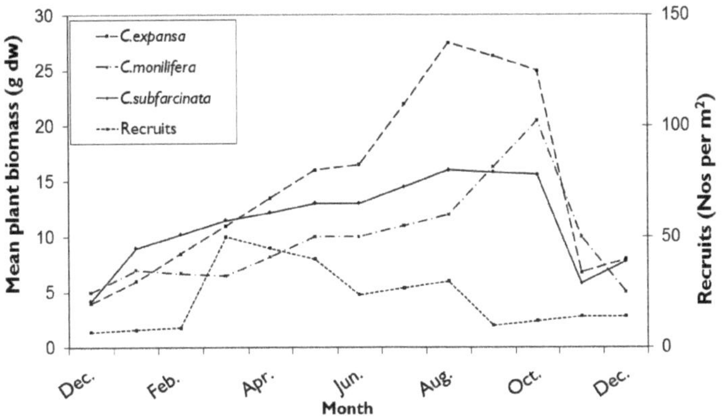

Figure 5.2: Change in mean plant biomass (g dry weight) of three species of *Cystophora* over a year at Cape Jervis, and change in density of recruits of the three species combined throughout the year (redrawn from Hotchkiss 1999, omitting standard errors).

term 'propagule' is often used for the product of the different types of reproductive cells. In the case of some species of *Sargassum*, fertilised eggs are larger and are retained on the parent plant for up to 2 days before release, whereupon they sink to the bottom and recruit locally. The life histories of several species of *Cystophora* and *Sargassum* are described below.

The genus *Cystophora*

The life histories of three common species, *Cystophora expansa, C. monilifera* and *C. subfarcinata*, were examined by Sarah Hotchkiss (1999) in a mixed *Cystophora* community at Cape Jervis, South Australia (Plate 6). Plants are perennial and live for several years, but their biomass changes greatly throughout the year. Plant biomass is low in late spring to summer as plants shed their laterals with ramuli and receptacles, leaving denuded stems. Plant biomass then increases again in winter, and reaches a peak in early spring, as ramuli and receptacles develop (Fig. 5.2). At full maturity, eggs and sperm are released epidemically, triggered by a temperature, light or lunar signal.

The reproductive strategies of the three species differ slightly, as shown in Table 5.1. *C. expansa* produces the most eggs, while *C. monilifera* produces fewer, but larger, ones. *C. subfarcinata* produces the smallest eggs, but is reproductive for longer than the other species. The benefit of a winter reproductive season for all the species is that winter storms tend to thin the canopy, creating gaps in which plant recruits can settle and grow faster in an environment richer in nutrients and with lengthening days. Fucoid algae are poor dispersers, and most spores settle near the parent plant. Hence, clearings caused by storms within a forest are quickly repopulated by spores from surrounding plants, or possibly from drift fertile plants transported for some distance.

The recruitment of *Cystophora* was examined by Emmerson and Collings (1998). In order to mimic storm effects, they made clearings of different sizes in

Table 5.1: Reproductive strategies and average growth rates of three species of *Cystophora* (after Hotchkiss 1999).

ww = wet weight.

Species	Egg release	No. eggs ($\times 10^6$)	Egg size ($\times 10^{-7}$g ww)	Growth rate (cm month^{-1})
C. expansa	Jun.–Oct.	6.6	3.8	4.9
C. monilifera	Late Aug.–Oct.	1.3	4.5	1.5
C. subfarcinata	Mid-Aug.–Oct.	2.2	2.2	1.6

a *Cystophora* forest near Sarah Hotchkiss' site. Initially, the largest clearings of 1 m^2 achieved the highest settlement of recruits, but after 5 months many of the recruits failed to survive, especially at the edges, apparently due to density-dependent mortality effects. Growth of recruits was faster and survival better in the middle of the clearings, where light reaching the bottom was greater. Recruitment was poor under a closed canopy. The authors concluded that gaps in the canopy, usually caused by storms, are soon recolonised, despite high mortality of recruits from herbivore grazing, scouring by adult plants and other factors.

At the end of the reproductive season, laterals and ramuli are shed in *C. expansa* and *C. monilifera*, but not in *C. subfarcinata*, which retains sterile ramuli during summer. In comparing the three species, Hotchkiss (1999) considered that *C. expansa* was the better competitor, with 'late-successional' features, whereas *C. subfarcinata* had opportunistic features – that is, it was able to quickly colonise any gap that appeared. Hence *C. expansa* often occurs in dense stands where it excludes other species, whereas *C. subfarcinata* can co-occur with other species over a range of wave energy environments. All three species allocate 87–91% of their biomass to reproduction.

In Sarah Hotchkiss' study, recruits (i.e. those < 5 cm high) appeared all year round, but were in highest numbers in March, when the canopy was at its thinnest (Fig. 5.2). She estimated that newly settled propagules took ~5 months to reach a size of 5 cm, and ~15 months to reach a canopy height of ~20 cm (Fig. 5.3). The average growth rate was derived from a growth model and did not show any seasonal changes in growth, which followed that of biomass increase. The three species have two stages of growth. Initially, they increase in length through summer–autumn, and then they increase in bulk as ramuli with reproductive tissue develop (the sub-canopy juvenile stage in Fig. 5.3). The minimum summer growth rates were likely due to nitrogen limitation.

The mortality of the species varied with season; in summer and early autumn it was 5–15% p.a., increasing to 20–35% in late autumn to winter, and falling to 10–15% in spring. The average length of lost primary axes (the stems) ranged from 25–35 cm in summer to 50–60 cm in winter.

The genus *Sargassum*

Although the 19 or so species of *Sargassum* are variously dominant on many semi- or fully sheltered reefs throughout southern Australia, their ecology is known mainly from studies of Gary Kendrick and associates at Rottnest I., Western Australia (Kendrick 1993, 1994; Kendrick and Walker 1991, 1994, 1995) and a study by Edgar (1983) in Tasmania (Plate 7). At Rottnest I., eight species of *Sargassum* comprise much of the algal biomass, and contribute ~20% of the annual primary production of the inshore reef system. At the Tasmanian site, two species of *Sargassum*, one *Cystophora* and one *Caulocystis* are present.

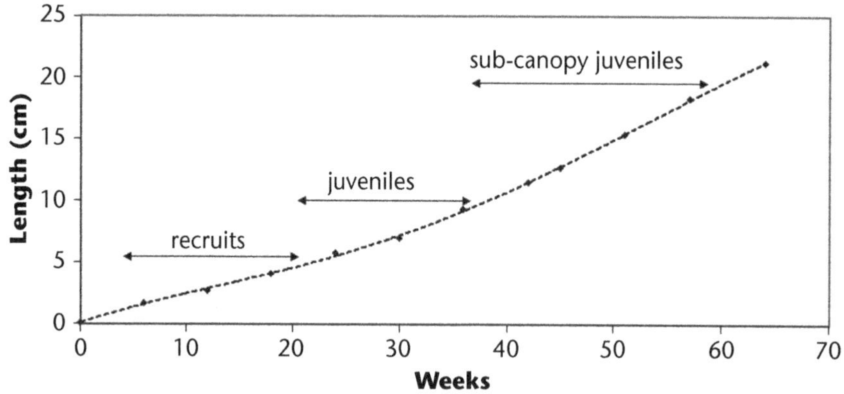

Figure 5.3: Modelled growth rate of recruits of three species of *Cystophora* at Cape Jervis, showing terminology used for three different growth phases (redrawn from Hotchkiss 1999).

Reproduction

The species of *Sargassum* have several primary branches that grow out of one or several central stems attached to a holdfast. From the primary branches lateral branches grow, and from these arise leaves, reproductive branches (the receptacles) and vesicles – the air bladders that give the plant buoyancy. Most species have a narrow seasonal reproductive season and, after eggs are released, the lateral branches, which make up the bulk of the algal biomass, are shed, leaving a stub or basal leaves, which can regenerate the following year (see below).

Like other fucoid algae, species of *Sargassum* have only one free-living generation. The sporophytes are unisexual or bisexual – that is, producing male and female gametes in separate or in the same individuals, respectively. After fertilisation, the fertilised egg (zygote) drifts to the bottom and grows into the next generation of sporophyte, as shown in Fig. 5.1. A detailed study of this process was carried out by May and Clayton (1991) on *Sargassum vestitum*, a common shallow-water species in Victorian and Tasmanian waters. The alga is fertile in summer, and when the eggs are released from the conceptacles they remain attached by stalks to the receptacles. Egg release is synchronised locally and occurs at low tide and over an 8-day period around the new moon. Fertilisation also occurs then, and the fertilised eggs remain attached to the parent plant for ~2 days as they grow rapidly to a size of ~350 µm, whereupon they are released into the water. It is thought that retention of the eggs by stalks to the plant's receptacles ensures 100% fertilisation of the eggs.

At Kendrick and Walker's (1994) study site in Nancy Cove, Rottnest I., three *Sargassum* species were present. Most (70%) of the individuals were *S. spinuligerum*, 26% were *S. distichum*, and the rest were *S. podacanthum*. Within a few days of the new moon during each month of the reproductive season, eggs were expelled from conceptacles in pulses, fertilised and then released as zygotes 1–2 days later. Life history information from the various studies is summarised in Table 5.2.

Dispersal

To determine the dispersal distance of propagules, Kendrick and Walker (1991) stained reproductive plants of *S. spinuligerum* and then measured how far the stained zygotes were carried by wave action. They found that the dispersal distance was very short, with 98% of zygotes settling within 1 m of the parent plant. Yet a single plant in a season could release up to 6 million zygotes. Although such localised dispersal ensured that a *Sargassum* patch would persist, it also raised the possibility that a genotype of a species could change from one place to another. This possibility was later confirmed. Kendrick (1993) found that a population of *S. spinuligerum* at North Point only 2 km distant from the Nancy Cove site was reproductive in winter, rather than in summer as at Nancy Cove (Table 5.2). He proposed two alternative explanations for the difference.

Table 5.2: Life history information for five species of *Sargassum*, with range of survival estimates of recruits from settlers (Rec) and from regenerating crusts (Cr) for different cohorts of the first three species combined (after Kendrick and Walker 1991, Kendrick 1993); May and Clayton 1991 for *S. vestitum*; and Edgar 1983 for *S. fallax* and *S. verruculosum*).

n.k. = not known; do = ditto.

Species	Egg release	Egg size (µm)	% survival (1 year)	% survival (2 years)
S. spinuligerum	Sept.–Jan.(mid-winter)[1]	120–280	17–35 (Rec)17–26 (Cr)	4–15 (Rec)4–9 (Cr)
S. distichum	Oct.–Dec.	do	do	do
S. podacanthum	Nov.–Dec.	do	do	do
S. tristichum	Mar.–Jun.	n.k.	n.k.	n.k.
S. vestitum	Oct.–Feb.	150–200	n.k.	n.k.
S. fallax	Oct.–Feb.	n.k.	n.k.	75[2]
S. verruculosum	All year	n.k.	n.k.	80[2]

[1]another deeper site at Rottnest Island 2 km from Kendrick and Walker's main site at Nancy Cove. [2]adult plants.

First, hormones released by another common species in the vicinity, *Sargassum tristichum*, which is reproductive in winter, could induce reproduction in *S. spinuligerum* (an unusual phenomenon known for other *Sargassum* spp.), or alternatively, the North Point population could have become established after a long-distance dispersal event from more northern populations via plants drifting south in the Leeuwin Current. In a later study specifically on the dispersal of propagules of *Sargassum* and *Cystophora*, Goldberg *et al.* (2004) put settlement plates in the sea in a systematic pattern at distances of up to 500 m from an island in Esperance Bay, and recorded the abundance of algal recruits of the above genera onto them over 8 months. The density of recruits declined with distance from the island. At 100 m distance, their density was about one-third to a half of that close to the island, and almost zero 500 m away. Hence, populations of both genera are likely to be largely recruited locally around each island of the Recherche Archipelago.

Given the very limited spore dispersal of *Sargassum* and *Cystophora*, favouring the persistence of local patches, and also the possibility of occasional, longer-distance dispersal by drifting plants, it is likely that the dynamics of metapopulations (i.e. local populations linked by propagule dispersal with other local populations to form a metapopulation) is quite complex, as it is in *Macrocystis* (Reed *et al.* 2006).

Recruitment

In the Kendrick and Walker study, recruitment of young plants < 10 cm size was measured over 2 years. Peaks occurred in February, but continued at low levels for the remainder of the year, suggesting that zygotes were either in the water column all year round or, more likely, that after settling, they remained dormant with subsequent delayed development. Recruits grew into adults at 50–90 cm size in 4–5 months after settlement, and then lived for up to 27 months (Table 5.2). However, mortality of recruits was high, declining from densities of ~100 adults m^{-2} in late summer to 15–18 adults m^{-2} some 6–7 months later (see below). But the adult population was derived not only from young recruits, but also from plants that regenerated vegetatively from crusts of the plant on the bottom. Kendrick and Walker (1994) estimated that 43% of the adult population came from recruits and 57% from vegetative regeneration. The life history details of five species are summarised in Table 5.2, and adult and recruit plant densities are shown in Fig. 5.4 for the period of the study.

Other fucoid algae

The genera *Acrocarpia, Seirococcus* and *Scytothalia*

Two species of *Acrocarpia* occur: *A. robusta* in SW Australia and *A. paniculata* east of the head of the Great Australian Bight (GAB) to SE Australia. In Greg Collings' (1996) study, biomass fluctuated extensively but showed no seasonality. *Scytothalia dorycarpa* and *Seirococcus axillaris* – two very similar strap-like species up to 2 m long – similarly showed no seasonality in abundance at Greg Collings' sites (Plate 6). The species are distributed on a water movement gradient – *Scytothalia* and *Acrocarpia paniculata* in

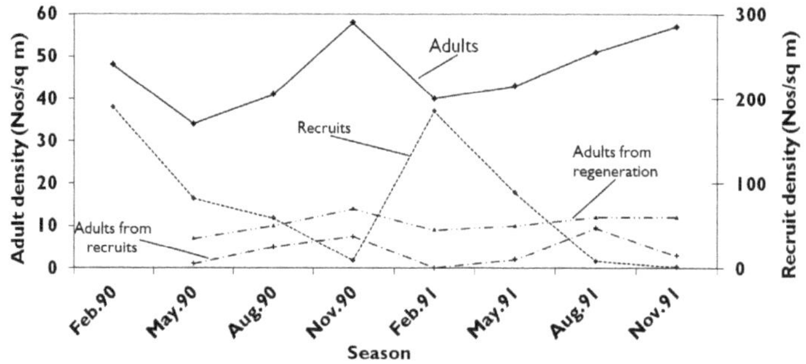

Figure 5.4: Trends in mean density (numbers m^{-2}) of adults and recruits of three species of *Sargassum* in permanent quadrats in Nancy Cove, Rottnest I., and trends in density of adults regenerating from holdfasts, and those growing from recruits (redrawn from Kendrick and Walker 1994, omitting standard errors).

extreme conditions, as also in Waterloo Bay (see later), and *Seirococcus* in moderate water movements. None of the species appear to dominate any sites, tending to occur in mixed fucoid assemblages. Their ecology is almost unknown.

The genus *Xiphophora*

Of the two southern species of *Xiphophora* – *X. chondrophylla* and *X. gladiata* – the latter was studied by Gillanders and Brown (1994) (Plate 8). This perennial species occurs in New Zealand and around Tasmania, although it is less common on Tasmania's north coast (Rice 1987). Gillanders and Brown found that in New Zealand it was first reproductive at a size of ~15 cm, and large plants (> 45 cm) were reproductive for 8 months of the year (May–December). The number of small recruits (< 15 cm) was highest from summer to early winter, and standing crop reached a maximum in autumn.

The crayweed *Phyllospora comosa*

Crayweed is abundant in the cooler waters of SE Australia and Tasmania, and also extends up the New South Wales coast as far north as Port Macquarie (Plate 9). It occurs to a depth of ~5 m on the east Australian coast, and further south below the *Durvillaea* zone to ~3 m deep, although on some exposed Tasmanian coasts it can go to 18 m depth. The species historically occurred around Sydney, but has disappeared there over some 70 km of coast (Coleman and Kelaher 2009).

The alga typically has a central main axis up to 3 m long, from which many branches arise, with shorter leaf-like laterals closely arranged along their length. Some of these short laterals bear conceptacles in which sperm- and egg-producing cells develop; the life cycle is similar to that of other fucoid algae. A genetic study by Coleman and Kelaher (2009) showed high connectivity between distant populations, suggesting that dispersal of propagules via floating, fertile drift plants was common.

Growth studies by Sanderson (1992) showed that growth is from apical cells and is confined to the distal 20–30 cm of the branches. It is also strongly seasonal, depending on the number of daylight hours. However, at temperatures > 14.5°C the growth rate peaks at ~6 h of sunshine, and then declines (Fig. 5.5) – a decline that may be due to the low level of nutrients at higher temperatures, rather than a direct temperature effect.

Craig Sanderson estimated that individuals 3 m long were at least 6 years old, although the laterals on the branches survived for only ~2 years. Studies of the survival of tagged individuals gave a half-life of ~2–3 years according to locality. *Phyllospora* forests are highly productive, and Sanderson estimated that the productivity of a forest at George III Reef was ~7.4 kg fresh weight m^{-2} $year^{-1}$.

The bull-kelp *Durvillaea*

Bull-kelp, once placed in the Order Fucales because of the similarity of reproduction, is now placed in its own order, Durvillaeales. *Durvillaea* dominates exposed to partly sheltered rocky shores in SE

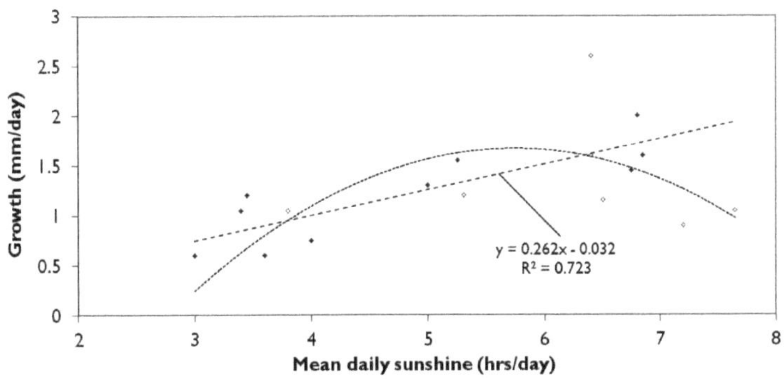

Figure 5.5: Plot of daily growth rate of branches of *Phyllospora comosa* versus mean number of daylight hours for sea temperatures ≤ 14.5°C (solid diamonds) and for those > 14.5°C (open diamonds). The regression equation is for data shown in solid diamonds and the curved dotted line is the curve of best fit for data shown in open diamonds (redrawn from Sanderson 1992).

Australia from Cape Jaffa to southern New South Wales, and around Tasmania. It is among the most massive of all algae, reaching a length of 8 m and a fresh weight of 75 kg. The ecology of the species is known from the studies of Cheshire and Hallam (1985; 1988a, b; 1989a, b), relating mainly to the morphology and distribution of the species.

Durvillaea is dioecious (i.e. male or female). The reproductive structures (conceptacles), with four eggs in each oogonium (the specialised egg-holding cell), are scattered over the majority of the frond in female plants, and antheridia (the sperm sacs) on branched filaments in male plants. The reproductive season runs from July to November each year (Cheshire *et al.* 1995; Phillips 2007).

Growth is spread diffusely over the blade, but is concentrated at the apex. Neither the growth rate nor the longevity of plants is known, although related New Zealand species live for up to 9 years (Hay 1978). Cheshire and Hallam (1989b) classified plants according to their vitality, based on plant weight, and 'age state', estimated by the circumference of the stipe. As plants aged, erosion of the blade exceeded growth, and plants gradually degenerated. In a survey of *Durvillaea* populations in SE Australia, the authors found that the oldest plants with highest vitality were in SE Tasmania and around King I. In general, the vitality of plants declined with decreasing exposure to swell.

Populations were classified as eastern and western according to slight differences in shape between them. These differences were thought to be due to past isolation of eastern and western populations when they were separated by the Bassian Isthmus during the past million years (see Chapter 2). Further, on an exposure gradient, there is a trend for plants to have long, thin stipes and short, wide blades in low wave exposure and short, thick stipes and long, narrow blades in extreme exposures. The high levels of alginates in the blades and stipes (up to 55%) provide strength and flexibility to the plants, and enable them to flex and bend in the extreme surge conditions. The holdfasts produce rhizoids that fill every tiny space in the substratum, and a force of up to 2.75 kN (= 280 kg) is required to remove them. Hence, *Durvillaea* can withstand strong swells, with water velocities of 14 m s^{-1}.

Alginates have many food and industrial applications, and, given the high standing crop of *Durvillaea*, ranging from 4 to 108 kg m^{-2} (average 22.5 kg m^{-2}), an industry developed at King I., in Bass Strait collecting drift plants washed ashore. Each year up to 5000 t (dry weight) of *Durvillaea* are harvested and exported.

FUCOID ASSEMBLAGES

Algal forests are a distinctive feature across the southern half of Australia in sheltered to moderately exposed waters on subtidal reefs, mainly to depths of 15–20 m. Whereas *Ecklonia* is a major dominant species on the west and east Australian coasts, with fucoid algae mainly limited to sheltered waters (Kendrick *et al.* 1999; Wernberg *et al.* 2003; Smale *et al.* 2010; Chapter 4), on the south coast fucoid algae become prominent over a wide range of exposures. The distribution of individual species is the outcome of their individual life histories interacting with temperature, depth, water movement, nutrients, competition, herbivory and other factors, very few of which have been examined. Here we describe the changing pattern of dominance of the many fucoid species across the south coast, from west to east.

SW Australia to the Coorong, central South Australia

Rottnest I

In sheltered waters, *Sargassum* is often the major dominant genus of canopy species, and Gary Kendrick (1993) studied a *Sargassum* assemblage at Rottnest I. Here, eight *Sargassum* species occur on shallow reefs in several multi-species associations, mainly at depths < 6m, although some species extend to 35 m depth. The component species in these associations varied with location and depth. The most common species was *S. spinuligerum* at most sites, switching to *S. tristichum* at some sites on the north coast, where it reached its highest densities at 15 m depth. At more exposed sites, *Sargassum* beds were mixed with *Ecklonia* and other species. In the very shallow sub- to intertidal zones, the rare species *S. ligulatum* was dominant. However, the variation in composition of species between sites and at different depths was so complex as to defy generalising about patterns, and Gary Kendrick suggested that every *Sargassum* bed was unique with its own combination of dominant species.

Recherche Archipelago

Subsequently, the subtidal assemblages of this region were described by Goldberg and Kendrick (2005). The islands of the Archipelago are exposed to strong SW Southern Ocean swells, and exposure gradients around islands and depth determined the presence and abundance of individual species. A schematic diagram of the major canopy species (Fig. 5.6) shows their approximate centres of abundance in terms of depth and water movement. However, it is notable that many species that prefer sheltered conditions in shallow water also occur in deeper water (> 20 m) on the exposed sides of islands. *Ecklonia* is a dominant only in the most exposed outer faces of islands and *Acrocarpia robusta* is found in calmer water than its eastern sister species, *A. paniculata*. Twelve species of *Sargassum* were recorded, but were too many to locate individually in the figure (Goldberg and Kendrick 2005).

Eastern Great Australian Bight

Several expeditions to islands of the Investigator Group and Nuyts Archipelago since the 1970s have examined their algal assemblages (reviewed by Baker *et al.* 2008). On the mainland coast, Shepherd and Womersley (1981) examined the distributional abundance of algae in Waterloo Bay. This bay is in a region of seasonal upwelling and faces the open swell of the eastern GAB, and is well known for its rich algal flora totalling > 364 species. Many reefs, from extreme exposure to sheltered, occur over a range of depths in different parts of the Bay, and 21 canopy species, each dominant at different depths and water movement conditions, were present. A mosaic chart (Fig. 5.7) shows the approximate centres of distribution, in terms of maximum biomass, of the main canopy species, and Fig. 5.8 shows a gradual replacement series of canopy species along a water movement gradient at

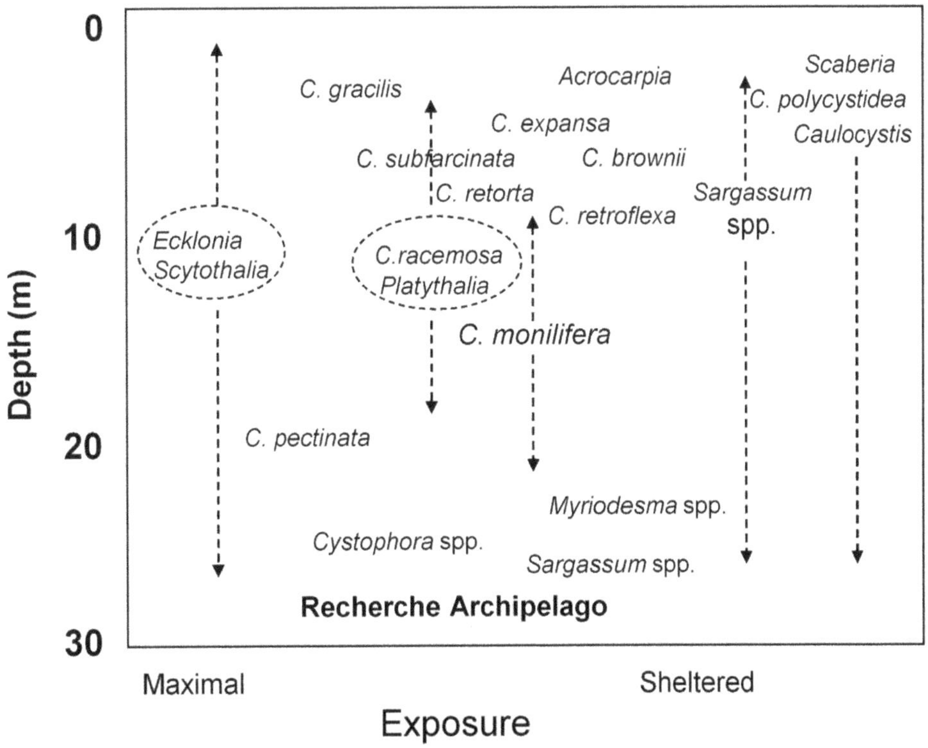

Figure 5.6: Schematic diagram showing approximate centres of distribution of dominant species and their depth range in the Recherche Archipelago, western Great Australian Bight (redrawn from Goldberg and Kendrick 2004).

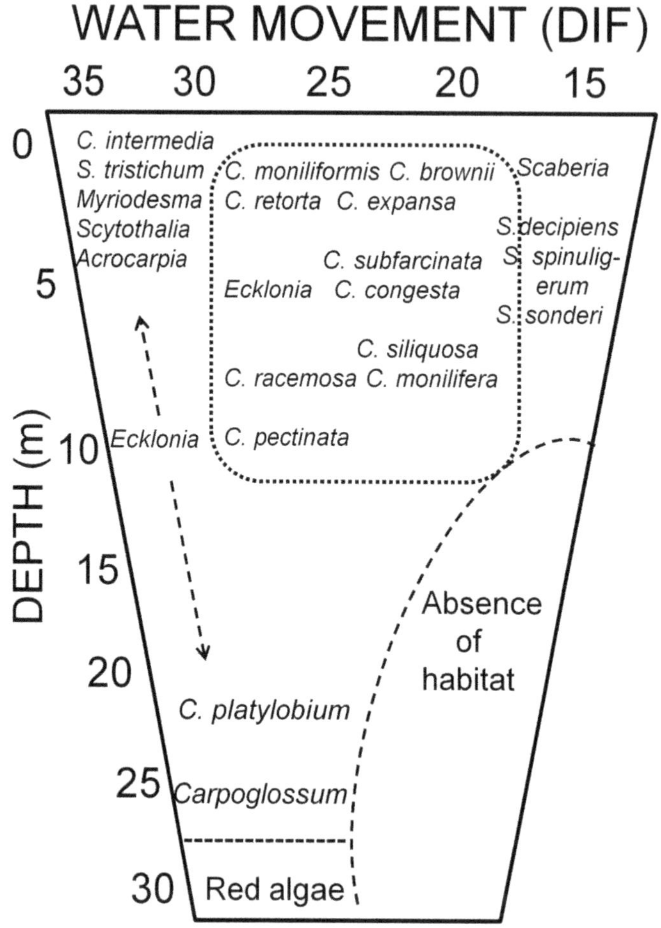

Figure 5.7: Depth–water movement mosaic showing approximate centres of distribution of canopy species according to depth and water movement at Waterloo Bay, eastern Great Australian Bight. Arrows show the broad ranges of *Ecklonia* and *Acrocarpia*. The boundaries of the assemblage dominated by *Cystophora* spp. are shown with a dotted line, and other assemblages are roughly separated by dotted lines (redrawn from Shepherd and Womersley 1981).

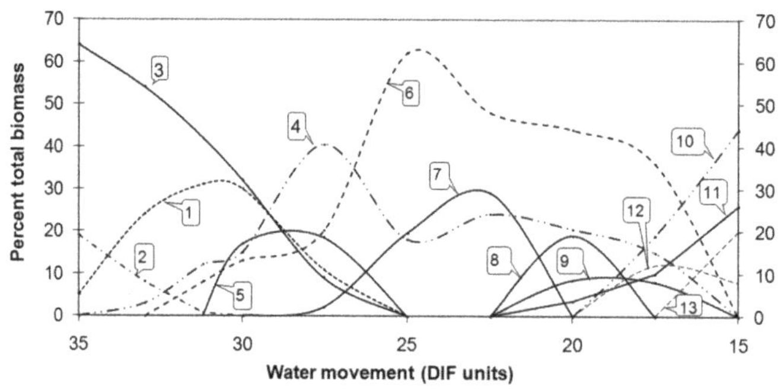

Figure 5.8: Distribution of percentage biomass (fresh weight) of canopy species in Waterloo Bay along a water movement gradient in DIF units at 5 m depth. Species are: 1, *Ecklonia* (dotted line); 2, *Myriodesma harveyana*; 3, *Acrocarpia paniculata*; 4, *Cystophora moniliformis*; 5, *C. racemosa*; 6, *C. siliquosa*; 7, *C. congesta*; 8, *C. subfarcinata*; 9, *C. monilifera*; 10, *Scaberia agardhii*; 11, *C. brownii*; (2–11 are continuous lines); two unnumbered species of *Sargassum* at far right (long dashes) are *S. decipiens* and *S. spinuligerum* (redrawn from Shepherd and Womersley 1981).

5 m depth, as an example of the replacement series that typically occurred along any gradient. Note that in these figures the unit of water movement strength used is the 'diffusion increase factor' (DIF), as measured by the rate of dissolution of plaster clod cards in Shepherd and Womersley (1981). At sites directly exposed to strong swell (30–35 DIF), fucoids, such as *Scytothalia* and *Acrocarpia*, and the kelp *Ecklonia* were dominant, but in moderate to sheltered conditions many species of *Cystophora* shared dominance (cf. Shepherd and Womersley 1971). Species of *Sargassum* tended to dominate in the calmest conditions, but these habitats were quite limited in the generally exposed Waterloo Bay.

Brief studies by Shepherd and Womersley (1976) and Turner and Cheshire (2003) at Nuyts Archipelago gave a similar picture, but the latter study used 20 m line intercept transects to estimate cover, mainly at depths < 12 m, and so was also able to examine the cover of the major patch-forming assemblages, and the scale of patchiness. By combining data for species in the same genera (*Cystophora* and *Sargassum*), Turner and Cheshire (2003) produced a picture of the shift in major assemblages along an exposure gradient, similar to that in Waterloo Bay (Fig. 5.9). But, in addition to the four assemblage groups, they recorded a fifth assemblage without any clear dominant canopy species, comprising many species in low abundance – some of them juveniles of canopy species. They called this the VLA (variable low abundance) group, and its presence led them to propose a model of how the stability of the four assemblages was maintained (discussed below).

The South Australian Gulfs

In contrast to the above studies, where long-period swells are the main water movement type, Collings and Cheshire (1998) examined the distributions of canopy species at less exposed sites subject mainly to wind-driven wave action near the entrances to the two South Australian gulfs. The biomass distributions of the individual species (Table 5.3) illustrate the gradual shift in dominance from species of *Cystophora* to species of *Sargassum*. When biomasses of species are combined at the generic level (Fig. 5.10), it can be seen that with declining water movement the biomass of *Cystophora* declines at about the same rate as that of *Sargassum* increases. Collings and Cheshire then derived an index of water movement from wind-driven waves based on average wind speed and fetch, and showed that the *Cystophora*:*Sargassum* biomass ratio (their 'fucoid index') was positively correlated with their water movement index. So, they proposed the fucoid index as a practical surrogate for water movement strength.

The acknowledged weakness of the index was that it only applied to communities dominated by *Cystophora* and *Sargassum*, and so was unlikely to be applicable at depths of > 10 m or to stronger water movements from swell, as is typical of oceanic coasts. For example, it would not apply to severe water

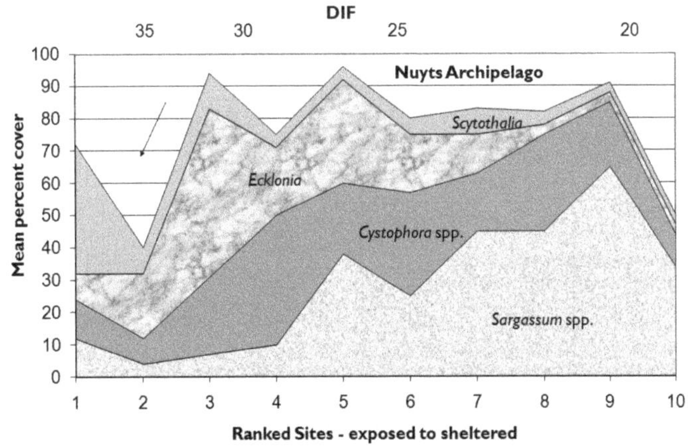

Figure 5.9: Change in mean percentage cover of patch-forming, dominant canopy species or groups of species in Nuyts Archipelago at 10 sites ranked according to declining exposure to swell (redrawn from Turner and Cheshire 2003). Approximate DIF values (see Fig. 5.7) are added for comparison with Waterloo Bay.

Table 5.3: Change in the composition of canopy species in terms of biomass (g dw m^{-2}) along the water movement gradient shown in Figs 5.7 and 5.9, using the same DIF units (redrawn from Collings and Cheshire 1998).

Canopy species	**Water movement (DIF units)**					
	25	**23**	**21**	**19**	**17**	**15**
Cystophora siliquosa	■	░	░	░		
Cystophora subfarcinata	▒					
Cystophora moniliformis	▒	▒	░	░		
Cystophora brownii			▒	░		
Cystophora expansa		░	▒	▒	▒	
Cystophora retorta		....	....	....		
Cystophora pectinata		....				
Cystophora polycystidea		....				░
Sargassum decipiens		▒	▒	▒	▒	░
Sargassum lacerifolium		░	░	....	░	░
Sargassum linearifolium			░	▒	▒	
Sargassum verruculosum		░	....	░		
Sargassum heteromorphum		░	░	░		
Seirococcus axillaris		▒	....	░		
Ecklonia radiata		░	░	░	░	
Acrocarpia paniculata		░	░	....	░	

Key: ■ > 500; ▒ 100–500; ░ 10–100; .... < 10. Numbers in g dw m^{-2}.

movement conditions, where *Cystophora* is replaced by *Ecklonia* (see Fig. 5.9). Further, there are exceptional species, such as *Cystophora siliquosa*, which occurs in quite strong water movement, *Cystophora polycystidea*, which is dominant in extreme shelter, and *Sargassum fallax*, which is common on moderately exposed coasts – all of which limit its application. Another problem is that a species (e.g. *Sargassum spinuligerum*) may prefer high exposure at one site (Marmion Lagoon) and low exposure elsewhere

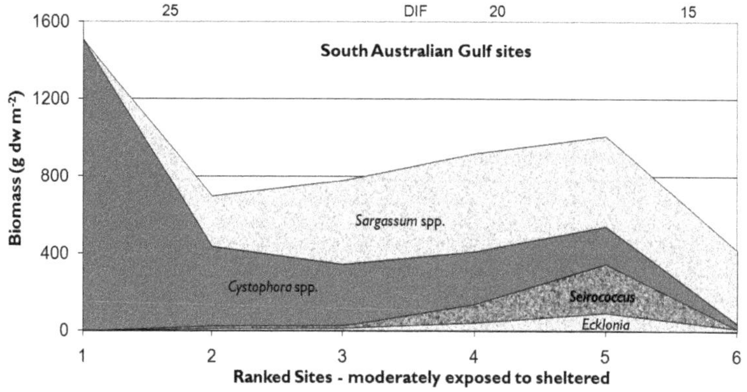

Figure 5.10: Change in biomass (g dry weight m^{-2}) along a water movement gradient of the canopy genera, *Cystophora*, *Sargassum*, *Seirococcus* and *Ecklonia*, for sites in lower Spencer and St Vincent Gulfs and Investigator Strait ranked according to decreasing water movement. Data for the last two genera were smoothed with running mean of three. Estimated DIF values are given for comparison with Figs. 5.7–5.10 (redrawn from Collings and Cheshire 1998).

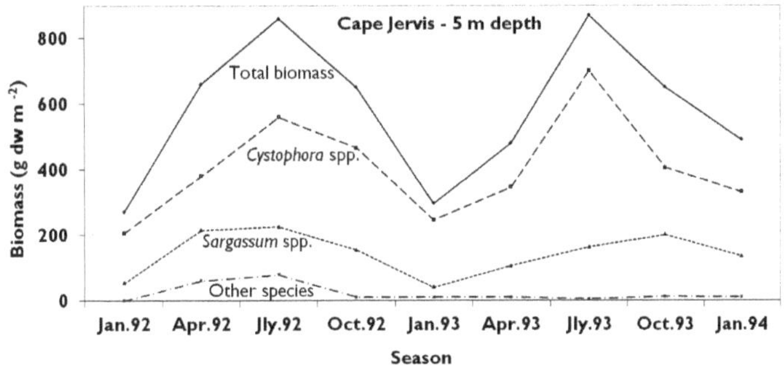

Figure 5.11: Changes in biomass (g dry weight m^{-2}) of a mixed fucoid community at Cape Jervis east (with a DIF value of 20–22) over 2 years. The species of *Cystophora* and *Sargassum* were: *C. brownii* (4%), *C. expansa* (28%), *C. monilifera* (9%), *C. moniliformis* (20%), *C. subfarcinata* (11%), and *S. fallax* (4%), *S. linearifolium* (9%), *S. varians* (8%, and *S. verruculosum* (5%). The other species of low biomass were: *Acrocarpia*, *Scytothalia* and *Seirococcus* (redrawn from Collings 1996).

(Waterloo Bay) for reasons not yet understood (Kendrick *et al.* 1999, Shepherd and Womersley 1981).

Greg Collings (1996) also examined the variation in space and over time of macroalgal assemblages at some of the above sites. At a semi-sheltered site (Cape Jervis east), the biomass of *Cystophora* spp. and *Sargassum* spp. fluctuated cyclically with maxima in winter and minima in summer (Fig. 5.11). These changes were due, in the case of *Cystophora*, to plants growing fertile laterals and ramuli in winter, and shedding them after their narrow reproductive period in spring. However, *C. subfarcinata* was exceptional because it reproduced all year and did not shed its ramuli. In the case of *Sargassum*, plants also shed their laterals, which make up the bulk of plant biomass, after reproduction in spring. In summer, all that is left of them is a stub or some basal leaves. Other species present – *Acrocarpia*, *Seirococcus* and *Scytothalia* – did not shed biomass after reproduction, and changes in biomass were due to other factors.

At more-exposed sites, *Acrocarpia*, *Scytothalia* and *Ecklonia* variously tended to be dominant, and the changes in biomass at one exposed site (Wright I., Victor Harbor) are typical of them (Fig. 5.12). Here, Greg Collings found modest fluctuations over time, with no evidence of seasonality. *Seirococcus*, *Cystophora* and *Sargassum* were only minor components of the canopy species at these sites.

SE Australia

Important floristic changes occur in the cooler waters of SE South Australia and Victoria. *Durvillaea potatorum* and *Phyllospora comosa*, together with *Ecklonia*, tend to dominate exposed shallow waters, while a mixed *Cystophora*–*Sargassum* assemblage is dominant in moderately exposed to sheltered waters (O'Hara 2001).

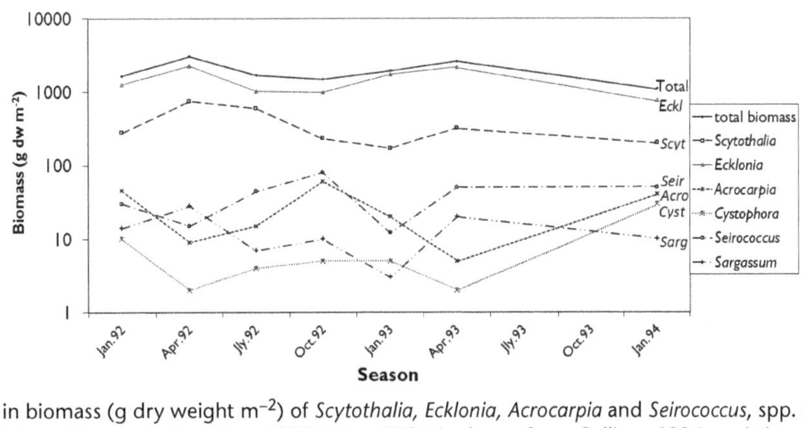

Figure 5.12: Trends in biomass (g dry weight m^{-2}) of *Scytothalia, Ecklonia, Acrocarpia* and *Seirococcus*, spp. at Wright I., Victor Harbor – an exposed site at 5 m depth from Jan. 1992 to Jan. 1994 (redrawn from Collings 1996, omitting standard errors). *C. moniliformis* and *S. fallax* were the main species of *Cystophora* and *Sargassum*, respectively.

In the still cooler waters of Tasmania, further changes occur as numerous canopy species reach the end of their southern range; for example, *Cystophora monilifera*, *C. polycystidea*, *Sargassum heteromorphum* and *Scaberia* rarely occur south of Bass Strait. Edgar (1984) described the general algal features of reef habitats around Tasmania, as shown in Fig. 5.13. Cold-water species become dominant, and the bull-kelp, *Durvillaea potatorum*, and the fucoids, *Lessonia corrugata*, *Xiphophora gladiata* and *Phyllospora*, dominate shallow waters, while deeper waters are dominated by *Ecklonia* and a mixed *Cystophora–Sargassum* assemblage. However, in extreme swell conditions, as at Eddystone Rock, some 25 km south of Tasmania, the *Durvillaea* zone goes to 18 m depth, and the other species listed above are restricted to even deep water, showing the powerful effect of surge on the depth range of species.

At sheltered sites, mixed communities of species of *Sargassum*, *Cystophora* and *Caulocystis* occur (Plate 10), and at one such site in eastern Tasmania Edgar (1983) found that *S. fallax* reached maximum biomass during the summer reproductive season, and then shed its reproductive branches until they regrew the following spring. In contrast, *S. verruculosum*, *Caulocystis* and *C. retroflexa* showed no seasonal change in biomass.

Productivity of fucoid assemblages

In the only study of the productivity of a fucoid assemblage for the region, Cheshire *et al.* (1996) examined the photosynthesis and respiration of an algal assemblage on boulders at 5 m depth. The assemblage comprised species of *Cystophora*, *Sargassum* and *Seirococcus* (48% of total biomass), and an understorey of erect corallines (30%) and turfing and foliose algae (22%), with a total biomass (wet weight) of ~3 kg m^{-2} (Plate 6). Net 24-h production was 6 g C m^{-2} $year^{-1}$ in summer and 2.6 g C m^{-2} $year^{-1}$ in winter, giving an annual production of ~19.2 kg wet weight m^{-2} $year^{-1}$. This production implied a turnover of plant biomass of six times a year, largely due to (a) the high production of reproductive fronds in spring and their later shedding in late summer, (b) exudate production and (c) detrital production from the turnover of vegetative fronds.

Patterns of community structure and their causes

In the above studies of fucoid assemblages, some authors have variously described them as 'mosaics of patches', 'a loosely ordered complexity of communities' and 'multi-species associations defying generalisation'. Others have emphasised niche separation of species or competition (e.g. Figs 5.6–5.8). However, these descriptions do not explain why so many species can co-exist with such extensive niche overlap, when competition theory suggests that inferior competing species should over time be excluded, resulting in the dominance of one or a few species. The application of disturbance theory and examination of factors affecting recruitment have provided some answers to these problems.

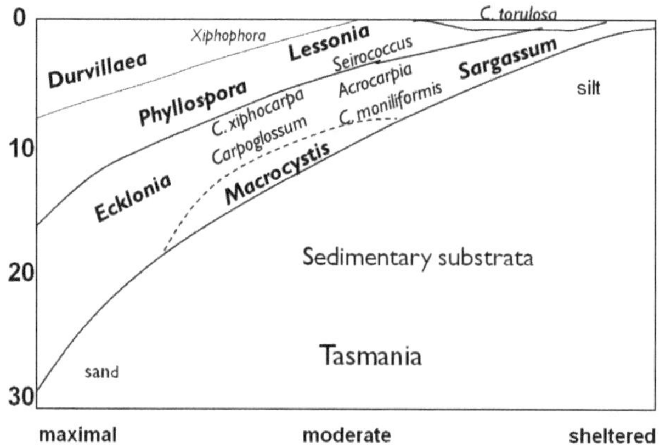

Figure 5.13: Schematic diagram showing the distributions of macroalgal assemblages defined by the canopy species, according to water movement and depth in Tasmanian waters (including the islands in Bass Strait). Major species are in bold, and located in the approximate centres of their distributions. Three species of *Cystophora* are shown – *C. xiphocarpa*, *C. torulosa*, and *C. moniliformis* (redrawn from Edgar 1984).

Disturbance

A disturbance can be described as a destructive event that removes biomass and, in the context of southern algal forests, is caused by storms, which remove plants, or by grazing (see Chapter 9). Storms are the major cause of disturbance on southern and western coasts of Australia, but both factors can be important on the east coast. The scale of a disturbance can range from the loss of single plants – essentially thinning a canopy – to creating clearings from a few hundred square centimetres to a few square metres. The severity of the effects of a disturbance depends on many factors, including substratum type, plant size and density, and life stage (Thomsen *et al.* 2004). For example, these authors found experimentally that most plants of *Ecklonia* broke away at the rock surface, and that this happened much more from a limestone rock surface than from granite.

Eventually a forest will recover from a disturbance. However, a disturbance is not simply a negative force, because it can have positive outcomes and may even be necessary for maintaining a stable community. Without it, a community might switch to a different kind of community. In this context the **intermediate disturbance hypothesis** explains the patchiness and diversity in communities on reefs. The idea is based on the view that communities rarely reach an equilibrium in which the competitively superior species eliminate others, because disturbances act to reverse the process by creating space for colonisation by less competitive species. If disturbances are too severe or frequent, few species will survive the harsh conditions. At the other extreme, if disturbances are rare, then monopolisation by the most competitively superior species will occur. The concept, although seldom tested, has intuitive appeal, although much remains to be learnt (reviewed by Minchinton 2007).

An understanding of the consequences of disturbance is essential to understanding the dynamics and ecology of algal forests, and one method widely used is to mimic the effects of a storm by creating experimental clearings in forests, and recording the species that recruit into them over the succeeding months. Seven studies have done this in different communities at sites across southern Australia, and they are each discussed in approximate order of sites on a water movement gradient, from strong surge to calm conditions.

Kendrick *et al.* (2004) cleared circular 10 m diameter plots dominated by *Ecklonia* on several reefs in Hamelin Bay, Western Australia, and followed recruitment in them over 11 months. The removal of *Ecklonia* saw the invasion of many other species, notable among them being the canopy-forming species in the genera, *Sargassum*, *Cystophora* and *Platythalia*, so that, over time, the clearings resembled the mosaic of small gaps that already existed in the *Ecklonia* forest. Reef topography, including rock type and relief, and depth also played a role in influencing the species that grew in the clearings. In summarising the many factors involved in recruitment, the authors concluded that 'local scale patterns reflect the interaction between local disturbance and colonisation history, localised dispersal and recruitment in macroalgae, and grazing'. Hence, storm-created gaps in *Ecklonia* forests were a natural mechanism that helped maintain a diversity of algae in the forest (and see Kendrick *et al.* 1999).

Goodsell and Connell (2005) did a similar experiment in *Ecklonia* forests at exposed sites in Encounter Bay, South Australia, but with smaller cleared plots of 1 m^2, and examined the plots 4 months later. Here, different species recruited into the clearings – *Sargassum fallax*, several *Cystophora* spp., *Acrocarpia* and *Scytothalia*. In this case, the disturbance facilitated a switch from an *Ecklonia* to a fucoid forest. In a later study in winter at nearby sites in Encounter Bay, Wernberg and Connell (2008) made small clearings in *Ecklonia* patches situated in a mixed forest, where several canopy species were present (*Ecklonia*, *Scytothalia*, *Sargassum* and *Cystophora*). A variety of recruits then colonised the patches over the next 8 months, depending on the species which formed the surrounding canopies, and presumably were the source of recruits. In this study, the data suggested that history (i.e. the species present from past recruitments) played an important role in determining recolonisation following disturbance. The authors also examined the scale of patch mosaics in the area and found that sheltered sites had many small patches, whereas exposed sites had large patches; further, exposed sites had more forest gaps than sheltered sites. They concluded that it was wave exposure that created the gaps, and that this process interacted with history to determine which species colonised the gaps.

The fourth study, also in an *Ecklonia* forest, was that of Steve Kennelly (see Chapter 4; Kennelly 1995). He cleared patches of the *Ecklonia* canopy in different seasons of the year at a moderately exposed site in Sydney Harbour. In all seasons except winter, the clearings led to a profuse growth, first of filamentous algae and then of turfing algae, until eventually the cleared patches were reinvaded from the edges by *Ecklonia* after 12–14 months. Only after the winter clearing did *Ecklonia* quickly re-establish a canopy within 3–4 months.

The last three studies examined the same question in less exposed fucoid assemblages. Emmerson and Collings (1998) made clearings of three different sizes (0.09–1 m^2) at a Cape Jervis site in a mixed assemblage of 14 canopy species in the genera *Cystophora*, *Sargassum*, *Seirococcus*, *Ecklonia* and *Acrocarpia*. The recruits in the clearings were mainly species of *Cystophora* and *Sargassum* that grew in the vicinity, but the authors were interested in the location of the recruits within the clearings, and their survival. The largest clearings had the most recruits at first, but clearing size became less important over time. Recruits in the centre of the clearings survived better than those around the edge or outside, likely due to better light conditions in the centre, enabling faster growth (Fig. 5.14). The authors considered that the two most important factors were the initial 'spore rain' providing the recruits, and then the density-dependent mortality, which greatly reduced their numbers.

Goldberg (2007) did her study for over a year on the sheltered side of an island in Esperance Bay. She made total clearings and partial clearings of canopy or understorey, and compared them with controls without clearings in a mixed *Sargassum*–*Cystophora*–*Caulocystis*–*Acrocarpia* assemblage of 18 canopy species. *Sargassum* recruits became established mainly in the shallow-water sites (< 10 m deep) under a *Sargassum* canopy or in partial clearings of understorey. In contrast, the recruits of the other species became established in all the treatments, but in greater density in total clearings. Nisse Goldberg concluded that the presence of a canopy layer and of an understorey did not present a barrier to recruitment for most of the canopy species, and especially *Sargassum*. Hence, at this site, disturbance played a minor role in promoting recruitment.

The seventh study by Gary Kendrick (1994) was at Nancy Cove, Rottnest I. Clearings (each 4 m^2) were made of the canopy or understorey in a *Sargassum* forest at 6 m depth, and recruitment was followed for a year on seeded and unseeded settlement plates. Recruits became established at high densities both in the clearings and in the natural forest – an unexpected result that mirrored the natural summer loss of the *Sargassum* canopy – and the forest canopy had little effect on recruit survival for the first 6 months. As the canopy re-established from perennial bases, the mortality of recruits increased steeply in the forest, but was much lower in the clearings (Fig. 5.15). This

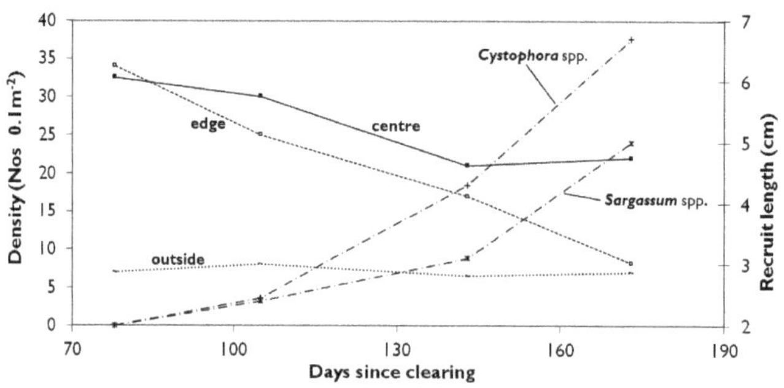

Figure 5.14: Density of recruits at the centre, edge and outside clearings in a mixed forest over 173 days at Cape Jervis. The growth of the recruits of *Cystophora* and *Sargassum* in the centres of the clearings is also shown over the period (scale on the right vertical axis). Growth of recruits at the edges of clearings was lower, but is not shown (redrawn from Emmerson and Collings 1998).

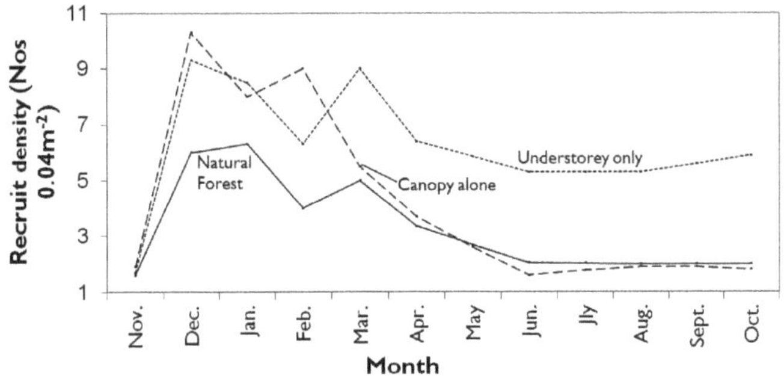

Figure 5.15: Mean survival of *Sargassum* recruits in the natural *Sargassum* forest (continuous heavy line), and in the two treatments – understorey alone without a canopy (dotted line), and canopy alone, without understorey (dashed line) (redrawn without error bars from Kendrick 1994).

showed that the canopy, probably by its intense shading effect, was largely responsible for the high mortality. Mortality was both density dependent (a within-species competition effect) and density independent (perhaps due to physical effects of available light or sediment/water movement, or to nutrient requirements. Kendrick concluded that the existence and persistence of the forest depended on history and the dispersal and survival of propagules. Competition, but not disturbance, was an important element in the recovery process.

The above seven studies have been considered in some detail, because they all show features unique to their particular site. Does this mean that the effects of disturbance are idiosyncratic and context-dependent, according to the species present and local environmental factors? Or can we deduce general principles that apply to recolonisation in algal forests after disturbance? First, we note that, in a comparison of studies done over a range of water movements, we might predict that disturbance from storms would decline in importance in calmer conditions.

The first four studies were at sites of strong water movement and all emphasised the role of disturbance, as did Turner and Cheshire (2003) and Turner and Collings (2008). These last two studies noted that in patches without dominants, their VLA groups discussed above, were 1–2 m in diameter, and they proposed a model to explain the diversity of species and their patchiness. In this model, patches resulting from past storms were initially colonised by many species and, over time, would become dominated by species of a main assemblage, and so merge with that assemblage (Fig. 5.16). Switches from one assemblage to an alternative assemblage might conceivably occur, but in the main the four assemblage types described by them would be relatively stable and would tend to persist even after a disturbance. However, the studies of Kendrick *et al.* (2004), Goodsell and Connell (2005) and Wernberg and Connell (2008) have shown that the situation is more complex than this simple model. Following a disturbance, numerous factors play a role in determining what recruits into clearings. These include the surrounding canopy species that provide the 'spore rain', water movement, the extent of spore dispersal and local topography.

In more benign environments, disturbances are rare and play little part in creating the patchiness that may still occur. Hence Collings (1996) and Collings and Cheshire (1998) emphasised small-scale differences, such as settlement and survival of propagules, grazing and local micro-topography. Over time, patches might expand and merge with others, or contract on meeting better competitors, but the end result was still a mosaic of patches. Kendrick (1993) also emphasised the role of local history. Once established, a species will release huge numbers of propagules around it, and so ensure its persistence. This, he said, coupled with rare long-distance dispersal from drifting plants, 'leads to genetic and ecological segregation of populations over very short distances'. Hence, patches of different species with ecologies that differed between locations were able to persist (Kendrick 1993). Goldberg (2007) similarly emphasised local factors.

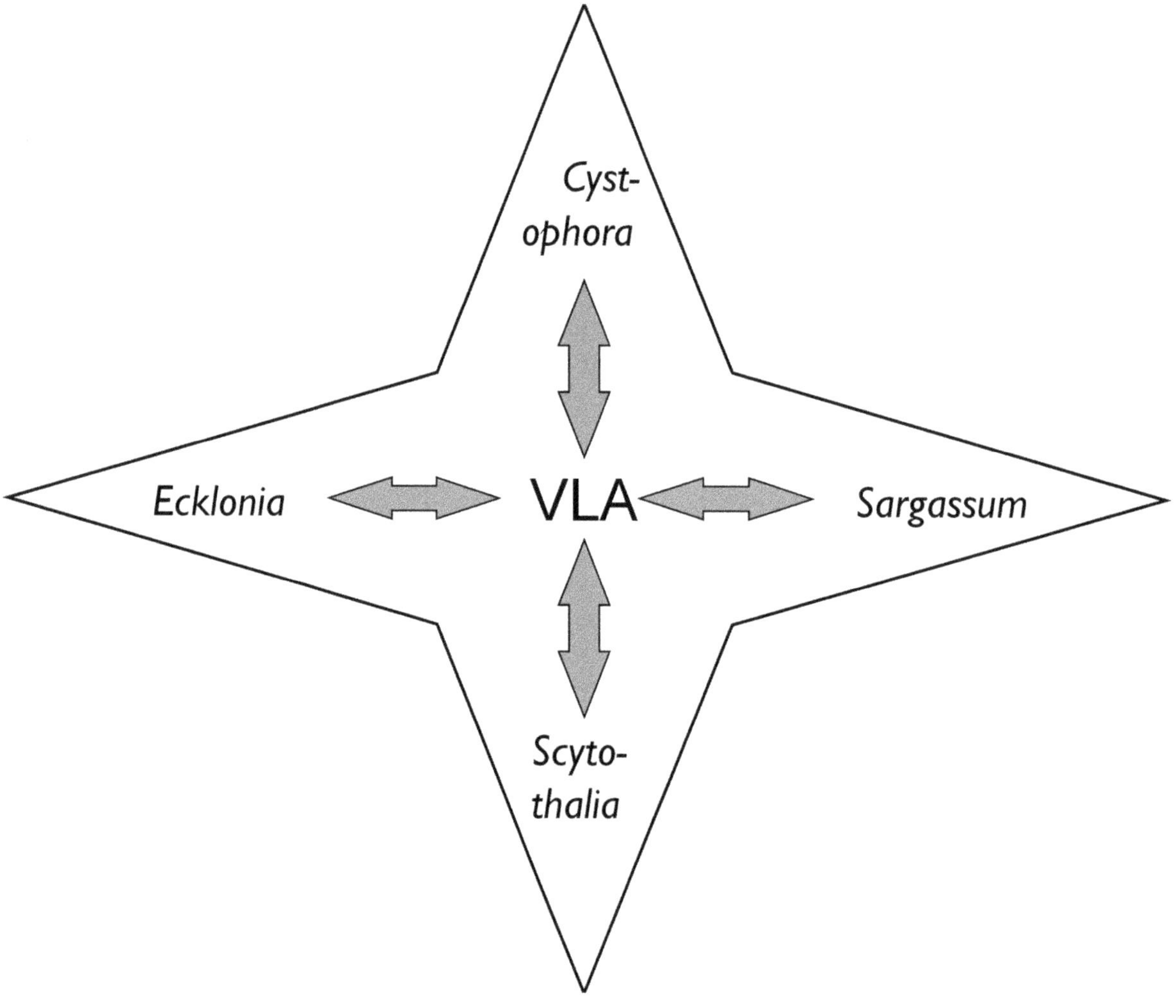

Figure 5.16: Schematic model of patch development. After a disturbance, a cleared patch – formerly in an assemblage at an apex of the star – goes to a variable low abundance state, and then reverts to a former state or to one of the other assemblages (redrawn from Turner and Cheshire 2003).

Competition

Competition is known to play a key role in structuring communities (Schiel 1988) and was inferred by some of the above authors to play a part. It can also be inferred from observations on local communities, such as the persistence of monospecific stands of canopy species over decades with virtually unchanging borders (Turner 2004). This may be due to:

- the suppression of other species from recruiting under them (allelopathy) or
- reduction of light below the canopy, such that propagules have insufficient light for growth, as noted by Kendrick *et al.* (2004).

Besides, in extensive surveys the abundance of adults and juveniles of species in the genera *Ecklonia, Cystophora* and *Sargassum* were strongly correlated (Turner 2004).

Conclusions

Natural storm disturbances over a wide range of exposures play a key role in maintaining the rich diversity of co-occurring canopy species along the southern Australian coast. In sheltered conditions, however, other factors, such as propagule dispersal from neighbouring species, predominate.

REFERENCES

Baker JL, Shepherd SA, Turner D, Edyvane K (2008) Investigator Group Expedition 2006: benthic macroalgal studies at islands in the eastern Great Australian Bight over three decades. *Transactions of the Royal Society of South Australia* **132**, 251–267.

Cheshire AC, Hallam ND (1985) The environmental role of alginates in *Durvillaea potatorum* (Fucales, Phaeophyta). *Phycologia* **24**, 147–153.

Cheshire AC, Hallam ND (1988a) Biomass and density of native stands of *Durvillaea potatorum* (southern bull-kelp) in south eastern Australia. *Marine Ecology Progress Series* **48**, 277–283.

Cheshire AC, Hallam ND (1988b) Morphology of the southern bull-kelp (*Durvillaea potatorum*, Durvillaeales, Phaeophyta) from King Island (Bass Strait, Australia). *Botanica Marina* **31**, 139–148.

Cheshire AC, Hallam ND (1989a) Morphological differences in the southern bull-kelp (*Durvillaea potatorum*) throughout south-eastern Australia. *Botanica Marina* **32**, 191–197.

Cheshire AC, Hallam ND (1989b) Methods for assessing the age composition of native stands of subtidal macro-algae: a case study on *Durvillaea potatorum*. *Botanica Marina* **32**, 199–204.

Cheshire AC, Conran JG, Hallam ND (1995) A cladistic analysis of the evolution and biogeography of *Durvillaea* (Phaephyta). *Journal of Phycology* **31**, 644–655.

Cheshire AC, Westphalen G, Wenden A, Scriven LJ *et al.* (1996). Photosynthesis and respiration of phaeophycean-dominated macroalgal communities in summer and winter. *Aquatic Botany* **55**, 159–170.

Clayton M (1990) The adaptive significance of life history characters in selected orders of marine brown macroalgae. *Australian Journal of Ecology* **15**, 439–452.

Coleman MA, Kelaher BP (2009) Connectivity among fragmented populations of a habitat-forming alga, *Phyllospora comosa* (Phaeophyceae, Fucales) on an urbanised coast. *Marine Ecology Progress Series* **381**, 63–70.

Collings GJ (1996) Spatiotemporal variation of macroalgal communities of southern Fleurieu Peninsula, South Australia. PhD thesis. University of Adelaide, Adelaide.

Collings GJ, Cheshire AC (1998) Composition of subtidal macroalgal communities of the lower Gulf waters of South Australia, with reference to water movement and geographical separation. *Australian Journal of Botany* **46**, 657–669.

Edgar GJ (1983) The ecology of south-east Tasmanian phytal animal communities. II. Seasonal change in plant and animal populations. *Journal of Experimental Marine Biology and Ecology* **70**, 159–179.

Edgar GJ (1984) General features of the ecology and biogeography of Tasmanian subtidal rocky shore communities. *Papers and Proceedings of the Royal Society of Tasmania* **118**, 173–186.

Emmerson LM, Collings GJ (1998) Macroalgal recruitment in artificially disturbed areas: interactive effects of temporal and spatial scale. *Marine and Freshwater Research* **49**, 541–546.

Gillanders BM, Brown MT (1994) Seasonal variation in standing crop, reproduction and population structure of *Xiphophora gladiata* (Phaeophyceae: Fucales). *Botanica Marina* **37**, 35–41.

Goldberg N (2007) Colonization of subtidal macroalgae in a fucalean-dominated algal assemblage, southwestern Australia. *Hydrobiologia* **575**, 423–432.

Goldberg N, Kendrick GA (2004) Effects of island groups, depth, and exposure to ocean waves on subtidal macroalgal assemblages in the Recherche Archipelago, Western Australia. *Journal of Phycology* **40**, 631–641.

Goldberg N, Kendrick GA (2005) A catalogue of the marine macroalgae found in the western islands of the Recherche Archipelago, Western Australia, with notes on their distribution in relation to island location, depth and exposure to wave energy. In *The Marine Fauna and Flora of Esperance, Western Australia. Proceedings of the Twelfth International Marine Biological Workshop* (Eds FE Wells, DI Jones and GA Kendrick) pp. 25–75. Western Australian Museum, Perth.

Goldberg N, Kendrick GA (2007) Recruitment ecology of marine macroalgae. In: *Marine Ecology* (Eds SD Connell and BM Gillanders) pp. 47–71. Oxford University Press, Melbourne.

Goldberg N, Kendrick GA, Heine J (2004) Highway or country road: algal recruitment with distance from an island reef. *Journal of the Marine Biological Association of the United Kingdom* **84**, 879–882.

Goodsell PJ, Connell SD (2005) Disturbance initiates diversity in recruitment of canopy forming algae: interactive effects of canopy-thinning on substratum availability. *Phycologia* **44**, 632–639.

Hay C (1978) Growth, mortality, longevity and standing crop of *Durvillaea antarctica* (Phaeophyceae) in New Zealand. *Proceedings of the 9th International Seaweed Symposium*. 20–27 August 1977, Santa Barbara, California (Eds A Jensen and JR Stein) pp. 97–103. Science Press, Princeton.

Hotchkiss SL (1999) Life history strategies of three species of *Cystophora* (Phaephyta, Fucales) from a shallow subtidal community in South Australia. PhD thesis. University of Adelaide, Adelaide.

Kendrick GA (1993) *Sargassum* beds at Rottnest Island: species composition and abundance. In *The Flora and Fauna of Rottnest Island, Western Australia* (Eds FE Wells, DI Walker, H Kirkman and R Lethbridge) pp. 455–472. Western Australian Museum, Perth.

Kendrick GA (1994) Effects of propagule settlement density and adult canopy on survival of recruits of *Sargassum* spp. (Sargassaceae: Phaeophyta). *Marine Ecology Progress Series* **103**, 129–140.

Kendrick GA, Walker DI (1991) Dispersal distances for propagules of *Sargassum spinuligerum* (Sargassaceae, Phaeophyta) measured directly by vital staining and venturi suction sampling. *Marine Ecology Progress Series* **79**, 133–138.

Kendrick GA, Walker DI (1994) Role of recruitment in structuring beds of *Sargassum* spp. (Phaeophyta) at Rottnest Island, Western Australia. *Journal of Phycology* **30**, 200–208.

Kendrick GA, Walker DI (1995) Dispersal of propagules of *Sargassum* spp. (Sargassaceae, Phaeophyta): observations of local patterns of dispersal and consequences for recruitment and population structure. *Journal of Experimental Marine Biology and Ecology* **192**, 273–288.

Kendrick GA, Lavery PS, Phillips JC (1999) Influence of *Ecklonia radiata* kelp canopy on structure of macro-algal assemblages in Marmion Lagoon, Western Australia. *Hydrobiologia* **398/399**, 275–283.

Kendrick GA, Harvey ES, Wernberg T, Harman N *et al.* (2004) The role of disturbance in maintaining diversity of benthic macroalgal assemblages in south-western Australia. *Japanese Journal of Phycology* **52**, Supplement, 5–9.

Kennelly SJ (1995) Kelp beds. In *Coastal Marine Ecology of Temperate Australia* (Eds AJ Underwood and MG Chapman) pp. 106–120. University of New South Wales Press, Sydney.

May DI, Clayton MN (1991) Oogenesis, the formation of oogonial stalks and fertilization in *Sargassum vestitum* (Fucales, Phaeophyta) from southern Australia. *Phycologia* **30**, 243–256.

Minchinton TE (2007) Natural disturbance and regeneration of marine benthic communities. In *Marine Ecology* (Eds SD Connell and BM Gillanders) pp. 138–174. Oxford University Press, Melbourne.

O'Hara T (2001) Consistency of faunal and floral assemblages within temperate subtidal rocky reef habitats. *Marine and Freshwater Research* **52**, 853–863.

Phillips J (2007) Phaeophyceae. In *Algae of Australia: Introduction* (Eds PM McCarthy and AE Orchard) pp. 264–287. ABRS, Canberra; CSIRO PUBLISHING, Melbourne.

Reed DC, Kinlan BP, Raimondi PT, Washburn L, *et al.* (2006) A metapopulation perspective on the patch dynamics of giant kelp in southern California. In *Marine Metapopulations* (Eds JP Kritzer and PF Sale) pp. 353–386. Elsevier Academic, Amsterdam.

Rice EL (1987) Small-scale distribution pattern in *Xiphophora gladiata* (Labillardiere) Montagne ex Kjellman (Fucales, Phaeophyta). *Phycologia* **26**, 247–251.

Sanderson CJ (1992) Subtidal macroalgal studies in east and south eastern Tasmanian coastal waters. MSc thesis. University of Tasmania, Hobart.

Schiel DR (1988) Algal interactions on shallow subtidal reefs in northern New Zealand: a review. *New Zealand Journal of Marine and Freshwater Research* **22**, 481–489.

Shepherd SA, Womersley HBS (1971) Pearson Island expedition 1969. The subtidal ecology of benthic algae. *Transactions of the Royal Society of South Australia* **95**, 155–167.

Shepherd SA, Womersley HBS (1976) The subtidal algal and seagrass ecology of St Francis Island, South Australia. *Transactions of the Royal Society of South Australia* **100**, 177–191.

Shepherd SA, Womersley HBS (1981) The algal and seagrass ecology of Waterloo Bay, South Australia. *Aquatic Botany* **11**, 305–371.

Smale DA, Kendrick GA, Waddington KI, Van Niel KP *et al.* (2010) Benthic assemblage composition on subtidal reefs along a latitudinal gradient in Western Australia. *Estuarine, Coastal and Shelf Science* **86**, 83–92.

Thomsen MS, Wernberg T, Kendrick GA (2004) The effect of thallus, size, life stage, aggregation, wave exposure and substratum conditions on the forces required to dislodge the small kelp *Ecklonia radiata* in temperate Australasia. *Botanica Marina* **47**, 454–460.

Turner DJ (2004) Effects of sedimentation on the structure of a phaeophycean dominated macro-algal community. PhD thesis. University of Adelaide, Adelaide.

Turner DJ, Cheshire AC (2003) Encounter 2002 Expedition to the Isles of St Francis: benthic macroalgal communities. *Transactions of the Royal Society of South Australia* **127**, 153–166.

Turner DJ, Collings GJ (2008) Subtidal macroalgal communities of Gulf St Vincent. In *Natural History of Gulf St Vincent* (Eds SA Shepherd, S Bryars, IR Kirkegaard, P Harbison, *et al.*) pp. 264–278. Royal Society of South Australia, Adelaide.

Wernberg T, Connell SD (2008) Physical disturbance and subtidal habitat structure on open rocky coasts: effects of wave exposure, extent and intensity. *Journal of Sea Research* **59**, 237–248.

Wernberg T, Kendrick GA, Phillips JC (2003) Regional differences in kelp-associated algal assemblages on temperate limestone reefs in south-western Australia. *Diversity & Distributions* **9**, 427–441.

6 Ecology of non-geniculate coralline algae

OVERVIEW

Non-geniculate coralline algae (NCA) – the group of calcified red algae lacking a jointed structure – are conspicuous on reefs from the low intertidal to 135 m depth, and can have high cover in shallow kelp habitats, urchin barrens and as rhodoliths in deep water. Their taxonomy is complex, in part because individual species can take different growth forms. Their growth rates are slow, but they readily colonise clean surfaces, where competitive interactions are strong. NCA are well adapted to low light conditions under kelp canopies, which facilitates their survival in various ways. They are morphologically defended against some grazers, and adapted to grazing pressure by others, particularly gastropods and chitons. Some NCA species share a mutualism with grazing molluscs (e.g. abalone) in which the coralline benefits from removal of epiphytes, and the mollusc by having an optimal settlement site for their larvae. In such relationships, the NCA surface micro-environment (i.e. oxygen levels, microalgal growth, surface topography, water flow and pH) is critical for larval settlement and survival.

Rhodoliths are free-living NCAs that provide a soft-bottom habitat with hard-bottom features. They are slow-growing and long-lived, and can form deep beds with a rich diversity of fauna within them – all features that endow them with high conservation value. Rhodolith beds are patchily distributed in southern Australia from shallow to deep water. However, they are sensitive to disturbance, either by trawling, sedimentation or excess nutrients, and can quickly degrade.

INTRODUCTION

NCA are among the most abundant organisms on coastal reefs (including coral reefs). They can take many different forms, including: (a) forms that encrust rocky substrata, often called crustose coralline algae; (b) forms that are free-living, and known as rhodoliths or sometimes maerl, the original French term; and (c) epibionts – those that live on other plants or animals. For all these forms, the strictly correct term is non-geniculate (meaning not jointed) coralline algae (NCA): a term that excludes erect, articulated or jointed forms of coralline algae, such as the genera *Corallina* and *Amphiroa*. Here we variously use both terms – NCA and crustose corallines – to refer to these calcified plants.

NCA are abundant in benthic habitats worldwide from the tropics to polar regions, and extend to > 130 m deep in the clear waters of the Great Australian Bight (GAB). They are common in shaded habitats, such as crevices and caves, and frequently cover most of the substratum in habitats, such as under kelp canopies. In southern Australia, their average cover is 70–85% under *Ecklonia* from southern Western Australia to central New South Wales (Fowler-Walker and Connell 2002; Connell 2005). Where sea urchin grazing is intense, their cover can be close to 100% (Chapters 11, 12).

In this chapter, our focus is mainly on the ecology of southern NCA, but because few studies have been undertaken here, we refer to other relevant studies in temperate waters, as well as global reviews (e.g. Johansen 1981; Steneck 1986; Woelkerling 1988; Foster 2001; Nelson 2009).

NCA are enigmatic because they combine very different ecological strategies, which can be summarised as follows:

- They have high plasticity in their external appearance and morphology, and are considered among the most challenging of algae to identify.
- They are opportunistic species, often being the first algae to colonise disturbed or bare areas, and are reproductive throughout the year (Ayling 1981; Davis and Wilce 1987).
- They are stress-tolerant, and can persist high in the intertidal where they may dry out in hot conditions, and in disturbed habitats, such as areas with intense urchin grazing (Chapter 12).
- Crustose forms can dominate rocky bottoms, such as urchin barrens or isoyake areas (the Japanese term) in many parts of the world (Chapter 12), and rhodoliths can dominate soft bottom habitats.
- They can induce settlement of a variety of invertebrate larvae, in some cases forming a mutualistic relation with the larvae, which, after metamorphosis and settlement, graze on their surfaces, thereby preventing them becoming overgrown.
- They are important in cementing some reefs together, so making them more robust.

ENCRUSTING CORALLINE ALGAE

Sabine Daume and SA Shepherd

Growth forms, species and life history

Crustose forms are conspicuous and important components of inter- and subtidal rocky habitats, especially on boulders and under kelp canopies, where they are often dominant. Yet little is known of the number of species that live in such habitats, and in ecological studies they are usually lumped together into a single category – crustose corallines.

Another complexity is that different growth-forms of crustose corallines occur; some of them are characteristic of a species, but in other cases several growth-forms can occur within the same species, even at the same locality. Woelkerling *et al.* (1993) recognised 10 different types of NCA growth-forms, and here we illustrate five of the commonest forms on rocky substrata (Plate 11):

1. **Encrusting:** flattened forms, which can be smooth or uneven
2. **Warty:** plants with warty projections, usually < 3 mm high
3. **Lumpy:** plants with many contiguous round swellings to 2–5 mm across
4. **Fruticose:** plants with round or compressed outgrowths ≥ 3 mm long
5. **Foliose:** plants with many flattened, sometimes curved, branches.

Less common forms are disc-shaped, layered, or with ribbon-like branches.

Often several species live together at the same site. In a study of three boulder habitats, Daume *et al.* (1999a) recorded nine species at a moderately sheltered site at Taylor I., South Australia, 10 species at a semi-exposed site at West I., but only three species in upper Port Phillip Bay, Victoria. At the two South Australian sites, the average cover of NCA on boulders was ~25%, with the smaller boulders tending to have a higher cover of NCA at Taylor I. At Rottnest I., Western Australia, Sim and Townsend (1999) recorded a total of 10 species in macroalgal communities at several sheltered and exposed sites. To illustrate the variety of forms present and their frequency of occurrence, Table 6.1 lists the five most common

Table 6.1: Species–growth form relations and percentage frequencies of occurrence of the five commonest species on 60 boulders in the NCA community at West I. (after Daume *et al.* 1999a).

Frequency of occurrence of species was obtained by dividing the number of boulders on which the species occurred by the total number of boulders with identified species. (Non-reproductive thalli could not be identified). The frequency of growth forms was obtained by dividing the number of boulders with the growth form by the total number of boulders.

	Growth form					
Species	**Crustose**	**Warty**	**Lumpy**	**Fruticose**	**Foliose**	**% frequency of occurrence**
Hydrolithon rupestre	20					24
Hydrolithon samoense	18	3				27
Mesophyllum engelhartii	10	7	7	7		32
Mesophyllum macroblastum		2		2	2	7
Sporolithon durum		12	5			19
Frequency of occurrence (%)	98	30	13	30	8	

species and their growth forms (Daume *et al.* (1999a). The remaining five species present were uncommon, occurring at 2–5% of sites.

Growth rates of temperate crustose corallines are not well known. Steneck (1986) noted that crustose coralline algae grow both laterally and vertically, with thinner crusts generally growing laterally faster than thicker ones. Lateral growth rates vary greatly, but typically grow faster in warmer water. Recorded lateral growth rates of temperate species are only 1–4 mm year^{-1}, while vertical accretion rates are exceedingly slow, often only ~0.3 mm year^{-1} (Littler and Littler (1984).

The recruitment and persistence of crustose corallines are poorly known. At a deep water (15 m) site off Cape Northumberland, South Australia, recruitment of crustose corallines (*Pneophyllum* sp.) on bare substratum was high in summer, but very low in winter (Shepherd 1979; Chapter 7). Conversely, at Rottnest I., NCA recruitment on experimental tiles in shallow water was high in winter and low in summer (Sim 2001). In algal forest, he found that the *Ecklonia* canopy did not affect recruitment or lateral growth of the corallines below them, whereas competing foliose and turf algae over 3 months had a negative effect on them. He suggested that crustose corallines can rapidly monopolise free space, but their later persistence depends on competition with other species, until a dense kelp canopy develops and eliminates competitors. Irving and Connell (2006b) similarly found that a kelp canopy prevents establishment of algal turfs that would overgrow crustose species. These and other studies (review of Steneck *et al.* 1991; Morcom *et al.* 2005) propose that encrusting algae compete for space mainly by overgrowing one another where their edges come into contact. Suggested principles governing the growth of, and competition between, corallines can be summarised as:

- thicker crusts generally overgrow thinner ones
- crusts of equal thickness often reach a competitive stand-off, in which neither wins
- crusts with raised edges will outgrow flat, adherent crusts, but
- sometimes thin crusts can stop thicker crusts by regenerating a secondary margin that can even grow over the thicker crust
- thinner, faster-growing species can dominate new substratum, although they may be competitively

inferior, so supporting the 'fugitive species' concept (i.e. that a competitively inferior species can co-exist with a dominant competitor, if it can colonise vacant space faster).

Canopy effects on crustose corallines

The kelp *Ecklonia radiata*

The algal canopy has a strong effect on the NCA community below it. At West I., a greater diversity of species and growth forms was present under the *Ecklonia* canopy than on nearby boulders without a canopy, but the percentage cover was lower under *Ecklonia* (Daume *et al.* 1999a).

In another study at the same site, Melville and Connell (2001) noted that, at low densities of < 7 *Ecklonia* plants m^{-2}, the kelp canopy had little effect on crustose corallines below them. However, at higher kelp densities the percentage cover of crustose corallines was greater below than outside the kelp canopy. The clearing of kelp from experimental patches, and reciprocal transplants of boulders between kelp patches and open patches, together showed that kelp canopies facilitated the growth of crustose corallines, but at the expense of erect coralline algae.

If the kelp canopy is removed or lost, light reaching the substratum increases, and bleaching of the NCA community occurs. Irving *et al.* (2004) examined this effect and found that light intensity on the bottom increased ~6 times (to ~300 μmol quanta m^{-2} s^{-1}) after canopy removal. The crustose species became bleached and lost much of their photosynthetic activity. However, the corallines did not die, but appeared to acclimate to the higher light levels, probably by reducing their level of light-harvesting pigments. A resulting photosynthetic activity-irradiance curve (Fig. 6.1) of the encrusting species *Lithophyllum* sp. (possibly *Hydrolithon* sp.) shows that maximum light absorption was at ~150 μmol quanta m^{-2} s^{-1} (i.e. at about half the level of light reaching the substratum after clearing of the kelp). For comparison with that curve, we show a photosynthetic activity-irradiance curve for *Lithothamnion* sp. from Danish waters, with a maximum at slightly lower light intensity levels (Kaspar 1992).

When kelp canopies are restored over areas previously cleared, or the area is covered with artificial shading, the crustose corallines recover to their former natural abundance (Connell 2003a, b). However, kelps not only provide shade, thereby facilitating crustose coralline growth, they also abrade the substratum and reduce sedimentation. In Sean Connell's (2003a) study, light reduction accounted for ~52% of the canopy effect on NCA percentage cover, but the reduction of sediment by the kelp accounted for only a few per cent. The abrasive effect of kelp also removes erect understorey algae, such as erect corallines and algal turf (Irving and Connell (2006a), and it is this algal turf that often accumulates sediment, and smothers crustose corallines (see Chapter 7). So the kelp's abrasive effect on the substratum may indirectly facilitate crustose species, both by excluding their competitors for space, as noted by the above authors, and also by reducing the likelihood of their being smothered. The strength of these indirect positive effects is yet to be studied.

Another complexity is that the morphology (relative lengths of stipe and frond, and surface area) of *Ecklonia* varies between eastern and southern Australia (Kennelly 1989; Fowler-Walker *et al.* 2005). In eastern Australia, stipes are longer and fronds smaller than on southern coasts, so that fronds are less able to sweep the substratum, resulting in less abrasion.

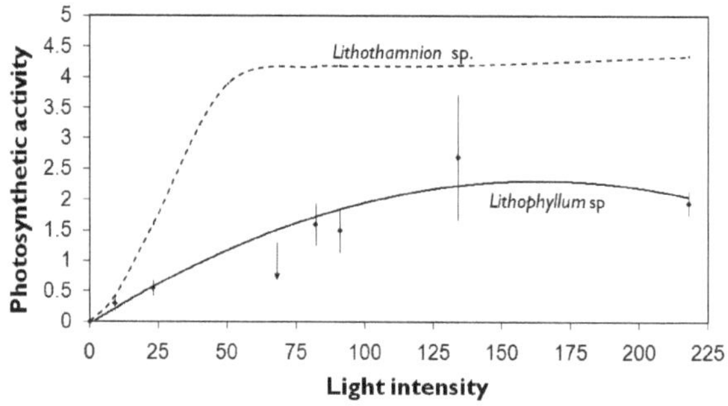

Figure 6.1: Photosynthetic activity-irradiance curves for the crustose coralline *Lithophyllum* sp. at West I. (redrawn from Irving *et al.* 2004), and for the cool temperate *Lithothamnion* sp. (redrawn from Kaspar 1992). Units for both axes are μmol quanta m^{-2} s^{-1}. Arrow indicates natural incident light level falling on boulders under the *Ecklonia* canopy at West I. Bars are standard errors.

Furthermore, more light can reach the substratum on eastern than southern coasts. These factors have favoured encrusting corallines in southern waters (Irving and Connell 2006a, b), and their additive effects can be summarised as shown in Fig 6.2.

Single and mixed species canopies

Throughout southern waters, macroalgal forests can occur as patches of single species, mixed stands of several canopy species or, less commonly, as open stands without canopy species. Several studies have examined the abundance of NCA under these three algal community types. The results (Fig. 6.2) show that Western and South Australia are very similar to each other, with the highest cover of NCA under a single-species canopy, and progressively less under a mixed-species canopy and no canopy. Other localised studies have confirmed the same pattern (Irving *et al.* 2003). However, it is notable that New South Wales deviates markedly from this pattern, with substantially higher cover of crustose corallines under a mixed canopy or no canopy. No similar comparison is available for Tasmania or Victoria, but we note that Ling's (2008) study in an eastern Tasmanian macroalgal forest recorded an NCA cover of ~40% under a mixed kelp canopy of *Ecklonia* and *Phyllospora*, which had a combined cover of ~80%. But what causes these understorey differences between canopy types and between regions?

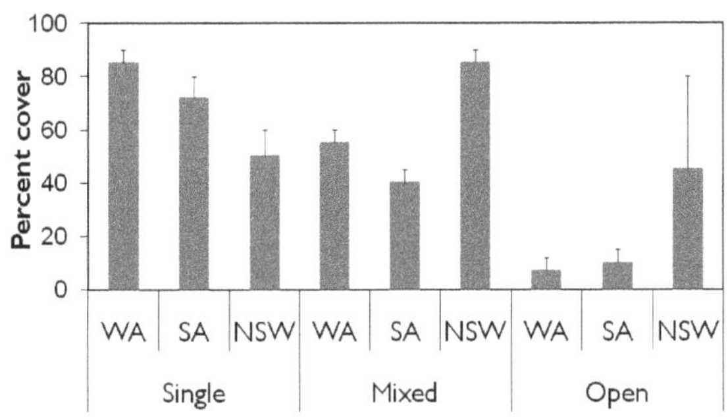

Figure 6.2: Average percentage cover of encrusting corallines recorded in three states: (1) under a single kelp species' canopy of *Ecklonia*; (2) under a mixed canopy of *Ecklonia* and the fucoids *Cystophora* and/or *Sargassum*; and (3) in open habitats without canopy species. Data extracted variously from Fowler-Walker and Connell (2002), Irving *et al.* (2004), Irving and Connell (2006a, b). Vertical bars are approximate range in published values.

Canopies of mixed kelps and fucoids tend to have more erect corallines and foliose understorey algae growing under them than *Ecklonia* does (Irving and Connell 2006a, b), leaving less space for NCA. These authors suggested that the greater habitat complexity of mixed forests, and especially their effect on physical variables (light, sedimentation and abrasion), enabled many algal species with different demands for light to become established in them. And, in the complete absence of a canopy, foliose algae can dominate the habitat.

The differences between regions are numerous and more complex, and specific aspects are considered in more detail elsewhere in this book. It is sufficient here to note that, in contrast to southern coasts, eastern Australian algal forests are greatly modified by intense herbivory by sea urchins (see Chapters 11 and 12). This major difference, coupled with the richness and abundance of fucoid algae, and the different morphology of the kelp *Ecklonia* in southern Australia, together mean that varied processes are operating at regional scales (Connell 2007; Connell and Irving 2008). The dominance or presence of the massive brown algae, *Phyllospora comosa*, *Macrocystis pyrifera*, *Lessonia corrugata* and *Durvillaea potatorum*, in the cooler Victorian and Tasmanian waters suggests that yet other interactions and processes are likely to operate there.

Effects of grazing and other disturbances

Any event or process that removes organisms or creates space on the substratum is a disturbance. It can be: (a) physical, as in storm waves that remove canopy and understorey algae, overturn boulders and make space for other organisms; (b) abrasion of the substratum by a kelp canopy; or (c) a biotic factor, as in grazing. A form of disturbance that keeps the encrusting surface coralline clean is critical for the long-term survival of many species of subtidal NCA. Grazing by herbivores, such as limpets, chitons, gastropods and sea urchins, removes filamentous algae, diatoms and bacteria from the coralline surface, and keeps it clean. Daume *et al.* (1997) described the diatom assemblage on the surface of the coralline red alga *Phymatolithon repandum* and the effect of postlarval abalone grazing on this microflora.

In several experiments examining the effect of grazing, the grazers were removed from a coralline habitat, whereupon the NCA became overgrown and eventually 'died' (review of Paine 1980; Steneck 1982). At West I., in the boulder habitat studied by Sabine Daume and colleagues, nine species of gastropod and 13 species of chiton, with a total density of up to 160 individuals m^{-2}, grazed on NCA and its epiphytes, besides other encrusting organisms. When they were removed, a filamentous algal turf ~5 mm high grew over the NCA over the next 65 days and accumulated a layer of sediment a few millimetres thick. When the grazers were restored to the boulders, they grazed the filamentous algae off the boulders within 7 days, exposing the now dead coralline, which never recovered (Kangas and Shepherd 1984; Clarkson and Shepherd 1985).

Paradoxically, NCA are well defended against herbivory because they have morphological and anatomical features that render them resistant to grazing (Steneck 1986). Firstly, they incorporate calcium carbonate (as calcite) into their cell walls, so providing a very hard framework that is a deterrent to some, but not all, grazers; some herbivores, such as chitons and gastropods require less force with their radulae to remove small pieces of the NCA thallus than pieces of macroalgae (Padilla 1985). Secondly, some species of corallines have multiple numbers of cell layers in their surface cells (the epithallium); the coralline tolerates grazing of the upper layers, while the deeper cell layers containing the meristem are protected. Herbivores, such as chitons, gastropods, limpets, some urchins and parrotfishes, are all able to excavate into NCA to varying extents, and these have an ecological and evolutionary impact on the crustose algae. Chitons have a bite depth of ~9 μm, limpets ~14 μm, and urchins up to 90 μm, while the average thickness of many corallines is 50–120 μm (Steneck 1986). Hence, most of the coralline remains intact despite molluscan grazing, and can regenerate. In this way, a stable coexistence between plant and herbivore is facilitated. Urchins, however, are an exception, and, depending on their density, can have a great impact on crustose corallines (Chapter 12).

Some corallines do not have to rely on grazers to keep them clean because they have an intrinsic mechanism to control fouling of their surface. Such species shed the surface layers of the thallus, and this is thought to be an effective mechanism against overgrowth (Johansen 1981; Breitburg 1984; Johnson and Mann 1986). In particular, some growth forms with protuberances accumulate diatoms and bacteria in small crevices and depressions, and these species are more prone to slough their surface cells to reduce fouling (Daume *et al.* 1997; Keats *et al.* 1997).

At a community level, where numerous species of crustose corallines co-exist and compete for space, the degree of disturbance to the habitat is important. If disturbances are rare, no free space becomes available and a few species that are the best long-term competitors will monopolise the habitat. At the other extreme, with continual disturbance, the slow-growing species will never become established and only a few thin, fast-growing species will survive. Disturbance rates at some intermediate range of intensity would prevent one species from becoming dominant, and allow many species with different growth and overgrowth strategies to co-exist. This principle – the intermediate disturbance hypothesis (Connell 1978) – has been applied to sessile invertebrates on boulders (Osman 1977) and panels (Svensson *et al.* 2007), and to algal communities (Sousa 1980; Davis and Wilce 1987).

The role of NCA in invertebrate larval settlement

NCA globally are known to induce the settlement of a variety of marine invertebrate larvae, such as soft octocorals, scleractinian corals (Morse *et al.* 1988), polychaetes (Gee 1965), limpets (Steneck 1982), chitons (Rumrill and Cameron 1983), asteroids (Barker 1977; Johnson and Sutton 1994), sea urchins (Pearce and Scheibling 1990) and abalone (see below). For many marine invertebrates, selection of an initial settlement site is a crucial and active process, with subsequent survival depending on correct selection of specific benthic substrata. These interactions have sometimes evolved into species-specific relationships or even mutualisms, in which both the grazer and the host plant benefit. Mutualisms are more likely to evolve among animals with low mobility (e.g. molluscs) and with a long-lived and predictable plant food source, such as corallines (Steneck 1982, 1992).

Herbivore–plant mutualisms

One of the best known herbivore–plant mutualisms is between small abalone and crustose corallines, and, so far, > 17 species of abalone worldwide are known to share this relationship (see Morse 1990; Kitting and Morse 1997). First, the abalone larva, when ready to settle, sinks passively to the bottom, and, if it falls onto an encrusting coralline, it receives from the coralline a chemical signal, which binds onto the larva's specialised receptors and induces metamorphosis and settlement. Over the next 3–6 months, the juvenile abalone lives, feeds and grows on the coralline (Fig. 6.3A). The benefits to the animal of this remarkable mutualism are:

- The smooth, extensive coralline surface is ideal for the firm attachment of the abalone's foot – something critically important in surgy conditions.
- The abalone feeds on the alga's surface mucus and diatoms initially (Fig. 6.3B), and after a week or so starts to rasp more deeply into the coralline's outer cell layers – all nutritious foods.

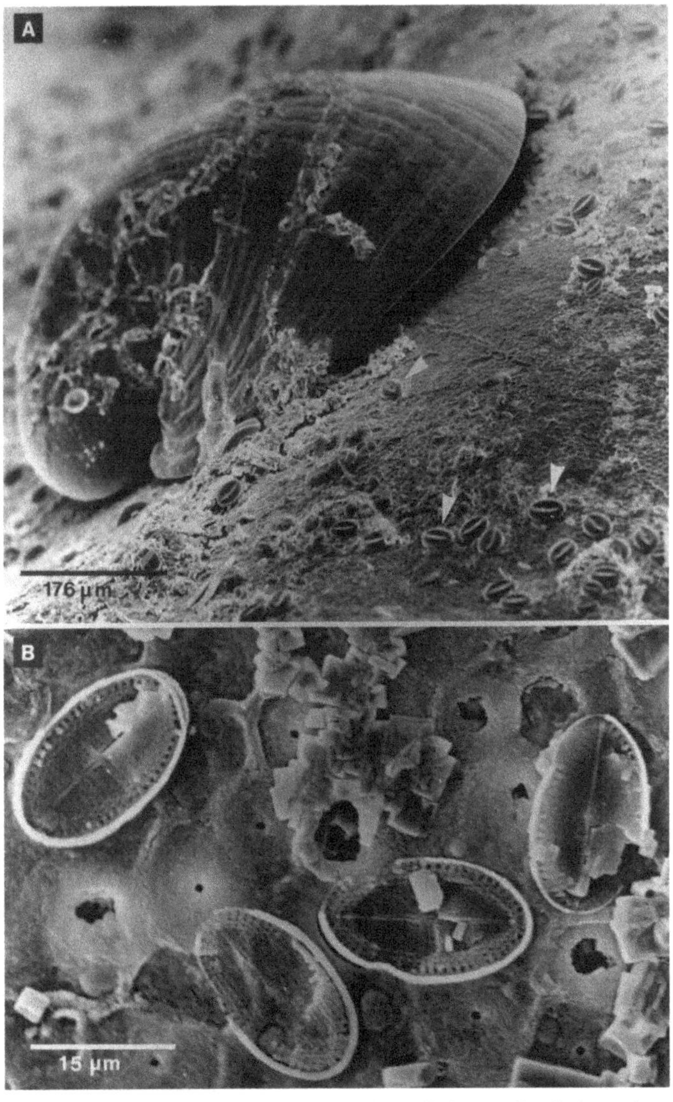

Figure 6.3: Scanning electron microscope pictures of (A) a recently settled greenlip abalone about 1 week old on crustose coralline algae, and (B) diatom and mucus growth on crustose coralline algal surface (photos S Daume).

- The tiny abalone acquires the cryptic, pink colour of its host, and becomes nearly invisible to searching predators, so enjoying an increased likelihood of survival.

Thus, the abalone has evolved to use the plant's unique chemical inducer to find the optimal microhabitat for it at a crucial phase of its life history.

The reciprocal benefits to the plant are equally impressive:

- The grazing juvenile abalone reduces epiphytic overgrowth of the host alga.
- The micro-grazing also appears to enhance growth of the host alga onto any neighbouring free surfaces (analogous to the pruning effect on bushes), as in the experiments of Kitting and Morse (1997) described below.
- The absence of epiphytic growth on the coralline surface prevents sediment becoming trapped on it, and allows more light to reach the coralline, which is important for the coralline's growth in shaded habitats, where the light is dim.

The above mutualism was described for the Californian red abalone *Haliotis rufescens* and its host corallines, *Lithothamnion californicum* and *Lithophyllum lichenare*, by Kitting and Morse (1997). The positive effect of grazing on the corallines by newly settled abalone, maintained at a density of 7 cm^{-2}, is illustrated in Fig. 6.4, which compares the growth of filamentous algae on the corallines over 9 weeks between grazed surfaces, ungrazed surfaces and dead corallines (the control). It is evident that grazing keeps the cover of filamentous algae to about half that on ungrazed surfaces.

Australian studies have shown similar mutualisms between two abalone species and NCA (Shepherd and Daume 1996). In experiments with the greenlip abalone *Haliotis laevigata*, most (85%) settling abalone larvae preferred the lumpy coralline *Sporolithon durum* and the remainder crustose forms of species of *Mesophyllum* or *Hydrolithon* (Daume *et al.* 1999a). When the pink photosynthetic pigment was removed from the *Sporolithon*, very few larvae settled on it, indicating that the chemical signal inducing settlement was associated with the coralline's photosynthetic pigment. Similarly, blacklip abalone larvae *Haliotis rubra* preferred to settle on crustose corallines, this time on *Phymatolithon repandum*, which was the dominant coralline in its habitat (Daume *et al.* 1999b; Huggett *et al.* 2005), although it did settle in response to a variety of other macroalgae. After settlement, the small blacklip feeds on the diatom film on the coralline, and in a few days starts to rasp the cuticle and outer cell layer of the coralline up to 3 µm deep – which the coralline soon regenerates (Garland *et al.* 1985; Daume *et al.* 1997, 2000).

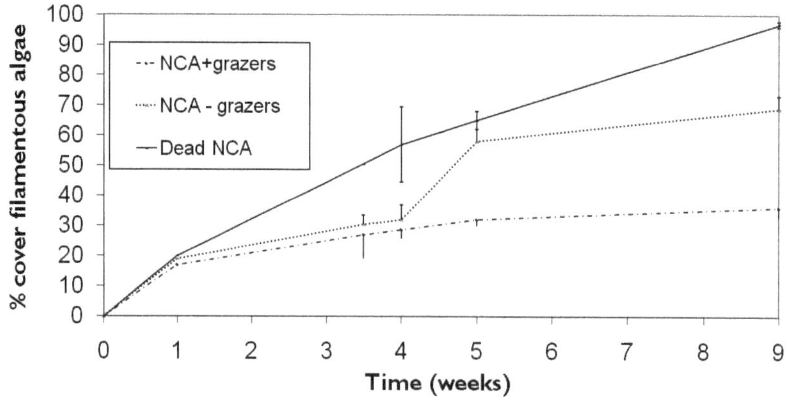

Figure 6.4: Plot of average cover of filamentous algae on crustose corallines (NCA) over 9 weeks, with and without grazing abalone. The control was dead NCA, without any grazers (simplified and redrawn from Kitting and Morse 1997). Vertical bars show the range of values in replicate experiments.

The micro-environment of the coralline surface

Studies of larval settlement have shown that oxygen concentrations on coralline surfaces are critical. These can vary greatly, depending on the growth form of the coralline (e.g. smooth, warty or lumpy), the presence of diatom films, bacteria or other fouling organisms, the extent of grazing, and water movement (Kaspar 1992). Here, the diffusive boundary layer (DBL) with low O_2 concentrations – defined as the distance between the coralline surface and the point at which oxygen concentration is at equilibrium with the water above – is critical for post-larvae. Another important factor is the demand for oxygen by the coralline itself at night. The thickness of the DBL under various conditions was as follows (Kaspar 1992):

- Under grazed conditions with slight water flow of ~2 cm s^{-1} above the coralline, the DBL was 0.1–0.2 mm thick, and O_2 saturation was more than twice air saturation just above it under light – both suitable conditions for larvae, whose diameters are often up to 0.5 mm.
- As water flow declined, O_2 saturation rapidly declined to low levels.
- In the absence of grazing for 2 weeks, the coralline surface developed a biofilm of detritus, microalgae, protozoans and bacteria, and the DBL increased to ~0.5 mm.
- a thin layer of sediment accumulating on the biofilm can increase the DBL to > 1 mm, and sheltered pockets in corallines, such as behind protuberances, may have the same effect – both creating oxygen stress for recently settled larvae.

The growth-forms of NCA with protuberances (e.g. lumpy forms) can also have an indirect effect on the settlement of invertebrate larvae. In Daume *et al.*'s (1999b) study, larvae preferred to settle on the lumpy form to the flat form of *Sporolithon*, probably because of differences in the oxygen supply (and/or pH) in the DBL. In another study, settlement rates of abalone larvae were tested in conditions of high and low water flow rates. In high flows, the larvae behaved as passive particles and few of them settled, because they had little chance to test the substratum for settlement cues. In low flows, a high proportion of the larvae could settle on the bottom and respond to the settlement cues (Boxshall 2000). So, different micro-environments, created by lumpy or warty coralline surfaces or water flows, can influence larval settlement (see Roberts *et al.* 2004). The pH at the surface, and in the cytoplasm of the cells, also has an effect on calcification rate, and thus on the growth of the algae (Digby 1977).

Diatoms and bacteria rapidly grow on the surface of NCA and accumulate in small crevices and behind protuberances. They can alter the oxygen supply and pH by taking up nutrients, and by photosynthesis, and thereby provide a suitable micro-environment for the precipitation of calcium carbonate in the cell walls of the algae, as well as for shell growth of newly settled abalone and other molluscs. Because bacteria are patchily distributed on the algal surface, the pH might rise only in areas with a high abundance of bacteria and diatoms, so that an optimal environment exists only in localised areas. Thus the suitability of the micro-environment for survival of larvae may vary significantly on a very small scale. Furthermore, as fouling communities develop, their topography becomes more complex, and the range of water-flow conditions on the spatial scale of larvae increases. Hence water-borne settlement cues occur patchily, and larvae encounter pulses of high cue concentration interspersed with patches of low concentration or no inducer at all (Koehl 2007; Koehl *et al.* 2007).

RHODOLITHS

Rhodoliths are free-living forms of NCA that live on sediment and sometimes among seagrass beds of *Amphibolis* or *Posidonia*, and form a habitat that has features of both hard and soft bottoms; areas of sea bed dominated by them are called rhodolith beds. Some 26 genera of NCA are recognised, and at least eight of them contain species that form rhodoliths (Woelkerling 1996; Harvey and Woelkerling 2007; Harvey and Bird 2008). In southern Australia, five main rhodolith-forming genera have been recorded – *Lithothamnion, Hydrolithon, Mesophyllum, Neogoniolithon* and *Sporolithon* – and they have a similar variety of shapes as the crustose forms – warty, lumpy, fruticose, and so on. They often occur in high densities to form deposits of living and dead aggregations. Sometimes a cobblestone or gravel core becomes overgrown with encrusting coralline, and these are termed nucleated rhodoliths. Rhodoliths are rolled about by water currents and swell, and Foster (2001) has colourfully called them the 'calcareous tumbleweeds of the sea' that form 'reefs which rock and roll'. Less imaginative prawn fishers call them 'popcorn'.

Rhodolith beds are found sparsely throughout southern Australia, but this scarcity may partly be an artefact of the patchiness of bottom surveying, especially in deep water offshore. Published records from west to east are listed below:

- on the Rottnest Shelf at 35–60 m depth – beds of *Sporolithon* (Collins 1988; James *et al.* 1999).
- in the western GAB at 27–65 m depth, among islands of the Recherche Archipelago, and in Esperance Bay – four species in the above genera, with three main forms – warty, lumpy, and fruticose (Goldberg 2006).
- off Twilight Cove 430 km NE of Esperance Bay, and generally on the Roe Shelf extending for 100 km offshore from the Baxter Cliffs (Harvey and Kendrick 2005) – cobble-sized rhodoliths of ~5 cm size; they were described as 'compact to rounded, branching or dendritic' (James *et al.* 2001).
- in the NE GAB at 60–135 m depth (James *et al.* 1994, 2001).
- in Spencer Gulf at 20–25 m depth in places of moderate to strong tidal current, and also in Investigator Strait – *Lithophyllum* beds (Svane *et al.* 2009).
- in Proper Bay, Port Lincoln - *Lithophyllum* beds (Fig. 6.5); and in Pelican Lagoon, Kangaroo I. – *Mesophyllum erubescens*; both sites are on rubbly bottoms at 1–4 m depth (Womersley 1956; Shepherd 1975).
- on the Lacepede Shelf, SE of Kangaroo I., at 60–80 m depths (James *et al.* 1994).
- in sheltered waters off Port MacDonnell at ~5 m depth – *Sporolithon* beds; and off Ocean Grove, western Bass Strait at depths of 30–35 m (S. Chidgey *pers. comm.*).

Figure 6.5: Rhodolith bed of *Lithophyllum* sp. in sparse *Posidonia australis* bed at 3–4 m depth off Billy Lights Point, Proper Bay, Port Lincoln recorded in 1975 (photo Kevin Branden).

- at the entrance to Western Port at a site 1–4 m deep, with moderate tidal currents – extensive beds of *Lithothamnion, Hydrolithon, Mesophyllum* and *Neogoniolithon* (foliose and fructose forms) over ~1 km^2 at average densities of ~500 rhodoliths m^{-2} of which up to two thirds are dead (Harvey and Bird 2008; Plate 12).
- in numerous bays in eastern Australia, from Gabo I. north to Byron Bay (Harvey *et al.* 2002), and in deeper shelf waters (Marshall and Davies 1978).

Ecology of rhodoliths

Habitat

Rhodoliths have strict habitat requirements. They are found mainly on sediments with a high calcareous content, and usually where shell, gravel or cobbles also occur. Carbonate production is high on the southern Australian shelf (see James *et al.* 1992, 1994, 1999), favouring rhodolith development. In Esperance Bay, for example, calcium carbonate comprises 83% of the sediment (Ryan *et al.* 2007), and rhodolith beds cover 14% of > 1000 km^2 of mapped bottom habitats (Baxter *et al.* 2005). Rhodoliths also require moderate water movement. If water movement is too low, they become buried by sediment, and if too high, they are rolled or carried away. The exposed southern Australian coasts are subject to prevailing swells with a period of 10–16 s and wavelengths of up to 200 m. These swells penetrate to > 100 m depth (rarely to 160 m depth), and can produce bottom orbital velocities of 50 cm s^{-1} at 30 m depth (see Goldberg 2006) – an oscillatory motion more than enough to rework and sort the sediments. Hence, deeper shelf bottoms throughout the region have been termed a 'shaved shelf', because the constant swell scours the bottom and prevents sediment accumulation (James *et al.* 1994). The depth range of 60–80 m on much of the southern exposed coast, and slightly shallower depths where swell is attenuated by coastal topography, are apparently optimal for rhodolith beds. However, fragments of living rhodoliths are likely to be found in depths down to 240 m, as on the Lacepede Shelf, SE of Kangaroo I. (James *et al.* 1992). One study on the east Australian coast found that rhodolith size declined substantially below ~60 m depth (Lund *et al.* 2000). The degree of water movement affects the shape and branching patterns of rhodoliths, with spherical shapes favoured by moderate water movement, and irregular shapes by lower water movement (Foster 2001).

In bays and gulfs, the distribution of rhodoliths seems to be controlled by tidal movement or by wind-driven waves. Field studies, sometimes employing underwater video, show that short-period waves 1 m high are able to move rhodoliths at ~5 m depth, and that tidal currents of ~30 cm s^{-1} can roll or rock them, depending on the complexity of their shape (Marrack 1999; unpublished observations). Hence, rhodolith beds in Spencer Gulf and in Western Port are both located in places of strong tidal current. Fish and some invertebrates can also move rhodoliths as they search for food, and may be important in some areas. Periodic rotation of nodules appears necessary for light to reach all sides of the nodule, as well as to prevent burial or fouling.

Growth and longevity

Rhodolith beds develop from crustose algal spores settling onto small grains of sand or gravel, or from broken pieces of coralline nodules. After a bed is established, recruitment is probably mainly by breakage and overgrowth. Rhodoliths grow extremely slowly, according to the sparse data available, and growth rates in temperate waters are typically 0.2–0.6 mm $year^{-1}$ in depths < 20 m, and much lower (0.01–0.1 mm $year^{-1}$) in deeper water (Foster 2001; Steller *et al.* 2007). Yet, a few other temperate studies have shown quite fast growth rates compared with the above. For example, in the northern Mediterranean at < 1 m depth, Boudouresque *et al.* (1972) showed that nodules of the much branched *Lithophyllum tortuosum* extended their margins at a rate of 6–11 mm $year^{-1}$, so that nodules 6 cm diameter were 2–3 years old, and reached a maximum size of ~10 cm.

The longevity of rhodoliths is known for two species in southern Australia. Shallow-water nodules of *Sporolithon durum*, 7–9 cm diameter and living at 1–3 m depth at Rottnest I., were estimated by radiocarbon dating to be < 60 years old (Goldberg and Heine 2008). Deeper-water forms at 38 m depth, with a size range of 2–6 cm, from a rhodolith bed in Esperance Bay, were found to range widely in age from modern (< 50 years old) to ~960 years. Because rhodoliths can grow uninterruptedly for > 100 years, it is likely that the older ones became buried, died, and later exposed and recolonised (Goldberg 2006). The growth and longevity information available suggest that established beds, especially in deeper water, are likely to be decades to many centuries old.

The production rates of rhodolith beds in temperate waters have been shown to be surprisingly similar to those for coral reefs. Typical rates are in the range 200–1200 g $CaC0_3$ m^{-2} $year^{-1}$ (review of Bosence and Wilson 2003). Such rates can cause beds to accumulate at a rate of ~1 mm $year^{-1}$, though thick beds have so far only been recorded in the western Great Australian Bight.

Fauna associated with rhodolith beds

Rhodoliths have been called foundation species or 'bio-engineers', because they modify benthic habitats by providing hard surfaces for some species, and shelter in their interstices and between nodules for others (Steller *et al.* 2003). Larger nodules, and those with higher branch density, support greater faunal densities than smaller or simpler ones (Steller *et al.* 2007).

Harvey and Bird (2008) measured the cryptic fauna living within the branches of the rhodolith beds in Western Port, but not other mobile fauna (e.g. predators associated with the beds, or animals living among the nodules). The beds had an average density of 400 individuals L^{-1} (of rhodolith volume) living in their interstices, comprising 89% polychaete worms, 8% bivalves and the remainder echinoderms and crustaceans. The two growth forms present, fruticose and foliose, had similar abundances of associated fauna.

In the Esperance rhodolith beds, the average density of fauna living on the rhodoliths (the epifauna), and those nestling in the interstices between the warts or lumps of nodules, as well as those living in tubes or galleries within the nodules (the endofauna), averaged 1250 individuals L^{-1} of rhodolith volume, excluding protists; 62% of these comprised the endofauna, and the rest epifauna. Polychaetes, with 59% of the total, were the most common group, followed by bryozoans (16%), arthropods (9%), sponges (6%) and ascidians (5%), with small numbers of hydroids, echiuran and sipunculid worms, and bivalve molluscs (Mathis *et al.* 2005).

The above studies did not compare the abundance of fauna living in rhodolith beds with those on nearby sandy bottoms. However, other studies (e.g. Steller *et al.* 2003) have shown that the richness of species is almost twice as high, and density a thousand times higher, in rhodolith beds compared with sandy bottom. The reason for such a huge difference is that the fauna of rhodolith beds includes: (a) many species requiring a hard substrate, such as sponges and ascidians; (b) species that live in the spaces between rhodoliths; (c) those that are interstitial in branching rhodoliths; and (d) predators. But is the species' richness of beds dependent on frequent disturbance by wave action?

To answer this, Hinojosa-Arango *et al.* (2009) artificially stabilised nodules in a rhodolith bed. The researchers found, not surprisingly, that more algae grew on the nodules and that the animals in the bed were sensitive to changes in the movement of nodules, with the result that some species moved out and others moved in. The biodiversity of rhodolith beds seems to be especially dependent on the fauna in adjacent habitats.

Disturbance

Rhodolith beds have very low resilience to fishing, specifically trawling, due to their slow growth rates and the negative and fatal impact of burial. Trawling is the severest form of human disturbance that initially reduces rhodolith density and size, and ultimately degrades them structurally to a gravel bottom (Kamenos *et al.* 2003; Steller *et al.* 2003). Other threats to rhodolith beds are: turbidity and sedimentation from terrestrial run-off, as in Western Port (Harvey and Bird 2008); organic enrichment from fish farms; and storm-water or effluent outfalls. In particular, high phosphate levels reduce growth and calcification of corallines (Bjork *et al.* 1995).

Disturbance effects on the fauna living in the beds are as severe as those on the rhodoliths themselves. Besides direct effects resulting from loss of habitat structure, indirect effects, such as declines in fragile and omnivorous species, increase in soft-sediment species and scavengers, and declines in predatory species, also take place. The outcome can be a high loss of species and functional diversity, and resource monopolisation by a few dominant species (Grall and Glémarec 1997). Fish farms, in particular, can cause a serious loss of biodiversity in the fauna of rhodolith beds, notably in the abundance of small crustaceans, such as ostracods, isopods, tanaids and cumaceans (Hall-Spencer *et al.* 2006). The effects of acidification on corallines in the context of global warming are discussed in Chapter 1.

A consequence of the increasing threats to rhodolith beds globally, and the recognition of their high scientific and conservation value, is that many of them are protected in European waters (Council of European Communities 1992), and also in New Zealand. Such protection is urgently needed for the still poorly known southern Australian rhodoliths.

REFERENCES

Ayling AM (1981) The role of biological disturbance in temperate subtidal encrusting communities. *Ecology* **62**, 830–847.

Barker MF (1977) Observations on the settlement of the Brachiolaria larvae of *Stichaster australia* (Verrill) and *Coscinasterias calamaria* (Gray) (Echinodermata: Asteroidea) in the laboratory and on the shore. *Journal of Experimental Marine Biology and Ecology* **30**, 95–108.

Baxter K, Bickers A, Kendrick GA, Shortis M (2005) Benthic habitat mapping in the Recherche Archipelago. In *Characterising the Fish habitats of the Recherche*

Archipelago. FRDC Final Report 2001/060. (Eds GA Kendrick, E Harvey, J McDonald, C Pattiaratchi *et al.*) pp. 29–83. University of Western Australia, Perth.

Bjork M, Mohammed SM, Bjorklund M, Semesi A (1995) Research and capacity building for sustainable coastal management. *Ambio* **24**, 502–505.

Bosence D, Wilson J (2003) Maerl growth, carbonate production rates and accumulation rates in the northeast Atlantic. *Aquatic Conservation: Marine and Freshwater Ecosystems* **13**, S21–S31.

Boudouresque C-F, Augier H, Guénoun Y-C (1972) Végétation marine de l'Ile de Port-Cros (Parc Nacionale) VIII. Premiers résultats de l'etude de la croissance *in situ* de *Lithophyllum tortuosum* (Rhodophycées, Corallinacées). *Bulletin de Muséum D'Histoire Naturelle de Marseille* **32**, 197–215.

Boxshall AJ (2000) The importance of flow and settlement cues to larvae of the abalone, *Haliotis rufescens* Swainson. *Journal of Experimental Marine Biology and Ecology* **254**, 143–167.

Breitburg DL (1984) Residual effects of grazing: inhibition of competitor recruitment by encrusting coralline algae. *Ecology* **65**, 1136–1143.

Clarkson PS, Shepherd SA (1985) Distribution and feeding of archaeogastropods in a boulder habitat at West Island, South Australia. *Journal of the Malacological Society of Australia* **7**, 35–44.

Collins L (1988) Sediments and history of the Rottnest Shelf, southwest Australia: a swell-dominated, non-tropical carbonate margin. *Sedimentary Geology* **60**, 15–49.

Connell JH (1978) Diversity in tropical rain forests and coral reefs. *Science* **199**, 1302–1310.

Connell SD (2003a) Negative effects overpower the positive of kelp to exclude invertebrates from the understorey community. *Oecologia* **137**, 97–103.

Connell SD (2003b) The monopolization of understorey habitat by subtidal encrusting coralline algae: a test of the combined effects of canopy-mediated light and sediment. *Marine Biology* **142**, 1065–1071.

Connell SD (2005) Assembly and maintenance of subtidal habitat heterogeneity: synergistic effects of light penetration and sedimentation. *Marine Ecology Progress Series* **289**, 53–61.

Connell SD (2007) Subtidal temperate rocky habitats: habitat heterogeneity at local to continental scales. In *Marine Ecology* (Eds SD Connell and BM Gillanders) pp. 378–401. Oxford University Press, Melbourne.

Connell SD, Irving AD (2008) Integrating ecology with biogeography using landscape characteristics: a case study of subtidal habitat across continental Australia. *Journal of Biogeography* **35**, 1608–1621.

Council of European Communities. (1992) Council Directive 92/43/EEC. Conservation of natural habitats and of wild flora and fauna. *International Journal of European Communities* **L206**, 7–49.

Daume S, Brand S, Woelkerling WJ (1997) Effects of post-larval abalone (*Haliotis rubra*) grazing on the epiphytic diatom assemblage of coralline red algae. *Molluscan Research* **18**, 119–130.

Daume S, Brand-Gardner S, Woelkerling WJ (1999a) Community structure of non-geniculate coralline red algae (Corallinales, Rhodophyta) in three boulder habitats in southern Australia. *Phycologia* **38**, 138–148.

Daume S, Brand-Gardner S, Woelkerling WJ (1999b) Settlement of abalone larvae (*Haliotis laevigata* Donovan) in response to non-geniculate coralline red algae (Corallinales, Rhodophyta). *Journal of Experimental Marine Biology and Ecology* **234**, 125–143.

Daume S, Krsinich A, Farrell S, Gervis M (2000) Settlement, early growth and survival of *Haliotis rubra* in response to different algal species. *Journal of Applied Phycology* **12**, 479–488.

Davis A, Wilce R (1987) Algal diversity in relation to physical disturbance: a mosaic of successional stages in a subtidal cobble habitat. *Marine Ecology Progress Series* **37**, 229–237.

Digby PSB (1977) Growth and calcification in the coralline algae, *Clathromorphum circumscriptum* and *Corallina officinalis*, and the significance of pH in relation to precipitation. *Journal of the Marine Biological Association of the United Kingdom* **57**, 1095–1109.

Foster MS (2001) Rhodoliths: between rocks and a hard place. *Journal of Phycology* **37**, 659–667.

Fowler-Walker MJ, Connell SD (2002) Opposing states of subtidal habitat across temperate Australia: consistency and predictability in kelp canopy-benthic associations. *Marine Ecology Progress Series* **240**, 49–56.

Fowler-Walker MJ, Gillanders BM, Connell SD, Irving AD (2005) Patterns of association between canopy-morphology and understorey assemblages across temperate Australia. *Estuarine, Coastal and Shelf Science* **63**, 133–141.

Garland CD, Cooke SL, Grant JF, McMeekin TA (1985) Ingestion of the bacteria on and the cuticle of crustose (non-articulated) coralline algae by post-larval and juvenile abalone (*Haliotis ruber* Leach) from Tasmanian waters. *Journal of Experimental Marine Biology and Ecology* **91**, 137–149.

Gee JM (1965) Chemical stimulation of settlement in larvae of *Spirorbis rupestris* (Serpulidae). *Animal Behaviour* **13**, 181–186.

Goldberg N (2006) Age estimates and description of rhodoliths from Esperance Bay, Western Australia. *Journal of the Marine Biological Association of the United Kingdom* **86**, 1291–1296.

Goldberg N, Heine JN (2008) Age estimates of *Sporolithon durum* (Corallinales, Rhodophyta) from Rottnest

Island, Western Australia, based on radiocarbon dating methods. *Journal of the Royal Society of Western Australia* **91**, 27–30.

Grall J, Glémarec M (1997) Biodiversité des fonds de maerl en Bretagne: approche fonctionelle et impacts anthropiques. *Vie et Milieu* **47**, 339–349.

Hall-Spencer J, White N, Gillespie E, Gillham K *et al.* (2006) Impact of fish farms on maerl beds in strongly tidal areas. *Marine Ecology Progress Series* **326**, 1–9.

Harvey AS, Bird FL (2008) Community structure of a rhodolith bed from cold-temperate waters (southern Australia). *Australian Journal of Botany* **56**, 437–450.

Harvey AS, Woelkerling WJ (2007) A guide to non-geniculate coralline red algal (Corallinales, Rhodophyta) identification. *Ciencias Marinas* **33**, 411–426.

Harvey AS, Woelkerling WJ, Millar AJK (2002) The Sporolithaceae (Corallinales, Rhodophyta) in south-eastern Australia. *Phycologia* **41**, 207–227.

Harvey E, Kendrick G (2005) The influence of reef isolation, exposure and depth on the structure and relative abundance of the shallow water reef fish assemblages of the Recherche Archipelago, South Western Australia. In *Characterising the Fish Habitats of the Recherche Archipelago: Appendix 6*. (Eds G Kendrick, E Harvey, J Hill, JI McDonald, *et al.*) pp. 143–167. FRDC Project 2001/060. University of Western Australia, Perth.

Hinojosa-Arango G, Magos CA, Johnson MP (2009) Like a rolling stone: the mobility of maerl (Corallinaceae) and the neutrality of the associated assemblages. *Ecology* **90**, 517–528.

Huggett MJ, de Nys R, Williamson JE, Heasman M *et al.* (2005) Settlement of larval blacklip abalone, *Haliotis rubra*, in response to green and red macroalgae. *Marine Biology* **147**, 1155–1163.

Irving AD, Connell SD (2006a) Physical disturbance by kelp abrades erect algae from the understorey. *Marine Ecology Progress Series* **324**, 127–137.

Irving AD, Connell SD (2006b) Predicting understorey structure from the presence and composition of canopies: an assembly rule for marine algae. *Oecologia* **148**, 491–502.

Irving A, Fowler-Walker MJ, Connell SD (2003) Associations of forest-type with morphology of *Ecklonia radiata* and abundance of understorey algae. *Transactions of the Royal Society of South Australia* **127**, 167–175.

Irving AD, Connell SD, Elsdon TS (2004) Effects of kelp canopies on bleaching and photosynthetic activity of encrusting coralline algae. *Journal of Experimental Marine Biology and Ecology* **310**, 1–12.

James NP, Bone Y, Von Der Borch CC, Gostin VA (1992) Modern carbonate and terrigenous clastic sediments on a cool water, high energy, mid-latitude shelf: Lacepede, South Australia. *Sedimentology* **39**, 877–903.

James NP, Borreen TD, Bone Y, Feary DA (1994) Holocene carbonate sedimentation on the weest Eucla Shelf, Great Australian Bight: a shaved shelf. *Sedimentary Geology* **90**, 161–177.

James NP, Collins LB, Bone Y, Hallcock P (1999) Subtropical carbonates in a temperate realm: modern sediments on the southwest Australian shelf. *Journal of Sedimentary Research* **69**, 1297–1321.

James NP, Bone Y, Collins LB, Kyser TK (2001) Surficial sediments of the Great Australian Bight: facies dynamics and oceanography on a vast cool-water carbonate shelf. *Journal of Sedimentary Research* **71**, 549–567.

Johansen HW (1981) *Coralline Algae, a First Synthesis*. CRC Press, Boca Raton, Florida.

Johnson CR, Mann KH (1986) The crustose coralline alga, *Phymatolithon* Foslie, inhibits the overgrowth of seaweeds without relying on herbivores. *Journal of Experimental Marine Biology and Ecology* **96**, 127–146.

Johnson CR, Sutton DC (1994) Bacteria on the surface of crustose coralline algae induce metamorphosis of the crown of thorns starfish *Acanthaster planci*. *Marine Biology* **120**, 305–310.

Kamenos NA, Moore PG, Hall-Spencer JM (2003) Substratum heterogeneity of dredged vs undredged maerl grounds. *Journal of the Marine Biological Association of the United Kingdom* **83**, 411–413.

Kangas M, Shepherd SA (1984) Distribution and feeding of chitons in a boulder habitat at West Island, South Australia. *Journal of the Malacological Society of Australia* **6**, 101–111.

Kaspar HF (1992) Oxygen conditions on surfaces of coralline red algae. *Marine Ecology Progress Series* **81**, 97–100.

Keats DW, Knight MA, Pueschel CM (1997) Antifouling effects of epithallial shedding in three crustose coralline algae (Rhodophyta, Corallinales) on a coral reef. *Journal of Experimental Marine Biology and Ecology* **213**, 281–293.

Kennelly SJ (1989) Effects of kelp canopies on understorey species due to shad and scour. *Marine Ecology Progress Series* **50**, 215–224.

Kitting CL, Morse DE (1997) Feeding effects of postlarval red abalone, *Haliotis rufescens* (Mollusca: Gastropoda) on encrusting coralline algae. *Molluscan Research* **18**, 183–196.

Koehl MAR (2007) Mini review: Hydrodynamics of larval settlement into fouling communities. *Biofouling* **23**, 357–368.

Koehl MAR, Reidenbach MA, Koseff JR, Hadfield MG (2007) Individual-based model of larval transport to coral reefs in turbulent, wave-driven flow: behavioral responses to dissolved settlement inducer. *Marine Ecology Progress Series* **335**, 1–18.

Ling SD (2008) Range expansion of a habitat-modifying species leads to loss of taxonomic diversity: a new and impoverished reef state. *Oecologia* **156**, 883–894.

Littler MM, Littler DS (1984) Models of tropical reef biogenesis: the contribution of algae. In *Progress in Phycological Research. Vol. 3* (Eds FE Round and VJ Chapman) pp. 323–364. Biopress, London.

Lund M, Davies PJ, Braga JC (2000) Coralline algal nodules off Fraser island, eastern Australia. *Facies* **42**, 25–34.

Marrack EC (1999) The relationship between water motion and living rhodolith beds in the southwestern Gulf of California, Mexico. *Palaios* **14**, 159–171.

Marshall JP, Davies PJ (1978) Skeletal carbonate variation on the continental shelf of eastern Australia. Australian Bureau of Mineral Resources. *Journal of Geology and Geophysics* **3**, 85–92.

Mathis BJ, Kohn AJ, Goldberg N (2005) Rhodoliths: the inside story. In *The Marine Flora and Fauna of Esperance, Western Australia.* (Eds FE Wells, DI, Walker and GA Kendrick) pp. 147–157. Western Australian Museum, Perth.

Melville AJ, Connell SD (2001) Experimental effects of kelp canopies on subtidal coralline algae. *Austral Ecology* **26**, 102–108.

Morcom NF, Woelkerling WJ, Ward SA (2005) Epiphytic nongeniculate corallines found on *Laurencia filiformis* f. *heteroclada* (Rhodomelaceae, Ceramiales). *Phycologia* **44**, 202–211.

Morse DE (1990) Recent progress in larval settlement: closing the gaps between molecular biology and ecology. *Bulletin of Marine Science* **46**, 465–483.

Morse DE, Hooker N, Morse ANC, Jensen RA (1988) Control of larval metamorphosis and recruitment in sympatric agariciid corals. *Journal of Experimental Marine Biology and Ecology* **116**, 193–217.

Nelson WA (2009) Calcified macroalgae – critical to coastal ecosystems and vulnerable to change. *Marine and Freshwater Research* **60**, 787–801.

Osman R (1977) The establishment and development of a marine epifaunal community. *Ecological Monographs* **47**, 37–63.

Padilla DK (1985) Structural resistance of algae to herbivores. *Marine Biology* **90**, 103–109.

Paine R (1980) Food webs: linkage, interaction strength and community infrastructure. *Journal of Animal Ecology* **49**, 666–685.

Pearce CM, Scheibling RE (1990) Induction of metamorphosis of larvae of the green sea urchin, *Strongylocentrotus droebachiensis*, by coralline red algae. *The Biological Bulletin* **179**, 304–311.

Roberts RD, Kaspar HF, Barker MF (2004) Settlement of abalone (*Haliotis iris*) larvae in response to five species of coralline algae. *Journal of Shellfish Research* **23**, 975–987.

Rumrill SS, Cameron RA (1983) Effects of γ-aminobutyric acid on settlement of larvae of the black chiton, *Katherina tunicata. Marine Biology* **72**, 243–247.

Ryan DA, Brooke BP, Collins LB, Kendrick GA *et al.* (2007) The influence of geomorphology and sedimentary processes on shallow-water benthic habitat distribution: Esperance Bay, Western Australia. *Estuarine, Coastal and Shelf Science* **72**, 379–386.

Shepherd SA (1975) 'An underwater survey of two sites receiving effluent discharge in Proper Bay, Port Lincoln'. Report No. 42. Department of Agriculture and Fisheries, Adelaide.

Shepherd SA (1979) The ecology of a deep water red algal community off Cape Northumberland, South Australia. MEnvSt thesis. University of Adelaide, Adelaide.

Shepherd SA, Daume S (1996) Ecology and survival of juvenile abalone in a crustose coralline habitat in South Australia. In *Survival Strategies in Early Life Stages of Marine Resources.* (Eds Y Watanabe, Y Yamashita and Y Oozeki) pp. 297–313. Balkema, Rotterdam.

Sim CB (2001) Recruitment ecology of crustose coralline algae (Corallinales, Rhodophyta) at Rottnest Island, Western Australia. *Australian Biologist* **14**, (1), 23–24.

Sim CB, Townsend RA (1999) An account of common crustose coralline algae (Corallinales, Rhodophyta) from macrophyte communities at Rottnest Island, Western Australia. In *The Seagrass Flora and Fauna of Rottnest Island, Western Australia.* (Eds DI Walker and FE Wells) pp. 395–408. Western Australian Museum, Perth.

Sousa W (1980) The responses of a community to disturbance: The importance of successional age and species' life histories. *Oecologia* **45**, 72–81.

Steller DL, Riosmena-Rodríguez R, Foster MS, Roberts CA (2003) Rhodolith diversity in the Gulf of California: the importance of rhodolith structure and consequences of disturbance. *Aquatic Conservation Marine and Freshwater Ecosystems* **13**, S5–S20.

Steller DL, Hernández-Ayón JM, Riosmena-Rodríguez R, Cabello-Pasini A (2007) Effect of temperature on photosynthesis, growth and calcification rates of the free-living coralline alga *Lithophyllum margaritae. Ciencias Marinas* **33**, 441–456.

Steneck RS (1982) A limpet-coralline alga association adaptation and defenses between a selective herbivore and its prey. *Ecology* **63**, 507–522.

Steneck RS (1986) The ecology of coralline algal crusts: convergent patterns and adaptive strategies. *Annual Review of Ecology and Systematics* **17**, 273–303.

Steneck RS (1992) Plant-herbivore coevolution: a reappraisal from the marine realm and its fossil record. In *Plant-Animal Interactions in the Marine Benthos.* (Eds DM John, SJ Hawkins and JH Price) pp. 477–491. Clarendon Press, Oxford.

Steneck RS, Hacker SD, Dethier M (1991) Mechanisms of competitive dominance between crustose coralline algae: an herbivore-mediated competitive reversal. *Ecology* **72**, 938–950.

Svane I, Hammett Z, Lauer P (2009) Impacts of trawling on benthic macro-fauna and -flora of the Spencer Gulf fishing grounds. *Estuarine, Coastal and Shelf Science* **82**, 621–631.

Svensson JR, Lindegarth M, Siccha M, Lenz M *et al.* (2007) Maximum species richness at intermediate frequencies of disturbance: consistency among levels of productivity. *Ecology* **88**, 830–838.

Woelkerling WJ (1988) *The Coralline Red Algae: An Analysis of the Genera and Subfamilies of Non-geniculate Corallinaceae.* Oxford University Press, Oxford.

Woelkerling WJ (1996) Non-geniculate Corallinales. In *The Marine Benthic Flora of Southern Australia: Rhodophyta—Part IIIB* (Ed HBS Womersley) pp. 146–282. ABRS, Canberra.

Woelkerling WJ, Irvine L, Harvey AS (1993) Growth-forms in non-geniculate coralline red algae (Corallinales, Rhodophyta). *Australian Systematic Botany* **6**, 277–293.

Womersley HBS (1956) The marine algae of Kangaroo Island. IV. The algal ecology of American River Inlet. *Australian Journal of Marine and Freshwater Research* **7**, 64–87.

7 Middle and lower stratum algae

OVERVIEW

The rich southern algal flora is considered here in three categories according to their height: prostrate species other than calcified crusts; algal turfs (to 1–2 cm high); and middle stratum species to 30–40 cm high. The growth rate, reproduction, longevity and productivity are considered for selected species or groups of species in each category. Species with a prostrate growth habit mainly prefer low light and tend to be found on shaded surfaces or in deeper water. Turf-forming algae comprise many species of filamentous algae, as well as early stages of other species. They are an ephemeral, early successional group, and tend to colonise newly cleared substratum. They have a rapid turnover and high productivity. The species–rich green algal genus, *Caulerpa*, is common in subtidal habitats, and spreads by sexual reproduction and by lateral growth of stolons attached to the rock. Some of the species have proved to be invasive and substantially expanded their range in southern Australia. Another group of erect algae common in shallow water is the corallines, which are calcified, and able to tolerate strong light conditions. Lastly, the foliose species of brown and red algae are considered. Red algae have a complex, usually three-phase life history, and sometimes two or three morphological forms. They become dominant on shaded surfaces and in deeper water. Their life history strategies are equally complex and, in one model applied to a deep-water red algal community, three strategies were recognised – early and late successional species, and stress-tolerant species. The succession over 2 years in this community is then described, whereby early successional species were gradually replaced by late successional dominants and stress-tolerant (low-light) species. Lastly, the effects on algae of excess nutrients and sedimentation – disturbances associated with urban environs –are described. Nutrients favour early successional algal turfs and are deleterious to canopy species. Sedimentation is also deleterious to most algae, except turfs and a few species with robust thalli. The two effects together are synergistic and induce an alternative stable state of early successional species.

INTRODUCTION

A rich flora of green, brown and red algae live under the canopy of brown algae, in canopy-gaps, on shaded vertical surfaces, and in open reef habitats to depths of ~30–70 m, depending on light availability. In this chapter, we are concerned with the three lower strata of algae: prostrate species (other than crustose corallines); low algal turfs of mainly filamentous species and small recruits of other species; and larger middle stratum species usually 5–40 cm high. Where known, the life history and ecology of one or more species in each group are described. First, we consider the shallow water assemblages, where algae often live below the canopy of larger brown algae, or in patches between canopies, or on steeper rock faces, and then we look at the poorly known deeper water flora.

SHALLOW-WATER FLORA

Non-calcified prostrate algae

Besides the crustose corallines discussed in Chapter 6, there are many species of non-calcified crusts that have a prostrate growth habit. This group includes taxa in the green algal genera *Dictyosphaeria* and *Rhipiliopsis*, the brown algal genera *Padina*, *Pseudolithoderma*, *Distromium* and *Lobophora*, and the red algal genera *Peyssonnelia* (Plate 13), *Sonderopelta* and *Hildenbrandia*. Except for the first three genera, which can tolerate high light conditions, these genera tend to prefer low light, and live in crevices, on shaded surfaces, such as on the sides of boulders, and sometimes under kelp canopies. They may extend into deep water to > 70 m depth at the lower limit of the photic zone. If exposed to high light levels, as when the kelp canopy is lost through storms, their cover declines (Kennelly 1987a, c), but within a kelp forest, abrasion by kelp has no effect (Kennelly 1989). The coverage of prostrate species is variable; on New South Wales coasts it can reach 40%, but elsewhere it is generally low (< 5%), although reaching 20% or more on boulders (Shepherd and Womersley 1981).

Little else is known about the ecology of non-calcified crusts. *Peyssonnelia novaehollandiae* and *Sonderopelta coriacea* were monitored for several years at Cape Northumberland (Shepherd 1979). The cover of *Peyssonnelia* increased rapidly in spring and autumn, with cover increase rates of 23–115 cm^2 $month^{-1}$, and declined after 3–6 months. Plants became fertile in winter and spring. Some plants disappeared after a single growth phase, but others persisted for up to 3 years, attaining a maximum cover of ~400 cm^2, whereupon the thallus divided at the margins and finally sloughed off the rock. The few plants of *Sonderopelta* had cover increase rates of ~29 cm^2 $month^{-1}$, and a maximum longevity of 5 months.

The brown alga *Pseudolithoderma* (*sub nom. Petroderma*) is grazed extensively by some small gastropods in the genera *Astralium*, *Phasianella* and *Euchelus*, while the red algae *Peyssonnelia* and *Sonderopelta* are grazed by the chiton *Callochiton crocinus* (Clarkson and Shepherd 1985; Chapter 8). However, the grazing by these molluscs may sometimes have positive effects on prostrate species by removing filamentous algae fouling their surfaces.

Algal turfs

Turf-forming algae comprise a complex and compact assemblage of small filamentous algae and short, early-stage foliose algae to ~1–2 cm high. The terms 'turf' and 'turfing species' have sometimes been used in Australian studies to refer to erect understorey species (not only filamentous species) to a few centimetres high (e.g. Underwood and Kennelly 1990), but are here restricted to the filamentous group up to ~1–2 cm high for consistency with more recent studies. Algal turfs form an important, though inconspicuous, part of the benthic algae in temperate Australian waters, and their ecology and productivity are known largely from the studies of Irving and Connell (2002, 2006a, b) and Copertino *et al.* (2005, 2006, 2009). These and other studies have examined their interactions with canopy species, and the effects of nutrients and sediment.

Turf assemblages are ephemeral and occur at the earliest successional stage, colonising newly cleared substrate or boulders rolled over by a storm. Hence, they are found in areas impacted by disturbances of various kinds. Eventually, they are replaced by larger macroalgal species, such as kelps or fucoids. An exception to this was in New South Wales in Kennelly's (1987a) study, where 'turfing' algae (in which he included some foliose brown algae such as *Zonaria*, *Lobophora* and *Padina*) inhibited the recruitment of *Ecklonia* for several years and created what he called 'an alternate stable state'.

On open coasts, the cover of turf algae is low, usually < 10%, especially under kelp or fucoid canopies, but where the canopy is absent turfs can become the main occupiers of space, with ~70% cover. In systems impacted by nutrients and sediments, coverage may increase to even higher levels. Hence, turfs form a significant component of reef systems. Furthermore, they are rich in number of species, and Copertino *et al.* (2005) recorded a total of 37 algal genera on settlement plates. Of these, the commonest turf species were in the genera *Ectocarpus, Feldmannia, Ceramium* and *Gelidium*. The brown filamentous *Hincksia* (formerly called *Giffordia*) was uncommon in Margareth Copertino's study, but is a dominant species in eastern Australia, where it has spring and autumn blooms (Kennelly and Larkum 1983; Kennelly 1987b). In addition to these species, a group of green filamentous species is seasonally common in shallow sheltered waters, where they are typically associated with high concentrations of nutrients that promote their rapid growth. This group includes species of the genera *Cladophora, Chaetomorpha, Enteromorpha* and *Cladophoropsis*.

Life history and productivity

Asexual reproduction is very common in the brown Ectocarpales, which include *Ectocarpus* and *Feldmannia*. Vegetative fragments form spontaneously, and these are viable and regenerate. Species may spread by means of stolon-like filaments emerging from a plant, and by spores, released from adult plants, which often settle and grow to form adult clones (Clayton 1974). Sexual reproduction, although it can occur year round, seems of minor importance, but when it does happen, female gametes secrete pheromones that attract swimming male gametes in the alternate gametophyte phase of the life history (Phillips 2007). The capacity for rapid asexual and sexual reproduction enables turfs to quickly colonise and monopolise bare space, and contributes to their 'early successional' character.

The productivity of turfs, growing on settlement plates and then placed in incubation chambers has been measured throughout the year (Copertino *et al.* 2005). Their net productivity was 1.3–2.9 g C m^{-2} d^{-1} (= 0.05–3.2 kg dry weight m^{-2} d^{-1}, assuming an ash content of 20% and that 1 g organic matter = 0.47 g C (see Larkum 1981). The authors then compared these values with those of kelp and fucoid canopy-forming assemblages, and found these differences:

- Although turf-forming algae are much more productive than canopy-forming species on a biomass basis, their annual net production per unit area is a fraction of that for canopy-forming species.
- Whereas the large macroalgae lose up to 40% of production from their high exudation rates, turfs lose only ~1% as exudate.
- Annual macroalgal turnover of biomass is on average 2.6 times, but in turfs it is ~29 times at 4 m depth, and ~79 times at 10 m depth.

The authors concluded that the contribution of turfs to the total carbon produced in a temperate reef system was extremely high – comparable, for example, to up to 71% of the carbon incorporated into the biomass of kelps. Copertino *et al.* (2006) then went on to show that in deeper water (10 m) turfs acclimated to low light by decreasing their photosynthetic capacity, light saturation and compensation point, and by accompanying increases in absorption of light by light-harvesting pigments. Hence, turfs maximise their photosynthetic efficiency throughout their depth range. Moreover, with almost no structural tissue for carbon storage, virtually all photosynthates are diverted to growth, so ensuring rapid growth rates.

Effects of canopy algae, sediments and micro-topography

Several studies by Andrew Irving, Thomas Wernberg and others have examined the relation between canopy species (kelps and fucoids) and the abundance of algal turfs. Generally, turf cover is low (< 5%) or zero under canopies, whether the canopy species are *Ecklonia*, fucoids or mixtures of them, but high in the gaps between them. In the latter habitats, turf cover has been found to be surprisingly varied according to site, ranging from ~18 to 40% in Western and South Australia (to > 50% on offshore islands in the eastern Great Australian Bight (GAB) – see Irving *et al.* 2003), but only ~5% in New South Wales due to intense grazing pressure (Irving *et al.* 2004; Irving and Connell 2006a). However, Fowler-Walker and Connell (2002) recorded double the above values on mainland sites across southern Australia (and see Fowler-Walker et al. 2005). Some causes of the absence of turfs under canopy algae were examined by Irving and Connell (2006a, b).

Firstly, light reduction under an *Ecklonia* canopy can be 95–99%, depending on the density of the kelp, and this is sufficient to inhibit the growth of turfs. Secondly, the sweeping of kelp fronds on the substrate (termed 'the abrasion effect') abrades the surface, and removes turf and other understorey species. This latter effect has been measured experimentally, and found to remove 60% of the turfs, with light reduction accounting for the remaining 40%.

Another effect relates to the process of sedimentation under canopies, and in gaps between them. Sediment levels are greatly reduced under canopies due to the abrasion effect of the sweeping blades, and the low abundance of understorey species that trap sediment (Kennelly 1989; Wernberg *et al.* 2005). Sediments tend to accumulate in gaps, especially among the compact filaments of turf algae, but their effect on algal turfs is unclear. At one urban site where natural sedimentation and nutrient levels were high, the experimental reduction of sediment led to twice the cover of algal turfs on settlement plates after 50 days (Irving and Connell 2002). In a later study over a similar time period, but at an oceanic site where natural sedimentation and nutrient levels were low, the experimental addition of sediments to similar levels as the previous study, had positive effects and increased the cover of turfs by 12% (Gorgula and Connell 2004). The reasons for the differences may have been due, in the case of the urban site, to a negative effect of high sedimentation in a habitat already saturated with high nutrient levels.

Irving and Connell's (2002) recruitment study also examined the effect of increased topographic complexity, by comparing smooth settlement plates with ones roughened with grooves ~1 mm deep). The latter increased the cover of algal turfs by ~8%, but only when sediment levels had been reduced. When sediment levels remained high, the sediments obliterated any differences in micro-topography, leading to less turf cover. The authors considered that a complex micro-topographic substratum increased turbulence at a micro-scale, and so enhanced spore settlement.

The middle stratum

We now consider the erect algae to ~30 (rarely 40) cm high, excluding the large canopy-forming brown algae, and use the term 'understorey' to describe this group of species growing under canopies or in cleared patches between canopies. First, we consider the ecologically distinctive green algal genus *Caulerpa*, then the equally distinctive, jointed calcareous red algae, and lastly other middle stratum algal assemblages that dominate shaded surfaces and deeper water beyond the depth limit of canopy species.

The genus *Caulerpa*

Caulerpa is mainly a tropical genus, which has diversified greatly in the southern half of Australia, where it arguably originated (Calvert *et al.* 1976). Here, some 27 species occur (Sanderson 1997), representing 32% of the world's total. The genus has two unique features:

- Its species are siphonous – that is, each plant is essentially a single cell that develops into an elaborate system of branching siphons
- The fronds, which are branched in various ways with ramuli, arise from a mat of creeping stolons, anchored to rock or sand by rhizomes.

The mats of stolons can be quite dense, and accumulate sediment between them, excluding all other algae. The mattes often form patches of up to several square metres, and have been observed to persist in the same locations for decades (unpublished observations). In some species, such as *C. brownii* and *C. flexilis*, the fronds die off annually in mid-summer leaving only stolons on the rock face and sprout from the stolons in autumn–winter (Shepherd and Womersley 1970). However, in southern Tasmania, Sanderson (1992) noted a quite different seasonal pattern of die-back; there, the fronds of *C. flexilis* disappeared in autumn–winter and recovered spectacularly in early spring.

Distribution in relation to depth and water movement

Each species of *Caulerpa* has a distinctive morphology and growth form, and some species vary in morphology according to ecological conditions. Studies elsewhere have shown that, as the morphology of the plant becomes more complex, photosynthetic rates decline (e.g. Gacia *et al.* 1996). Besides, more complex shapes are less resilient in strong water movements. Hence, morphology is expected to play a part in the distribution of species in relation to depth and water movement.

At two well studied sites, Waterloo Bay (Shepherd and Womersley 1981) and Crawfish Rock, in upper Western Port (Shepherd *et al.* 2009), numerous species of *Caulerpa* co-existed with broadly overlapping ranges. However, the species showed peak abundances in different parts of the depth–water movement mosaic, with several species showing multiple peaks (Fig. 7.1). Some correlations are apparent between the morphology of the species and their location on the mosaic. Small, compact species, with short densely clumped ramuli, such as *C. brownii*, *C. papillosa* and *C simpliciuscula*, and species with simple, flattened fronds, such as *C. scalpelliformis*, were most abundant in areas with strong water movement. On the other hand, species, whose fronds are armed with loose arrays of ramuli (*C. trifaria*), or with bulbous ramuli (*C. cactoides* and *C. vesiculifera*) occurred in areas with reduced water movement. Where species have two different growth forms (*C. flexilis* and *C. simpliciuscula*), the variety with looser ramuli was in deeper water with low light at their lower depth limit; here, self-shading of the ramuli is less and photosynthetic efficiency likely higher (cf. Gacia *et al.* 1996). In places with strong currents, *Caulerpa* fronds may be very long – up to 1 m. From the above observations, two questions – the relations between functional morphology and water movement on the one hand, and photosynthesis in the genus on the other – would well repay detailed investigation.

Invasive species

Some species of *Caulerpa* have proved to be highly invasive, and several have either become established in Australia, or have so extended their Australian range that they are considered pest species (Davis *et al.* 1997). In these cases, the aquarium trade or

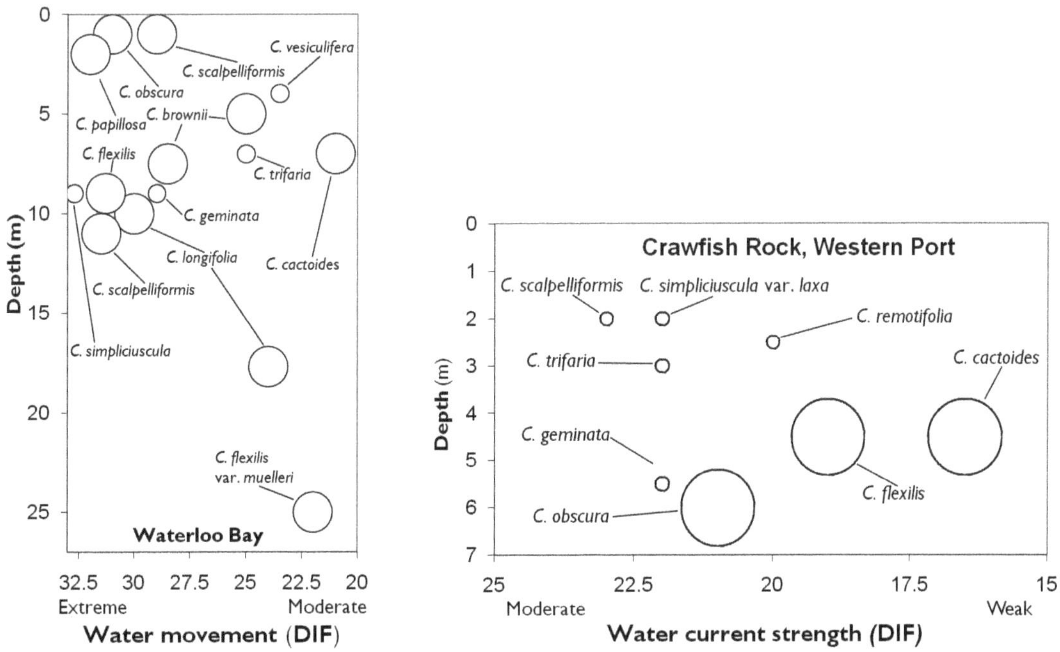

Figure 7.1: Locations of peak abundances in fresh weight biomass of species of *Caulerpa* on depth–water movement mosaic at (a) Waterloo Bay, South Australia, and (b) Crawfish Rock, Victoria. Small circles indicate < 100 g m^{-2} biomass in each figure, and large circles indicate 0.1–1 kg m^{-2} at Waterloo Bay and 0.1–0.3 kg m^{-2} at Crawfish Rock (data from Shepherd and Womersley 1981, and from Shepherd *et al.* 2009).

shipping has been the most likely vector. The species are described below.

- *Caulerpa taxifolia* is a notorious, recent introduction from Queensland into southern waters, but it has not spread from bays or ports in South Australia or New South Wales onto open coasts. Being strictly a species of sandy bottoms, it is not discussed further here;
- *C. scalpelliformis*, a southern species, suddenly appeared in Botany Bay, a leap of 150 km beyond its northern limit in southern New South Wales, in ~1994, and soon achieved a cover of 50–90% on deep reef habitats at 20 m depth (Davis *et al.* 1997). As the plant spread with its stolonic mat, the dense fronds above accumulated sediment, and led over time to 'a shift to a less diverse, more silt-tolerant sponge assemblage' (Davis *et al.* 1997);
- *C. filiformis* was first seen in Botany Bay in the 1920s, and now extends for > 200 km on the New South Wales coast. Normally a species of the shallow subtidal zone, it has extended its depth range to 25 m. The species also occurs in South Africa, but whether it originated there or in eastern Australia is still in doubt (Pillmann *et al.* 1997);
- *C. racemosa* f. *cylindracea* is a tropical Indian Ocean species extending into SW Australia and east to the Recherche Archipelago (Goldberg and Kendrick 2005). During the late 1990s, it jumped to Port Adelaide, South Australia, and then to nearby Gulf waters, where patches of it now occur, but are quite rare (Westphalen 2008). It also arrived in the 1990s in the Mediterranean, where it proved highly invasive (reviewed by Klein and Verlaque 2008).

Life history

Not surprisingly, more is known about invasive species of *Caulerpa* than any others, and here we describe the life history of the native Australian 'invader', *C. racemosa*, which is probably typical of many other species in the genus. *C. racemosa* reproduces sexually and vegetatively. Plants are both male and female. Sexual reproduction involves the movement of cytoplasm into specialised structures (the gametangia), where it is transformed into male and female gametes. Gametes are released into the water synchronously, often in a mass spawning event. For example, in one study, *C. racemosa* spawned on only one summer day a year, 14 min before dawn, producing a green cloud of gametes in the water. Plants released their entire cell contents in the single spawning event, leaving only 'white, ghost fronds' (Panayotidis and Zuljević 2001). Other species have similar epidemic spawning patterns (Clifton 1997). In vegetative reproduction, stolons grow at their apices and spread over the bottom. Fronds that are fragmented during a storm event can survive for several days in the drift, and then re-establish themselves elsewhere on suitable substrate (Ceccherelli and Piazzi 2001).

The fronds of *C. racemosa* grow and die within a year, but stolons are perennial and form mats that persist for years; the growth rate of stolons has been measured at rates of 0.8–1.3 cm day^{-1}. The maximum growth of fronds and stolons occurs in summer months (Klein and Verlaque 2008).

Chemical defences of Caulerpa

Caulerpa species contain a suite of biologically active secondary metabolites, notably sesquiterpenes, that make them unpalatable to herbivores and hence provide them with a defence against most large grazers. But is this chemical defence a good deterrent against common temperate Australian herbivores? Davis *et al.* (2005) examined this question by testing extracts from *C. scalpelliformis, C. filiformis* and *C. taxifolia* on local turbinid snails, a sea hare and sea urchins. Overall, temperate grazers were relatively undeterred from feeding on two species of *Caulerpa*, but extracts from *C. filiformis* did deter the turbinid snail *Turbo torquatus*, the sea urchin *Heliocidaris tuberculata*, and the sea hare *Aplysia sydneyensis.* When offered a choice of various other algal species, *Turbo undulatus* showed the lowest preference for *C. filiformis.* So the study gave limited support for the view that these invasive species were under lower consumer pressure in the invaded habitat (the enemy-release hypothesis).

At least six species of sacoglossan nudibranchs on southern coasts are specialist herbivores, feeding on species of *Caulerpa* (Williams and Walker 1999). Two examples are *Edenttelina typica*, an obligate feeder on *C. brownii*, and *Midorigai australis*, an obligate feeder on *C. simpliciuscula* (Burn 1989). See Table 9.2 for other species.

Middle stratum diversity

The diversity of middle stratum/understorey macroalgae, excluding canopy species and crustose or prostrate species, is high throughout southern Australia, but declines up the west and east coasts. As examples of this richness in kelp or fucoid habitats at a regional scale of a few square kilometres, Phillips *et al.* (1997) recorded 78 species belonging to this group in Marmion Lagoon, with similar numbers in Hamelin Bay and Fitzgerald River, Western Australia (Wernberg *et al.* 2003). In Recherche Archipelago in the western GAB, Goldberg and Kendrick (2005) recorded 203 species in the group, while in Waterloo Bay in the eastern Bight Shepherd and Womersley (1981) recorded 324 species. Further east, 119 species were recorded at West I. (Shepherd and Womersley 1970), and 188 species at Cape Northumberland, SE South Australia (Shepherd 1979). In the Bunurong Marine Park, Victoria 161 species were recorded (Ferns and Hough 2002). Diversity appears to decline markedly in New South Wales with decreasing latitude and as grazing intensity increases. May and Larkum (1981) recorded 50 species in the group at a site in Jervis Bay, SE New South Wales, Van der Velde and King (1984) some 65 species at Bare I., Botany Bay, but Kennelly (1987c) only found 27 species at a site in Sydney Harbour.

Erect coralline algae

This group of algae is distinguished from the crustose forms by their multi-axial, jointed construction – hence the terms 'articulated' or 'geniculate' commonly applied to them (except *Metamastophora*). The joints or genicula are uncalcified areas that give flexibility, and the heavy calcification of the rest of the plant provides rigidity and the ability to tolerate strong light. Together, these characteristics enable the plant to inhabit shallow habitats with moderate to slight surge (Johansen 1974). The erect forms are closely related to the crustose forms (Chapter 6) from which they have evolved, and common southern genera are *Amphiroa*, *Corallina*, *Cheilosporum* (Plate 14), *Haliptilon*, *Jania*, *Metagoniolithon* and *Metamastophora*. Little is known about the life history of these genera, except for three species of *Corallina*, *Metagoniolithon* and *Metamastophora* studied at 15 m depth off Cape Northumberland (see Table 7.2). In mid-summer, *Corallina* and other corallines lose their fronds and branchlets, leaving only bare primary and secondary axes (Shepherd and Womersley 1970).

Coralline communities of single or mixed species often dominate habitats exposed to slight to moderate swell, but their coverage varies according to habitat. Early studies (e.g. Shepherd and Womersley 1970) showed that species may have differing preferences for depth and wave energy. For example, on a wave energy gradient the upper depth limit of two species, *Metamastophora flabellata* and *Cheilosporum elegans*, is elevated with declining exposure to swell. Later studies from western to eastern Australia have explored numerous other factors that influence the abundance of corallines, as discussed below.

In Western Australia *Corallina*, *Amphiroa*, *Metagoniolithon*, *Metamastophora* and *Jania* are common corallines, and their cumulative cover ranges from a few per cent in kelp forests, up to 20% in mixed fucoid forests, and up to 40% in gaps or open habitats (e.g. Fowler-Walker and Connell 2002). Canopy abrasion has little effect on the understorey, but light reduction has a major effect, substantially reducing both cover and diversity of species. Hence, species such as *Amphiroa anceps*, which prefer strong water movement, are found mainly in canopy gaps created by storm events (Kendrick *et al.* 1999, 2004; Wernberg *et al.* 2003; Toohey *et al.* 2004; Wernberg and Goldberg 2008).

In South Australia, Sean Connell (2005) tested the effects of the physical factors, light and sediment, on erect corallines. He found that their growth and survival depended on an absence of shade (i.e. they needed strong light), and were favoured by an accumulation of fine sediment. These experiments were done in cleared plots, where canopy abrasion of the understorey was absent. To test this latter effect, Irving and Connell (2006b) manipulated kelp/fucoid canopies, and found that under *Ecklonia* the additive effect of shade and abrasion removed 59% of the coralline, with two-thirds of it due to abrasion; however, under a mixed kelp/fucoid canopy only ~33% of the coralline was removed, partly because fucoids have less abrasive effect. Where kelp densities were $< 7\ m^{-2}$, the reduction of light was inconsequential (Melville and Connell 2001). Here, it was also observed that kelp abrasion removed the lateral branchlets of *Haliptilon*, so causing its decline.

On the SE coast of New South Wales, major coralline species are *Amphiroa*, *Haliptilon* and *Corallina*. Here, coralline cover under kelp is much higher (up to 40% under kelp canopies and up to 90% in open habitats) compared with southern and western coasts. The reasons for these differences on the east coast are the high grazing pressure by sea urchins and the taller stipes on *Ecklonia* that minimise its abrasion there (see Chapters 4, 12; Kennelly and Underwood 1992; Fowler-Walker and Connell 2002; Irving *et al.* 2004).

From all these studies, we conclude that corallines are mainly high-light species, which decline under kelp and fucoid canopies due to their shading effect. Abrasion by fronds of canopy species also reduces coralline cover, although this effect is slight in New South Wales, where *Ecklonia* stipes are longer, and abrasion less. For these reasons, erect corallines reach their highest cover in open habitats or in gaps in algal forests throughout southern Australia.

Foliose algae

While green algae and corallines dominate shallow water, red algae increase in abundance and diversity with depth, so that at depths of 20 m or more, as canopy species disappear, red algal assemblages become dominant. Here we summarise the studies in different States of Australia, which have considered particular questions of relevance to their environments.

Western Australia

Some studies of the under-canopy assemblages have experimentally manipulated kelp cover or amount of sediment, while others have compared abundances in different habitats on depth and water movement gradients. A range of factors influencing the distribution or abundance of particular species have been noted (Table 7.1). These relate to kelp density and its effects on available light, sediment and water movement.

Table 7.1: Understorey algae and the environmental factors that increase or decrease their abundance, indicating the specific factor and its effect.

Full species names are given in the references at foot of table. +ve = positive effect; –ve = negative effect; water movement refers to exposure to swell.

Region	Genera/Species	Factor (reference)
WA – Marmion Lagoon	*Heterodoxia, Dictyomenia, Lobophora, Jeannerettia,Pterocladia, Rhodymenia*	+ve: mod-low kelp densities[1]
	Pterocladia, Dictyomenia	–ve: increased water movement[1]
	Heterodoxia, Ulva, Chauviniella,	–ve: more sediment[3]
	Rhodymenia, Dictyomenia	+ve: more sediment[3]
	Pterocladia, Rhodymenia, Heterodoxia, Jeannerettia,Hypnea, Ulva, Chauviniella, Dictyomenia	+ve: more light (i.e. –ve effect of canopy)[3]
	Pterocladia, Rhodymenia	+ve: reduced water movement[3]
	Jeannerettia	–ve: reduced water movement[3]
	Pterocladia, Rhodymenia	Tolerant to presence/absence kelp cover[4]
	Pterocladia, Rhodymenia, Chauviniella	+ve: adaptation to low light[6]
Hamelin Bay	*Lobospira, Glossophora, Pterocladia, Codium duthiae*	+ve: clearings in *Ecklonia*[2]
Recherche Archipelago	*Botryocladia, Osmundaria.Codium* spp., *Rhodymenia* spp., *Thamnoclonium*	+ve: sheltered sites, deep exposed sites[5]+ve: deeper water (10–20 m)[5]

[1] Kendrick *et al.* (1999); [2] Kendrick *et al.* (2004); [3] Wernberg *et al.* (2005); [4] Wernberg and Goldberg (2008); [5] Goldberg and Kendrick (2004); [6] Toohey *et al.* (2004)

Concerning the first of these, Wernberg and Goldberg (2008) found that the canopy cover of *Ecklonia* had a strong influence on the understorey. During winter–spring, kelp biomass declined by 80% compared with late summer, due to high winter swells, and the number of understorey species increased on average from 10 to 25 per 0.25 m^2 quadrat. The turnover of species between sampling periods was also high, ranging from 45 to 103%. The authors concluded that the kelp *Ecklonia* was a powerful habitat-former (an ecosystem engineer) and had a strong seasonal influence on the understorey, by its control of available light reaching the habitat.

The authors categorised the understorey algae according to their light requirements– that is, obligate, low-light species that require shade; obligate 'gap' species that do best outside a canopy; and tolerant species that cope with variable light conditions. This last group, the light-tolerant species, was also examined by Toohey *et al.* (2004). They found that three species of red algae, *Pterocladia lucida, Rhodymenia sonderi* and *Chauviniella coriifolia*, were able to photo-acclimate to low light by up to 11% (i.e. they were able to increase their levels of photosynthetic pigments and so absorb a greater amount of ambient light). This gave them a competitive advantage in low-light conditions and explained their increased abundance in such conditions, while species with little ability to photo-acclimate declined or were absent.

Another recurrent theme in many studies has been the emphasis on the patchy occurrence of understorey species, due to the combination of the many factors that influence their distribution. Localised dispersal of propagules is one cause of clumped distributions, and was emphasised by Kendrick *et al.* (1999) as an effective competitive strategy for pre-empting space via local persistence and spreading. At a larger scale, Wernberg *et al.* (2003) compared the understorey algae at sites spread over 1000 km of coast and found a gradual turnover of species at a regional scale, with only 11% of the 152 species common to the whole region. There were also large differences at a local scale, as noted above. Similarly, in their study of islands in the Recherche Archipelago, Goldberg and Kendrick (2004) found high diversity in the understorey, with a high turnover of species between replicate quadrats. Many of the species were uncommon, with ~60% of them having a low biomass of < 5 g 0.25 m^{-2}.

South Australia and Tasmania

The two detailed studies of the algal vegetation on central South Australian coasts (Shepherd and Womersley 1970, 1981) showed an extraordinarily rich flora of middle stratum species of brown and red algae. The studies plotted distributions of biomass on depth and water movement gradients, and showed that many species are widely dispersed on these gradients. Many of the species also occur on vertical faces or on other semi-cryptic surfaces. In the second of these studies, water movement was measured by the rate of dissolution of Doty clod cards (Doty 1971), which are made of calcium sulphate blocks that slowly dissolve according to water movement intensity. Hence, water movement could be calibrated in terms of a diffusion increase factor (DIF), as shown in Figs 7.1 and 7.2.

Although it is convenient to plot distributions on a depth–water movement mosaic, it is likely, as noted by the authors, that factors other than light and water movement also help determine distributions. These include sediment in calmer conditions, competition for space, disturbance, aspect and differences in the substratum – factors that are still largely unexplored. Depth–water movement mosaics, showing the locations of peak abundances of the most common understorey species for West Island and Waterloo Bay, are shown in Fig. 7.2.

In Tasmania, surveys have also been carried out by Edgar (1984), Sanderson and Thomas (1987) and Sanderson (1992). Here, the flora has cold temperate elements and is as rich as that in South Australia. The most common genera of middle stratum brown and red algae, as recorded in those papers, are shown in Fig. 7.3.

New South Wales

Two systematic algal surveys have been done in New South Wales (see review of Kennelly 1995). May and Larkum (1981) in a study of a proposed sewage outfall site at Jervis Bay, SE New South Wales, recorded ~50 erect understorey species, of which a surprisingly high proportion (48%) were seasonal, with most of the commoner species found only from winter to

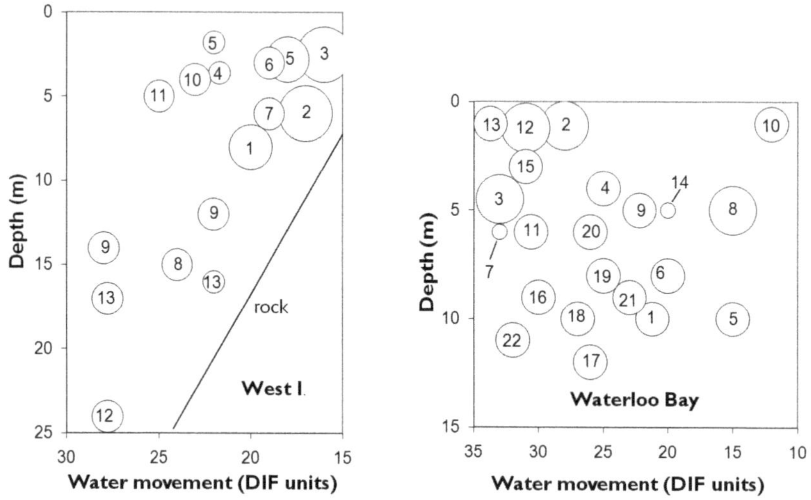

Figure 7.2: Locations of approximate peak abundances of common brown and red understorey algae on a depth–water movement mosaic at (a) West I. and (b) Waterloo Bay, South Australia.
At West I., the species/species groups are: (1) *Halopteris* spp.; (2) Zonaria spp.; (3) *Amphiroa anceps*; (4) *Asparagopsis armata*; (5) *Corallina cuvieri*; (6) *Cheilosporum elegans*; (7) *Metagoniolithon radiatum*; (8) *Osmundaria prolifera*; (9) *Phacelocarpus labillardieri*; (10) *Phacelocarpus apodus*; (11) *Pterocladia lucida*; (12) Rhodophyllis *membranacea*; and (13) *Rhodophyllis multipartita*. Four circle sizes indicate fresh weight biomass ranges as follows: smallest = < 1 kg m^{-2}; next = 1–2 kg m^{-2}; next = 2–3 kg m^{-2}; largest = > 3 kg m^{-2}. (Redrawn from Shepherd and Womersley 1970).
At Waterloo Bay the species are: (1) *Areschougia congesta*; (2) *Callophyllis lambertii*; (3) *Callophyllis rangiferinus*; (4) *Corallina cuvieri*; (5) *Gelidium australis*; (6) *Haloplegma preissi*; (7) *Hypnea episcopalis*; (8) *Osmundaria prolifera*; (9) *Metagoniolithon radiatum*; (10) *Jania fastigiata*; (11) *Laurencia elata*; (12) *Laurencia filiformis*; (13) *Laurencia shepherdi*; (14) *Lobospira bicuspidata*; (15) *Phacelocarpus labillardieri*; (16) *Phacelocarpus apodus*; (17) *Plocamium angustum*; (18) *Plocamium costatum*; (19) *Plocamium cartilagineum*; (20) *Plocamium mertensii*; (21) *Plocamium preissianum*; and (22) *Sarcomenia delesserioides*. Three circle sizes indicate fresh weight biomass ranges as follows: smallest = < 100 g m^{-2}; next = 0.1–1 kg m^{-2}; largest = > 1 kg m^{-2}. (Redrawn from Shepherd and Womersley 1981).

mid-summer. The commonest species were in the brown algal genera *Lobophora* and *Zonaria*, and the red algal genera, *Acrosorium, Fosliella, Gelidium, Heterosiphonia, Laurencia, Plocamium* and *Polysiphonia*. Subsequently, detailed studies of the dynamics of kelp forests and their relations with the understorey were done by Steve Kennelly around Sydney and further south.

A notable feature of the algal vegetation in New South Wales is its patchiness, with patches of *Ecklonia* forest interspersed with small clearings of understorey species, and with other patches of white rock grazed by the black urchin *Centrostephanus* (Chapter 12). To understand the dynamics of the clearings without urchins, Kennelly carried out experimental clearings and other experiments at 4–5 m depth in the kelp forest at various sites on the New South Wales coast. In a series of papers (Kennelly 1983, 1987a–c, 1989; Kennelly and Larkum 1983; Kennelly and Underwood 1992, 1993), Steve Kennelly found the following.

- The main underflora species in the clearings were variously species in the genera *Padina, Zonaria, Lobophora, Polysiphonia, Gelidium* and *Rhodymenia* spp., which also occurred sparsely under the kelp canopy. Some of the species were seasonal in occurrence, such as *Lobophora* (autumn–winter), *Colpomenia* (summer), *Zonaria* (autumn and spring), *Gelidium* (winter–summer) and *Polysiphonia* (spring).
- Clearings with the above underflora formed a dense and persistent assemblage that inhibited the recruitment of kelp within them.
- In fresh clearings a succession occurred. An initial, fast-growing filamentous turf appeared in the first few weeks, and was later followed by the above underflora. The order in which species appeared,

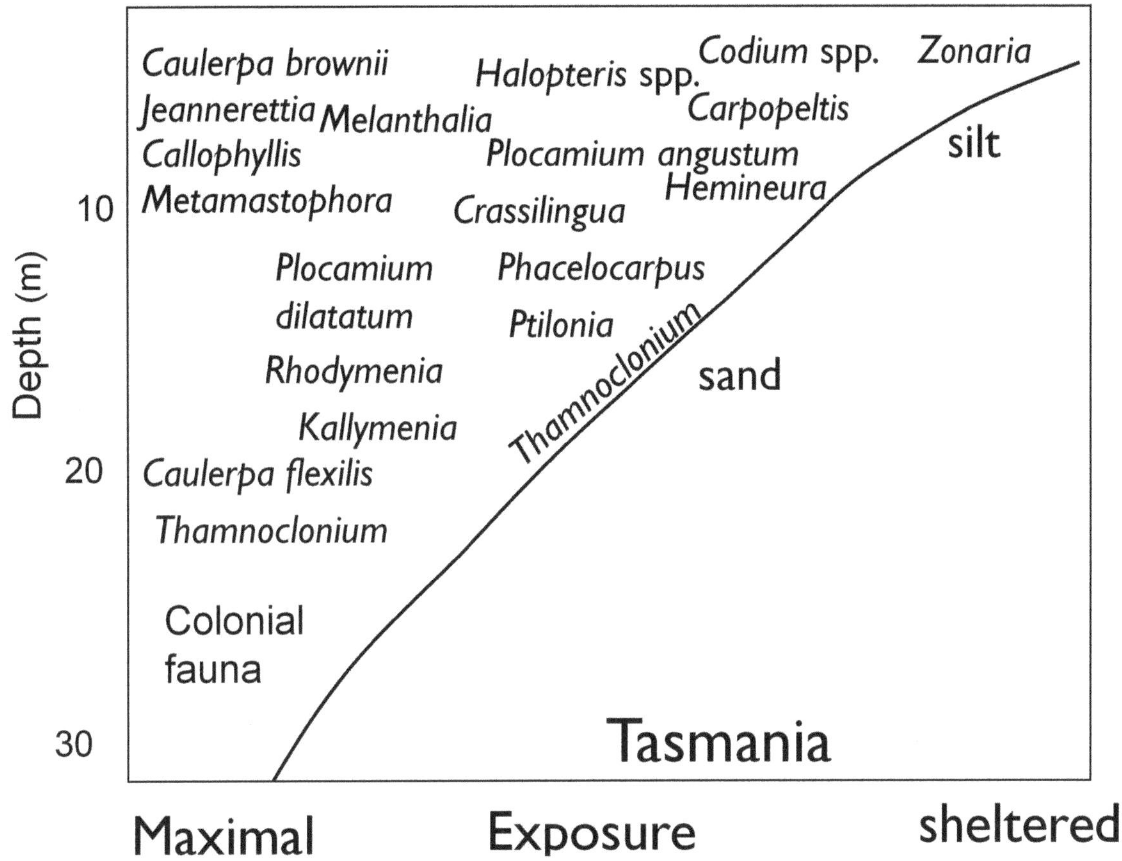

Figure 7.3: Qualitative depth–water movement mosaic for southern Tasmania, showing location of approximate peak abundances of some common understorey genera (after Edgar 1984, Sanderson 1992). The curved line indicates depth of bottom substratum according to exposure.

and their cover increase, depended on the season of clearing. For example, *Zonaria/Lobophora* reached 40% cover in 8–12 weeks in autumn–spring, but took almost twice as long to reach this level of cover in other seasons.

- In canopy-removal experiments, the cover of *Dictyota, Zonaria, Lobophora* and *Rhodymenia* increased, showing that light was the limiting factor.
- Grazing by fish and invertebrates also affected the composition of species in clearings. Some species were avoided (e.g. *Zonaria, Lobophora*), and others preferred (e.g. *Polysiphonia, Ulva*).

In a final review, Kennelly and Underwood (1993) concluded that the patch dynamics within kelp forests were largely determined by local factors. Although some generalisations emerged as above, the time scales and composition of species varied from place to place, a not surprising conclusion bearing in mind the likely variability in algal spore dispersal, grazing and local environmental factors.

Red algal life history

The genus *Delisea*, with four species, is moderately common throughout southern Australia, and the complexity and variability among life histories of red algae are well illustrated by this genus. *D. pulchra* is very common in the shallow subtidal in SE Australia and its ecology is well known. The species has three separate life-history stages with distinct forms (termed triphasic), as shown in Fig. 7.4 (Bonin and Hawkes 1988). Gametophytes produce male and female gametes that upon fertilisation form a carposporophyte stage that remains attached to the female gametophyte. The carposporophyte produces carpospores that drift off and settle on the bottom and grow into the tetrasporophyte, which is identical in appearance to the gametophyte. The tetrasporophytes then produce tetraspores that are liberated and settle on the bottom, where they germinate into a gametophyte plant, so completing the cycle. This triphasic cycle, however, differs among the species. In *D. elegans* and *D. hypneoides*, tetrasporophyte plants similar in morphology to gametophytes are unknown, but possibly occur as small crusts bearing tetraspores on filaments. Similarly, the gametophyte stage is not always present, as described below.

Studies on *D. pulchra* in Botany Bay by Wright *et al.* (2000 a, b) and Wright and Steinberg (2001) showed that the gametophyte stage was absent in > 90% of populations studied, indicating that most populations were maintained by asexual reproduction of tetrasporophytes, in which tetraspores settled on the bottom without meiosis (i.e. nuclear division in which chromosome number is reduced by one half), and grew as tetrasporophytes. Tetraspores lack flagella – fine threads that beat and give motility – and sink when released from the plant, so that dispersal should be quite localised. In genetic studies, Wright *et al.* (2000b) found that at a local scale there was some genetic divergence, indicating that local water currents or fine-scale selection occurred, and spores were well dispersed locally at a scale of a few metres. At a larger scale of hundreds of metres, however, populations were genetically distinct, indicating a lack of dispersal over these larger distances.

Wright and Steinberg (2001) examined the recruitment and survival of *D. pulchra* at 3 and 8 m depths. Recruits appeared all year round at the shallow sites at rates of 15–50 m^{-2} a month, but in smaller numbers at the deeper sites. At the shallow sites, their survival varied with density of the purple urchin *Heliocidaris*, which moved around grazing the bottom. At urchin densities of up to ~80 m^{-2}, survival of recruits was 66–80% a month, but was much lower (to 40% a month) at higher urchin densities. At the deeper site, survival was ~5% after 2 months, and plants lived for

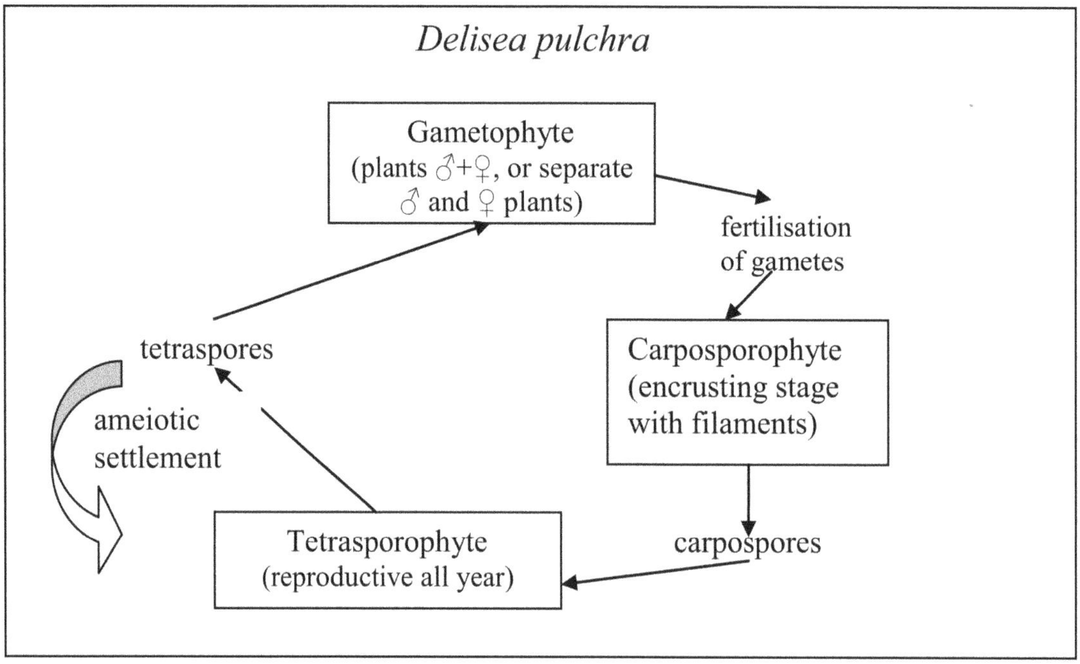

Figure 7.4: Life history of *Delisea pulchra* showing sexual gametophyte and non-reproductive tetrasporophyte stages, and also the occurrence of asexual cycling of ameiotic tetraspores (i.e. without nuclear division and genetic segregation).

only ~5 months. Urchins were rare here, and the cause of the poor plant survival unknown.

Plants grew to ~15 cm high, reaching adulthood and reproductive maturity in ~6 months, but their survival was only ~35% a year. Few were still alive after 18 months. The significant findings of the study were the shifting importance of recruitment and survival to the plant's population size, depending on urchin density. At high urchin densities, recruitment was the most significant factor, but at low urchin densities survival was the most important.

Delisea pulchra produces secondary metabolites (furanones) that

- deter some herbivores
- deter epiphytes from settling on the blades,
- reduce the attachment of bacteria on the plant (de Nys *et al.* 1995).

However, the chemical deterrent effect is apparently partial, because the urchin *Holopneustes purpurascens* and the gastropod *Turbo torquatus*, which were abundant in *Delisea* habitat, had a low preference for the alga (Wright *et al.* 2000a). Amphipods were also abundant, and avoided *Delisea*. Curiously, the sea-hare *Aplysia* did consume the chemically rich tips of *Delisea* (Chapter 9), which it incorporated into its surface tissues to provide protection against its own predators.

DEEP-WATER ALGAE

In the deeper lower sublittoral zone, canopy species are replaced by red algal middle and lower strata, and sessile animals increase. This zone generally starts at 15–30 m depth (depending on light penetration) on upward-facing rock, and may go to 70–80 m depth in clear, oceanic water (e.g. Shepherd 1979; Smale *et al.* 2010). However, on vertical surfaces or in turbid coastal waters, the zone may be as shallow as a few metres deep (e.g. Shepherd *et al.* 2009). The numerous deep (20–70 m), offshore reef platforms off the SE coast of South Australia – remnants of ancient shorelines – all have extensive, species-rich red algal assemblages (e.g. Haig *et al.* 2006).

Red algal assemblages are often rich in species, but only two studies have so far examined them. In this section, we compare the results of the two studies – Shepherd (1979) and Sanderson (1992) – undertaken on exposed coasts off Cape Northumberland, South Australia at 15 m depth and George III Rock, southern Tasmania at 12 m depth, respectively. Both studies started in a *Macrocystis* forest, but in each case the *Macrocystis* canopy was torn out during a storm, leaving red algal assemblages. In all, 114 species were recorded at the Cape Northumberland site and 69 species at George III Reef.

Firstly, species–area relations were obtained at both sites, by replicate sampling of the habitat at one time, but at Cape Northumberland sampling continued over 3 years to give a species–area–time curve. The results (Fig. 7.5) show almost identical species–area relations for both sites at the initial sample size of 4 m^2; however, at Cape Northumberland with repeated later sampling at 3-monthly intervals, rare species kept accumulating in an apparently linear manner over time, suggesting that the total richness of species was still far from being realised even after 3 years of sampling.

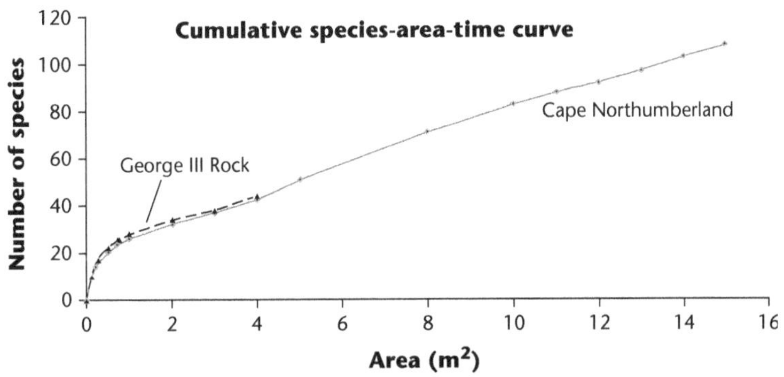

Figure 7.5: Species–area curves (up to 4 m^2 area) for George III Rock, Tasmania (redrawn from Sanderson 1992), and species–area curve, extended into a species–area–time curve, for Cape Northumberland (redrawn from Shepherd 1979). Vertical bars are standard deviations for the Cape Northumberland species–area curve.

Both studies examined the algal biomass–environment relations over time at fixed sites. At Cape Northumberland, algal biomass (*B*) (in g fresh weight 2400 cm^{-2}) was linearly related to a storm index (*SI*) and algal production, *P*, (g 2400 cm^{-2}) for the preceding 3-month quarter, as shown by the equation:

$$B = 648.7 - 5.7\ SI + 0.7\ P\ (r^2 = 0.86;\ P < 0.001)$$

With an average quarterly SI value of 14 (range 3–49), the average quarterly biomass loss over 3 years was ~275 g 2400 cm^{-2} (i.e. a quarterly loss of 36% of the average algal standing crop). This amount of loss was clear evidence of the major role of storms and swell in the community dynamics. At the less exposed George III Rock, algal biomass (*B*) in a 2400 cm^2 quadrat was related only to the number of sunshine hours (*SH*) over the preceding 2 months, as shown by the equation:

$$B = 393.7 + 363.9\ SH\ (r^2 = 0.79;\ P = 0.003)$$

However, 'the action of swell can strip the rocks over the winter-spring period ...' (Sanderson 1992), indicating that rare storm events could still reduce algal biomass.

Life history and productivity

Growth, survival, recruitment and longevity

The growth, survival, recruitment and maximum age of 32 red algal species were measured in permanent quadrats at Cape Northumberland, and the results for the 17 most common species are given in Table 7.2. Growth rates were usually 2–6 cm $month^{-1}$, and

Table 7.2: Growth (elongation) rates, half-life,[1] longevity and recruitment rates of 14 common species of red algae in permanent quadrats at 15 m depth off Cape Northumberland (data from Shepherd 1979).

Note: for Plocamium spp., growth rates are for fronds, and half-life and maximum age data are for colonies.

Species	Growth rate (cm $month^{-1}$)	Half-life (months)	Max. age (months)	R (Nos m^{-2} $year^{-1}$)
Ballia callitricha	5.8	3–7	13	3.1
Callophyllis lambertii	2.6	4–5	34	1.5
Callophyllis rangiferina	5.8	2–3	9	2.6
Carpopeltis phyllophora	2.1	3–4	12	2.6
Champia viridis	3.7	4–14	14	0.5
Corallina officinalis	2.8	4–5	11	1.5
Hymenena affinis	3.2	6–8	33	15.3
Melanthalia concinna	2.7	9–11	31	2.1
Metagoniolithon radiatum	1.6	3–5	9	2.1
Metamastophora flabellata	2.1	4–7	20	2.5
Myriogramme gunniana	4.6	~2	6	0.5
Nitospinosa pristoidea	3.9	1–3	10	4.1
Nizymenia australis	2.2	12–34	34	1.5
Phacelocarpus peperocarpos	2.1	8–10	33	4.1
Plocamium angustum	4.1	5–10	34	2.0
Plocamium dilatatum	3.9	7–9	36	6.6
Rhodophyllis multipartita	5.7	2–3	12	11.7
Rhodymenia australis	1.6	~2	3	0.5
Thuretia quercifolia	2.7	3–6	18	2.6

[1]The instantaneous mortality coefficient, Z, and survival, S, can be calculated from the half-life, with the relations: Z = half-life/ln2 and $S = e^{-z}$.

maximum ages recorded (the ecological longevity) ranged from a few months to several years. The range of half-lives (i.e. the time for a population to be reduced by one half), recorded in the three permanent quadrats over 3 years, is also given.

The species of *Plocamium* are of especial interest because they have two types of growth. The species form discrete colonies of individual fronds arising from a mat of basal stolons attached to the rock. First, colonies can expand laterally as the basal stolons spread over the rocks and, second, fronds elongate by apical growth. Colony expansion takes place mainly in spring–early summer, and in autumn, and decline of the colony occurs some 4–7 months later, as fronds, and sometimes the stolonic matte, slough off the rock. The fronds also pass through a period of growth and decline. After reaching 10–20 cm high, the fronds become reproductively mature before yellowing and sloughing off. Maximum recorded growth rates of stolonic expansion were 65 cm^2 $month^{-1}$, and the maximum size of colonies were ~200 cm^2 for *Plocamium angustum*, and 500 cm^2 for *Plocamium dilatatum*.

Productivity

From information on recruitment, growth and survival of plants in permanent quadrats, the net annual production of a red algal assemblage can be estimated. At the Cape Northumberland site, Shepherd (1979) estimated this to average 5.0 kg fresh weight m^{-2} (range 4.9–5.4), with epiphyte production of 80 g m^{-2}, giving an average dry weight annual production of ~1.0 kg m^{-2}. At George III Reef, Sanderson (1992) found an average annual production of 3.0 kg fresh weight m^{-2}. Some of Craig Sanderson's quadrats were under a *Macrocystis* canopy and some outside. Those under the canopy averaged 2.5 kg m^{-2}, and those outside averaged 4.4 kg m^{-2}. The *Macrocystis* canopy reduced the light reaching the bottom by 25%, and when part of the forest disappeared, some species (e.g. *Phacelocarpus* spp. and *Delisea*) decreased, while others (e.g. *Crassilingua and Thamnoclonium*) increased in abundance. Hence, in this assemblage at 12 m depth, changes in available light affected both the overall productivity and the assemblage composition.

Reproduction

Regular sampling at Cape Northumberland and George III Reef enabled seasonal reproductivity to be determined. The results for the most common species (Table 7.3) show that most species were reproductive for more than half the year, and almost always during winter. Tetrasporophytes were generally more common than male/female plants and, in three species, *B. callitricha*, *Inkyuleea mariana* (formerly *Ballia mariana*) and *C. rangiferina*, no sexual plants were ever recorded. This suggested that either their life history was asexual, as occurs among some algae, or that gametophytes differed morphologically from tetrasporophytes. In the case of *B. callitricha*, vegetative reproduction sometimes occurred, as algal fragments, anchored by discs of the epiphytic coralline alga *Synarthrophyton patena*, which grew only on that species, could become established on the reef. Four species were present at both sites, enabling a comparison of their reproductivity. Although some similarity exists, it was noticeable that the species in Tasmania had longer reproductive seasons and higher reproductivity than those at Cape Northumberland. Winter sea temperatures are marginally cooler in southern Tasmania (Table 4.1), and possibly contributed to these differences.

Succession

Succession is the process of development of a community, and has been much studied in theory and practice. Three principles governing succession are given below (Odum 1969).

- It is an orderly, directional and predictable process.
- The developing community modifies the environment, so that to some extent the succession is community-controlled.
- The end point, often termed the climax, is a stabilised system in which biomass and productivity tend to a maximum.

Because storm events are a major force at Cape Northumberland, episodically tearing algae off the

Table 7.3: Seasonal changes in reproductivity and distribution of reproductive plants of 17 species of red algae.

Data are for Cape Northumberland (Cape N.), except where indicated for George III Reef (George III). A heavy line indicates high abundance, a fine line moderate abundance, and a dotted line rare occurrence, of reproductive organs. θ = presence of tetrasporophytes. Male and female symbols indicate the presence of male and female gametophytes, respectively. A line adjacent to the symbol indicates the temporal extent of reproductive plants in a season.

Species	Summer	Autumn	Winter	Spring
Ballia callitricha, Inkyuleea mariana			— θ—	
Callophyllis lambertii	— θ —♀		♀— θ—♀	
Callophyllis rangiferina	— θ —	— θ—	— θ—	
Carpopeltis phyllophora	——	——	——	——
Crassilingua marginifera		♀	♀— θ—	♀— θ—
Curdiea laciniata	θ—♀	♀	♀- θ—	- ♀ -
Hymenena affinis	— θ—♀	♂♀— θ—	— θ—	- θ - ♂♀
Metagoniolithon radiatum	——		—	
Myriogramme gunniana	♂-♀		— θ—	θ—♀—
Nizymenia australis	—	— θ—		
Peyssonnelia capensis	——	—	——	— - -
Phacelocarpus peperocarpos	——		—	———
Plocamium angustum (Cape N.) (George III)	- - ♂♀ -	- -♂♀ θ— θ—	♂♀ —θ— ♂♀ —θ—	♂♀—θ—
Plocamium dilatatum (Cape N.) (George III)	♂♀ θ — ♂♀	♀ - —θ——	—θ— —θ—♂♀	—θ
Plocamium mertensii (Cape N.) (George III)	θ—	- θ ♀- - - -θ—♀	—θ —♀	- - θ
Ptilonia australasica (Cape N.) (George III)	— —♀—	— ♀—	—♀— θ	—— ——
Rhodophyllis multipartita	- ♂♀ -		♀—θ—♂	—θ
Rhodymenia australis	- θ ♀ -		♀—	- - θ

bottom, the process of community recovery from such disturbances can be studied by two kinds of experiment. First, placement of concrete blocks on the seabed will show the species that first colonise a bare substratum – called the primary succession. Second, by scraping clean the rocky substratum of all erect algae, except for the calcareous encrusting forms, the order in which colonising species appear in the clearing will indicate the secondary succession – with the proviso that scraping the substratum does not completely remove all resting spores.

Concrete blocks were placed out in summer and winter and after 3 months virtually the only colonisers were the encrusting coralline *Pneophyllum* sp. and a turf of epiphytic, filamentous *Ectocarpus*, mainly in winter, and the green alga *Chaetomorpha* in summer (see Table 7.4). After 3 months, a few other colonisers began to appear, with a total of six species in the first 6 months. The settlement of *Ecklonia* and *Macrocystis* on the blocks was curious, because the former was rare in the habitat and the latter absent in the mature community over the whole time of the study. In many

Table 7.4: Temporal succession of red algae on concrete blocks (–6–0) i.e. 6 months before clearing, and on cleared substrate over 2 years at the Cape Northumberland site.

Key: — = > 10% of individual plants; → = 5–10% of plants; — = 2–5% of plants; ····· = < 2% of plants.

Species	Time Blocks –6 - 0	Time (months) Cleared substratum → 0–3	3–6	6–9	9–12	12–15	15–18	18–21	21–24
Ecklonia radiata	—- - - - -								
Macrocystis f. *angustifolia*		—- -							
1.*Glaphyrymenia pustulosa*	—- - - - -								
2. *Myriogramme gunniana*		— - - - -	——	——	——	——	——	——	——
3. *Chaetomorpha darwinii*		——-							
4. *Halopteris* spp.		- - - —	— -	— -	— -	— -	— -	- - - -	
5. *Corallina officinalis*	- -		— -	— -	- - - -	- - - -	- - -		
6. *Hymenena affinis*		- -	— -	—-	—-	—-	—-	—-	— -
7. *Rhodophyllis multipartita*		-	——	——	— -	— -	— -	— -	— -
8. *Rhodymenia australis*			—	— -	—-	— -	——	——	——
9. *Ballia callitricha*		-	——	——	——	——			
10. *Sarcodia marginata*			—	——	— -	— -	— -	— -	—-
11. *Curdiea laciniata*				——	——	— -			
12. *Shepleya wattsi*						-	——	——	— -
13. *Carpopeltis phyllophora*								-	—-
14. *Phacelocarpus peperocarpos*				······	······	······	······	······	——
15. *Melanthalia concinna*				······	——	——	——	——	——
16. *Plocamium angustum*			— -	— -	——	——	——	——	——
17. *Plocamium dilatatum*				······	······	——	— -	—-	—-
18. *Metagoniolithon radiatum*				······		······	······		······
19. *Metamastophora flabellata*				······	······	······		······	
20. *Myriogramme gunniana*						— -	——	——	— -
21. *Niżymenia australis*			······	······	— -	— -	— -	—-	—-
22. *Callophyllis lambertii*			——	——	——	——	——	——	——
23. *Callophyllis rangiferina*			······	······					
24. *Peyssonnelia novaehollandiae*					······	······	—-	······	······
25. *Sonderopelta coriacea*								······	
26. *Thuretia quercifolia*				······					

respects, the results were similar to those of Reimers and Branden (1994), who placed out modules in deeper water off Glenelg, Gulf St Vincent (see Fig. 7.8). There, they found that crustose corallines were the first colonists in winter–spring, but that algal turfs, comprising 56 species of mainly red algae, were the initial colonists at other times of the year.

In the Cape Northumberland clearing, recolonisation was much more rapid than on the blocks, and the following sequence of events took place.

- In the first 3–4 months, depending on the season of clearing, 200–300 individual plants m^{-2}, comprising 34–52 species, recolonised the area, totalling 99 species in all. Among the new colonists, *Hymenena affinis* and *Plocamium angustum* were the most prolific. Of the other species, 71 were rare in the mature community. These species had the characteristics of short-lived, pioneer or early successional species, because they were mostly small and < 5 cm high, and rapidly became reproductively mature, and then disappeared.
- the dominants, *Callophyllis* spp., *Melanthalia* spp., *Phacelocarpus* spp. and *Nizymenia*, became established in the first 6 months, and gradually increased in biomass, so that after 17 months they were large enough to exclude early colonising species. After 2 years they comprised most of the algal biomass.
- Most middle-to-late successional species appeared after ~6 months and persisted over the 2 years; however, a few genera/species (e.g. *Halopteris* spp., *Callophyllis rangiferina, Corallina, Curdiea* and *Ballia*) persisted for only a year or so, or were even more transitory.
- The colonisation behaviour of the two species of *Plocamium* was markedly different. *P. angustum* colonised the clearing extensively in the first 3 months, then declined, but persisted at a low level after that. In contrast, *P. dilatatum* did the reverse, colonising at a low level early in the succession and reaching maximum biomass after 20 months, as it spread vegetatively.
- Diversity of species was highest midway through the successional process, and then declined as early successional species disappeared.
- Most species recruited into the clearing throughout the year, and only a few minor species showed evidence of seasonality of recruitment.

The abundance and succession of the main species over the 2 years are shown in Table 7.4, and changes in algal biomass and dominant species are shown in Fig. 7.6.

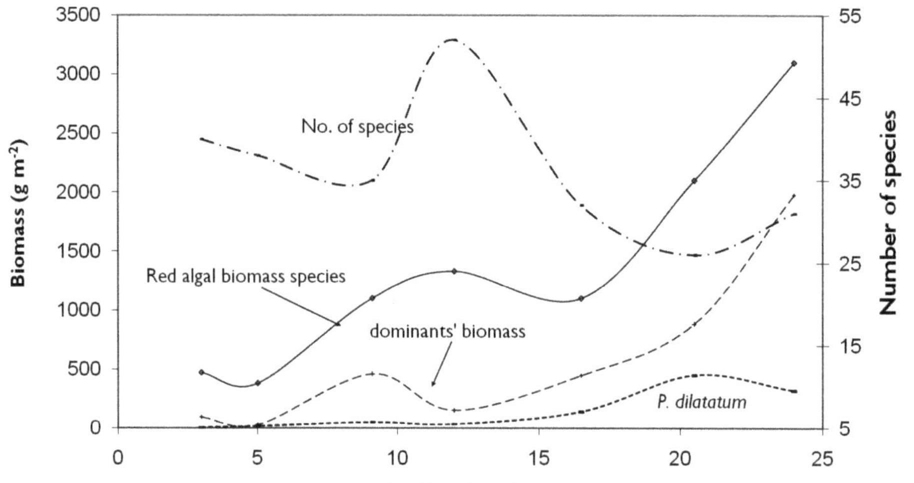

Figure 7.6: Changes in total numbers of species, red algal biomass (fresh weight), biomass of the dominants combined (*Callophyllis, Melanthalia* spp., *Phacelocarpus* spp. and *Nizymenia*), and biomass of *Plocamium dilatatum*, over 2 years after clearing (redrawn from Shepherd 1979).

We now return to Odum's three principles cited above. The succession had some order, in that initially many early successional (pioneer) species invaded the clearing, followed by other longer lived species, including those that eventually achieved dominance in terms of biomass. But, as to Odum's second principle, resolution of the question whether early colonists facilitated or inhibited later colonists would require careful experimentation (see Foster and Sousa 1985). It is possible that the very early establishment of crustose corallines was a prerequisite for the initial turfing species, which in turn may have facilitated the recruitment of *Ecklonia* and some other species (e.g. Harger and Tustin 1973). The dominants invaded the clearing early in the succession, and their recruitment rates differed little between those in the clearing and those in the mature forest. This evidence suggests that they arrived according to their usual recruitment pattern, and then grew slowly to achieve eventual dominance. On this view, the sequence of species was determined by their morphology and life history characteristics, more in accordance with the tolerance model discussed by Connell and Slatyer (1977). On Odum's last principle – whether a stable end point was reached after 2 years – Shepherd (1979) concluded that it had not, because complete dominance by a few species had not yet been attained. However, given the frequency of disturbance by storms in this community, creating gaps for the recruitment of early successional species, the community would likely comprise many patches at various successional stages, so that the whole community would never fully reach a stable end point.

Successional studies of deeper water red algal communities are rare, and the only other one known is that of Boudouresque (1973) in the French Mediterranean. Here, a very similar sequence occurred: first crustose corallines, then an algal turf, then development of a species-rich community of 43 species, and lastly after 12 months the dominance of a few species, which began to arrive after 3 months.

Ecological strategies of algae

To promote an understanding of the processes by which many algae can co-exist, despite the tendency for a few larger dominants to out-compete other species, models have been elaborated to illustrate the underlying principles. A well known one is that of Grime (1977), which was applied to the algal community at Cape Northumberland (Shepherd 1981). The model proposes three primary strategies of algae – dominance, ephemerality and shade-tolerance:

- **Dominance:** by this strategy, plants with larger canopies in terms of spread, height and density can out-compete other species for light and space – the main limiting resources. They can achieve this in several ways: by being long-lived; by having tough stems that are resistant to breakage from swell; or by rapid regeneration from the more persistent holdfasts.
- **Ephemerality:** removal of plants by storms, grazing and canopy whiplash is a major disturbance affecting algae. Ephemeral (i.e. early-successional or pioneer) species are the first to grow in gaps created by disturbances, and they can achieve this by having short life histories (i.e. rapid growth and rapid completion of the life cycle).
- **Shade tolerance:** low light levels in shaded environments, deeper water or under canopies are a stress that limits growth, and only shade-tolerant species adapted to such conditions can survive. Other stresses are nutrient shortages or excess, and scour from sand or other plants.

These three strategies are similar to those suggested by Wernberg and Goldberg (2008) above, and suggest that life history traits are decisive determinants of the primary strategies of algae in a community, at least regionally (see Grime 1988).

In the Cape Northumberland community, a dominance index was obtained from data on plant density, height and spread, and an ephemerality index from data on the half-life, and time of appearance in the succession. A plot of these two indices (Fig. 7.7) for 26 common species shows that some species conform primarily to one strategy, while others represent a compromise between several strategies. But significantly, the plot also shows the emergence of a third shade-tolerant strategy in addition to the first two. Omitted from the plot are the 70-odd rare pioneer species that appeared very early in the clearing and then disappeared; these would all cluster at the disturbance maximum. As the succession proceeded, the

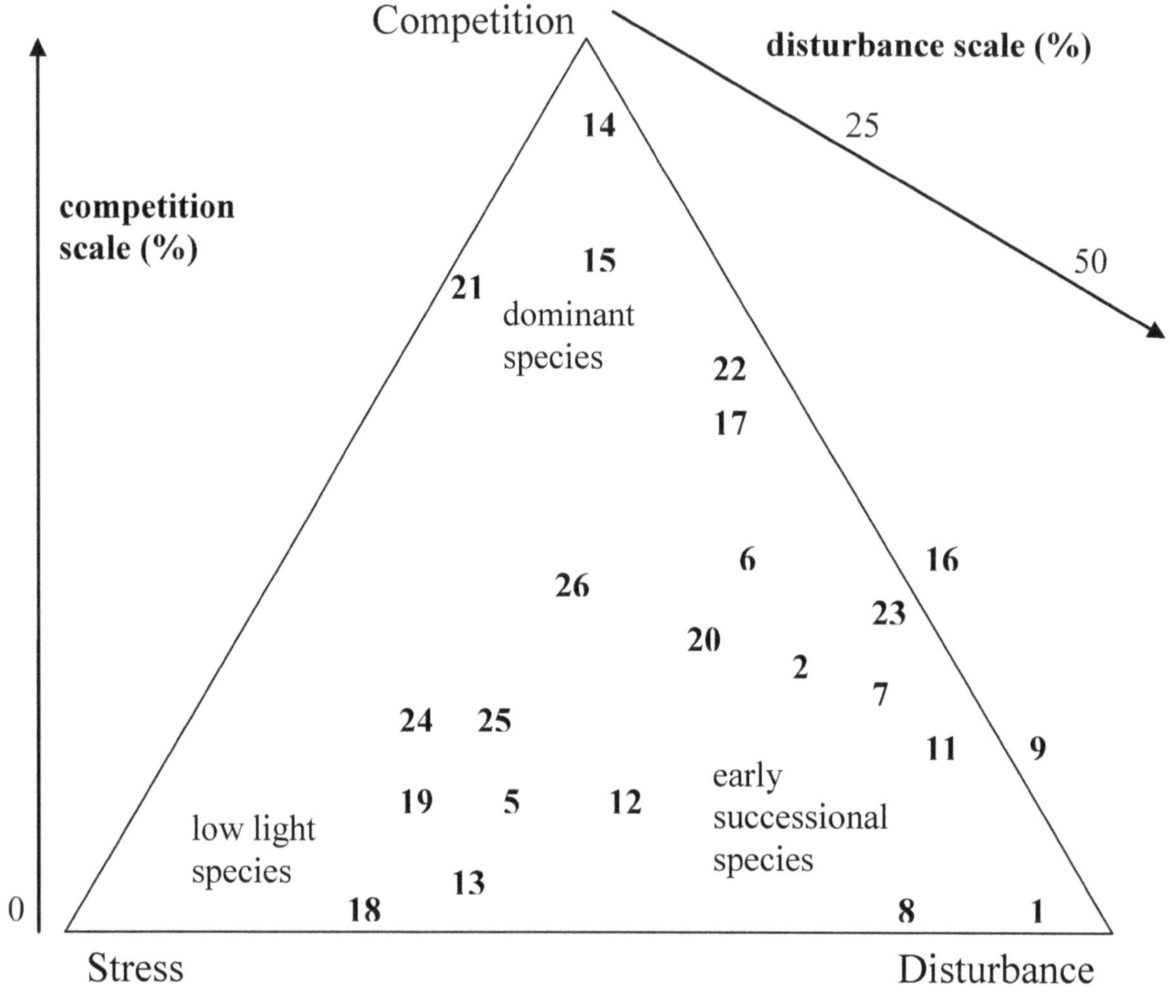

Figure 7.7: A triangular ordination of algae in the mature red algal community at the Cape Northumberland site, showing the location of mainly middle and late successional species, omitting ~70 early successional species located at the disturbance apex. The apices of the triangle represent maxima in competition, disturbance and stress. Per cent scales are shown. Key to species plotted: 1–13 are the species as numbered in Table 7.4. Other species are: (14) *Phacelocarpus peperocarpos*; (15) *Melanthalia concinna*; (16) *Plocamium angustum*; (17) *Plocamium dilatatum*; (18) *Metagoniolithon radiatum*; (19) *Metamastophora canaliculata*; (20) *Myriogramme gunniana*; (21) *Nizymenia australis*; (22) *Callophyllis lambertii*; (23) *Callophyllis rangiferina*; (24) *Peyssonnelia novaehollandiae*; (25) *Sonderopelta coriacea*; and (26) *Thuretia quercifolia* (redrawn, with additional data added, from Shepherd 1981).

community consisted of clusters of species – some towards the competition maximum, and others towards the stress maximum, as low-light conditions developed in the shade of larger, red algal dominants.

A problem with the ecological-strategy model is that algae have complex life histories, with two or three alternate phases and morphology, and may have alternative reproductive pathways, as in sexual or asexual reproduction. This means that algae may have dual or other strategies in alternative phases, which cannot be accommodated by the model, or may have unusual adaptive abilities (Russell 1986; Toohey *et al.* 2004; Craine 2005).

EFFECTS OF NUTRIENTS AND SEDIMENTATION

As human populations have expanded in coastal areas, land use and clearing have intensified, and storm-water run-off and effluent discharge from industry and sewage treatment into coastal waters have increased. Offshore dredging and coastal construction have caused increasing sedimentation. The above activities have more than doubled soluble nutrient levels in coastal waters (Gabric and Bell 1993) and injected suspended material into near-shore waters, creating eutrophication and sedimentation problems. In this section, we summarise their effects on algal communities and illustrate them with two case studies.

Nutrients

Sewage discharges off the east and SE Australian coasts have given rise to many studies of their effects on inter- to subtidal algae (e.g. Bellgrove *et al.* 1997; Fairweather 1990). Toxicological studies (see Burridge *et al.* 1996 and references) have shown that algal macrophytes are sensitive to chronic effluent discharges at their germination stage, and this sufficiently explains the widespread decline and disappearance of the larger brown algae close to outfalls. However, in other places kelps and fucoid algae have disappeared over large areas more distant from outfalls, and here more complex explanations have been invoked, related to natural algal loss by storms and the replacement of those species by diverse algal turfing species (Connell 2008 and see below).

With the loss of canopy species, a range of ephemeral species, favoured by the elevated nutrients, rapidly colonise the substratum. Typical early colonists of bare rock are *Ulva* and *Enteromorpha*, often followed by turf-forming, filamentous species, *Ceramium, Chaetomorpha, Ralfsia*, and others (Brown *et al.* 1990). Curiously, *Corallina* is rare near outfalls, likely because elevated phosphate levels inhibit its calcification (Brown *et al.* 1977; reviewed by Nelson 2009). Globally, the replacement of kelps or fucoids by algal turfs has become epidemic near urban centres (Airoldi and Beck 2007) and the problem has been explored in southern Australia by Russell et al. (2005) and Russell and Connell (2007).

First, the authors showed experimentally that increasing nutrients by 89% over 3 months caused a three-fold increase in an algal turf cover of *Feldmannia* spp., using *Ecklonia* blades as a standardised substrate. They then showed that algal turf cover over 5000 km of the southern coast was directly related to nutrient levels. On the central New South Wales coast, nutrient levels are far higher than elsewhere in southern Australia, where waters are typically low in nutrients (oligotrophic). Interestingly, their experiments suggested that increases in nutrients in oligotrophic waters will promote algal turf growth much more than in eutrophic waters.

It is rare that anthropogenic changes only affect nutrient levels near urban centres, and often other physical factors change as well. Gorgula and Connell (2004) found that, while nutrients and sediments each increased growth of algal turfs, both together had a dramatic additive effect, increasing turf cover by 77%, with the nutrient component having six times the effect as sediment alone. Hence, urban coasts, where both sediments and nutrients are enhanced, are likely to suffer a serious expansion of turfs on rocky bottom. However, the situation can be more complicated where grazers (e.g. herbivorous molluscs) are present because they can in some situations consume excess algal turf and ameliorate the effect (Russell and Connell 2007).

Another complexity explored by Russell *et al.* (2009) was the relation between projected increasing acidification from climate change and increased nutrient levels in coastal waters. Here, the authors found that both factors favoured growth of turfs, but together they had a synergistic effect, increasing turf cover by an extra 34%. Hence, the replacement of algal forests by turfs would be exacerbated under future climate scenarios.

Sedimentation

Deposition of sediment on rocky bottoms is not uniform, and depends on the wave energy environment. In moderate to high wave energy, sediments are rapidly dispersed and persist only where they are trapped in crevices or depressions or by sessile fauna or algal mats. Sedimentation on reefs has therefore become a problem mainly in sheltered waters near urban centres (reviewed by Airoldi 2003), but there are few reports of its effects in southern waters.

The main effects of sedimentation on algae can be summarised as:

- **smothering**, which reduces light, oxygen and nutrients reaching the plant, and may favour the accumulation of hydrogen sulphide

- **scour**, which can abrade, damage or remove organisms
- **a switch from a stable to an unstable substratum**, so blurring the distinction between reef and soft-bottom habitats.

Essentially, sedimentation can be viewed as a stress or a disturbance, which inhibits settlement/recruitment of algal spores, causes the decline of some species, but favours a few others, changes species composition and reduces diversity. Airoldi (2003) suggested that several common patterns have emerged from the literature:

- Sexually reproducing species are more vulnerable than those that reproduce vegetatively.
- 'Sand-tolerant' species adapted in various ways to withstand sediment accumulation (see examples below) are favoured.
- Many grazers, notably molluscs and sea urchins, may decline, so indirectly favouring the survival of sand-tolerant species.
- Filamentous algal turfs tend to increase in abundance in sediment-affected areas, favoured by their ability to grow rapidly, spread vegetatively and pre-empt vacant space – features consistent with their early successional character, noted above. They trap sediment, causing it to accumulate, and so prevent colonisation by other species. Hence, sediment can be considered a structural constituent of the modified habitat, causing a drastic shift of the community to another state of uncertain stability.

Two case studies

Reef habitats of lower Gulf St Vincent, South Australia

Discontinuous reefs of low relief, remnants of ancient shorelines, lie on the eastern side of the Gulf at depths of 5–10 m, and were pristinely dominated by *Ecklonia* or fucoid canopy species. Since the 1950s, with population expansion, assaults on the Gulf have been numerous. They include: discharges from numerous sewage outfalls after secondary treatment; many non-point, stormwater discharges; and periodic dredging of sand from 1997 for beach replenishment. As a result, nutrient levels doubled compared with natural levels, and sediments impacted near-shore reefs.

In consequence, the reefs in the region began undergoing change in the 1970s, so that 40 years later their pristine state was forgotten – an example of an (almost) lost baseline. Canopy species largely disappeared, and turf algal and mussel communities increased, depending on site location and proximity to particular discharges (Table 7.5, Fig. 7.8; Connell *et al.* 2008). However, although sites close to urban outfalls suffered the most serious effects, more distant sites remained unchanged (Table 7.5).

Table 7.5: Percentage cover of algal canopy species and algal turfs in a pristine state (1960s) and in 2005–07, with information on proximity to impact source(s).

Data extracted from Connell *et al.* (2008), Turner *et al.* (2007) and Collings *et al.* (2008).

	% Canopy cover		% Turf		
Site	Pristine	2005–7	Pristine	2005–7	Site location
1. Aldinga control	> 90	> 90	~5	~5	Reserve located 18 km south of nearest outfall
2. Point Noarlunga Reef	> 90	20	~5	30	2 km south of outfall
3. Horseshoe Reef	~80	~15	~5	~30	1 km south of outfall
4. Broken Bottom	20–30[1]	0	< 5	~42	Offshore from major outfall
5. Semaphore	20–30[1]	0	~10[1]	~10	Offshore from major outfall
6. North Reef (control)	27[2]	27	45[2]	45	40 km NW of major outfall

[1]estimates (S.A. Shepherd unpublished observations in the 1960s); [2]control sites estimated, assuming no change.

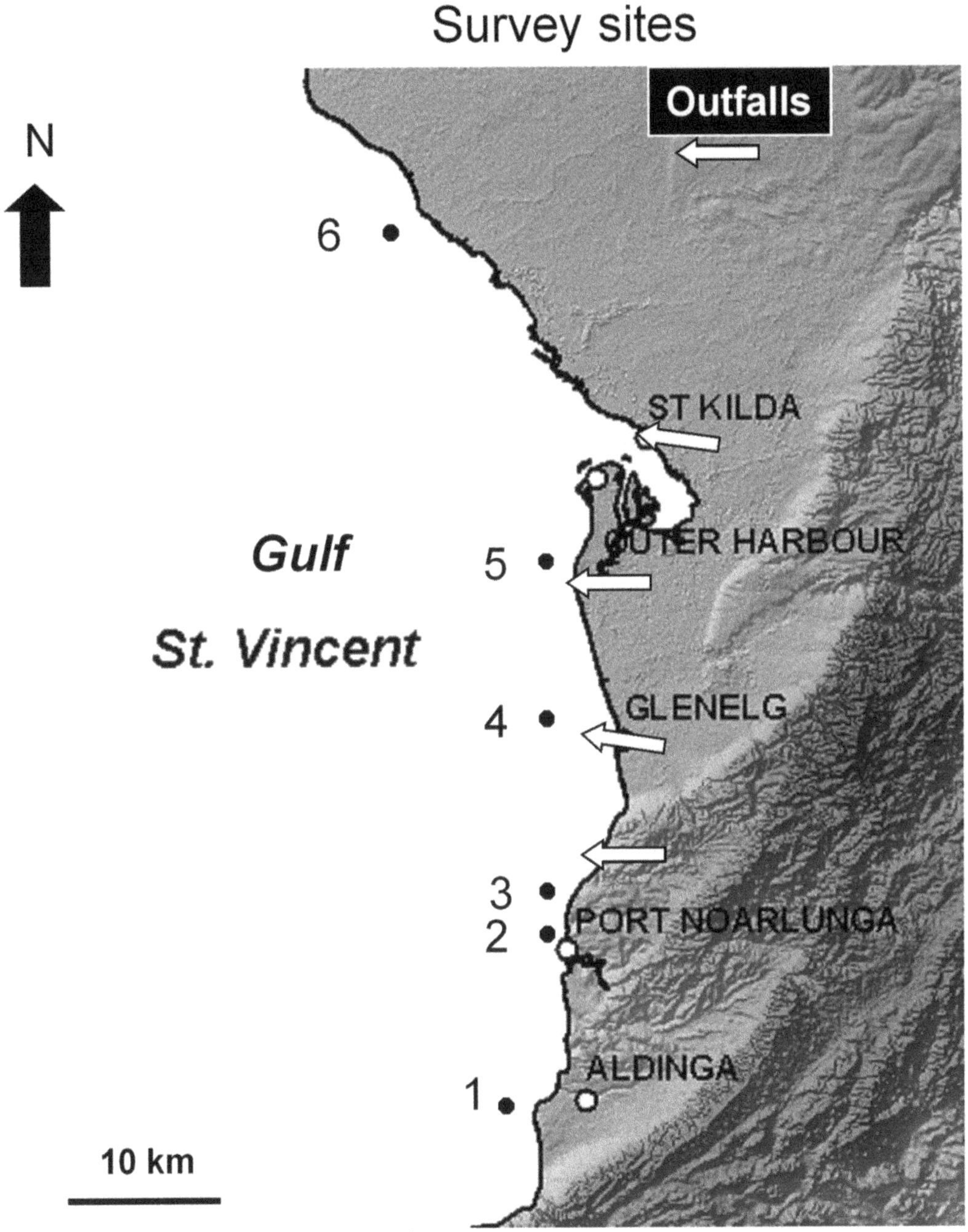

Figure 7.8: Map of eastern Gulf St Vincent off Adelaide, showing six survey sites, numbered 1 to 6, as in Table 7.5, and the location of four major sewage effluent outfalls during the period of decline.

Mussels are common organisms on temperate reefs, and often settle in algal turfs (Chapter 9; Turner *et al.* 2007). At Point Noarlunga Reef and the adjacent Horseshoe Reef, both near an outfall where discharges increased during the 1970s, cover of the mussel *Brachidontes rostratus* increased dramatically from near zero in 1975 (Shepherd and Sprigg 1976) to 40–80% some 15 years later as canopy cover declined.

The four changes – the decline of canopy species, and the increases in turf algae, entrained sediment and mussel cover – suggest that positive feedbacks were driving the reef community to an alternative state (Smith 2000; Turner *et al.* 2007). These were:

- Increased nutrient levels, which favoured a spreading algal turf.
- The expanding turf promoted mussel settlement.
- An increased planktonic food supply from the higher nutrient levels favoured mussel growth.
- The growing matte of mussels increasingly filtered algal spores from the water.

The theory underlying regime shift or alternative stable states in a community is as follows. Many natural feedbacks maintain a community in one state, but a catastrophe or a longer term disturbance can cause the community to switch to another state. This new state then stabilises through newly operating positive feedbacks, as discussed above for nutrient and/or sedimentation disturbances (see Connell 2008).

Western Port, Victoria

During the 1970s, an environmental catastrophe occurred, in which almost 700 km^2 of seagrass were lost, exposing mudflats to severe tidal erosion. Thereupon, sedimentation and turbidity increased dramatically throughout the Bay; nutrient levels, however, increased only slightly. Fortunately, a pre-catastrophe algal study at Crawfish Rock in the upper Bay, repeated some 35 years later, enabled a before-and-after evaluation of the sedimentation effects on the original rich, benthic algal flora of 138 species (Shepherd *et al.* 2009).

In all, 66% of algal species were lost as the algal community became fragmented, and the photic zone was elevated from ~13 m to 4 m depth. Foliose red algae were the most seriously affected, presumably smothered by the layers of sediment. However, a few species were unaffected by the greater sedimentation. The green algae, *Caulerpa flexilis* and *C. obscura*, increased in cover at the expense of other *Caulerpa* species, which persisted with low cover (see Fig. 7.1). Resistant red algae that survived were species of *Areschougia* and *Thamnoclonium*; both have robust, wiry thalli – a known adaptation to sandy habitats. Curiously, the kelp *Ecklonia* was little affected, except for an elevation of its lower depth limit, probably because the strong tidal currents continually swept clean the sloping rock surfaces favoured by *Ecklonia*.

Conclusions

Excess nutrients and sediments constitute serious assaults on the integrity of algal communities. The former cause the decline of canopy species, and favour their replacement by algal turfs, while sediments are deleterious to algae, and cause their decline by several mechanisms. Together the two assaults are synergistic and can cause a switch from a species-rich productive ecosystem to an impoverished alternative stable state.

REFERENCES

Airoldi L (2003) The effects of sedimentation on rocky coast assemblages. *Oceanography and Marine Biology - an Annual Review* **41**, 161–236.

Airoldi L, Beck MW (2007) Loss, status and trends for coastal marine habitats of Europe. *Oceanography and Marine Biology - an Annual Review* **45**, 345–405.

Bellgrove A, Clayton MN, Quinn GP (1997) Effects of secondarily treated sewage effluent on intertidal macroalgal recruitment processes. *Marine and Freshwater Research* **48**, 137–146.

Bonin DR, Hawkes MW (1988) Systematics and life histories of New Zealand Bonnemaisoniaceae (Bonnemaisoniales, Rhodophyta): II. The genus *Delisea*. *New Zealand Journal of Botany* **26**, 619–632.

Boudouresque C-F (1973) Etude in situ de la reinstallation d'un peuplement sciaphile de mode battu après sa destruction expérimentale, en Mediterranée. *Helgoländer Wissenschaftliche Meeresuntersuchungen* **24**, 202–218.

Brown V, Ducker SC, Rowan KS (1977) The effect of orthophosphate concentration on the growth of articulated coralline algae (Rhodophyta). *Phycologia* **16**, 125–131.

Brown VB, Davies SA, Synnot RN (1990) Long-term monitoring of the effect of treated sewage effluent on the intertidal macroalgae community near Cape Schank, Victoria, Australia. *Botanica Marina* **33**, 85–98.

Burn R (1989) Opisthobranchs (Subclass Opisthibranchia) In *Marine Invertebrates of Southern Australia. Part II.* (Eds SA Shepherd and IM Thomas) pp. 725–788. South Australian Government Printer, Adelaide.

Burridge TR, Portelli T, Ashton P (1996) Effect of sewage effluents on germination of three marine brown algal macrophytes. *Marine and Freshwater Research* **47**, 1009–1014.

Calvert HE, Dawes CJ, Borowitzka MA (1976) Phylogenetic relationships of *Caulerpa* (Chlorophyta) based on comparative chloroplast ultrastructure. *Journal of Phycology* **12**, 149–162.

Ceccherelli G, Piazzi I (2001) Dispersal of *Caulerpa racemosa* in the Mediterranean: lack of attachment time effect on establishment. *Botanica Marina* **44**, 209–213.

Clarkson PS, Shepherd SA (1985) Distribution and feeding of archaeogastropods in a boulder habitat at West Island, South Australia. *Journal of the Malacological Society of Australia* **7**, 35–44.

Clayton MN (1974) Studies on the development, life history and taxonomy of the Ectocarpales (Phaeophyta) in southern Australia. *Australian Journal of Botany* **22**, 743–813.

Clifton KE (1997) Mass spawning by green algae on coral reefs. *Science* **275**, 1116–1118.

Collings G, Bryars S, Turner D, Brook J *et al.* (2008) *Examining The Health of Subtidal Reef Environments in South Australia. 4. Assessment of Community Reef Monitoring and Status of Selected Reefs Based on the Results of the 2007 Surveys.* SARDI Publication No. F2008/000511-1. SARDI, Adelaide.

Connell JH, Slatyer RO (1977) Mechanisms of succession in natural communities and their role in community stability and organization. *American Naturalist* **111**, 1119–1144.

Connell SD (2005) Assembly and maintenance of subtidal habitat heterogeneity: synergistic effects of light penetration and sedimentation. *Marine Ecology Progress Series* **289**, 53–61.

Connell SD (2008) Water quality and the loss of coral reefs and kelp forests: alternative states and the influence of fishing. In *Marine Ecology* (Eds SD Connell and BM Gillanders). Oxford University Press, Melbourne.

Connell SD, Russell BD, Turner DJ, Shepherd SA *et al.* (2008). Recovering a lost baseline: missing kelp forests from a metropolitan coast. *Marine Ecology Progress Series* **360**, 63–72.

Copertino M, Connell SD, Cheshire A (2005) The prevalence and production of turf-forming algae on a temperate subtidal coast. *Phycologia* **44**, 241–248.

Copertino M, Cheshire A, Watling J (2006) Photoinhibition and photo-acclimation of turf algal communities on a temperate reef, after *in situ* transplantation experiments. *Journal of Phycology* **42**, 580–592.

Copertino M, Cheshire A, Kildea T (2009) Photophysiology of a turf algal community: integrated responses to ambient light and standing biomass. *Journal of Phycology* **45**, 324–336.

Craine JM (2005) Reconciling plant strategy theories of Grime and Tilman. *Journal of Ecology* **93**, 1041–1052.

Davis AR, Roberts DE, Cummins SP (1997) Rapid invasion of a sponge-dominated deep-reef by *Caulerpa scalpelliformis* (Chlorophyta) in Botany Bay, New South Wales. *Australian Journal of Ecology* **22**, 146–150.

Davis AR, Benkendorf K, Ward DW (2005) Responses of common SE Australian herbivores to three suspected invasive *Caulerpa* spp. *Marine Biology* **146**, 859–868.

de Nys R, Steinberg PD, Willemsen P, Dworjanyn SA *et al.* (1995) Broad spectrum effects of secondary metabolites from the red alga *Delisea pulchra* in antifouling assays. *Biofouling* **8**, 259–271.

Doty MS (1971) Measurement of water movement in reference to benthic algal growth. *Botanica Marina* **14**, 32–35.

Edgar GJ (1984) General features of the ecology and biogeography of Tasmanian subtidal rocky shore communities. *Papers and Proceedings of the Royal Society of Tasmania* **118**, 173–186.

Fairweather PG (1990) Sewage and the biota on seashores: assessment of impact in relation to natural variability. *Environmental Monitoring and Assessment* **14**, 197–210.

Ferns LW, Hough D (Eds) (2002) *High Resolution Marine Habitat Mapping of the Bunurong Coast (Victoria) – Including the Bunurong Marine and Coastal Park.* Flora and fauna Division, Dept Natural Resources and Environment, Melbourne.

Foster MS, Sousa WP (1985) Succession. In *Handbook of Phycological Methods: Ecological field Methods – Macro-algae.* (Eds MM Littler and DS Littler) pp. 269–290. Cambridge University Press, Cambridge.

Fowler-Walker MJ, Connell SD (2002) Opposing states of subtidal habitat across temperate Australia: consistency and predictability in kelp canopy-benthic associations. *Marine Ecology Progress Series* **240**, 49–56.

Fowler-Walker MJ, Gillanders BM, Connell SD, Irving AD (2005) Patterns of association between canopy-morphology and understorey assemblages across temperate Australia. *Estuarine, Coastal and Shelf Science* **63**, 133–141.

Gabric AJ, Bell PRF (1993) Review of the effects of non-point nutrient loading on coastal ecosystems. *Australian Journal of Marine and Freshwater Research* **44**, 261–283.

Gacia E, Littler MM, Littler DS (1996) The relationships between morphology and photosynthetic parameters within the polymorphic genus *Caulerpa*. *Journal of Experimental Marine Biology and Ecology* **204**, 209–224.

Goldberg NA, Kendrick GA (2004) Effects of island groups, depth, and exposure to ocean waves on subtidal macroalgal assemblages in the Recherche Archipelago, Western Australia. *Journal of Phycology* **40**, 631–641.

Goldberg NA, Kendrick GA (2005) A catalogue of the marine macroalgae found in the western islands of the Recherche Archipelago, Western Australia, with notes on their distribution in relation to island location, depth and exposure to wave energy. In *The Marine Flora and Fauna of Esperance, Western Australia.* (Eds FE Wells, DI Walker and GA Kendrick) pp. 25–89. Western Australian Museum, Perth.

Gorgula SK, Connell SD (2004) Expansive covers of turf-forming algae on human-dominated coast: the relative effects of increasing nutrient and sediment loads. *Marine Biology* **143**, 613–619.

Grime JP (1977) Evidence for the existence of three primary strategies in plants and its relevance to ecological and evolutionary theory. *American Naturalist* **111**, 1169–1194.

Grime JP (1988) A comment on Loehle's critique of the triangular model of primary plant strategies. *Ecology* **69**, 1618–1620.

Haig J, Russell B, Murray-Jones S (2006) *Offshore Marine Habitat Mapping and Near-Shore Marine Biodiversity within the Coorong Bioregion.* Report to Murray–Darling Basin NRM Board. Department for Environment and Heritage, Adelaide.

Harger JRE, Tustin K (1973) Succession and stability in biological communities. I. Diversity. *The International Journal of Environmental Studies* **5**, 117–130.

Irving AD, Connell SD (2002) Interactive effects of sedimentation and microtopography on the abundance of sub-tidal turf-forming algae. *Phycologia* **41**, 517–522.

Irving AD, Connell SD (2006a) Predicting understorey structure from the presence and composition of canopies: an assembly rule for marine algae. *Oecologia* **148**, 491–502.

Irving AD, Connell SD (2006b) Physical disturbance by kelp abrades erect algae from the understorey. *Marine Ecology Progress Series* **324**, 127–137.

Irving AD, Fowler-Walker MJ, Connell SD (2003) Associations of forest-type with morphology of *Ecklonia radiata* and abundance of understorey algae. *Transactions of the Royal Society of South Australia* **127**, 167–175.

Irving AD, Connell SD, Gillanders BM (2004) Local complexity in patterns of canopy-benthos associations produces regional patterns across temperate Australia. *Marine Biology* **144**, 361–368.

Johansen HW (1974) Articulated coralline algae. *Oceanography and Marine Biology - an Annual Review* **12**, 77–127.

Kendrick GA, Lavery PS, Phillips JC (1999) Influence of *Ecklonia radiata* kelp canopy on structure of macro-algal assemblages in Marmion lagoon, Western Australia. *Hydrobiologia* **398/399**, 275–283.

Kendrick GA, Harvey ES, Wernberg T, Harman N *et al.* (2004) The role of disturbance in maintaining diversity of benthic macroalgal assemblages in southwestern Australia. *Japanese Journal of Phycology* **52** (Suppl.), 5–9.

Kennelly SJ (1983) A preliminary study of temporal colonisation of subtidal algae in an *Ecklonia radiata* community. *Aquatic Botany* **17**, 275–282.

Kennelly SJ (1987a) Inhibition of kelp recruitment by turfing algae and the consequences for an Australian kelp community. *Journal of Experimental Marine Biology and Ecology* **112**, 49–60.

Kennelly SJ (1987b) Physical disturbances in an Australian kelp community. I. Temporal effects. *Marine Ecology Progress Series* **40**, 145–153.

Kennelly SJ (1987c) Physical disturbances in an Australian kelp community. II. Effects on understorey species due to differences in kelp cover. *Marine Ecology Progress Series* **40**, 155–165.

Kennelly SJ (1989) Effects of kelp canopies on understorey species due to shade and scour. *Marine Ecology Progress Series* **50**, 215–224.

Kennelly SJ (1995) Kelp beds. In *Coastal Marine ecology of Temperate Australia* (Eds AJ Underwood and MG Chapman) pp. 106–120. University of New South Wales Press, Sydney.

Kennelly SJ, Larkum AWD (1983) A preliminary study of temporal variation in the colonization of subtidal algae in an *Ecklonia radiata* community. *Aquatic Botany* **17**, 275–282.

Kennelly SJ, Underwood AJ (1992) Fluctuations in the distributions and abundances of species in sublittoral forests in New South Wales. *Australian Journal of Ecology* **17**, 367–382.

Kennelly SJ, Underwood AJ (1993) Geographic consistencies of effects of experimental physical disturbance on understorey species in sublittoral kelp forests in central New South Wales. *Journal of Experimental Marine Biology and Ecology* **168**, 35–58.

Klein J, Verlaque M (2008) The *Caulerpa racemosa* invasion: a critical review. *Marine Pollution Bulletin* **56**, 205–225.

Larkum AWD (1981) Marine primary productivity. In *Marine Botany: an Australasian Perspective.* (Eds MN Clayton and RJ King) pp. 370–385. Longman Cheshire, Melbourne.

May V, Larkum AWD (1981) A subtidal transect in Jervis Bay, New South Wales. *Australian Journal of Ecology* **6**, 439–457.

Melville AJ, Connell SD (2001) Experimental effects of kelp canopies on subtidal coralline algae. *Austral Ecology* **26**, 102–108.

Nelson WA (2009) Calcified macro-algae - critical to coastal ecosystems and vulnerable to change: a review. *Marine and Freshwater Research* **60**, 787–801.

Odum EP (1969) The strategy of ecosystem development. *Science* **164**, 262–270.

Panayotidis P, Zuljević A (2001) Sexual reproduction of the invasive green alga *Caulerpa racemosa* var. *occidentalis* in the Mediterranean Sea. *Oceanologica Acta* **24**, 199–203.

Phillips JA (2007) Heterokontophyta: Phaeophyceae. In *Algae of Australia: Introduction*. (Eds PM McCarthy and AE Orchard) pp. 264–287. ABRS, Canberra; CSIRO PUBLISHING, Melbourne.

Phillips JC, Kendrick GA, Lavery PS (1997) A test of a functional group approach to detecting shifts in macroalgal communities along a disturbance gradient. *Marine Ecology Progress Series* **153**, 125–138.

Pillmann A, Woolcott GW, Olsen JL, Stam WT *et al.* (1997). Inter- and intra-specific genetic variation in *Caulerpa* (Chlorophyta) based on nuclear rDNA ITS sequences. *European Journal of Phycology* **33**, 379–386.

Reimers H, Branden K (1994) Algal colonisation of a tyre reef – influence of placement date. *Bulletin of Marine Science* **55**, 460–469.

Russell G (1986) Variation and natural selection in marine macroalgae. *Oceanography and Marine Biology - an Annual Review* **24**, 309–377.

Russell BD, Connell SD (2007) Response of grazers to sudden nutrient pulses in oligotrophic versus eutrophic conditions. *Marine Ecology Progress Series* **349**, 73–80.

Russell BD, Elsdon TS, Gillanders BM, Connell SD (2005) Nutrients increase epiphyte loads: broad-scale observations and an experimental assessment. *Marine Biology* **147**, 551–558.

Russell BD, Thompson JA, Falkenberg LJ, Connell SD (2009) Synergistic effects of climate change and local stressors: CO_2 and nutrient-driven change in subtidal rocky habitats. *Global Change Biology* **15**, 2153–2162.

Sanderson JC (1992) Subtidal macroalgal studies in east and south eastern Tasmanian coastal waters. MSc thesis. University of Tasmania, Hobart.

Sanderson JC (1997) 'Subtidal macroalgal assemblages in temperate Australian coastal waters. Australia: State of the Environment'. Technical Paper Series (Estuaries and the Sea) Report. Department of the Environment, Canberra.

Sanderson JC, Thomas DP (1987) Subtidal macroalgal communities in the D'Entrecasteaux Channel, Tasmania. *Australian Journal of Ecology* **12**, 41–51.

Shepherd SA (1979) The ecology of a deep water red algal community off Cape Northumberland, South Australia. MEnvSt thesis, University of Adelaide; Fisheries Internal Report No. 4. Department of Fisheries, Adelaide.

Shepherd SA (1981) Ecological strategies in a deep water red algal community. *Botanica Marina* **24**, 457–463.

Shepherd SA, Sprigg RC (1976) Substrate, sediments and subtidal ecology of Gulf St Vincent and Investigator Strait. In *Natural History of the Adelaide Region.* (Eds CR Twidale, MJ Tyler and BP Webb) pp. 161–174. Royal Society of South Australia, Adelaide.

Shepherd SA, Womersley HBS (1970) The sublittoral ecology of West Island, South Australia. I. Environmental features and the algal ecology. *Transactions of the Royal Society of South Australia* **94**, 105–138.

Shepherd SA, Womersley HBS (1981) The algal and seagrass ecology of Waterloo Bay, South Australia. *Aquatic Botany* **11**, 305–371.

Shepherd SA, Watson JE, Womersley HBS, Carey J (2009) Long-term changes in macroalgal assemblages after increased sedimentation and turbidity in Western Port, Victoria, Australia. *Botanica Marina* **52**, 195–206.

Smale DA, Kendrick GA, Waddington KI, Van Niel KP *et al.* (2010) Benthic assemblage composition on subtidal reefs along a latitudinal gradient in Western Australia. *Estuarine, Coastal and Shelf Science* **86**, 83–92.

Smith N (2000) The impacts of the mussel *Xenostrobus pulex* on subtidal South Australian macroalgal systems. BSc (Hons). University of Adelaide, Adelaide.

Toohey B, Kendrick GA, Wernberg T, Phillips JC *et al.* (2004) The effects of light and thallus scour from *Ecklonia radiata* canopy on an associated foliose algal assemblage: the importance of photoacclimation. *Marine Biology* **144**, 1019–1027.

Turner D, Kildea T, Westphalen G (2007) *Examining the Health of Subtidal Reef Environments In South Australia. 2. Status of Selected Reefs Based on the Results of the 2005 Surveys.* SARDI Publication No. RD 03/0252-6. SARDI, Adelaide.

Underwood AJ, Kennelly SJ (1990) Ecology of marine algae on rocky shores and subtidal reefs in temperate Australia. *Hydrobiologia* **192**, 3–20.

Van der Velde JT, King RJ (1984) The subtidal seaweed communities of Bare Island, Botany Bay. *Wetlands* **4**, 7–22.

Wernberg T, Goldberg N (2008) Short-term temporal dynamics of algal species in a subtidal kelp bed in relation to changes in environmental conditions and canopy biomass. *Estuarine, Coastal and Shelf Science* **76**, 265–272.

Wernberg T, Kendrick GA, Phillips JC (2003) Regional differences in kelp-associated algal assemblages on temperate limestone reefs in south-western Australia. *Diversity & Distributions* **9**, 427–441.

Wernberg T, Kendrick GA, Toohey BD (2005) Modifications of the physical environment by an *Ecklonia radiata* (Laminariales) canopy and implications for associated foliose algae. *Aquatic Ecology* **39**, 419–430.

Westphalen G (2008) Exotic species in Gulf St Vincent. In *Natural History of Gulf St Vincent* (Eds SA Shepherd,

S Bryars, IR Kirkegaard, P Harbison *et al.*) pp. 224–237. Royal Society of SA, Adelaide.

Williams SI, Walker DI (1999) Meso-herbivore – macroalgal interactions: feeding ecology of sacoglossan sea slugs (Mollusca, Opisthobranchia) and their effects on their algal food. *Oceanography and Marine Biology - an Annual Review* **37**, 87–128.

Wright JT, Steinberg PD (2001) Effect of variable recruitment and post-recruitment herbivory on local abundance of a marine alga. *Ecology* **82**, 2200–2215.

Wright JT, de Nys R, Steinberg PD (2000a) Geographic variation in halogenated furanones from the red alga *Delisea pulchra* and associated herbivores and epiphytes. *Marine Ecology Progress Series* **207**, 227–241.

Wright JT, Zuccarello GC, Steinberg PD (2000b) Genetic structure of the subtidal red alga *Delisea pulchra*. *Marine Biology* **136**, 439–448.

PART 3

THE INVERTEBRATES

After looking at the large-scale patterns in Part I and the algae in Part II, we now move on to the fine-scale reef picture. Part III describes the ecology of some of the key invertebrate players, which are significant from a food web point of view because they transfer food and energy from the algae to higher levels. They are the meio- and macrofauna (Chapter 8), molluscan herbivores, predators and filter-feeders (Chapters 9 and 10), and the echinoderms (Chapters 11 and 12), which are striking for their pentamerous radial symmetry. Then follows in Chapter 13 the sessile filter-feeders (i.e. sponges, ascidians and bryozoans), which are notable for their competition for limited space on reefs, and in Chapter 14 the cnidarians, remarkable for their alternating generations of sessile and pelagic forms, although one group, the anthozoans, has only a polyp form. Finally, in Chapter 15 we treat the larger crustaceans, representing an impressive evolutionary jump, as they have evolved an astonishing diversity of species that fill every conceivable niche. They are equipped with an armour-like external skeleton and appendages to facilitate movement, and sometimes even migrations.

8 Ecology of meiofauna and macrofauna

OVERVIEW

The motile meiofauna and macrofauna are very small forms of animal life < 10–20 mm size that are critically important in the food web of reef systems. The meiofauna (< 1–2 mm size) is poorly known and only briefly described. The macrofauna, in which we include animals up to a size of ~20 mm, mainly comprises small crustaceans, molluscs and polychaetes, although many other marine animal phyla are represented. Their range of feeding is broad and they include herbivores, carnivores, suspension feeders, and detritivores. A few herbivorous species feed on the macroalgal host plants, but most feed on epiphytes, diatoms, detritus and organic sediment among the algal fronds or turf on the reef. We describe the feeding ecology and life history of amphipods (including caprellids), isopods, shrimp, mysids, krill, worms (polychaetes, flatworms and nemerteans), and some common grazing and predatory gastropods. Environmental factors influence the relative abundances of the different feeding types. Suspension feeders tend to be dominant in stronger water movements, and detritivores in calmer conditions where organic sediment accumulates. The abundance of the motile macrofauna living in algae also varies according to host plant density and architecture, season and numerous environmental factors such as depth, water movement, light and salinity. Kelp holdfasts also provide a micro-habitat for an abundant and diverse macrofauna, which varies according to kelp density, forest type and matrix, depth, amount of detritus and sediment, and other factors.

Predation can be important controlling abundance, particularly in more open habitats.

The macrofauna has an important and positive ecosystem role, controlling epiphyte abundance and filamentous algal growth. Overall, the productivity of the macrofauna is high and provides the major pathway by which energy is transferred from the primary producers (algae) to higher trophic levels. Caenogastropods have a much more diverse trophic role than herbivorous vetigastropods, and feed on a wide variety of sessile and sedentary organisms. The ecology of several species is discussed in detail.

INTRODUCTION

We noted in Chapter 1 that algae, whether as microscopic forms (phytoplankton) or as benthic macroalgae, are at the base of the food web and produce the energy needed to support all marine life, apart from hydrothermal vent and cold seep chemosynthetic communities. Similarly, algal consumers range in size from the tiny **meiofauna** to larger **macrofauna** to even larger **megafauna**; the term **epifauna** is used loosely to cover all or any one of them, depending on the context.

The meiofauna is often loosely defined as the small, mobile fauna that passes through a 1 mm mesh sieve, but is caught on a 0.1 mm mesh sieve; however, the definition is rather vague, because different workers have used different mesh sizes according to their specific purposes, so the boundary can range from ~0.5 mm to 2 mm at the upper end and 50 μm to 100 μm at the lower end. The macrofauna is then roughly defined as the mobile animals that are > 1–2 mm and < 10–20 mm size, and the megafauna as all the larger invertebrates. Here again the boundary between macro- and megafauna is vague, and the treatment of marginal groups in terms of size may be based on convenience. In this chapter, we shall mainly discuss the first two groups, drawing on a New Zealand study by Hicks (1986) for the meiofauna, and many Australian studies for the macrofauna. For convenience, we include small shrimps in the discussion because they are intimately associated with the macrofauna, although they are strictly part of the megafauna.

The macrofauna strictly includes motile and sessile forms. Motile forms are amphipods, isopods, tanaidaceans, small decapods, free-living polychaetes, gastropods, chitons, platyhelminths, pycnogonids and nemerteans, while sessile forms include hydroids, bivalves, bryozoans, ascidians, tubicolous polychaetes (e.g. serpulids) and small sponges. The sessile forms are considered in Chapter 13.

MEIOFAUNA

The meiofauna is mainly associated with algae, both on fronds and among holdfasts, but also occurs in sediments and in detritus in algal beds, on algal mats, and among sessile animals such as mussels and oysters. It includes diverse groups, such as nematodes, harpacticoid copepods, ostracods, rotifers, mites, tardigrades, polychaetes and a few other minor groups.

The meiofauna is important as food for many benthic macrofauna, and densities probably average 10^6 m^{-2} on many bottom habitats, so they are a dominant trophic pathway linking microbial life with the macrofauna. Their food, the microbiota (sometimes called periphyton), includes bacteria, diatoms and blue-green algae. Microalgae form a filamentous coating on rock and algae, and are also abundant in associated sediment. The meiofauna feed on this microbiota, variously by scraping, grazing, sucking or licking the surfaces on which they lie.

In Hicks' (1986) New Zealand study, harpacticoid copepods comprised 37%, ostracods 24%, and nematodes 18% of the meiofauna. Algal species with dense fronds or complex architecture tended to accumulate fine particles, and had the highest densities of meiofauna. In contrast, flat algal blades, such as those in *Ecklonia*, offered only flat surfaces from which to feed, and gave little protection from turbulence or predators; on these surfaces densities were much lower.

The major predators of the meiofauna include some macrofaunal species (e.g. flatworms, nematodes, polychaetes and copepods) and also small fishes (blennies, gobies and clingfish). Hence, the complex architecture of algae is important in reducing predation. Coralline and *Caulerpa* algal turfs and finely branched algae, as in some *Ulva* species (formerly *Enteromorpha*), are favoured by many forms of meiofauna.

Turbidity and fine sediment deposited in algal habitats may favour a few species of nematodes, copepods and ostracods, but they are generally detrimental to the meiofauna, as they are to larger forms, and reduce the productivity of the system. Nearly all meiofauna lack a larval dispersive stage, so swimming by adults from nearby habitats is the main way in which they colonise a new substrate. Nematodes are often the most abundant of the meiofauna.

CRUSTACEAN MACROFAUNA

The crustaceans of the macrofauna have sometimes been dubbed 'marine insects', because of their high abundance, small size relative to their hosts, and because they use their host plants as sources of food and as habitat (Hay *et al.* 1987) (Plate 15). The macrofauna in algal forests and turfs can reach densities of 23×10^6 m^{-2} for animals > 0.5 mm (Taylor 1998). In Graham Edgar's (1983a, b, c, d) Tasmanian studies: gammarid amphipods were the most abundant group, with 63% of all individuals present; caprellid amphipods were next with 21%, then polychaetes with 10%, and isopods with 4%; the remainder comprised < 3%. In Alastair Hirst's (2003) study in the eastern Great Australian Bight (GAB), gammarid amphipods were dominant, comprising 78% of the small crustaceans, followed by isopods 11%, tanaidaceans 8% and caprellids 3%.

Australian studies have shown that macroalgal beds and seagrass beds have many similarities as habitat for the macrofauna. Indeed, some groups of macrofauna – motile species such as amphipods – can move extensively (up to 30 m) between algal beds and from algae to seagrass. Even sedentary animals, such as molluscs, float or drift quite long distances (Edgar and Moore 1986; Edgar 1992). In all, the two habitats harbour epifaunas that are prolific in numbers and in species diversity. In this section, we concentrate on the epifauna of algae, but draw on studies in seagrass, where relevant, to illustrate particular points (reviewed by Edgar and Moore 1986).

Diet of macrofauna

Amphipods

The food of macrofaunal species is as varied as the habitats in which they live, and collectively they consume a wide variety of living and detrital plant material (Nagle 1968). Macroalgae are eaten by some gammarids, and epiphytes are favoured by many species. Edgar (1991a, b) found that artificial algae colonised by epiphytic (often filamentous) algae and diatoms attracted many more species and individuals than those without epiphytes. Some amphipods comb detritus or diatoms from epiphytes with the hairs on their antennae, and then remove them with their mouthparts; others pick up algal fragments with their gnathopods (anterior claws), and then eat the detritus. A few amphipods have very specific food preferences. *Bircenna* bores into fucoid algae while *Tethygeneia* and *Hyale rubra* eat the tissue of a *Zonaria*, but specifically choose parts of the plant that are not defended by deterrent secondary metabolites (Poore 1994). In preference experiments: the amphipods *Exampithoe kutti* and *Ampithoe ngana* preferred the brown alga *Dictyopteris acrostichoides*; *Hyale rubra* preferred a red alga *Pterocladia capillacea*; and *Hyale media* preferred sea lettuce *Ulva australis*, a green alga. The amphipod *Peramphithoe parmerong* also had a specific preference for particular host algae, mainly *Sargassum linearifolium*, on which it fed (Poore and Steinberg 1999; Taylor and Steinberg 2005).

Chemical deterrents are present in some algae and influence the diets and choice of host algae by herbivorous amphipods in the family Ampithoidae (reviewed by Poore *et al.* 2007). The 27 known Australian species in this family are relatively unspecialised in their diet, and many of them feed on several hosts. The amphipod genus *Peramphithoe* (discussed below) had the most restricted host range, and was the least likely to live on chemically rich hosts. Overall, chemically defended host algae were colonised by more amphipod species than chemically poor algae, suggesting that tolerance of such defended algae was an evolutionary advantage to the amphipod, because it enabled species to have a broader range of hosts. However, secondary metabolites, such as bromines in the red alga *Asparagopsis*, can affect choice of preferred hosts (Paul *et al.* 2006).

Caprellid amphipods

Caprellids are opportunistic feeders, mainly consuming detritus, but a few species scrape diatoms and dinoflagellates from the substratum or capture live prey (Keith 1969; Guerra-García and de Figueroa 2009). Some are suspension feeders, and are abundant in places of strong current, or in turbulent water, where they can capture the most food (Fenwick 1976). Hence, they suspend themselves high up on plants where water movement is strongest, and are virtually absent lower down towards the plant's holdfast, because the algal canopy baffles water movement and reduces the amount of available food there.

Isopods

Isopods have diets as varied as amphipods. They may be carnivorous, herbivorous, scavengers and even cannibalistic at high densities. The herbivorous

Haswellia carnea is a generalist, eating a range of green and brown algae, but its top three preferred species are the green algae *Ulva, Codium* and *Caulerpa* (Taylor and Steinberg 2005). Other isopod species are obligately associated with a seagrass species; for example, several boring species of *Lynseia* with *Posidonia* or *Heterozostera*, and a *Limnoria* with the seagrass *Amphibolis* (Brearly and Walker 1993). These species burrow into seagrass blades, form burrows and feed on the leaf tissue, ultimately leading to leaf loss. Similarly, two isopods *Phycolimnoria* spp. bore into the holdfasts and stipes of the kelp *Macrocystis* causing plants to become detached from the substratum (Edgar 1987).

Shrimps

Caridean shrimps, which are at the margin between macro- and megafauna, are significant as the main predators of macrofauna, but they also consume detritus. In Howard's (1984) study of four species in three genera, *Macrobrachium*, *Hippolyte* and *Chlorotocella*, polychaetes comprised 10–70% of their diet (depending on the shrimp species), amphipods, ostracods and mysids comprised 12–20% of the diet, small gastropods (to 16%), and plant detritus (5–70%). Hence, these shrimps have a key role as predators in algal and seagrass systems, strongly influencing the structure of macrofauna and transferring their productivity to higher consumers.

Spatial and seasonal patterns

The patterns and species of epifauna vary greatly both spatially from fine scales (e.g. location on an alga) to large scales (e.g. depth and exposure), and over time (hourly to seasonally). We will look at these patterns and then examine the interactions between the epifauna and their habitats.

Macroalgae, especially perennial canopy species such as the fucoids, can be colonised by epiphytes, which are mainly ephemeral algal species with a rapid turnover; these species settle, grow and reproduce within a season. Hence, the overall architecture of a plant community can be quite complex, with a canopy of upper stratum species with their own epiphytes, a middle stratum of green, brown or red algae (sometimes called an algal turf) to ~10–15 cm high, and a lower stratum of filamentous, prostrate or encrusting algae a few mm high. Macroalgae have been classified into many life forms according to their structure and architecture (Steneck and Watling 1982; Baker *et al.* 2008). In noting the structural differences relevant for macrofaunal colonisation, Edgar (1983a) recognised four basic forms – finely branched species (e.g. *Halopteris, Caulocystis, Acrocarpia* and *Caulerpa*); species with broader, flattened branches (e.g. *Zonaria, Seirococcus, Pollexfenia* and *Plocamium*), and the wide-bladed kelp species (e.g. *Ecklonia* and *Macrocystis*). Kelp holdfasts are unique because they provide complex habitat for a wide size range of macrofauna. Of course, species with intermediate structural forms also occur.

At the scale of a single plant, epifaunal size patterns differ according to algal architecture and distance along the length of the alga. In Graham Edgar's study, finely branched algae harboured a predominance of the smallest fauna, while algae with wider thalli harboured larger animals (Fig. 8.1). Algal species with intermediate architectural features had a broader range of faunal sizes. *Ecklonia* blades had very few larger faunal sizes, but curiously a large number of the smallest sizes – meiofaunal harpacticoid copepods, which swarmed among the blades. In contrast, *Ecklonia* holdfasts harboured a range of faunal sizes, due to their more complex form, providing a cryptic micro-habitat. Edgar (1983a) also found that all amphipods showed preference for living on particular host algae, although the same species occurred on most host plants.

Epiphyte abundance on all canopy species varied seasonally, being highest in late summer to autumn. This is illustrated by data for *Sargassum verruculosum* showing how macrofaunal numbers tracked epiphyte abundance (Fig. 8.2). Most of the common species of amphipods, molluscs, isopods and polychaetes graze epiphytes, and Edgar (1983b) attributed the increased abundances of macrofauna on plants with many epiphytes to the greater food supply. Similarly, the numbers of epifaunal predators, such as crabs, shrimps and nemerteans, also peaked in February–April, as they took advantage of the increased abundance of prey species. The summer increase in abundance of most animals was likely due to rapid breeding in the presence of

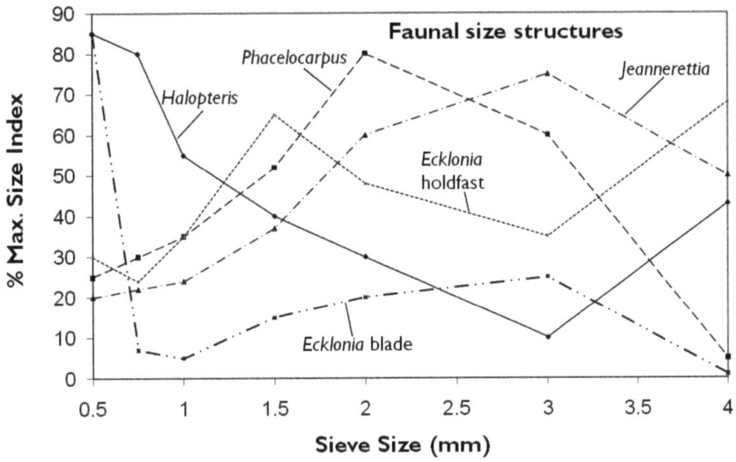

Figure 8.1: Size structures of the macrofaunas associated with different algal forms.

increased resources – smaller species can double population size every month – and, for larger species, migration from areas without epiphytes. The overall densities of the macrofauna peaked at > 30 000 m^{-2} in early autumn, due mainly to high abundances of amphipods and gastropods; polychaetes peaked a little earlier during summer (Fig. 8.3). The equally rapid decline in macrofaunal numbers in late autumn was apparently due to intense predation by carnivorous or omnivorous fish, notably little weed whiting (*Neoodax balteatus*), Castelnau's goby (*Nesogobius pulchellus*) and bridled leatherjacket (*Acanthaluteres spilomelanurus*) (see Edgar 1983b).

Environmental gradients and habitat features

The abundance and composition of the epifauna vary along environmental gradients and according to specific habitat features. Here we discuss some that have been reported in temperate Australasian waters.

Estuarine gradient

Edgar (1991b) used a standardised imitation 'habitat' (rope fibre) to examine the abundance of epifauna along the 25 km estuary at Port Davey, Tasmania. The gradient was quite complex, because numerous factors – salinity, nutrients, light reaching the habitat and water movement – all declined with distance up the estuary. Of these factors, light was critical for the

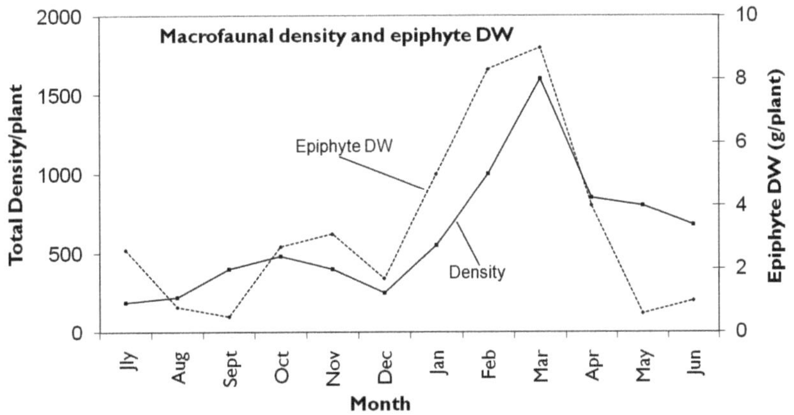

Figure 8.2: Mean seasonal abundance of epiphyte dry weight (DW), comprising mainly *Ceramium, Polysiphonia* and *Ectocarpus*, on *Sargassum verruculosum*, and mean seasonal distribution of macrofaunal density per adult *Sargassum* plant (average length 60–80 cm, but 80–120 cm March–August) (redrawn from Edgar 1983b).

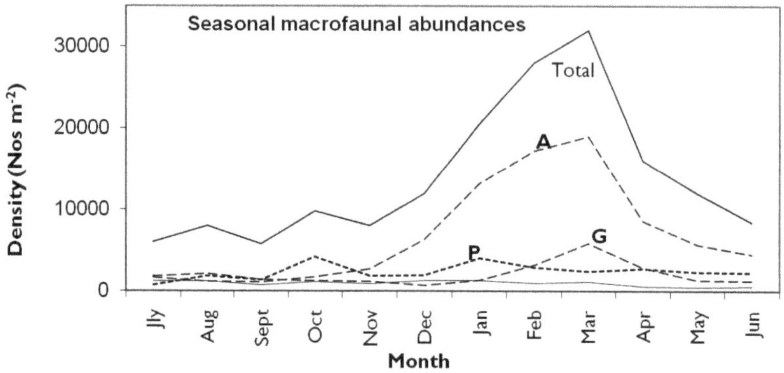

Figure 8.3: Seasonal changes in density of total macrofauna, and its components – amphipods (A), gastropods (G), polychaetes (P) and isopods, tanaids, mysids and cumaceans (bottom continuous line) (redrawn from Edgar 1983b).

epifauna because freshwater entering the estuary was darkly stained with tannins. The epifauna comprised mainly amphipods and gastropods, and the study found that the number of species at the head was only ~11% of that at the entrance (243 species). Productivity (see below) also declined steeply up the estuary, due mainly to the decline in available light reducing the growth of filamentous algal epiphytes and diatoms, on which the epifauna mainly fed.

Water movement and depth

Both of these factors are important but are often difficult to separate due to their interaction and because they entail other consequences, such as the amount of organic detritus, which increases with decreasing water movement (Edgar and Shaw 1993), and changes in algal assemblages along gradients (Shepherd and Womersley 1981). In exposed conditions, algae erode and liberate particulate matter and also dissolved organic carbon, so providing food for suspension feeders and detritivores (Edgar and Moore 1986).

Hirst (2003) examined the crustacean epifauna at a constant depth along exposure gradients in the eastern GAB. Epifaunal densities declined from ~1200 m^{-2} at sheltered sites to ~400 m^{-2} at the most exposed sites, due in part to declines in tanaids and algal-grazing amphipods. However, the algal assemblages also changed, so it was unclear whether these changes were due to changes in canopy species or available food. The former is more likely because canopy species with wide blades (e.g. *Ecklonia* and *Scytothalia*) tended to have relatively more isopods and fewer amphipods. The flattened body shape of isopods probably better enabled them to cling to broad algal blades, which dominated more exposed conditions.

In a study of the amphipod epifauna of *Caulerpa brownii* along an exposure gradient, Fenwick (1976) found the opposite trend. Amphipod abundances were highest (~225 000 m^{-2}) at the most exposed site, and lowest (~9000 m^{-2}) at sheltered sites (calculated on the assumption that *Caulerpa* biomass averaged 3750 g m^{-2}). *Caulerpa* patches are unusually rich in epifauna because they incorporate two microhabitats: a thick, basal, rhizomatous mat, which provides plenty of cryptic space among the rhizomes; and erect fronds above, which form a quite dense array and trap detritus. Nestling, tube-dwelling and inquiline species (those harmlessly associated with animals) were most common at the exposed sites. Fenwick attributed this abundance pattern to the rich, suspended food supply at exposed sites, and sediment deposition at sheltered sites (a negative effect). The wealth of epifauna living in *Caulerpa* patches often makes them a prime target for foraging by western blue groper (see Chapter 16).

Edgar (1983a) specifically examined changes in the epifauna along a depth gradient in SE Tasmania and found a 75% decline of amphipods from the intertidal to 4 m depth, and an increase in molluscs and isopods. These changes were related to declining water movement with depth, and the accompanying changes in algal species and structure (see above). Caprellid abundance was a good indicator of water movement because they preferred strong water movement conditions (see also Nagle 1968). In other interesting experiments simulating strong swells on the epifauna in *Amphibolis* beds, Edgar and Robertson (1992) found that free-living amphipod genera, such as *Tethygeneia*, *Paradexamine* and *Hyale*, were nearly all dislodged from the seagrass with strong water movement, whereas tube-dwelling groups, such as *Ampithoe* and *Cymadusa*, were little affected. Overall, water movement and depth gradients are very complex because they can also involve light and sediment gradients; hence, predictions are difficult, especially as species and predators vary – indicating the strong context-dependence of epifaunal abundances.

Habitat features – species matrix and plant density

In addition to the architecture of different species of algae, plant density and spatial arrangement (called the species matrix) may play a role in influencing epifaunal abundance. Poor quality plants or unsuitable habitat may affect the ability of herbivores to find a preferred host, and Poore (2004) found that one amphipod species was much less likely to find a good host if the host was surrounded by unpalatable algal species, but would settle on the latter if preferred hosts were rare.

Seagrasses, with a relatively simple structure, have been used in various studies to illustrate how plant density affects the macrofauna (for a global seagrass review see Jernakoff *et al.* 1996). Edgar and Robertson (1992) found that reducing leaf densities by half in one experiment, and removing epiphytes in another, reduced epifaunal densities by 61% and 31%, respectively. When plants were isolated and placed in the open, similar steep declines of ~60% in abundance occurred, suggesting that the epifauna actively selected dense habitats to avoid predation and possibly also intense sunlight. A similar study by Bell and Westoby (1986) gave similar results. Both studies also found that fish predators were not responsible for the decline in epifauna in the 'reduced leaf density' treatments. Although plant density per se seemed to be the key factor controlling macrofaunal abundance, could the greater food supply in denser seagrass have also been important? To test this, Edgar (1999) manipulated the detritus level by adding seagrass debris, and found that most epifaunal species increased substantially in biomass and productivity – so much so that he concluded that detritus far outweighed the structural importance of seagrass leaves and the trophic role of leaf epiphytes in terms of enhancing epifaunal abundance and productivity.

In another *Posidonia* seagrass study, Tanner (2005) found a strong 'edge effect' in which crustacean (especially amphipod and tanaid) densities were highest at the sand–seagrass boundary, declining with distance both inside and outside the boundary – a pattern he attributed to predation. If the *Posidonia* was surrounded by the much finer and denser seagrass *Heterozostera*, the results were different and, further, densities of amphipods and polychaetes in *Posidonia* were more than double those in *Posidonia* surrounded by sand (Tanner 2006). Moreover, the epifauna in *Heterozostera* was different from that in *Posidonia*, and also varied with patch size of either seagrass. Thus, the landscape context – the species mix – is an important determinant of the species that live in such habitats. These sorts of landscape studies on epifauna in algal habitats are yet to be done, but, as shown by studies of the holdfast fauna (see next section), they are certain to be equally complex.

The holdfast micro-habitat

Holdfasts of some kelps, notably *Macrocystis* and to a lesser extent *Lessonia* and *Ecklonia*, provide refuges for many small, motile animals among their root systems (known as haptera), which grow down from the stipe to give a conical shape to the holdfast. Some sessile animals, such as sponges, hydroids, bryozoans, mussels and ascidians, may grow on the haptera, and sediments may accumulate there, so that the holdfast structure over time develops many interstices of different sizes and forms a complex micro-habitat.

To test whether drifting kelps could raft species living in their holdfasts across oceans, Edgar (1987) examined the holdfast fauna of *Macrocystis pyrifera*, and suspended some of the kelps with holdfasts in the water column for 6 months. He recorded ~1200 individuals in 172 species per holdfast, with the most abundant individuals being amphipods (74%) and isopods (14%), with tanaids, echinoderms, polychaetes and gastropods each contributing ~2%.

Graham Edgar found that many species stayed in the floating holdfasts for the 6 months and concluded that drifting *Macrocystis* plants were a possible means of dispersal by rafting in the Southern Ocean, provided boring isopods did not eat up the kelp on the way!

Another study by Smith *et al.* (1996) on the smaller and less complex *Ecklonia* holdfasts at exposed headlands recorded a total of 386 species, and an average of ~490 individuals per holdfast (maxima of ~1900 individuals in the largest holdfasts). The proportional abundances of different groups were similar to those in *Macrocystis* holdfasts, with the amphipods and polychaetes together comprising 78% of the individuals and 43% of all the species. The trophic composition of this rich fauna was: omnivores 48%; carnivores 26%; suspension feeders 12%; deposit feeders 10%; and herbivores 3%. Most of the holdfast species also lived in surrounding habitats, indicating that the selection of species was not unique, even if the microhabitat had unique features. The average size of animals and number of species increased with increasing holdfast size and age. The overall successional trend in holdfasts – as the fauna aged and grew larger – was from dominance by small polychaetes and bivalves to a community of amphipods, decapods, large polychaetes, and some ophiuroids and sea cucumbers. In trophic terms, this meant an increase in omnivores and a decrease in deposit feeders. In another study, Smith (1996) examined the effect of trapped sediment and sponges, which accumulated or grew over time. He found that carnivores (mostly polychaetes) had a positive relation with the amount of sediment, omnivores (mostly amphipods) a negative relation, and the five common polychaete deposit feeders a positive relation. Carnivores had a positive relation with sponges, but deposit feeders and omnivores a generally negative relation.

In another study, Smith and Simpson (1993) examined the effect of sewage effluent on the holdfast fauna of *Ecklonia*, and noted an increase in suspension feeders (mainly bivalves) and a decrease in crustaceans, polychaetes and ophiuroids on holdfasts near the outfall compared with controls. Hence, monitoring the fauna of *Ecklonia* holdfasts was a cost-effective surrogate for measuring pollution effects in coastal waters.

In other studies at a calm-water South Australian site, faunal densities in *Ecklonia* holdfasts averaged 90–160 individuals per holdfast (depending on the experiment), with a total species diversity of 81 species (Goodsell and Connell 2002, 2005; Goodsell *et al.* 2004) – much lower than those in Smith and colleagues' study in New South Wales. Crustaceans comprised 58% and polychaetes 36% of individuals, perhaps due to the higher levels of detritus; hence, there were more deposit-feeding polychaetes at the South Australian site. Like Poore's (2004) study on a mobile amphipod, these studies set out to examine the effect that (a) different distribution patterns of *Ecklonia* forests (pure stands; *Ecklonia* clumps among fucoid algae; and solitary plants), and (b) disturbance (i.e. reducing plant density) had on the number of holdfast species.

They found that in general more species and more individuals lived in holdfasts in dense *Ecklonia* forests than in sparse ones, and that the removal of *Ecklonia* plants resulted in more species crowding into the remaining holdfasts. The latter effects were stronger in naturally sparse forests than in dense ones, suggesting that the effects of disturbance depended on the history of the algal forest and how severe the disturbance was. Because the species in holdfasts appear to be quite mobile, it seems, perhaps not surprisingly, that the responses shown were due to changes in movement patterns of individuals among holdfasts, and that rare species were more affected than common ones, as suggested by the authors.

In a follow-up study, Goodsell and Connell (2008) experimentally eliminated all macrofauna from holdfasts and then measured rates of recolonisation in different situations – sparse and dense *Ecklonia* forest, and forests manipulated by 'planting' large *Sargassum* and crushed *Sargassum* (to detect algal deterrent effect) in the habitat. The low-density holdfasts attracted more colonists, supporting the 'nearest refuge' model, which predicts that mobile, dispersive species will concentrate more in isolated habitats than in abundant ones. Adding *Sargassum* plants, crushed or whole, to the forest, however, made little difference.

In another study in SW Australia, the macrofaunas of shallow (~5 m) and deep (~20 m) *Ecklonia* holdfasts were compared on a high-energy coast (Coleman *et al.* 2007). The shallow holdfasts averaged 60–90 individuals per holdfast – mainly amphipods, isopods and polychaetes – which was three to six times the number in deep holdfasts. In all, there were 160 species, with little overlap of species between shallow and deep water.

The above studies show that the large *Macrocystis* holdfasts have a much richer and more diverse macrofauna than those of *Ecklonia*, which in turn differ greatly between sites and regions. The macrofauna appear to be as mobile as the epifauna on algal blades, so may vary according to forest type and density, depth, amount of detritus and trapped sediment, and probably other factors as well.

Algal drift

Storms and wave action tear macroalgae from their substrata, and the detached algae so produced generally lie on the seabed until moved along by currents or wave action. Some are washed ashore, producing the wrack commonly seen on the beach, but most drift about on the sea bed until, over time, they break down into detritus. Yet, some species of deeper water red algae continue to photosynthesise and seem to thrive as drift in bays. Examples are the beds of *Hennedya crispa* in bays in the eastern GAB (Shepherd and Womersley 1976), and the *Gracilaria* beds, with some *Ecklonia*, in Jervis Bay, New South Wales (Langtry and Jacoby 1996). These beds no doubt have an epifauna, as evidenced by their resident, predatory fish fauna. Close to shore, drift algae often accumulate, and here the amphipod, *Allorchestes compressa* is very common, showing a strong preference for decaying *Ecklonia* tips, although it also feeds at slower rates on *Sargassum* and other algae (Robertson and Lucas 1983). Generally, these habitats largely remain unstudied.

Predator effects

Some of the above studies have implicated, in one way or another, the role of predators in affecting macrofaunal abundances. The main predators in these algal or seagrass systems are fishes, small crabs and shrimps. In some systems, fishes have a strong effect (Edgar and Moore 1986), whereas in others, shrimps appear more important (Howard 1984). But is there evidence for a dominant role of predators in structuring the epifauna?

In dense algal beds, predators have low prey-capture efficiency, as shown by studies of Bell and Westoby (1986) and Edgar and Robertson (1992), but, conversely, they can have high efficiency in open habitats (Kennelly 1983). Edgar (1983d) showed, for example, that more (and larger) amphipods were captured by fish in the open branches of *Cystophora* compared with the more compact *Zonaria* plants. This finding was confirmed in microcosm experiments in which the size structures of the epifauna showed that most of the largest individuals were eaten, while the smallest remained intact (Edgar and Aoki 1993; Edgar 1994). In other experiments, Irving *et al.* (2007) showed that predatory shrimps had marked effects on the epifauna as well as on epiphytic plants and animals.

Grazer effects on the habitat

Another question that sometimes arises is whether macrofaunal grazers have the potential to modify benthic community structure or reduce algal abundance. To examine this question, Poore *et al.* (2009) mixed an insecticide in plaster of Paris, so creating a slow-release insecticide as the plaster gradually dissolved; this successfully kept out almost all crustacean macrofauna from an algal bed. Despite the steep reduction in epifaunal abundance, the growth of macroalgae, epiphyte cover and the algal assemblage were unaffected – in contrast to the potential effect of sea-urchins (see Chapter 12). However, this study does not exclude the possibility that in other contexts higher background densities of epifauna could have an effect, as reported by Howard (1982). In the study of Alastair Poore and colleagues in Sydney Harbour, fish predation on epifauna was low, so natural epifaunal densities were unlikely to ever be high enough to have an effect on the algal vegetation.

A case study on the ecology of an amphipod

A series of studies on the amphipod *Peramphithoe parmerong* illustrates the fascinating complexities of behaviour that can occur in a small herbivore (Poore and Steinberg 1999). This amphipod is common around Sydney Harbour in shallow algal beds, where it builds nests with algal fragments using a sticky secretion as adhesive. It lives in the nest and feeds on algae around its entrance. The sexes are distinct and the female becomes mature when about 1 month old and, straight after moulting, lays eggs, which are then fertilised in her brood pouch by the male. When the eggs hatch, the juveniles stay in the pouch for some days, to be eventually evicted when the mother moults and lays more eggs. However, the juveniles tend to remain close to their mothers for life. She is reproductive all year, and this rapid life history facilitates the study of the growth, survival and fecundity of the species on different diets.

The amphipod has marked preferences for building its nest in particular algal habitats. The order of preference for algal hosts as micro-habitat was:

Sargassum linearifolium, S. vestitum > *Colpomenia, Padina* > *Dictyopteris, Dictyota, Dilophus, Zonaria* (the last group of Dictyotales being largely avoided).

Feeding preferences were very similar to this preference order for hosts, and in algal preference trials in which eight species were on offer, the amphipod ate *Sargassum linearifolium* and almost nothing else (Taylor and Steinberg 2005). Avoidance of the Dictyotales group was due to the presence in them of chemical deterrents (non-polar secondary metabolites), but the amphipod was not deterred by the presence of another chemical deterrent, phlorotannins (a polyphenolic), present in *S. vestitum*. Poore and Steinberg (1999) then measured the performance of individuals fed the different species, by rearing their offspring. They found that the amphipod was not simply choosing the most palatable algae (i.e. those that were low in chemical deterrents), but those that gave the highest growth and survival to their young. Hence, population fitness is maximised if the parents choose the best algal host as habitat for their young. Survival, growth and fecundity (number of eggs produced) of this amphipod followed the same order as food preference, and the number of grandchildren (as eggs) derived from a single brood could be readily calculated for different diets. The preferred food in decreasing order of preference (with number of grandchildren per brood) was:

Sargassum linearifolium (104) > *S. vestitum* (81) > *Colpomenia* (44) > *Padina* (29)

These startling results indicated the strong effect of host plant diet on subsequent generations. But was the preference for a particular host plant a heritable trait? Poore and Steinberg (2001) answered this by separately rearing the brood of *Peramphithoe* on the high-quality *Sargassum* and the low-quality *Padina*.

They found that the genetic traits of survival and growth were heritable (with heritability estimated at 0.23), but, despite this, there was no evidence of local adaptation to a particular host alga. Gene flow between populations was probably enough to prevent local adaptations arising, especially as *Padina* became seasonally very rare and colonists must have then migrated to other algae. Despite the amphipod's strong preference for its maternal host plant, it showed a surprising ability to migrate for distances of > 1 m (Poore 2005). But a puzzle still remained as to why some *Peramphithoe* persisted in living and feeding on *Padina*, despite its poor quality as food. Poore and Hill (2006) then went on to show that the amphipod actively sought a mixed diet, and had a strong tendency to switch hosts from time to time. Hence, a few of them would always be found on poor-quality hosts. In summary, this herbivore, like many other generalist herbivores, performs best on a mixed diet.

Macrofaunal productivity

A significant advance in macrofaunal research in the last two decades has been the development of a way to estimate macrofaunal productivity, which is the rate at which biomass is produced in a group of animals (Edgar 1990). This method requires macrofaunal samples to be sieved with different mesh sizes, and numbers and weight of each sieve-size group determined. Generalised equations relating production, biomass and temperature (derived from 41 published studies) can then be applied to any dataset to estimate daily production.

The measure has turned out to be much more useful than other community measures because it can be applied generally to investigate food web processes in community ecology. Edgar (1993) applied the measure to many shallow-water habitats and showed that epifaunal production was constant for many algal and seagrass habitats worldwide, indicating that food resource ceilings are a major constraint limiting epifaunal abundances. In New Zealand Taylor (1998) also used the same measure to compare the productivity of four habitats – a structurally complex fucoid forest, an *Ecklonia* forest, sea-urchin barrens and a coralline algal turf (Table 8.1). Complex habitats were manifestly far more productive than simple ones, and the epifauna was by far the most important contributor to the secondary production of rocky reef habitats, except in urchin barrens, where sea-urchins were relatively more important. In that study, consumption of epifauna by fish amounted to ~26% of the epifaunal production. This example illustrates the power of the technique to estimate the productivity of different habitats in coastal food webs.

Animal-associated epifauna

Many epifaunal and other species use larger sessile marine invertebrates, such as sponges, echinoderms, large crustaceans, cnidarians and ascidians, either solely as habitat – termed commensalism (i.e. 'messmates') – or as both habitat and food – termed parasitism. The copepods, amphipods and others that live in these poorly known associations are, as stated by Gotto (1979), 'mainly unobtrusive forms, their ecology clandestine, and their ecologic significance limited … curious types generated in odd moments of evolutionary whimsy'.

Amphipod tenants may live in almost any hiding place – under the shells of molluscs, such as abalone; within hermit crab shells (see Chapter 15); in the mantle cavities of bivalves; in the siphons of ascidians; within sponge tissue; or with soft corals and cnidarians. Caprellids may live on sponges and bryozoans, and are especially abundant on hydroids on which they predate. In one study, Poore *et al.* (2000) recorded 56 amphipod species on sponge and soft coral hosts in Botany Bay. Some amphipods were specific to one host, whereas others were less specialised in their host and feeding preferences. Some tenants may even use the host as protection against predators or, as in the study of Underwood and Verstegen (1988), just for shelter and to avoid desiccation. In that study, up to 20 amphipods, all *Hyale media*, crowded together under a large limpet *Cellana tramoserica* during low tide.

Some species live in the pharynx (oral siphon) of ascidians where they eat particles from the mucus sheets lining the walls, while others are more committed to parasitism, feeding on mucus and host tissue. In a study of the symbionts of the ascidian *Pyura praeputialis* (formerly *P. stolonifera* – see Chapter 13) in Bass Strait, Dalby (1996) found a copepod (*Doropygus*), an amphipod (*Paraleucothoe*) and a nemertean worm (*Gononemertes*) all living and breeding inside the host. The same or a related species of *Paraleucothoe* also lived in this ascidian elsewhere in SE Australia. Similarly, related species of copepod and nemertean live in the ascidian *Pyura pachydermatina* in New South Wales (Egan 1984a,b). Nemerteans (see below) are voracious carnivores or scavengers, feeding on protozoans, diatoms, mucus, phytoplankton, and even eggs attached to the lobster *Jasus edwardsii* (Gibson 1997).

Other invertebrates, such as polychaetes and molluscs, may also be obligately associated with other animals. The marginellids (Family Marginellidae;

Table 8.1: Mean epifaunal density (no. > 0.5 mm sieve size m^{-2}), biomass (g ash-free dry weight (= AFDW) m^{-2}) and productivity (mg AFDW m^{-2} d^{-1}), and percentage contribution to productivity by epifauna, mega-invertebrates and fish in four habitats (extracted from Taylor 1998).

	Epifauna			Productivity (%)		
Habitat	**Density**	**Biomass**	**Productivity**	**Epifauna**	**Megafauna**	**Fish**
Fucoid	20 301	27.2	472.0	97.9	1.2	0.8
Ecklonia	71 122	11.2	174.2	83.9	11.5	4.5
Urchin barrens	24 428	4.7	63.9	54.5	35.2	10.3
Coralline turf	159 080	25.3	413.6	98.5	0.7	0.9

Gastropoda) are notable in this respect. Some are parasitic on bryozoans, such as *Microginella*, and others are associated with ascidians, bivalves and even sleeping fish (Ponder 1998a).

Mysids (Order Mysidacea)

Mysids are highly mobile, shrimp-like crustaceans up to ~25 mm long that live in vast numbers in southern coastal waters. Some 20–30 species are found associated with reef habitats, while others live in estuaries or on sandy bottoms, or are wholly pelagic in oceanic waters (Mauchline 1980; Keable *et al.* 2003). They are closely related to amphipods, isopods and cumaceans, but differ markedly from them in morphology and social behaviour.

Most species aggregate for at least part of their life cycle into shoals (defined as very large groups) or swarms (defined as smaller constituent parts of shoals). The primary function of this social behaviour is protection against predators; a secondary function is thought to be that it helps to keep them in a preferred zone. Some species are carnivorous, some browse on the epifauna of macroalgae, some feed on detritus, and others are filter-feeders, consuming diatoms and dinoflagellates.

Females carry embryos in a pouch (called a marsupium) and brood them there for 1–6 weeks through their several larval stages, depending on temperature. Eventually the larvae emerge as miniature adults, without having any pelagic larval stage.

Studies on mysids in Australia have been nearly all done in Tasmania by David Ritz and his students. Here, we describe the ecology of three species common in the waters of Tasmania and Bass Strait to 82 m depth, and studied at Bruny I., Tasmania by O'Brien and Ritz (1988), O'Brien (1987, 1988a, b, 1989), Johnston *et al.* (1997), Fenton (1992, 1996a, b), Metillo and Ritz (1993, 2003) and Metillo (2007).

Three Tasmanian mysids

Life history

The three most common species at the Bruny I. study site, out of 14 species present, were *Paramesopodopsis rufa, Anisomysis mixta australis* and *Tenagomysis tasmaniae*, each occupying a slightly different habitat. *P. rufa*, an orange-red species, swarms among and above the kelp, and also in sand patches. The second species *A. mixta australis* lives at the rock–sand interface, and the third species *T. tasmaniae* also lives at the interface between rock and sand, but in areas where disintegrating drift algae collect on the bottom. All three species occur in swarms within half a metre of the bottom, but *P. rufa* is most abundant in spring, with densities of up to 400 m^{-2}, *T. tasmaniae* in early summer and winter (peak densities 900 m^{-3}), and *A. m. australis* peaks in autumn with densities of 600 m^{-3}. The first two species breed throughout the year, but the third *A. m. australis* breeds in all seasons except winter. Each species appears to produce three generations a year. Brooding of larvae by all three species lasts from 14 to 28 days, with *P. rufa* taking 4–6 days longer than the others. Summer broods by all three species take about a week less to emerge than winter broods. Life history and dietary data for the three species are compared in Table 8.2.

Diet

Substantial niche differences in diet were evident between the species (Table 8.2). *P. rufa* was more carnivorous than the other two species, and fed partly on copepods and other small crustaceans. In capturing prey, *P. rufa* suddenly shoots out its short frontal legs (the endopods) at speeds of 6–16 mm sec^{-1}, which is fast enough to capture small, slower-swimming prey. All three species consumed large amounts of particulate matter (algal detritus), either suspended in the water or on the surface of the seabed.

Behaviour

Studies of mysid schooling behaviour and response to predators (O'Brien 1988a; O'Brien and Ritz 1988) showed that the above species tended to be polarised in one direction, especially during increased current flow when they oriented into the current. At the approach of a predator, the swarm moved away from the source of danger and became compressed as individuals packed more closely together. If the predator was within striking range, the whole swarm swam faster or split into smaller groups and, on an attack, individuals use a tail-flip avoidance reflex or sought

Table 8.2: Average life history, productivity and dietary data for three species of mysid at Bruny I., Tasmania.

The Production:Biomass ratio is the annual turnover rate.

	Tanagomysis tasmaniae	***Anisomysis m. australis***	***Paramesopodopsis rufa***	**Ref.**
Growth rate: larvae/ juvenilesSub-adult/adult	2.15 0.85	2.02 0.97	2.02 1.87	Fenton (1996a)
Life span of largest size class (months)	9.4	7.7	7.3	Fenton (1996a)
Swarm biomass (mg m^{-3})	~14	!5.5	~2.8	Fenton (1996a)
Production (mg m^{-3} $year^{-1}$)	~75	~22	~29.5	Fenton (1996a)
Production:Biomass $year^{-1}$)	5.4	7.6	5.4	Fenton (1996a)
Diet (% composition)	Detritus ~90	Detritus ~92	Detritus ~75 Crustaceans ~20	Fenton (1996a, b) Metillo and Ritz (1993)
Food selectivity	*P* > D; Z > D	*P* > D > Z	P ≥ D > Z	Metillo *et al.* (2007); Metillo and Ritz (2003)
Feeding activity	Nocturnal	Uncertain	Daytime	Fenton (1996b)

Food selectivity is for not-starved mysids, and symbols are: *P* = phytoplankton; D = algal detritus; Z = zooplankton.

cover by clamping down onto the substrate or an algal frond. However, stragglers are more often attacked than those in a swarm, so in general it is better for the individual to remain in the swarm.

Later Ritz *et al.* (1997) examined the response of mysids to two different styles of predator attack. These are 'lunging', when a fish, such as salmon, swims rapidly at a school, and 'ambush', when, for example, a slow moving seahorse selects and flicks its head to capture a mysid. In the former type, an energetic flight response ensued as mysids sought to escape, with tail–flipping a common response (see below). In the latter type, the seahorse elicited no response, and the mysid swarm simply parted to allow the seahorse through. In all, the array of mysid responses to predators seems as complex as those of schooling fishes.

Associations

A curious association between mysid swarms and larval fishes is sometimes seen on coastal reefs. Sandfish (Family Leptoscopidae) and clingfish larvae have been observed by Barry Bruce (*pers. comm.*) swimming with mysid swarms. In one case, the sandfish larvae swam with and mimicked the colour of a reddish mysid species. Bruce also found that some sandfish larvae consumed small mysids, indicating that the association was in part predatory and of no apparent benefit to the mysids.

Predators

Fish are the main predators of mysids, and at the Tasmanian site four fish species out of 20 present consumed mysids. These were Australian salmon, yellow-eyed mullet, cardinal fish and the purple wrasse. Curiously, the majority of prey mysids were not the three discussed above, but a fourth less common species *Tenagomysis australis.* Elsewhere, the scorpionfish *Centropogon australis*, the rock flathead *Platycephalus laevigatus*, many cryptic fishes such as three-fins, gobies, blennies and weedfishes, and pipefishes have been recorded as predators (Fenton 1996b).

In reviewing the role of mysids in coastal ecosystems, Gwen Fenton concluded that they were highly important for two reasons:

- Coastal mysid species form a substantial part of the diet of many exploited fish species, and especially of juvenile fish that use shallow coastal waters and estuaries as nurseries.
- Mysids have a significant role in the turnover of macroalgal biomass, and on the microorganisms coating disintegrating algae. This is indicated by

their very high abundance, their feeding preference for algal detritus and their high productivity as a consequence of rapid growth rates and short lifespan.

Krill or euphausiids (Order Euphausicaea)

Krill are small crustaceans resembling mysids, but are more closely related to decapods. Most of the > 80 species worldwide live in oceanic waters, but *Nyctiphanes australis* occurs in extremely high abundance in SE Australian coastal waters from the eastern Great Australian Bight to northern New South Wales, and around Tasmania. It prefers cool temperate conditions with an optimal temperature range of 12–18°C. Subsequent to the early work of Sheard (1953), this krill species has been much studied by David Ritz and his students (Morgan and Ritz 1982; Ritz and Hosie 1982; Hosie and Ritz 1983; O'Brien 1986, 1987, 1988a, b, 1989; Dalley and McClatchie 1989; Ritz *et al.* 1990).

Life history of *Nyctiphanes australis*

This species has a 1:1 sex ratio. The female retains eggs in egg sacs until they emerge as free-swimming larvae, which pass through six stages before transforming into juveniles. After ~3 months juveniles reach adulthood and start breeding. During the breeding season, which lasts all year except from late autumn to winter, the female continually produces eggs totalling up to ~1000 during her life. Individuals grow to a maximum size of ~20 mm, and moult every few days. In southern Tasmania, three generations are produced per year: the first larval batch emerges in late August, and these produce a second generation by early December, with a third generation produced in March–April. Longevity is likely 1–2 years.

Diet

Krill are opportunistic omnivores/detritivores. During the larval stages they feed mainly on detritus, gradually shifting to a diet that includes sponge spicules, diatoms, dinoflagellates and crustaceans (mainly exuviae from moulted shells). The krill gather detritus during benthic feeding by day, and at night rise above the bottom, where they swim in shoals with their feeding baskets open to take up more detritus and other suspended particles.

Behaviour

The swarming capacity of krill is notable, with vast schools of up to 1 million m^{-3} recorded in Tasmania. Swarming near the surface is common in all seasons along the SE coast of Australia from Bass Strait to Robe, South Australia, and individual swarms covering an area of up to 1 ha are common. Swarms tend to be smaller off Tasmania and are most visible during calm seas (Gill and Morrice 2003). During swarming, all individuals swim polarised in the same direction at speeds of up to 10 cm sec^{-1}. Swarming occurs sometimes by day and often at night, and appears to have multiple reasons. Sub-adults appear to swarm for feeding, while adults swarm sometimes for feeding, but mainly for reproduction (i.e. to prepare for mating to aid in mate recognition, and then for mating itself). One intriguing aspect of swarming behaviour is this krill's unique strategy of 'matting' in which masses of krill lie together on the bottom, whether on reef or sand, under-side up, forming a red carpet that may last for many hours. O'Brien *et al.* (1986) thought the odd behaviour was probably functional, perhaps to prevent the whole school stranding on the shore as occasionally happened when the school found itself in too shallow water.

The social swarming behaviour of krill has many similarities to schooling in fish, indicating that similar selective pressures have shaped their evolution. Ritz (1994) reviewed krill swarming behaviour, and showed that:

- Swarms had higher feeding success because they were more successful at finding patchy food.
- When threatened by a predator, the swarm becomes more tightly cohesive, and this increases survival through increased vigilance (the 'safety in numbers' principle), although food capture suffers in the process.

Yet, the benefits and costs may conflict, and Ritz (1994) summarised the trade-offs as: 'Swarms and schools must be viewed as groups of social but potentially selfish individuals which continually monitor the costs and benefits of remaining within the group … If the swarm is so large that the returns per individual fall below some threshold level, it would pay some individuals to leave and for the swarm to split.'

Predators

Nyctiphanes has many predators, including marine mammals, birds, fish, squid and other invertebrates. Filter-feeding whales have long been known to feed on krill since the early descriptions of the whaler and naturalist William Scoresby (1820):

> *'When the whale feeds, it swims with considerable velocity below the surface, with its jaws widely extended. A stream of water consequently enters its capacious mouth, and along with it large quantities of water insects; ...the food is entangled and sifted, as it were by the whalebone, which from its compact arrangement, and the thick internal covering of hair, does not allow a particle the size of the smallest grain to escape.'*

For a full discussion of whale feeding see Chapter 18 and Gill and Morrice (2003).

The many small predators of krill include the short-tailed shearwater or mutton bird, which can dive to 20 m depth and visually select swarms of prey krill. The shearwater follows a zigzag course, darting to the left and right with beak open to capture krill with its oral filtering apparatus. Other bird predators are the little penguin and fairy prion. Fish predators range in size from pelagic southern bluefin tuna, barracouta, salmon and tiger flathead, down to pilchards and bottom-living hardyheads, yellow-eyed mullet, gobies, snake blennies and weedfishes. Barracouta are known to work together in groups and herd the krill into a packed mass.

In a study on the influence of swarming on capture success, Morgan and Ritz (1984) found that a salmon predator had a higher capture success at low krill densities than at higher densities – so supporting the view that swarming increases the likelihood of avoiding a chasing predator. However, they noted that this model may not apply to predators that have a filter-feeding mode of krill capture.

Krill have several specific escape responses from predators:

- A primary response for a school is a 'flash expansion' in which individuals spread out, or alternatively 'splitting' where they divide into smaller swarms. These can be effective against whales, seals and penguins, especially if the prey detects the predator first, because then detection of a small swarm can be very difficult against a hazy background.
- Another response is a coordinated tail flip by those nearest to the predator – which may confuse it.
- An individual under attack can tail-flip several times, and later re-join the swarm in a concerted escape.
- Bioluminescence is another defence, and momentary flashes of light emitted from the krills' photophores may startle, confuse or even temporarily blind the predator (see Ritz 1994)
- Swarms continually change shape, direction and depth, so making it difficult for a fish to track their targets before a strike.

One documented effect of a changing climate in SE Tasmania is the decline in krill, as the East Australia Current has extended further south in recent decades. This has caused cascade effects in the ecosystem and the near collapse of some fisheries such as jack mackerel (Johnson *et al.* 2011).

WORMS

We now return to a benthic group, briefly mentioned above in discussing *Ecklonia* holdfast fauna. Several common groups of invertebrates, loosely described as worms, live in reef habitats, either in crevice or cryptic sites, such as holdfasts, and under rocks. They occur in many shapes and sizes, sessile and mobile, and are classified into 10 or more phyla. The ecology of very few species is known.

Flatworms (Phylum Platyhelminthes)

Polyclad flatworms are the most primitive and simply organised of marine worms. They are flattened animals, oval shaped and up to a few centimetres long and have mucus glands that secrete mucus, enabling them to lay down a slime trail and glide over the bottom; they usually live in cryptic habitats, such as under boulders or among sessile fauna (Prudhoe 1982; Newman and Cannon 2003). Flatworms are all hermaphroditic with both male testes and female ovaries scattered over the body. They are armed with a stylet or spine, and during copulation some species inseminate their partner by stabbing the stylet through the partner's skin and injecting sperm; they may even engage in preliminary sparring with their stylets! Other species copulate more conventionally tail to tail, each delivering sperm to the other's female genital system. The sperm circulate through the body and fertilise eggs, which are then aggregated in capsules and laid on a firm surface.

Life history

The life history of only one species is known – that of the southern flatworm *Notoplana australis* – examined by Anderson (1977). This is a common species on reefs to ~15 m depth on the south and eastern Australian coast. It lays egg masses all year round, but mainly in spring and early summer. The egg mass is a coiled, gelatinous string of hundreds of egg capsules, each containing a single egg. The egg hatches after 12–14 days into a tiny, flattened larva that swims near the surface for a week, then near the bottom for 2 weeks, and metamorphoses and settles on reef.

Food and feeding

Polyclads are all carnivorous and feed on sessile animals such as bryozoans, soft corals, compound ascidians and shellfish. Several species feed on cultured oysters, causing much economic loss, hence their common name 'oyster leeches'. For example, in Coffin Bay, South Australia the polyclad *Enterogonia orbicularis* commonly predates on cultivated Japanese oysters (Prudhoe 1982).

In Port Phillip Bay, the flatworm *Stylochus pygmaeus* is a common species that commonly feeds on the barnacles *Balanus anphitrite* and *Elminius modestus* (Merory and Newman 2005). To enter the barnacle, the flatworm first releases copious amounts of mucus, which apparently paralyse or suffocate the barnacle, as well as aiding in digesting the tissue. The flatworm then inserts its pharynx between the opercular plates and sucks out the contents. This species engages in reciprocal mating with its partner, and the resulting egg masses are laid in dead barnacle shells. The eggs hatch after 5–6 days into larvae, which metamorphose into juvenile flatworms after another week or so.

Polyclads in the genus *Imogine* are common in Botany Bay and Port Stephens, New South Wales, where they variously predate at night on scallops, mussels, pearl oysters and barnacles (O'Connor and Newman 2001; Lee *et al.* 2006). One of the species *I. mcgrathi* consumed pearl oysters at a rate of ~1 a month, with large (> 6 cm) flatworms preferring larger oysters, whereas *I. lateotentare* ate barnacles and, like *S. pygmaeus*, laid thousands of eggs in the empty barnacle shells.

Most temperate flatworms are dull in colour, but a few species have striking colours or patterns. Such bright designs suggest that these species may be toxic, and advertise their toxicity with their warning colouration (termed aposematic), like many of their tropical–warm temperate relatives (Ang and Newman 1998).

Polychaetes

The features distinguishing polychaetes from other worms are their segmented body, typically with a well defined head, and sensory or feeding appendages, called parapodia, on each segment. Some are free living and others sessile, while some live in tubes made of sand, shell fragments or mud–calcium carbonate in the case of the serpulids. Tubes may be attached to hard substrates or stand erect within the sediment. The magisterial work edited by Beesley *et al.* (2000) summarises knowledge of Australian polychaetes. The ecology of very few species is known, and here we discuss two sessile species – a fan worm and a coral worm – and a group of burrowing mudworms.

The European fanworm *Sabella spallanzanii*

This species is common in the Mediterranean, although there it never reaches the densities that have been reported from Australia (Plate 16). It was introduced to southern Australian ports, probably on the hulls of ships, during the 1960s–80s, and is now a dominant species in Cockburn Sound, Western Australia, and Port Phillip Bay, Victoria (Carey and Watson 1992; Clapin and Evans 1995; Currie *et al.* 1999, 2000). In Port Phillip, it first appeared in Corio Bay (although populations have since declined), and then spread clockwise around Port Phillip Bay, as larvae were transported by the current and adults were dispersed by trawlers. The fanworm is also present in numerous harbours or marinas in the South Australian gulfs (Wiltshire *et al.* 2010).

The fanworm colonises both reefs and soft bottoms in sheltered waters to ~30 m deep, and forms clustered colonies of tubes sometimes at densities of up to 250 m^{-2}. Distinctive, spiralling tentacles emerge from the tube to capture suspended matter and plankton in the size range 2–40 µm. The worms are male or female with a 1:1 sex ratio, and reach sexual maturity at a body size of ~5 cm. Free spawning in Port Phillip Bay occurs in March–May annually, with large females shedding > 50 000 eggs into the water. Larvae are non-feeding and settle after ~2 weeks (NIMPIS 2011).

Sabella grows ~15 cm a year and reaches a length of up to 40 cm high, so maximum longevity would appear to be ~3 years. A possible predator of the worm is leatherjackets (at least in aquaria), and another bizarre one is the water rat, which dives off jetty piles to capture them in Port Phillip Bay (DR Currie *pers. comm.*); there are no other known predators. However, most animals avoid this species, because the body and the gametes contain saponins, which provide some protection from predation.

An interesting question about *Sabella* is whether or not it is a serious pest (i.e. adversely affects the ecosystem). Holloway and Keough (2002a, b) first noted that worm colonies reduced water movement under the canopy, and reduced the amount of larvae and micro-plankton that might otherwise colonise the bottom or provide food for filter-feeders there. Their experiments on the effects of *Sabella* colonies on hard and soft bottoms found that in the short term *Sabella* affects recruitment of many epifaunal species – some were inhibited, but others favoured. Growth rates and survival of some species were also reduced. There was no effect on polychaete worms or other infauna, but the epifauna of small crustaceans (< 1 mm size) was reduced in abundance (O'Brien *et al.* 2006). Overall, in the short term (a few weeks) effects were mainly negative, but in the longer term (> 6 months) the differences in reef and soft bottom habitats with and without *Sabella* became minimal.

On the other hand, Carey and Watson (1992) and O'Brien *et al.* (2006) noted many positive features following *Sabella* invasion. The worm tubes provided habitat for many sessile species, such as filamentous red algae, hydroids and molluscs (e.g. *Electroma georgiana*), and mobile species, such as echinoderms and larger crustaceans. In particular, the abundance of red algal epiphytes would increase the productivity of the bottom and provide food for grazers. A curious consequence of *Sabella*'s invasion of Port Phillip Bay was the marked increase in abundance of the little weed whiting *Neoodax balteatus* (see Hobday *et al.* 1999; Neira and Sporcic 2002). This whiting feeds on algae and invertebrates (Edgar *et al.* 1982); hence, clumps of worm tubes would provide shelter, and the epifauna on the tubes would provide food.

Coral worm *Galeolaria caespitosa*

This tube-building polychaete forms a distinct zone in the intertidal and on jetty piles on wave-exposed coasts of south and SE Australia. In fact, this *Galeolaria* is a species complex, and two cryptic species are now recognised – *G. caespitosa* on the east coast and *G. gemineoa* on the southern coast (Halt *et al.* 2009). These authors considered that the species diverged 5–50 million years ago, possibly as late as the Pliocene during a glacial maximum when the Bassian Isthmus may have created a barrier to gene flow.

The coral worm lives in mixed-sex clusters, and at high densities the tubes intertwine as they grow vertically from the substratum and form continuous calcareous mats some 5–10 cm high. This contrasts with a sister species *Galeolaria hystrix*, which usually grows subtidally as single tubes over the rock face, but can sometimes form colonies in deeper water with strong current flow.

The coral worm appears to be a protogynous hermaphrodite in which some females later become males, because individuals are sometimes found with both eggs and sperm. The worm is fertile almost all year round, but most spawning happens in spring. Females have a fecundity of up to 20 000 eggs. Eggs and sperm are released into the water column where fertilisation takes place. In this species, polyandry (multiple male sperm) is beneficial because females spawning in the presence of several males increase the likelihood that their eggs will be fertilised by the most compatible males (Marshall and Evans 2005).

The resulting larvae are plankton-feeders and larval life is ~14 days (longer if starved). They have a pelagic swimming stage of ~7 days and a benthic stage of another 7 days, after which metamorphosis is complete and the larvae settle and transform into the adult form (Marsden and Anderson 1981; Kupriyanova *et al.* 2001; Kupriyanova 2006). The subtidal species *G. hystrix* has a similar life history to its sister species. The major predators appear to be crabs and fishes. Another predator is the pycnogonid *Ammothea australiensis*, and Staples (1997) noted that during feeding the sea spider inserted its claws behind the worm's crown to prevent it from withdrawing into its tube.

Larval settlement, at least in New South Wales, takes place mostly in spring and summer. In a study of larval settlement in artificial clearings in *Galeolaria* mats, Minchinton (1997) found that larvae settled in response to a combination of chemical cues from adult worms and also shade, thus ensuring the development of dense colonies on wave-washed shores. The long spawning season and the ease of laboratory larval culture have facilitated many other studies beyond the scope of this book on fertilisation, larval behaviour and the use of the larvae in toxicity testing.

Mudworms (Family Spionidae)

The *Polydora*-complex is a large group of small species – many of them in the genera *Polydora, Dipolydora* and *Boccardia*, often only 1–2 cm long – that bore into encrusting corallines, barnacles, corals and mollusc shells. Five of the 37 species known from southern Australia are notorious for the damage they do to cultured scallops, oysters or abalone (Table 8.3) and. for this reason, have been extensively studied. They are all cosmopolitan, and some have likely been translocated round the world with the development of oyster culture since the 19th century (Walker 2011).

Species in the *Polydora*-complex are known as interface feeders, feeding on surface deposits and/or on the plankton. The larvae invade host shells in several ways. They may settle on the outside of a shell, often in an indented part of the shell sculpture, or they swim into the mantle cavity or burrow between the shell edge and the mantle (Blake and Evans 1972). Some species, such as *Polydora websteri* and *Dipolydora giardi*, bore U-shaped tubes, whereas other species bore W-shaped, S-shaped or more complex tubes. Burrowing by the worm is not a simple mechanical process because the worm secretes a viscous fluid that

Table 8.3: Life history of five mudworm species in the *Polydora*-complex and *Boccardia*, with cultured host species, type of egg development, larval type and time planktonic.

Species	Host	Development	Eggs	Planktonic period	Australian distribution
Polydora hoplura	Sydney rock oyster, Tasmanian oyster	Brooded; nurse eggs; plankton feeder	Strings of ~50 capsules, each with ~3 larvae, + 60 nurse eggs	not known	NSW to Tasmania
Polydora websteri	Sydney rock oyster	Brooded; no nurse eggs; plankton feeder	Strings of ~10 capsules, each with ~50 larvae	40 days at 16°C; 60 days at 10°C	NSW
Dipolydora giardi	abalone	Brooded; no nurse eggs; plankton feeder	Hollow cylinder; eggs attached by mucous filaments	24 days. (Can delay metamorphosis)	Southern Australia
Boccardia chilensis	abalone	Brooded; no nurse eggs; plankton feeder	Strings of 20 capsules, each with ~65 eggs	45 days +	NSW, Victoria Tasmania and SA
Boccardia knoxi	abalone	Brooded with nurse eggs	?	?	Tasmania

Data extracted from Zottoli and Carriker (1974), Blake and Arnofsky (1999), Sato–Okoshi (1999), Lleonart *et al.* (2003), McDiarmid *et al.* (2004), and Walker (2011).

dissolves the organic matrix and the calcareous shell crystals. However, they rarely penetrate the inner surface of the host, probably because burrowing induces the host to secrete a protective layer of conchiolin (a protein shell matrix), and then a layer of nacre at the point of entry. Severe worm infestation can cause high mortality of the host, and sublethal effects are a significant reduction in host flesh weight and growth (Lleonart *et al.* 2003).

Polydorid worms are male or female, and internal fertilisation occurs after copulation. The female then lays her eggs in capsules attached to the inner walls of her tube, and the eggs develop within the capsule, until they are sufficiently advanced to leave the tube and become planktonic. In some species, unfertilised nurse eggs provide nutrition for developing larvae. Asexual reproduction can also occur in some species by fission of the body into segments, which then regenerate into new individuals.

The known life histories of five polydorids are summarised in Table 8.3. *P. websteri* is also known to have a life span of ~1 year, with major spawning occurring in early summer and early winter, and larval settlement 1–2 months later. Considerable taxonomic confusion still exists in *Polydora* and *Boccardia* (see Walker 2011).

Ribbon worms (Phylum Nemertea)

Nemerteans are unsegmented worms with elongate bodies and a long proboscis, which are well adapted to capture active prey, such as polychaetes, flatworms, small molluscs and crustaceans (Gibson 1997). They range in length from a few millimetres to > 1 m. Many species are free-living in cryptic reef habitats, but a few are parasitic or commensal. Two genera, *Carcinonemertes* and *Gononemertes*, are parasitic or symbiotic on crustaceans and ascidians, and the latter genus has been studied by Gibson and Egan (1976), Egan and Anderson (1979) and Egan (1984a, b). Here we discuss the life history of well-studied species of both genera.

The symbiotic *Gononemertes australiensis*

This nemertean lives deep inside the ascidian *Pyura gibbosa* (previously called *P. pachydermatina*) among liver or gonad tissues, but does not appear to feed on its host's tissues – it apparently emerges and feeds on protozoans and animal material on the ascidian's external test. So, this ribbon worm must be considered a symbiont rather than a parasite.

This species is dioecious (with two sexes) and spawns continuously all year. Mature worms have ~10 000 gonads, each with one mature egg at any one time, closely packed along the full length of the body on both sides. Both eggs and sperm are thought to be released into the water where fertilisation takes place. The egg hatches ~22 h later and is free-swimming for the next 3–5 days, when it sinks to the bottom in search of a host ascidian. On finding a host, it grows a proboscis for feeding and develops into a miniature adult over ~2 weeks to a size of ~0.7 mm.

The host *Pyura* is an annual species breeding in winter, with settlement of the next generation occurring in late-winter to spring. Soon after spawning, the adult ascidian dies, with only a few weeks overlap between generations. A major problem for the worm is to ensure that its offspring will successfully find a host, and the success of this nemertean as a symbiont was found by Egan (1984b) to be due to three factors:

- A winter peak occurs during the host's reproductive period, so that a large number of the newly settled hosts are on the bottom ready to receive a maximum number of worm larvae. A peak in worm larvae also occurs because large worms breed in the full-grown 1-year old ascidians at the end of their life.
- The worm reproduces throughout the year, thereby ensuring maintenance of the population all year.
- During the peak breeding season of both worm and host, a remarkable phenomenon of precocious maturation of newly settled worms in their young hosts occurs. Some newly settled worms become sexually mature within 2–3 weeks of settling at a length of only 1–7 mm, compared with 4–6 cm for normal adults. The new settlers live in a different part of the host – the lobes of the digestive gland – and soon start breeding. The population of precociously mature worms occurs at a critical time when the old ascidian population and the large breeding worms living in them are dying off, so ensuring continuation of the population.

As concluded by Roe (1988) 'The life cycle of *Gononemertes australiensis* is tightly linked to that of its host, and it apparently receives cues from its host to drive its life cycle.'

The parasitic genus *Carcinonemertes*

Several species in this genus are parasitic on crabs and lobsters. The basic life history of the genus is well known. Juvenile worms feed on yolk and developing embryos of the host and, when they reach maturity, lay egg strings among the host's embryos. Like *Gononemertes*, species of *Carcinonemertes* produce huge numbers of eggs and reproduction is also timed to synchronise with crab or lobster egg-laying so that worm and crustacean larvae all hatch at the same time. In some species, the hatched larvae can start feeding at once on the host's eggs, but in other species the hatchlings leave the host and swim towards the water surface in search of another host.

Another adaptation of *Carcinonemertes* is facultative parthenogenesis – that is, females can produce egg strings which hatch into larvae without being fertilised by male sperm when males are not available (Roe 1988). This is a distinct advantage for a species when it is only sparsely present or in a marginal habitat – as could easily happen in some host crab or lobster populations.

One nemertean species *Carcinonemertes australiensis* occurs in the eggs of the western rock lobster, and an undescribed species of *Carcinonemertes* infests the southern rock lobster, but the infection rates are unknown (Shields *et al.* 2006). Yet another undescribed species of *Carcinonemertes* infests the smooth pebble crab *Philyra laevis*, and Bell and Hickman (1985) found that large crabs harboured ~25 or so immature worms in their gills within the branchial chamber.

MOLLUSCAN MACROFAUNA

The diversity and abundance of small molluscs in rocky subtidal habitats are extremely high – far too high to consider in any detail. Their taxonomy is poorly known, with many as yet undescribed species, and their ecology virtually unknown, except for Edgar's studies outlined above, which included some molluscs. In this section, we consider only the most common families, genera and species of these inconspicuous gastropods, up to a maximum size of 2–3 cm. Table 8.4 lists the most common groups of macrofaunal gastropods found in temperate reef habitats. Some species live among algal fronds, but many others are cryptic, and live within crevices, under boulders or in sediment under boulders.

Small species in the Super-order Vetigastropoda, comprising mainly trochids, turbinids, fissurellids, and phasianellids, are almost all herbivorous and are quite catholic in their diet. Many of them are cryptic by day to escape predatory fish, emerging at night to graze. In contrast, the caenogastropods have a vast array of feeding strategies and live in many microhabitats; some are herbivorous, but the majority are carnivorous. The three herbivorous groups – Rissoideans, Cerithioideans and the Eatoniellids – feed largely on microalgae, or detritus, or scrape bacteria and epiphytes from algae. Some also live on seagrass and graze epiphytes from the blades. The remaining caenogastropod groups vary widely in the degree of specialisation of their diet. Some species are highly specialised – those parasitic on particular host species – and others less so, specialising on many species within a phylum, or even across phyla; these species usually live in close association with their prey. A few are omnivorous and one species *Sabia australis* (Family Hipponicidae), discussed below, lives on the shell of its gastropod host and feeds on its faecal pellets.

To a large extent, the feeding capabilities of herbivorous gastropods can be predicted from the kind of feeding apparatus (the radula) that they possess and the shape and toughness of the various forms of algae available for food (Steneck and Watling 1982). All vetigastropods have rhipidoglossan radulae ('brooms'), with rows of many teeth, which enable wide grazing strokes, with limited penetration. Species with this kind of radula have difficulty eating leathery macroalgae, and calcareous algae. Many caenogastropods have taenioglossate radulae ('rakes'), with fewer teeth, which can exert greater force. This type of radula indicates a scraping or rasping mode of feeding, which can more easily cope with tough, leathery algae or articulated

Table 8.4: Diet and habitat of common genera or species, listed according to Family or Superfamily, of small gastropods, 1–20, (rarely 30) mm size, that live subtidally on southern temperate reefs, with approximate maximum size.

Superorder, family	Common genera/species (no. of temperate species)	Size (mm)	Diet (references)	Habitat
Vetigastropoda				
F. Trochidae	*Granata imbricata*	30	Ab, Cor, Seag[1]	Cryptic under rocks
topshells, periwinkles	*Cantharidus(3), Prothalotia*	18	Mic, Alg[3]	On algae, seagrass
	Diloma(4), D. odontis	25	Alg	On fucoids/*Ecklonia*
	Phasianotrochus(4), Thalotia(3)	30	Alg (b,r), Mic.[2,3]	Fucoids, kelps, seagrass
	Herpetopoma aspersa	10	Alg (b,r), Cor[1]	Cryptic under rocks
	Clanculus (21), Notogibbula(4)	10–30	Alg (g,b,r), Cor, Seag[1]	Cryptic under rocks
F. Fissurellidae Keyhole-,	*Notomella (6), Montfortula*	25	Alg	Cryptic under rocks
slit- limpets	*rugosa*	40	Spo / Spo, bry	Sponges/sponge,
	Diodora lincolnensis, Hemitoma	35	Spo	bryozoans. Sponge
	subemarginata, Tugali	25	Alg	Cryptic under rocks
	cicatricosa	25	Col. Asc.	Ascidians
	Amblychilepas(4),Cosmetalepas			
	concatenata			
F. Turbinidae	*Astralium squamiferum*	20	Alg (b,r), Cor, Seag[1]	Cryptic under rocks On
F. Phasianellidae	*Turbo(3), T. undulatus,*	35	Alg (g,b,r), Cor, Seag[1]	*Ecklonia, Caulerpa*
turbans, pheasant shells	*Phasianella (4), P. ventricosa*	30	Alg (g,b,r),Cor[1]	Cryptic or in kelp
Caenogastropoda	Many undescribed species	10	Mic, forams, detritus	In algae, cryptic
Superf. Rissooidea	*Pisinna (60), Rissoina*	20	Mucous trap feeding	Large colonies on rock
Superf. Vermetoidea	*Serpulorbis sipho*	30/5	Detritus/Mic, Cor.	Cryptic/in algae
Superf. Cerithioidea	*Cacozeliana granaria, Diala(6)*	<	Nk/Alg	In algae, seagrass
Superf. Cingulopsoidea	*Alaba* spp., *Plesiotrochus* (4)	5/25	Mic	In algae, or cryptic
F. Eatoniellidae	*Eatoniella, Crassitoniella, (20)*	2		
Superf. Vanikoroidea	*Sabia(2), S. australis*	25	Faecal pellets of host	Commensal on shells of
F. Hipponicidae				many molluscs
Superf. Cypraeoidea	*Notocypraea(6), Ellatrivia*	25/10	Spo, bry, detritus/bry	Cryptic under rocks
	merces			
Superf. Velutinoidea	*Lamellaria(5), Trivia (10)*	30/10	Col. Asc./bry	Cryptic under rocks
F. Ovulidae	*Primovula heleneae, P. cruenta*	30	*Euplexaura/Echino-*	Gorgonians, soft corals
	Phenacovolva, Prosimnia		*gorgia/Mopsella*	
Superf. Tonnoidea	*Sassia(10)*	30	Col Asc, Spo	Cryptic under rocks
Superf. Triphoroidea	27 genera, > 68 species	10	Spo	Cryptic or on sponge
F. Triphoridae	> 100 spp. e.g. *Cerithiopsis* spp.	10	Spo	On sponge
F. Cerithiopsidae				
Superf. Janthinoidea	*Epitonium*(14+)	25	Anemones	Sediments, under rocks
Superf. Eulimoidea	*Pelseneeria brunnea*	5	Parasitic – *Heliocidaris*	On sea-urchin
	Curveulima cornuta	5	Parasitic – *Comatulella*	On crinoid
	Sabinella munita	8	Parasitic – *Goniocidaris*	On sea-urchin
	Stilapex lactarius	8	Parasitic – ophiuroids	On ophiuroids
Superf. Muricoidea	*Bedeva paivae, Lepsiella(3)*			
F. Muricidae	*Haustrum baileyanum,*	30	Oysters, mussels, etc.	Rocky bottom
	Prototyphis angasi,	30	Parasitic – abalone/nk	On abalone/cryptic
	Tripterotyphis robustus	20	*Spondylus tenellus*	Rock and sand

(Continued)

Table 8.4: (Continued)

Superorder, family	Common genera/species (no. of temperate species)	Size (mm)	Diet (references)	Habitat
F. Buccinidae	*Cominella lineolata*	25	Scavenger	Rock and sand
F. Columbellidae	*Anachis(6), Mitrella(12)*	20	Carnivorous, herbivorous	Rock
F. Volutomitridae	*Waimatea obscura*	20	Bry	On rock, in algae
F. Costellariidae	*Austromitra(4),Vexillum(6)*	20	Molluscs, Asc/ molluscs	On rock
F. Mitridae	*Mitra(2)*	20	Sipunculans, invertebrates	Cryptic, on rock
F. Marginellidae	*Austroginella(9), Cystiscus(17)* *Mesoginella(14), Pillarginella(4)*	20 20	Carnivorous e.g. Bry, Asc	Cryptic, on rock or in sand near rock
Superf. Cancellarioidea	*Cancellaria(8)*	30	Bivalves, gastropods	Cryptic under rocks
Superf. Conoidea	*Conus clarus*	30	Carnivore e.g. Polychaetes	In sand under rocks

Data for caenogastropods and fissurellids variously extracted from Beesley *et al.* (1998), Ludbrook and Gowlett-Holmes (1989), Wilson (1993–1994) and Gowlett-Holmes (2008), and data sources for remaining vetigastropods given in the table. Common genera with numerous species are listed as genera only, with the number of known temperate species in brackets. Where several species are listed in a group, the dietary information for each, if different, is separated by a slash. Alg = algae (macroalgae; b = brown, r = red, g = green algae); Asc = ascidians; Bry = bryozoan; Col Asc = colonial ascidian; Cor. = erect or crustose corallines; Mic = microalgae (diatoms, dinoflagellates, bacteria, etc.); Seag = seagrass; Spo = sponge; nk = not known.
[1]Clarkson and Shepherd (1985); [2]Nielson and Lethbridge (1989); [3]Hickman (2005)

corallines. Other kinds of radulae, such as 'shovel' types and 'all-purpose tools', also occur in some species.

The ecology of subtidal macrofaunal species is known only for a few, mostly larger, forms and examples of some herbivorous and carnivorous species are given here.

Ecology of four species of subtidal macrofaunal molluscs

The marginellid *Microginella minutissima*

This tiny marginellid, < 3 mm size, lives on the curly fronds of the bryozoan *Amathia biseriata*. The host species lives attached to rock or shell at depths below ~10 m in places of moderate to strong current. *M. minutissima* is dioecious (male and female) and lives for up to 6 months. After mating (which can last for several hours), the female lays an egg every second day, and the eggs hatch into crawling juveniles after 6–7 weeks (Murray 1970). This marginellid lives and feeds on its host bryozoan, and the relationship is thought to be obligate, but nothing is known about how dispersal of the marginellid is achieved. It may depend on rafting on host colonies torn off the substratum by water movement and carried by the current.

The bonnet limpet *Sabia australis*

This limpet (formerly called *Hipponix conicus* – see Ponder 1998b) is commonly found commensal with other molluscs, notably abalone shells, the whelk, *Pleuroploca*, murex shells, such as *Pterynotus*, the turbinid *Turbo torquatus*, and many others. The biology of the species was studied by Laws (1970) and Knudsen (1991). It feeds to some extent on detritus, but often on faecal pellets of its host, using its long and mobile proboscis. To facilitate such feeding, it usually locates itself on the part of the shell where it has direct access to its host's excrement.

The female produces up to 10 egg capsules, mostly during the winter, each with 9–24 embryos that hatch into crawling juveniles. The mode of dispersal is uncertain, but it is likely that the juveniles can drift or swim in the water and then seek out a host on which to settle. The sexual development of the young appears to be influenced by the proximity of conspecifics. If it settles alone on a host mollusc, it quickly develops into a female – the male phase being very short or even absent – but if it settles near a female, it will develop into a male and live on the back of the female.

Large males can change sex to become females, and transitional individuals have been recorded with a penis and also well-developed eggs. This unusual sexual mode, involving male-to-female sex change is termed protandric hermaphroditism. As a male, it remains much smaller than the female and erodes a groove at the base of her shell to secure easy entry of its penis. This point of entry is very close to the opening of the female oviduct, so is optimal for fertilisation. However, some have suggested that the male may also use the point of entry to feed on the female's faecal pellets. Large females of 16–25 mm size commonly have two males 5–15 mm size on their backs, each with its own entry groove. In SW Australia, however, Knudsen (1991) found that the bonnet limpet often lived on many host shells without access to their faeces, and so depended on sediment and detritus for food. The longevity of this limpet is unknown.

The trochid *Cantharidus pulcherrimus*

Species of *Cantharidus* and other related cantharidine genera, such as *Thalotia, Prothalotia, Phasianotrochus* and *Strigosella*, live on macroalgae and seagrass in the shallow subtidal zone (Hickman 2005). The life cycle and ecology of one species *C. pulcherrimus* was examined by Wells and Keesing (1987) on a reef in SW Western Australia where it reached extraordinarily high densities of up to 600 m^{-2}. This species is dioecious and a broadcast spawner, spawning annually in March–April. After a brief larval life of a few days, larvae settle and juveniles are detectable at a size of ~1 mm by June. They reach ~8 mm in 1 year, but very few survive to reach 2 years of age at ~10 mm size.

This trochid fed mainly on *Sargassum* and its epiphytes at Wells and Keesing's site, but, in general, cantharidines feed on microbial films and epiphytes, using their broad radulae to sweep clean the algal or seagrass substratum (Hickman 2005).

Two weeks in the life of the whelk *Haustrum baileyanum*

This small whelk is a common subtidal species that appears to be obligately associated with abalone, mainly blacklip and greenlip. The parasitic relationship between the whelk and abalone was explored in a fascinating study by Thomas and Day (1995). The whelk seeks out abalone shells by chemoreception – probably by detection of ammonia released as waste by the abalone – and then climbs on to the back of the shell, ready for drilling into the shell. But the abalone is not defenceless, and can violently twist its shell, push the whelk away with the shell's edge, or crawl onto the whelk to prevent it climbing onto its shell. Once the whelk is aboard the shell, it must first locate the best spot to drill into the shell: this is nearly always in the middle of the shell where the muscular foot is attached. It probably orients itself on the shell by locating the spire and then moving forwards to about the centre where it starts drilling. The drilling mechanism is itself quite a complex two-step process. The whelk alternately employs its boring organ to drill the hole in the abalone shell, secreting an acidic solution to soften the shell, and then scraping off the exposed crystals with its radula. In this study, the whelk on average, spent 14 days on an abalone and drilled at a rate of 0.23 mm per day. The whelk usually took 7 days to drill through the shell, and then spent another 7 days sucking nutrients from the muscle tissue. The abalone's defence to such assaults is to first secrete a black protein layer, and later nacre, around the hole, but this takes several weeks – long after the whelk has stopped feeding and left. So it seems that the whelk sucks out food from the abalone tissue till it is full, and then moves on to drill and suck another abalone another day. A curious feature is that the whelk prefers to drill over the thickest part of the shell, right where the muscle is, taking ~25% more drilling time than over a thinner part of the shell, where, for example, the viscera are. The reason for this appears to be that muscle tissue has about one third more energy by volume than visceral tissue. Hence, in this example, the benefits of drilling over the thicker part of the shell outweighed the extra cost involved.

Molluscan grazer effects on the habitat

Often many species of small grazing gastropods (termed guilds) co-exist in crevices or boulder habitats, where they find protection from their major predators – wrasses and crabs. In these habitats, they emerge at night to graze algae on the boulders' upper surface. Where drift algae are available and are trapped among the boulders, they do not even need to emerge from cryptic habitat to feed on the drift.

Cumulatively, molluscan grazers can have an important effect in controlling the growth of filamentous algae. In a study in a boulder habitat at West I., 17 herbivorous gastropods were present and the diet of 11 of them examined (Clarkson and Shepherd 1985). The main genera involved were *Clanculus* (2 spp.), *Chlorodiloma, Euchelus, Turbo* (2 spp.), *Astralium* (formerly *Bellastraea*), *Phasianella, Granata, Gibbula* and *Scutus*. In all the species, except the last two, the diet was 30–60% of crustose coralline algae and/or erect coralline algae, with most of the remainder macroalgae or, in a few cases, drift seagrass. Animal matter (bryozoans and colonial ascidians) and filamentous algae comprised a few per cent in each. The largest species, the elephant snail *Scutus*, fed almost exclusively on drift algae. But what effect did this guild, with a combined density of > 150 individuals per m^2 at 4–5 m depth, have in the ecosystem?

To answer this, all grazing gastropods and chitons were removed from experimental boulders, which had a natural cover of crustose corallines with some macroalgae (mainly erect corallines). After 2 months, a fine, filamentous algal turf of *Feldmannia* spp. had grown all over the upper surfaces of the experimental boulders and sediment had then settled among the filamentous algae. The crustose corallines were smothered and dead. Adjoining controls remained healthy and free of the covering of filamentous algae. Some of the experimental boulders were then returned to their original location, and within a week grazing molluscs had completely grazed the filamentous algal covering on the boulders, exposing the dead crustose corallines. Crustose corallines have an important functional role as a micro-habitat on which a variety of molluscan larvae settle and feed (see Chapter 6), so their demise can have serious flow-on effects on the system.

In an elaboration of the above study at the same site, Russell and Connell (2005) examined the interaction between nutrient addition, which favours algal growth, and the grazing effect of the same guild of small molluscs as that studied by Clarkson and Shepherd (1985). In the presence of nutrients, both the fine filamentous algae and a more robust and taller species *Cladosiphon filum* grew on the rock, but the grazing molluscs (called bulldozers by Russell and Connell) consumed only the fine filamentous species. This selective 'bulldozing' gave space for *Cladosiphon* to increase substantially in abundance.

ECOSYSTEM EFFECTS OF MACROFAUNA

From all the above studies in algal communities, some general principles can be stated, although their widespread applicability to temperate Australian reefs in most cases is yet to be tested.

1. Epifaunal grazers can reduce epiphyte density on algae and settled detritus by up to a half, and thereby increase algal turnover and the primary production in the system. This has been better quantified in seagrass than in algal communities, but the same principle likely applies in both systems.
2. Gastropods are more efficient, but less selective, than amphipods at consuming epiphytes, but amphipods still may have significant effects.
3. In complex reef habitats, macrofaunal gastropods can control filamentous algal growth and prevent smothering of the crustose coralline micro-habitat.
4. Epifaunal densities tend to be highest in dense vegetation. This is partly because mobile species prefer dense habitats, and partly because their predators are less efficient in dense habitats than in more open ones. But little-examined effects of landscape, such as patch size and proximity of sand or other habitats, are also likely to be important.
5. Macrofaunal diversity is often highest at intermediate water movements (although exceptions occur). Possibly the 'intermediate disturbance hypothesis', which predicts higher algal diversity in habitats with intermediate levels of disturbance, may play a role promoting epifaunal diversity at moderate water movements.
6. Predation on the epifauna tends to be higher in patchy habitats, patch edges and more open habitats than in dense vegetation. Nevertheless, predators have marked effects on all algal communities in several ways. Fish predators often preferentially take the larger epifaunal invertebrate predators (inducing a cascade effect), and all predators reduce epifaunal abundances and so transfer energy to higher trophic

levels. This emphasises the importance of predator–prey interactions at this level in reef systems.

7. Despite predation effects, epifaunal densities may be constrained by resource ceilings (i.e. food abundance) in some systems.
8. Overall, the epifauna is a major contributor to the productivity of reef habitats, transferring energy from primary producers to higher trophic levels.
9. Both macrofaunal gastropods and crustaceans have evolved a broader array of feeding niches than we have considered. Some are grazers or carnivores, but many have diversified into parasitism, commensalism or detritivory.

REFERENCES

Anderson DT (1977) The embryonic and larval development of the turbellarian *Notoplana australis* (Schmarda, 1859) (Polycladida: Leptoplanidae). *Australian Journal of Marine and Freshwater Research* **28**, 303–310.

Ang HP, Newman LJ (1998) Warning colouration in pseudoceratid flatworms (Platyhelminthes, Polycladida). A preliminary study. *Hydrobiologia* **383**, 29–33.

Baker JL, Shepherd SA, Turner D, Edyvane K (2008) Investigator Expedition 2006: benthic macroalgal studies at islands in the eastern Great Australian Bight over three decades. *Transactions of the Royal Society of South Australia* **132**, 251–267.

Beesley PL, Ross GJB, Wells A (1998) *Mollusca: The Southern Synthesis. Fauna of Australia. Vol. 5. Parts A, B.* CSIRO PUBLISHING, Melbourne.

Beesley PL, Ross GJB, Glasby CJ (2000) *Polychaetes and Allies: The Southern Synthesis. Fauna of Australia. Vol. 4A Polychaeta, Myzostomida, Pogonophora, Echiura, Sipuncula.* CSIRO PUBLISHING, Melbourne.

Bell PJ, Hickman JL (1985) Observations on *Carcinonemertes* (Nemertea: Carcinonemertidae) associated with the smooth pebble crab, *Philyra laevis. Papers and Proceedings of the Royal Society of Tasmania* **119**, 65–68.

Bell JD, Westoby M (1986) Abundance of macrofauna in dense seagrass is due to habitat preference, not predation. *Oecologia* **68**, 205–209.

Blake JA Arnofsky PL (1999) Reproduction and larval development of the spioniform Polychaeta with application to systematics and phylogeny. *Hydrobiologia* **402**, 57–106.

Blake JA, Evans JW (1972) *Polydora* and related genera as borers in mollusc shells and other calcareous substrates. *The Veliger* **15**, 235–249.

Brearly A, Walker DI (1993) Isopod borers in seagrass species from south-western Australia. In *Proceedings of the 5th International Marine Biology Workshop: the Marine Flora and Fauna of Rottnest Island, Western Australia.* January 1991, Rottnest Island (Eds FE Wells, DI Walker, H Kirkman and R Lethbridge) pp. 415–428. Western Australian Museum, Perth.

Carey JM, Watson JE (1992) Benthos of the muddy bottom habitat of the Geelong arm of Port Phillip Bay, Victoria, Australia. *Victorian Naturalist* **109**, 196–202.

Clapin G, Evans DR (1995) 'The status of the introduced marine fanworm *Sabella spallanzanii* in Western Australia: a preliminary investigation. Centre for research on Introduced Marine pests'. CSIRO. Technical Report No. 2, CSIRO Marine Research, Hobart.

Clarkson PS, Shepherd SA (1985) Distribution and feeding of archaeogastropods in a boulder habitat at West Island, South Australia. *Journal of the Malacological Society of Australia* **7**, 35–44.

Coleman MA, Vytopil E, Goodsell PJ, Gillanders BM *et al.* (2007) Diversity and depth-related patterns of mobile invertebrates. *Marine and Freshwater Research* **58**, 589–595.

Currie DR, McArthur MA, Cohen BF (1999) Exotic marine pests in the port of Geelong, Victoria. In *Marine Biological Invasions of Port Phillip Bay, Victoria.* (Eds CL Hewitt, ML Campbell, RE Thresher and RB Martin) pp. 227–246. CRIMP Technical Report No. 20. CSIRO Marine Research, Hobart.

Currie DR, McArthur MA, Cohen BF (2000) Reproduction and distribution of the invasive European fanworm *Sabella spallanzanii* (Polychaeta: Sabellidae) in Port Phillip Bay, Victoria, Australia. *Marine Biology* **136**, 645–656.

Dalby JE Jr (1996) Nemertean, copepod, and amphipod symbionts of the dimorphic ascidian *Pyura stolonifera* near Melbourne, Australia: specificities to host morphs, and factors affecting prevalence. *Marine Biology* **126**, 231–243.

Dalley DD, McClatchie S (1989) Functional feeding morphology of the euphausiid *Nyctophanes australis. Marine Biology* **101**, 195–203.

Edgar GJ (1983a) The ecology of south-east Tasmanian phytal animal communities. I. Spatial organization on a local scale. *Journal of Experimental Marine Biology and Ecology* **70**, 129–157.

Edgar GJ (1983b) The ecology of south-east Tasmanian phytal animal communities. II. Seasonal change in plant and animal populations. *Journal of Experimental Marine Biology and Ecology* **70**, 159–179.

Edgar GJ (1983c) The ecology of south-east Tasmanian phytal animal communities. III. Patterns of species diversity. *Journal of Experimental Marine Biology and Ecology* **70**, 181–203.

Edgar GJ (1983d) The ecology of south-east Tasmanian phytal animal communities. IV. Factors affecting the distribution of ampithoid amphipods among algae.

Journal of Experimental Marine Biology and Ecology **70**, 205–225.

Edgar GJ (1987) Dispersal of faunal and floral propagules associated with drifting *Macrocystis pyrifera* plants. *Marine Biology* **95**, 599–610.

Edgar GJ (1990) The use of the size structure of benthic macrofaunal communities to estimate faunal biomass and secondary production. *Journal of Experimental Marine Biology and Ecology* **137**, 195–214.

Edgar GJ (1991a) Artificial algae as habitats for mobile epifauna: factors affecting colonization in a Japanese *Sargassum* bed. *Hydrobiologia* **226**, 111–118.

Edgar GJ (1991b) Distribution patterns of mobile epifauna associated with rope fibre habitats within the Bathurst Harbour estuary, south-western Tasmania. *Estuarine, Coastal and Shelf Science* **33**, 589–604.

Edgar GJ (1992) Patterns of colonization of mobile epifauna in a Western Australian seagrass bed. *Journal of Experimental Marine Biology and Ecology* **157**, 225–246.

Edgar GJ (1993) Measurement of the carrying capacity of benthic habitats using a metabolic-based index. *Oecologia* **95**, 115–121.

Edgar GJ (1994) Observations on the size-structure of macrofaunal assemblages. *Journal of Experimental Marine Biology and Ecology* **176**, 227–243.

Edgar GJ (1999) Experimental analysis of structural versus trophic importance of seagrass beds. I. Effects on macrofaunal and meiofaunal invertebrates. *Vie et Milieu* **49**, 239–248.

Edgar GJ, Aoki M (1993) Resource limitation and fish predation: their importance to mobile epifauna associated with Japanese *Sargassum*. *Oecologia* **95**, 122–133.

Edgar GJ, Moore PG (1986) Macro-algae as habitats for motile macrofauna. *Monografias Biologicas* **4**, 255–277.

Edgar GJ, Robertson AI (1992) The influence of seagrass structure on the distribution and abundance of mobile epifauna: pattern and process in a Western Australian *Amphibolis* bed. *Journal of Experimental Marine Biology and Ecology* **160**, 13–31.

Edgar GJ, Shaw C (1993) Inter-relationships between sediments, seagrasses, benthic invertebrates and fishes in shallow marine habitats off south-western Australia. In: *Proceedings of the 5th International Marine Biology Workshop: the Marine Flora and Fauna of Rottnest Island, Western Australia.* January 1991, Rottnest Island (Eds FE Wells, DI Walker, H Kirkman and R Lethbridge) pp. 429–442. Western Australian Museum, Perth.

Edgar GJ, Last PR, Wells MW (1982) *Coastal Fishes of Tasmania and Bass Strait.* Cat and Fiddle Press, Hobart.

Egan EA (1984a) The seasonal occurrence of the copepod, *Pachypygus australis* Gotto (Notodelphyidae) in its host *Pyura pachydermatina* (Herdman). *Journal of Experimental Marine Biology and Ecology* **76**, 247–262.

Egan EA (1984b) The seasonal reproductive cycle of the nemertean *Gononemertes australiensis* Gibson in relation to that of its ascidian host *Pyura pachydermatina* (Herdman). *Journal of Experimental Marine Biology and Ecology* **76**, 225–246.

Egan EA, Anderson DT (1979) The reproduction of the entozoic nemertean *Gononemertes australiensis* Gibson (Nemertea: Hoplonemertea: Monostylifera) – gonads, gametes, embryonic development and larval development. *Australian Journal of Marine and Freshwater Research* **30**, 661–681.

Fenton GE (1992) Population dynamics of *Tenagomysis tasmaniae* Fenton, *Anisomysis mixta australis* (Zimmer) and *Paramesopodopsis rufa* Fenton from south-eastern Tasmania (Crustacea: Mysidacea). *Hydrobiologia* **246**, 173–193.

Fenton GE (1996a) Production and biomass of *Tenagomysis tasmaniae* Fenton, *Anisomysis mixta australis* (Zimmer) and *Paramesopodopsis rufa* Fenton from south-eastern Tasmania (Crustacea: Mysidacea). *Hydrobiologia* **323**, 23–30.

Fenton GE (1996b) Diet and predators of *Tenagomysis tasmaniae* Fenton, *Anisomysis mixta australis* (Zimmer) and *Paramesopodopsis rufa* Fenton from south-eastern Tasmania (Crustacea: Mysidacea). *Hydrobiologia* **323**, 31–44.

Fenwick GD (1976) The effect of wave exposure on the amphipod fauna of the alga *Caulerpa brownii*. *Journal of Experimental Marine Biology and Ecology* **25**, 1–18.

Gibson R (1997) Nemerteans. In *Marine Invertebrates of Southern Australia* (Eds SA Shepherd and M Davies) pp. 905–974. SARDI and Fauna and Flora of South Australia Handbooks Committee, Adelaide.

Gibson R, Egan E (1976) Some histochemical observations on digestive and other enzymes of the entozoic hoplonemertean *Gononemertes australiensis* Gibson, with comments on its possible feeding behaviour. *Journal of Experimental Marine Biology and Ecology* **24**, 285–296.

Gill PC, Morrice MG (2003) 'Blue whale research in the Bonney upwelling, south-east Australia – current information'. Deakin University, School of Ecology and Environment Technical Paper 2003/1, Geelong.

Goodsell PJ, Connell SD (2002) Can habitat loss be treated independently of habitat configuration? Implications for rare and common taxa in fragmented landscapes. *Marine Ecology Progress Series* **239**, 37–44.

Goodsell PJ, Connell SD (2005) Historical configuration of habitat influences the effects of disturbance on mobile invertebrates. *Marine Ecology Progress Series* **299**, 79–87.

Goodsell PJ, Connell SD (2008) Complexity in the relationship between matrix composition and inter-patch distance in fragmented habitats. *Marine Biology* **154**, 117–125.

Goodsell PJ, Fowler-Walker MJ, Gillanders BM, Connell SD (2004) Variations in the configuration of algae in subtidal forests: implications for invertebrate assemblages. *Austral Ecology* **29**, 350–357.

Gotto RV (1979) The association of copepods with marine invertebrates. *Advances in Marine Biology* **16**, 1–109.

Gowlett-Holmes K (2008) *A Field Guide to the Marine Invertebrates of South Australia*. Notomares, Hobart.

Guerra-García JM, de Figueroa JMT (2009) What do caprellids (Crustacea: Amphipoda) feed on? *Marine Biology* **156**, 1881–1890.

Halt MN, Kupriyanova EK, Cooper SJB, Rouse GW. (2009) Naming species with no morphological indicators: species status of *Galeolaria caespitosa* (Annelida: Serpulidae) inferred from nuclear and mitochondrial gene sequences and morphology. *Invertebrate Systematics* **23**, 205–222.

Hay ME, Duffy JE, Pfister CA, Fenical W (1987) Chemical defense against different marine herbivores: are amphipods insect equivalents? *Ecology* **68**, 1567–1580.

Hickman CS (2005) Seagrass fauna of the temperate southern coast of Australia I: the Cantharidine trochid gastropods. In *Proceedings of the 12th International Marine Biological Workshop: the Marine Flora and Fauna of Esperance, Western Australia*. 3–21 February 2003, Esperance (Eds FE Wells, DI Walker and GA Kendrick) Vol. 1 pp. 199–220. Western Australian Museum, Perth.

Hicks GRF (1986) Meiofauna associated with rocky shore algae. In *The Ecology of Rocky Coasts* (Eds PG Moore and R Seed) pp. 36–56. Columbia University Press, New York.

Hirst AJ (2003) Encounter 2002 expedition to the Isles of St Francis, South Australia: peracarid crustacean epifauna of subtidal macroalgal canopies. *Transactions of the Royal Society of South Australia* **127**, 189–203.

Hobday DK, Officer RA, Parry GD (1999) Changes to demersal fish communities in Port Phillip Bay, Australia, over two decades, 1970–1991. *Marine and Freshwater Research* **50**, 397–407.

Holloway MG, Keough MJ (2002a) Effects of an introduced polychaete, *Sabella spallanzanii*, on the development of epifaunal assemblages. *Marine Ecology Progress Series* **236**, 137–154.

Holloway MG, Keough MJ (2002b) An introduced polychaete, *Sabella spallanzanii*, affects recruitment and larval abundance of sessile invertebrates. *Ecological Applications* **12**, 1803–1823.

Hosie GW, Ritz DA (1983) Contribution of moulting and eggs to secondary production in *Nyctiphanes australis* (Crustacaea: Euphausiacea). *Marine Biology* **77**, 215–220.

Howard RK (1982) Impact of feeding activities of epibenthic amphipods on surface-fouling of eelgrass leaves. *Aquatic Botany* **14**, 91–97.

Howard RK (1984) The trophic ecology of a caridean shrimp in an eelgrass community. *Aquatic Botany* **18**, 155–174.

Irving AD, Tanner JE, McDonald BK (2007) Priority effects on faunal assemblages within artificial seagrass. *Journal of Experimental Marine Biology and Ecology* **340**, 40–49.

Jernakoff P, Brearly A, Nielsen J (1996) Factors affecting grazer-epiphyte interactions in temperate seagrass meadows. *Oceanography and Marine Biology - an Annual Review* **34**, 109–162.

Johnson CR, Banks SC, Barrett NS, Cazassus F *et al.* (2011) Climate change cascades: shifts in oceanography, species' ranges and subtidal marine community dynamics in eastern Tasmania. *Journal of Experimental Marine Biology and Ecology* **400**, 17–32.

Johnston NM, Ritz DA, Fenton GE (1997) Larval development in the Tasmanian mysids *Anisomysis mixta australis, Tenagomysis tasmaniae* and *Paramesopodopsis rufa* (Crustacea: Mysidacea). *Marine Biology* **130**, 93–99.

Keable SJ, Fenton GF, Lowry JK (2003) Mysidacea. In *Crustacea: Malacostraca: Peracarida: Amphipoda, Cumacea, Mysidacea*. (Eds JK Lowry and HE Stoddart) pp. 419–469. *Zoological Catalogue of Australia. Vol. 19.2B*. CSIRO PUBLISHING, Melbourne.

Keith DE (1969) Aspects of feeding in *Caprella californica* Stimpson and *Caprella equilibra* Say (Amphipoda). *Crustaceana* **16**, 119–124.

Kennelly S (1983) An experimental approach to the study of factors affecting algal colonization in a sublittoral kelp forest. *Journal of Experimental Marine Biology and Ecology* **68**, 257–276.

Knudsen J (1991) Observations on *Hipponix australis* (Lamarck, 1919) (Mollusca, Gastropoda, Prosobranchia) from the Albany area, Western Australia. In *Proceedings of the International Marine Biological Workshop: The Marine Flora and Fauna of Albany, Western Australia*. (Eds FE Wells, DI Walker, H Kirkman and R Lethbridge) pp. 641–660. Western Australian Museum, Perth.

Kupriyanova EK (2006) Fertilization success in *Galeolaria caespitosa* (Polychaeta: Serpulidae): gamete characteristics, role of sperm dilution, gamete age, and contact time. *Scientia Marina* **70**, S3, 309–317.

Kupriyanova EK, Nishi E, Ten Hove HA, Rzhavsky AV (2001) Life-history patterns in serpulimorph polychaetes: ecological and evolutionary perspectives.

Oceanography and Marine Biology - an Annual Review **39**, 1-101.

Langtry SK, Jacoby CA (1996) Fish and decapod crustaceans inhabiting drifting algae in Jervis Bay, New South Wales. *Australian Journal of Ecology* **21**, 264–271.

Laws HM (1970) Reproductive biology and shell site preference in *Hipponix conicus* (Schumacher). *The Veliger* **13**, 115–121.

Lee K-M, Beal AM, Johnston EL (2006) A new predatory flatworm (Platyhelminthes, Polycladidae) from Botany Bay, New South Wales, Australia. *Journal of Natural History* **39**, (47), 3987–3995.

Lleonart M, Handlinger J, Powell M (2003) Spionid mudworm infestation of farmed abalone (*Haliotis* spp.). *Aquaculture* **221**, 85–96.

Ludbrook N, Gowlett-Holmes K (1989) Chitons, gastropods and bivalves. In *Marine Invertebrates of Southern Australia. Part II.* (Eds SA Shepherd and IM Thomas) pp. 504–724. South Australian Government Printer, Adelaide.

Marsden JR, Anderson DT (1981) Larval development and metamorphosis of the serpulid *Galeolaria caespitosa* Lamarck. *Australian Journal of Marine and Freshwater Research* **32**, 667–680.

Marshall DJ, Evans JP (2005) The benefits of polyandry in the free-spawning polychaete *Galeolaria caespitosa. Journal of Evolutionary Biology* **18**, 735–741.

Mauchline J (1980) The biology of mysids and euphausiids. *Advances in Marine Biology* **18**, 3–637.

McDiarmid H, Day R, Wilson R (2004) The ecology of polychaetes that infest abalone shells in Victoria, Australia. *Journal of Shellfish Research* **23**, 1179–1188.

Merory M, Newman LJ (2005) A new stylochid flatworm (Platyhelminthes, Polycladida) from Victoria, Australia and observations on its biology. *Journal of Natural History* **39**, (28), 2581–2589.

Metillo EB (2007) In situ feeding rate and diet selectivity in Tasmanian mysid species (Crustacea: Mysidacea). *Hydrobiologia* **589**, 207–218.

Metillo EB, Ritz DA (1993) Predatory feeding behaviour in *Paramesopodopsis rufa* Fenton (Crustacea: Mysidacea). *Journal of Experimental Marine Biology and Ecology* **170**, 127–141.

Metillo EB, Ritz DA (2003) Differential chemosensory feeding behaviour by three co-occurring mysids (Crustacea: Mysidacea) from southeastern Tasmania. *Comparative Biochemistry and Physiology Part A* **134**, 399–408.

Metillo EB, Ritz DA, Swadling KM (2007) In situ feeding rate and diet selectivity in Tasmanian mysid species (Crustacea, Mysidacea). *Hydrobiologia* **589**, 207–218.

Minchinton TE (1997) Life on the edge: conspecific attraction and recruitment of populations to disturbed habitats. *Oecologia* **111**, 45–52.

Morgan WL, Ritz DA (1982) Comparison of the feeding apparatus in the muttonbird, *Puffinus tenuirostris* (Temminck) and the fairy prion, *Pachyptila turtur* (Kohl) in relation to the capture of the krill, *Nyctiphanes australis. Journal of Experimental Marine Biology and Ecology* **59**, 61–75.

Morgan WL, Ritz DA (1984) Effect of prey density and hunger state on capture of krill, *Nyctiphanes australis* Sars, by Australian salmon, *Arripis trutta* (Bloch & Schneider). *Journal of Fish Biology* **24**, 51–58.

Murray FV (1970) The reproduction and life history of *Microginella minutissima* (Tenison-Woods, 1876) (Gastropoda: Marginellidae). *Memoirs of the National Museum of Victoria* **31**, 31–37.

Nagle JS (1968) Distribution of the epibiota of macroepibenthic plants. *Contributions in Marine Science* **15**, 105–144.

Neira FJ, Sporcic MI (2002) Use of ichthyoplankton ecology to evaluate ecosystem changes: a case study in a large, semi-enclosed Australian bay. *Marine and Freshwater Research* **53**, 339–354.

Newman L, Cannon L (2003) *Marine Flatworms: the World of Polyclads.* CSIRO PUBLISHING, Melbourne.

Nielson J, Lethbridge R (1989). Feeding and the epiphyte food resources of gastropods living on leaves of the seagrass *Amphibolis griffithii* in south-western Australia. *Journal of the Malacological Society of Australia* **10**, 47–58.

NIMPIS (2011) *Sabella spallanzanii* feeding and predators. National Introduced Marine Pest Information System, Canberra. <http://www. marinepests.gov.au/nimpis> (Accessed February 2013).

O'Brien DP (1986) Stranding and mating behaviour in *Nyctiphanes australis* (Euphausiidae: Crustacaea). *Marine Biology* **93**, 465–473.

O'Brien DP (1987) Description of escape responses of krill (Crustacea: Euphausiacea), with particular reference to swarming behaviour and the size and proximity of the predator. *Journal of Crustacean Biology* **7**, 449–457.

O'Brien DP (1988a) Direct observations of clustering (schooling and swarming) behaviour in mysids (Crustacea: Mysidacea). *Marine Ecology Progress Series* **42**, 235–246.

O'Brien DP (1988b) Surface schooling behaviour of the coastal shrimp *Nyctiphanes australis* (Crustacaea: Euphausiacea) off Tasmania. *Marine Ecology Progress Series* **42**, 219–233.

O'Brien DP (1989) Analysis of the internal arrangement of individuals within crustacean aggregations (Euphausiacea, Mysidacea). *Journal of Experimental Marine Biology and Ecology* **128**, 1–30.

O'Brien DP, Ritz DA (1988) Escape responses of gregarious mysids (Crustacea: Mysidacea): towards a general classification of escape responses in aggregated crustaceans.

Journal of Experimental Marine Biology and Ecology **116**, 257–272.

O'Brien DP, Ritz DA, Kirkwood RM (1986) Stranding and matting behaviour in *Nyctiphanes australis* (Euphausiidae: Crustacea). *Marine Biology* **93**, 465–473.

O'Brien AL, Ross DJ, Keough MJ (2006) Effects of *Sabella spallanzanii* physical structure on soft sediment macrofaunal assemblages. *Marine and Freshwater Research* **57**, 363–371.

O'Connor WA, Newman LJ (2001) Halotolerance of the oyster predator, *Imogine mcgrathi*, a stylochid flatworm from Port Stephens, New South Wales, Australia. *Hydrobiologia* **459**, 157–163.

Paul NA, de Nys R, Steinberg PD (2006) Seaweed-herbivore interactions at a small scale: direct tests of feeding deterrence by filamentous algae. *Marine Ecology Progress Series* **323**, 1–9.

Ponder WF (1998a) Family Vanikoroidea. In *Mollusca: the Southern Synthesis, Part B. Fauna of Australia. Vol. 5.* (Eds PL Beeseley, GJB Ross and A Wells) pp. 769–772. CSIRO PUBLISHING, Melbourne.

Ponder WF (1998b) Family Marginellidae. In *Mollusca: the Southern Synthesis, Part B. Fauna of Australia. Volume 5.* (Eds PL Beeseley, GJB Ross and A Wells) pp. 838–41. CSIRO PUBLISHING, Melbourne.

Poore AGB (1994) Selective herbivory by amphipods inhabiting the brown alga *Zonaria angustata*. *Marine Ecology Progress Series* **107**, 113–123.

Poore AGB (2004) Spatial associations among algae affect host use in a herbivorous marine amphipod. *Oecologia* **140**, 104–112.

Poore AGB (2005) Scales of dispersal among hosts in a herbivorous marine amphipod. *Austral Ecology* **30**, 219–228.

Poore AGB, Hill NA (2006) Sources of variation in herbivore preference: among-individual and past diet effects on amphipod host choice. *Marine Biology* **149**, 1403–1410.

Poore AGB, Steinberg PD (1999) Preference-performance relationships and effects of host plant choice in an herbivorous marine amphipod. *Ecological Monographs* **69**, 443–464.

Poore AGB, Steinberg PD (2001) Host-plant adaptation in an herbivorous marine amphipod: genetic potential not realised in field populations. *Evolution* **55**, 68–80.

Poore AGB, Watson MJ, de Nys R, Lowry JK *et al.* (2000) Patterns of host use among alga- and sponge-associated amphipods. *Marine Ecology Progress Series* **208**, 183–196.

Poore AGB, Hill NA, Sotka EE (2007) Phylogenetic and geographic variation in host breadth and composition by herbivorous amphipods in the family Ampithoidae. *Evolution* **62–1**, 21–38.

Poore AGB, Campbell AH, Steinberg PD (2009) Natural densities of mesograzers fail to limit growth of macroalgae or their epiphytes in a temperate algal bed. *Journal of Ecology* **97**, 164–175.

Prudhoe S (1982) Flatworms. In *Marine Invertebrates of Southern Australia. Part I.* (Eds SA Shepherd and IM Thomas) pp. 220–227. Government Printer, Adelaide.

Ritz DA (1994) Social aggregation in pelagic invertebrates. *Advances in Marine Biology* **30**, 155–216.

Ritz DA, Hosie GW (1982) Production of the Euphausiid *Nyctophanes australis* in Storm Bay, south-eastern Tasmania. *Marine Biology* **68**, 103–108.

Ritz DA, Hosie GW, Kirkwood RJ (1990) Diet of *Nyctophanes australis* (Crustacea: Euphausiacea). *Australian Journal of Marine and Freshwater Research* **41**, 365–374.

Ritz DA, Osborn JE, Ocken AEJ (1997) Influence of food and predatory attack on mysid swarm dynamics. *Journal of the Marine Biological Station of the United Kingdom* **77**, 31–42.

Robertson AI, Lucas JS (1983) Food choice, feeding rates, and the turnover of macrophyte biomass by a surf-zone inhabiting amphipod. *Journal of Experimental Marine Biology and Ecology* **72**, 99–124.

Roe P (1988) Ecological implications of the reproductive biology of symbiotic nemerteans. *Hydrobiologia* **156**, 13–22.

Russell BD, Connell SD (2005) A novel interaction between nutrients and grazers alters relative dominance of marine habitats. *Marine Ecology Progress Series* **289**, 5–11.

Sato-Okoshi W (1999) Polydorid species (Polychaeta: Spionidae) in Japan, with descriptions of morphology, ecology and burrow structure. 1. Boring species *Journal of the Marine Biological Station of the United Kingdom* **79**, 831–848.

Scoresby W (1820) *An Account of the Arctic Regions with a History and Description of the Northern Whale Fishery.* Archibald Constable & Co., Edinburgh.

Sheard K (1953) Taxonomy, distribution and development of the Euphausiacea (Crustacea). *Report of B.A.N.Z Antarctic Research Expedition Ser. B* **8**(1), 1–72.

Shepherd SA, Womersley HBS (1976) The subtidal algal and seagrass ecology of St Francis Island, South Australia. *Transactions of the Royal Society of South Australia* **100**, 177–191.

Shepherd SA, Womersley HBS (1981) The algal and seagrass ecology of Waterloo Bay, South Australia. *Aquatic Botany* **11**, (4), 305–371.

Shields JD, Stephens FJ, Jones B (2006) Pathogens, parasites and other symbionts. In *Lobsters: Biology, Management, Aquaculture and Fisheries.* (Ed. BF Phillips) pp. 146–205. Blackwell Publishing, Oxford.

Smith SD (1996) The macrofaunal community of *Ecklonia radiata* holdfasts: variation associated with sediment regime, sponge cover and depth. *Australian Journal of Ecology* **21**, 144–153.

Smith SD, Simpson RD (1993) Effects of pollution on the macrofaunal community of the kelp *Ecklonia radiata*: discrimination at different taxonomic levels. *Marine Ecology Progress Series* **96**, 199–208.

Smith SD, Simpson RD, Cairns SC (1996) The macrofaunal community of *Ecklonia radiata* holdfasts: description of the faunal assemblage and variation associated with holdfast volume. *Australian Journal of Ecology* **21**, 81–95.

Staples DA (1997) Sea spiders or pycnogonids (Phylum Arthropoda). In *Marine Invertebrates of Southern Australia. Part III.* (Eds SA Shepherd and M Davies) pp. 1040–1072. SARDI (Aquatic Sciences) and Flora and Fauna of South Australia Handbooks Committee, Adelaide.

Steneck RS, Watling L (1982) Feeding capabilities and limitation of herbivorous molluscs: a functional group approach. *Marine Biology* **68**, 299–319.

Tanner JE (2005) Edge effects on fauna in fragmented seagrass meadows. *Austral Ecology* **30**, 210–218.

Tanner JE (2006) Landscape ecology of interactions between seagrass and mobile epifauna: the matrix matters. *Estuarine, Coastal and Shelf Science* **68**, 404–412.

Taylor RB (1998) Density, biomass and productivity of animals in four subtidal rocky reef habitats: the importance of small mobile invertebrates. *Marine Ecology Progress Series* **172**, 37–51.

Taylor RB, Steinberg PD (2005) Host use by Australasian seaweed mesograzers in relation to feeding preferences of larger grazers. *Ecology* **86**, 2955–2967.

Thomas M, Day RW (1995) Site selection by a small drilling predator: why does the gastropod *Haustrum baileyanum* drill over muscle tissue of the abalone *Haliotis rubra? Marine and Freshwater Research* **46**, 647–656.

Underwood AJ, Verstegen PH (1988) Experiments on the association between the intertidal amphipod *Hyale media* Dana and the limpet *Cellana tramoserica* (Sowerby). *Journal of Experimental Marine Biology and Ecology* **119**, 83–98.

Walker LM (2011) A review of the current status of the *Polydora*-complex (Polychaeta: Spionidae) in Australia and a checklist of recorded species. *Zootaxa* **2751**, 40–62.

Wells FE, Keesing JK (1987) Population characteristics of the gastropod *Cantharidus pulcherrimus* on intertidal platforms in the Perth area of Western Australia. *Journal of the Malacological Society of Australia* **8**, 23–35.

Wilson B (1993–1994) *Australian Marine Shells. Prosobranch Gastropods. Parts 1,2.* Odyssey Publishing, Perth.

Wiltshire K, Rowling K, Deveney M (2010) 'Introduced marine species in South Australia. A review of records and distribution mapping'. Report No. F2010/000305-1. SARDI Research Report Series No. 468, Adelaide.

Zottoli RA, Carriker MR (1974) Burrow morphology, tube formation and micro-architecture of shell dissolution by the spionid polychaete *Polydora websteri*. *Marine Biology* **27**, 307–316.

9 The molluscan megafauna: herbivores, carnivores and filter-feeders

'Some have a mouth and teeth, as the snail, which has small, sharp, and smooth teeth. They also have a proboscis . . . and this organ is like a tongue. In the whelk this organ is hard, and powerful in strength, for they can pierce through the shells of the baits.'

Aristotle, *Historia Animalium*, Bk. IV, Ch. IV

OVERVIEW

The molluscan megafauna discussed in this chapter comprises molluscs from ~2 cm size and above. The important representative groups discussed are limpets, chitons, shelled gastropods, opisthobranchs, mussels and a scallop. They include grazers, consumers and filter feeders. We describe the general aspects of the ecology of each group and give examples of southern reef-associated species whose ecology has been studied. First we consider two species of limpet that extend into the subtidal zone, and then the five major species of abalone: *Haliotis laevigata*, *H. rubra*, *H. scalaris*, *H. roei* and *H. cyclobates*. We describe some predators of abalone and the prey responses to those predators. An example of the process of collapse of an exploited population under intense fishing is given to illustrate a major threat to their survival, followed by a summary of the fisheries management principles. We also describe the ecology of two commonly exploited turbinid whelks, *Turbo undulatus* and *Turbo torquatus*. The general ecology of chitons is then briefly discussed, followed by an account of the diet and ecology of opisthobranchs and notably of two herbivores: the sea hare *Aplysia parvula* and the sacoglossan *Placida dendritica*, with an account of their symbioses with chloroplasts and zooxanthellae. This is followed by a description of the carnivorous nudibranchs, their defences against predators and one notable defence armament: the ability to sequester nematocysts from certain prey.

We then consider several carnivorous molluscs, most of them generalist predators, and describe in more detail the ecology of three well-known predators: *Dicathais orbita*, *Cabestana spengleri* and *Conus anemone*. We consider the life history and feeding ecology of the large cowries in the genera *Zoila* and *Umbilia*, both much prized by shell collectors. Lastly, we describe the life histories of filter-feeding molluscs, specifically mussels in the genera *Mytilus*, *Limnoperna* and *Trichoma*, and finally the doughboy scallop *Mimachlamys asperrima*.

INTRODUCTION

The diversity of habitats on southern Australian shallow reefs, variously comprising limestone, dolerite, sandstone, basalt and granitic substrata, and their dominance by algal forests, provides an abundant food source for a remarkably diverse fauna of invertebrate grazers and detritivores, such as crustaceans, molluscs, echinoderms and several groups of worms (Edgar 2001; Connell and Vanderklift 2008). Here, we consider one extremely diverse group – the molluscan megafauna – ranging in size from a few centimetres to > 20 cm for the largest abalone. They include limpets, chitons and grazing gastropods. The ecology of most is poorly understood; hence, their role in benthic systems is yet to be properly assessed.

GRAZERS AND BROWSERS

Although limpets and some grazing molluscs often dominate the intertidal zone, very few limpets live subtidally, and small grazing molluscs are inconspicuous among the kelp forests and understorey algae, tending to remain hidden within algal fronds or in crevices and caves, or under boulders. Only a few larger species have been studied in any detail, and here we summarise what is known about them.

Firstly, two kinds of algal feeding should be distinguished – grazing and browsing. The former refers to species that consume algal plants indiscriminately (whether filamentous algal turf or larger species, and whether attached or not), and the term 'browser' refers to species that feed on specific parts of attached algae, such as ramuli or blades. The term 'detritivore' refers to those species that feed on algal drift in various stages of decomposition (Connell and Vanderklift 2008).

Limpets

Although most limpets live intertidally at locations largely free of macroalgae, where they can cling more easily to the rock and graze microalgae, several species extend into the subtidal zone, especially in areas grazed bare by sea-urchins in SE Australia. The ecology of two well-studied species, *Patelloida alticostata* and *Scutellastra peroni*, is described.

The lottiid limpet *Patelloida alticostata*

The life history of this widely distributed, southern limpet has been studied by Anderson (1965), Parry (1982b) and Fletcher (1987). The species is dioecious (with two sexes), and eggs are apparently laid as an egg mass during spawning events. These occur from October to March in New South Wales, but only in February in Victoria (Parry 1982a). Eggs hatch into trochophores, an early planktonic larval stage, 18 h after fertilisation and, after 4 more days, the larvae metamorphose and settle on the bottom. In New South Wales, recruiting young appear 6 months after spawning (i.e. in November–January and March–May) and grow to ~20 mm in the first year, ~32 mm in 2 years, and to 35 mm in the third. They can reach a maximum size of ~48 mm in 7 years, which is about their longevity. Growth rates in Victoria are much lower: 1-year olds average 7 mm, 2-year olds 12 mm and 3-year olds are 15 mm and sexually mature. Maximum size is ~30 mm and mortality rates range from 40–60% a year.

This limpet feeds by rasping microalgae from the rock with its radula at rates exceeding 1 rasp per second (as do other limpets in experiments where feeding has been detected using underwater hydrophones – Petraitis 1992). Consequently, limpet abundance can be reduced by intense competition with other limpets feeding on the same food resource (Fletcher and Creese 1985; Fletcher and Underwood 1987). However, in sea-urchin barrens, where this limpet can be very common, its diet is > 90% crustose coralline algae, and the rest diatoms, while in porous limestone habitats it rasps deep into the rock to feed on endolithic microalgae (Black *et al.* 1988). On other limestone reefs in Western Australia, it goes on feeding excursions, always returning to its home scar by following its mucus trail (Black 1977). At one intertidal site, ~1/3 of deaths were caused by the drilling whelk *Dicathais orbita*, with heat and desiccation likely causing much of the remainder. In New South Wales, the limpet's mortality rate was estimated to be 36–39% a year subtidally, which was less than half the mortality rates of intertidal populations. Major predators are whelks (e.g. *Dicathais orbita*) and wrasses (Ludbrook and Gowlett-Holmes 1989).

The patellid limpet *Scutellastra peroni*

This limpet ranges around the southern half of Australia to ~20 m depth, and is common in sea-urchin barrens where densities can reach > 50 m^{-2} (see Fig. 12.4). It spawns twice a year (June and February–April) in New South Wales, but only in late summer (February–March) in Victoria (Parry 1982a). Like other limpets, this limpet has a short, lecithotrophic (i.e. they feed on their egg yolks) larval stage, as shown by the appearance of juveniles a month or so after spawning (Underwood 1974). In Victoria, this limpet grows to ~20 mm in year 1, to 25 mm in year 2 (the size of sexual maturity), to 30 mm in year 3, and to 35 mm in year 4. Maximum size is ~47 mm at ~7 years of age. Mortality rates were estimated to be ~50% per year. The major predators on this and other limpets are the purple wrasse *Notolabrus fucicola* and the blue-throated wrasse *Notolabrus tetricus*; minor predators include the oystercatcher, which feeds on limpets at the lowest spring tides, and whelks such as *Dicathais orbita* and *Agnewia tritoniformis* (Parry 1982b). In an overall comparison of life histories of four limpets, Greg Parry showed that species with higher mortality rates evolved to invest relatively more in reproductive effort than species with lower mortality rates – an example of the familiar *r*- and *k*-selection theory.

Turbinid gastropods

Trochid and turbinid snails, followed perhaps by abalone, are the most widespread and abundant subtidal herbivores, ranging in size from very small species (Chapter 8) to the very large, such as greenlip and blacklip abalone growing to > 20 cm, as discussed below. Here we consider the ecology of several larger species.

Two turbinids, *Turbo undulatus* and *Turbo torquatus* (Plates 17, 18), have been studied in the most detail. A third species, Jourdan's turban shell *Turbo jourdani*, is one of the largest southern gastropods, growing to ~25 cm size. This *Turbo* is the only mollusc that can create mini-barrens and is sometimes seen at the edge of its bared patch of up to a square metre in size, which it has grazed clean of macroalgae. It ranges south from Geraldton, Western Australia and east through the Great Australian Bight (GAB), mainly at depths of > 10 m.

The warrener or periwinkle *Turbo undulatus*

Although *Turbo undulatus* reaches a maximum size of ~4 cm over most of its range throughout southern Australia and New South Wales, it can grow to almost 7 cm where food is abundant. It is the smallest turbinid to be studied in Australia – mainly because it is fished commercially in Bass Strait and SE Australia.

The warrener ranges from the intertidal zone to a depth of ~10 m and can occur in densities up to ~100 m^{-2}; hence, it can have a substantial effect on algal communities. In Tasmania, the shell width can grow ~8–9 mm a year, so that it reaches a maximum width (68 mm) in ~6 years (Zacharin 1992). In New South Wales, growth rates appear to be much faster (Steinberg and van Altena 1992). The species is dioecious (two sexes) with a 1:1 sex ratio, and reaches sexual maturity at ~27 mm shell width. During spawning, eggs and sperm are released into the water where fertilisation occurs. Epidemic spawning may be induced by rough seas, as occurs in many vetigastropods, but this has yet to be verified for this species (Ward and Davis 2002). The eggs hatch into non-feeding larvae that drift in the water column for a few days before settlement. The spawning season extends from October to May in New South Wales (Underwood 1974), but starts 2–3 months later on the east and west coasts of Tasmania, and even later, in March–April, at Recherche Bay in southern Tasmania (Zacharin 1992).

The warrener is a generalist grazer. In South Australia, it often lives and feeds on the blades of the kelp *Ecklonia radiata*, while green algae *Caulerpa* spp. are also favoured at some sites (Clarkson and Shepherd 1985). In New South Wales, Worthington and Fairweather (1989) found that, in the lower intertidal zone, it preferred to live in a coralline turf habitat, probably because more shelter was available there. Here, a third of its diet was coralline algae, with the remainder filamentous brown and green algae. Curiously, *T. undulatus* has a broader diet than might be predicted from its rhipidoglossan radula. In explanation of this, Worthington and Fairweather (1989) showed that it has unusually large central teeth in its radula, enabling it to rasp more deeply into tough brown algae or corallines.

In another New South Wales study, Steinberg and van Altena (1992) examined the role of chemical deterrents present in different algae in affecting the food preferences of this *Turbo*. They found that phlorotannins – tannins present in *Ecklonia* and species of *Sargassum* – were only a mild deterrent to its feeding, and that it fed equally on both algal genera, each of which had substantial levels of these deterrents in their tissues. A surprisingly subtle outcome of their study was that body weight, but not shell weight, increased on a diet of only *Ecklonia*, implying that a mixed algal diet is necessary for this species.

The turban *Turbo torquatus*

This large turban snail grows to ~12 cm width. It has a disjunct distribution in southern Australia, being absent from Victorian and Tasmanian waters (though fossils have been preserved there in quaternary deposits, indicating past warmer conditions). In New South Wales, the turban is exploited commercially in a small fishery that is concentrated in shallow waters to ~10 m depth. On the southern coasts of Western Australia and South Australia, the turban is much less abundant than on the east and west coasts of Australia, rarely reaching densities of 0.1–0.2 m^{-2} (see references in Wernberg *et al.* 2008).

Like its congener, this species is dioecious, with spawning occurring from October–January and May to July on both the east and west coasts of Australia (Joll 1980; Ward and Davis 2002). This turban inhabits kelp forests, usually < 20 m depth, at densities of up to 3 m^{-2}, and prefers resting sites under the algal canopy or in crevices (Ettinger-Epstein and Kingsford 2008; Wernberg *et al.* 2008). The former study lasted over 18 years and found that the kelp habitat was important, providing both shelter from swell and a food source. In a 10-year cycle of changing kelp abundance, under the influence of the Pacific Decadal Oscillation, turban abundance followed the same oscillation. Another finding of the study was that tagged individuals moved extensively within and between kelp patches, up to a distance of ~40 m per month. Juveniles apparently recruited into shallow, complex coralline habitat, and then moved into deeper water as they grew.

The diet of the turban has been examined at several sites. In Western Australia, it consumes a wide range of macroalgae (e.g. *Ulva, Hypnea, Hennedya* and *Ecklonia*), at least in aquarium experiments. In South Australia, in a fucoid/*Ecklonia* habitat, it browses mainly on the erect coralline *Haliptilon roseum* (50%) and juvenile *Cystophora* plants (22%), while additionally ingesting filamentous algae, red and green macroalgae, and seagrass (Clarkson and Shepherd 1985). In New South Wales, the turban consumed a wide range of algae in this order of preference:

Ulva > *Pterocladiella* > *Dictyopteris* > *Delisea* > *Ecklonia*

Chemical deterrents, such as phlorotannins, apparently played only a minor role in influencing food preferences (Taylor and Steinberg 2005). Hence, both the turban shell and the smaller warrener can be considered generalist herbivores, but moderately selective of algal food species, depending on availability and palatability.

Abalone *Haliotis* spp.

Abalone occur on most coasts of the world and have been known since the 4th Century BC when Aristotle called them 'sea ears', which is the meaning of their generic name *Haliotis.* Abalone have a single, flattened, cup-shaped and spirally curved shell with a row of pore-holes, that marks the passage of an exhalant current carrying wastes from the body. The shell protects the muscular foot, the gut and gonad. On high-energy coasts, some species have a smooth shell, which reduces drag from water movement, while other species have an elaborately sculptured shell that increases protection against shell-crushing predators. Hence shell shape is a trade-off between the various forces acting on it. The abalone shell is structurally one of the strongest natural products known, due to its fine laminated micro-structure, and has inspired materials scientists to develop mimicking synthetic materials for spacecraft (Sellinger *et al.* 1998).

Abalone belong to a primitive group of gastropods that evolved in the Tethys Sea or the Indo-West Pacific, and radiated from there (see Chapter 3). Of the many tropical species, five extend down the east and/or west Australian coasts, of which two species, *H. coccoradiata* and *H. brazieri*, are limited to eastern Australia; five temperate species occur on the southern coasts (Geiger 1999; Geiger and Owen 2012). Three of these five are exploited commercially and comprise substantial fisheries in the southern States.

As noted in Chapter 3, two of the these three species, greenlip and blacklip abalone, grow to > 20 cm long, and both likely evolved to that size when the giant kelp *Macrocystis* arrived from the Northern Hemisphere and a rich red algal flora evolved in the region. The exploited species have been studied in Australia more than any other mollusc, and this brief account of their ecology emphasises some little known curiosities, and is drawn from Shepherd *et al.* (1992), unpublished reports and later reviews, such as McShane (1999), Shepherd (1999, 2008) and Morgan and Shepherd (2006).

Ecology of five southern abalone species

Abalone live attached to a firm substratum and each species has a distinctive habitat. The small *Haliotis cyclobates* lives in sheltered waters from Esperance, Western Australia to the central Victorian coasts, and is especially common on the razor shell *Pinna bicolor* among seagrass beds of the South Australian Gulfs to depths of ~20 m. *Haliotis scalaris* is a small cryptic species found under rocks from the subtidal fringe to depths of ~50 m, and Roe's abalone *Haliotis roei* lives in narrow crevices or on intertidal reef platforms on exposed coasts to a depth of 3–5 m. The habitat of greenlip abalone *Haliotis laevigata* is on rock of low relief near the sand-line at depths of 5–50 m, on moderate- to high-energy coasts or in places of strong tidal currents. In contrast, the habitat of blacklip abalone *Haliotis rubra* is in crevices, caves or vertical walls, usually on reefs of high relief on low- to high-energy coasts to depths of ~25 m. Another supposed species, brownlip, common in Western Australia, is a subspecies of blacklip, and named *Haliotis rubra conicopora* (Geiger 1999).

All five abalone species are herbivores and show a strong preference for red algae, rejecting many tough and leathery brown algal species with high phenolic content (a chemical deterrent to grazing). However, *H. roei* is exceptional because it does feed on the kelp *Ecklonia* and the fucoids *Sargassum* spp., as well as on red and green algae (Wells and Keesing 1989; Shepherd and Steinberg 1992). In Tasmania, Guest *et al.* (2008) found that blacklip abalone fed substantially on brown algae, probably *Macrocystis* and *Durvillaea*, both of which are low in phenolics. But perhaps the most interesting aspect of abalone feeding is the strong preference that all species have for catching drift algae, rather than foraging for it, probably due to the abundance of algal drift on open southern coasts. To achieve this feat, even during a strong swell, the abalone elevates the front of the foot, and adeptly seizes a piece of fast drifting alga floating past. At first sight, this specialised way of feeding seems too uncertain to satisfy an abalone's needs, but there are three distinct advantages:

- Foraging, with a high risk of being taken by a predator, is avoided.
- A diversity of algal species, often carried from distant sites or from seagrass, is available in the drift.
- The abalone can rest in a protected or cryptic site, where drift algae accumulate or are carried, and can feed night and day.

Life history

Abalone have two sexes and generally become sexually mature at 2–4 years of age. Each species has a specific spawning season in which eggs and sperm are released into the water (Table 9.1). Spawning is synchronous in a population and is usually induced by an environmental trigger, such as a lunar cycle or a sharp change in swell conditions. Greenlip abalone spawn around the time of the new moon, and *H. scalaris* around the full moon, while blacklip spawn epidemically in oily calm, low-swell conditions (Nash *et al.* 1995; Shepherd and Daume 1996; Morgan and Shepherd 2006). Abalone naturally aggregate together in clusters before the spawning season, and Babcock and Keesing (1999) showed in a series of field experiments that the fertilisation success rate of eggs falls sharply with increasing distance between abalone (Fig. 9.1). Hence, intense fishing, by reducing population densities to a low level (termed the Allee effect), seriously reduces fertilisation success, and subsequent recruitment to the population.

Abalone are essentially sedentary all their life. Juveniles move about in the local vicinity, often to catch a piece of drift algae, while adults tend to move even less, especially when living at a site with abundant drift algae. Hence, a greenlip abalone on one reef

Table 9.1: Life history features of five temperate Australian abalone species, after Day and Fleming (1992), Shepherd and Breen (1992) and Shepherd (2008).

SL = shell length; the parameters K and L_∞ relate to the growth curves fitted to abalone. Legal size limits may vary with region, and whether fished by recreational or commercial fishers.

Feature	*H. laevigata*	*H. rubra*	*H. roei*	*H. scalaris*	*H. cyclobates*
Growth rate (linear) (mm/year) (curvilinear phase). K/year	20–28 0.4–0.6	~20 0.15–0.4	~20 0.3–0.44	14–20	~18 0.32
Maximum age (years)	15–30	15–20	8–10	~8	~6
Max. size (L_∞) (mm)	130–150	115–175	80–86	~100	~70
Sexual maturity (mm SL)	75–115	70–110	40–50	50–70	40–50
Spawning season	Oct.–Dec.	Oct.–Dec. Mar.–Jun.	All year [1] July–Dec.[2]	Nov.–July	Oct.–Dec.
Annual mortality (%) Juveniles (< 1 year) 1–2 years > 2 years	 97–99 ~50 20–45	 90–99 50–80 10–40	 No data	 67	 51
Legal size limits (mm SL)	130–145	130–145	60–70	–	–

[1] In South Australia (Shepherd and Laws 1974); [2] in Western Australia (Wells and Keesing 1989);

may remain on the same rock the whole of its life, while on another reef exposed to swell it may move many metres over time, usually into the direction of the approaching swell until it finds an optimal site to catch drift algae. Roe's abalone may move between the shallow subtidal and intertidal zones, and blacklip movement is usually restricted to the same cave or crevice. *H. cyclobates* moves about locally, often in nearby seagrass beds, to graze on epiphytic algae before returning to firm substrate, whether on rock or a razor shell.

Abalone larvae have a planktonic stage of ~5 days and during that time can drift for 20–30 km, but may be retained within the same bay or even in the same reef system, depending on coastal topography, larval behaviour and local currents. Larvae of shallow-water species, such as blacklip, tend to be philopatric (settling on or near their natal reefs), while larvae of

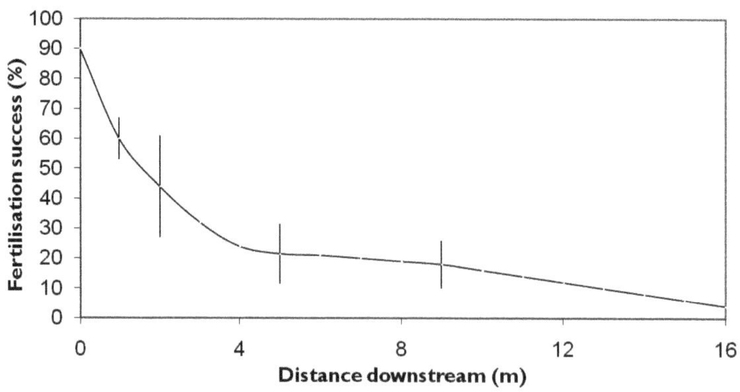

Figure 9.1: Fertilisation success rate of greenlip abalone eggs measured from field experiments in which sperm were released at one given point and eggs were released at six points at increasing distances (0–16 m) downstream (redrawn from Babcock and Keesing 1999). Vertical lines are standard deviations.

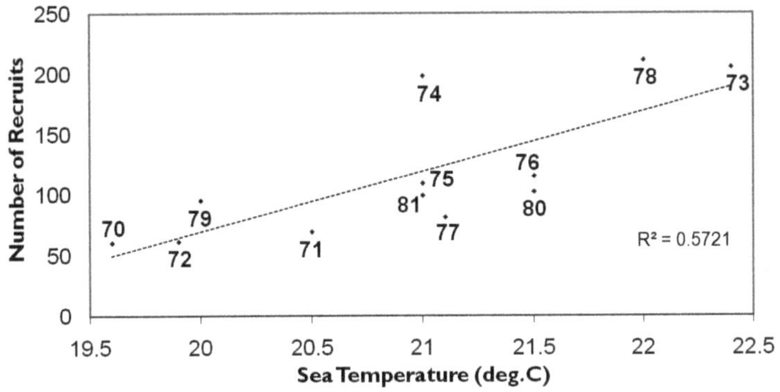

Figure 9.2: Plot of recruitment strength of 1-year-old greenlip abalone versus mean October–December sea temperature for the previous year (i.e. during the pelagic larval stage), measured over 12 years at a study site at West I. (after Shepherd 1984).

deep-water species tend to be dispersed more widely (reviewed by Morgan and Shepherd 2006; Piggott *et al.* 2008). In the last few days of its pelagic life, the late veliger larva of greenlip abalone alternately swims up and down in the water column, and eventually on the fifth day stays near the bottom searching for a suitable crustose coralline micro-habitat on which to settle. On finding this substratum, the veliger is induced to settle and undergoes metamorphosis. The abalone's relation with coralline algae is an example of a mutualism, which benefits both host and settler (see Chapter 6). Another cue that can induce abalone larvae to settle is the mucous trails of large abalone – termed the 'slime trail hypothesis' – a mechanism whereby abalone larvae tend to settle near others of the same species. This is clearly adaptive for the species, and also potentially has serious implications for a fishery, because if a local population is depleted in an area, veligers are less likely to settle there, and continued recruitment failures will inevitably lead to that population's local extinction. An interesting exception to the abalone veliger's preference for corallines is *H. cyclobates*, which can also settle on seagrass blades (Stevenson and Melville 1999). For the first month or so after settlement, the tiny abalone feeds on bacteria, diatoms and mucus, and later filamentous algae on the coralline surface (Chapter 6).

The rate of settlement, and later recruitment, to an abalone population is critical for its persistence, and this is especially so for exploited species subject to continual fishing pressure. In a study of recruitment strength of one greenlip population in a reserve at West I., South Australia, Shepherd (1984) found that in warm summers recruitment was up to four times higher than in cool summers (Fig. 9.2). The same pattern of recruitment was reflected in the catch 6 years after larval settlement for the whole of the greenlip abalone catch in the western zone of South Australia during the 1970–80s before catch quotas were introduced. Hence, the relationship between recruitment strength and sea temperature appears to be a general one, and likely due to the fact that cooler summers are windier, so that larvae are more likely to be transported away from their parent population in surface currents, as discussed below.

Population structure

Abalone populations, like most benthic species, typically occur patchily within local populations on reefs (Morgan and Shepherd 2006). These patches are separated by unsuitable habitat, but the local populations are linked by larval dispersal, and together the cluster of local populations form what are termed metapopulations, which themselves are genetically distinct, and separated from other metapopulations. In blacklip abalone, metapopulations typically comprise several local populations close to shore (Fig. 9.3), forming a stepping stone series, with each local population connected to its neighbours by larval dispersal. In Roe's abalone, reefs 10 km apart have distinct metapopulations (Hancock 2003). However, in greenlip abalone, metapopulations usually occur on expanses of

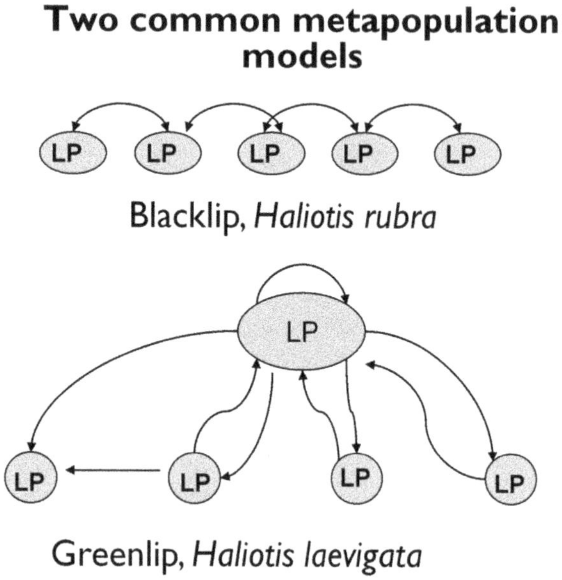

Figure 9.3: Models illustrating two common types of metapopulation structure in blacklip and greenlip abalone. Local populations (LP) are represented by circles or ellipses, and their inter-connections with other LPs by larval dispersal shown by arrows. The blacklip model (upper) is common along exposed coasts, where LPs are in series and inter-connected by local currents. The model may also apply to Roe's abalone. The greenlip model (lower) is typical of the model shown by greenlip on deeper reefs, depending on local currents (modified from Morgan and Shepherd 2006).

moderately deep reefs of low relief, and are typically separated by 30–50 km of unsuitable habitat (e.g. sand) or unfavourable currents preventing larval transport between them. Hence, metapopulations of an abalone species are essentially independent stocks scattered along a coastline, separated by headlands, stretches of sandy beach or adverse currents.

Metapopulation structure can be visualised as a group of circles, each representing a local population as shown in Fig. 9.3. Under intense fishing, metapopulations can spatially contract as the likelihood of larvae being transported to more distant reefs declines. Eventually, the metapopulation will collapse and, if overfishing is widespread, serial collapse of a fishery can occur. One monitored greenlip metapopulation in South Australia (Fig. 9.4; Shepherd *et al.* 2001) conformed partly to the blacklip stepping-stone model of a metapopulation (Fig. 9.3), and partly to a source-sink model, as the major source of larvae was the upstream 'source' population. Under intense fishing, the local populations contracted spatially, recruitment declined (Fig. 9.5), likely aided by two mechanisms – the Allee effect and the slime trail hypothesis – and then the local populations collapsed serially from upstream to downstream over the succeeding years. The arrow in Fig. 9.5 shows the approximate threshold adult density critical for the maintenance of this stock.

Genetic studies have commonly been used to establish or confirm the metapopulation structure of abalone, as described above, and recent studies on blacklip have thrown more light on the frequency and scale of dispersal in blacklip larvae in different conditions, such as in both healthy and overexploited stocks, and in exposed and sheltered habitats (Temby *et al.* 2007; Appleyard *et al.* 2008; Miller *et al.* 2009). In SE Tasmania, genetic analyses suggest bimodality of larval dispersal, whereby most larvae settle in or close to their natal habitat, while a few are dispersed more widely by coastal currents. No differences were found between embayment populations and those on more open coasts. On the other hand, in SE South Australia the picture was rather one of more extensive larval dispersal along the coast, possibly because of the strong prevailing swells and coastal currents.

Growth

The growth rate of most species is well known from tagging or ageing studies, and varies widely from place to place depending on algal food supply, water movement and other habitat factors (Day and Fleming 1992). Examples of growth curves for the five southern species at specific sites (Fig. 9.6) show the large differences between species. Most species have an initial linear phase of growth for the first few years and then a curvilinear phase as growth slows down in the reproductive years and the abalone approaches its maximum size (Table 9.1). Descriptions of the curvilinear phase of growth are usually given in terms of the von Bertalanffy growth equation fitted to the data, and described by two growth parameters, K and $L\infty$, where K is the rate of approach to $L\infty$, the average

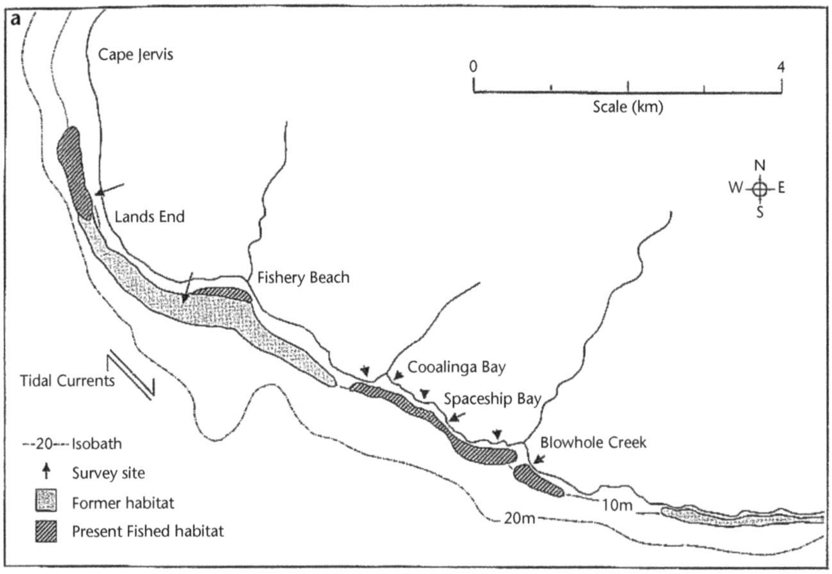

Figure 9.4: Map of Backstairs Passage, South Australia, showing original extent of greenlip local populations in 1970, and their gradual spatial contraction and eventual collapse under intense fishing. Source populations are to the NW, because net water outflow is to the SE. Under fishing pressure, the source LP collapsed first and the others later collapsed serially downstream (after Shepherd *et al.* 2001).

maximum length reached by the local population. The growth rate of Roe's abalone in the first year (20 mm) found by Hancock (2003) is only half that recorded by Keesing and Wells (1989), suggesting that early growth is highly variable between sites.

In one 30-year long-term study of a heavily fished greenlip abalone population in Waterloo Bay facing the eastern GAB (SA Shepherd unpublished data), the growth rate of the 3+ and older age classes (i.e. from ~70 mm shell length) slowed down by almost

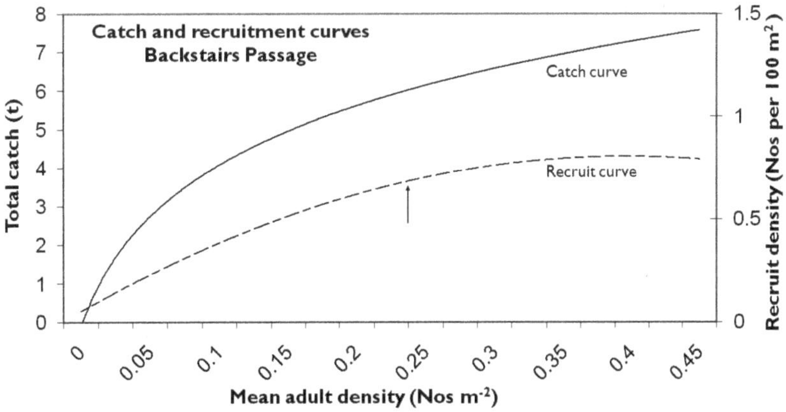

Figure 9.5: Smoothed plots of greenlip abalone in Backstairs Passage (see Fig. 9.4), showing (1) a catch curve, i.e. averaged total annual catch (t) versus average adult density (numbers of 4+ age class m^{-2}), and (2) averaged recruit density (numbers of the 3+ age-class 100 m^{-2}) versus average adult density 3 years earlier, as the metapopulation gradually collapsed from 1979 to 1999. The arrow indicates approximate threshold density for a sustainable metapopulation (redrawn from Shepherd *et al.* 2001).

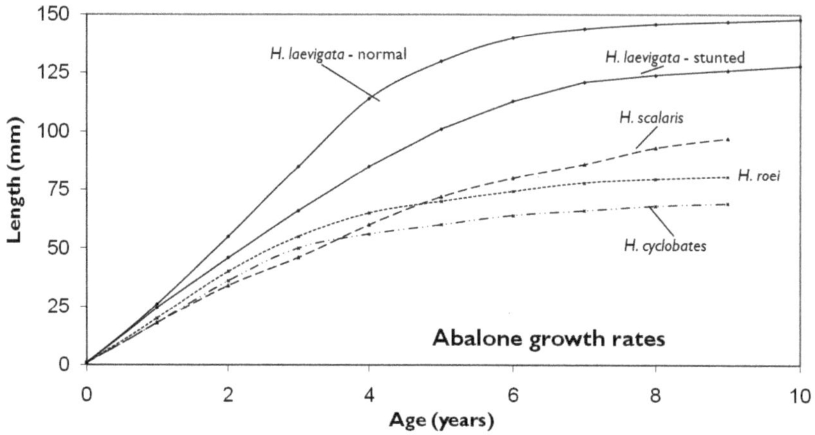

Figure 9.6: Generalised growth curves for five species of abalone. The two curves for greenlip *H. laevigata* in Waterloo Bay are: 'normal' in 1967–70 in early fishing days; and 'stunted' in 1996, after the fishery collapse (Shepherd and Hearn 1983; Shepherd unpublished data). The curves for *H. scalaris* and *H. cyclobates* are derived from Shepherd 2008, and that for *H. roei* from Hancock 2003.

50% from 30 mm year^{-1} to 18 mm year^{-1} between 1969, when fishing began, and 1997 after the fishery collapsed; the bay was then closed to fishing (see Fig. 9.6). Furthermore, the size of sexual maturity declined from 102 mm SL to 75 mm SL, indicating that the abalone was diverting energy from growth into reproduction. A complexity of the abalone's growth in this population was that females grew ~25% faster than males (Shepherd and Hearn 1983). This meant that when the size limit was increased from 130 mm to 145 mm SL in 1986 to enable a recovery of the population, males rarely reached the size limit and only females were fished. This led to a severely biased male:female sex ratio of ~17:1, a failure of recruitment and a population collapse. The study is an example of how fishers, fishing above a legal minimum size, select the faster growing individuals in a population, leaving mainly stunted abalone to reproduce. In this small metapopulation with no refuge to maintain genetic diversity, rapid evolutionary change in critical life history traits occurred in about four generations, and precipitated the collapse (e.g. Miethe *et al.* 2010).

Predators and abalone responses to them

The predators of abalone vary with size. Very small abalone < 5 mm SL are taken by predators that share their coralline habitat. They include anemones (which mainly capture settling veligers), flatworms, nemerteans, sea-stars and crabs. With increasing size abalone become more obvious, and vulnerable to wrasses, crabs, whelks and sea-stars. Eventually, those surviving to 50–100 mm SL gradually emerge from their crevice habitat into the open, where they may be taken by rays and blue groper, particularly when their shells are weakened by boring polychaete worms or sponges. In these attacks on adults, the ray attempts to crush the shell, and the groper uses a ram-and-bite technique to dislodge or crush the prey (Chapters 16, 17; Shepherd 2006). Overall, abalone mortality rates decline with size, as the shell thickens and strengthens (Table 9.1).

Yet, abalone are not without defences. For example, small abalone have well-developed, stereotyped responses to slow moving predators such as the 11-armed sea-star *Coscinasterias muricata* and whelks (Shepherd 1978). Parsons and MacMillan (1979) described four sequential components to the flight response as follows:

1. When first approached by the sea-star, the abalone protrudes its tentacles and waves them about.
2. It then 'mushrooms' by elevating the shell up to 1 cm above the bottom.
3. It then engages in either shell-twisting or thrusting behaviour. In the former, used against a sea-star, it

violently twists the shell back and forth through an arc of up to 90°, and in the latter, used against a whelk, it thrusts the shell vigorously forward to push away the predator.

4. Finally, it elevates the forepart of the foot and flees, as shown in Plate 19.

However, Day *et al.* (1995) showed that not all of the four sequences of responses were rigidly followed, because the blacklip abalone used in their experiments often took off in flight as soon as the predator was detected. These authors also made an interesting discovery. As the small abalone fled, they ejected viscous white mucus through their 4th shell pore from the edge – sufficient to cause the sea-star to move off in another direction. The mucus is produced by many tiny goblet cells in the tissue surrounding the foot and appears to be an acidic defensive deterrent against predators.

The escape response of abalone, however, can be dangerous, especially in the presence of wrasse predators, which sometimes follow an 11-armed sea-star, and seize any escaping prey. Besides, in very cramped cave and crevice habitats, an escape response may not be possible. In these situations, and for large abalone in open habitats, a more appropriate response is to clamp down on the substrate, where dislodgement is impossible for most predators. In this response, the abalone typically spreads its foot over the rock to increase adhesion and extends all its sensory tentacles. In a study of the variable responses of abalone to the sea-star, Shepherd (1978) found that four species switched from a flight to a clamping response at a size and manner unique to each species, depending on habitat and size of emergence (Fig. 9.7). Some blacklip seem to retain an escape response at all sizes, and have been observed fleeing up a vertical face with a sea-star in hot pursuit, and then letting go and falling to the bottom in a very effective escape from its slower moving predator. In Roe's abalone, both responses occurred, with a gradual switch to clamping with increasing size. However, *H. scalaris* has a response that varies with size and habitat. At a small size, it is cryptic in narrow crevices, where sea-stars cannot enter, so clamping tends to be the best response. As it grows and enters larger crevices, flight becomes more common, and then decreases again at a large size. The reason for this latter decline is unclear, but may be due to the high risk of fleeing in a habitat where predatory wrasses dominate. The role that mucus discharge plays in the clamping response of abalone has yet to be investigated.

A common parasite of abalone in SE Australia is the muricid snail *Haustrum baileyanum*, which bores through the abalone shell and feeds on the tissue below (Chapter 8). The abalone does not passively endure the snail's assault, and has some defences (Thomas and Day 1995). If the abalone detects the snail's approach, it either uses a shell-twisting or thrusting behaviour to push away the snail, or it

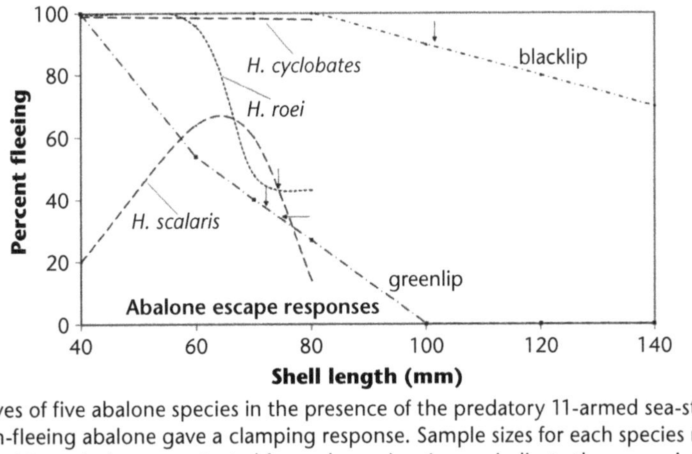

Figure 9.7: Response curves of five abalone species in the presence of the predatory 11-armed sea-star *Coscinasterias muricata*. In each trial, non-fleeing abalone gave a clamping response. Sample sizes for each species ranged from 47 to 134. About 20 animals were used in each size group tested for each species. Arrows indicate the approximate size of emergence of each species from cryptic habitat.

crawls onto the snail. If neither defence succeeds, the abalone may then, if resting in a crevice, rub its shell dorsally against the rock above, and dislodge the snail (unpublished observations). If all else fails and the snail succeeds in drilling into shell, the abalone secretes a black proteinaceous layer at the drilling site, and forms a black nodule inside the shell – a process that repairs the damage. It is not clear from the above study whether the snail finally withdraws because it is satiated or because of the abalone's secretion.

Diseases of abalone

Over the history of abalone fishing, diseases that have decimated abalone populations have been occasionally been reported and investigated. A serious parasite infesting blacklip and greenlip is the protozoan *Perkinsus*, which produces soft, yellow pustules in the foot. The disease decimated abalone populations in Gulf St Vincent and affected blacklip populations elsewhere in SE Australia (Goggin and Lester 1995; Shepherd 2008). More recently in 2005, a ganglioneuritis infection of blacklip abalone escaped from an onshore hatchery and invaded wild populations along the western Victorian coast, inducing a mortality rate of 40–90% (Gorfine *et al.* 2008). This herpes-like disease damages the nerve tissues and generally results in paralysis and death. Another, but less common, disease is 'Blue Spot', which discolours the foot, and occurs sporadically in Tasmanian coastal waters when sea temperatures rise above ~18°C. This happens when the East Australia Current penetrates strongly to the south in SE Australia (Thrower 1977). A similar 'marine heatwave' occurred on the SW Australian coast during an extreme La Niña event, when sea temperatures increased 3–5°C in February 2011 and caused a total mortality of Roe's abalone over hundreds of kilometres (Pearce *et al.* 2011). Such extreme events may well recur in the future, as sea temperatures continue to rise on the east and west Australian coasts (Chapter 3).

Management of abalone fisheries

Throughout southern Australia, fishing has been controlled since its early days (later in New South Wales) by individual quotas issued to a limited number of licence holders, with divers observing minimum size limits, and allotted to specific fishing zones, and in some States subject to other restrictions such as daily catch limits, local closures or local fishing caps. However, these management tools have been applied at spatial scales that have provided only indirect formal control over harvesting discrete abalone metapopulations, so that localised stock declines or collapses have occurred in each State. In New South Wales, declines in the northern areas and around population centres due to recreational fishing have been common, and *Perkinus* infection has also played a role in these population collapses (Worthington *et al.* 2001). In Victoria, a serial decline in productive reefs has occurred, as well as severe declines in western Victoria from ganglioneuritis disease (H Gorfine *pers. comm.*). Tasmania has the largest fishery, with an annual catch of ~2500 t, but localised declines have occurred, notably on the east coast, as urchin barrens have extended south and expanded in area (see Chapter 12). In South Australia and Western Australia, despite some serious metapopulation depletions (Shepherd and Rodda 2001; Shepherd *et al.* 2001), the fisheries have generally been maintained (see Mayfield *et al.* 2011). Western Australia is the only State with a small fishery (~110 t annually) for Roe's abalone.

Current trends in management in all States are to apply fishing controls (e.g. caps on catch, local closures and local size limits at capture) at finer spatial scales, so that abalone metapopulations can be maintained and restored. Fine-scale management was pioneered by Jeremy Prince in Victoria (see Prince *et al.* 2008), and adopted for management of adjacent stunted and normal blacklip sub-populations in SE South Australia (Saunders *et al.* 2008).

Chitons (Class Polyplacophora)

Unlike other molluscs, chitons are remarkably uniform in structure, but more varied in their ecology. The account below is extracted from the reviews of Boyle (1977), Ludbrook and Gowlett-Holmes (1989) and Kaas *et al.* (1998), as well as other studies referred to specifically. As yet, nothing is known about the growth or longevity of subtidal chitons, and little about intertidal species, although Wells and Sellers (1987) suggested longevity of at least 2 years for *Acanthopleura hirtosa* (and see Otway 1994 below).

Chitons have eight overlapping plates embedded in a tough girdle enclosing the foot by which they adhere to the rocky substratum. Their radulae comprise a strong ribbon with transverse rows of teeth, some hardened with magnetite, which are used to scrape food from the substratum. An intriguing feature of chitons is the presence in the calcareous plates of pigmented cavities and/or aesthetes – channels, which penetrate through the various shell layers of the plates (Currie 1989, 1992). These structures are thought to have multiple functions, such as light receptors to indicate when the chiton is exposed to the light, and statocysts to enable the chiton to perceive gravity for orientation on the bottom.

Chitons have separate sexes that release sperm and eggs into the water where fertilisation takes place. In some species, the eggs are clustered in strings or masses. The spawning season of chitons varies, with no clear pattern except for some slight evidence of longer spawning seasons at lower latitudes (Currie 1990). Hence, some warm temperate New South Wales species, such as *Ischnochiton australis* and *Chiton pelliserpentis* (both autumn and spring), and *Cryptoplax mystica* (autumn–winter) have longer spawning seasons than *Acanthopleura hirtosa* (autumn only) on reef platforms off Perth (Sakker 1986; Wells and Sellers 1987). After fertilisation, the larvae drift in the water column and settle, usually after a period of ~1–4 days.

In the colder Tasmanian waters, four species are known to brood their eggs in the mantle cavity (Turner 1978; Edgar 2008), where they are retained until they metamorphose into crawling juveniles. Females of one species *Ischnochiton mayi* retain eggs for up to 2 weeks after fertilisation, and up to 36 juveniles have been found under one female.

Intertidal chitons are subject to predation by wrasses, especially the western blue groper at high tide, and to oystercatchers at low tide. In one oystercatcher 'midden' examined by the author, the chiton *Plaxiphora alba* was the second most common prey item, after the limpet *Cellana tramoserica*. Subtidal chitons are well hidden by day under boulders or deep within narrow crevices to avoid their main predators – crabs and wrasses. When exposed by overturning the boulder under which they are resting, they rapidly move away from the light towards a cryptic position. At night, herbivorous and omnivorous chitons may emerge to forage, while carnivorous species variously seek their prey in their particular micro-habitat.

The diet of chitons varies greatly between species. In a study in a boulder habitat at 1–5 m, Kangas and Shepherd (1984) examined the diet of 16 species of chiton. Six species were herbivorous, eating crustose or geniculate coralline algae, filamentous and macroalgae. Two of the six species were more strictly browsers: the brooding red-marked chiton *Callochiton crocinus* selected the algal crust *Petroderma*; and Torr's chiton *Ischnochiton torri* ate only seagrass and its epiphytes. Seven species were omnivorous: eating algae, sponges and bryozoans, while another three species were carnivorous. Of these last three, one *Notoplax speciosa* specialised on sponge tissue and another *Loricella angasi* trapped amphipods under its broad mantle. In addition, two other rarer species, *Acanthochitona pilsbryi* and *A. suerii*, lived in sponges on which they apparently fed. Overall, chitons played an important role in boulder and cryptic habitats by preventing the overgrowth of filamentous algae, compound ascidians and bryozoans on crustose corallines – an important habitat for juvenile gastropods (Clarkson and Shepherd 1985; unpublished data).

Population ecology of two low intertidal species

The eyed chiton *Onithochiton quercinus* and the giant chiton *Plaxiphora albida* inhabit the low intertidal cunjevoi zone at Cape Banks, Botany Bay, New South Wales. Otway (1994) found that both species reached reproductive maturity at an age of 2 years, and spawned in early autumn. Later in autumn, 2–3 mm sized juveniles of both species appeared among the cunjevoi (the ascidian *Pyura praeputialis*). The two species grew to 14 mm after 1 year, 23–24 mm after 2 years, and to 32 mm (for *O. quercinus*) and to 41 mm (for *P. albida*) after 4 years. Longevity was ~8 years. The mortality rate of a cohort of the eyed chiton was ~24% a year.

Sea slugs or opisthobranchs (Subclass Opisthobranchia)

Sea slugs are a gastropod subclass with species that have, in most cases, lost their shells or retained only a vestige of one. They comprise four main groups: the bubble shells (Order Cephalaspidea); the sea-hares (Order Anaspidea); the sap-sucking slugs (Order Sacoglossa); and the largest group, the nudibranchs (Order Nudibranchia). The loss of a shell in some of these groups has enabled them to develop the most bizarre body shapes, and to evolve protection against predators by chemical means, aided sometimes by brilliant warning colouration.

Sea slugs are hermaphrodites with internal fertilisation (i.e. each has a penis and vagina). During copulation, which can last up to a day, each partner inserts its penis into the other's vagina, and they exchange sperm. Some sacoglossans lack a vaginal opening, and the penis injects sperm into the partner's body cavity. Eggs are usually encased in a capsule and laid on the substratum in strings or spiral coils. Egg capsules of some common southern species have been illustrated by Smith *et al.* (1989). Egg size differs greatly between sea slug species. Small eggs hatch into plankton-feeding larvae that drift in the water column for days to weeks, medium-sized eggs hatch into larvae that feed on their internal egg reserves and usually settle within a few days, while very large eggs develop directly into crawling juveniles (Rose 1985). Sea slugs can live for up to a year, although very small species may complete their life cycle within a few weeks. They have few predators. Some fish eat bubble shells in the Order Cephalaspidea, and a burrowing anemone eats *Akera*, but the most important predators are other sea slugs, notably *Philinopsis* and *Pleurobrachaea* (Burn 1989).

The life cycle of sea slugs is closely tied to their diets. Some groups are specialised herbivores and spend their life cycle in the micro-habitats of their food species. The majority of species are carnivorous, preying variously on sponges, hydroids, anemones, bryozoans and other sea slugs. Rudman *et al.* (1998) noted that 'species that feed on transient food organisms such as short-lived hydroids, are usually small and have a short life cycle . . . Conversely, species which feed on large, long-living colonies such as sponges or alcyonarians are often larger and have longer life cycles'. Hence, small aeolids may undergo several generations in a year, whereas sponge feeders often have an annual cycle (Todd 1981).

Here we provide examples of the ecology of some southern sea slugs, drawn mainly from the reviews of Todd (1981), Burn (1989), Rudman *et al.* (1998) and Williams and Walker (1999). We focus on several notable features – the specialised diet of some species and some remarkable abilities of others. These abilities are:

- the capacity of some herbivorous species to take up chloroplasts from their algal diet and incorporate them into their own bodies (termed kleptoplasty)
- the various defences of carnivorous species, including the ability of some carnivorous species to sequester the nematocysts (stinging cells) of hydroids and other cnidarians, and incorporate them into their own cerata (specialised projections on the body where the gills are located) for protection against predators
- the ability to take up zooxanthellae from their cnidarian prey.

Food and feeding of herbivorous slugs

Because most species of sea slug have evolved a preference for one prey species or group of species, they live in close association with that species and the diet is readily determined. The known diets of some southern Australian herbivorous opisthobranchs are given in Table 9.2, with the caveat that in a few cases the diet has been inferred from the association alone, without examination of gut contents. It is evident that many sacoglossans have adapted to specialise on green algae, notably species of *Caulerpa*, a green algal genus, which, with some 22 southern species and > 30% of the world's species, has likely evolved in southern waters (Womersley 1984). In a study of the foraging behaviour of two *Elysia* species, Jensen (1991) found that the specialist species *E. filicaula* and the generalist feeder *E. australis* moved around randomly in search of food patches. The specialist species was more efficient at locating a food patch than the generalist feeder, and both species, having located a high-value food patch, stayed on it till the patch was exhausted.

The other main group of herbivorous slugs is the sea-hares, so named by Pliny nearly 2000 years ago, who mistakenly thought that they were venomous and 'cause terror in the sea' (Magnus 1555), because of the

Table 9.2: Diets of some herbivorous opisthobranchs in reef habitats.

Numbered references are listed at bottom of table.

Order, Family, Species	Diet of sea slug
Sacoglossa: Volvatellidae *Ascobulla fischeri*	*Caulerpa brownii, C.geminata, C. racemosa*[1]
Juliidae: *Berthelinia australis*	*C. simpliciuscula*[1]
Berthelinia babai	*C. cactoides, C. geminata, C. simpliciuscula*[1]
Berthelinia rottnesti, Midorigai australis	*C. racemosa, C. simplicuscula*[2,3]
Berthelinia typica	*C. brownii*[1]
Oxynoidae: *Oxynoe viridis*	*C. cactoides, C. racemosa, Caulerpa* spp.[1,2]
Roburnella wilsoni	*C. brownii, C. remotifolia, C. scalpelliformis*[1]
Placobranchidae (= Elysiidae) *Elysia australis, E. filicauda*	*Chaetomorpha, Enteromorpha, Rhizoclonium, Polyphysa peniculus*[1,2,4,7]
Elysia ornata	*Bryopsis plumosa*[2]
Elysia maoria	*Codium*[4]
Polybranchiidae *Polybranchia pallens*	*C. longifolia*[1]
Limapontiidae: *Placida dendritica*	*Codium fragile*[1]
Ercolania margaritae	Filamentous green algae; *Bryopsis*[3]
Ercolania sp.1	*Apjohnia laetevirens*[3]
Stiliger smaragdinus	*Caulerpa* spp. often *C. cactoides*[5,6]
Anaspidea: Aplysiidae *Aplysia parvula* + five other species of *Aplysia*	Red/green algae; *Laurencia, Plocamium, Ulva*[1]

[1]Burn (1989); [2]Williams and Walker (1999); [3]Burn (1974, 2006); [4]Rudman *et al.* (1998); [5]Gowlett-Holmes (2008); [6]Edgar (2008); [7]Jensen (1991)

ink they discharge. But Steinbeck's description of them in his book 'The Sea of Cortez' as 'shell-less flabby sea-slugs, somewhat resembling a rabbit crouched over' was more colourful and accurate.

Six southern sea-hare species are known, many of them common in seagrass habitats. Here we discuss the ecology of a well-studied cosmopolitan species *Aplysia parvula* (also the smallest), which occurs principally in red algal habitats around southern coasts to a depth of ~20 m (Shepherd 1979).

The sea-hare Aplysia parvula

Recent studies on this sea-hare (Plate 20) indicate an animal remarkably well adapted to its environment (Carefoot 1987; Rogers *et al.* 2003). The life span of the species is up to ~12 months. It matures after ~1–2 months and then spends its life eating, laying eggs and copulating. After a spawning peak at its maximum size (80–90 mm), the sea-hare senesces and dies. Eggs hatch after ~7–10 days at 18–24°C and then the veliger larvae swim upward in the water column where they can be transported long distances, depending on the strength of the current. After 20–30 days, the larvae are ready to settle and they then sink to the bottom and settle on a host red alga, being induced to metamorphose by substances in the alga.

In the New South Wales study, the sea-hare lived mainly on one of two host red algae – *Delisea* or *Laurencia* – and its activity varied with time of day. At dusk, sea-hares emerged onto the algal canopy, and some began to feed, while others copulated. Just before dawn, they moved back down to the base of the plants where they stayed till the following afternoon. Their eggs were usually found at the base of red algae and apparently were laid during the day because no egg laying was observed at night.

Although the sea-hare's life history was played out on one or other of the two algae, *Delisea* and *Laurencia*, according to site, it did eat other species of red algae. However, dietary studies revealed a complex picture (Rogers *et al.*1995, 2002). In *Delisea* habitat, sea-hares ate mainly the host species, which is chemically rich in secondary metabolites and thus a strong deterrent to grazing. But feeding on *Delisea* came with both costs and benefits. The costs were that the sea-hare ate less, grew more slowly, and laid fewer eggs, than when fed *Laurencia*. The benefit was that on a *Delisea* diet, the sea-hare incorporated the alga's secondary metabolites into its body and surface mucus.

Secondary metabolites are a powerful deterrent to some predators. This was evident in experiments in which sea-hares were offered to the most common fish predators, wrasses and white ears (*Parma microlepis*). Some fish repeatedly bit and spat out the sea-hare to dissipate the secretions, but eventually ate the animal, whereas others rejected the sea-hare after an initial bite – both reactions indicating a strong distaste. Overall, sea-hares were considered to benefit by using the *Delisea* habitat as a refuge from fishes, even though predators might take some exposed animals. Interestingly, these sea-hares are also eaten by a pycnogonid (Rogers *et al.* 2000), although in this case the sea-hare could escape if of sufficient size (see Chapter 15).

Another deterrent is the well-known discharge of ink, which apparently deters predatory crabs as well as some fishes (DiMatteo 1982).

The sea slug Placida dendritica

This sacoglossan, probably a species complex, is common in warm temperate regions of the world and occurs in eastern Australia south to Tasmania. The species feeds on *Codium* and *Bryopsis* over much of its range. In Oregon, studies by Trowbridge (1992) showed that the species was a specialist feeder on species in these two algal genera, although it also feeds on the plants' epiphytes. Adult densities were highest in spring and declined in late summer–autumn to very low winter levels. Egg masses were laid by adults for most of the year, except in winter, and the larvae, after hatching and passing through a planktotrophic stage, settled on their preferred algae.

Settlement was induced by substances leaching from their preferred host alga (*Codium*), especially from damaged plants. Recruitment occurred most strongly in spring and summer, but persisted through most of the year. Growth was rapid and the slugs took only a month to reach sexual maturity after settlement. Hence, populations of the slug consisted of many overlapping generations. Young slugs tended to settle and feed on stressed plants, but *Codium* was largely restricted to physically disturbed locations (e.g. with strong water movement and burial by sand) that the slugs could not tolerate so well. So, Trowbridge (1998) proposed that a prey-stress model applied to the interaction between the alga and its consumer. The model, which is rarely applied to marine situations, predicted that prey (i.e. *Codium*) defences are weakened under the added stress of grazing, so that the consumer contributes to the restricted prey distribution pattern.

Solar-powered sea slugs (kleptoplasty)

A remarkable feature of herbivorous sacoglossans is that they can exploit chloroplasts – the specialised chlorophyll-containing structures in a living algal cell where photosynthesis takes place – taken from the food algae and retain them internally for periods of up to several months. What makes this so remarkable is that a 'naked' foreign body can be sustained within the host slug. The chloroplasts then continue to photosynthesise in the sea slugs, and release photosynthetic products (polysaccharides and amino acids); hence, the apt title 'solar-powered' given to host slugs by Rudman (1998). This summary of the ecology of the symbiosis is taken from Williams and Walker (1999) and Rumpho *et al.* (2000).

So far, three Australian genera, *Elysia* (four species), *Oxynoe* and *Placida* have been surveyed and found to contain chloroplasts. These sea slugs are able to suck out the cell contents of siphonaceous algal cells, mainly from green algae, and retain the chloroplasts, which are then distributed as 'photosynthetic factories' just a cell layer thick under the sea slug's skin. In sunny conditions, the sea slug unfolds its parapodia and sunbathes.

In a notable study of one of the *Elysia* species in Botany Bay, Brandley (1984) followed the species over its lifetime of 1 year. Egg masses laid in late November led to the recruitment of young ones at ~2 mm size the following March–April; these grew to adults in a few months. In spring, adults fed on *Codium* and took up their chloroplasts, so imparting a green colour to the host, whereas in winter the young adults fed on red algal epiphytes, and then incorporated their rhodoplasts into their body tissue, which gave them a red or brown colour. This was the first record of a species which could take up two types of chloroplasts (green and red).

Sea slugs gain important ecological benefits through this symbiosis. Depending on species, 40–50% of the carbon fixed by the chloroplasts is taken up by the slug and, in the three genera referred to above, this can amount to 50–60% of the total food ingested by animal (Raven *et al.* 2001). Hence, if food is generally or seasonally scarce, the sea slug can still comfortably survive on 'solar power'.

The other significant benefit for the slug is protection from predators. Firstly, it acquires camouflage, because it adopts the green colour of its algal food and blends perfectly with its algal micro-habitat. Secondly, it can sequester toxins produced by the alga (especially in *Caulerpa* species) and this can provide yet further protection against predators (Paul and Van Alstyne 1988).

Nudibranchs (Order Nudibranchia)

The nudibranchs are by far the largest group of sea slugs and have invaded a wide range of habitats. They also 'rank first among the gastropods in evolutionary enterprise' (Morton 1967). Lacking a protective shell, they have evolved elaborate, and sometimes subtle, defence mechanisms against predators. The brilliant colours of many species, especially chromodorids, not only 'enliven the solitudes of the ocean' (Adams and Reeve 1850), but warn potential predators that they are distasteful. Most species are carnivorous, and have diversified to feed on many sessile animal groups, including sponges, bryozoans, cnidarians (anemones, hydroids, alcyonarians and zoanthids), and even mobile crustaceans, on soft and hard bottoms. There is also a pelagic genus (*Glaucus*) that predates on jellyfishes (Thompson and Bennett 1970). Table 9.3 lists the known diets of many common reef-associated southern nudibranchs.

It is evident that the carnivorous nudibranchs have the broadest diversity of diet, feeding on sessile invertebrates from many phyla. Many species have an obligate relation with one prey species, but for others the obligate relation is still uncertain. In a few nudibranchs, the diet changes as they mature into adulthood (R Burn *pers. comm.*) and, at the other extreme, a few species have a broad diet. The same variability is shown by species that feed on sponges – many feed on only one species, while others consume a range of related species. Furthermore, Rudman and Bergquist (2007) found a distinct pattern in the sponge-feeding preferences of different groups. *Chromodoris* and related genera preferred darwinellid sponges, *Hypselodoris* and related genera preferred dysideid sponges, and *Glossodoris* fed only on thorectid sponges. These authors suggested that the availability of strong, distasteful chemicals in some of the sponges enabled the chromodorid group to evolve the astonishing colour patterns that they now possess.

Zooxanthellae capture

In addition to their 'normal' diet, some slugs have evolved the ability to take up certain species of microscopic, unicellular photosynthetic dinoflagellates – called zooxanthellae – from their cnidarian prey, such as anemones, zoanthids, alcyonarians (soft corals) and hydroids. They then transfer the zooxanthellae to their cerata (the dorsally projecting, leaf-like appendages that serve as gills) where they can photosynthesise and provide food for the host – analogous to the stealing of chloroplasts by sacoglossans. This

Table 9.3: Diet of some common carnivorous cephalaspideans, umbraculids and nudibranchs in reef habitats.

Numbers refer to the references listed at bottom of table.

Order, Family, Species	Diet of sea slug
Cephalaspidea: Gastropteridae: *Sagaminopteron ornatum*	Grey sponge (*Dysidea*?)[1]
Aglajidae: *Melanochlamys queritor*	On coralline algae: polychaetes[1]
Umbraculida: Tylodinidae: *Tylodina corticalis*	Aplysinellid sponges, Thorectidae (?)[2]
Umbrachidae: *Umbraculum umbraculum*	Spirastrellid sponges: *Tethya, Aaptos, Ancorina, Plakina*[2,4]
Pleurobranchidae: *Berthellina citrina*	Encrusting sponges[1]
Berthella medietas	Sponges[1]
Pleurobranchus peronii	Ascidians[5,6]
Pleurobranchaea maculata	Anemones, *Actinia tenebrosa*, invertebrates[1,3]
Nudibranchia: Chromodorididae:	
Noumea haliclona	*Darwinella* spp., *Aplysilla* sp.[10]
Noumea closei	Sponges *Darwinella* sp., *Dendrilla rosea, Dictyodendrilla digitata*[10]
Verconia verconis	Sponges *Dendrilla* sp., *Darwinella gardineri*[10]
Chromodoris alternata, C. epicuria, C. tasmaniensis	Sponge *Darwinella gardineri* [1,10]
Chromodoris ambigua, C. thompsoni	Sponge *Chelonaplysilla violacea*[10]
Ceratosoma amoenum, C. brevicaudatum,	Sponge *Dysidea* sp.[10] Soft coral *Drifa gaboensis*[13]
Ceratosoma splendida	Sponges *Darwinella* (3 spp.), *Dictyodendrilla dendyi*[10]
Hypselodoris bennetti	Sponge *Psammocinia* sp.
Mexichromis macropus	Sponge *Euryspongia* sp.[1,7,10]
Dorididae: Discodorididae: Dendrodorididae: Phyllidiidae: many genera and > 100 species	Sponges[1,6,7,8,10]
Aegiridae: *Aegires villosus*	Calcareous sponges[7]
Goniodorididae: *Trapania brunnea*	Kamptozoans on orange sponges (*Mycale* sp.)[9]
Goniodoris meracula	Colonial ascidians (*Botrylloides* spp.)[9]
Okenia zoobotryon	Soft bryozoan *Zoobotryon verticillatum*[7]
Onkidorididae: *Corambe* sp.	Encrusting bryozoan *Electra pilosa*[9]
Gymnodorididae: *Gymnodoris arnoldi Paliolla cooki*	Crustose bryozoans e.g. *Beania magellanica*[1]
Polyceridae: *Tambja verconis*	Soft bryozoan *Bugula dentata*[1]
Polycera hedgpethi	Soft bryozoans *Bugula* spp.[1]
Vayssiereidae: *Vayssierea caledonica*	Serpulid polychaete *Salmacena dysteri*[9]
Madrellidae: *Madrella sanguinea*	Red crustose bryozoan *Mucropetraliella elleri*[1,9,12]
Janolidae: *Calduckia affinis*	Crustose bryozoan *Beania magellanica*[1]
Tethydidae: *Melibe australis*	Traps small crustaceans[7]

(Continued)

Table 9.3: Continued

Numbers refer to the references listed at bottom of table.

Order, Family, Species	Diet of sea slug
Tritoniidae: *Tritonia* sp.	Gorgonian *Mopsella zimmeri*[7]
Marianina rosea	Hydroids[7]
Zephyrinidae: *Janolus* sp.	Crustose bryozoans[7]
Aeolidiidae: *Anteaeolidiella foulisi, Baeolidia australis, Spurilla macleayi*	Sea anemones[7]
Austraeolis ornata, Cerberilla sp.	Hydroids[7,8]
Dotidae: *Doto ostenta*	Hydroids[7]
Doto pita	Hydroids *Sertularia tenuis, Amphisbetia minima*14
Doto sp. 2	Hydroid *Gymnangium* sp.[14]
Facelinidae: *Austraeolis ornata*	Hydroids, nudibranchs, carrion[1,7]
Phyllodesmium serratum	Alcyonarians, *Clavularia* sp., *Carijoa* sp., *Junceda* sp.[7,8,11]
Phyllodesmium poindimiei	Alcyonarian *Carijoa* sp.[11]
Pteraeolidia ianthina	Alcyonarians, hydroids[7,8]
Flabellinidae: *Flabellina rubropurpurata*	Hydroids[7]
Flabellina rubrolineata	Hydroids, especially *Halocordyle disticha*[8]
Arminidae: *Armina* sp.	Sea pen *Sarcoptilus grandis*[8]
Tergipedidae: *Trinchesia kuiterorum*	Hydroid *Zyzzyzus spongicolus*[14]
Trinchesia sp. 2; sp. 3; sp.4	Hydroid *Lobataria newtoni*[14]
Trinchesia sp. 5	Hydroid *Obelia dichotoma*[14]
Tergipes sp.	Hydroid *Pennaria wilsoni*[14]

[1]Burn (1989); [2]Willan (1984); [3]Ottaway (1977); [4]Watson (1976); [5]Thompson (1970); [6]Rudman *et al.* (1998); [7]Gowlett-Holmes (2008); [8]Edgar (2008); [9] Burn (2006), R. Burn *pers. comm.*; [10]Rudman and Bergquist (2007); [11]Rudman (1991); [12]Keough (1999); [13]Farrant (1987); [14]J.E. Watson and R. Burn *pers. comm.*

plant–animal association ranges from the primitive, where zooxanthellae only last in the host for a few days, to the advanced, where the host has adaptive features that allow the guest zooxanthellae to remain in their tissues indefinitely, and photosynthesise (Rudman 1981b).

This ability has been studied for one tropical–temperate Australian species common in southern Australia (except Victoria and Tasmania). This is the aeolid *Pteraeolidia ianthina*: the largest (to 15 cm) slug on open coasts. This aeolid preys on alcyonarians (soft corals) and appears to have a relatively 'advanced' association with alcyonarian zooxanthellae. Hoegh-Guldberg *et al.* (1986) found that when zooxanthellae were in high densities in the aeolid, they supplied ~80% of its carbon needs in winter and more than twice as much carbon in summer. The aeolid is usually brown, due to the zooxanthellae, but a rare white form exists that lacks zooxanthellae. Gowlett-Holmes (2008) noted that in South Australia the species is predominantly blue, without zooxanthellae, and is much smaller than those living in SE Australia. The species produces ~5000 eggs in a long egg-string in late summer to autumn. As a rare example of a brooding species, the slug sits on the egg mass, which then hatch in ~11 days to produce short-lived, non-feeding larvae (Rose and Hoegh-Guldberg 1982; Smith *et al.* 1989).

Because slugs with high densities of zooxanthellae were the only ones to brood, Rose and Hoegh-Guldberg suggested that maturation of the slug depended on its nutritionally advantageous association with zooxanthellae.

Nudibranch defences

Nudibranchs possess three main types of defence:

- **Behavioural** – i.e. avoiding detection by crypsis
- **Shape and colour** – Many slugs display brilliant colours, and there are two reasons for this, which have sometimes been confused. First, the slug may be colourful because it has taken up the bright colour of its prey, and has become virtually indistinguishable from its background host or prey species (e.g. some sponges). In other cases, it may be translucent and simply matches its background. Alternatively, the disruptive colours and shape of the cerata may effectively camouflage the slug when present on its prey. Yet another adaptation is warning colouration (called aposematism), which advertises that the slug is unpalatable. This defence is only useful if the slug has taken up a deterrent chemical described below, or mimics a chemically defended species.
- **Chemical** – Many sponges produce distasteful substances that are highly toxic to fish, and the fish soon learn to avoid them. Some chromodorids are able to isolate these chemicals and incorporate them into specific glands on the body. If attacked, a few species can discharge the chemicals as a sticky white substance, which quickly deters the would-be predator. Other species can store the chemicals in the mantle glands, which explode or release their contents when bitten by a would-be predator. Such species often have the well-known, brilliant colours referred to above, and this warning colouration is a defence strategy evolved to deter predators. Even more elaborately, some slugs can take up the prey's stinging cells, called nematocysts, which we now discuss.

Sequestration of nematocysts

Some slugs, notably aeolids that prey on cnidarians (see Table 9.3), can take up the nematocysts from hydroids and anemones on which they feed. The nematocysts then migrate to tiny sacs at the tip of the slug's cerata, where they remain functional. When an aeolid is attacked or aggravated, its cerata typically bristle, perhaps as a threat display or to confuse the predator (Todd 1981). This sequestration ability of aeolids has intrigued biologists, who wonder how they can do so without discharging the stinging cells. Some consider that the secretion of toxic mucus by the slug might inactivate the cell, and Conklin and Mariscal (1977) observed that some nematocysts do discharge initially, but that the slug quickly becomes acclimated and can then consume its prey with impunity. In other cases, the slug takes up immature nematocysts, which later mature in the cerata (Greenwood and Mariscal 1984). In one unusual example, an aeolid *Cuthona kuiteri* that occurs in SE Australia, not only takes up the nematocysts of its specialised prey hydroid *Zyzzyzus spongicola*, but has evolved cerata that perfectly mimic in colour and shape the polyps of the hydroid – a good example of camouflage (Rudman 1981a).

PREDATORY GASTROPODS

Carnivorous gastropods occur in several families, mostly within the neogastropods. Some are scavengers and others are specialist or generalist predators. Laxton (1971) distinguished three kinds of carnivorous gastropods: hunting predators such as *Charonia* equipped with a toxin to immobilise prey; 'grazing' predators such as *Ranella* that feed on sessile prey; and scavengers that also feed on carrion. Hunters tends to move at random at night, using chemoreception to detect their prey, whereas 'grazing' predators tends to move up-current by day or night.

For many carnivores, the diet is known in general terms, but very few species have been studied in detail. In Table 9.4 we list some commoner reef-dwelling species with notes on their diet. Note that, firstly, in contrast to predatory opisthobranchs, most of the gastropods listed are generalist hunters that search for their prey. Secondly, the prey species are diverse and spread among many phyla. *Cassis fimbriata* is one of the few specialists feeding on the purple sea-urchin

Table 9.4: Diet of some common carnivorous gastropods in reef habitats.

Numbers refer to the references listed below the table.

Infra-order, Family, Species	Diet of gastropod
Littorinimorpha Ranellidae: *Charonia lampas rubicunda*	Sea-urchins, ophiuroids, sea-stars e.g. *Meridiastra calcar, Coscinasterias muricata*[1,2,3]
Argobuccinum pustulosum	Solitary ascidians[2]
Ranella australasia	Solitary ascidians e.g. *Pyura stolonifera*, carrion[1]
Cabestana tabulata	Solitary ascidians e.g. *Microcosmus*[1]
Sassia sexcostata	Colonial ascidians[2]
Cassidae: *Cassis fimbriata*	Sea-urchin *Heliocidaris erythrogramma*[2,3]
Neogastropoda Buccinidae: *Cominella lineolata, C. tasmanica*	Herbivorous gastropods e.g. *Siphonaria, Clanculus*[4]
Pleuroploca australasiae	Gastropods, bivalves, ascidians, carrion[2]
Muricidae: *Lepsiella flindersi*	Mussels *Xenostrobus* spp.; limpet *Patelloida alticostata*; serpulids, carrion[2,5,6]
Lepsiella vinosa	Mussels, oysters, gastropods, barnacles, serpulids[2,5]
Mitridae: *Mitra glabra*	Sipunculan *Themiste cymodoceae*; polychaetes[3,7]
Mitra badia	Sipunculans: *Apiosoma, Aspidosiphon, Phascolosoma*[8]

[1]Laxton (1971); [2]Gowlett-Holmes (2008); [3]Edgar (2008); [4]Kohn *et al.* (2005); [5]Wells and Keesing (2005); [6]Morton (1999); [7]Ludbrook and Gowlett-Holmes (1989); [8]Taylor (1993).

Heliocidaris erythrogramma. *Cassis* emerges at night onto reefs to hunt for sea-urchins, and on finding one, holds it with the foot. It penetrates the test with its radula, drilling at a rate of ~0.1 mm $min.^{-1}$, and then sucks out the internal tissue for the next 1–2 h (Hughes and Hughes (1981).

We now consider the life history and ecology of three well-studied generalist predators and one group, the cowries, which are specialist sponge feeders, except for one genus, which are detritivores.

The whelk *Dicathais orbita* (previously called *Thais orbita*)

Dicathais orbita is a common and widespread whelk that occurs from the intertidal to depths of ~10 m on coasts of the southern half of Australia. It has been intensively studied by Bruce Phillips and colleagues (Phillips 1969; Phillips and Campbell 1974). The species has a 1:1 sex ratio and after mating, females lay egg capsules, mainly in September and October. One female can lay 40 or more capsules, each containing ~300–3000 eggs, depending on female size. After ~45 days, the eggs hatch into a free-swimming veliger stage, which lasts ~20 days, and the veliger then settles in shallow subtidal areas where it grows and gradually disperses into surrounding areas. Sexual maturity is reached by both sexes at ~38 mm when they are 4–5 years old. The annual mortality averages ~46% per year, giving an average life expectancy of ~5 years and a (rarely attained) maximum age of ~20 years at a size of ~75 mm. The growth rate is ~4 mm $year^{-1}$ for much of its life, but slows markedly when above ~50 mm size. Mortality was found to be due largely to predation by a muricid *Agnewia* and also octopus.

This whelk is a generalist predator with a varied diet, depending on location. It has a variety of attack methods, including shell drilling, valve wedging and anaesthetising prey with a blue secretion from the hypobranchial gland. Where mussels are abundant, it feeds largely on them, but elsewhere its food can include other bivalves, the hammer oyster *Malleus*, limpets and barnacles (Butler 1979; Fairweather 1988). Feeding in the intertidal zone, however, can impose heat stress on the whelk. Hence, at Rottnest I., in the adverse intertidal conditions where they lived, Morton and Britton (1993) found them to be time-optimisers rather than energy-maximisers in terms of their foraging strategy. Yet, food was plentiful and their diet comprised > 30 species, mainly gastropods, as well as carrion. Feeding occurred about every 5 days. At times, they even drilled through the thick operculum of the turban *Turbo torquatus*, probably because they feed opportunistically and attack the first prey they encounter (Taylor and Glover 1999). In one study, the limpet *Patelloida alticostata* comprised ~23% of their diet (Black 1978, and see below), and in another the diet was barnacles, *Galeolaria* tube worms, limpets and mussels (Morton 1999). In a fourth study, Fairweather (1988) found that *Dicathais* crawled on average 1.2 m day^{-1}, with maximum daily recorded movements of ~6 m, in order to exploit intertidal prey and then retreated to the subtidal zone or into crevices for shelter.

Black (1978) examined in detail the way in which the whelk ate the limpet *Patella alticostata*. The whelk chose to drill holes in the thickest part of the shell near the apex – the part of the shell lying over the gonad and digestive gland – and took 12–24 h to drill through 1.5 mm of shell. Robert Black suggested that it chose that precise location to drill because the visceral mass underneath contained energy-rich food, which could be eaten quickly. Quick consumption was a good tactic, because if the whelk drilled into the limpet's muscle tissue there was a high risk of the limpet dying, and then being swept off the rock, before the whelk could obtain a good meal. In that field study, the time and energy spent by *Dicathais* drilling the limpet were well rewarded, because 65% of drill holes gave direct access to gonad tissue, and unfinished drill holes on limpets amounted to only 2%. It is interesting to note how the optimal drilling strategy of a shell-drilling species may vary according to the prey. In the previous, Chapter we noted how the best strategy for *Haustrum baileyanum* was to drill into the abalone muscle rather than into gonad tissue, because in that case there was no risk of the abalone dying during drilling and being swept away. Removal of the risk of failure completely changes the optimal drilling strategy.

The triton *Cabestana spengleri*

This triton occurs in SE Australia and as far west as the South Australian gulfs, as well as in New Zealand. The species is common on shallow reefs near Sydney where it feeds on cunjevoi. Elsewhere it feeds on a range of solitary ascidians and polychaete worms. Its life history and diet have been variously described by Anderson (1959), Laxton (1969, 1970, 1971) and Riedel (1992). The sexes are separate and, after a sometimes prolonged courtship period, copulation between breeding pairs occurs in spring (October–November). The female then lays her egg mass, comprising ~250 tough, protective capsules. Each capsule contains many yolked eggs floating in a nutritious medium – an attractive target for asteroid predators and microorganisms! To protect them from such predators, the female broods the eggs until they hatch; defence against microorganisms is provided chemically by antibacterial substances (Benkendorff *et al.* 2001). After ~1 month, the eggs hatch into veligers, which swim away, although in Anderson's (1959) laboratory experiments they hatched directly into crawling young – a result possibly due to laboratory artefacts.

In a New Zealand population of this triton, courtship between males and females lasted for most of the year even though eggs were laid only in late spring. Courtship involved pairing whereby the male mounted the female, sometimes with and sometimes without copulation. John Laxton (1969) suggested that pairing and copulation apparently had become a ritual, independent of the purpose of reproduction – behaviour far more complex than anything ever before suspected.

The triton reaches sexual maturity in about 1 year, and full size of 10–15 cm in ~3 years. Longevity is not known, but is at least 5 years.

The cone shells *Conus* spp.

The genus *Conus* is essentially a tropical group, but has three temperate southern species whose ecology is known from the work of Kohn (1993), Kohn and Almasi (1993) and Kohn *et al.* (2005). Cone shells have two sexes and the females of each species lay distinctive clusters of egg capsules on stones or under rocks. In *C. anemone*, the most common southern cone, the capsule may contain up to 60 eggs, with often 10 or more capsules in each mass (Smith *et al.* 1989). The egg masses are laid in summer and the eggs hatch directly into crawling juveniles after a few weeks. In the other southern species, however, the eggs hatch after 2–3 weeks into plankton-feeding veliger larvae.

Cones are quiescent by day, resting under rocks or in sand pockets in reefs, but emerge at night to hunt for prey. They differ from other gastropods in having a powerful venom apparatus and a harpoon-like radula specialised for hypodermic venomous injections. The prey is swallowed whole after being harpooned and paralysed by the venom. The diet of southern cones is almost entirely polychaete worms, mainly nereids and eunicids, eaten at a rate of about one worm every second night. The longevity of some tropical species is ~10 years, but may be more for temperate species (Perron 1983).

The cowries (Superfamily Cypraeoidea)

This group contains two families: the true cowries, Cypraeidae; and the spindle cowries, Ovulidae. In the Cypraeidae, the genus *Zoila* has four southern species and the genus *Umbilia* two species. The spindle cowries are mainly tropical with only four southern species, two of which appear to be relict in upper Spencer Gulf. The ecology of *Zoila* and *Umbilia* is known from a remarkable field study of Peter Clarkson (Wilson 1998; Wilson and Clarkson 2004) and the little knowledge of spindle cowries from the work of Gowlett-Holmes and Holmes (1989).

The genera *Zoila, Umbilia* and *Primovula*

Life history

All six species of the first two genera lay an egg mass containing up to 300 capsules, each containing many eggs, while *Primovula heleneae* lays a clutch of egg capsules on its host gorgonian. The season of egg laying for the cowries varies between, and even within, species over their geographic range (Table 9.5). After ~15 days, the eggs develop into a veliger, and one veliger then feeds on the remaining eggs or emerging veligers within the capsule. Hatching and emergence from the capsule occur after 6–8 weeks and the juvenile cowry crawls out to live an independent life.

In the Armenian cowry *U. armeniaca*, the mother protects the hatchlings, often within the cup of the

Table 9.5: The southern species of cowry showing their depth range and distribution, time of brooding of eggs, and principal food.

Species	Depth (m)	Distribution	Egg brooding	Diet (major sponge genera) or other food
Zoila friendii	To 200	Shark Bay, WA to central Vic.	Summer (WA), all year (SA)	*Caulospongia, Chondrilla, Geodia, Halichondria, Spirastrella, Tedania*
Zoila marginata	25– > 60	Shark Bay, WA to central S.A.	Spring–autumn (WA)	*Geodia, Trachycladus*
Zoila rosselli	8–150	Shark Bay, WA to central GAB	All year (WA); Spring–summer (SA)	*Spongoserites, Spirastrella, Sigmaxinella, Calyx*
Zoila venusta	0–150	Shark Bay, WA to NE GAB	Spring–autumn	*Chondrilla, Corticium, Phyllospongia, Sarcotragus*
Umbilia armeniaca	25–160	SW WA to central SA	Summer–autumn	Detritus, microorganisms, *Pericharax*
Umbilia hesitata	12–200	SE SA to SE Qld	Unknown	Detritus

sponge *Phoriospongia*, until they are almost adults. This direct-development strategy has the advantage that larval mortality is low, but the disadvantage is that the species has limited dispersal potential. Perhaps in partial compensation for this restricted dispersal, Peter Clarkson observed that 'hatchlings show surprising skill and agility even when trailing their redundant egg capsules behind them'. He noted that hatchlings were neutrally buoyant and he observed them to crawl up the sides of a host cup-sponge, then retract the foot, and launch themselves into the current, so enabling advection over a greater distance.

Overall, the limited dispersal of the various species implies limited genetic mixing and gives rise to localised populations, each with distinctive features in terms of shell shape and colour patterns – a diversity that endows the group with endless fascination to shell-collectors.

The growth and longevity of cowries are known for only two species. Ludbrook and Gowlett-Holmes (1989) recorded that the subspecies *Z. friendii thersites* in Gulf St Vincent built the final lip to its shell after 2years, and matured sexually in the following year. However, *Z. marginata* and *Z. venusta* appear to reach sexual maturity in 2 years. Longevity is 10–15 years (Wilson and Clarkson 2004).

Habitat and diet

The habitat of cowries is largely determined by the habitat of their preferred host sponge species (whether one or several) and, as noted by Peter Clarkson, a cowry rarely forages by day, and usually lives more or less permanently on its host. For example, over 10 years of observations, Clarkson noted that a particular cup-shaped *Spirastrella* sponge was a favoured host for egg-laying, as well as feeding, by the black cowry *Z. friendii thersites* (Plate 21).

On exposed coasts, sponges are generally restricted to cryptic habitats, such as caves and crevices in shallow water, but become increasingly dominant in open habitats below 20–25 m depth (Chapter 13). Similarly, these cowries live in caves and crevices at shallow depths, but in open habitats with increasing depth.

Each cowry subspecies grazes a range of sponge species, each of which varies with geography and depth. Peter Clarkson recorded some species of cowry feeding on up to 20 different sponge species, but here we list (Table 9.5) the most common hosts. In a few instances, the end of the geographic range of a cowry species or subspecies coincides with that of its preferred host, but, in most cases, the cowry species switches its preferred host both geographically and with depth, independently of the range of a particular host. For detailed information on the ecology of each cowry species and its preferred host see the magisterial work of Wilson and Clarkson (2004).

In contrast to *Zoila*, the Armenian cowry *Umbilia armeniaca* lives in the open on low-relief limestone among algal beds dominated by the red alga *Osmundaria* and sponges. It feeds mainly on detritus, but occasionally on the sponge *Phoriospongia*. The food of the wonder cowry *U. hesitata* is always detritus from sandy bottoms.

The gorgonian hosts of the two relict ovulids are *Echinogorgia* sp. for *Primovula cruenta* and *Euplexaura* sp. for *P. heleneae* (Plate 22); both host species also appear to be relict in upper Spencer Gulf in places of strong current.

Predators

The main predators of these cowries are fish, octopuses, crabs and carnivorous molluscs and, notably, large labrids (e.g. baldchin groper and western blue groper in SW Australia), stingrays, blue morwong, and the two large octopuses *Octopus maorum* and *Octopus* cf. *tetricus*. The buccinid whelk *Pleuroploca australasia* also attacks cowries, as does the giant spider crab *Leptomithrax gaimardii*, which prefers juveniles (Wilson and Clarkson 2004). Peter Clarkson observed that sometimes the claws of this spider crab became wedged in the cowry shell aperture and wrote: 'the crabs are a comical sight as they lumber along burdened by the weight of a food hamper they cannot open or discard.'

FILTER-FEEDING MOLLUSCS

Southern reefs host a diverse group of filter-feeding molluscs, which includes some gastropods such as worm shells (e.g. vermetids) and bivalves such as mussels, oysters, a rock-dwelling scallop, rock-borers that drill into soft rock, and shipworms that bore into wood. In addition, two species, the hammer oyster and the razor shell, widely occur in sheltered waters. Hammer oysters can form reef-like aggregations of dead shells up to 2 m high in places of strong current, as in upper Spencer Gulf, and the large shells of oysters also form mini-reefs for reef-associated benthic species. Subtidally, most filter-feeding molluscs are inconspicuous or rare on reefs, except mussels, which occasionally form extensive beds at the sublittoral fringe, and the doughboy scallop, which can be common in sheltered waters. Here, we discuss the ecology of two mussel species and the doughboy scallop *Mimachlamys asperrima*.

Mat-forming mussels

The biology of mussels in Australia is known mainly from studies of Wilson and Hodgkin (1967), Dix and Ferguson (1984) and Ompy (2003) (and see Gosling 2003, who reviews the extensive European work on closely related species). The upper limit of mussels is about the mid-intertidal zone, because they can only feed, and their enlarged gills can only function, in water. The gills remove suspended particles, such as bacteria, phytoplankton and detritus, from the water, as well as dissolved organic matter (DOM). Mussels are usually aggregated in patches, which provide the adaptive benefits of increasing stability and a reduction in predation risk, but at the cost of greater competition for food (see Svane and Ompi 1993).

The largest southern mussel is the blue mussel *Mytilus galloprovincialis*, which is often found on jetty piles in harbours and occasionally in clumps on rocky coasts. The species has apparently migrated several times to Australia from the northern hemisphere, first during the Pleistocene (~1.2 Ma) and possibly twice recently via shipping (Hilbish *et al.* 2000; Gérard *et al.* 2008; Svane 2011).

The blue mussel's spawning season varies with latitude across southern Australia. Off Sydney, it spawns from June to December, off Fremantle from July to September, and off Port Lincoln from July to October, whereas in Tasmania a double spawning occurs from August to October and January to March. Spatfall occurs about a month after spawning events, after a planktonic larval life of 2–5 weeks, depending on temperature. Montelli (2009) recorded settlement from June to October, peaking in August, in Fremantle Harbour. Ompy (2003) found that settling larvae preferred to settle on conspecific byssus threads, a behaviour that explains the formation of mussel patches.

Growth of the mussel is almost linear, with sexual maturity reached at ~4 cm length in one year, and a length of ~10cm in ~2½ years. Longevity is 10–15 years. The mussel has sometimes been used as a bio-indicator of pollution. If animals are suspended in the water column, their uptake over time of heavy metals, such as copper, cadmium and zinc, provides a measure of water contamination (Harris *et al.* 1979).

In a closely related species studied in the Northern Hemisphere, mussels subject to high predation by crabs grew thicker shells and stronger byssus threads as a protection against predation. The thicker shells were induced by water-borne cues from the crab predator (Leonard *et al.* 1999). A similar response in the blue mussel is yet to be tested.

Of the numerous smaller mussel species, the flea mussel *Limnoperna pulex* (formerly *Xenostrobus pulex*) is possibly the best known because it forms dense mats in the mid to low intertidal zone on rock platforms on many semi-exposed southern shores. Its cover is often 20–30%, especially near nutrient sources such as estuaries and upwelling waters. The spawning season is from July to January, with spatfall occurring about a month after spawning. The shell grows at ~2 mm month^{-1}, becoming sexually mature in ~11 months and reaching a maximum length of ~25 mm. Longevity is ~2 years. The lower limit of this mussel is likely determined by predation by whelks, crabs and wrasses, which forage intertidally at high water in southern Australia as in New Zealand (Rilov and Schiel 2006; unpublished observations).

Another subtidal species, the hairy mussel *Trichomya hirsuta* is commoner in the warmer waters of New South Wales and upper Spencer Gulf than elsewhere. It grows on reefs to ~20 m depth, often attached to algal holdfasts. This species spawns from November to January and the planktonic larvae settle after about 2 weeks, preferably on conspecific byssus threads (Ompy 2003).

The doughboy scallop *Mimachlamys asperrima*

Most scallops live on soft sediments, but the doughboy scallop attaches itself to firm substrata with byssus threads (Plate 23). It is common on southern sheltered reefs, as well as on razor shells, rubbly or shelly bottoms, and on jetty piles. When attacked by its main predator, the 11-armed sea-star *Coscinasterias muricata*, the scallop breaks its byssus and swims away. It then settles in another site, where it secretes a new byssus.

The biology of the doughboy scallop is reasonably well known because it was exploited during the 1970s and early 1980s in D'Entrecasteaux Channel before the fishery's sudden collapse in 1982 (Young and Martin 1989). Reproduction of the scallop is known from several parts of its range (O'Connor and Heasman 1996; Styan and Butler 2003; Butler 2008). In Jervis Bay, New South Wales and Gulf St Vincent, South Australia, spawning generally occurs from June to September, but occurs only from June to July in southern Tasmania. The species is dioecious (with males and females) and a serial spawner (i.e. after a spawning episode, the gonad builds up rapidly for another spawning episode). After fertilisation of the eggs, the larvae drift for 20–22 days and then settle as spat (Rose and Dix 1984). The longevity of this scallop is not known, although other related scallops live for up to 14 years.

Some interesting aspects of the doughboy scallop's fertilisation ecology have been described. Free spawning, in which sperm and eggs are released into the water column, is a risky business, because sperm may never encounter an egg if the spawners are too far apart, or alternatively too many sperm may encounter the same egg (called polyspermy). Styan and Butler (2000) examined this problem and found that males and females had to be < 50 cm apart to overcome the dilution effect of distance for successful fertilisation. Polyspermy, however, was not a problem, because eggs possessed an intrinsic block inhibiting the penetration of more than one sperm.

Another interesting feature is that the doughboy scallop is typically encrusted by either a yellow crellid, or a pink myxillid, sponge. Pitcher and Butler (1987) found that a mutualism existed between the scallop and the sponge. Scallops with sponges on them grew faster, probably because sponge-less scallops moved about more often to escape predators. Scallops with sponges also had better survival. This was because their major predator, the sea-star *Coscinasterias*, could not grasp the scallop with its tube feet, so enabling the scallop to make good its escape. The benefits to the sponge were less clear, but apparently the scallop facilitated its recruitment. Young scallops recruited near adult scallops, which passed the sponge on directly to the juveniles. Another possible benefit to the sponge is that it gained protection against grazing by dorid nudibranch predators. Helen Chernoff (1987) found that the doughboy scallop survived better when attached to an elevated substratum, probably because of higher levels of siltation and predation on the sea bed – hence, their preference for living on razor shells, jetty piles or on reefs. She also found that they moved on average ~1 m a day (maximum 4 m a day) and that growth rates were slow (< 5 mm a year).

REFERENCES

Adams A, Reeve L (1850) Mollusca. Part 3. In *The Zoology of the Voyage of the H.M.S. Samarang under the Command of Captain Sir Edward Belcher C.B., F.R.A.S. F.G.S. during the Years 1843–1846.* (Ed. A Adams) pp. 45–87. Reeve, Bentham and Reeve, London.

Anderson DT (1959) The reproduction and early life history of the gastropod *Cymatilesta spengleri* (Perry) (Fam. Cymatidae). *Proceedings of the Linnean Society of New South Wales* **84**, 232–237.

Anderson DT (1965) The reproduction and early life histories of the gastropods *Notoacmaea petterdi* (Ten.-Woods), *Chiazacmaea flammea* (Quoy and Gaimard) and *Patelloida alticostata* (Angas) (Fam. Acmaeidae). *Proceedings of the Linnean Society of New South Wales* **90**, 106–114.

Appleyard SA, Carr NA, Elliott NG (2008) Molecular analysis of blacklip abalone population structure. In *Towards Optimising the Spatial Scale of Abalone Fishery Management.* (Eds S Mayfield and TM Saunders) pp. 61–77. FRCD Project No. 2004/019. SARDI Research Report Series No. 273. SARDI Aquatic Sciences, Adelaide.

Babcock R, Keesing JK (1999) Fertilisation biology of the abalone *Haliotis laevigata*: laboratory and field studies. *Canadian Journal of Fisheries and Aquatic Sciences* **50**, 1668–1678.

Benkendorff K, Davis AR, Bremner JB (2001) Chemical defense in the egg masses of benthic invertebrates: an assessment of antibacterial activity in 39 molluscs and 4 polychaetes. *Journal of Invertebrate Pathology* **78**, 109–118.

Black R (1977) Population regulation in the intertidal limpet *Patelloida alticostata* (Angas, 1865). *Oecologia* **30**, 9–22.

Black R (1978) Tactics of whelks preying on limpets. *Marine Biology* **46**, 157–162.

Black R, Lymbery A, Hill A (1988) Form and function: size of radular teeth and inorganic content of faeces in a guild of grazing molluscs at Rottnest Island, Western Australia. *Journal of Experimental Marine Biology and Ecology* **121**, 23–35.

Boyle PR (1977) The physiology and behaviour of chitons (Mollusca: Polyplacophora). *Oceanography and Marine Biology - an Annual Review* **15**, 461–509.

Brandley BK (1984) Aspects of the ecology and physiology of *Elysia* cf. *furvacoida* (Mollusca: Sacoglossa). *Bulletin of Marine Science* **34**, 207–219.

Burn R (1974) Notes on some benthonic opisthobranchs from Port Phillip Bay, Victoria. *Journal of the Malacological Society of Australia* **3**, 43–57.

Burn R (1989) Opisthobranchs (Subclass Opisthobranchia). In *Marine Invertebrates of Southern Australia. Part II.* (Eds SA Shepherd and IM Thomas) pp. 725–788. South Australian Government Printer, Adelaide.

Burn R (2006) A checklist and bibliography of the Opisthobranchia (Mollusca: Gastropoda) in Victoria and Bass Strait area, south-eastern Australia. *Museum Victoria Reports* **10**, 1–42.

Butler AJ (1979) Relationships between height on the shore and size distributions of *Thais* spp. (Gastropoda: Muricidae). *Journal of Experimental Marine Biology and Ecology* **41**, 163–194.

Butler AJ (2008) Razor fish and scallops. In *Natural History of Gulf St Vincent* (Eds SA Shepherd, S Bryars, IR Kirkegaard, P Harbison, *et al.*) pp. 238–247. Royal Society of South Australia, Adelaide.

Carefoot TH (1987) *Aplysia*: its biology and ecology. *Oceanography and Marine Biology - an Annual Review* **25**, 167–284.

Chernoff H (1987) Factors affecting mortality of the scallop, *Chlamys asperrima* (Lamarck) and its epizoic sponges in South Australian waters. *Journal of Experimental Marine Biology and Ecology* **109**, 155–171.

Clarkson PS, Shepherd SA (1985) Distribution and feeding of archaeogastropods in a boulder habitat at West Island, South Australia. *Journal of the Malacological Society of Australia* **7**, 35–44.

Conklin EJ, Mariscal RN (1977) Feeding behaviour, ceras structure, and nematocyst storage in the aeolid nudibranch, *Spurilla neapolitana* (Mollusca). *Bulletin of Marine Science* **27**, 658–667.

Connell SD, Vanderklift MA (2008) Negative interactions: the influence of predators and herbivores on prey and ecological systems. In *Marine Ecology* (Eds SD Connell and B Gillanders) pp. 72–99. Oxford University Press, Melbourne.

Currie DR (1989) Valve structuring and aesthete distributions in four species of Australian chitons (Mollusca: Polyplacophora). *Journal of the Malacological Society of Australia* **10**, 69–86.

Currie DR (1990) Reproductive periodicity of three species of chitons at a site in New South Wales, Australia (Mollusca: Polyplacophora). *Invertebrate Reproduction & Development* **17**, 25–32.

Currie DR (1992) Photoreceptor or statocyst? The ultrastructure and function of a unique sensory organ embedded in the shell valves of *Cryptoplax mystica* Iredale & Hull, 1925 (Mollusca: Polyplacophora). *Journal of the Malacological Society of Australia* **13**, 15–25.

Day RW, Fleming AE (1992) The determinants and measurement of abalone growth. In *Abalone of the World: Biology, Fisheries and Culture* (Eds SA Shepherd, MJ Tegner and SA Guzmán del Próo) pp. 141–168. Fishing News Books, Oxford.

Day RW, Dowell A, Sant G, Klemke J *et al.* (1995) Patchy predation: foraging behaviour of *Coscinasterias calamaria* and escape responses of *Haliotis rubra. Marine and Freshwater Behaviour and Physiology* **26**, 11–33.

DiMatteo T (1982) The ink of *Aplysia dactylomela* (Rang, 1828) (Gastropoda: Opisthobranchia) and it role as a defense mechanism. *Journal of Experimental Marine Biology and Ecology* **57**, 169–180.

Dix TG, Ferguson A (1984) Cycles of reproduction and condition in Tasmanian blue mussels, (*Mytilus edulis planulatus*) in Tasmania. *Australian Journal of Marine and Freshwater Research* **35**, 307–313.

Edgar GJ (2001) *Australian Marine Habitats in Temperate Waters.* New Holland Publishers, Sydney.

Edgar GJ (2008) *Australian Marine Life: the Plants and Animals of Temperate Waters.* 2nd Edition. New Holland Publishers, Sydney.

Ettinger-Epstein P, Kingsford MJ (2008) Effects of the El Niño southern oscillation on *Turbo torquatus* (Gastropoda) and their kelp habitat. *Austral Ecology* **33**, 594–606.

Fairweather PG (1988) Movements of intertidal whelks (*Morula marginalba* and *Thais orbita*) in relation to availability of prey and shelter. *Marine Biology* **100**, 63–68.

Farrant P (1987) Population dynamics of the temperate Australian soft coral *Capnella gaboensis*. *Marine Biology* **96**, 401–407.

Fletcher WJ (1987) Life history dynamics of the limpet *Patelloida alticostata* in intertidal and subtidal environments. *Marine Ecology Progress Series* **39**, 115–127.

Fletcher WJ, Creese RG (1985) Competitive interactions between co-occurring herbivorous gastropods. *Marine Biology* **86**, 183–191.

Fletcher WJ, Underwood AJ (1987) Interspecific competition among subtidal limpets: effect of substratum heterogeneity. *Ecology* **68**, 387–400.

Geiger DL (1999) Distribution and biogeography of the recent Haliotidae (Gastropoda: Vetigastropoda) worldwide. *BolletinoMalacologico* **35**, 57–120.

Geiger DL, Owen B (2012) *Abalone: World-wide Haliotidae*. Conch Books, Hackenheim, Germany.

Gérard K, Bierne N, Borsa P, Chenuil A *et al.* (2008) Pleistocene separation of mitochondrial lineages of *Mytilus* spp. mussels from Northern and Southern Hemispheres and strong genetic differentiation among southern populations. *Molecular Phylogenetics and Evolution* **49**, 84–91.

Goggin CL, Lester RJG (1995) *Perkinsus*, a protistan parasite of abalone in Australia: a review. In *Progress in Abalone Fisheries Research* (Eds SA Shepherd, RW Day and AJ Butler) *Marine and Freshwater Research* **46**, 639–646.

Gorfine H, Day R, Bardos D, Taylor B *et al.* (2008) 'Rapid response to abalone virus depletion in Western Victoria: information acquisition and reef code assessment'. FRDC Project 2007/066. University of Melbourne, Melbourne.

Gosling E (2003) *Bivalve Molluscs: Biology, Ecology and Culture*. Fishing News Books, Oxford.

Gowlett-Holmes K (2008) *A Field Guide to the Marine Invertebrates of South Australia*. Notomares, Hobart.

Gowlett-Holmes K, Holmes NJC (1989) Rediscovery of *Primovula (P) heleneae* Cate and description of a new species of *Crenavola* from South Australia (Mollusca: Gastropoda: Ovulidae). *Transactions of the Royal Society of South Australia* **113**, 205–214.

Greenwood PG, Mariscal RN (1984) Immature nematocyst incorporation by the aeolid nudibranch *Spurilla neapolitana*. *Marine Biology* **80**, 35–38.

Guest MA, Nichols PD, Frusher SD, Hirst AJ (2008) Evidence of abalone (*Haliotis rubra*) diet from combined fatty acid and stable isotope analyses. *Marine Biology* **153**, 579–588.

Hancock AT (2003) The biology and fishery of Roe's abalone *Haliotis roei* Gray in south-western Australia, with emphasis on the Perth fishery. PhD thesis. University of Western Australia, Perth.

Harris JE, Fabris GJ, Statham PJ, Tawfik F (1979) Biogeochemistry of selected heavy metals in Western Port, Victoria, and use of invertebrates as indicators with emphasis on *Mytilus edulis planulatus*. *Australian Journal of Marine and Freshwater Research* **30**, 159–178.

Hilbish TJ, Mullinas A, Dolven SL, Meyer A *et al.* (2000) Origin of the antitropical distribution pattern of marine mussels (*Mytilus* spp.): routes and timing of transequatorial migration. *Marine Biology* **136**, 69–77.

Hoegh-Guldberg O, Hinde R, Muscatine L (1986) Studies on a nudibranch that contains zooxanthellae. II. Contribution of zooxanthellae to animal respiration (CZAR) in *Pteraeolidia ianthina* with high and low densities of zooxanthellae. *Proceedings of the Royal Society of London. Series B, Biological Sciences* **228**, 511–521.

Hughes RN, Hughes HPI (1981) Morphological and behavioural aspects of feeding in the Cassidae (Tonnacea, Mesogastropoda). *Malacologia* **20**, 385–402.

Jensen KR (1991) Foraging behaviour of two Australian species of *Elysia* (Mollusca, Episthobranchia). In *The Marine Flora and Fauna of Albany, Western Australia. Proceedings of the 3rd International Marine Biology Workshop.*11–28 January 1988, Albany (Eds FE Wells, DI Walker, H Kirkman and R Lethbridge) pp. 541–552. Western Australian Museum, Perth.

Joll LM (1980) Reproductive biology of two species of Turbinidae (Mollusca: Gastropoda). *Australian Journal of Marine and Freshwater Research* **31**, 319–336.

Kaas P, Jones AM, Gowlett-Holmes KL (1998) Class Polyplacophora. In *Mollusca: The Southern Synthesis. Fauna of Australia. Vol. 5. Part A.* (Eds PL Beasley, GJB Ross and A Wells) pp. 161–177. CSIRO PUBLISHING, Melbourne.

Kangas M, Shepherd SA (1984) Distribution and feeding of chitons in a boulder habitat at West Island, South Australia. *Journal of the Malacological Society of Australia* **6**, 101–111.

Keesing JK, Wells FE (1989) Growth of the abalone *Haliotis roei* Gray. *Australian Journal of Marine and Freshwater Research* **40**, 199–204.

Keough M (1999) Sessile animals. In *Under Southern Seas: the Ecology of Australia's Rocky Reefs*. (Ed. N Andrew) pp. 136–147. University of New South Wales Press, Sydney.

Kohn AJ (1993) Development and life history of three temperate Australian species of *Conus* (Mollusca: Gastropoda). In *The Marine Flora and Fauna of Rottnest Island, Western Australia. Vol. 2. Proceedings of the 5th International Marine Biological Workshop.* January 1991, Rottnest Island (Eds FE Wells, DI Walker, H Kirkman and R Lethbridge) pp. 509–521. Western Australian Museum, Perth.

Kohn AJ, Almasi KN (1993) Comparative ecology of a biogeographically heterogeneous *Conus* assemblage. In *The*

Marine Flora and Fauna of Rottnest Island, Western Australia. Vol. 2. Proceedings of the 5th International Marine Biological Workshop. January 1991, Rottnest Island.Eds FE Wells, DI Walker, H Kirkman and R Lethbridge) pp. 523–538. Western Australian Museum, Perth.

Kohn AJ, Curran KM, Mathis BJ (2005) Diets of the predatory gastropods *Cominella* and *Conus* at Esperance, Western Australia. In *The Marine Flora and Fauna of Esperance, Western Australia. Proceedings of the 12th International Marine Biological Workshop.* 3–21 February 2003, Esperance (Eds FE Wells, DI Walker and GA Kendrick) pp. 235–244. Western Australian Museum, Perth.

Laxton JH (1969) Reproduction in some New Zealand Cymatiidae (Gastropoda: Prosobranchia). *Zoological Journal of the Linnean Society* **48**, 237–253.

Laxton JH (1970) Shell growth in some New Zealand Cymatiidae (Gastropoda: Prosobranchia). *Journal of Experimental Marine Biology and Ecology* **4**, 250–260.

Laxton JH (1971) Feeding in some Australasian Cymatiidae (Gastropoda: Prosobranchia). *Zoological Journal of the Linnean Society* **50**, 1–9.

Leonard GH, Bertness MD, Yund PO (1999) Crab predation, waterborne cues, and inducible defenses in the blue mussel, *Mytilus edulis*. *Ecology* **80**, 1–14.

Ludbrook N, Gowlett-Holmes K (1989) Chitons, gastropods and bivalves. In *Marine Invertebrates of Southern Australia. Part II.* (Eds SA Shepherd and IM Thomas) pp. 504–724. South Australian Government Printer, Adelaide.

Magnus O (1555) *Historia de Gentibus Septentrionalibus.* Rosenkilde and Bagger, Copenhagen, 1972.

Mayfield S, Chick RC, Carlson IJ, Ward TM (2011) Invertebrate dive fisheries can be sustainable: forty years of production from a greenlip fishery off southern Australia. *Reviews in Fisheries Science* **18**, 218–230.

McShane P (1999) Blacklip abalone. In *Under Southern Seas: the Ecology of Australian Rocky Reefs* (Ed. N Andrew) pp. 72–77. University of New South Wales, Sydney.

Miethe T, Dytham C, Dieckmann U, Pitchford JW (2010) Marine reserves and the evolutionary effects of fishing on size at maturation. *ICES Journal of Marine Science* **67**, 412–425.

Miller KJ, Maynard BT, Mundy CN (2009) Genetic diversity and gene flow in collapsed and healthy abalone fisheries. *Molecular Ecology* **18**, 200–211.

Montelli L (2009) Seasonal variation in mussel spatfall at Cockburn Sound. Poster Abstract. In *Australian Marine Sciences Association 46th Annual Conference: AMSA 2009; Program and Abstract Handbook.* 5-9 July 2009, Adelaide (Eds S Dittmann, N Hall and M Loo) p. 232. Australian Marine Sciences Association, Adelaide.

Morgan LE, Shepherd SA (2006) Population and spatial structure of two common temperate reef herbivores: abalone and sea urchins. In *Marine Metapopulations* (Eds JP Kritzer and PF Sale) pp. 209–246. Elsevier, London.

Morton J (1967) *Molluscs.* Hutchinson Press, London.

Morton B (1999) Competitive grazers and the predatory whelk *Lepsiella flindersi* (Gastropoda: Muricidae) structure a mussel bed (*Xenostrobus pulex*) on a southwest Australian shore. *The Journal of Molluscan Studies* **65**, 435–452.

Morton B, Britton JC (1993) The ecology, diet and foraging strategy of *Thais orbita* (Gastropoda: Muricidae) on a rocky shore of Rottnest island, Western Australia. In *The Marine Flora and Fauna of Rottnest Island, Western Australia. Vol. 2. Proceedings of the 5th International Marine Biological Workshop.* January 1991, Rottnest Island (Eds FE Wells, DI Walker, H Kirkman and R Lethbridge). pp. 539–564. Western Australian Museum, Perth.

Nash WJ, Anderson JC, Bridley J, Wideley D *et al.* (1995) Larval settlement of blacklip abalone (*Haliotis rubra*) on artificial collectors in southern Tasmania. *Marine and Freshwater Research* **46**, 531–538.

O'Connor WA, Heasman MP (1996) Temporal patterns of reproductive condition in the doughboy scallop, *Chlamys (Mimachlamys) asperrima* Lamarck, in Jervis Bay, Australia. *Journal of Shellfish Research* **15**, 237–244.

Ompy M (2003) The reproductive ecology of four mytilid bivalves: *Mytilus galloprovincialis* Lamarck, 1819, *Brachidontes rostratus* (Dunker, 1857), *Brachidontes erosus* (Lamarck, 1819), and *Trichomya hirsutus* (Lamarck, 1819). PhD thesis. Flinders University, Adelaide.

Ottaway JR (1977) *Pleurobranchaea novaezelandiae* preying on *Actinia tenebrosa. New Zealand Journal of Marine and Freshwater Research* **11**, 125–130.

Otway NM (1994) Population ecology of the low-shore chitons *Onithochiton quercinus* and *Plaxiphora albida. Marine Biology* **121**, 105–116.

Parry GD (1982a) Reproductive effort in four species of intertidal limpets. *Marine Biology* **67**, 267–282.

Parry GD (1982b) The evolution of the life histories of four species of intertidal limpets. *Ecological Monographs* **52**, 65–91.

Parsons DW, MacMillan DL (1979) The escape response of abalone (Mollusca, Prosobranchia, Haliotidae) to predatory gastropods. *Marine Behaviour and Physiology* **6**, 65–82.

Paul VJ, Van Alstyne KL (1988) Use of ingested algal diterpenoids by *Elysia halimedae* Macnae (Opisthobranchia: Ascoglossa) as anti-predator defenses. *Journal of Experimental Marine Biology and Ecology* **119**, 15–29.

Pearce A, Lenanton R, Jackson G, Moore J *et al.* (2011) *The 'Marine Heat Wave' off Western Australia During the Summer of 2010/2011.* Fisheries Research Report No. 222. Department of Fisheries, Perth.

Perron FE (1983) Growth, fecundity and mortality of *Conus pennaceus* in Hawaii. *Ecology* **64**, 53–62.

Petraitis PS (1992) Effects of body size and water temperature on grazing rates of four intertidal gastropods. *Australian Journal of Ecology* **17**, 409–414.

Phillips BF (1969) Population ecology of the whelk *Dicathais aegrota. Australian Journal of Marine and Freshwater Research* **20**, 225–265.

Phillips BF, Campbell NA (1974) Mortality and longevity in the whelk *Dicathais orbita* (Gmelin). *Australian Journal of Marine and Freshwater Research* **25**, 25–33.

Piggott MP, Banks S, Tung P, Beheregaray LB (2008) Genetic evidence for different scales of connectivity in a marine mollusc. *Marine Ecology Progress Series* **365**, 127–136.

Pitcher CR, Butler AJ (1987) Predation by asteroids, escape response and morphometrics of scallops with epizootic sponges. *Journal of Experimental Marine Biology and Ecology* **112**, 233–249.

Prince JD, Peeters H, Gorfine H, Day RW (2008) The novel use of harvest policies and rapid visual assessment to manage spatially complex abalone resources (Genus *Haliotis*). *Fisheries Research* **94**, 330–338.

Raven JA, Walker DI and Jensen KR. (2001). What fraction of the organic carbon in sacoglossans is obtained from photosynthesis by kleptoplastids? An investigation using the natural abundance of stable carbon isotopes. *Marine Biology* **138**, 537–545.

Riedel F (1992) A re-evaluation of the ontogeny of *Cabestana spengleri* (Perry, 1811) (Gastropoda: Tonnoidea: Ranellidae) *The Veliger* **35**, 117–121.

Rilov G, Schiel DR (2006) Trophic linkages across seascapes: subtidal predators limit effective mussel recruitment in rocky intertidal communities. *Marine Ecology Progress Series* **327**, 83–93.

Rogers CN, Steinberg PD, de Nys R (1995) Factors associated with oligophagy in two species of sea hares (Mollusca: Anaspidea). *Journal of Experimental Marine Biology and Ecology* **192**, 47–73.

Rogers CN, de Nys R, Steinberg PD (2000) Predation on juvenile *Aplysia parvula* and other small anaspidean, ascoglossan and nudibranch gastropods by pycnogonids. *The Veliger* **43**, 330–337.

Rogers CN, de Nys R and Steinberg PD (2002) Effects of algal diet on the performance and susceptibility to predation of the sea hare *Aplysia parvula. Marine Ecology Progress Series* **236**, 241–254.

Rogers CN, de Nys R, Steinberg PD (2003) Ecology of the sea hare *Aplysia parvula* (Opisthobranchia) in New South Wales, Australia. *Molluscan Research* **23**, 185–198.

Rose RA (1985) The spawn and development of 29 New South Wales opisthobranchs (Mollusca: Gastropoda). *Proceedings of the Linnean Society of New South Wales* **108**, 23–36.

Rose RA, Dix TG (1984) Larval and juvenile development of the doughboy scallop, *Chlamys (Mimachlamys) asperrima* (Lamarck) (Mollusca: Pectinidae). *Australian Journal of Marine and Freshwater Research* **35**, 315–323.

Rose RA, Hoegh-Guldberg O (1982) A brood-protecting nudibranch with pelagic lecithotrophic development. *The Journal of Molluscan Studies* **48**, 231–232.

Rudman WB (1981a) Polyp mimicry in a new species of aaeolid nudibranch mollusc. *Journal of Zoology* **193**, 421–427.

Rudman WB (1981b) The anatomy and biology of alcyonarian-feeding aeolid opisthobranch molluscs and their development of symbioses with zooxanthellae. *Zoological Journal of the Linnean Society* **72**, 219–262.

Rudman WB. (1991). Further studies on the taxonomy and biology of the octocoral-feeding genus *Phyllodesmum* Ehrenberg, 1831 (Nudibranchia: Aeolidoidea). *The Journal of Molluscan Studies* **57**, 167–203.

Rudman WB (1998) Sea slug forum. <www.austmus.gov.au/seaslugs/solarpow> (Accessed March 2009).

Rudman WB, Bergquist PR (2007) A review of feeding specificity in the sponge-feeding Chromodorididae (Nudibranchia: Mollusca). *Molluscan Research* **27**, 60–88.

Rudman WB, Willan RC, Burn R, Thompson TE *et al.* (1998) Opisthobranchia. In *Mollusca: The Southern Synthesis. Fauna of Australia. Vol. 5. Part B.* (Eds PL Beasley, GJB Ross and A Wells) pp. 915–1035. CSIRO PUBLISHING, Melbourne.

Rumpho ME, Summer EJ, Manhart JR (2000) Solar-powered sea slugs. Mollusc/algal chloroplast symbiosis. *Plant Physiology* **123**, 29–38.

Sakker ER (1986) Seasonal reproductive cycles of three Australian species of chitons (Mollusca: Polyplacophora). *International Journal of Invertebrate Reproduction and Development* **10**, 1–16.

Saunders TM, Mayfield S, Hogg AA (2008) A simple, cost-effective, morphometric marker for characterising abalone populations at multiple scales. *Marine and Freshwater Research* **59**, 32–40.

Sellinger A, Weiss PM, Nguyen A, Lu Y *et al.* (1998). Continuous self-assembly of organic-inorganic nanocomposite coatings that mimic nature. *Nature* **394**, 256–260.

Shepherd SA (1978) 'Studies on southern Australian abalone (Genus *Haliotis*): ecological significance of variable defensive reactions'. Internal Report. Department of Fisheries, Adelaide.

Shepherd SA (1979) The ecology of a deep water red algal community off Cape Northumberland. South Australia. MEnvSt. thesis. 255 p. University of Adelaide, Adelaide.

Shepherd SA (1984) 'Evidence for a summer sea temperature effect on the recruitment of greenlip abalone'. Internal Report No. 113. Department of Fisheries, Adelaide.

Shepherd SA (1999) Greenlip abalone. In *Under Southern Seas: the Ecology of Australian Rocky Reefs* (Ed. N Andrew) pp. 78–85. University of New South Wales, Sydney.

Shepherd SA (2006) Ontogenetic changes in diet, feeding behaviour and activity of the western blue groper, *Achoerodus gouldii*. In *The Marine Fauna and Flora of Esperance, Western Australia Vol. 2. Proceedings of the 12th International Marine Biology Workshop.* 3–21 February 2003, Esperance. (Eds FE Wells, DI Walker and GA Kendrick) pp. 477–494. Western Australian Museum, Perth.

Shepherd SA (2008) Abalone of Gulf St Vincent. In *Natural History of Gulf St Vincent* (Eds SA Shepherd, S Bryars, IR Kirkegaard, P Harbison, *et al.*) pp. 448–455. Royal Society of South Australia, Adelaide.

Shepherd SA, Breen PA (1992) Mortality in abalone: its estimation, variability and causes. In *Abalone of the World: Biology, Fisheries and Culture* (Eds SA Shepherd, MJ Tegner and SA Guzmán del Próo) pp. 253–275. Fishing News Books, Oxford.

Shepherd SA, Daume S (1996) Ecology and survival of juvenile abalone in a crustose coralline habitat in South Australia. In *Survival Strategies in Early Life History Stages of Marine Resources. Proceedings of an International Workshop.* 11–14 October 1994, Yokohama, Japan (Eds Y Watanabe, Y Yamashita and Y Oozeki) pp. 297–313. Balkema, Rotterdam.

Shepherd SA, Hearn WS (1983) Studies on southern Australian abalone (genus *Haliotis*). IV. Growth of *H. laevigata* and *H. ruber. Australian Journal of Marine and Freshwater Research* **34**, 461–475.

Shepherd SA, Laws HM (1974) Studies on southern Australian abalone (genus *Haliotis*). II. Reproduction of five species. *Australian Journal of Marine and Freshwater Research* **25**, 49–62.

Shepherd SA, Rodda KR (2001) Sustainability demands vigilance: evidence for serial decline of the South Australian greenlip abalone fishery and a review of management. *Journal of Shellfish Research* **20**, 829–841.

Shepherd SA, Steinberg PD (1992) Food preferences of three Australian abalone species with a review of the algal food of abalone. In *Abalone of the World: Biology, Fisheries and Culture* (Eds SA Shepherd, MJ Tegner and SA Guzmán del Próo) pp. 169–181. Fishing News Books, Oxford.

Shepherd SA, Tegner MJ, Guzmán del Próo SA (Eds) (1992) *Abalone of the World: Biology, Fisheries and Culture.* Fishing News Books, Oxford.

Shepherd SA, Rodda KR, Vargas KM (2001) A chronicle of collapse in two abalone stocks with proposals for precautionary management. *Journal of Shellfish Research* **20**, 843–856.

Smith BJ, Black JH, Shepherd SA (1989) Molluscan egg masses and capsules. In *Marine Invertebrates of Southern Australia. Part II* (Eds SA Shepherd and IM Thomas) pp. 841–891. South Australian Government Printer, Adelaide.

Steinberg PD, van Altena I (1992) Tolerance of marine invertebrate herbivores to brown algal phlorotannins in temperate Australasia. *Ecological Monographs* **62**, 189–222.

Stevenson J, Melville A (1999) Settlement and recruitment of the abalone *Haliotis cyclobates* Péron 1816. *Marine and Freshwater Research* **50**, 229–234.

Styan C and Butler AJ. (2000). Fitting fertilization kinetics models for free-spawning marine invertebrates. *Marine Biology* **137**, 943–951.

Styan C, Butler AJ (2003) Asynchronous patterns of reproduction for the sympatric scallops *Chlamys bifrons* and *Chlamys asperrima* (Bivalvia: Pectinidae) in South Australia. *Marine and Freshwater Research* **54**, 77–86.

Svane I, (2011) An overview of the blue mussel in southern Australia – a serial invader, a blind passenger, or just a welcome addition to the menu? *Transactions of the Royal Society of South Australia* **135**, 134–139.

Svane I, Ompi M (1993) Patch dynamics in beds of the blue mussel *Mytilus edulis* L.: effects of site, patch size, and position within a patch. *Ophelia* **37**, 187–202.

Taylor JD (1993) Dietary and anatomical specialization of mitrid gastropods (Mitridae) at Rottnest Island, Western Australia. In *The Marine Flora and Fauna of Rottnest Island, Western Australia. Vol. 2. Proceedings of the 9th International Marine Biological Workshop.* January 1991, Rottnest Island. (Eds DI Walker and FE Wells) pp. 583–599. Western Australian Museum, Perth.

Taylor JD, Glover EA (1999) Penetrating the defences: opercular drilling by *Dicathais orbita* on the turbinid gastropod *Ninella torquata*. In *The Marine Flora and Fauna of Rottnest Island, Western Australia. Vol. 2. Proceedings of the 9th International Marine Biological Workshop.* January 1991, *Rottnest Island.* (Eds DI Walker and FE Wells) pp. 177–198. Western Australian Museum, Perth.

Taylor R, Steinberg PD (2005) Host use by Australasian seaweed mesograzers in relation to feeding preferences of larger grazers. *Ecology* **86**, 2955–2967.

Temby N, Miller K, Mundy C (2007) Evidence of genetic subdivision among populations of blacklip abalone (*Haliotis rubra* Leach) in Tasmania. *Marine and Freshwater Research* **58**, 733–742.

Thomas M, Day RW (1995) Site selection by a small drilling predator: why does the gastropod *Haustrum baileyanum* drill over muscle tissue of the abalone *Haliotis rubra? Marine and Freshwater Research* **46**, 647–655.

Thompson TE (1970) Eastern Australian Pleurobranchomorpha (Gastropoda: Opisthobranchia). *Journal of Zoology* **160**, 173–198.

Thompson TE, Bennett I (1970) Observations on Australian Glaucidae (Mollusca: Opisthobranchia). *Zoological Journal of the Linnean Society* **49**, 187–197.

Thrower SJ (1977) Blue spot in abalone – possible causes and solutions. *Australian Fisheries* **36**, (10), 6–7.

Todd CD (1981) The ecology of nudibranch molluscs. *Oceanography and Marine Biology - an Annual Review* **19**, 141–234.

Trowbridge CD (1992) Phenology and demography of a marine specialist herbivore: *Placida dendritica* (Gastropoda: Opisthobranchia) on the central coast of Oregon. *Marine Biology* **114**, 443–452.

Trowbridge CD (1998) Stenophagous, herbivorous sea slugs attack dessication-prone green algal hosts (*Codium* spp.): indirect evidence of prey-stress models (PSMs)? *Journal of Experimental Marine Biology and Ecology* **230**, 31–53.

Turner E (1978) Brooding of chitons in Tasmania. *Journal of the Malacological Society of Australia* **4**, 43–47.

Underwood AJ (1974) The reproductive cycles and geographical distribution of some common eastern Australian prosobranchs (Mollusca: Gastropoda). *Australian Journal of Marine and Freshwater Research* **25**, 63–88.

Ward DW, Davis AR (2002) Reproduction of the turban shell *Turbo torquatus* Gmelin 1791 (Mollusca: Gastropoda) in New South Wales, Australia. *Marine and Freshwater Research* **53**, 85–91.

Watson AL (1976) Preliminary observations on the influence of water movement on population structure in *Ancorina corticata* (Carter) (Choristida: Demospongiae). *Proceedings of the New Zealand Ecological Society* **23**, 45–50.

Wells FE, Keesing JK (1989) Reproduction and feeding of the abalone *Haliotis roei* Gray. *Australian Journal of Marine and Freshwater Research* **40**, 187–197.

Wells FE, Keesing JK (2005) Feeding of *Lepsiella flindersi* (Adams & Angas, 1863) on the limpet *Patelloida alticostata* (Angas, 1865), near Esperance, Western Australia. In *The Marine Flora and Fauna of Esperance, Western Australia. Proceedings of the 12th International Marine Biological Workshop.* 3–21 February 2003 *Esperance.* (Eds FE Wells, DI Walker and GA Kendrick) pp. 315–323. Western Australian Museum, Perth.

Wells FE, Sellers RJ (1987) Reproductive periodicity of the chiton *Acanthopleura hirtosa* on intertidal platforms in the Perth area of Western Australia. *Journal of the Malacological Society of Australia* **8**, 87–92.

Wernberg T, White M, Vanderclift MA (2008) Population structure of turbinid gastropods on wave-exposed subtidal reefs: effects of density, body size and algae on grazing behaviour. *Marine Ecology Progress Series* **362**, 169–179.

Willan RC (1984) A review of the diets in Notaspidea (Mollusca: Opisthobranchia). *Journal of the Malacological Society of Australia* **6**, 125–142.

Williams SI, Walker DI (1999) Meso-herbivore – macroalgal interactions: feeding ecology of sacoglossan sea slugs (Mollusca, Opisthobranchia) and their effects on their algal food. *Oceanography and Marine Biology - an Annual Review* **37**, 87–128.

Wilson BR (1998) Superfamily Cypraeoidea. In *Mollusca: The Southern Synthesis. Fauna of Australia. Vol. 5. Part B.* (Eds PL Beasley, GJB Ross and A Wells) pp. 780–786. CSIRO PUBLISHING, Melbourne.

Wilson BR, Clarkson P (2004) *Australia's Spectacular Cowries. A Review and Field Study of Two Endemic Genera:* Zoila *and* Umbilia. Odyssey Publishing, El Cajon.

Wilson BR, Hodgkin EP (1967) A comparative account of the reproductive cycles of five species of marine mussels (Bivalvia: Mytilidae) in the vicinity of Fremantle, Western Australia. *Australian Journal of Marine and Freshwater Research* **18**, 175–203.

Womersley HBS (1984) *The Marine Benthic Flora of Southern Australia. Part I.* Government Printer, Adelaide.

Worthington DG, Fairweather PG (1989) Shelter and food: interactions between *Turbo undulatus* (Archaeogastropoda: Turbinidae) and coralline algae on rocky seashores in New South Wales. *Journal of Experimental Marine Biology and Ecology* **129**, 61–79.

Worthington DG, Blount C, Chick RC, Gibson PT *et al.* (2001) *An Assessment of the Performance Indicators of the Abalone Share Management Plan in 2000.* Fisheries Resource Assessment Series No. 11. NSW Fisheries Centre, Cronulla.

Young PC, Martin RB (1989) The scallop fisheries of Australia and their management. *Critical Reviews in Aquatic Science* **1**, 615–638.

Zacharin W (1992) 'Report to Demersal Mollusc Research Group on the commercial dive fishery for the common wavy turbo, *Turbo undulatus*'. Unpublished Report November 1992. Department of Primary Industry and Fisheries, Hobart.

10 Cephalopods

OVERVIEW

Cephalopods are a carnivorous and highly mobile group of molluscs. The ecology and life history of the smallest squid species *Idiosepius notoides* is described, followed by an account of the life history, behaviour and ecology of one of the larger cephalopods, the giant cuttlefish *Sepia apama*, which forms unique, vast spawning aggregations in upper Spencer Gulf. Lastly, the ecology of four common octopods is discussed. These are: the two small, venomous octopuses, *Hapalochlaena maculosa* and *Hapalochlaena fasciata*; the western gloomy octopus *Octopus* cf. *tetricus*; and the Maori octopus *Octopus maorum*. These octopuses are voracious carnivores, and the last two species are significant predators of lobsters; their interactions with lobster fishing are briefly discussed.

INTRODUCTION

This chapter pursues the theme of the last chapter on molluscs and considers a specialised class, the Cephalopoda, which contain three major Orders – cuttlefish, squid and octopus. In the last two Orders the shell is completely lost or is vestigial.

CEPHALOPODS

Cephalopods are an ancient and highly evolved group of molluscs. Aristotle described aspects of their life history in ~330 BC (cited in Boyle and Rodhouse 2005) in these terms: 'These molluscs are all carnivorous; and … the calamari and the sepia are more than a match for fishes of even the largest species. The octopus for the most part gathers shellfish, extracts the flesh, and feeds on that; in fact fishermen recognise their holes by the number of shells lying about'. The name cephalopod, which means 'head-foot', was first applied because of the fusion between the two. The head, which contains the eyes and brain, also has attached to it a circle of tentacles or arms, equivalent to the foot on other molluscs. The trunk is surrounded by the mantle, and this has evolved to enable the mantle cavity to fill with water, which is ejected forcibly through a funnel. In this way, the animal can jet propel itself through the water, forwards or backwards, with impressive speed. However, jet propulsion is generally used only for escape from predators. Normal swimming in cuttlefish and squid is achieved by undulating their narrow lateral fins, whereas octopuses crawl over the seabed.

Cephalopods have a high growth rate and short life span, rarely living for more than ~2 years. This live-fast-die-young life history strategy has unusual implications for them. They must eat large amounts of food to fuel growth, for energy for hunting prey and to escape from predators, and for reproduction.

Cephalopods, notably the octopus, are considered to be the most intelligent of all invertebrates, and have very large brains, which are comparable with those of fish, earning them the title of 'honorary vertebrates' (Hanlon and Messenger 1996). They have evolved complex behaviours to aid in prey capture, predator avoidance and mating.

Sometimes also called the 'chameleons of the sea', cephalopods can instantly change their colour by expanding or contracting pigment cells, called chromatophores, to match their background. One moment they can be white and smooth to match a sandy background when swimming over sand, and then switch instantly to a mottled brown and wrinkled form when over a reef. Their unpredictable pattern-changing behaviour, which is used to escape from predators (e.g. seals, dolphins and sharks), is termed 'Protean' after the mythical Proteus, who avoided capture by continually changing shape.

Cephalopods also have a large ink sac from which they can eject ink to deter predators. In some species, the ink emerges as a cloud and acts as a smoke screen. In other species, the spurts of ink are held together with mucus, whose shape may resemble prey and act as a distracting decoy to confuse the predator and facilitate the cephalopod's escape. The ink may also affect the predator's chemical senses, making the prey distasteful – which in itself may lead to learned avoidance of the prey in later encounters (Derby 2007). The sexes of cephalopods are separate, and the males have a specialised tentacle adapted for placing a sperm package in the female's mantle or buccal cavity, or in a specialised sac at the base of her arms.

Of the major groups in the cephalopods, the Nautilus and the argonauts are pelagic, and squid mainly so: some living in the open ocean and others in bays and gulfs. Hence, they are not discussed in this book, except for the pygmy squid *Idiosepius notoides*, which commonly lives in macroalgae or seagrass on the southern and eastern Australian coasts. The cuttlefish and octopuses discussed are associated with reefs for at least part of their life cycle.

The larger octopuses, squid and cuttlefish are exploited commercially in southern Australia. Octopuses are taken as by-catch by lobster fishers, and are sometimes caught by setting earthen or plastic pots. Boyle and Rodhouse (2005) recount an unusual variant of fishing, in which octopuses are used to recover items from the bottom in salvage operations. They are tethered and lowered to the seabed, whereupon they seize the lost object, and are then hauled back to the surface by the salvage team! Squid, and sometimes cuttlefish, are taken by jig fishing, where the barbless hook resembles a prawn, and the jig's jerking movement imitates that of the prawn, so that the squid seizes it and is caught.

The pygmy squid *Idiosepius notoides*

This tiny squid (Plate 24) is common on the east and south coasts of Australia on reefs and in seagrass beds to ~10 m depth. It is notable for its dorsal suction disc on the mantle. This account is taken from the notes of English (1980) in the Sydney region and the study of Tracey *et al.* (2003) in Tasmania.

In Tasmania, the lifespan differs between the sexes. Females reach sexual maturity at ~12 weeks and live for 17–18 weeks, while males mature at ~9 weeks and live for up to 13 weeks. Mature females carry eggs in their ovaries at all stages of development, indicating that they can spawn repeatedly for the duration of their short adult life. Females reach a mantle length of 20 mm, and grow at an exponential rate during life (Fig. 10.1), whereas males grow almost as fast, but scarcely reach half the weight of females. For both sexes, growth is strongly influenced by temperature. Individuals hatching during warmer seasons grow much faster, and mature younger and at a larger size, than those hatching in cooler seasons. Unlike almost all other squids, the hatchlings of pygmy squid live continually in the plankton before settling into reef or seagrass habitat (Boletzky 2003).

The pygmy squid mimics a piece of alga or seagrass by attaching itself to the blade with its adhesive pad, as it sits in wait for prey. When a mysid, shrimp or other

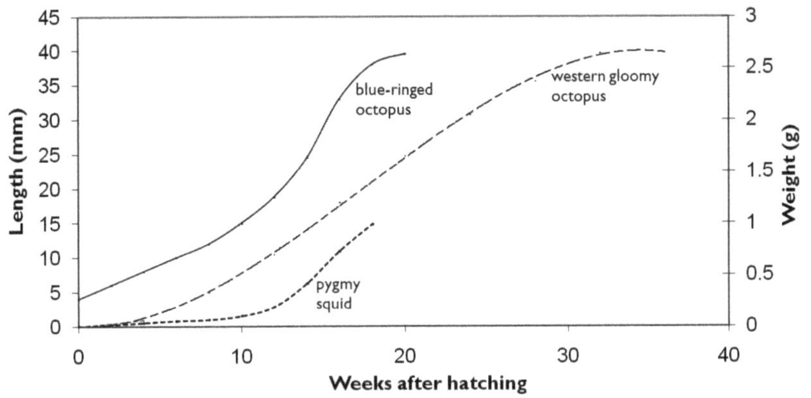

Figure 10.1: Generalised growth curves for three cephalopods: that is, length (from the eyes to the rear end of the body) versus age for the blue-ringed octopus *Hapalochlaena maculosa* (redrawn from Tranter and Augustine 1973); weight (kg) versus age for the western gloomy octopus *Octopus cf. tetricus* (redrawn from Joll 1977b); and weight (g) versus age for the pygmy squid *Idiosepius notoides* (redrawn from Tracey *et al.* (2003).

small crustacean approaches to within 1 cm or so, the squid seizes it with its tentacles, inserts its beak through the exoskeleton and eats the flesh.

Many fishes cryptic in reef and seagrass habitats, including young western Australian salmon, are predators of the pygmy squid (e.g. Hindell *et al.* 2001).

The giant cuttlefish *Sepia apama*

Sepia apama (Plate 25) appears to be a species-complex, with separate species in south-west (SW) and south-east (SE) Australia, as well as one or more in central South Australia (S Donellan *pers. comm.*). Species within the complex extend to 300 m deep and aggregate during spawning. The east Australian species is the largest cuttlefish in the world, reaching a length of ~1 m and a weight of 12 kg, but less in cooler southern waters. This cuttlefish also supports small fisheries.

In the upper Spencer Gulf, massive aggregations have been recorded in the spawning season, giving rise to ecotourism ventures in the region and many studies. This section draws on the works of Norman *et al.* (1999), Hall and Hanlon (2002), Hall and Fowler (2003), Naud *et al.* (2004, 2005), Hall *et al.* (2007), the review of Gillanders and Payne (2013), as well as other studies.

Life history

Juvenile cuttlefish hatch at a size of up to 15 mm from eggs laid at their parents' spawning sites during spring, and grow during the following summer. They then mate, lay batches of single eggs – like large white marbles attached to the underside of rock overhangs – and die at the end of autumn–early winter at an age of 1 (or 2 years at one site). This life cycle requires a very large investment in food for growth and for a final 'big bang' spawning period, after which the exhausted cuttlefish dies.

Except during the spawning season, cuttlefish live solitary lives, centred in a cave or deep crevice or on sandy bottoms, and range locally in their search for food. The size of spawning aggregations varies over the range of the species. Besides the massive one in upper Spencer Gulf, small aggregations are known in the eastern Great Australian Bight (GAB) off Anxious Bay, lower Gulf St Vincent (Rowling 1994), and bays along the east coast, notably Jervis Bay.

Food and foraging

The daily activity of cuttlefish in southern Spencer Gulf was monitored over several weeks using acoustic tags (Aitken *et al.* 2005). Here, the cuttlefish lived in dens from which they foraged for food in the surrounding area. Their home ranges varied from 15 000 to 24 000 m^2, and they swam up to 400 m a day during foraging activities. Some foraged entirely in reef habitat, while others searched for food in seagrass beds. For most of the time, the cuttlefish were inactive in their dens and, on average, they foraged for only 30 min by day and 15 min at night.

In upper Spencer Gulf, the movement of the cuttlefish at its egg-laying site is still a puzzle, due to ambiguities arising from the use of acoustic telemetry to follow their behaviour. Payne *et al.* (2010) conjectured that they rested at night in rock crevices or alternatively moved offshore to rest or forage.

Cuttlefish are active carnivores and efficient predators, as suggested by their brief daily foraging period, preying primarily on crabs, prawns, bivalves and small benthic fish (such as weedy whiting). A cuttlefish typically 'approaches its prey slowly, with slight fluttering motions of the lateral fins, and seizes the prey by suddenly protruding the pair of powerful tentacles' (Saunders 2009). A common pattern is for juveniles to feed on small crustaceans, and larger ones on small fish (Aitken *et al.* 2005; Boyle and Rodhouse 2005).

Spawning

Low-density aggregations

Low densities of cuttlefish are typical of the species over its geographic range. At one such site on a 500 m^2 calcrete reef at Edithburgh, South Australia, Rowling (1994) recorded up to 50 individuals living in dens for periods up to ~3 months, while they foraged locally. Here, the males guarded their dens (small crevices) that closely matched their body size, and at times competed with other males for dens, as well as for females. After successful matings in August, the females deposited eggs deep inside the male dens and died very soon after. Keith Rowling observed that, in contests with other males, a stereotyped threat behaviour involving six sequential stages occurred. These began with a rapid colour change to dark purple on initial contact (Stage 1), followed by a zebra stripe display (Stage 2), then side-by-side body alignment (Stage 3), with successive waves of dark colour (Stage 4), escalating to a head-on display with arms spread wide (Stage 5), and finally to physical attack (Stage 6).

High-density aggregation

In the only known large aggregation of the species: > 180 000 cuttlefish once gathered annually from May to August in upper Spencer Gulf west of Point Lowly on an extensive shallow reef ~7 km long by ~80 m wide. Here, densities increased to 1.05 m^{-2}, with males arriving first and staying the full period, but females staying for only about half that time. This gave rise to an initial male:female sex ratio of 11:1, declining to 4:1 at the height of the spawning season, and then to an equal sex ratio by late winter.

At this site, the cuttlefish followed one of two developmental patterns, inferred from the growth increments of the internal cuttle-bone, during this growth period:

- One group grew rapidly in their first summer and autumn, reaching sexual maturity at 7–8 months at the beginning of winter, and then in its first year joined the spawning aggregation as a component of small individuals.
- The other group grew much more slowly during the first summer, and did not reach sexual maturity until the second year, and finally joined the spawning aggregation at 2 years of age as a component of large individuals.

Spawning uses up to 30% of the cuttlefish's body weight as fuel for reproduction, and the emaciated cuttlefish then dies – often to enter the food web as food for scavenging crabs, dolphins and sharks.

This dual life history, as yet recorded only in upper Spencer Gulf for this cuttlefish, raises the question whether it is genetically determined, or a response to its unique upper Gulf environment (Hall and Fowler 2003; Hall *et al.* 2007; Gillanders and Payne 2013). Genetic studies in progress (see above) suggest that genetic differentiation has occurred and that the Spencer Gulf population has evolved into a distinct species.

The most plausible explanation for the collective spawning aggregation involves two factors. First, the upper half of the Gulf has an almost entirely sandy bottom, and the cuttlefish possibly spends much of its life in this habitat, as shown by its frequent capture as by-catch during prawn trawling over the deeper sandy bottom of the region. Second, virtually the only reef habitat available for attachment of eggs in the region is along the shallow reef system to the west and north of Point Lowly and so the cuttlefish come here to spawn and deposit their eggs. Hence, in this unique region, both environmental and genetic factors seem to have determined the life history.

An adaptive consequence of this life history is that populations become more resilient to adverse conditions, such as a recruitment failure in a particular year. Hence, a dual strategy serves to spread the risk in a variable environment (Hall and Fowler 2003).

The mid- to upper Spencer Gulf population has an uncertain future (Hall 2012). It has been exploited by fishers and observed by researchers since the 1960s, and studied intensely from 1997. Over this period, the population appears to have varied its egg-laying site, while the Point Lowly aggregation has declined almost linearly since 1999 to critically low levels in 2012 (a 93% decline) as shown by the monitoring reports of Karina Hall after fishing was prohibited near the egg-laying sites in 1998 (Fig. 10.2). Moreover, the most recent surveys have shown only one year class, instead of the previous two, implying a total mortality after 1 year. The cause(s) are still speculative and include several human-caused contributing factors, such as:

- cuttlefish by-catch in the prawn fishery, estimated to be 5000–15 000 a year from average cuttlefish densities on trawling grounds (see Currie *et al.* 2009), as well as substantial continuing fishing mortality
- acoustic trauma, to which the cuttlefish is sensitive, caused by the nearby port development and ship traffic (André *et al.* 2011)
- the annual input of ~200 t of industrial ammonia causing *Hincksia* blooms on reefs, as well as contaminated groundwater, has affected egg laying.

Spawning behaviour

The aggregation of cuttlefish at spawning sites has led to the evolution of flamboyant and remarkable behaviours (Norman *et al.* 1999; Hall and Hanlon 2002). Initially, a male courts a female with an elaborate colour display of zebra stripes pulsing across the upper mantle. But courtship is short-lived, lasting a few seconds, unless prolonged by interference from other males. Mating then occurs in a head-to-head position as the male spreads his arms around the female's head. She accepts him by opening her arms and laying them over his. Then follow three stages of mating, as vividly described by Hall and Hanlon (2002).

1. Initially, the male spends time jetting water from his funnel over the female's mouth area, as he moves his second and third arms towards her mouth. This is called flushing behaviour, whose purpose is explained below.
2. Next, with his fourth arm, he quickly transfers sperm packages to the female's mouth area, where her sperm receptacles are located.
3. He breaks open the sperm packages around her mouth with his fourth arm.

The whole mating episode lasts for 2–3 min, with small males mating faster than large ones, unless a challenging male interrupts the affair, before he can open his sperm package. This happened up to a third of the time for small males, but much less often for large ones.

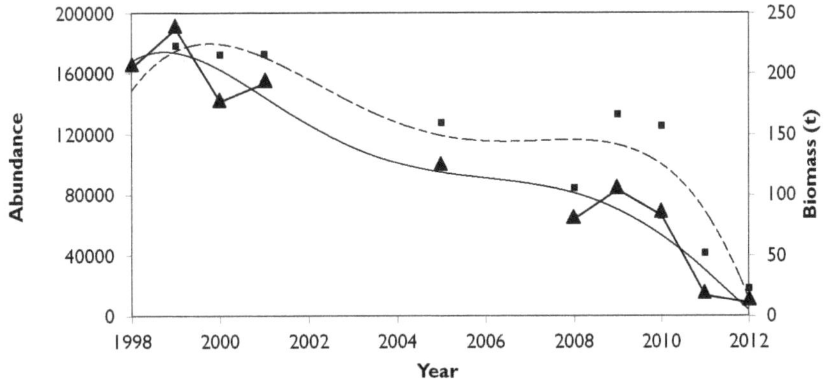

Figure 10.2: Decline in estimated total biomass (■) and numbers (▲) of giant cuttlefish *Sepia apama* at annual spawning grounds in upper Spencer Gulf since 1998, with curves of best fit (redrawn from Hall 2012, omitting standard errors). Biomass estimates include commercial catches in 1997–1998.

After mating – sometimes repeated at ~10 min intervals with different males – the female starts to lay her eggs one at a time – which takes up to 10 min per egg if she is undisturbed. First, she holds the egg in her arms, when it is fertilised by sperm deposited around her mouth, and then transfers it down the arms, which are all cupped around it. She now looks for the underside of a crevice in which to attach it. She does not 'put all her eggs in the same basket', but places them in different crevices, often where eggs of other females have been laid, resulting in clusters of eggs from different females accumulating in crevices. After laying an egg, she flares her arms and cleans them of the gelatinous secretions from the previous egg and prepares to lay the next one.

Male behaviour

With many unmated males on watch for females and ready to sneak in for a quick copulation, the male guards her vigilantly during her egg laying by hovering above her, or by placing himself between her and other males. Male challengers are met by threat displays, as described by Rowling (1994) – usually an adequate deterrent – but occasionally, ~1% of the time, the contest escalates into biting, with resulting injury to the challenger or defender. Usually the victory is decided by size. Sometimes a paired male is approached by a lone female, and he promptly includes (and defends) her in his 'mini-harem'.

In the competition for females, small males are the underdogs, but, like other cuttlefish, they are masters of disguise and use it in elaborate displays – and to good purpose. They may adopt various tactics to overcome their low status:

- **sneaking:** with this strategy, they hover round, watching a mating pair, until the large male partner is distracted, and then sneak in for a quick transfer of sperm. Alternatively, they might hide under or behind a rock, and attempt to sneak a mating with a female looking for an egg-laying site.
- **transvestism:** this is where a small male impersonates a female by adopting her colouration and posture. By his cross-dressing tactic, he outwits, and avoids, aggression from the larger male partner. He then cuckolds the larger male and copulates with the female.

An even more sophisticated deception has been recorded for the mourning cuttlefish *Sepia plangon* in New South Wales (Brown *et al.* 2012). A male can display courtship markings to a female on one side of his body and female markings on the other side to avoid being perceived as a rival to a nearby male.

Female behaviour

Females spend much of their time at the spawning site searching for suitable cryptic places in which to affix their eggs. During this time, they reject many males, sometimes because they have an egg in their arms ready to deposit, as happens ~29% of the time. In these instances, they may use an 'arching' posture (arching themselves dorsally) or a lateral white stripe display to warn the male to back off. At other times, they may use a welcoming 'white banner' display (all arms white and waving), which induces a prompt mating; this happens in ~27% of male advances, irrespective of their size, while another 3% are forced copulations. There is some evidence from other species of *Sepia* that females prefer males that have recently mated (Boal 1997).

Sexual selection

We now summarise the basis of the sophisticated and elaborate behaviours of cuttlefish at a spawning site. The term 'sexual selection' was coined by Darwin in his attempt to resolve the puzzling problem why animals evolve conspicuous traits with no evident survival value. What he had in mind was selection arising out of competition for mates or mating. The abundant evidence for this is in this cuttlefish (Hall and Hanlon 2002; Naud *et al.* 2004, 2005) is as follows:

- Contests between males are very common, partly due to the extreme male bias in the sex ratio.
- Large males adopt specific strategies to assert their dominance, such as guarding females and flushing. The latter behaviour is when a male jets water across her mouth while copulating – an action that is likely to remove fresh, competing ejaculate.
- Small males have developed effective tactics – that is, transvestism and sneaking – to overcome their low fighting status.

- Females exhibit a range of behaviours related to choosing a mate, with 97% of their matings being of their own choosing. Their movement around the spawning area in search of good egg-laying sites opens them to encounters with large numbers of males, many of whom are rejected and a few chosen, possibly based on persistence or fitness. There is some evidence that the most recent partner is the one that sires the next egg to be laid.
- The scope for sperm competition is high, with the following features: multiple matings in which sperm is deposited in two locations (around her mouth and on a sperm receptacle under the beak); accumulation by the female of ejaculate of up to five different males; and lastly the intriguing possibility that a female, by appropriate muscular movement, can choose where to store sperm and when to use it for fertilisation.

Population structure

Ongoing genetic studies of the giant cuttlefish show three distinct stocks over its range in southern Australia: a western stock on Australia's western and SW coast extends east to Esperance; a South Australian and Victorian stock extends to eastern Victoria; and a New South Wales stock extends up the east coast to Queensland. The last two stocks are well differentiated, suggesting isolation during the Pleistocene when the Bassian Isthmus separated them, with a small hybrid zone of overlap forming as sea level subsequently rose. Within South Australia, further subdivision of stocks has occurred, with the eastern GAB, upper and lower Spencer Gulf and Gulf St Vincent populations all differing from each other. The inferred population structure, supported by anecdotal and some tagging information, is illustrated in Fig. 10.3. Although some of these populations may yet prove to be separate species, they are at present considered genetically distinct sub-stocks within a single species complex (Kassahn *et al.* 2003). Moreover, the genetic differences between populations imply that this cuttlefish is either a poor disperser and/or returns to its natal site to spawn.

The blue-ringed octopus and the blue-lined octopus

SA Shepherd and Craig Proctor

Once thought to be a single species, two species are now known to occur in southern Australia – the blue-ringed octopus *Hapalochlaena maculosa* in SE Australia and the blue-lined octopus *Hapalochlaena fasciata* along the New South Wales coast to southern Queensland (Lu 1998). The ecology and life history of the blue-ringed octopus is known only from an unpublished study by Craig Proctor at Edithburgh, South Australia in 1980–1987. The life history of the blue-lined octopus was studied by Tranter and Augustine (1973) on individuals reared in a Sydney aquarium.

Blue-ringed octopus *Hapalochlaena maculosa*

The blue-ringed octopus (Plate 26) is a small species whose arm spread rarely exceeds ~12 cm, and ranges from the intertidal to a depth of ~30 m. It is a cryptic species, hiding by day in crevices under rocks or in rubble, and often in empty razor-shells or bivalves, which it can clap shut if danger threatens. This octopus feeds on small crustaceans, mainly shrimps and crabs, which it hunts and paralyses with its toxic venom. Normally, it is a pale sandy colour with pale blue markings, but, if provoked, it instantly flashes brilliant blue iridescent rings as a warning. The octopus bites the prey with its beak and radula, and releases venomous saliva from salivary glands into the mouth, which flows into the wound and the prey quickly dies.

Off Edithburgh in SW Gulf St Vincent, many individuals were individually marked by injecting dye spots into an arm, enabling their movements to be followed for over a year. They lived solitary lives, except for pairing with another individual of either sex occasionally, and were strictly nocturnal: emerging after dusk to forage, and returning home, usually inside an empty razor-shell, before dawn. However, during the warmer months (November–April), they became nomadic, frequently shifting their home base.

Over the 7 years of the study, the octopus followed a consistent annual pattern of spawning and development. Mating occurred at night in March and April

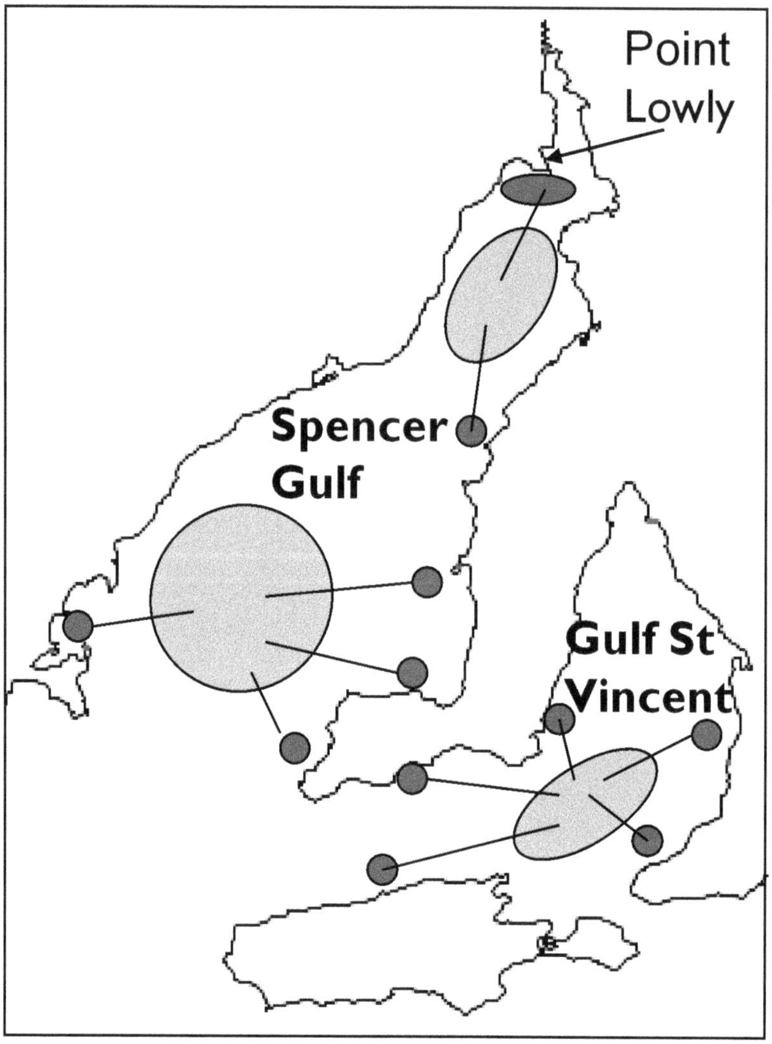

Figure 10.3: Inferred population structure of the giant cuttlefish *Sepia apama* in the South Australian gulfs. Light coloured circles or ellipses represents mixing areas and dark-coloured circles represent spawning areas, based on research and anecdotal information. Lines represent the net direction of movement to and from spawning areas (redrawn from data from Kassahn *et al.* 2003).

and the female then laid on average 50–60 eggs (maximum 177), usually over about 2 weeks. She brooded the eggs for 6–7 months till they hatched in late October to early November and, during all this, time stayed in her shelter day and night. Continuous monitoring of brooding females showed that ~40% of their eggs disappeared during the period of brooding. This loss also happened in accompanying aquarium studies, so predation was an unlikely cause. After removing the few diseased or unfertilised eggs, females probably consumed the remaining proportion of 'lost eggs' to maintain their body condition during the long period of brooding.

By the time of hatching, the female had lost body condition and soon died. The decaying remains of females and their empty egg capsules were sometimes found in their shelters. At the end of the mating season, exhausted males were obviously senescing and died within another month. Hence, females lived for 12–13 months and males for 6–8 months. At this site, winter temperatures were 12–14°C during much of the period of embryonic development and the life cycle was almost twice as long as that of the blue-lined octopus described below.

The blue-lined octopus *Hapalochlaena fasciata*

After laying a batch of up to 50 eggs, the female broods them for ~2 months and, for this purpose, she forms a brood pouch by cradling eggs between her raised skirt and body. She holds the eggs firmly with suckers while she moves about on her remaining two arms. After hatching at ~3–4 mm size, the young octopus darts away with its yolk sac, which it consumes in the next few hours, but stays with its mother for several days. Within 3 days, it is able to catch and feed on small amphipods and a day or so later may even cannibalise weaker siblings. By now the juvenile is able to eject ink, and within another month it can stalk, capture and eat small crabs. It pierces the prey crab's carapace, injects venom from its salivary gland and then sucks out the partly pre-digested tissue.

Juveniles grow quickly, reaching sexual maturity in ~4 months at ~4 cm size (Fig. 10.1) and eggs, attached to the ovarian wall, develop and become ripe over the next month. The eggs then break free and are stored in the ovisac, and later laid in clusters. Mating starts at about this stage. On finding a female, the male mounts her by wrapping his six front arms around her neck while transferring sperm packages with his third arm from the penis into her mantle cavity. He stays on her back for an hour, and then withdraws the third arm. He often tries to mate with other males, but soon abandons those enterprises. The male dies soon after mating. After the eggs hatch, the female never eats again and soon dies. So, in all, the life cycle is completed in ~7 months – 4 months to maturity, 1 month for egg laying and 2 months till hatching.

Venom of blue-ringed and blue-lined octopuses

These octopuses are infamous for their potent toxin – a salivary venom, designed to immobilise and kill prey within a few seconds. The toxin is also present in their eggs to protect them from predation. Human victims have also been paralysed by the toxin, and pharmacological studies (e.g. Sheumack *et al.* 1978; Hwang *et al.* 1989) have shown that the venom contains maculotoxin, which is closely related to tetrodotoxin present in puffer fish, xanthid crabs and some other marine species. The active agent in the venom is produced by various bacteria – *Bacillus, Pseudomonas* and *Vibrio.* In humans, this potent neurotoxin causes a rapid painless paralysis, nausea, respiratory failure, unconsciousness and death if untreated (Edmonds 1976). Emergency treatment includes irrigation of the bite, application of a pressure bandage and mouth-to-mouth resuscitation until the patient receives skilled medical support. Almost five centuries ago, Li Shih-Chen (1596) wrote that the related tetrodotoxin was 'the poisoning no remedy can relieve' and even now no antivenine for these toxins is available.

Western gloomy octopus *Octopus cf. tetricus*

This large octopus is closely related to the eastern gloomy octopus *Octopus tetricus* of eastern Australia (Kailola *et al.* 1993), but has not yet been formally described as a separate species. It is common in SW Australia, living on reef and in seagrass to > 50 m depth. After aquarium and field studies by Joll (1976, 1977a, b, 1978, 1983), exploitation of this octopus developed as a small fishery. It is also notorious as a pest species, predating on lobsters in lobster pots.

Life history

Egg laying by the female takes place at night. In one observed episode, a large female laid ~100 strings of eggs over several days, with up to 150 000 eggs per string. The strings of eggs were fastened to the roof of the octopus's lair. During development of the eggs, the female guarded them closely, continually bathing them by directing her exhalant siphon onto them. She vigorously attacked any attempt by a predator to remove eggs. The eggs hatched after ~35 days and planktonic hatchlings ~2.5 mm long emerged, each with a large internal yolk sac, which sustained them for a week or so. After that they were able to hunt and capture mysids. The length of the planktonic hatchling stage is not known, but probably lasts 30–50 days in temperate waters.

A generalised growth curve for octopus reared in aquarium conditions and fed to excess (Fig. 10.1) shows rapid growth to a maximum size in ~32 weeks, with animals losing weight after that, either due to egg laying by females or senility in males. However, Lindsay Joll (1983) found that the growth rate is linear with the feeding rate, and in field trials he found that the growth rate in the wild was much slower, and probably only ~40% of that recorded in the aquarium. Sexual maturity is reached at weights of 100–150 g in males, while field samples showed that females > 350 g weight begin to store sperm packages. In one case, a female stored sperm for 114 days before laying fertilised eggs.

In mating, the male projects his specialised third arm into the female's mantle cavity where he deposits sperm packages, all the while pumping water from his mantle cavity – apparently normal behaviour for these octopuses. Joll recorded that mating behaviour continued for up to 7 hours, with frequent resting intervals of up to 10 min. Egg-laying followed over the next 2–15 weeks. Females died within a few weeks of the young hatching from the eggs and males died of senility at about the same time. Hence, the life span of this octopus is about a year. Small individuals are found throughout the year, but are most common in late spring to early summer, reflecting greater larval settlement at that time. It is likely that some breeding goes on year round.

Maori octopus *Macroctopus maorum* and pale octopus *Octopus pallidus*

The Maori octopus occurs throughout SE Australian waters, the subantarctic and New Zealand to depths of > 500 m, and the pale octopus is limited to SE Australia; both species are taken commercially. The Maori octopus is the largest southern species, reaching 15 kg in weight, and > 3 m long. It has a similar life history to the gloomy octopus, producing planktonic hatchlings that settle on the sea floor and adopt a benthic life. The species lives for up to a year in reef or sand habitats or the interface between them, and matures sexually at 3–4 kg weight. It is a major predator of the southern rock lobster and is fished commercially in the SE Australian States.

The pale octopus matures sexually at 400–500 g, but its longevity is still uncertain, and it may live in cooler Tasmanian waters for possibly 2 years (Leporati *et al.* 2007, 2008a,b).

Like many other octopuses, the Maori octopus lives solitarily in a den and forages in the vicinity for food at night. In one brief tagging experiment, individuals occupied their own dens for 3 months and accumulated prey remains in them to form a midden. During this period, several of the tagged animals barricaded the entrances to their dens, and 2–3 weeks later laid egg strings on their den roofs. The eggs hatched in ~11 weeks (Anderson and Babcock 1999).

According to Anderson (1999), the spawning season of the Maori octopus extends over the warmer half of the year (October–March). In aquarium observations of mating behaviour, the male first pinned down the female underneath and, if she struggled, he subdued her by placing his web over her eyes. Mating lasted from 20 min to 3.5 h, and the female laid 5000–7000 eggs some 2–3 months later. She stopped feeding 2 weeks before egg laying, and after the eggs were laid she constantly cleaned and guarded them for the next 6 weeks till they hatched over several days. The hatching period was much longer than in the field studies mentioned above, likely due to temperature differences. The hatchlings were planktonic and swam actively towards the surface for several days before ultimately settling on the bottom.

It is curious that occasionally large aggregations of some thousands of mature Maori octopus form in Eaglehawk Bay, SE Tasmania. It is not known where they come from, nor the reason for aggregating (Doubleday *et al.* 2008). This study used the elemental structure of the octopus stylet to examine the population structure of the octopus. The stylet is a pair of fine, mainly organic, rods embedded in the mantle muscle system that incorporate numerous traceable elements (e.g. Mn, Sr, Co, Ni, Zn and Cu), which are useful for distinguishing stocks. Samples from northern, eastern and southern Tasmania were different from each other and from samples elsewhere in SE Australia, implying that populations are distinct and larval dispersal is localised. Genetic studies confirmed this picture, but, curiously, showed that South Australian populations were homogeneous with southern Tasmanian populations, apparently because the paralarvae could be transported from west to east by the Leeuwin Current around southern Tasmania (Doubleday *et al.* 2009).

Growth

Growth of octopus appears to be a two-phase process, consisting of an early exponential phase of rapid growth followed by a slowing growth phase as the octopus invests energy in reproduction. Temperature and food ration strongly influence the initial, but not the later, growth phase (review of Semmens *et al.* (2004). The generalised growth curves for the western gloomy octopus and the blue-ringed octopus (Fig. 10.1) are examples of this composite two-phase growth pattern. The pale octopus grows much more slowly in its initial phase, reaching on average ~1.5 g in 5 months in cool conditions (14–15°C) and ~2 g in warm conditions (17–18°C) (André *et al.* 2008).

Recent studies of the pale octopus have further developed the use of stylets, which produce daily growth rings, to estimate age and growth (e.g. Leporati *et al.* 2008a; André *et al.* 2009). This species shows great plasticity in growth, which depends on sea temperature at time of hatching, diet and other factors. Hence, the growth rate of individuals hatching in different seasons can amplify during life.

The growth rate of the Maori octopus has proved to be problematic using stylets, but may be feasible using lipofuscin – a fluorescent bio-molecule that accumulates with age (Doubleday *et al.* (2011). It is yet to be examined, but, being an annual species, may be similar to that of the western gloomy octopus.

Foraging behaviour and food

The Maori octopus uses several methods to capture prey – ambush, direct attack and searching. It crawls or swims over the bottom and searches in crevices with its arms. It can jet-propel through the water and pounce on and cover the prey with its membranous arms. It is armed with suckers that can tightly grasp the prey, and can use its toxic saliva, containing cephalotoxin, to paralyse captured animals. Although crabs can be paralysed in seconds, lobsters take longer – up to 20 min – to become paralysed (Joll 1977a). To capture a lobster, the octopus first entangles it in its arms, and then envelops it with the web.

To consume molluscs, the octopus uses a variety of means (Steer and Semmens 2003). It can pull bivalves apart by force or alternatively drill through the shell and insert its paralysing secretions. The choice between the two methods is a trade-off between the cost to the octopus in terms of handling time and effort, and the benefit derived from the food. Gastropods usually have to be drilled; for example, the warrener *Turbo undulatus* is drilled on the body whorl just above the aperture and toxin injected through the drill hole (Lock 1988). Other drilled gastropods recorded in middens are cowries, volutes, murex shells and harp shells (Wilson and Clarkson 2004). In the case of the greenlip abalone's thicker shell, the octopus simply spreads its web over the prey and releases its secretions, which localises the toxin, and gradually paralyses the prey, so that within 5–10 min the octopus can overturn the abalone and carry it off (unpublished observations). The accumulation of shells and crab carapaces in octopus middens provides good evidence of the diet, although it can be subject to bias, because, for example, gastropod shells may well be taken by hermit crabs, and crab carapaces may break up.

The diet of the Maori octopus was examined by Grubert *et al.* (1999) in a bay in SE Tasmania. They found that, in numerical terms, the diet was: crustaceans (crabs and shrimp) 78%; fishes (mainly gobies) 18%; cephalopods 3%; and gastropods 1%. As the octopus digests bivalves (e.g. mussels and scallops) externally, and discards the shells, these items if present would be virtually undetectable in the gut. The authors also recognised that octopuses feeding on rocky reefs, with a wider variety of benthic organisms present, would take different prey such as lobster and abalone.

The Maori octopus is a major predator of lobsters captured in pots in the lobster fisheries of southern Australia and 4–5% of the Tasmanian and South Australian lobster catches are lost to them. In the latter State alone, this amounts to an annual consumption of ~250 000 lobsters by octopuses. This loss has given rise to several studies designed to find ways of reducing this high cost to the fishery (Brock and Ward 2004; Brock *et al.* 2006a, b; Harrington *et al.* 2006). These studies have also shed some light on octopus behaviour. Firstly, octopuses are more active at night than in the day, depending on their hunger, although they do enter traps in the day as well as at night. Secondly, an octopus locates lobster traps by olfaction and is primarily attracted by the bait, rather than by the lobster. Thirdly, octopus abundance varies spatially, suggesting that some bottom habitats are more attractive than others. A new design of two-chambered pot by Danny Brock was found to reduce octopus predation by 70%, because the octopus could enter and leave the pot by a side entrance, whereas the lobster entered from the top and was trapped. Unfortunately, the new design of pot reduced the lobster catch and was never adopted. Overall, in whatever habitat the octopus forages, it is a voracious predator and may well act as a keystone species in the benthic ecosystem, when present in sufficient densities.

Predators

The Maori octopus has many of the same predators as the western gloomy octopus. They include sharks (e.g. wobbegongs, school and gummy sharks), mulloway, queen snapper and, perhaps most of all, dolphins and seals (Joll 1983; Coleman and Mobley 1984). For example, recent analysis of the scat of Australian sea lions in South Australia gave on average 15% content of this octopus, as well as substantial amounts of other octopus species (K. Peters *pers. comm.*). It is plausible that overfishing of sharks (see Chapter 17) has caused a substantial increase in octopus abundance, as often claimed by fishers.

REFERENCES

Aitken JP, O'Dor RK, Jackson GD (2005) The secret life of the giant Australian cuttlefish *Sepia apama* (Cephalopoda): behaviour and energetics revealed through radio acoustic positioning and telemetry (RAPT). *Journal of Experimental Marine Biology and Ecology* **320**, 77–91.

Anderson T (1999) Morphology and biology of *Octopus maorum* Hutton 1880 in northern New Zealand. *Bulletin of Marine Science* **65**, 657–676.

Anderson T, Babcock RC (1999) Subcutaneous electromagnetic tagging of benthic octopus: a preliminary evaluation. *Marine and Freshwater Research* **50**, 225–227.

André J, Pecl GT, Semmens JM, Grist EPM (2008) Early life history processes in benthic octopus: relationships between temperature, feeding, food conversion and growth in juvenile *Octopus pallidus*. *Journal of Experimental Marine Biology and Ecology* **354**, 81–92.

André J, Pecl GT, Grist EPM, Semmens JM *et al.* (2009) Modelling size-at-age in wild immature female octopus: a bioenergetic approach. *Marine Ecology Progress Series* **384**, 159–174.

André J, Solé M, Durfort M, Quero C *et al.* (2011) Low-frequency sounds induce acoustic trauma in cephalopods. *Frontiers in Ecology and the Environment* **9**, 489–493.

Boal JG (1997) Female choice of males in cuttlefish (Mollusca: Cephalopoda). *Behaviour* **134**, 975–988.

Boletzky Sv (2003) Biology of early life stages in cephalopod molluscs. *Advances in Marine Biology* **44**, 143–203.

Boyle P, Rodhouse P (2005) *Cephalopods: Ecology and Fisheries*. Blackwell Science Ltd, Oxford.

Brock DJ, Ward TM (2004) Octopus (*Octopus maorum*) bycatch and lobster (*Jasus edwardsii*) mortality in the South Australian rock lobster fishery. *Fishery Bulletin* **102**, 430–440.

Brock DJ, Saunders TM, Ward TM, Linnane AJ (2006a) Effectiveness of a two-chambered trap in reducing within-trap predation by octopus on southern spiny rock lobster. *Fisheries Research* **77**, 348–355.

Brock DJ, Saunders TM, Ward TM, Linnane AJ (2006b) A two-chambered trap reduces within-trap predation by octopus on rock lobsters in aquarium trials. *Fisheries Research* **80**, 129–135.

Brown C, Garwood MP, Williamson JE (2012) It pays to cheat: tactical deception in a cephalopod social signalling system. *Biology Letters* **8**, 729–732.

Coleman N, Mobley M (1984) Diets of commercially exploited fish from Bass Strait and adjacent Victorian waters, south-eastern Australia. *Australian Journal of Marine and Freshwater Research* **35**, 549–560.

Currie DR, Dixon CD, Roberts SD, Hooper GE *et al.* (2009) *Fishery-Independent By-Catch Survey to Inform Risk Assessment of the Spencer Gulf Prawn Trawl Fishery.* Publication No. F2009/000369.1 SARDI Research Report Series No. 390. SARDI, Adelaide.

Derby CD (2007) Escape by inking and secreting: marine molluscs avoid predators through a rich array of chemicals and mechanisms. *The Biological Bulletin* **213**, 274–289.

Doubleday ZA, Pecl GT, Semmens JM, Danyushevsky L. (2008). Using stylet elemental signatures to determine the population structure of *Octopus maorum*. *Marine Ecology Progress Series* **360**, 125–133.

Doubleday ZA, Semmens JM, Smolenski AJ, Shaw PW (2009) Microsatellite DNA markers and morphometrics reveal a complex population structure in a merobenthic octopus species (*Octopus maorum*) in south-east Australia and New Zealand. *Marine Biology* **156**, 1183–1192.

Doubleday ZA, White J, Pecl GT, Semmens JM (2011) Age determination in merobenthic octopuses using stylet increment analysis: assessing future challenges using *Macroctopus maorum* as a model. *ICES Journal of Marine Science* **68**, 2059–2063.

Edmonds C (1976) *Dangerous Marine Animals of the Indo-Pacific Region.* 2nd edn. Wedneil Publications, Newport.

English S (1980) Aspects of the biology of *Idiosepius notoides* (Mollusca: Cephalopoda). *Journal of the Malacological Society of Australia* **4**, 256.

Gillanders BM, Payne NL (2013) Giant Australian cuttlefish. In *Natural History of Spencer Gulf* (Eds S Madigan, SA Shepherd, S Murray-Jones and D Wiltshire). Royal Society of South Australia, Adelaide (in press).

Grubert MA, Wadley VA, White RWG (1999) Diet and feeding strategy of *Octopus maorum* in southeast Tasmania. *Bulletin of Marine Science* **65**, 441–451.

Hall KC (2012) 'Estimated abundance and biomass of giant Australian cuttlefish *Sepia apama* in the spawning area of northern Spencer Gulf'. Report to Olympic Dam EIS Project, Adelaide.

Hall KC, Fowler AJ (Eds) (2003) 'The fisheries biology of the cuttlefish *Sepia apama* Gray in South Australian Waters'. FRDC Report No. 98/151. SARDI Aquatic Sciences, Adelaide.

Hall KC, Hanlon RT (2002) Principal features of the mating system of a large spawning aggregation of the giant Australian cuttlefish *Sepia apama* (Mollusca: Cephalopoda). *Marine Biology* **140**, 533–545.

Hall KC, Fowler AJ, Geddes MC (2007) Evidence for multiple year classes of the giant Australian cuttlefish *Sepia apama* in northern Spencer Gulf, South Australia. *Reviews in Fish Biology and Fisheries* **17**, 367–384.

Hanlon RT, Messenger JB (1996) *Cephalopod Behaviour.* Cambridge University Press, Cambridge.

Harrington JJ, Semmens JM, Gardner C, Frusher SD (2006) Predation of trap-caught southern rock lobsters, *Jasus edwardsii* (Hutton, 1875), in Tasmanian waters by the Maori octopus, *Octopus maorum* (Hutton, 1880): spatial and temporal trends. *Fisheries Research* **77**, 10–16.

Hindell JS, Jenkins GP, Keough MJ (2001) Spatial and temporal variability in the effects of fish predation on macrofauna in relation to habitat complexity and cage effects. *Marine Ecology Progress Series* **224**, 231–250.

Hwang DF, Arakawa O, Saito T, Simidu U *et al.* (1989) Tetrodotoxin-producing bacteria from the blue-ringed octopus *Octopus maculosus. Marine Biology* **100**, 327–332.

Joll LM (1976) Mating, egg-laying and hatching of *Octopus tetricus* (Mollusca: Cephalopoda) in the laboratory. *Marine Biology* **36**, 327–333.

Joll LM (1977a) 'The predation of pot caught Western Rock Lobster (*Panulirus longipes cygnus*) by octopus (Western Australia)'. Western Australian Department of Fisheries and Wildlife. Report No. 29, Perth.

Joll LM (1977b) Growth and food intake of *Octopus tetricus* (Mollusca: Cephalopoda) in aquaria. *Australian Journal of Marine and Freshwater Research* **28**, 45–56.

Joll LM (1978) Observations on the embryonic development of *Octopus tetricus* (Mollusca: Cephalopoda). *Australian Journal of Marine and Freshwater Research* **29**, 19–30.

Joll LM (1983) *Octopus tetricus.* In *Cephalopod Life Cycles. Vol. 1.* (Ed. PR Boyle) pp. 325–334. Academic Press, London.

Kailola PJ, Williams MJ, Stewart PC, Reichelt RE *et al.* (1993) *Australian Fisheries Resources.* Department of Primary Industries and Energy, Canberra.

Kassahn KS, Donnellan SC, Fowler AJ, Hall KC, Adams M, Shaw PW (2003) Molecular and morphological analyses of the cuttlefish *Sepia apama* indicate a complex population structure. *Marine Biology* **143**, 947–962.

Leporati SC, Pecl GT, Semmens JM (2007) Cephalopod hatchling growth: the effects of initial size and seasonal temperatures. *Marine Biology* **151**, 1375–1383.

Leporati SC, Semmens JM, Pecl GT (2008a) Determining the age and growth of wild octopus using stylet increment analysis. *Marine Ecology Progress Series* **367**, 213–222.

Leporati SC, Pecl GT, Semmens JM (2008b) Reproductive status of *Octopus pallidus*, and its relationship to age and size. *Marine Biology* **155**, 375–385.

Li S (1596) *The Great Herbal.* Reprint, Cultural Library Co., Taipei, 1956.

Lock I (1988) *Octopus cyaneus* blues. *Australian Shell News* **64**, 9.

Lu CC (1998) Order Octopoda. In *Mollusca: the Southern Synthesis. Vol. 5.* (Eds Beesley PL, Ross GJB and Wells A) pp. 545–555. CSIRO PUBLISHING, Melbourne.

Naud M-J, Hanlon RT, Hall KC, Shaw PW *et al.* (2004) Behavioural and gene tic assessment of reproductive success in a spawning aggregation of the Australian giant cuttlefish, *Sepia apama. Animal Behaviour* **67**, 1043–1050.

Naud M-J, Shaw PW, Hanlon RT, Havenhand JN (2005) Evidence for biased use of sperm sources in wild female giant cuttlefish (*Sepia apama*). *Proceedings. Biological Sciences* **272**, 1047–1051.

Norman MD, Finn J, Tregenza T (1999) Female impersonation as an alternative reproductive strategy. *Proceedings. Biological Sciences* **266**, 1347–1349.

Payne NL, Gillanders BM, Webber DM, Semmens JM (2010) Interpreting diel activity patterns from acoustic telemetry: the need for controls. *Marine Ecology Progress Series* **419**, 295–301.

Rowling KP (1994) Den ecology and behavioural interactions of the Australian giant cuttlefish *Sepia apama.* BSc (Hons) thesis. Flinders University, Adelaide.

Saunders B (2009) *Shores and Shallows of Coffin Bay.* Australian Printing Specialists, Adelaide.

Semmens JM, Pecl GT, Villanueva R, Jouffre D *et al.* (2004) Understanding octopus growth: patterns, variability and physiology. *Marine and Freshwater Research* **55**, 367–377.

Sheumack DD, Howden MEH, Spence I, Quinn RJ (1978) Maculotoxin: a neurotoxin from the venom glands of the octopus *Hapalochlaena maculosa* identified as tetrodotoxin. *Science* **199**, 188–189.

Steer MA, Semmens J (2003) Pulling or drilling, does size or species matter? An experimental study of prey handling in *Octopus dierythraeus. Journal of Experimental Marine Biology and Ecology* **290**, 165–178.

Tracey SR, Steer MA, Pecl GT (2003) Life history traits of the temperate mini-maximalist *Idiosepius notoides*, (Cephalopoda: Sepioidea). *Journal of the Marine Biological Association of the United Kingdom* **83**, 1297–1300.

Tranter DJ, Augustine O (1973) Observations on the life history of the blue-ringed octopus *Hapalochlaena maculosa. Marine Biology* **18**, 115–128.

Wilson B, Clarkson P (2004) *Australia's Spectacular Cowries. A Review and Field Study of Two Endemic Genera:* Zoila *and* Umbilia. Odyssey Publishing, El Cajon, California.

11 Echinoderms

> 'Non coelo tantum, sed et mari suae stellae sunt.'
> *(Not only the heavens, but the seas also have their stars.)*
>
> JH Linck, *De Stellis Marinis*, 1733
> (The first book written on sea-stars)

OVERVIEW

The echinoderms comprise five classes: feather-stars, sea-stars, brittle-stars, sea-urchins and sea cucumbers. The ecology of one feather-star, *Aporometra wilsoni*, is described. The diets of 19 species of sea-star are also described, with species categorised as browsers, omnivores or predators. The effects of predatory sea-stars in temperate Australian reef systems are examined and concluded to be minor compared with the effects of competition for space and other factors. The early life history of some species of the family Asterinidae is reviewed. Many species have evolved in this family in southern Australia, together with many different life histories, ranging from planktonic, lecithotrophic larvae to directly developing benthic egg masses, to viviparous, intra-gonadal larval development. The ecology of a predatory sea-star *Coscinasterias muricata* is also outlined. Among the brittle-stars, the life history of one cosmopolitan species *Amphipholis squamata*, and four other common brittle-stars, is described. Finally, the ecology and life history of sea-urchins – and, in particular, three species, the arboreal *Holopneustes purpurascens*, the purple urchin *Heliocidaris erythrogramma* and the black urchin *Centrostephanus rodgersii* – are described in detail. Lastly, the ecology of a holothuroid *Australostichopus mollis* is briefly described.

INTRODUCTION

The echinoderms form a group of animals that are remarkably different from all others. Firstly, their body pattern has a symmetry of five (pentametry), that is to say that almost all of the group structures are in fives – five arms, five sets of spines, and so on. Secondly, the animals all possess tube feet, which are extensions of the coelom, the main body cavity, and work on the hydraulic principle. Internal pressure is maintained by water being sucked in by the madreporite (a safety valve) and water pressure is used to control the tube feet. A third feature of increasing relevance in modern times (given the increasing ocean acidification from climate change – see Chapter 1) is the ability to lay down a strong, internal calcareous skeleton.

The echinoderms are divided into five classes: feather-stars (Crinoidea); sea-stars (Asteroidea);

brittle-stars and basket-stars (Ophiuroidea); sea-urchins (Echinoidea) and sea cucumbers (Holothuroidea). In the most primitive group, the feather-stars, the mouth and anus point upwards; in the sea-stars, brittle-stars and sea-urchins the mouth points downwards; and in the sea cucumbers it points forwards. The mouth position (called the oral side) reflects the feeding habits of the different groups.

FEATHER-STARS

Feather-stars or crinoids have a central disc housing the mouth and body system, with a surrounding array of feather-like arms, each of which carries many pinnules, branching alternately from the sides. Each pinnule has a groove leading to another groove in the arm, which then leads to the mouth. Tube feet line the grooves and, when food touches them, they rapidly close over it and eject a series of mucous strings from mucus glands at their base. The food becomes trapped and is then pushed into the groove and on towards the mouth (Byrne and Fontaine 1981, 1983).

Crinoids rely on water movement to carry suspended food particles to their pinnules or tube feet. In places where water currents are uni- or bidirectional, species such as *Comanthus* spp. (Plate 27) respond by aligning their arms so as to form a filtration fan, but in turbulent water the arms are oriented in different planes to increase their exposure to multi-directional flows. Many crinoids are inactive by day, but extend their arms at night for food capture, probably to take advantage of small crustaceans that emerge from the substrate after dusk to feed in the water column. Examination of a crinoid at night often reveals an amphipod or copepod trapped in a groove, and being transported to the mouth. Curiously, the shrimp *Periclimenes carinidactylus* has a commensal association with the crinoid *Comanthus trichoptera* and possibly steals food from its crinoid host (Bruce 1980).

The southern feather-star *Aporometra wilsoni*

Of the eight or so crinoids in southern Australia (Shepherd *et al.* 1982), only one, *Aporometra wilsoni*, has been at all studied (Haig and Rouse 2008). This species is the smallest of all southern crinoids, with a maximum arm length of only ~25 mm. It lives by attaching itself to algal fronds where there is moderate water movement and up to 200 individuals may be found on one alga. The species has separate sexes and internal fertilisation of the lecithotrophic (i.e. yolked) eggs. Females then brood their larvae within their pinnules – a type of brooding called ovoviviparity. The larvae pass through two stages of development before they are liberated from the pinnules at a non-swimming, late larval stage (called doliolaria), when they fall off their parent onto another part of the plant or onto the bottom. From there, they pass through later sessile larval stages and, after detachment from the stalk, swim as miniature adults to another alga. At the height of the reproductive season in September–October, Haig and Rouse (2008) estimated that 2000–3000 doliolariae were released by one individual, although the breeding season extended to the end of summer (February) for many individuals in a population. Adults die at the end of summer, leaving only small juveniles in the population (J. Haig *pers. comm.*). This abbreviated larval stage and the limited swimming ability of juveniles help explain the patchy distribution of the species and the large aggregations that sometimes occur on an alga. To explain the surprisingly broad geographic range of the species, Helgen and Rouse (2006) suggested that dispersal of the species was aided by rafting of juveniles or adults on drift algae.

SEA-STARS

Sea-stars or asteroids have long symbolised marine life in popular culture and even appeared in Minoan frescos some 4000 years ago in ancient Crete. They have a semi-rigid network of interlocking plates, which provide support and flexibility, allowing them to use their tube feet for walking. In reef-dwelling species, each tube foot can form a suction cap. Sets of the tube feet are controlled by the nervous system, enabling individual sea-stars to climb vertical surfaces and walk at rates of up to 15–20 cm a minute. Even so, sea-stars are relatively sedentary and generally do not move far. Nevertheless, some sea-stars, such as *Coscinasterias muricata*, are voracious predators, foraging over many square metres of reef or sandy bottom (see below).

Any animal with a rigid surface runs the risk of becoming fouled by larvae settling on it and covering its surface. A unique feature shared by sea-stars and sea-urchins is the presence of pincer-like organs, called pedicellariae, often armed with teeth, which remove settling larvae on exposed surfaces and keep the sea-stars clean.

The ecology of only a few of ~40 species of sea-stars that live on southern Australian reefs is known. Here, we describe the feeding of some common species and their role in reef systems, with more detailed information provided on the early life history of the asterinids. This well-studied family has diversified in southern Australia and has been intensively studied by Maria Byrne and her colleagues (reviewed by Byrne (2006). We discuss the ecology and early life history of five species: namely, the diminutive *Parvulastra exigua*; the gastric-brooder *Smilasterias multipara*; the introduced *Patiriella regularis*; and two voracious predators, *Coscinasterias muricata* and the introduced *Asterias amurensis.*

Food and feeding

Sea-stars feed by extruding their thin-walled stomach and digesting their food externally – a behaviour, which serendipitously for the biologist, enables the food being eaten at any one time to be readily identified. Reef-dwelling species feed mainly on sessile fauna or detritus, while 'hunters' such as the eleven-armed sea-star *Coscinasterias* feed on molluscs, such as bivalve mussels, which they can open by using the tube feet to pull the shells apart with a force as high as 20–30 N (i.e. ~2–3 kg).

The diet of 19 species that commonly occur in reef or boulder habitats, or on jetty piles, on southern Australian coasts is summarised in Table 11.1. Some of the species are browsers, some are omnivores, and some are hunting predators. Twelve species browse on sessile forms, such as encrusting bryozoans, compound ascidians and sponges, with some of the species showing apparent preference for one or more sessile animal groups. Three species of *Meridiastra* are omnivorous, consuming red algae and seagrasses and their epiphytes, mainly as drift, as well as sessile animals. The feeding of two of these species, and also of *Petricia vernicina*, was examined at West I., South Australia in more detail (see Table 11.1). *Meridiastra gunnii* browsed mainly on drift algae and seagrass epiphytes trapped among boulders; *Meridiastra media* showed a strong preference for encrusting bryozoans and compound ascidians, although it also fed substantially on bare rock and attached spirorbids; and *Petricia vernicina* showed a strong preference for encrusting sponges, but also fed on bryozoans and ascidians more or less in accordance with their abundance (Table 11.1).

The diet of the diminutive species *Parvulastra exigua* has been examined over much of its southern range. Several authors have noted that the stomach was often extended over bare rock, and thereby assumed the sea-star grazed on microalgae. This question was clarified by Jackson *et al.* (2009) who verified that assumption, and further showed that one individual could remove up to 60% of the microalgae from a small area in one feeding event lasting ~20 min. In a study on the Victorian coast, Stevenson (1992) noted that *Meridiastra media* predated on *P. exigua* and suggested that, at places where both species occurred, *M. media* may control the lower limit of *P. exigua.* When the latter was experimentally put in contact with the predator, it showed an escape response by fleeing at speeds of up to 11 cm min.$^{-1}$, which is more than double its usual crawling speed.

The last four species in Table 11.1 are hunters that feed mainly on molluscs. The feeding ecology of one of the most voracious of them, *Coscinasterias muricata*, is described in more detail below. But what role do asteroids play in reef systems?

This question was examined by Keough and Butler (1979) on jetty piles where four species of sea-star in the genera *Tosia, Meridiastra, Petricia* and *Coscinasterias* lived and fed on the sessile fauna. They found that while sea-star predation measurably reduced the abundance of sponges and compound ascidians, the cumulative effect of the four species was minor compared with other forces at work: namely predation by crabs and fish and, more importantly, seasonal growth of the sessile animals and competition for space between them. Although jetty piles are artificial habitats, they mostly contain the same species as in reef systems, and Butler (2008) regarded them as 'windows' that may give insights into the ecology of shaded reef habitats, such as caves and vertical walls. This is borne out by the strong similarity in diet of the sea-star *Petricia* on jetty piles and in boulder habitat (Table 11.1).

Table 11.1: Diets and water movement preferences (WM) of 19 southern Australian sea-stars from the various sources indicated.

The three water movement categories are: exposed (Exp), moderate exposure (Md) and sheltered (sh). Col. ascid. = colonial ascidian. Unid. = Unidentified. Percentages given are the frequency of sea-stars with stomach found everted over the organism indicated, as extracted from dietary studies at specific sites. References are listed at foot of table.

Sea-star	WM	Diet
Oreasteridae		
Nectria macrobrachia	Exp	Sponges (30%), Bryozoa (70%).[1,3]
Nectria ocellata	Md–sh	Sponges, Bryozoa, encrusting algae, Col. ascid., detritus.[5]
Nectria saoria	Exp	Bryozoa (36%), coralline algae (18%), Col. ascid. (46%).[1,3]
Goniasteridae		
Pentagonaster dubeni	Exp–sh	Sponges (43%), Bryozoa (36%) col. ascid.(20%).[2,3]
Tosia australis[a]	Md–sh	Sponges (95%), Bryozoa (5%).[4]
Asteropseidae		
Petricia vernicina	Exp–sh	Boulder habitat – sponges (64%), Bryozoa (11%), Col. ascid.(11%), *Petroderma* (4%), CCA (7%) detritus (3%).[3b] Sponges (76%), Bryozoa (12%), Col. ascid. (6%), CCA (6%).[4]
Ophidiasteridae		
Fromia polypora	Exp	Sponges, bryozoa, Col. ascid.[2]
Asterinidae		
Meridiastra calcar	Md-sh	Detritus, algae (e.g. *Laurencia, Gelidium*), molluscs.[2,3,6] Detritus (~45%), macroalgae (~38%), live prey (~4%).[9]
Meridiastra gunnii	Sh	Carrion, detritus, red algae, molluscs[2]. Boulder habitat – red algae (48%), seagrass (24%), bryozoa (14%), spirorbids (10%), Col. ascid. (5%).[3b] Jetty – sponge (22%) carrion (38%), Col. ascid. (9%), Unid. (28%).[4]
Meridiastra media	Md-sh	Boulder habitat – Corallines (9%), bryozoa (45%), bare rock with spirorbids (21%), Col. ascid.(24%).[3b]
Parvulastra exigua	Md-sh	Bare rock+*Hildenbrandia* (70%), macroalgae (e.g. *Ulva*) 16%, shells (16%), detritus (3%).[7] Microalgae.[8] Microalgae (~80%), macroalgae (~10%), detritus (~9%).[9] Macroalgae (91%), shells and gastropods (9%)[10]
Paranepanthia grandis	Sh	Col. ascid., bryozoa, gastropods.[2]
Pseudonepanthia troughtoni	Exp	Coralline algae (25%), bryozoa (40%), sponges (20%), Col. ascid. (15%).[3]
Echinasteridae		
Plectaster decanus	Sh	Sponges (55%), Col. ascid. (18%), detritus (27%).[3]
Asteriidae		
Uniophora granifera	Md–sh	Solitary and col. ascidians e.g. *Pyura, Microcosmus* (85%), detritus (15%).[3]
Coscinasterias muricata	Md–sh	Gastropods (43%), crabs (28%), ascid. (3%) carrion (13%).[4]
Allostichaster polyplax	Md–sh	Gastropods (70%), chitons (15%), bivalves (15%).[3]
Smilasterias irregularis	Md–sh	Gastropods (57%), chitons (28%), Col. ascid. (15%).[2,3]
Astrostole scabra	Md–Sh	Crustacea, molluscs (> 60 spp.).[6]

[1]Shepherd (1967); [2]Shepherd (1968); [3] S.A. Shepherd unpublished data; [4]Keough and Butler (1979); [5]Coleman (1977); [6]Sloan (1980); [7]Arrontes and Underwood (1991); [8]Jackson *et al.* (2009); [9]Stevenson (1992); [10]Duyverman (1976).

[a] *Tosia australis stricto sensu* – see Naughton and O'Hara (2009) regarding distribution of cryptic sister species *Tosia neossia*; [b]Data obtained at West I., South Australia, with sample sizes of 20–35 collected over a year. In each case, feeding differed significantly from that expected according to the cover of crustose organisms as shown in Kangas and Shepherd (1985) ($\chi^2_5 > 33$ for each; $P < 0.001$)

Asterinid life histories

Asterinid sea-stars have especial interest because they have evolved the greatest diversity of life histories of any group of echinoderms (McHugh and Rouse 1998; Byrne 2006). These histories are thought to have contributed to the evolution of new species in southern Australia. Eight species are moderately common, and two have interest because they are the smallest sea-stars in the world. There are nine species of *Meridiastra*, three species of the tiny *Parvulastra* and also *Paranepanthia grandis*. The *Meridiastra* species are broadcast spawners and shed their yolked eggs into the water column. Some have buoyant eggs and others have heavier eggs that sink to the bottom. The former type drifts about for some days until they metamorphose and settle on the bottom, whereas the latter type undergo development and attach themselves to the bottom. The tiny *Parvulastra* species have very small eggs, which develop within the mother's gonad and, upon release, quickly settle as miniature adults. In *Parvulastra vivipara* the juveniles within the gonad cannibalise their clutch-mates to support their rapid growth, and then they emerge as large, almost sexually mature, sea-stars. The early life histories of some of these southern asterinids are summarised in Table 11.2.

Table 11.2: Early life history of eight asterinids and one asteriid.

Egg diameter (µm) is given with information on whether eggs have yolks (lecithotrophic) and sink or float. The spawning season, method of fertilisation and sex ratio (female:male) are given, with comments on egg development, larval life and larval settlement, where known. References are listed at foot of table.

Species	Eggs (µm)	Spawning season, fertilisation	Sex ratio	Comments
Meridiastra calcar	415; yolk. Eggs sink	Sept.–Feb. Broadcast.	1:1	Eggs develop over 2 years. Planktonic larvae settle in ~10 days.[1,2]
Meridiastra gunnii	430; yolk Buoyant	Broadcast.		Planktonic larvae.[1]
Meridiastra occidens	~380;yolk. Buoyant	Mar.–Apr. Broadcast.	1:1	Planktonic larvae. Sexual maturity at ~7 months, spawning at ~1 year old.[3]
Meridiastra oriens	400; yolk Eggs sink	Aug.–Jan. Broadcast.	1:1	Eggs develop over 2 years. Planktonic larvae.[1]
Patiriella regularis	150;	Dec.–Feb. Broadcast		Planktonic larvae for 9–10 weeks.[6]
Parvulastra exigua	390; yolk. Benthic egg mass	Jun.–Feb. (NS.) Males fertilise egg mass Oct.–Dec. (SA)	8:1	Protandrous i.e. over life some males change to female. Eggs develop directly into juveniles in ~3 weeks[1] Recruitment in summer[5]
Parvulastra vivipara	145; yolk	Internal fertilisation; development within gonad.		Hermaphroditic, with testes and ovaries. Intra-gonadal cannibalism occurs, and juvenile crawls out as tiny adult.[4]
Parvulastra parvivipara	135; yolk	As above		As above
Smilasterias multipara	1000; yolk Eggs sink	Aug.–Nov. External fertilisation.		Gastric brooding; eggs develop in 8 weeks, and emerge from mother's mouth.[7]

[1]Byrne (2006); [2]Lawson-Kerr and Anderson (1978); [3]Grice and Lethbridge (1988); [4]Byrne and Cerra (1996), Byrne (1996); [5]Duyverman (1976); [6]Byrne and Barker (1991); [7]Komatsu *et al.* (2006).

It is now widely thought that a feeding, planktonic larva was the ancestral state and that, over the last few million years, multiple transformations occurred in which some species evolved a planktonic, yolk-feeding larva and others the brooding of eggs. In the latter case, the transformation was accompanied by a much reduced adult size of sea-star, which brooded non-dispersing offspring within the gonad and released them onto the bottom. In one species, the offspring when released are only slightly smaller than their parent. Yet there is another twist to our story. The two tiniest species, *P. vivipara* and *P. parvivipara*, have gone further, with the evolution of cannibalism among brooded larvae to support the growth of a few survivors. Byrne (1996) described the battles that occur within the gonad, with often a large juvenile eating smaller ones or, as in one case, four smaller ones ganging up on a large one and eating it. Sometimes large juveniles faced each other mouth-to-mouth in a 'stomach duel' to avoid falling victim to the other. Thus, over evolutionary time, the viviparous *P. vivipara* likely had a *P. exigua*-like ancestor (Byrne 2006) whose precursor would have been a species that produced feeding, planktonic larvae.

A notable feature of the last four species in Table 11.2 is their small size and their non-pelagic development by brooding, viviparity or within egg capsules – the last commonly seen in cold-water or Antarctic species. Although many explanations for this have been proposed and much debated, the most plausible theory for these four species seems to be that their small size and restricted habitat mean that, had they produced planktonic larvae, there would have been too few of them to have had much chance of surviving and settling in the right habitat. By producing offspring that develop directly and settle near their parents, the likelihood of the population surviving is much higher. This life history feature itself has favoured speciation, as is evident in the two genera involved: *Parvulastra* and *Smilasterias* (see Pearse *et al.* 2009).

The diminutive *Parvulastra exigua*

This sea-star ranges across southern Australia in intertidal to shallow subtidal areas and lives on rocky, sandy or muddy substrata, but with a strong preference for rock. Spawning occurs from June to February near Sydney, but over a shorter season (October to December) in South Australia. The species is unique among asterinids in that it is protandrous; that is, some individuals develop into males and then change sex with increasing size, so that adults have a strongly biased sex ratio towards females (Table 11.2). The females lay their large sticky egg masses onto rocky surfaces and the males then fertilise them directly, ensuring high fertilisation success. The eggs develop directly into crawling juveniles in about 3 weeks.

The gastric brooder *Smilasterias multipara*

This species grows to a size (arm radius) of ~4 cm and is found only in western Victoria and northern and eastern Tasmania. It is remarkable for being one of only two sea-stars known globally to brood its eggs in the stomach (Table 11.2). Eggs are laid on the sea bed in spring, fertilised externally by the male and then transferred by the female, one by one with her tube feet to the mouth, and thence to the stomach. She holds them there for 2 months, while they develop through several larval stages, and they eventually emerge from her mouth at a size of ~3 mm.

The New Zealand sea-star *Patiriella regularis*

This sea-star is native to New Zealand and was introduced accidentally to southern Tasmania, probably with shipments of oysters. It has not spread to other parts of southern Australia. The species has separate sexes that spawn eggs and sperm into the water column during summer months. Spawning occurs during slack or calm water conditions to maximise fertilisation success (Byrne and Barker 1991). Unlike other species of asterinids discussed, the eggs are small and hatch into larvae that feed on plankton for 9–10 weeks before settling (Table 11.2).

The predatory sea-star *Coscinasterias muricata*

The eleven-armed sea-star, *Coscinasterias muricata* (formerly called *C. calamaria*) is a common and voracious predator, ranging across southern Australia and New Zealand. The species is notable for the fact that it can reproduce both sexually and by fission (i.e. by a spontaneous splitting of the animal into two parts, each of which can regrow a complete disc and a further suite of arms). In a New Zealand study of this process, Crump and Barker (1985) found that individuals tore themselves into two parts by their tube feet, each half keeping a piece of the disc. The wound healed after ~14 days and the arms on each part had fully regenerated after 24–48 weeks. The researchers noted that subtidal individuals divided less than intertidal ones and proposed that fission occurred as a response to stress. When an individual was subjected to physical stress, such as a lack of food, it was more likely to undergo fission – a behaviour recorded more often in small than in large sea-stars. Yet, some populations apparently never reproduced asexually, and Mladenov (1996) proposed a model by which external environmental factors might regulate fission-inducing substances in the sea-star and switch on asexual reproduction (see also Sköld *et al.* (2002). From genetic studies on the sea-star, both clonal and sexual reproduction seemed to contribute to the genetic diversity of the species (Johnson and Threlfall 1987).

Spawning in *Coscinasterias* occurs from July–January (Barker 1978; Crump and Barker 1985) and fertilisation occurs externally in the water column. The resulting larvae then drift in the plankton for ~30 days and finally settle on the bottom. Experiments showed that they settled on most hard surfaces, often on the undersides of rocks, providing a primary surface film was present (Barker 1977). Juveniles live among algal fronds, but little is known of their ecology at a small size (Klemke 1985).

Much more is known about the foraging ecology of adult *Coscinasterias* and the defences of its prey species. Keough and Butler (1979) found that on jetty piles a majority of the diet was molluscs (mainly bivalves, but also some gastropods, such as *Phasianella australis* and *Thais orbita*), ~28% was crabs (mainly hermit crabs and *Plagusia chabrus*), and the rest was scavenged dead animals (Table 11.1). Elsewhere, McShane and Smith (1986) noted that they 'fed on each other and almost every other epibenthic species, including algae, sponges, tunicates … and sea-urchins'. In a pioneering study using archival tags, Lamare *et al.* (2009) found that *Coscinasterias* ate on average a blue mussel a day, moving ~6 m up into the intertidal zone daily to feed and then down again into the subtidal zone for shelter.

Day *et al.* (1995) carried out extensive and detailed studies on the foraging behaviour of the sea-star in Port Phillip Bay. It foraged by day and the most commonly eaten prey were mussels (48%), followed by other bivalves (31%) and small gastropods (7%), with the rest a variety of other small species. But the most interesting questions arose when the mussels had been eaten out or were absent through recruitment failure, leaving voracious and starving sea-stars. In this situation, the sea-star pursued small abalone – a risky venture, because small abalone have a well-developed escape response (see Chapter 9), and also produce a sticky chemical substance that acts as some deterrent to the sea-star. Despite these defences, starving sea-stars still managed to catch small abalone when they were cornered without a crevice into which to retreat. Rob Day and colleagues were especially interested in the sea-star's predation on abalone and the circumstances under which it happened. They proposed a 'patchy predation' hypothesis that predicted that abalone would be eaten only when preferred prey species, such as mussels and gastropods, were eaten out or absent; in such situations, small abalone could be eaten in large numbers.

Other studies in South Australia (S.A. Shepherd unpublished data) have cast more light on Rob Day and colleague's patchy foraging hypothesis. During the 1970–1980s, abalone divers reported seeing on several occasions feeding fronts of the sea-star. In these accounts, large numbers of the sea-star emerged from deeper sandy habitats onto reefs occupied by abalone and attacked both adults and juveniles, severely reducing their numbers. Similar events or their aftermath have been observed by researchers several times in South Australia and once in Tasmania (GJ Edgar *pers. comm.*), and data obtained on two of them. After one such event in outer Coffin Bay in the eastern GAB, when large numbers of sea-stars moved in from a sandy bottom, tagged abalone suffered 99% mortality over a few months (Rodda *et al.* 2002). After another invasive event near Point Marsden on Kangaroo I., Dixon (2011) similarly recorded the almost total mortality (99%) of ~880 tagged abalone.

One puzzling question emerging from the above studies was why the invading sea-stars sometimes ate adult abalone and at other times ate only small ones. To answer this question, the author (unpublished data) set up six cages under Cape Jervis jetty, in which a *Coscinasterias* was caged either with a large greenlip abalone, or with a large greenlip together with some other gastropods (*Phasianella* and *Dicathais orbita*), with three replicates of each experiment. After 3 days, the sea-stars caged with only a large abalone had attacked and overturned it, and eaten the abalone's soft parts (but not the muscle) in two of the cages, whereas in the cages with abalone and small gastropods the large abalone were left untouched, and only the small gastropods attacked. Elsewhere at the site, numerous abalone and *Coscinasterias* sea-stars co-existed, with no evidence of the latter's predation on abalone, presumably because other gastropod prey (*Phasianella*) were plentiful. Laboratory experiments by Day *et al.* (1995) confirmed that starved sea-stars, but not well-fed ones, attacked adult abalone. The varied escape responses of abalone to this predator are discussed further in Chapter 9.

The cause of outbreaks of *Coscinasterias* is not known, but the echinoderms are known to be a boom–bust phylum in which outbreaks occur in some species (Uthicke *et al.* 2009), as in the notorious crown-of-thorns sea-star. The outbreaks may result from enhanced larval survival through nutrient enhancement or from multiple successive recruitment events.

The northern Pacific sea-star *Asterias amurensis*

This East Asian species (Plate 28), apparently introduced into Australia in ballast water in the mid-1980s, is present in the Derwent Estuary, Tasmania, and Port Phillip Bay, where it arrived from Tasmania in the mid-1990s (Byrne *et al.* 1997, 2013; NIMPIS 2002). The species has been claimed to be the most damaging marine pest in Australia, because of the damage it can do to native shellfisheries. In the Derwent Estuary, the sea-star became the dominant invertebrate predator, with a preference for bivalves at or near the sediment surface, which were soon decimated (Ross *et al.* 2006). In Port Phillip Bay, the effects were even more devastating and by 2000 the pest was estimated to number ~75 million or 2800 t, concentrated in the deeper parts of the Bay; later, numbers declined somewhat (Hutson *et al.* 2005). The decline of deeper demersal fish populations is considered to be partly due to *Asterias* (Hirst 2012).

The sexes are separate, with females releasing up to 25 million eggs during the June to October spawning season, with spawning apparently controlled by photoperiod. After fertilisation, larvae remain in the water column for 2–4 months, depending on sea temperature, before settling onto soft sediments or reefs. They reach sexual maturity at a size of 4–5.5 cm and are ready to reproduce in the following winter. The species grows to ~7 cm in 1 year, and after that more slowly at up to 2.4 cm a year. They mostly live for 2–3 years, with a maximum longevity of ~5 years. Asexual reproduction can occur by fission of an arm with part of the central disk attached.

Asterias amurensis is a voracious, generalist predator that attacks molluscs, crabs, sea-urchins and ascidians, and scavenges dead prey. It is especially damaging to native bivalves, such as scallops, because these have not evolved any escape response to them, as they have to other native predatory sea-stars, such as *Coscinasterias muricata*. For example, Hutson *et al.* (2005) found that the doughboy scallop *Mimachlamys asperrima* escaped rapidly from *Coscinasterias* by clapping its valves rapidly and jetting away from the predator high in the water column. In contrast, it did not respond to *Asterias amurensis* at all, and fell an easy prey.

The known predators of *Asterias* are *Coscinasterias muricata* and the giant spider crab *Leptomithrax gaimardii*.

BRITTLE-STARS AND BASKET-STARS

Ecology and life histories

Ophiuroids (i.e. brittle-stars and basket-stars) play an important role in the ecology of coastal waters. They are preyed upon by fish and they scavenge organic detritus or filter feed. Alan Baker (1982) recorded 73 southern Australian species, most of which live on reefs within crevices, under boulders or in algae. Brittle-stars are cryptic in behaviour to avoid their predators, such as wrasses, which readily take them. If a brittle-star is disturbed, it will sacrifice an arm to the predator and attempt to escape under a boulder. Hence, brittle-stars are sometimes seen with a smaller regenerating arm, as witness to a past encounter. In deeper water, the filter-feeding basket-stars usually live above the bottom on sponges, bryozoans or gorgonians, where they spread out their arms at night to trap plankton. The brittle-stars have no anus and undigested food is expelled from the mouth on the underside of the disk. They feed on detritus and microalgae, catching food in sticky mucus secreted along the arms and then transporting it to the mouth. Some are carnivorous and catch small worms and crustaceans.

The brittle-star's arms are fragile (hence their name) and composed of calcite vertebra-like plates, with the tube feet running along their underside. They use their tube feet for movement over the bottom, but can also swim just above the bottom by sinuous flexure of the arms, when the central disk and one arm are pushed forwards by oar-like movements of the other four arms. The basket-stars are distinguished by their articulated arm joints, which enable them to coil up, extend or climb up vertical faces. In some species of basket stars, the arms branch repeatedly, enabling them to form a fan for more efficient food capture.

The sexes are separate in most species and the gonads are located on the disk open to the outside on each side of the arms. There is some evidence that species may aggregate for spawning to enhance fertilisation success (Falkner and Byrne 2004). Eggs and sperm are liberated into the water where fertilisation takes place, and the larvae subsequently pass through several larval stages before metamorphosing and settling on the bottom. However, a few species brood their eggs, while others are hermaphrodites. We now describe the life history of one remarkable species, *Amphipholis squamata* (Plate 29), and the reproduction and early life history of four other common southern species.

The viviparous brittle-star *Amphipholis squamata*

This is a tiny species with a disc diameter of rarely more than 4 mm and arms up to 15 mm long (Plate 29). It is the only viviparous echinoderm occurring worldwide on tropical and temperate coasts, but is now believed to be a complex of species (Le Gac *et al.* 2004). In southern waters, it lives among corallines, mussels, bryozoans and sponges, from the intertidal zone in rock pools to > 2000 m depth. The species has long been known for its luminescent properties, being first called *Asterias noctiluca* (Viviani 1805), and has been studied in many parts of the world, but not Australia. The following account is taken from various temperate studies (Fell 1946; Jones and Smaldon 1989; Emson and Whitfield 1989; Deheyn and Jangoux 1999).

Amphipholis squamata is a simultaneous hermaphrodite (i.e. both male and female) and capable of self-fertilisation, although cross fertilisation between pairs is likely the usual way of reproduction, when sperm shed from one individual enters the genital sac (called a bursa) on the disc between the arms of another close by. The fertilised egg (one per bursa) is brooded by the parent as it goes through several larval stages before eventual release at a disc diameter of ~1 mm. After ~8 months, and at a size of 1.8 mm, the diminutive brittle-star starts breeding and is capable of breeding for the rest of its brief life of ~18 months. Although

Table 11.3: Reproduction and early life history of four southern Australian brittle-stars, after Selvakumaraswamy and Byrne (2000, 2004), and Falkner and Byrne (2003).

Species	Spawning season	Egg size (μm)	Fecundity	Larval feeding type; time to larval settlement
Ophiactis resiliens	May–Nov.	83	29 000	Planktotrophic; 3 months
Ophiothrix caespitosa	July–?	105	?	Planktotrophic; > 26 days
Ophiothrix spongicola	Mar.–Jun.	128	59 000	Planktotrophic; 30 days
Ophionereis schayeri	Dec.–July	241; yolk	6000	Non-feeding planktonic larva; 7 days

breeding is mostly during summer at the temperate sites studied, it is capable of continuing year round, probably at pulsed intervals rather than continuously. The number of embryos being brooded at any one time depends on the brittle-star's size, with large animals (3.5 mm disc diameter) brooding up to nine embryos at a time.

This brittle-star has been variously described as a suspension-feeder, deposit-feeder and omnivore. Gut contents have shown the presence of detritus, diatoms, macroalgae, crustacean appendages, polychaete fragments, bryozoan polyps, sponge spicules, faecal pellets, foram tests and gastropod shells. To capture such diverse prey, which includes highly mobile forms, it uses a mucous net as well as its tube feet and spines.

A behavioural curiosity of *A. squamata* is its arm-coiling response to disturbance. If attacked by a predator, or if it falls off the frond of an alga, it folds its arms up over the disc to adopt a spherical shape and falls to the bottom. The response enables an animal to avoid predators by dropping rapidly out of sight into dense algal fronds (Emson and Wilkie 1982). Another notable feature is its bioluminescent arms – something which has long puzzled biologists. Several colour varieties of the brittle-star occur and these show different intensities of luminescence, which can also vary seasonally. Furthermore, the different colour varieties of the brittle-star do not cross-breed, suggesting that body colour is inherited, and that the several varieties may be distinct sibling species (Deheyn and Jangoux 1999). But what is the benefit of luminescence to the brittle-star? Various suggestions have been made (Deheyn *et al.* 1997, 2000).

The luminescence of the arms peaks during the summer breeding season, so it may serve to promote courtship of sexual partners or the aggregation of many (called the 'group effect' and known for some ophiuroids). It may also function to deter predators, and studies have shown that the main predators, fish and crabs, reject luminescent brittle-stars in favour of non-luminescent ones (Grober 1988). This is, in effect, another example of warning colouration that has evolved to mislead or deter a predator. Alternatively, luminescence may be a sacrificial lure, as the arm can readily be severed when a predator does attack. Many of these ideas are yet to be tested.

Early life history of four southern brittle-stars

The spawning and development of four common brittle-stars have been studied off Sydney by Selvakumaraswamy and Byrne (2000, 2004) and by Falkner *et al.* (2006). The four species illustrate well the evolutionary change from species whose larvae feed in the plankton to those that produce yolked eggs and non-feeding, short-lived larvae (Table 11.3). Another brooding species is *Amphiura constricta*, which spends its life in coralline algal turf, but its life history is not known (Falkner and Byrne 2006).

Ophionereis schayeri is the most common of the above four species and lives under boulders in shallow water. Like *Amphipholis squamata*, its arms are luminescent and the phenomenon may serve a similar function as in that species. Although spawning may extend over 8 months, it is concentrated in December–January and becomes intermittent after that. In this species, the eggs are quite large, and the larvae non-feeding, with settlement occurring after a short planktonic life of only 6–7 days (Table 11.3). Hence, this species exemplifies an evolutionary shift to the production of non-feeding larvae, which requires a much higher level of maternal provisioning per egg, as shown by the large egg size. The other three brittle-stars listed in Table 11.3 have a different strategy. They produce a greater number of small eggs, which hatch into plankton-feeding larvae, and settle after 1–3 months, often in coralline algae – a favoured habitat for juvenile brittle-stars. With increasing size, they move under boulders, into crevices or among algae. Their longevity is not known.

SEA-URCHINS

Ecology and life histories

Sea-urchins, or echinoids, are the most conspicuous of all the echinoderms in subtidal habitats. There are 29 southern species of regular sea-urchins that live in reef habitats, and another 20 that burrow into sandy or muddy bottoms. Here, we consider only the former group and discuss in detail the ecology of several species, some of which can modify algal habitats by their grazing.

Sea-urchins have an external skeleton, or test, made up of an array of interlocking, calcite plates that surround the internal organs. The plates carry articulated spines – large primary ones and much smaller secondary ones – which serve a variety of purposes, such as movement, digging and protection from predators. The mouth on the underside of the test is a framework of skeletal bars, supporting five large vertically aligned teeth, and described by Hyman (1955) as 'an admirable complex of muscles and calcareous pieces that serve to grasp and chew up food'.

The sea-urchin oral framework was first described by Aristotle, who compared it with a lantern, and so now it is known as Aristotle's lantern. The teeth gradually wear away with use and need continual sharpening by the urchin before its nocturnal grazing excursions. Indeed, in one fascinating unpublished study, a dentist researching bruxism (nocturnal grinding of the teeth by humans) recorded with an underwater microphone the orchestra of many urchins grinding their teeth in anticipation of feeding! As a tooth wears down at the lower end, the whole tooth moves downwards and continues to grow at the upper end.

Like the sea-stars, echinoids have well developed pincers, the pedicellariae, over their entire surface; these can remove fouling organisms that might settle on the test and also transport food to the mouth.

The sexes of sea-urchins are separate, and eggs and sperm are shed into the water, where fertilisation takes place. Some species have planktotrophic (plankton-feeding) larvae that drift in the sea for weeks, while others have short-lived, yolked larvae (lecithotrophs) that settle on the sea bed after a few days. Planktotrophy in echinoid larvae is believed to be the ancestral mode (i.e. the most ancient) and after 500 million years two-thirds of known species have the derived lecithotrophic larval type, which evolved recently (< 4 Ma) (Byrne and Voltzow 2004; Uthicke *et al.* 2009). The roe of several sea-urchin species is edible, and these species are harvested in fisheries.

Table 11.4: The feeding of seven species of southern Australian sea-urchins from various sources as indicated.

For some species the mean is given for several sites. CCA = crustose coralline algae; green = green algae; brown = brown algae; red = red algae; corallines = articulated coralline algae; turfs = filamentous algae; P = preference experiments. References are listed at foot of table.

Sea-urchin	Diet
Heliocidaris erythrogramma	Four sites: Drift algae (seagrass 30%, green 10%, brown 3%, red 5%); CCA, turfs 53%.[1] Preferred alga, *Macrocystis pyrifera.*[2] Drift *Ecklonia.*[3] Drift algae (green 7%, brown 56%, red 35%), animal 1%.[4a] Drift algae (*Ecklonia, Scytothalia dorycarpa*).[5] Four sites: algal drift (brown 34%, foliose 38%,); grazed (turfs 3%, corallines 11%, CCA 6%[6]. **P**: *Pterocladiella* > *Ecklonia* > *Caulerpa, Amphiroa, Codium* > *Dictyopteris, Sargassum* > *Delisea* > *Lobophora.*[7]
Heliocidaris tuberculata	Macroalgae, mainly *Corallina, Haliptilon*, CCA[11]
Holopneustes purpurascens	**P**: *Ecklonia* > *Sargassum* > *Dictyopteris* > *Lobophora* > *Ulva*[7]
Phyllacanthus irregularis	Drift algae (green 15%, brown 43%, red 31%), animal 9%.[4a]
Phyllacanthus parvispinus	Solitary Is: carnivorous (coral, barnacles, bivalves, polychaetes) ; southern NSW herbivorous.[12] Mainly calcareous algae.[11]
Centrostephanus tenuispinus	Drift algae (green 5%, brown 37%, red 40%), animal 11%.[4a]
Centrostephanus rodgersii	**P**: *Ulva* > *Codium, Pterocladiella* > *Sargassum, Caulerpa* > *Dictyopteris, Delisea* > *Lobophora*[7] **P**: *Ecklonia* > *Sargassum linearifolium, Zonaria* > *S. vestitum*[8] Algal forests: macroalgae, invertebrates. Barrens: filamentous algae, CCA[9] Diatoms (16%), CCA (12%), filamentous spp. (18%), macroalgae (green (2%), brown (*Ecklonia, Cystophora, Acrocarpia*) (20%), red (14%), sponge (7%), sand, etc. (11%)[10]

[1]Connolly (1986); [2]Sanderson *et al.* (1996); [3]Vanderklift and Kendrick (2005); Vanderklift *et al.* (2009); [4]Vanderklift *et al.* (2006), [5]Vanderklift and Wernberg (2008); [6]Naumann (2007); [7]Taylor and Steinberg (2005) **P**; [8]Hill *et al.* (2003); [9]Ling (2008), Ling and Johnson (2009); [10]SA Shepherd unpublished data from incipient barrens, Deal I., Tasmania in 1981(see Johnson *et al.* 2005); [11]Mortensen (1928, 1943) cited in De Ridder and Lawrence (1982); [12]O'Connor (1976); O'Connor *et al.* (1978).

[a] animal matter = sponge, ascidians, hydroids, bryozoans, crustaceans

The ecology of three southern species, *Holopneustes purpurascens*, *Heliocidaris erythrogramma* and *Centrostephanus rodgersii*, has been studied in much detail and their life histories are discussed below. In Chapter 12, we consider the ecological effects of the last two species, *Heliocidaris* and *Centrostephanus*, on reefs.

The eastern egg urchin *Holopneustes purpurascens*

This species (Plate 30) ranges down the east Australian coast and is closely related to *Holopneustes inflatus*, a south-eastern species, and also to an undescribed red species of *Holopneustes* that ranges across southern Australia (Edgar 2008). *Holopneustes* species have an unusual 'arboreal' habitat in *Ecklonia* and other algal forests, where they wrap themselves in kelp blades and feed on them; as yet, only the ecology of *H. purpurascens* is known.

H. purpurascens has separate sexes, with a 1:1 sex ratio, and reaches sexual maturity at a size of 11–25 mm test diameter. The urchin has a well defined breeding season from September to December, although some sporadic spawning may occur at other times of the year – itself an odd feature for echinoids (Williamson and Steinberg 2002). The females produce large, floating, yolked eggs, which, upon external fertilisation, drift in the plankton for ~6 days, when they are ready to metamorphose and settle (Morris 1995; Williamson *et al.* 2000). It is at the point of settlement that the urchin larvae have evolved a remarkable behaviour. They are induced to metamorphose and settle on the red alga *Delisea* (as well as on coralline turf) in response to histamines (a red algal sugar) released into the water by the plant, which has high concentrations of the sugar (Swanson *et al.* 2006). However, if they do not find a suitable settlement habitat, larvae can postpone settlement for up to 18 days. Larval settlement on *Ecklonia* plants is very rare because *Ecklonia* has virtually none of this algal sugar. At first sight, the avoidance of *Ecklonia* seems paradoxical, because studies have shown that juvenile urchins have much poorer growth and survival living in *Delisea* than in *Ecklonia*. The explanation is that, soon after settlement, most juveniles move from *Delisea* plants to *Ecklonia*, and remain in that habitat for the rest of their lives, where they grow faster, live longer and produce more eggs than if they had stayed on *Delisea* (Williamson *et al.* 2004). Additional laboratory experiments confirmed this, and showed that urchins much preferred to live among *Ecklonia* blades. Provocatively, Williamson *et al.* (2004) suggested that the larval preference for *Delisea* was a 'mistake', because other red algae also induced settlement of the urchin larvae, albeit at a lower rate. They argued that urchin larvae were simply 'responding to the highest local concentration of a more widespread chemical cue that in this habitat resulted in larvae metamorphosing on a host plant that is not consumed …'. However, another possibility not considered by the authors is that settling larvae would receive much better protection against predators searching visually in the dense canopy of the finely branched *Delisea* plants than in the more open canopy of *Ecklonia* (see Rogers *et al.* 1998, and Chapter 5). Hence, the increased survival of post-larvae in *Delisea* would tend to select for a cue that favoured settlement on it, even though it was unpalatable. Yet other questions remained unanswered. Why, for example, do some larvae settle in coralline algae? In response to this, Swanson *et al.* (2006) suggested that secondary inducing agents might operate, such as microalgal biofilms on the corallines.

Although the primary adult habitat of *H. purpurascens* is in *Ecklonia*, it also lives less commonly in the fucoid *Sargassum vestitum* and in the red alga *Delisea pulchra*. This is perhaps not surprising because multiple choice food preference studies (Taylor and Steinberg 2005) found that several algal species were eaten and that the urchin showed the following order of preference:

Ecklonia > *Sargassum* > *Dictyopteris*, *Lobophora* > all other species

H. purpurascens has the curious behaviour of hiding low near the base of its host plant by day. This location is safe against predation by the damsel fish, *Parma microlepis*, which will voraciously bite off the urchin's tube feet and spines if it is exposed (Williamson *et al.* 2004). At night, the urchin climbs up the host plant and feeds on the blade (Rogers *et al.* 1998). But does the urchin have an impact on *Ecklonia* forests? To test this, Peter Steinberg (1995) caged up to two large *Holopneustes* per kelp plant (an unnaturally high density) for 14 weeks, and found that, even though the urchins ate up to 80% of the plants' secondary laminae (the lateral blades), the plants survived and their growth rates were unaffected. These case studies illustrate the complexities and puzzles that can emerge about an animal's behaviour or life history even after intensive studies – matters that can only be resolved by painstaking field or laboratory experiments.

The purple urchin *Heliocidaris erythrogramma*

This is the most common urchin subtidally in western and southern Australia (Plate 31), and is also abundant in eastern Australia, although in lower densities than the black urchin, *Centrostephanus*, discussed below. Nevertheless, high densities of up to 60 m^{-2} can occur on the east and SW coast, and densities up to 20 m^{-2} sometimes occur in Spencer Gulf, but densities elsewhere rarely exceed a few per square metre. The common name, purple urchin, is not entirely accurate, because greenish-blue and white variants can occur. The colour appears to be influenced both by the environment and genetics (Growns and Ritz 1994; Beck and Styan 2010), with the whitish variety being common on exposed coasts and the dark purple variety in sheltered waters. The ecology of the purple urchin has been well reviewed by John Keesing (2006), and this account draws on that review, supplemented by more recent work.

Reproduction

This sea-urchin has an approximate 1:1 sex ratio and males and females spawn freely, with fertilisation of eggs taking place in the water column. The spawning season in eastern Australia is in summer–autumn (December to March), but with evidence of earlier sporadic spawning in November (Laegdsgaard *et al.* 1991). In Tasmania, spawning is from November to January, possibly extending into February (Dix 1977). Spawning appears to start earlier in the eastern GAB, where epidemic spawning events have been witnessed in October (Table 11.1 in Pennington 1985). In this region, oceanic swell is perennial and oily calm days with very low swell are uncommon. Two epidemic spawning events were witnessed in such conditions, and they produced vast rafts of floating orange eggs (unpublished observations). Such calms with very low swell also trigger spawning events in some other sea-urchin species (Pennington 1985). The evolution of this trigger may be due to the greater fertilisation success of sea-urchins in slight water movement compared with rapid water movement, as illustrated by Fig. 11.1, which shows the results of Craig Styan's (1997) experiments measuring the fertilisation success for *H. erythrogramma* in slow currents, and the

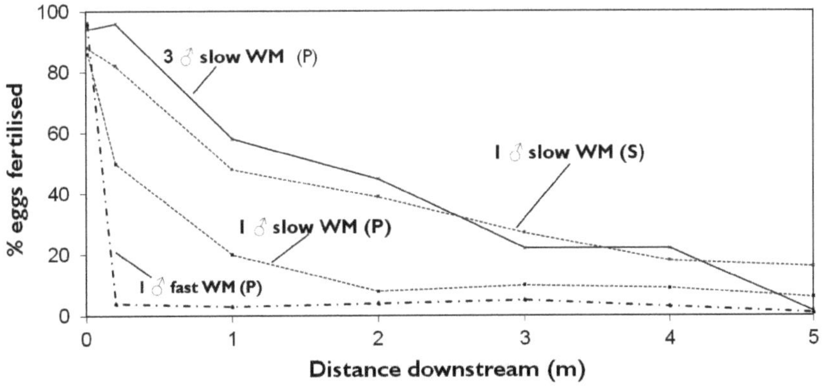

Fig. 11.1: Fertilisation success of sea-urchins experimentally measured under different water movement (WM) conditions: '1 slow WM (S)' refers to a male purple urchin *H. erythrogramma* in slow water movement (WM); remaining graphs refer to experiments in fast and slow WM, with one or three males. Slow WM is ~0.1 m s^{-1}, and fast WM is > 0.2 m s^{-1}. *P* = experiments by Pennington, and S = those by Styan. (Redrawn from Pennington 1985 and Styan 1997, respectively).

comparable data of Pennington (1985) for another sea-urchin species in both slow and rapid currents. The comparisons illustrate how higher densities of urchins or reduced water currents dramatically reduce the dilution effect (i.e. the rate of decline in the percentage of eggs fertilised with increasing distance downstream).

Settlement and growth

The large eggs (430 μm diameter) of the purple urchin are buoyant and emerging larvae have a life of ~5 days. The settling larvae then seek a bottom habitat and are induced to settle on coralline algae, such as *Corallina* spp. and *Amphiroa*, by the bacterial biofilm on them (Huggett *et al.* 2006, 2008). Here, they use up their yolk reserves over the next 1–2 months and then start feeding on the bottom. Surprisingly, the authors recorded new recruits in low numbers among corallines almost every month of the year, despite a short summer–autumn spawning season. Newly settled urchins have also been recorded in corallines, such as *Haliptilon*, but not in other algae, at West I., South Australia (unpublished data). In Tasmania, recruitment was estimated to be from 0.3–1.0 urchins m^{-2} $year^{-1}$ – enough to permit fishing once every 4–5 years (Sanderson *et al.* 1996). Overall, recruitment appears to be low for this species.

The growth of *H. erythrogramma*, although examined by several workers, is still poorly understood, due to the problems inherent in measuring the growth of an animal that is difficult to tag effectively, and where numbers vary from place to place and over time according to food availability. Several methods have been used – marking with tetracycline, physical tagging of the test, reading growth rings in Aristotle's lantern and interpreting size-frequency distributions – and all have their problems. The available information on growth and methods of its measurement were critically examined by Keesing (2006) and here we present growth curves (Fig. 11.2) for: (a) purple urchins living in barrens and those in algal forests in eastern Tasmania (after Pederson and Johnson 2008); (b) for slow and fast growing urchins, covering the range in growth rates obtained by Sanderson *et al.* (1996) in various habitats in southern Tasmania; and (c) data of Ebert (1982) from Point Peron in SW Australia. Of these curves, those of Pederson and Johnson (2008) seem the most reliable. Ebert's suggested growth of ~40 mm in 1 year is rather high, and it is possibly 2 years old at that size, while the remainder of the curve indicating a slow growth rate, with an average maximum test diameter (D∞) of 76 mm (Fig. 11.2) seems more reliable. Ebert (1982) was the only study that estimated annual survival rates, which

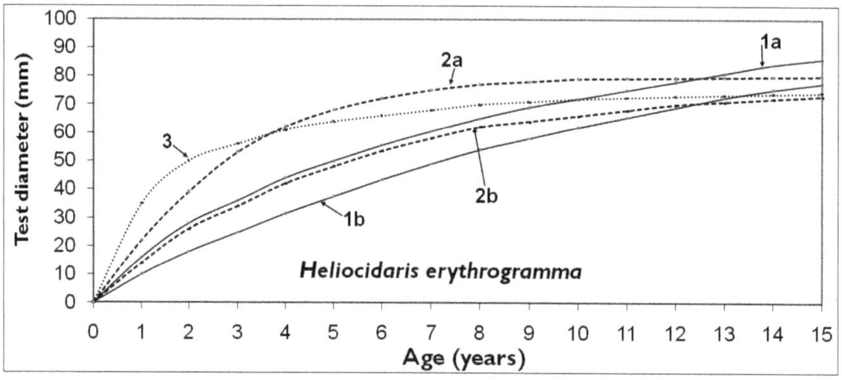

Fig. 11.2: Growth curves for the purple urchin *Heliocidaris erythrogramma* in different southern habitats or regions. Curves 1a and 1b (continuous line) are for algal forest and barrens, respectively, in SE Tasmania (redrawn from Pederson and Johnson 2008); Curves 2a and 2b (dotted lines) are for fast and slow growing areas (after Sanderson *et al.* 1996) in southern Tasmania; Curve 3 is for Point Peron, Western Australia (after Ebert 1982). Maximum test diameters (D_{∞}) from growth curves are: 1a – 101 mm; 1b – 94 mm; 2a – 80 mm; 2b – 75 mm; 3 – 76 mm.

were 91–96% per year for populations at Point Peron and Sydney Harbour. Interestingly, that study also found that survival declined with increasing exposure to swell. From his several studies, Ebert concluded that the purple urchin was a slow-growing and long-lived species, with a longevity of > 20 years, given the low estimated mortality. But Pederson and Johnson (2008) found that SE Tasmanian populations were much older, with average ages of 10–15 years and a maximum age of 37 years.

Food and feeding

The purple urchin feeds by grazing the rocky substratum or by catching drift algae. In the former case it moves about at night as it scrapes the rock, and in the latter case it has a 'sit-and-wait' behaviour, remaining in a crevice for prolonged periods and catching algal drift trapped there (Vanderklift *et al.* 2009). In other habitats, such as those adjoining seagrass, it will similarly remain in a location where drift algae or seagrass pass by or accumulate. The diet of this urchin has been examined at many southern sites (Table 11.4).

At most exposed to slightly exposed sites on the southern Australian coast from SW Australia to Tasmania and eastern Victoria, algal forests are dominant, and algae are periodically torn off the substratum by swell and storms and transported to sites of lower wave energy. On these coasts, the purple urchin uses the sit-and-wait strategy to catch algal drift on its spines. In SW Australia the purple urchin catches mainly *Ecklonia* or *Scytothalia*, with a little other algae and seagrass, apparently preferring drift macroalgae to drift seagrass. Stable isotope studies confirmed that the drift consumed was almost entirely *Ecklonia* throughout the region (Vanderklift and Kendrick 2005; Vanderklift and Wernberg 2008, 2010). In Victoria, Constable (1989) compared different habitats occupied by the purple urchin and recorded that drift algae were available to it in the following order of habitats:

kelp habitats > barrens habitats > seagrass habitats

He also found that macroalgae were more nutritious than seagrass and that when food was in short supply the gonad was resorbed and the test shrank in size.

In South Australia, Livore and Connell (2012a) also examined the food quality and its source, and remarkably found that, although the kelp *Ecklonia* had lower nitrate levels on exposed than on sheltered coasts, the urchin ate less kelp from exposed coasts but had better gonad condition there. The enigma was explicable in two ways. Either

- Some of the nitrogen was bound up in forms (e.g. tannins) that could not be assimilated.
- Urchin reproductive output might peak at intermediate qualities of food, (i.e. 'more is not necessarily better').

In a different approach Sanderson *et al.* (1996), who were interested in gonad quality and size from a fishery point of view, found that, where urchin densities were high and food conditions were poorer in consequence, gonads were smaller and roe quality poorer. Similarly, on exposed coasts in South Australia, high urchin densities led to a reduced per capita food supply and poorer urchin condition (Livore and Connell 2012b).

Within the calmer waters of the South Australian gulfs, where algal drift is largely replaced by seagrass drift, the purple urchin has a dual strategy. It catches drift, including seagrass when available, and at other times grazes on crustose corallines, filamentous algae and some macroalgae. Rod Connolly (1986) examined the urchin's feeding at four sites from exposed to sheltered coasts in mid and upper Spencer Gulf and concluded that it was a generalist and opportunist feeder – on average half its food was grazed and the other half was captured as drift (Table 11.4). The urchin preferred to aggregate on higher and larger rocks, presumably to graze on filamentous algae, as well as to catch drift algae.

In another Gulf study, Naumann (2007) found that urchin density (range 0.01–1.3 m^{-2}) strongly affected the food eaten. The urchins ate 10 times as much crustose corallines at the higher density sites, compared with the lower density sites, suggesting more intense scraping activity at sites where there was less available food. At two other sites with even higher densities (range 6.5–20 m^{-2}), the urchins grazed only on nearby turfing algae or, if that was lacking, on surrounding seagrass, as evident from small haloes of surrounding bare bottom (Shepherd 1974; Naumann 2007). Elsewhere, they were able to adopt a sit-and-wait behaviour and feed on the abundant drift algae and seagrass.

Feeding fronts of the purple urchin have occasionally been seen in New South Wales and Victoria. A vast aggregation of up to 120 urchins m^{-2} formed a front in *Posidonia* seagrass beds in Botany Bay, and gradually consumed many hectares of seagrass (unpublished observations; Keesing 2006). At Bare I. in northern Botany Bay, another feeding front of similar density occurred in 1995. The aggregation persisted for 3 years, and its impact was studied as it slowly moved over ~3 months (Wright and Steinberg 2001;Wright *et al.* 2005). In Corio Bay, Victoria, a stationary feeding front of even higher density (to 180 m^{-2}) occurred at the edge of seagrass beds and lasted for several years, implying that the urchins were surviving on algal drift (Andrew and Constable (1999). The basic cause of the above feeding fronts is probably enhanced local recruitment events, creating local barrens and a shortage of food. The phenomenon has been studied elsewhere in the world, but not in Australia.

In SE Tasmania, Pederson and Johnson (2008) examined the size of the purple urchin's jaw within its Aristotle's lantern in different habitats and found that urchins living in barrens possessed larger jaws than urchins that lived within nearby algal forests. In sea-urchins, larger jaws are better able to rasp the substratum for food and so larger jaws are typically associated with habitats with very limited resources, as in barrens. The study is significant because it suggests that purple urchins living in barrens can adapt to the limited food supply there and can change their diet and rasp the substratum for crustose corallines more effectively than urchins living in algal forest habitats.

Movement

The purple urchin does not move during daylight hours, but at night it may move about to graze. At Deal I. in eastern Bass Strait, tagged urchins started grazing at dusk and moved randomly (χ^2 = 2.0; ns) within mini-barrens at a rate of 0.5–1 m per 2 h (Fig. 11.4), but ceased moving after midnight. In a study in Spencer Gulf, comparing movement between urchins placed on rocks cleared of macroalgae, and ones on natural rocks, Connolly (1986) found that movement was only 5–6 cm a night and did not differ between the two experimental habitats. (Here, 55% of its food was grazed and the rest was drift). Similarly, Livore and Connell (2012c) recorded localised grazing movements of only a few centimetres.

Sanderson *et al.* (1996) cleared 100 m^2 squares of urchins and followed the rate of 'diffusion' by urchins into the cleared area. He found that they gradually moved in over a year, with the greatest movement from January to June and very little for the rest of the year. Net distance moved by urchins was mainly in the range 10–25 m over the full year.

Conclusions

These various studies of the purple urchin indicate that its movement and grazing are interlinked, and largely depend on food supply. In the presence of drift algae, the urchin ceases moving altogether and feeds on the drift, and in the absence or shortage of drift it grazes on the surrounding turf, and occasionally forms feeding fronts. On exposed coasts, drift algae are more plentiful, whereas in sheltered bays or gulfs drift seagrass is often more abundant. So it is not surprising that on open and on sheltered coasts, the extent of feeding on drift is similar. In summary, the studies above suggest that the purple urchin has the following general preference in terms of food and the availability of the preferred food-type determines its foraging behaviour:

Drift algae > algal turf > seagrass > crustose corallines

Urchin density is another factor that clearly affects the per capita availability of food, but its causes have not been studied. Suffice it to say that moderate to high densities of > 2–3 m^{-2} tend to occur in bays and gulfs, suggesting that coastal topography plays an important role restricting larval dispersal and increasing recruitment strength.

The behavioural flexibility of this urchin is very similar to that of urchins elsewhere (Dix 1970) and even abalone (Shepherd 1973; Chapter 9). The effect of urchin aggregations on benthic ecosystems is discussed in detail in Chapter 12.

Predators

The predators of the purple urchin include wrasses, snapper, crabs, stingrays, dog sharks (Shepherd 1974) and notably Port Jackson sharks, whose lips may sometimes be seen stained with the purple colour of consumed urchins at the sites where they lay their eggs. Other common predators are the common octopus and the southern rock lobster. In Tasmania, Guest *et al.* (2009) found that sea-urchins comprised 10–65% of the diet of lobsters, according to location. Pederson and Johnson (2006) noted a negative correlation between the purple urchin and rock lobsters, a phenomenon most clearly evident within marine reserves where populations of purple urchins declined following a massive increase in biomass of rock lobsters (Barrett *et al.* 2009; Chapter 20). Pederson and Johnson (2006) carried out numerous field experiments to determine the relative predation of lobsters, – as compared with fishes, such as wrasses, banded morwongs and bastard trumpeters – on small, medium-sized and large urchins. Lobsters proved to be a far more serious predator than fish and their effects depended on size of both prey and predator. For example, small urchins were more successful in sheltering in cryptic micro-habitats from their predators and small lobsters could consume medium-sized, but not large, urchins. Large lobsters could readily capture large urchins and the mortality rates of urchins were three to nine times greater in marine reserves than in adjacent areas where lobsters were fished. Thus lobsters essentially controlled the abundance of the purple urchin and the fishing of lobsters accordingly had a strong cascading effect in shallow coastal systems.

The black urchin *Centrostephanus rodgersii*

The black urchin is the dominant herbivore subtidally on eastern Australian coasts and is notorious for the extensive urchin barrens, devoid of macroalgae, that it creates on shallow reefs among macroalgal forests. Hence, the species has been called an ecosystem engineer by some, because it can profoundly modify a habitat, and a keystone species by others, because its effect is out of proportion to its size or abundance, as described in Chapter 12. Its life history and ecology have been studied extensively over the past two decades and are well reviewed by Andrew and Byrne (2007).

Reproduction

The sexes are separate with a 1:1 sex ratio and individuals become sexually mature at 40–60 mm diameter at an age of ~2 years. The species has a very short winter

Table 11.5: Early life history of some southern Australian sea-urchins.

Egg size is given, and whether the eggs are yolked (lecithrotrophic), or plankton-feeding (planktotrophic). n.k. = not known. References are listed at foot of table.

Species	Spawning season	Egg size (μm)	Larval feeding type; time to settlement;
Heliocidaris erythrogramma	Nov.–Feb. Broadcast	450; yolked; buoyant	Non-feeding, planktonic; 5 days[1]
Heliocidaris tuberculata	Jan.–Feb.; Jun.–July Feb.–Oct.	95; eggs sink	Planktotrophic; planktonic 21–30 days[3,4]
Phyllacanthus parvispinus	All year (north NSW); Dec. (central NSW)	700; yolked; buoyant	Non-feeding, planktonic; 5 days[2,4]
Centrostephanus rodgersii	Jun. (Solitary Is); Jun.–Nov. (Eden) July–Aug. (Tas.)	110; eggs sink	Planktotrophic; planktonic ~100 days[4,5,6] Planktonic for 3–5 months[7]
Holopneustes purpurascens	n.k.	580; yolked; Buoyant	Non-feeding, demersal larva[8]

[1]Keesing (2006); [2]Parks *et al.* (1989); [3]Laegdsgaard *et al.* (1991); [4]O'Connor *et al.* (1978); [5]King *et al.* (1994); Huggett *et al.* (2005); [6]Ling *et al.* (2008); [7]Andrew and Byrne (2007); Morris (1995).

spawning season for about a month from June to July in northern New South Wales, extending to ~5 months (June–October) at Eden in the south (Table 11.5). The black urchin lives both in barrens it creates and in fringe habitats, where algae are abundant. Byrne *et al.* (1998) examined its reproductive output in both habitats and found that in fringe habitats the urchin grew larger and, for equivalent sizes, also had two to three times the size of gonads. It also spawned over a longer period in fringe habitats than in barrens habitats. The gonads of urchins are their major store for nutrients and the marked differences in reproductive output clearly showed the poverty of food in the barrens habitat. Further south in eastern Tasmania, where the urchin has expanded its range in the 1980s, the species apparently has a shorter 2–3 month spawning season compared with that in Eden (Ling *et al.* 2008).

Recruitment, growth and mortality

After fertilisation, larvae have a long planktonic life. In contrast to other species with a planktonic larval stage, the larvae are not able to respond to poor food conditions by growing longer arms to enhance food capture, as can larvae of *Heliocidaris tuberculata* (Soars *et al.* 2009). After about 3 months the larvae metamorphose and settle in reef habitat (Table 11.5). The larvae are widely dispersed by the East Australian Current (EAC). Some are caught in eddies when the EAC parts from the coast, some travel north in an inshore counter-current, while others are carried southwards towards Tasmania. Coastal topography also plays a role, because headlands form eddies downstream from them, and bays can trap larvae within them. Hence, not surprisingly, minor genetic differences in populations are apparent at a fine scale, but not over the geographic range of the species (Johnson *et al.* 2005; Banks *et al.* 2007). As expected for a long-lived larva, recruitment to local populations is patchy and highly irregular.

The age of the urchin can be estimated directly by growth checks on Aristotle's lantern, and Neil Andrew (1991, 1993) suggested that they grew to ~3–4 cm in 1 year, and lived for up to 20 years, reaching a maximum size of 11 cm test diameter off Sydney. In SE Tasmania, where growth is slower than in central New South Wales, they can reach a size of 12.5 cm over 20 years (Fig. 11.3; Ling *et al.* 2009a). These last authors examined growth rates of individuals marked with tetracycline in barrens and macroalgal habitats, and showed faster growth and larger maximum size in the latter habitat, although variability was high, as indicated by the broad 95% confidence intervals (Fig. 11.3). The tests and length of spines also differed between the urchins living in barrens and those living in macroalgae. The barrens-dwellers grow longer spines and lighter tests, together increasing their resistance to

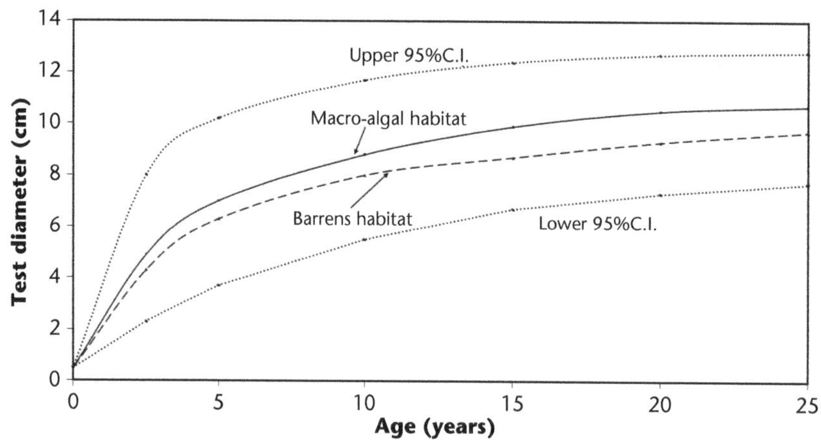

Fig. 11.3: Growth rates of the black urchin *Centrostephanus rodgersii* in macroalgal habitat, with upper and lower 95% confidence intervals (C.I.), and in barrens habitat (redrawn from Ling and Johnson 2009).

predators, whereas the forest-dwellers have shorter spines and thicker and heavier tests, apparently better suited to whiplash from fronds of algal canopy species and surge (Ling and Johnson 2009). The diets of urchins also differed between the two habitats as shown in Table 11.4.

The only estimate of the mortality rate for New South Wales is from Ebert (1982) who estimated a rate of 6.2% $year^{-1}$ within Sydney Harbour. However, this estimate was for a very sheltered site and mortality would be greater at more exposed sites, as he also noted. In eastern Tasmania, the mean mortality rate was estimated to be 11–17% $year^{-1}$, except in the extreme south (Ling *et al.* 2009a).

Another factor to consider in understanding the growth and mortality of urchins is their unique physiology. They can go for long periods without food, and during food shortages shrinkage, and possibly resorption, of the test can occur, as in *Heliocidaris* and other species (Constable 1993). So, their survival is not greatly affected by food shortages, nor do they suffer much competition for food with their own or other species (Andrew 1989).

Habitat and predators

Young recruits are found under boulders and in crevices and, as they grow, they move to larger crevices, but retain through life an obligate association with crevices, at least in New South Wales (Andrew and Underwood 1989). This sheltering behaviour has several benefits because it provides protection against predators, shelter against dislodgement by wave action and also serves to circumscribe the area of grazing, because the urchin never moves far from its daytime shelter. In eastern Victoria and Tasmania, the urchin is often found in the open, especially when at high densities (Andrew and O'Neill 2000). Whether predation is less in those cooler waters is possible, but has not been examined. The major predators of the black urchin are the eastern blue groper as well as smaller wrasses, the Port Jackson shark, rays, eastern rock lobster and octopus. Of these, the Port Jackson shark and the groper are believed to be the most important (Andrew 1993; Davis *et al.* 2003), both studies noting that urchins not in crevices are very quickly taken by groper. The strong inference that historical overfishing in New South Wales of important predators, such as the lobster, eastern blue groper and snapper, led to an explosion of black urchin populations has often been drawn by ecologists (e.g. Connell 2008). If so, it must have occurred before 1965, when urchins were already in high abundance in extensive barrens along the New South Wales coast from Sydney to Green Cape (unpublished observations).

In eastern Tasmania, the main predator of the black urchin is the southern rock lobster. Other less important predators are wrasses, banded morwong and bastard trumpeter. In a study of predation on black urchins inside and outside marine reserves, Ling *et al.* (2009b) found that predation was much higher within the reserve where lobsters were larger and more abundant than at sites outside the reserve, where lobsters are fished. Only larger, legal-sized lobsters were able to predate on the black urchin, a result similar to the findings of Pederson and Johnson (2006) for lobster predation on the purple urchin (see above). The authors concluded that intense fishing of rock lobster for well over a century had reduced the resilience of macroalgal beds to invasion of the region by the black urchin (see Chapter 12). Hence, intense fishing had contributed to a catastrophic shift to urchin barrens – a consequence that calls for a strong ecosystem management approach to reverse the severe loss of predators.

Movement, food and feeding

At night-time the black urchin leaves its crevice and goes on grazing excursions, sometimes moving 5–10 m in a night, although mostly it forages in a small area around its shelter. In larger barren areas with higher densities, urchins tend to forage further than those in crevices with only a few individuals (Jones and Andrew 1990). Average distances moved over 2 h periods in two barrens off Cape Banks, NSW and in incipient barrens at Deal I. are shown in Fig. 11.4. At Deal I., the black urchins grazed within their barren, and by dawn they had all returned to a crevice abutting or within the barren.

The black urchin diet includes a variety of macroalgae, seagrasses, encrusting algae and invertebrates, and is strongly influenced by the algae and

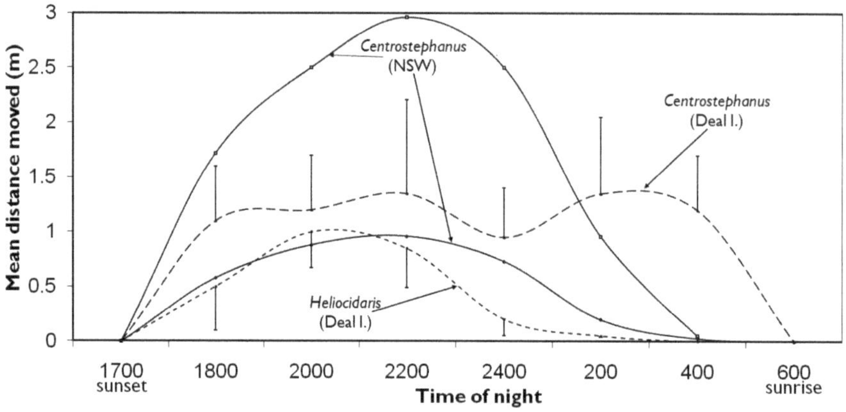

Fig. 11.4: Mean distances moved per 2 hour periods between sunset and dawn (a) by the black urchin *Centrostephanus rodgersii* at two barrens sites in central New South Wales (redrawn without standard errors from Jones and Andrew 1990), and (b) by the black urchin and the purple urchin *Heliocidaris erythrogramma* at Deal I., Bass Strait, in 1981, with standard errors (SA Shepherd, unpublished data).

invertebrates available (Jones and Andrew 1990; Andrew and Byrne 2007). In two choice experiments, in which black urchins were offered species of macroalgae (Table 11.4), they showed distinct preferences for particular macroalgae, but these experiments were designed to test how their feeding might affect the adjoining macroalgal vegetation, not whether the urchins preferred encrusting species. In fact, gut analyses suggest that much of the diet is filamentous and encrusting algae. For example, at Deal I., where macroalgal forest surrounded small barrens, much of the food eaten was taken by grazing. In the gut content analysis for that site, given in Table 11.4 (Ref. 10), crustose corallines comprised 12% by volume of the food eaten, but 70% by dry weight.

SEA CUCUMBERS

The holothuroids are a Class of cylindrical echinoderms with the mouth at one end and the anus at the other. Their pentamerous symmetry, typical for the Class, can be seen in the radial arrangement of the tube feet in many species, although others, adapted for crawling on the bottom, have a superimposed bilateral symmetry. Holothuroids are sluggish animals, and live on the bottom, some under rocks or in crevices, and others in sediment or seagrass, or in several of these habitats. The ecology of very few species is known, and here we describe the life history of *Australostichopus mollis* (Plate 32), which is common in SE Australia in reef and soft bottom habitats. Its ecology is known largely from studies on its culture (Sewell 1990; Morgan 2008, 2009, 2011, 2012; Slater and Jeffs 2010). Recent studies show that what was thought to be one species *A. mollis* is in fact a species-complex, with closely related cryptic species yet to be described (M Byrne *pers. comm.*).

Southern sea cucumber *Australostichopus mollis*

This sea cucumber lives on reefs, often with a boulder bottom, as well as on sandy bottoms, on semi-exposed to sheltered shores to depths of ~100 m.

Life history

Larvae are thought to settle on algae or on a substrate with high benthic diatom cover. After settlement, juveniles are cryptic and found usually under boulders, shell and rubble in shallow water of 2–5 m depth in or near kelp and fucoid forests, which are also a source of detrital food. After settlement, juveniles may take up to a year to grow to 15–20 g. Juveniles then grow rapidly and at 2 years of age (3 years post-settlement) reach a maximum length of ~25 cm and an average weight of ~116 g. Average population longevity is ~3–4 years. Sexual maturity is reached after a year as a juvenile and at a weight approaching 75 g, with a 1:1 sex ratio. The annual spawning season extends from November to February, when eggs and sperm are shed into the water, where fertilisation takes place. Larval life stages take 3–4 weeks to settlement and the early juvenile period is 4–12 months according to food availability and other seasonal influences. In Andrew Morgan's (2012) study, the natural mortality rate varied seasonally, reaching a maximum rate of 40–44% a year in late autumn to winter and falling to zero for the rest of the year.

Ecology

Like other holothuroids, the southern sea cucumber feeds on detrital deposits by extending the branched tentacles, sweeping them over the sediment surface, and transferring the adherent particulate matter to the mouth. Feeding is mostly at night and adults move up to a few metres a night over the bottom as they gather detritus.

Sea cucumbers have developed several defences against predation, including a thick body wall containing toxins, nocturnal activity, cryptic behaviour and evisceration and the discharge of sticky threads (called Cuvierian tubules) – factors which together are thought to explain the few known predators (Francour 1997). For the southern sea cucumber, the only verified predators are the sea-star *Luidia* and grubfishes (*Parapercis* spp.).

When a sea cucumber eviscerates, it ejects its alimentary canal, part of the respiratory tree and gonad – a phenomenon that seems to occur when an animal is subjected to severe stress. As yet, there is no satisfactory explanation for evisceration and it may simply reflect a pathological condition. If the sea cucumber survives, regeneration is a slow process, and Dawbin (1949) found that it was only partially completed after some 5 months, although animals were able to ingest food after 3 months. Some species, but not *Australostichopus*, have Cuvierian tubules. The distal ends of these tubules float freely inside the anus, so that if a sea cucumber is irritated, it opens its anus, and emits the tubules which lengthen, become very sticky and immobilise the cause of irritation.

REFERENCES

Andrew NL (1989) Contrasting ecological implications of food limitation in sea urchins and herbivorous gastropods. *Marine Ecology Progress Series* **51**, 189–193.

Andrew NL (1991) Changes in subtidal habitat following mass mortality of sea urchins in Botany Bay, New South Wales. *Australian Journal of Ecology* **16**, 353–362.

Andrew NL (1993) Spatial heterogeneity, sea urchin grazing, and habitat structure on reefs in temperate Australia. *Ecology* **74**, 292–302.

Andrew NL, Byrne M (2007) Ecology of *Centrostephanus*. In *Edible Sea Urchins: Biology and Ecology. Developments in Aquaculture and Fisheries Science. Vol. 37* (2nd edn) (Ed. JM Lawrence) pp. 191–204. Elsevier, Amsterdam.

Andrew NL, Constable A (1999) Sea urchins. In *Under Southern Seas: the Ecology of Australia's Rocky Shores*. (Ed. N Andrew) pp. 127–136. University of New South Wales Press, Sydney.

Andrew NL, O'Neill T (2000) Large-scale patterns in habitat structure on subtidal rocky reefs in New South Wales. *Marine and Freshwater Research* **51**, 255–263.

Andrew NL, Underwood AJ (1989) Patterns of abundance of the sea urchin *Centrostephanus rodgersii* (Agassiz) on the central coast of New South Wales, Australia. *Journal of Experimental Marine Biology and Ecology* **131**, 61–80.

Arrontes J, Underwood AJ (1991) Experimental studies on some aspects of the feeding ecology of the intertidal starfish *Patiriella exigua*. *Journal of Experimental Marine Biology and Ecology* **148**, 255–269.

Baker AN (1982) Brittle-stars (Class Ophiuroidea). In *Marine Invertebrates of Southern Australia. Part I.* (Eds SA Shepherd and IM Thomas) pp. 418–436. Government Printer, Adelaide.

Banks S, Piggott MP, Williamson JE, Boye U *et al.* (2007) Oceanic variability and coastal topography shape genetic structure in a long-dispersing sea urchin. *Ecol ogy* **88**, 3055–3064.

Barker MF (1977) Observations on the settlement of the brachiolaria larvae of *Stichaster australis* (Verrill) and *Coscinasterias calamaria* (Gray) (Echinodermata: Asteroidea) in the laboratory and on the shore. *Journal of Experimental Marine Biology and Ecology* **30**, 95–108.

Barker MF (1978) Descriptions of the larvae of *Stichaster australis* (Verrill) and *Coscinasterias calamaria* (Gray) (Echinodermata: Asteroidea) from New Zealand obtained from laboratory culture. *The Biological Bulletin* **154**, 32–46.

Barrett NS, Buxton CD, Edgar GJ (2009) Changes in invertebrate and macroalgal populations in Tasmanian marine reserves in the decade following protection. *Journal of Experimental Marine Biology and Ecology* **370**, 104–119.

Beck HJ, Styan CA (2010) Colour patterns in the sea urchin, *Heliocidaris erythrogramma*, suggest limited connectivity across the southern and Pacific Ocean coastlines of Australia. *Marine and Freshwater Research* **61**, 143–152.

Bruce AJ (1980) *Periclimenes carinidactylus* Bruce, a crinoid-associated Pontoniine shrimp, from South Australia. *Crustaceana* **38**, 295–305.

Butler AJ (2008) Fauna of jetty piles, artificial reefs and biogenic surfaces. In *Natural History of Gulf St Vincent* (Eds SA Shepherd, S Bryars, IR Kirkegaard, P Harbison and JT Jennings) pp. 279–296. Royal Society of South Australia, Adelaide.

Byrne M (1996) Viviparity and intragonadal cannibalism in the diminutive asterinid sea stars *Patiriella vivipara* and *P. parvivipara*. *Marine Biology* **125**, 551–567.

Byrne M (2006) Life history diversity and evolution in the Asterinidae. *Integrative and Comparative Biology* **46**, 243–254.

Byrne M, O'Hara TD, Lawrence JM (2013) *Asterias amurensis*. In *Starfish: Biology and Ecology of the Asteroidea*. (Ed. JM Lawrence) pp. 174–180. John Hopkins University Press, Baltimore.

Byrne M, Barker MF (1991) Embryogenesis and larval development of the asteroid *Patiriella regularis* viewed by light and electron microscopy. *The Biological Bulletin* **180**, 332–345.

Byrne M, Cerra A (1996) Evolution of intragonadal development in the diminutive asterinid sea stars *Patiriella vivipara* and *P. parvivipara* with an overview of development in the Asterinidae. *The Biological Bulletin* **191**, 17–26.

Byrne M, Fontaine AR (1981) The feeding behaviour of *Florometra serratissima* (Echinodermata: Crinoidea). *Canadian Journal of Zoology* **59**, 11–18.

Byrne M, Fontaine AR (1983) Morphology and function of the tube feet of *Florometra serratissima* (Echinodermata: Crinoidea). *Zoomorphology* **102**, 175–187.

Byrne M, Voltzow J (2004) Morphological evolution in sea urchin development: hybrids provide insights into the pace of evolution. *BioEssays* **26**, 343–347.

Byrne M, Morrice MG, Wolf B (1997) Introduction of the northern Pacific asteroid *Asterias amurensis* to Tasmania: reproduction and current distribution. *Marine Biology* **127**, 673–685.

Byrne M, Andrew NL, Worthington DG, Brett PA (1998) Reproduction in the diadematid sea urchin *Centrostephanus rodgersii* in contrasting habitats along the coast of New South Wales, Australia. *Marine Biology* **132**, 305–318.

Coleman N (1977) *A Field Guide of Australian Marine Life*. Rigby, Adelaide.

Connell SD (2008) Water quality and loss of coral reefs and kelp forests: alternative states and the influence of fishing. In *Marine Ecology* (Eds SD Connell and BM Gillanders) pp. 556–568. Oxford University Press, Melbourne.

Connolly R (1986) Behaviour and ecology of the sea urchin *Heliocidaris erythrogramma* (Valenciennes). BSc (Hons) thesis. University of Adelaide, Adelaide.

Constable AJ (1989) An investigation of resource allocation in the sea urchin *Heliocidaris erythrogramma* (Valenciennes). PhD thesis. University of Melbourne, Melbourne.

Constable AJ (1993) The role of sutures in shrinking the test in *Heliocidaris erythrogramma* (Echinoidea: Echinometridae). *Marine Biology* **117**, 423–430.

Crump RG, Barker MF (1985) Sexual and asexual reproduction in geographically separated populations of the fissiparous asteroid *Coscinasterias calamaria* (Gray). *Journal of Experimental Marine Biology and Ecology* **88**, 109–127.

Davis AR, Fyfe SK, Turon X, Uriz MJ (2003) Size matters sometimes: wall height and the structure of subtidal benthic invertebrate assemblages in south-eastern Australia and Mediterranean Spain. *Journal of Biogeography* **30**, 1797–1807.

Dawbin WH (1949) Auto-evisceration and the regeneration of viscera in the Holothurian *Stichopus mollis* (Hutton). *Transactions of the Royal Society of New Zealand* **77**, 497–528.

Day RW, Dowell A, Sant G, Klemke J *et al.* (1995) Patchy predation: foraging behaviour of *Coscinasterias calamaria* and escape responses of *Haliotis rubra*. *Marine and Freshwater Behaviour and Physiology* **26**, 11–33.

De Ridder C, Lawrence JM (1982) Food and feeding mechanisms: Echinoidea. In *Echinoderm Nutrition* (Eds M Jangoux and JM Lawrence) pp. 57–116. Balkema, Rotterdam.

Deheyn D, Jangoux M (1999) Colour varieties as sibling species in the polychromatic ophiuroid *Amphipholis squamata* (Echinodermata): evidence from inheritance of body colour and luminescence characters. *Journal of Experimental Marine Biology and Ecology* **234**, 219–234.

Deheyn D, Mallefet J, Jangoux M (1997) Intraspecific variations of bioluminescence in a polychromatic population of *Amphipholis squamata* (Echinodermata: Ophiuroidea). *Journal of the Marine Biological Association of the United Kingdom* **77**, 1213–1222.

Deheyn D, Mallefet J, Jangoux M (2000) Evidence of seasonal variation in bioluminescence of *Amphipholis squamata* (Echinodermata: Ophiuroidea): effects of environmental factors. *Journal of Experimental Marine Biology and Ecology* **245**, 245–264.

Dix TG (1970) Biology of *Evechinus chloroticus (Echinoidea: Echinometridae)* from different localities. 2. Movement. *New Zealand Journal of Marine and Freshwater Research* **4**, 267–277.

Dix TG (1977) Reproduction in Tasmanian populations of *Heliocidaris erythrogramma (Echinodermata: Echinometridae). Australian Journal of Marine and Freshwater Research* **28**, 509–520.

Dixon CD (2011) Experimental approach for understanding density dependence of greenlip abalone (*Haliotis laevigata*) populations. PhD thesis. University of Melbourne, Melbourne.

Duyverman H (1976) Factors influencing the local distribution and abundance of the sea star *Patiriella exigua* L. MSc thesis. Flinders University, Adelaide.

Ebert TA (1982) Longevity, life history, and relative body wall size in sea urchins. *Ecological Monographs* **52**, 353–394.

Edgar GJ (2008) *Australian Marine Life: the Plants and Animals of Temperate Waters*. 2nd edn. New Holland, Sydney.

Emson RH, Whitfield PJ (1989) Aspects of the life history of a tide pool population of *Amphipholis squamata* (Ophiuroidea) from south Devon. *Journal of the Marine Biological Association of the United Kingdom* **69**, 27–41.

Emson RH, Wilkie IC (1982) The arm-coiling response of *Amphipholis squamata* (Delle Chiaje). In *Echinoderms: Proceedings of the 4th International Echinoderm Conference*.14–17 September 1981, Tampa Bay. (Ed. JM Lawrence) pp. 11–18. A.A. Balkema, Rotterdam.

Falkner I, Byrne M (2003) Reproduction of *Ophiactis resiliens* (Echinodermata: Ophiuroidea) in New South Wales with observations on recruitment. *Marine Biology* **143**, 459–466.

Falkner I, Byrne M (2004) Seasonal trends in population density of *Ophiactis resiliens* (Echinodermata: Ophiuroidea) in the Sydney region, Australia. In *Proceedings of the 11th International Echinoderm Conference. Echinoderms*. 6–10 October 2003, Munich (Eds T Heinzeller and JH Nebelsick) pp. 183–185. Balkema Publishers, Leiden.

Falkner I, Byrne M (2006) Skeletal characters for identification of juvenile *Ophiactis resiliens* and *Amphiura constricta* (Echinodermata): cryptic ophiuroids in coralline turf habitat. *Journal of the Marine Biological Association of the United Kingdom* **86**, 1199–1207.

Falkner I, Byrne M, Sewell M (2006) Maternal provisioning in *Ophionereis fasciata* and *O. schayeri*: brittle stars with contrasting modes of development. *The Biological Bulletin* **211**, 204–207.

Fell HB (1946) The embryology of the viviparous ophiuroid *Aphipholis squamata* Delle Chiaje. *Transactions of the Royal Society of New Zealand* **75**, 419–464.

Francour P (1997) Predation on holothurians: a literature review. *Invertebrate Biology* **116**, 52–60.

Grice AJ, Lethbridge RC (1988) Reproductive studies on *Patiriella gunnii* (asteroidea: Asterinidae) in south-western Australia. *Australian Journal of Marine and Freshwater Research* **39**, 399–407.

Grober MS (1988) Brittle-star bioluminescence functions as an aposematic signal to deter crustacean predators. *Animal Behaviour* **36**, 493–501.

Growns JE, Ritz DA (1994) Colour variation in southern Tasmanian populations of *Heliocidaris erythrogramma* (Echinodermata: Echinoidea). *Australian Journal of Marine and Freshwater Research* **45**, 233–242.

Guest MA, Frusher SD, Nichols PD, Johnson CR *et al.* (2009) Trophic effects of fishing southern rock lobster *Jasus edwardsii* shown by combined fatty acid and stable isotope analyses. *Marine Ecology Progress Series* **388**, 169–184.

Haig JA, Rouse GW (2008) Larval development of the featherstar *Aporometra wilsoni* (Echinodermata: Crinoidea). *Invertebrate Biology* **127**, 460–469.

Helgen LE, Rouse GW (2006) Species delimitation and distribution in *Aporometra* (Crinoidea: Echinodermata): endemic Australian featherstars. *Invertebrate Systematics* **20**, 395–414.

Hill NA, Blount C, Poore AGB, Worthington D *et al.* (2003) Grazing effects of the sea urchin *Centrostephanus rodgersii* in two contrasting rocky reef habitats: effects of urchin density and its implications for the fishery. *Marine and Freshwater Research* **54**, 691–700.

Hirst A (2012) Decadal shifts in demersal fish assemblage structure in Port Phillip Bay correspond with drought and the introduction of an exotic starfish. Abstract in *Australian Society for Fish Biology and Oceania Chondrichthyan Society 2012 Joint Conference and Symposium*. April 2012, Adelaide.

Huggett MJ, King CK, Williamson JE, Steinberg PD (2005) Larval development and metamorphosis of the Australian diadematid sea urchin *Centrostephanus rodgersii. Invertebrate Reproduction & Development* **47**, 197–204.

Huggett MJ, Williamson JE, de Nys R, Kjelleberg S *et al.* (2006) Larval settlement of the common Australian sea urchin *Heliocidaris erythrogramma* in response to bacteria from the surface of coralline algae. *Oecologia* **149**, 604–619.

Huggett MJ, Crocetti GR, Kjelleberg S, Steinberg PD (2008) Recruitment of the sea urchin *Heliocidaris erythrogramma* and the distribution and abundance of inducing bacteria in the field. *Aquatic Microbial Ecology* **53**, 161–171.

Hutson KS, Ross DJ, Day RW, Ahern JJ (2005) Australian scallops do not recognise the introduced predatory seastar *Asterias amurensis. Marine Ecology Progress Series* **298**, 305–309.

Hyman LH (1955) *The Invertebrates. IV. Echinodermata*. McGraw-Hill, New York.

Jackson AC, Murphy RJ, Underwood AJ (2009) *Patiriella exigua*: grazing by a starfish in an overgrazed intertidal system. *Marine Ecology Progress Series* **376**, 153–163.

Johnson MS, Threlfall TJ (1987) Fissiparity and population genetics of *Coscinasterias calamaria*. *Marine Biology* **93**, 517–525.

Johnson C, Ling S, Ross J, Shepherd SA *et al.* (2005) *Establishment of the Long-Spined Sea Urchin (*Centrostephanus Rodgersii*) in Tasmania: First Assessment of Potential Threats to Fisheries.* FRDC Project 2001/044. TAFI, University of Tasmania, Hobart.

Jones GP, Andrew NL (1990) Herbivory and patch dynamics on rocky reefs in temperate Australasia: the roles of fish and sea urchins. *Australian Journal of Ecology* **15**, 505–520.

Jones MB, Smaldon G (1989) Aspects of the biology of the cosmopolitan brittlestar *Amphipholis squamata* (Echinodermata) from the Firth of Forth, Scotland. *Journal of Natural History* **23**, 613–625.

Kangas M, Shepherd SA (1985) Distribution and feeding of chitons in a boulder habitat at West Island, South Australia. *Journal of the Malacological Society of Australia* **6**, 101–111.

Keesing JK (2006) Ecology of *Heliocidaris erythrogramma*. In *Edible Sea Urchins: Biology and Ecology.* 2nd edn. Developments in Aquaculture and Fisheries Science **37**, 339–351.

Keough MJ, Butler AJ (1979) The role of asteroid predators in the organization of a sessile community on pier pilings. *Marine Biology* **51**, 167–177.

King CK, Hoegh-Guldberg O, Byrne M (1994) Reproductive cycle of *Centrostephanus rodgersii* (Echinoidea), with recommendations for the establishment of a sea urchin fishery in New South Wales. *Marine Biology* **120**, 95–106.

Klemke J (1985) Habitats and diets of adults and juveniles of the sea-star *Coscinasterias calamaria* (Gray) in Port Phillip Bay. BSc (Hons) thesis. University of Melbourne, Melbourne.

Komatsu M, O'Loughlin PM, Bruce B, Yoshizawa H *et al.* (2006) A gastric-brooding asteroid *Smilasterias multipara*. *Zoological Science* **23**, 699–705.

Laegdsgaard P, Byrne M, Anderson DT (1991) Reproduction of sympatric populations of *Heliocidaris erythrogramma* and *H. tuberculata* (Echinoidea) in New South Wales. *Marine Biology* **110**, 359–374.

Lamare MD, Channon T, Cornelisen C, Clarke M (2009) Archival electronic tagging of a predatory sea star—testing a new technique to study movement at the individual level. *Journal of Experimental Marine Biology and Ecology* **373**, 1–10.

Lawson-Kerr C, Anderson DT (1978) Reproduction, spawning and development of the starfish *Patiriella exigua* (Lamarck) (Asteroidea: Asterinidae) and some comparisons with *P. calcar*. *Australian Journal of Marine and Freshwater Research* **29**, 45–53.

Le Gac M, Féral J-P, Poulin E, Veyret M *et al.* (2004) Identification of allopatric clades in the cosmopolitan ophiuroid species complex *Amphipholis squamata* (Echinodermata). The end of a paradox? *Marine Ecology Progress Series* **278**, 171–178.

Ling SD (2008) Range expansion of a habitat-modifying species leads to loss of taxonomic diversity: a new and impoverished reef state. *Oecologia* **156**, 883–894.

Ling SD, Johnson CR (2009) Population dynamics of an ecologically important range-extender: kelp beds versus sea urchin barrens. *Marine Ecology Progress Series* **374**, 113–125.

Ling SD, Johnson CR, Frusher S, King CK (2008) Reproductive potential of a marine ecosystem engineer at the edge of a newly expanded range. *Global Change Biology* **14**, 907–915.

Ling SD, Johnson CR, Ridgway K, Hobday AJ *et al.* (2009a) Climate-driven range extension of a sea urchin: inferring future trends by analysis of recent population dynamics. *Global Change Biology* **15**, 719–731.

Ling SD, Johnson CR, Frusher SD, Ridgway KR (2009b) Overfishing reduces resilience of kelp beds to climate-driven catastrophic phase shift. *Proceedings of the National Academy of Science* **106**, 22341–22345.

Livore JP, Connell SD (2012a) Effects of food origin and availability on sea urchin condition and feeding behaviour. *Journal of Sea Research* **68**, 1–5.

Livore JP, Connell SD (2012b) Reducing per capita food supply alters urchin condition and habitat. *Marine Biology* **159**, 967–973.

Livore JP, Connell SD (2012c) Fine-scale effects of sedentary urchins on canopy and understory algae. *Journal of Experimental Marine Biology and Ecology* **411**, 66–69.

McHugh D, Rouse GW (1998) Life history evolution of marine invertebrates: new views from phylogenetic systematics. *Trends in Evolution and Ecology* **13**, 182–186.

McShane P, Smith M (1986) Starfish vs abalone in Port Phillip Bay. *Australian Fisheries* **45**, (4), 16–18.

Mladenov PV (1996) Environmental factors influencing asexual reproductive processes in echinoderms. *Oceanologica Acta* **19**, 227–235.

Morgan AD (2008) The effect of food availability on phenotypic plasticity in larvae of the temperate sea cucumber *Australostichopus mollis*. *Journal of Experimental Marine Biology and Ecology* **363**, 89–95.

Morgan AD (2009) Spawning of the temperate sea cucumber *Australostichopus mollis* (Levin). *Journal of the World Aquaculture Society* **40**, 363–373.

Morgan AD (2011) Patterns of distribution and abundance of the temperate sea cucumber *Australostichopus mollis* on a rocky subtidal reef. *New Zealand Journal of Zoology* **38**, 195–206.

Morgan AD (2012) Use of a growth model to estimate size at age in the temperate sea cucumber *Australostichopus mollis. SPC Beche-de-Mer Information Bulletin* **32**, 24–32.

Morris VB (1995) Apluteal development in the sea urchin *Holopneustes purpurescens* Agassiz (Echinodermata: Echinoidea: Euechinoidea) *Zoological Journal of the Linnean Society* **114**, 349–364.

Naughton KM, O'Hara TD (2009) A new brooding species of the biscuit star *Tosia* (Echinodermata: Asteroidea: Goniasteridae) distinguished by molecular, morphological and larval characters. *Invertebrate Systematics* **23**, 348–366.

Naumann JR (2007) Herbivory on rocky reefs: just drifting and scraping a living. BSc (Hons) thesis. University of Adelaide, Adelaide.

NIMPIS (2002) *National Introduced Marine Pest Information System.* Australian Government, Canberra. <www.marinepests.gov.au/nimpis> (Accessed February 2013).

O'Connor C (1976) A study of the reproductive periodicities and population dynamics of some New South Wales sea urchins with a view to open-sea culture. *Proceedings of the 10th European Symposium on Marine Biology in Ostend. Vol. 1.* 17–23 September 1975, Ostend, Belgium (Eds G Persoone and E Jaspers) pp. 285–301. IZWO, Wetteren.

O'Connor C, Riley G, Lefevre S, Bloom D (1978) Environmental influences on histological changes in the reproductive cycle of four New South Wales sea urchins. *Aquaculture* **15**, 1–17.

Parks AL, Bisgrove BW, Wray GA, Raff RA (1989) Direct development in the sea urchin *Phyllacanthus parvispinus* (Cidaroidea): phylogenetic history and functional modification. *The Biological Bulletin* **177**, 96–109.

Pearse JS, Mooi R, Lockhart SJ, Brandt A (2009) Brooding and species diversity in the Southern Ocean: selection for brooders or speciation within brooding clades? In *Smithsonian at the Poles: Contributions to International Polar Year Science.* (Eds I Krupnik, MA Lang and SE Miller) pp. 181–196. Smithsonian Institution Scholarly Press, Washington DC.

Pederson HG, Johnson CR (2006) Predation of the sea urchin *Heliocidaris erythrogramma* by southern rock lobsters (*Jasus edwardsii*) in no-take marine reserves. *Journal of Experimental Marine Biology and Ecology* **336**, 120–134.

Pederson HG, Johnson CR (2008) Growth and age structure of sea urchins (*Heliocidaris erythrogramma*) in complex barrens and native macroalgal forests in eastern Tasmania. *ICES Journal of Marine Science* **65**, 1–11.

Pennington JT (1985) The ecology of fertilization of echinoid eggs: the consequences of sperm dilution, adult aggregation, and synchronous spawning. *The Biological Bulletin* **169**, 417–430.

Rodda KR, Mayfield S, Shepherd SA (2002) 'Translocation of greenlip abalone in South Australia'. Report to PIRSA Fisheries Policy Group. SARDI Aquatic Sciences, Adelaide.

Rogers CN, Williamson JE, Carson DG, Steinberg PD (1998) Diel vertical movement by mesograzers on seaweeds. *Marine Ecology Progress Series* **166**, 301–306.

Ross DJ, Johnson CR, Hewitt CL (2006) Abundance of the introduced seastar, *Asterias amurensis*, and spatial variability in soft sediment assemblages in SE Tasmania: clear correlations but complex interpretation. *Estuarine, Coastal and Shelf Science* **67**, 695–707.

Sanderson JC, Le Rossignol M, James W (1996) 'A pilot program to maximise Tasmania's sea urchin (*Heliocidaris erythrogramma*) resource'. FRDC Report. Project 93/221. Tasmanian Sea Urchin Developments & Marine Environmental Systems, Hobart.

Selvakumaraswamy P, Byrne M (2000) Reproduction, spawning, and development of five ophiuroids from Australia and New Zealand. *Invertebrate Biology* **119**, 394–402.

Selvakumaraswamy P, Byrne M (2004) Metamorphosis and developmental evolution in *Ophionereis* (Echinodermata: Ophiuroidea). *Marine Biology* **145**, 87–99.

Sewell MA (1990) Aspects of the ecology of *Stichopus mollis* (Echinodermata: Holothuroidea) in north-east New Zealand. *New Zealand Journal of Marine and Freshwater Research* **24**, 97–103.

Shepherd SA (1967) A revision of the starfish genus *Uniophora* (Asteroidea, Asteriidae). *Transactions of the Royal Society of South Australia* **91**, 1–15.

Shepherd SA (1968) The shallow water echinoderm fauna of South Australia. 1. The asteroids. *Records of the South Australian Museum* **15**, 729–756.

Shepherd SA (1973) Studies on southern Australian abalone (genus *Haliotis*). I. Ecology of five sympatric species. *Australian Journal of Marine and Freshwater Research* **24**, 217–257.

Shepherd SA (1974) 'An underwater survey near Crag Point in upper Spencer Gulf'. Technical Report No. 1. South Australian Department of Fisheries, Adelaide.

Shepherd SA, Marshall J, Rowe FWE (1982) Feather-stars (Class Crinoidea). In *Marine Invertebrates of Southern Australia. Part 1.* (Eds SA Shepherd and IM Thomas) pp. 396–400. Government Printer, Adelaide.

Sköld M, Barker MF, Mladenov PV (2002) Spatial variability in sexual and asexual reproduction of the fissiparous seastar *Coscinasterias muricata*: the role of food and fluctuating temperature. *Marine Ecology Progress Series* **233**, 143–155.

Slater MJ, Jeffs AG (2010) Do benthic sediment characteristics explain the distribution of juveniles of the deposit-feeding sea cucumber *Stichopus mollis? Journal of Sea Research* **64**, 241–249.

Sloan NA (1980) Aspects of the feeding biology of asteroids. *Oceanography and Marine Biology - an Annual Review* **18**, 57–124.

Soars NA, Prowse TAA, Byrne M (2009) Overview of phenotypic plasticity in echinoid larvae, '*Echinopluteus transversus*' type vs typical echinoplutei. *Marine Ecology Progress Series* **383**, 113–125.

Steinberg PD (1995) Interaction between the canopy dwelling echinoid *Holopneustes purpurascens* and its host kelp *Ecklonia radiata. Marine Ecology Progress Series* **127**, 169–181.

Stevenson JP (1992) A possible modification of the distribution of the intertidal seastar *Patiriella exigua* (Lamarck) (Echinodermata: Asteroidea) by *Patirella calcar* (Lamarck). *Journal of Experimental Marine Biology and Ecology* **155**, 41–54.

Styan CA (1997) Inexpensive and portable sampler for collecting eggs of free-spawning marine invertebrates underwater. *Marine Ecology Progress Series* **150**, 293–296.

Swanson RL, de Nys R, Huggett MJ, Green JK *et al.* (2006) *In situ* quantification of a natural settlement cue and recruitment of the Australian sea urchin *Holopneustes purpurascens. Marine Ecology Progress Series* **314**, 1–14.

Taylor RB, Steinberg PD (2005) Host use by Australasian seaweed mesograzers in relation to feeding preferences of larger grazers. *Ecology* **86**, 2955–2967.

Uthicke S, Schaffelke B, Byrne M (2009) A boom-bust phylum? Ecological and evolutionary consequences of density variations in echinoderms. *Ecological Monographs* **79**, 3–24.

Vanderklift MA, Kendrick GA (2005) Contrasting influence of sea urchins on attached and drift macroalgae. *Marine Ecology Progress Series* **299**, 101–110.

Vanderklift MA, Wernberg T (2008) Detached kelps from distant sources are a food subsidy for sea urchins. *Oecologia* **157**, 327–335.

Vanderklift MA, Wernberg T (2010) Stable isotopes reveal a consistent consumer-diet relationship across hundreds of kilometres. *Marine Ecology Progress Series* **403**, 53–61.

Vanderklift MA, Kendrick GA, Smit AJ (2006) Differences in trophic position among sympatric sea urchin species. *Estuarine, Coastal and Shelf Science* **66**, 291–297.

Vanderklift MA, Lavery PS, Waddington KI (2009) Intensity of herbivory on kelp by fish and sea urchins differs between inshore and offshore reefs. *Marine Ecology Progress Series* **376**, 203–211.

Viviani D (1805) *Phosphorencia Maris Quattuordecim Lucescentium Animalculorum Novis Speciebus Illustrata*. Genova, Italy.

Williamson JE, Steinberg PD (2002) Reproductive cycle of the sea urchin *Holopneustes purpurascens* (Temnopleuridae: Echinodermata). *Marine Biology* **140**, 519–532.

Williamson JE, de Nys R, Kumar N, Carson DG *et al.* (2000) Induction of metamorphosis in the sea urchin *Holopneustes purpurascens* by a metabolite complex from the algal host *Delisea pulchra. The Biological Bulletin* **198**, 332–345.

Williamson JE, Carson DG, de Nys R, Steinberg PD (2004) Demographic consequences of an ontogenetic shift by a sea urchin in response to host plant chemistry. *Ecology* **85**, 1355–1371.

Wright JT, Steinberg PD (2001) Effects of variable recruitment and post-recruitment herbivory on local population size of a marine alga. *Ecology* **82**, 2200–2215.

Wright JT, Dworjanyn SA, Rogers CN, Steinberg PD *et al.* (2005) Density-dependent sea urchin grazing: differential removal of species, changes in community composition and alternative community states. *Marine Ecology Progress Series* **298**, 143–156.

12 Effects of sea-urchins on benthic habitats

'The sea urchin is a most voracious and promiscuous feeder, always ready to eat ...'
TR Jones, 1858

OVERVIEW

The black sea-urchin *Centrostephanus rodgersii* has had a dramatic effect on subtidal habitats in New South Wales and eastern Tasmania by grazing down macroalgal forests and creating extensive urchin barrens dominated by crustose corallines and a sparse encrusting fauna. In New South Wales, these barrens have existed at least since the establishment of the colony there, and studies have been directed towards description of the barrens and algal habitats, and their interrelations, and an understanding of the reverse process of recovery from barrens to algal forest. Since the early 1960s, the black urchin has gradually expanded its range southwards across Bass Strait and down the east Tasmanian coast, creating at first 'incipient barrens' and, over time, more extensive ones. The process of destruction of algal forests and the negative effects on the ecosystem and inshore fisheries, notably abalone, are described, and it is suggested that a mutualism exists between the black urchin and encrusting corallines, in which populations of both animal and plant benefit from the creation of barrens. The purple urchin *Heliocidaris erythrogramma* is also present in macroalgal forest habitats on exposed and sheltered coasts throughout southern Australia. The effect of the species, however, varies with habitat. On exposed coasts, where algal drift is abundant, the urchin traps the drift and has no effect on surrounding algal communities. But, on more sheltered coasts in eastern and southern Australian, where algal drift may be less abundant, the purple urchin will graze attached algae and produce 'haloes' or 'mini-barrens'. These are patches of partly cleared rock of varying area, in which some, but not all, algae are grazed. The sometimes complex, reverse transition from barrens back to algal forest is described, and may depend on other factors. In some conditions, algal forests and barrens may constitute alternative stable states.

INTRODUCTION

Sea-urchins have long been known to have a pivotal role in the structure of benthic communities and have caused the deforestation of vast areas of algal forests in many parts of the world. For this reason, they have been sometimes called ecosystem engineers, because they can seriously modify the structure of a benthic community.

The destruction of algal habitats and their replacement by encrusting corallines was first noted in Japanese waters in 1830, and termed 'isoyake' (meaning burnt rock) (Okamura 1915, cited in Ohmi 1951). Since then, Japanese and other workers have published many studies on the isoyake phenomenon, and at least 20 examples of similar deforestation due to urchins have been recorded on almost all temperate coasts of the world (reviews by Lawrence 1975; Harrold and Pearse 1987).

In Australia, isoyake areas have long been known on the coast of New South Wales, and variously called 'white rock', 'urchin barrens' or 'barrens habitat'; they are thought to have originated from overfishing of urchin predators (see Chapter 11). Here, we simply use the term 'barrens', although the term 'isoyake' has priority, and the term 'barrens' is strictly a misnomer, because many species of crustose corallines and a diverse, cryptic fauna are present in them (Chapter 6). But urchin barrens differ in size and in their characteristics, so we also use the terms 'incipient barrens' and 'haloes' to describe specific types and sizes of urchin-grazed patches.

THE BLACK URCHIN *CENTROSTEPHANUS RODGERSII*

Ecology in New South Wales

Underwood *et al.* (1991) described the main habitat types that occur along the NSW coast. Besides the *Ecklonia* forest habitat, these included:

- **a barrens habitat**, with its two basic features – a seasonally variable cover of crustose corallines and high densities of the black urchin (Plate 33)
- **a Fringe habitat**, comprising a mixed macroalgal forest, so termed because it surrounds or abuts the barrens; this habitat is colonised by fucoids, mainly *Sargassum* spp. and sometimes *Cystophora* spp. and among them plants of the kelp *Ecklonia*
- **a Turf habitat**, covered by erect coralline algae, mainly *Corallina, Haliptilon* and *Amphiroa*, up to 5–10 cm high, but without canopy species or encrusting corallines
- **a *Phyllospora* forest habitat**, which is mainly found in southern New South Wales, and limited to shallow depths. Here the black urchin sometimes occurs in densities of up to 20 urchins m^{-2}.

Another habitat, not referred to by Underwood *et al.* (1991), probably because it is transitional, is the filamentous algal (or turf) habitat, which arises after release of barrens from urchin grazing. It comprises filamentous red algae and foliose species, such as *Rhodymenia*, and soon disappears as canopy species develop or urchin grazing resumes.

Early studies described in detail the extent of the above habitats and later examined the processes involved in their persistence. Aerial surveys in 1996–97 between Port Stephens and the Victorian border (Andrew and O'Neill 2000) showed that the barrens habitat averaged ~50% cover of near-shore reefs and was strongly correlated with the density of the black urchin. Both declined near Cape Howe in southern New South Wales (Fig. 12.1). On the other hand, the Fringe habitat and the large kelp *Phyllospora* increased, while *Ecklonia* decreased, towards the south (Fig. 12.2).

The black urchin and the barrens it forms occur from a few metres depth to > 20 m deep, depending on swell, where they gradually become replaced by sponge communities. The barrens range in size from haloes of a few square metres surrounding a few crevice-living urchins to extensive areas exceeding several hectares. As discussed in Chapter 11, the black urchin is crevice-dependent, due to the presence of predators, although south of ~36°S. (near Eden), urchins are often seen outside their crevice shelters by day. Geographically, the extent of barrens declines towards northern New South Wales, and at the south near Cape Howe, but continues sporadically into eastern Victoria (e.g. Shepherd 1973). The deepest recorded urchin barrens are at 50 m depth on a high-relief granite bank (New Zealand Star Bank) in north-east (NE) Bass Strait (see Table 13.2).

The dramatic effect of the black urchin on near-shore reefs and their habitats has been examined by many authors (reviewed by Andrew and Byrne 2006). Examining the opposite process of habitat recovery after removal of urchins is one way of clarifying the role that urchins play in the system, and this method has been used by several authors. Fortuitously, a mass mortality of the black urchin occurred in Botany Bay

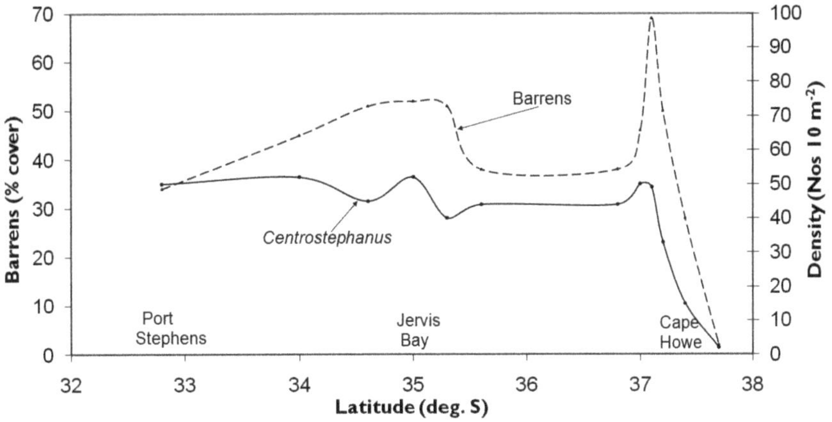

Figure 12.1: Trends in mean percentage cover of urchin barrens and density of the black urchin *Centrostephanus rodgersii* on coastal near-shore reefs of New South Wales, south from Port Stephens to Cape Howe (data extracted from Andrew and O'Neill 2000, using running means of 3).

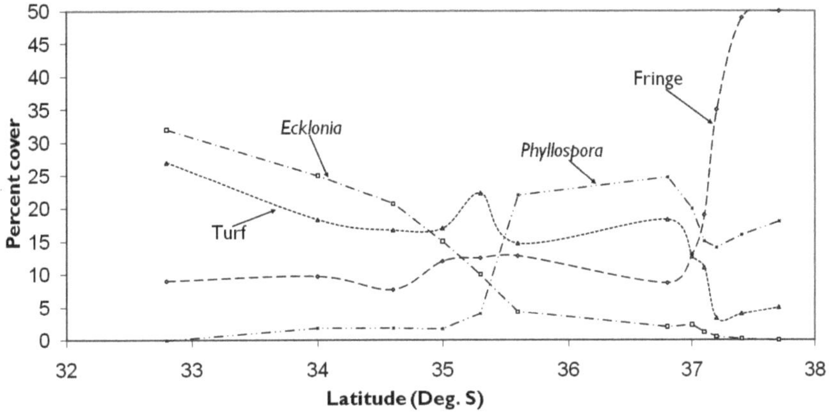

Figure 12.2: Latitudinal trends in percentage cover of the Fringe habitat, the kelps *Ecklonia* and *Phyllospora comosa* and the turf habitats at the same sites as in Fig. 12.1, on coastal near-shore reefs of New South Wales from Port Stephens to Cape Howe (data extracted from Andrew and O'Neill 2000, using running means of 3).

after floods in early August 1986 and Neil Andrew (1991) followed changes in the reef habitats over the next 2 years at several sites. The cover of crustose corallines, which was high in the barrens, fell sharply, as filamentous and foliose algal cover increased over the succeeding months, and *Sargassum* and *Ecklonia* densities increased dramatically some 6 months later. Limpets increased for a short time and then declined to near zero after 15 months. Then, in early 1987, a strong recruitment of *Centrostephanus* occurred, but these urchins died during a second flood event in April 1988. The succession of species/events is summarised in Fig. 12.3, and the sequential habitat changes following the initial urchin mortality can be summarised as follows:

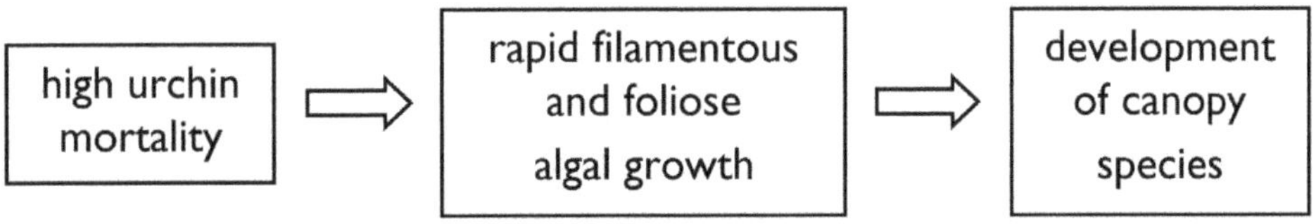

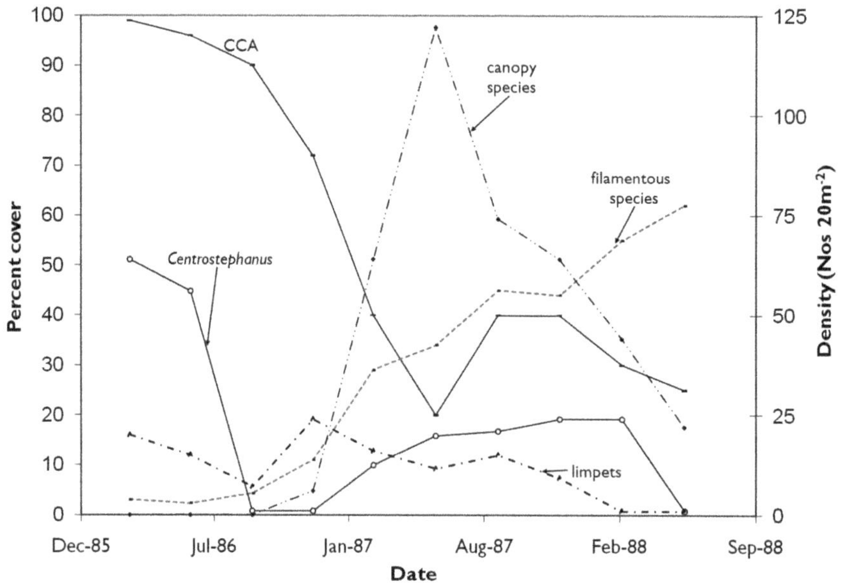

Figure 12.3: Changes over time from 1986 to April 1988 in: percentage cover of crustose corallines (CCA) (Δ); filamentous and foliose algae (-) (running mean); density of the black urchin, *Centrostephanus* (numbers 20 m^{-2}) (Δ); combined density of the canopy species, *Sargassum* and *Ecklonia* (numbers m^{-2}) (•); and the combined density (numbers 0.05 m^{-2}) of the three limpets, *Patelloida mufria*, *P. alticostata* and *Cellana tramoserica* (Δ) at Bare, I., Botany Bay. Flood events occurred in August 1986 and April 1988 (data extracted from Andrew 1991, omitting error bars).

The study actually followed the course of events at seven sites in Botany Bay, and Fig. 12.3 summarises those at Site 2, which roughly represented what happened at the others. However, there were minor differences between sites – at some sites, the kelp *Ecklonia* developed a canopy, and at others, species of *Sargassum* did, probably due to the different availability of algal spores. Because the recruitment of the black urchin resumed some time after the flood event, Neil Andrew suggested that the barrens habitat must be stable, and resilient to flood and other disturbances.

The 'natural' experiment described above confirmed a series of experiments previously carried out by Fletcher (1987) at the same site. Rick Fletcher removed *Centrostephanus* and/or limpets from urchin barrens of ~4 m^2 size surrounding crevices at two depths (~2 and 8 m) and, similar to Neil Andrew's findings, recorded a profusion of algal growth at both depths. At the shallow site, ~45% (by weight) of algal growth was erect corallines (*Haliptilon* and *Amphiroa*), 27% was red algae, mainly *Rhodymenia* and filamentous species, and almost all the rest was *Colpomenia*; at the deeper site, 93% was filamentous red algae and *Rhodymenia*, with a little *Zonaria*. There were slight differences between the inclusion and exclusion of limpets in the urchin-removal experiments. When both urchins and limpets were removed, algal cover reached ~100% in a year, but when only urchins were removed it took almost twice as long for algal cover to reach 100%, suggesting that limpets delayed the recovery process. Rick Fletcher concluded that black urchins played a keystone role in the system, and also played a facilitative role for the limpets because they modified the habitat and were ultimately necessary for limpet survival. These latter findings supported earlier suggestions by Underwood and Jernakoff (1981) that the presence of limpets depended on the grazing activity of black urchins, which maintained the crustose corallines and provided space for limpets to graze; otherwise they would be swamped by the growth of foliose algae.

Later, Andrew and Underwood (1993) examined what happened to other species and the surrounding habitats when urchin densities were reduced by varying percentages. Were the effects of urchin grazing related in a linear manner to their density? What they found was that whether urchin densities were reduced by one or two thirds, the cover of crustose corallines declined by only about one third and filamentous algae increased by ~20%, compared with ~40% with complete urchin removal. Foliose algae did not colonise patches when only some urchins were removed. These non-linearities appeared to be due to compensatory changes in the remaining urchins' foraging behaviour. Further, they suggested that either foliose algae became more susceptible to grazing, or the filamentous species grew and expanded faster. As for the limpets, removal of all black urchins saw their demise, as Fletcher (1987) had earlier found. Partial removal of urchins led to some reduction in the size of barrens patches, again not related linearly with density, but this was complicated by the extent and shape of the crevices. But what about the reverse process – the development of barrens in algal forests?

To examine this, Andrew (1993) transferred large boulders (1–2 m size) into algal forest, so creating a crevice habitat for urchins, where previously there had been none. As predicted, the black urchin recruited to the boulders, and lived in the newly created crevices, and over the next 2 years *Ecklonia* and *Sargassum* gradually disappeared, and crustose corallines increased in area, so creating small barrens around the boulders. The results showed unequivocally the central role of shelter (crevices) in the ecology of the black urchin. As a corollary to this, Andrew (1994) went on to see whether in natural situations the black urchin made forays into algal forests surrounding barrens to fell *Ecklonia* plants, or foraged only within established barrens. He found that annual mortality rates of *Ecklonia* plants were 31–59%, and mainly caused by herring cale (Chapter 16). Moreover, the mortality rates were similar in the centre and at the edge of *Ecklonia* forests, and because black urchins only grazed to the edge of the forest, they could not be a cause of the mortality of plants within the forest. So, he concluded that urchin barrens were stable over time and that the black urchins had little impact on *Ecklonia* forests outside their established barrens.

Hill *et al.* (2003) did a similar study to that of Andrew and Underwood (1993) and manipulated urchin densities in cages. In one experiment in Fringe habitat, they found that they had to increase urchin densities 2.5 fold above the natural barrens density of 4 urchins m^{-2} (i.e. to 10 m^{-2}), to convert the Fringe macroalgal forest into a barren in 20 weeks. They also confirmed that when urchins were reduced in density, only about one third of the natural barren density was sufficient to maintain the barren.

Interactions between urchins and invertebrates

Black urchin grazing can modify the abundance of both sessile invertebrates and algae, although some invertebrates can defend themselves chemically. Sponges, ascidians and bryozoans can survive in urchin barrens, although with low cover, and are particularly evident on vertical walls. At sites in southern New South Wales, Andy Davis and colleagues found that the presence of the above sessile invertebrate groups on walls depended on the wall height. Low walls < 2 m high had ~80% cover of crustose corallines, and invertebrate cover of ~20%, and high ones of 3–4 m had virtually no corallines and almost 100% cover of invertebrates. The low walls had mainly sponge species in the genera, *Darwinella, Pronax, and Mycale*, while high walls had mainly two sponges, *Euryspongia* and *Ircinia*, the bryozoan *Celleporaria*, and ascidians *Botrylloides, Didemnum* and *Sycozoa*. It seemed that the black urchins could incorporate low walls into their barrens, because they could climb and graze them at night, and return to shelter before dawn, whereas high walls were either inaccessible or lacked any shelter (or both) (Davis *et al.* 2003). Curiously, in some places, a large barnacle *Austrobalanus imperator*, up to 2.5 cm high and 4 cm across, reached densities of ~200 m^{-2} on walls, and interfered with the grazing of the black urchin, thereby creating refuges for many invertebrate species. Hence, the barnacles indirectly promoted recruitment of invertebrates around them, although recruitment rates were low, and took > 4 years to recruit a few species (Davis and Ward 2009).

Sponges were common in macroalgal forests, with a cover often of ~10%, but less common in barrens, with ~2% cover. Wright *et al.* (1997) examined whether sponges defended themselves chemically and deterred urchins from grazing on them. In imaginative experiments, they prepared agar disks with sponge extract mixed into them and fed them to black urchins (the control disks had only sea lettuce mixed into them). They found that the black urchin avoided eating disks that contained sponge extracts from the three main barrens species, but were not deterred from eating disks containing sponge extract from the four main forest-dwelling sponge species. The latter sponges contained high levels of sand in them (80–90%), and were different species from those that lived in barrens, which contained only 10–20% sand. Barrens sponges never lived in algal forests, nor forest sponges in barrens. It was evident that the sponges that lived in barrens were chemically defended, and so escaped being eaten. As an aside, the authors suggested that the sand incorporated into the forest sponges may help them resist abrasion from the whiplash of *Ecklonia* fronds.

Another once common species on southern New South Wales coasts is the blacklip abalone *Haliotis rubra*. It was early observed by Shepherd (1973) that this abalone was absent from barrens habitats, and he suggested that the black urchin out-competed the abalone for food, with the result that abalone were rarely found in barrens. Figure 12.4 shows the segregation of black urchins and abalone, as well as densities of other common herbivorous molluscs, in an urchin barren off Gabo I., far eastern Victoria. Andrew and Underwood (1992) later examined the segregation of the two herbivores at three sites south of Sydney and confirmed that they were invariably segregated – the abalone preferring the Fringe habitat or even deep habitat, as shown in Fig. 12.4, and the black urchin the barrens habitat. Subsequently, Andrew *et al.* (1998) cleared areas of the black urchin and measured the rate of increase of small blacklip abalone in those clearings over 4 years. They found that abalone numbers increased 10-fold over the period in the areas cleared of black urchins, but not in areas with urchins, confirming that black urchins excluded abalone from their barrens patches. The segregation of the two species has subsequently become of great relevance in eastern Tasmania, as discussed below.

Black urchin ecology in Bass Strait and Tasmania

As described in Chapter 2, the East Australian Current (EAC) flows southwards down the east coast of Australia, bringing warm, salty water south, and with it the larvae of the black urchin. The EAC is much stronger in summer and, when it reaches Tasmania, eddies peel off and flow westwards to collide with the coast (Ridgway and Dunn 2003). The flow of the EAC is dynamic, and historically has increased in strength since ~1750 (as estimated from the chemistry of deep-sea octocorals – see Thresher *et al.* 2004), and flowed further south, warming the coastal waters of south-east (SE) Australia. This warming trend is well shown by long-term CSIRO sea temperature data from Maria I. (Fig. 12.5). As waters have warmed, black urchin populations have gradually extended their range southwards down the New South Wales coast, probably since ~1750, and continued to Tasmania during the 1960s (Johnson *et al.* 2005, 2011).

The expansion of the black urchin's range and its implications – the resulting habitat changes – has been studied extensively by Tasmanian workers. The initial urchin invasion led to small haloes or grazed patches of a few square metres. These patches of bare substratum, surrounded by algal forest, soon became overgrown by white encrusting corallines – incipient barrens. As they gradually expanded in area, they coalesced and eventually formed extensive barrens over several hectares.

Invasion of the Kent Group of Islands, Bass Strait

The black urchin apparently first reached the Kent Group of Islands in mid Bass Strait during the 1960s. A large sample of the black urchin there was aged in 2000, and gave a range in ages of up to 18 years, with an average age of ~12 years, from which it was inferred that, by 1974, when they were well established in a few sites, they had been there for at least a decade (Johnson *et al.* 2005). This relatively early invasion of the Kent Group was unusual, but may have been due to the warm tongue of water from southern New South Wales that enters Bass Strait and bathes the islands. During the 1960s, as late winter sea temperatures began to exceed ~12°C – the development threshold of the black urchin larvae (Ling *et al.* 2009b) – the larvae would have arrived from southern New South Wales, settled there and survived.

The abundance of both the black urchin and the purple urchin and the extent of urchin barrens were measured in East Cove, Deal I. in the Kent Group

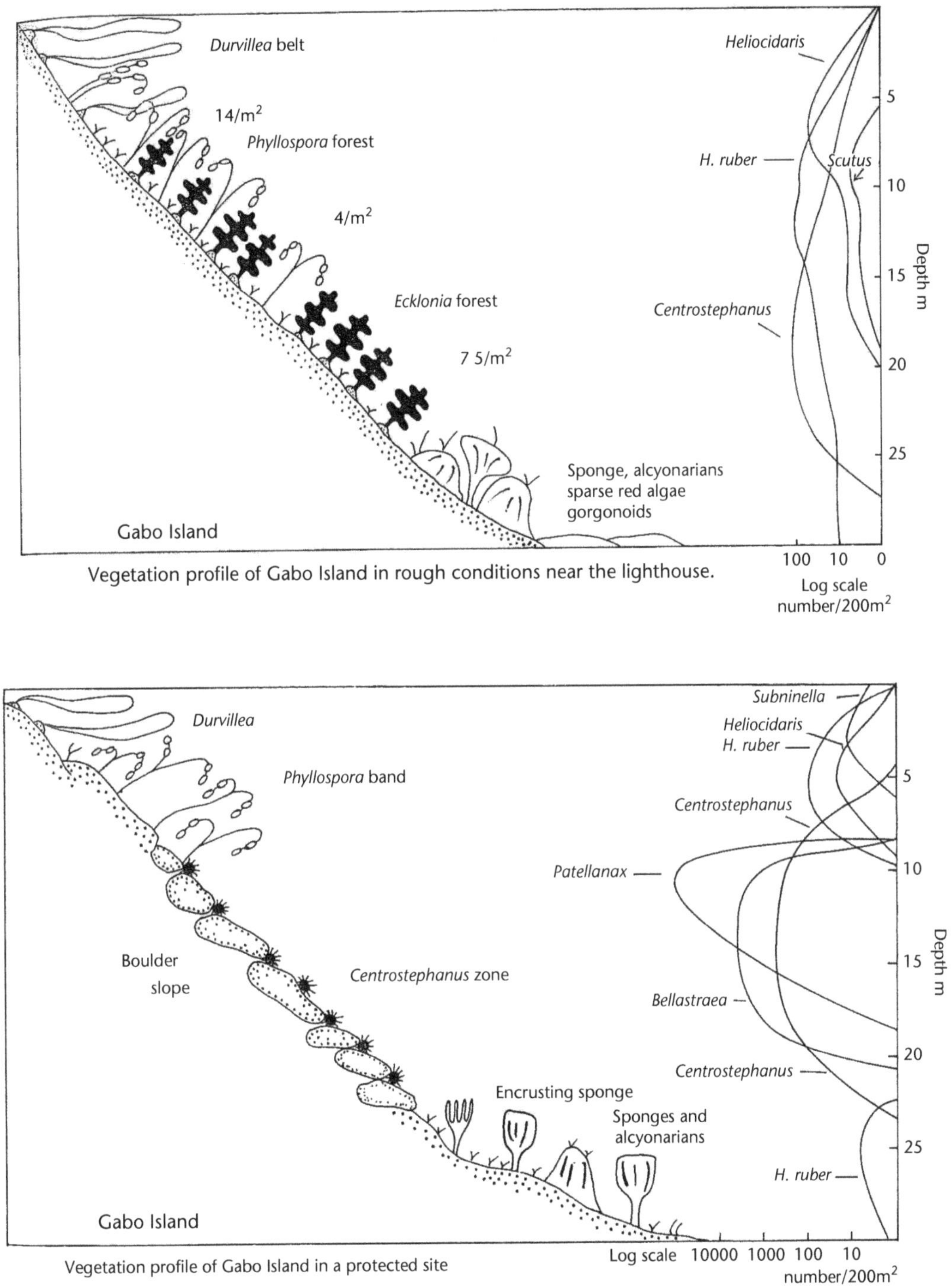

Figure 12.4: (Upper) Vegetation profile at exposed site at Gabo I. showing distribution of algal habitats on the reef slope and change in densities with depth of four herbivores in numbers per 200 m^2, the purple urchin *Heliocidaris erythrogramma*, blacklip abalone *Haliotis rubra* (formerly called *H. ruber*), elephant snail *Scutus antipodes* and the black urchin *Centrostephanus rodgersii*.
(Lower) Vegetation profile at sheltered site at Gabo I. showing distribution of habitats on the reef slope, and change in densities with depth in numbers per 200 m^2 of seven herbivores. Those species only at the sheltered site are: *Turbo undulatus* (formerly *Subninella*); *Scutellastra* (formerly *Patellanax*); and *Astralium* (formerly *Bellastraea*) (after Shepherd 1973).

during the Museum of Victoria expeditions there in 1974 and 1981, and in a later visit there in 2000 (Johnson *et al.* 2005). In East Cove, the substratum was continuous granite to a few metres depth and then comprised boulders of 0.5–2 m size to a depth of 15–20 m. In 1974, the black urchin was present in aggregations in small incipient barrens at densities of ~0.2–0.3 m^{-2}, often under a patchy canopy of *Ecklonia*, *Sargassum* and *Cystophora*, at depths of 8–11 m (Figs 12.6). By 1981, densities of the black urchin had increased below ~9 m depth and the habitat there was the reverse of that in 1974 (Fig. 12.7). The barrens, no longer in small patches, were almost continuous below ~9 m depth, and the algal forest was reduced to a few fragmented patches; and, even in 2000, remnants of the forest still persisted among the barrens. The mean densities of the black urchin over the 26-year period are shown in Fig. 12.8, and the percentage cover of barrens in Fig. 12.9. The apparent decline in cover of barrens between 1981 and 2000, despite an increase in *Centrostephanus* densities, may be an artefact of the different survey method used, or may be due to the coalescence of many small incipient barrens into a few larger ones. Elsewhere around the islands, the extent of

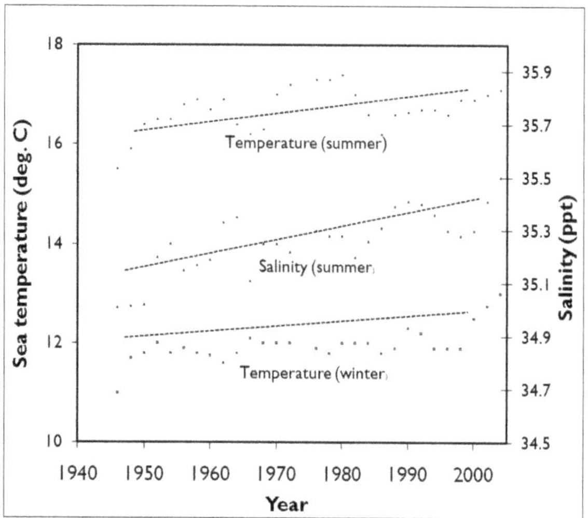

Figure 12.5: Trends in summer and winter sea water temperatures (°C), and in summer salinity (‰), with a 5 year running mean, off Maria I., SE Tasmania from 1944–2002 (redrawn from Johnson *et al.* 2005).

Figure 12.7: Expanded incipient barren cleared by the black urchin *Centrostephanus rodgersii* in East Cove, Deal I., at ~8 m depth in 1981 (photo Rob Lewis).

Figure 12.6: Small incipient barren grazed by the purple urchin *Heliocidaris erythrogramma* at ~5 m depth in East Cove, Deal I., Kent Group in 1974. Two purple urchins can be seen at right (photo Rob Lewis).

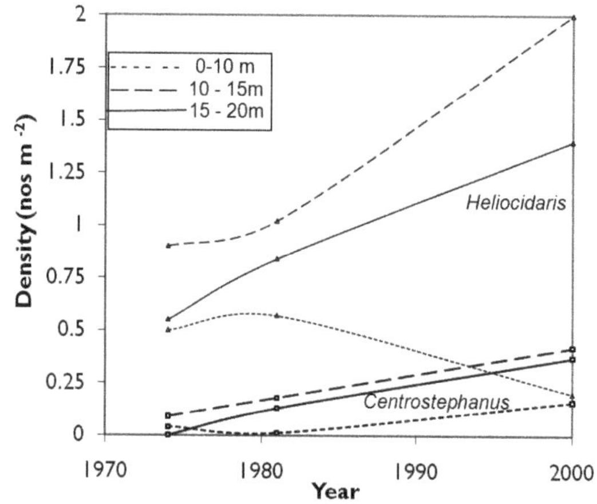

Figure 12.8: Mean density of the black urchin *Centrostephanus* (heavy lines) and the purple urchin *Heliocidaris* (light lines) over 26 years from 1974 to 2000 in East Cove, Deal I., the Kent Group, at three depths, 0–10 m, 10–15 m, and 15–20 m (redrawn from Johnson *et al.* 2005).

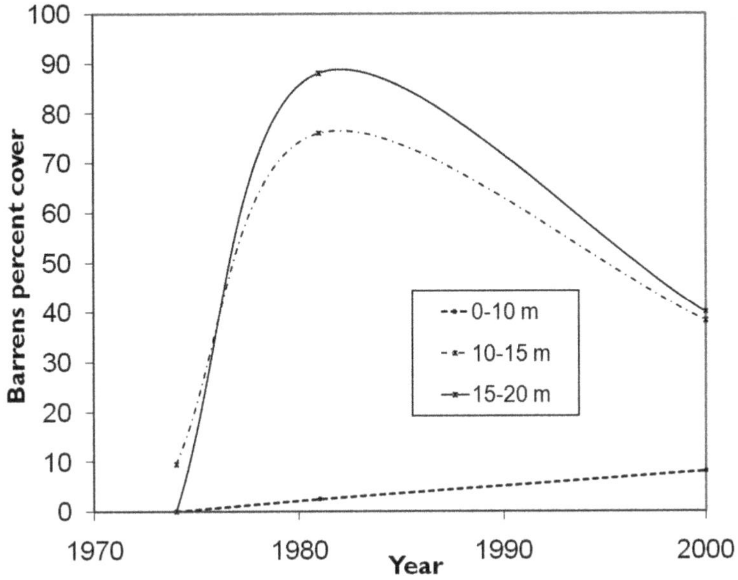

Figure 12.9: Percentage cover of urchin barrens at three depths in East Cove, Deal I., from 1974 to 2000 (redrawn from Johnson *et al.* 2005).

barrens had increased enormously since 1981, so that in 2000 the black urchin was the dominant invertebrate in the Kent Group. The extent of barrens in percentage cover (%B) was significantly correlated with the density of the black urchin (in numbers 10 m^{-2} (D_C) along transect lines, as shown by the equation:

$$\%B = 3.53D_C - 3.87\ (P < 0.0001)$$

In simple terms, this means that an area eventually became 100% barrens at a density of ~0.3 urchins m^{-2}. However, black urchin densities were much higher in incipient barrens. In a specific study of 14 such barrens over a range of sizes off NE Dover I. during the 2000 expedition, the average density of the black urchin in small barrens was ~1.2 m^{-2} (Fig. 12.10), suggesting that, over time, as incipient barrens coalesced the urchins spread out. In these barrens patches, purple urchin densities were very similar to those of the black urchin, but their inclusion lowered the slope of the regression line in Fig. 12.10, indicating that purple urchin presence did not affect the size of the barrens (see below).

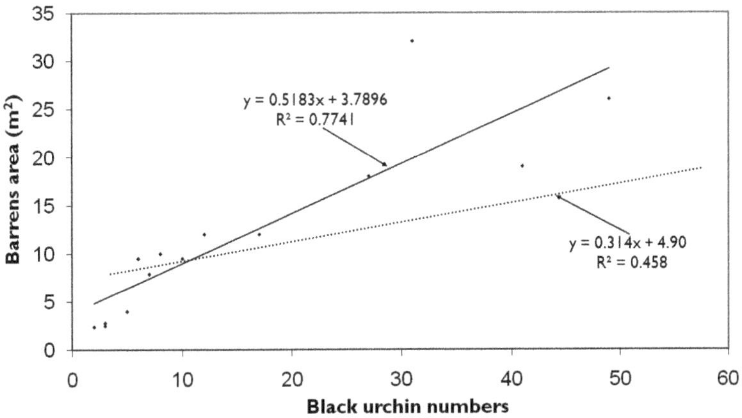

Figure 12.10: Plot of density of the black urchin *Centrostephanus* versus. area of barrens off Dover I., Kent Group in 2000, with regression (heavy line) and regression equation. The light, dotted line is the regression of the total number of black and purple urchins (*Centrostephanus* and *Heliocidaris*) present in the barrens, with regression equation (data points not shown) (unpublished data).

The consequences of the extensive barrens on the Islands' marine ecosystem were profound. Algal biomass data along a depth gradient, measured in 1974 and in 2000 (i.e. before and after the black urchin invasion) are shown in Fig. 12.11 for a site off NE Dover I. Assuming a uniform slope of ~30°, the total algal biomass of a strip 1 m wide from 0–50 m, where rock gave way to sand was calculated to be ~296 kg fresh weight in 1974 and ~197 kg in 2000: a loss of ~33%. If the barrens extend further into shallower or deeper water, as seems likely, the loss will increase substantially. Extending these calculations to the whole of the Kent Group of Islands, and conservatively estimating a coastline length of 45 km, the total loss of algal standing crop by 2000 was 4450 t – a serious loss of productivity.

The abundance of the purple urchin also increased over the period, 1974–2000 throughout the islands, but its density was unrelated to the percentage cover of barrens. However, purple urchin density, D_H, was inversely related to black urchin density, D_C (Johnson *et al.* 2005, 2011), as shown by the 95th quartile regression equation:

$$D_H = \exp(-0.14\, D_C + 2.75\ (P = 0.056)$$

This arose partly because the purple urchin was more abundant in shallower depths, where the black urchin was less common, and less abundant in deeper water, where the black urchin was dominant. At places where the purple urchin was locally abundant in 1974, it foraged under the canopy, then consisting mainly of *Ecklonia* and *Sargassum verruculosum*, and partially cleared haloes around its crevices of some of the foliose understorey algae, leaving some algal turf (Fig. 12.7). But the purple urchin, in its low densities, did not create barrens dominated by crustose corallines, as did the black urchin, nor did it increase their size.

The long-term data set for the Kent Group of Islands is especially instructive, because it shows the time course of the progressive effects of an invasive species on benthic habitats – from small haloes around crevices, to incipient barrens as densities increase, and finally to widespread and generalised barrens. The differences between the two urchin species, in terms of their effects on the habitat, seem to be largely due to the differences in their diet and grazing behaviour. As shown in Chapter 11, the black urchin is a non-selective generalist herbivore, and grazes extensively on crustose corallines by scraping the upper coralline layers, whereas these corallines form only a small fraction of the purple urchin's diet, which favours foliose or filamentous algae (see below). The significance of such a diet for the black urchin, in terms of population fitness, is discussed below.

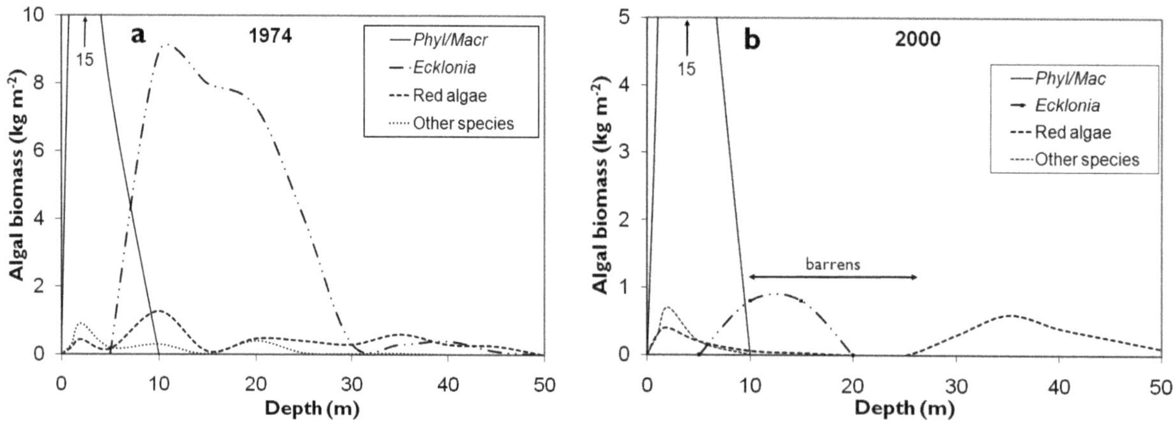

Figure 12.11: Comparison between fresh weight biomass of macroalgae *Phyllospora comosa* and *Macrocystis angustifolia* (plotted together), *Ecklonia radiata*, other species of green and brown algae combined, and red algae, measured at 1 m, 2 m and 5 m depth, and then at 5 m depth intervals to 50 m depth on the NE side of Dover I., Kent Group in (a) 1974, and (b) 2000. Biomass values in (b) are calculated from biomass values in (a) after deducting algal biomass from barrens area in 2000 unpublished data).

Invasion of eastern Tasmania by the black urchin

The rate of advance of the black urchin down the eastern Tasmanian coast, and the extent of urchin barrens, are known from diver and video surveys, and were summarised by Johnson *et al.* (2005). The urchin had reached the NE Tasmanian coast by 1978, Recherche Bay, southern Tasmania by 2000, and south-west (SW) Tasmania by 2005. However, barrens were increasingly rare south of ~43°S (Fig. 12.12). In 2000, the urchin barrens were largely incipient (i.e. in patches of a few square metres surrounded by macroalgal forests) except for sites at St Helens (at 41.4°S in Fig. 12.12), where almost a third of them were in patches of > 10 m size.

The expansion of barrens is an ongoing process in SE Tasmania. Over time, as populations of the urchin matured, the number of age classes increased, average age increased and the barrens kept expanding. Widespread barrens only occurred when the average population age reached ~18 years, as at St Helens at 41.4°S (Ling and Johnson 2009). When estimated in 2008, the average age of urchins declined down the east coast of Tasmania, from ~20 years in the NE to ~5 years in the SW. These authors suggested that the final transition from algal forest to barrens occurred when the large canopy species, their haptera (i.e. stems) and roots weakened by foraging urchins, senesced or were torn off the reef by storm events. So, gradually over time, the barrens extended further into shallow water.

The presence and extent of barrens also varied according to the substratum type and depth. Barrens usually started at 5–10 m depth, because they were limited in shallower depths by the whiplash effect of the large kelps, and continued to > 30 m depth. The percentage cover of barrens on boulders was more than twice that on flat rock, because the urchins prefer a crevice habitat, which provides shelter and some protection from predators (Fig. 12.13). The total percentage cover of barrens on rock of all types reached a maximum of ~17% at 20–25 m depth (Fig. 12.13), and extended beyond 33 m depth where the surveys stopped.

Other factors affecting the density of black urchins and the occurrence of barrens are the interactions between coastal topography, sea temperature and the EAC (see Ling *et al.* 2009b). Within bays, winter sea temperatures are often in the range 9.5–12°C, which is too cold for black urchin larvae to survive, whereas around headlands, where the EAC reaches the coast, sea temperatures are > 12°C (i.e. above the larval survival threshold of the black urchin) (Ling *et al.* 2008). Hence, populations can recruit and survive near headlands, but not in bays, and this contributes to the variability in the extent of barrens as can be seen in Fig. 12.12.

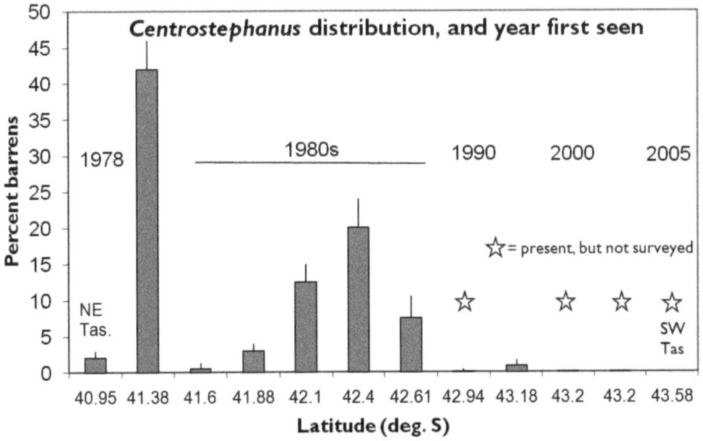

Figure 12.12: The southward advance of the black urchin *Centrostephanus* down the east coast of Tasmania showing percentage cover of barrens at 12 sites in 2000 (redrawn from Johnson *et al.* 2005), with information on year when first seen (extracted from Ling and Johnson 2009 and Ling *et al.* 2009b).

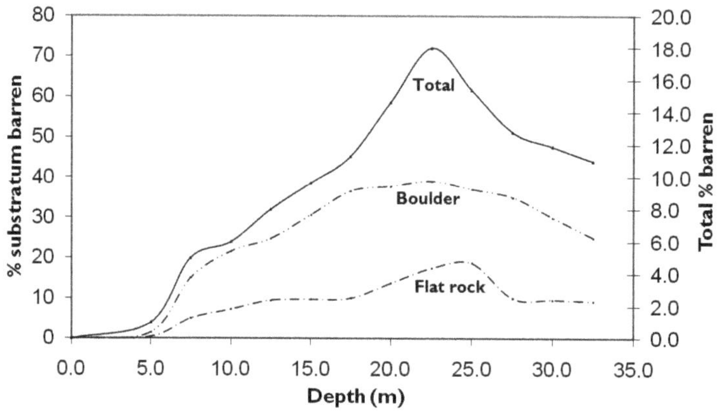

Figure 12.13: Percentage cover of black urchin barrens in boulder habitat and on flat (continuous) reef, and percentage cover of barrens in all reef habitats combined (redrawn, omitting error bars, from Johnson *et al.* 2005).

Lastly, Ling and Johnson (2009) suggested that the population fitness of the black urchin is maximised by its morphological, behavioural and feeding characteristics. The adaptation to grazing the substratum bare promotes the growth of crustose corallines and the formation of barrens, which in turn pave the way for larger black urchin population sizes by the following mechanisms:

- the recruitment rate increases
- the carrying capacity of the habitat increases, and this favours larger population sizes of urchins
- larger population sizes and the urchin's aggregative behaviour in turn favour increased fertilisation success in a free-spawning species in an environment of strong surge; hence, larval production is increased, and the risk of local extinction reduced
- the urchin's morphological plasticity (growth of longer spines in barrens and lighter tests) enables the species to reach a size where they are resistant to predators at ~15 years of age in barrens, as against ~21 years when they live in algal forests. This is an example of 'the predatory size refuge' concept.

One could go further and argue that the black urchin–crustose coralline relation is a type of animal–plant mutualism, as previously noted for the relation between grazing molluscs, notably small abalone, and corallines, which they also graze (see Chapter 6). The sea-urchin populations receive the benefits noted above, and the crustose corallines benefit from the removal of filamentous algae, sediment and fouling organisms, which would otherwise smother them. Some of these ideas have been proposed for the relation between other urchin species and crustose corallines, but have yet to be tested rigorously in southern Australia.

Effects of the black urchin invasion on habitats and fisheries

The advance of the black urchin into Tasmania has provided an opportunity to examine its effects in a cool temperate reef environment, in contrast to the warm temperate reefs of New South Wales. The now classical urchin removal experiments from incipient barrens have been done and the gradual recovery of macroalgal habitat monitored (Ling 2008). In that study, the algal succession followed more or less the same trajectory as in New South Wales (Fig. 12.3). The main canopy species in Tasmania were *Ecklonia* and *Phyllospora* and both species grew rapidly and reached a combined cover of ~60% after 15 months, while crustose coralline cover declined to low levels over ~9 months. Understorey foliose red algae gradually increased to natural forest levels of ~15% cover over 2 years, whereas filamentous algae rapidly increased from low levels to ~30% cover, and then declined to near zero. Of the main faunal components, bryozoans recovered to higher levels that in algal forests and sponges rather more slowly, reaching about one third of their cover in adjacent forests over 2 years.

The recovery process from barrens to algal forest in Tasmania proved to be slightly faster and more predictable than that in New South Wales. Notable differences between the two regions are that the limpet meso-grazers are absent from Tasmanian barrens, and the barrens are largely incipient (i.e. they are small and surrounded by algal forest), whereas they are much more extensive in New South Wales. Ling (2008) suggested that, in Tasmania, propagules from algal forests were likely to be more readily available to colonise the substratum, and less likely to be grazed, than in New South Wales.

Overall, the switch from algal forest to urchin barrens can be catastrophic for local biodiversity. About 150 invertebrate species (e.g. polychaetes, small crustaceans, gastropods, bivalves, echinoderms and brachiopods) and > 370 species of algae were missing in barrens, as compared with adjoining algal forest (Ling 2008). This loss implies a serious decline in primary and secondary productivity, by which energy is transmitted to higher levels of the food chain. So what is the implication of increasing barrens for inshore fisheries?

The blacklip abalone is an important fishery on the east Tasmanian coast and the negative impact of expanding urchin barrens there was examined by Strain and Johnson (2009). When the black urchin was added to experimental algal plots inhabited by the abalone, the abalone's body condition and survival were seriously compromised, as the urchin monopolised the food supply. But the converse did not happen (i.e. abalone did not have an adverse effect on the urchin). The reason underlying the superior competitive ability of the urchin relates to the food and feeding mode of the two species. The black urchin is a generalist herbivore that grazes non-selectively on virtually all algae, encrusting or erect, and also on encrusting animals (see Chapter 8). In contrast, the blacklip abalone feeds selectively largely on red algae, preferentially by trapping it as drift (Chapter 9). Nevertheless, the urchin cannot completely exclude abalone, because the latter inhabits shallow water < 5 m deep, which on open coasts is too exposed for the black urchin.

The same authors also tested whether the black urchin negatively affected the purple urchin, by removing the former from experimental plots containing the latter. They found that invariably both juvenile and adult purple urchins increased in abundance after removal of black urchins, suggesting that adults of the two species directly competed for shelter in crevices. However, the black urchin could not completely exclude the purple urchin because the latter, being a smaller species, can live in crevices too narrow for the black urchin to occupy. Regarding the increase in numbers of juvenile purple urchins after removal of black urchins, it seems that the habitat for recruitment of purple urchins was improved, possibly by the initial algal growth after black urchin grazing had terminated. Yet, in experiments where adult black and purple urchins were retained in the same plots, the purple urchins declined in numbers. So, Strain and Johnson (2009) suggested that, in low densities, the black urchin facilitated the purple urchin, but in high densities it had a negative impact.

THE PURPLE URCHIN *HELIOCIDARIS ERYTHROGRAMMA*

Effects of purple urchin grazing

The variable grazing behaviour of the purple urchin has long puzzled biologists in southern Australia. In some places they graze algal populations destructively, creating mini-barrens, and in other areas they have no obvious effect at all. We shall now discuss various studies on the effects of their grazing carried out across the purple urchin's southern range.

Western Australia

On the west-facing Australian coast, the purple urchin is abundant on the near- to offshore reefs running in a south–north direction for over 250 km of coast. Here, Matt Vanderklift and colleagues have extensively studied the distribution and abundance of the purple urchin. The reefs are typically dominated by *Ecklonia*, with plant densities of up to 30 m^{-2}. The density of the urchin can be quite high also, often ~8 m^{-2}, with maxima of > 40 m^{-2}, and occurring equally on flat limestone reef and on steep to vertical rock faces (Vanderklift and Kendrick 2004, 2005; Vanderklift and Wernberg 2008). Yet, despite such high densities, neither grazed patches of reef nor urchin barrens have ever been seen. It seems that drift algae are especially abundant on these reefs, and can be transported shorewards by the swell for several kilometres from distant reefs, as well as from local reefs. With such an abundant food supply, the purple urchin sits in a crevice or hollow, rarely moving, and adopts a sit-and-wait behaviour to catch algal drift, mainly *Ecklonia*. Hence, grazing on attached algae is extremely low (Vanderklift *et al.* 2009).

South Australian and Victorian coasts

On open coasts, the purple sea-urchin *Heliocidaris erythrogramma* occurs in low densities of 0.1–0.2 urchins m^{-2} (Connell and Irving 2008) and is confined to crevice habitats. Here, it is sedentary and feeds on drift algae, but it still can have a tiny, but measurable, effect on the algae around it. In one study on a semi-exposed shore with urchin densities of ~5 m^{-2}, the purple urchin ate some foliose algae within ~4 cm radius and virtually eliminated the fucoid canopy there, which was largely replaced by an algal turf. With increasing distance up to ~15 cm away, the canopy gradually increased to ~60% cover, while the turf declined to near zero, and foliose and erect coralline algae stabilised at ~30% cover (Livore and Connell 2012).

In many sheltered and semi-sheltered locations, as within gulfs and bays, densities can rise to 10–20 (rarely to 40) urchins m^{-2}. At these higher densities, the purple urchin grazes much of the macroalgae in its vicinity, and forms noticeable patches of partly cleared bottom around it. Such areas differ markedly from the isoyake areas described in Japan, and from the incipient or late-stage barrens typically formed by *Centrostephanus* grazing (Underwood *et al.* (1991). So, here we simply call them 'mini-barrens' (Plate 34), because they have their own characteristic features, and typically retain some cover of crustose corallines and foliose algae (Figs 12.7, 12.14; and see Fig. 8.5a;).

Figure 12.14: A mini-barren of the purple urchin *Heliocidaris erythrogramma* on rock patches among seagrass at Crag Point, upper Spencer Gulf, South Australia (photo K.L. Branden).

Such mini-barrens, ranging in size up to ~10 m^2 are occasionally seen in inshore waters and sheltered bays of the South Australian Gulfs (Fig. 12.14) (Shepherd 1974; Connolly 1986), and in Victorian bays (see Chapter 11).

Elsewhere in Port Phillip Bay, such as Portarlington and Avalon, local feeding aggregations of the purple urchin have sometimes appeared. In some cases, they have persisted for years and created mini-barrens, and in one event a strong recruitment suddenly appeared, and over the next 3 years ate out all algae on the reef (except for the Japanese kelp *Undaria*) before quickly disappearing (JE Watson *pers. comm.*).

South-East Australian coasts

In sheltered bays on these coasts, at least as far north as Sydney, densities of the purple urchin can rise to 50–100 m^{-2} (e.g. Wright and Steinberg 2001), and several studies there have shown the effects of higher densities on the habitat. At Bare I., Botany Bay, Wright *et al.* (2005) compared the grazing effect of different purple urchin densities on the benthic algae over 210 days in cage-plots. In the high-density plots (100 urchins m^{-2}) the urchins ate all the species present, including encrusting species, and reduced the plots to nearly all bare rock. In the lower density plots (40 urchins m^{-2}), some algal species were avoided or eaten much more slowly. Hence, the rate at which species were consumed became a measure of their preference/avoidance. The results for the two urchin density trials are given in Fig. 12.15. Preferred species rapidly consumed within 30–90 days were the brown algae *Sargassum linearifolium* and *Zonaria diesingiana* (neither shown in the figure), and the corallinaceous red alga *Amphiroa anceps*. The red algae *Delisea pulchra* and *Corallina officinalis* and the brown alga *Sargassum vestitum* were avoided at the lower density, but mostly eaten at the higher density level as food became scarce. Note that in the high density plots (Fig. 12.15b), nearly all the rock was eventually grazed bare, even of encrusting coralline algae, by the end of the trial when the urchins were essentially starving. At other sites nearby, where urchin densities averaged ~10 m^{-2} over 6 months, total macroalgal cover averaged ~60%, encrusting corallines ~8% and bare rock ~24%. In contrast, in other places where densities averaged ~80 m^{-2}, as in the plots in Fig. 12.15b, all foliose algae had been consumed leaving only bare rock and encrusting corallines, so confirming the effect of high urchin densities in algal communities.

The above study is significant, because it showed that the purple urchin is a selective grazer, and that a continuum exists from macroalgal forest to barrens, according to urchin density and the time during which grazing persists. This is illustrated below, with arrows showing the transitions that occur with increasing urchin densities. The illustration is an over-simplification, because it does not take time into account, and Wright *et al.* (2005) predicted that it could take years at a constant urchin density for some avoided species, such as *Corallina* and *Delisea* to be eliminated locally.

But under what circumstances does a community switch from an algal forest to a barren? Some researchers have suggested that massive urchin recruitment is sufficient to precipitate an algal forest in the direction shown by the arrows below, but evidence for this kind of event is largely anecdotal. Rather, Pederson and Johnson (2008) argued that a combination of small-scale recruitments, coupled with a change in urchin grazing behaviour and grazing intensity, are enough to initiate the transition to barrens. In support of their argument, they showed that, in places where algal abundance had declined, the purple urchin had developed larger jaws, which were better adapted for

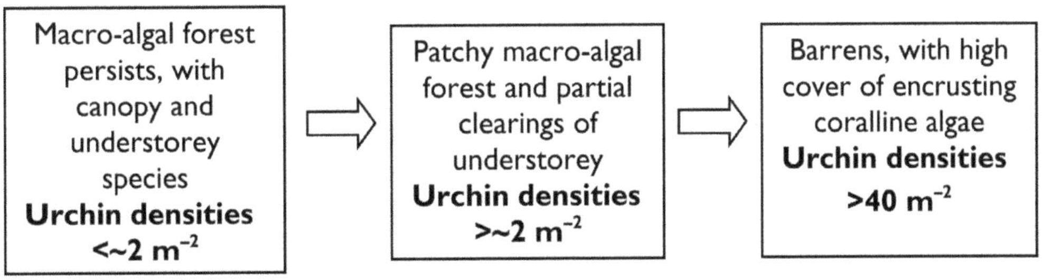

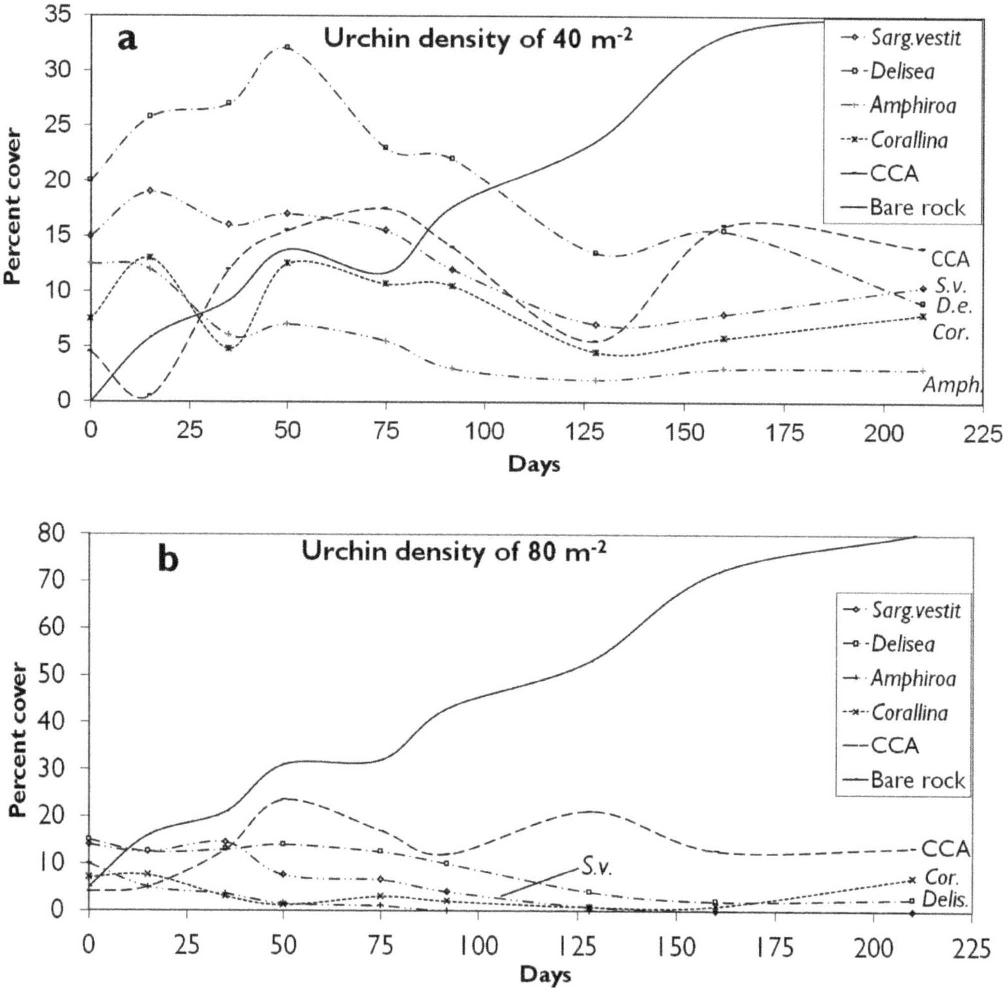

Figure 12.15: Trends in percentage cover over 210 days of various algal groups or species, and increase in bare rock under grazing by the purple urchin *Heliocidaris erythrogramma* in (a) at densities of 40 urchins m^{-2}, and (b) at densities of 80 urchins m^{-2} (redrawn from Wright *et al.* 2005, omitting error bars). To simplify the figures, bare rock is plotted as a running mean of 3. *Sarg.vestit* = *Sargassum vestitum*; *Delisea* = *Delisea elegans*; *Amphiroa* and *Corallina* = two genera of coralline algae; CCA = crustose coralline algae.

scraping crustose corallines. So, as macroalgae declined, they relied more on encrusting algae scraped from the rock (see Chapter 11).

For the reverse transition from a barren back to macroalgal forest, one might expect that the switch should occur with a return to lower urchin densities. But a switch back to forest can be much more complex and may require something more than merely reduced urchin densities. A catastrophic event, such as the flood-induced mass urchin mortality described by Andrew (1993), and shown in Fig. 12.3, did achieve the transition back to algal forest, but in the example below, no such reverse transition occurred.

In SE Tasmania at a sheltered site with mini-barrens at 5–10 m depth, Valentine and Johnson (2005b) cleared all purple urchins, which were at densities of 7–10 individuals m^{-2}, from experimental plots. At the beginning, the plots had ~20% cover of filamentous and foliose red algae, ~10% encrusting corallines and 1–5% brown algal turf and green algae. The algal succession over the following 2 years (Fig. 12.16) proved to be unique and unexpected. The

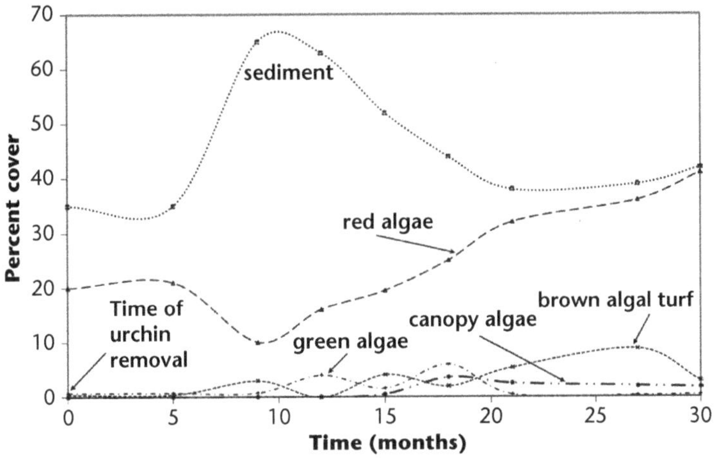

Figure 12.16: Changes over 30 months after removal of the purple urchin *Heliocidaris erythrogramma* from mini-barrens at Lords Bluff, Mercury Passage, SE Tasmania. Canopy algae refers to native species of *Cystophora* and *Sargassum*; red algae include filamentous and foliose species (mostly *Dasya ceramioides*); green algae were mainly *Codium* spp.; and brown algal turf were mainly species of *Zonaria* and *Dictyopteris* (redrawn from Valentine and Johnson 2005b).

canopy did not recover, and virtually only the red algae increased substantially in cover. However, sediment cover increased greatly and persisted for the whole 2.5 years of the study. It was thought that the original algal canopy of native species, which had once existed well before the study, had prevented sediment settling by the sweeping action of its fronds on the rock surface. After the canopy declined and disappeared, due to purple urchin grazing, sediment accumulated and inhibited the establishment of native canopy species, but not the Japanese kelp *Undaria pinnatifida* (Valentine and Johnson (2005a). Hence, it appeared that at this site a positive feedback was operating to maintain the mini-barrens habitat, as illustrated below:

A later study by Ling *et al.* (2010) at an exposed site in SE Tasmania gave quite different results to the above experiments and exemplified the enigmatic, context-dependent behaviour of this urchin. On wave-exposed reefs, the purple urchin was cryptic within crevices, and fed by catching the abundant drift algae, so even at densities of > 8 m^{-2}, the urchin had no effect on the surrounding algal community.

Urchin predators

A final question relates to the role of predators in controlling the abundance of urchins. We shall see (Chapter 20) how the mortality rates of sea-urchins is three to nine times higher inside marine reserves than outside them, due to lobster predation (Pederson and

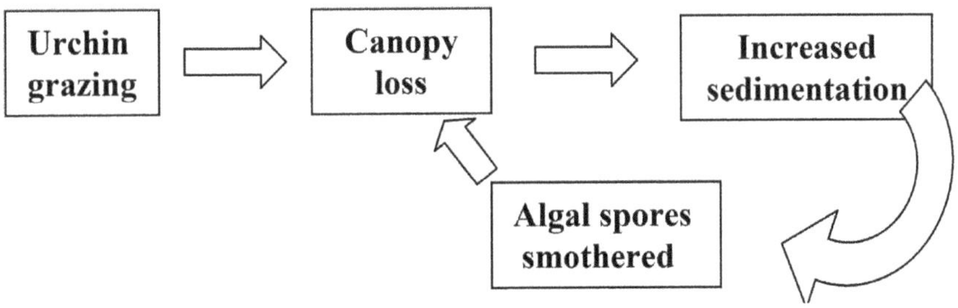

Johnson (2006), so it is plausible that the historic, high fishing intensity on lobsters has had a cascade effect, in which sea-urchins, released from predation, have increased greatly in abundance and overgrazed the algae to produce the barrens that are now prevalent. Evidence from existing reserves suggests that it may take several decades for lobsters to reduce urchin densities to low levels. Thus, in the Derwent Estuary Reserve established for > 30 years, sea-urchin densities are significantly lower than in nearby fished areas, and at the Maria I. Reserve, established for < 15 years, sea-urchin densities are much lower than those outside the reserve (Pederson and Johnson 2006, 2008; Ling *et al.* 2009a). Experiments in which lobsters are released into barrens habitat are currently underway to test the above ideas. If the results validate the cascade effects suggested, the evidence will clearly and doubly implicate humans in the advance of barrens in SE Tasmania – globally by causing climate change that is increasing sea temperatures, and locally by intense lobster fishing.

CONCLUSIONS

The grazing of the black urchin *Centrostephanus rodgersii* and the purple urchin *Heliocidaris erythrogramma* can seriously affect the composition and abundance of macroalgae over extensive areas of bottom habitats, and for this reason they have been called ecosystem engineers. The extent of their effects depends on the dietary breadth and food preferences of the urchin and their density. The black urchin is a much larger generalist forager that grazes and scrapes all algae non-selectively and can completely eliminate macroalgae in areas and create barrens. The purple urchin is more selective in its feeding. Where algal drift is abundant, it feeds exclusively on the drift that it traps, but, where it is not, the urchin browses selectively, and generally partially clears grazing patches – creating haloes or mini-barrens. Only in extremely high densities does it approach the foraging intensity of the black urchin.

Manipulations of urchin density show that the transition from algal forest to barrens is a continuum for both species. The causes that lead to sufficient densities of urchins to create barrens are not well known. A strong natural recruitment of urchins occasionally occurs and creates high sedentary densities or feeding fronts. Another factor is that, as algal availability in an area where urchins live declines, both species of urchin may adapt to the reduced food supply by developing larger feeding jaws. This enables them to scrape the bottom more effectively and promote the growth of encrusting coralline algae. There is some evidence that this modified 'barrens habitat' becomes an alternative stable state to the algal forest habitat, and may require some catastrophic event to precipitate a switch back to algal forest. Under climate change, the range of the black urchin has extended to southern Tasmania in recent decades. Evidence about the role of predatory lobsters in controlling urchin abundance, the role of fishing in reducing the numbers of lobsters in coastal waters and the role of marine protected areas in conserving natural ecosystems is still emerging, and firm conclusions about them would be premature.

REFERENCES

Andrew NL (1991) Changes in subtidal habitat following mass mortality of sea urchins in Botany Bay, New South Wales. *Australian Journal of Ecology* **16**, 353–362.

Andrew NL (1993) Spatial heterogeneity, sea urchin grazing, and habitat structure on reefs in temperate Australia. *Ecology* **74**, 292–302.

Andrew NL (1994) Survival of kelp adjacent to areas grazed by sea urchins in New South Wales, Australia. *Australian Journal of Ecology* **19**, 466–472.

Andrew NL, Byrne M (2007) Ecology of *Centrostephanus*. In *Edible Sea Urchins: Biology and Ecology Developments in Aquaculture and Fisheries Science. Vol. 37* (2nd edn) (Ed. JM Lawrence) pp. 191–204. Elsevier, Amsterdam.

Andrew NL, O'Neill AL (2000) Large-scale patterns in habitat structure on subtidal rocky reefs in New South Wales. *Marine and Freshwater Research* **51**, 255–263.

Andrew NL, Underwood AJ (1992) Associations and abundance of sea urchins and abalone on shallow subtidal reefs in southern New South Wales. *Australian Journal of Marine and Freshwater Research* **43**, 1547–1559.

Andrew NL, Underwood AJ (1993) Density-dependent foraging in the sea urchin *Centrostephanus rodgersii* on shallow subtidal reefs in New South Wales, Australia. *Marine Ecology Progress Series* **99**, 89–98.

Andrew NL, Worthington DG, Brett PA, Bentley N *et al.* (1998) 'Interactions between the abalone fishery and sea urchins in New South Wales'. FRDC Report No. 93/102. NSW Fisheries Research Institute, Cronulla.

Connell SD, Irving AD (2008) Integrating ecology with biogeography using landscape characteristics: a case study of subtidal habitat across continental Australia. *Journal of Biogeography* **35**, 1608–1621.

Connolly RM (1986) Behaviour and ecology of the sea urchin *Heliocidaris erythrogramma* (Valenciennes). BSc (Hons) thesis. University of Adelaide, Adelaide.

Davis AR, Ward DW (2009) Establishment and persistence of species-rich patches in a species-poor landscape: role of a structure-forming subtidal barnacle. *Marine Ecology Progress Series* **380**, 187–198.

Davis AR, Fyfe SK, Turon X, Uriz MJ (2003) Size matters sometimes: wall height and the structure of subtidal benthic invertebrate assemblages in south-eastern Australia and Mediterranean Spain. *Journal of Biogeography* **30**, 1797–1807.

Fletcher WJ (1987) Interactions among subtidal Australian sea urchins, gastropods, and algae: effects of experimental removals. *Ecological Monographs* **57**, 89–109.

Harrold C, Pearse JS (1987) The ecological role of echinoderms in kelp forests. *Echinoderm Studies* **2**, 137–233.

Hill NA, Blount C, Poore AGB, Worthington D *et al.* (2003) Grazing effects of the sea urchin *Centrostephanus rodgersii* in two contrasting reef habitats: effects of urchin density and its implications for the fishery. *Marine and Freshwater Research* **54**, 691–700.

Johnson C, Ling S, Ross J, Shepherd SA *et al.* (2005) 'Establishment of the long-spined sea urchin (*Centrostephanus rodgersii*) in Tasmania: first assessment of potential threats to fisheries'. FRDC Project 2001/044. Tasmanian Aquaculture and Fisheries Institute, Hobart.

Johnson CR, Banks SC, Barrett NS, Cazassus F *et al.* (2011) Climate changes cascades: shifts in oceanography, species' ranges and subtidal marine community dynamics in eastern Tasmania. *Journal of Experimental Marine Biology and Ecology* **400**, 17–32.

Lawrence JM (1975) On the relationships between marine plants and sea urchins. *Oceanography and Marine Biology - an Annual Review* **13**, 213–286.

Ling SD (2008) Range expansion of a habitat-modifying species leads to loss of taxonomic diversity: a new and impoverished reef state. *Oecologia* **156**, 883–894.

Ling SD, Johnson CR (2009) Population dynamics of an ecologically important range-extender: kelp beds versus sea urchin barrens. *Marine Ecology Progress Series* **374**, 113–125.

Ling SD, Johnson CR, Frusher S, King CK (2008) Reproductive potential of a marine ecosystem engineer at the edge of a newly expanded range. *Global Change Biology* **14**, 907–915.

Ling SD, Johnson CR, Frusher S, Ridgway KR (2009a) Overfishing reduces resilience of kelp beds to climate-driven catastrophic phase shift. *Proceedings of the National Academy of Sciences of the United States of America* **106**, 22341–22345.

Ling SD, Johnson CR, Ridgway K, Hobday AJ *et al.* (2009b) Climate-driven range extension of a sea urchin: inferring future trends by analysis of recent population dynamics. *Global Change Biology* **15**, 719–731.

Ling SD, Ibbott S, Sanderson JC (2010) Recovery of canopy-forming macroalgae following removal of the enigmatic grazing sea urchin *Heliocidaris erythro-gramma*. *Journal of Experimental Marine Biology and Ecology* **395**, 135–146.

Livore JP, Connell SD (2012) Fine-scale effects of sedentary urchins on canopy and understory algae. *Journal of Experimental Marine Biology and Ecology* **411**, 66–69.

Ohmi H (1951) Studies on *Isoyake* or 'decrease of seaweeds' along the coast of northern Japan. *Bulletin of the Faculty of Fisheries Hokkaido University* **2**, 110–117.

Pederson HG, Johnson CR (2006) Predation of the sea urchin *Heliocidaris erythrogramma* by rock lobsters (*Jasus edwardsii*) in no-take marine reserves. *Journal of Experimental Marine Biology and Ecology* **336**, 120–134.

Pederson HG, Johnson CR (2008) Growth and age structure of sea urchins (*Heliocidaris erythrogramma*) in complex barrens and native macroalgal beds in eastern Tasmania. *ICES Journal of Marine Science* **65**, 1–11.

Ridgway KR, Dunn JR (2003) Mesoscale structure of the mean East Australia Current system and its relationship with topography. *Progress in Oceanography* **56**, 189–222.

Shepherd SA (1973) Competition between sea urchins and abalone. *Australian Fisheries* **32**, (6), 4–7.

Shepherd SA (1974) 'An underwater survey near Crag Point in upper Spencer Gulf'. Technical Report No. 1. South Australian Department of Fisheries, Adelaide.

Strain E, Johnson JR (2009) Competition between an invasive urchin and commercially fished abalone: effect on body condition, reproduction and survivorship. *Marine Ecology Progress Series* **377**, 169–182.

Thresher R, Rintoul SR, Koslow A, Weidman C *et al.* (2004) Oceanic evidence of climate change in southern Australia over the last three centuries. *Geophysical Research Letters* **31**, L07212

Underwood AJ, Jernakoff P (1981) Effects of interactions between algae and grazing gastropods on the structure of a low shore intertidal algal community. *Oecologia* **48**, 221–233.

Underwood AJ, Kingsford MJ, Andrew NL (1991) Patterns in shallow subtidal marine assemblages along the New South Wales coast. *Australian Journal of Ecology* **6**, 231–249.

Valentine JP, Johnson CR (2005a) Persistence of the exotic kelp *Undaria pinnatifida* does not depend on sea urchin grazing. *Marine Ecology Progress Series* **285**, 43–55.

Valentine JP, Johnson CR (2005b) Persistence of the sea urchin (*Heliocidaris erythrogramma*) barrens on the east coast of Tasmania: inhibition of macro-algal recovery in the absence of high densities of sea urchins. *Botanica Marina* **48**, 106–115.

Vanderklift MA, Kendrick GA (2004) Variation in abundances of herbivorous invertebrates in temperate subtidal rocky reef habitats. *Marine and Freshwater Research* **55**, 93–103.

Vanderklift MA, Kendrick GA (2005) Contrasting influence of sea urchins on attached and drift macroalgae. *Marine Ecology Progress Series* **299**, 101–110.

Vanderklift MA, Wernberg T (2008) Detached kelps from distant sources are a food subsidy for sea urchins. *Oecologia* **157**, 327–335.

Vanderklift MA, Lavery PS, Waddington KI (2009) Intensity of herbivory on kelp by fish and sea urchins differs between inshore and offshore reefs. *Marine Ecology Progress Series* **376**, 203–211.

Wright JT, Steinberg PD (2001) Effects of variable recruitment and post-recruitment herbivory on local population size of a marine alga. *Ecology* **82**, 2200–2215.

Wright JT, Benkendorff K, Davis AR (1997) Habitat associated differences in temperate sponge assemblages: the importance of chemical defence. *Journal of Experimental Marine Biology and Ecology* **213**, 199–213.

Wright JT, Dworjanyn SA, Rogers CN, Steinberg PD *et al.* (2005) Density-dependent sea urchin grazing: differential removal of species, changes in community composition and alternative community states. *Marine Ecology Progress Series* **298**, 143–156.

13 Sessile fauna: sponges, ascidians and bryozoans

OVERVIEW

Sessile fauna – mainly sponges, ascidian and bryozoans (with some cnidarians, polychaetes and other invertebrates) – dominate shaded reef surfaces in shallow water and open reef habitats in deeper water. Sponges are filter-feeders, with great plasticity in form and function. Reproductively, they vary greatly and both asexual and sexual reproduction occurs. Development may be oviparous or viviparous and larval life is usually short with localised dispersal. Longevity ranges from < 1 year to decades. Associations with microalgae and other animals are common. Ascidians are either colonial animals, like sponges, or solitary, and are active filter-feeders. Solitary forms are broadcast spawners with wide larval dispersal, while colonial forms brood eggs in a pouch and produce emergent tadpole larvae with localised dispersal. The life history and ecology of two common species – 'bluebells', a clavelinid colonial species, and cunjevoi, a pyurid solitary form – are described. Bryozoans are filter-feeding colonial animals and their form may be encrusting, bushy or erect. In the photic zone, they live in shaded habitats. Reproduction is sexual and asexual, while larvae settle in hours to days for brooding species, and in up to 2 months for non-brooding species. The life histories of three species with different forms – encrusting, bushy and erect – are described in detail. In shaded habitats, sponges, ascidians, bryozoans and a few species of other phyla compete for space. Sponges and colonial ascidians are good competitors, but poor recruiters, and bryozoans, and serpulid and sirorbid polychaetes, are the reverse. Trade-offs operate between these life history features and determine the occupants of a particular habitat. The succession of organisms on firm substrata is complex and varies according to their recruitment and competitive ability, their longevity, predation, and the intensity and frequency of disturbances. This is illustrated by three examples of succession on jetty pilings, level-bottom reefs and razor shells. Lastly, the fauna of deep-water communities, described by remote sensing equipment, is mentioned. Sessile cnidarians – hydroids, sea anemones and corals – are discussed in Chapter 14.

INTRODUCTION

Below the photic zone dominated by algae is the so-called circa-littoral zone. This domain is semi-dark and extends to the edge of the continental shelf and beyond, but knowledge of it is limited to benthic habitats on open coasts at 20–50 m depth towards the upper limit of the zone, and to the much shallower, near-shore, shaded habitats under overhangs or in crevices and caves. On all these low-light reef surfaces, algae (mostly encrusting forms) contribute very little to the biomass, and the substratum is dominated by sessile fauna – principally sponges, ascidians, bryozoans and cnidarians (such as soft or stony corals, anemones, hydroids and gorgonians), and a few polychaetes (such as galeolarian tube-worms and feather-duster worms). The polychaetes are considered in Chapter 8 and the cnidarians in Chapter 14. Here, we consider the ecology of sessile representatives of three major phyla – sponges, ascidians and bryozoans – and then the ecology of their shared habitats. A good example of such a habitat is jetty piles, which have proved to be convenient places to do extensive observations and experiments. These habitats have been the subject of many studies by Alan Butler and his students over > 20 years (see reviews of Butler 1995, 2008). Although jetty piles are strictly artificial habitats, they contain species that occur widely on similarly shaded surfaces and Alan Butler (2008) viewed these artificial habitats as 'a window on the whole system'. Hence, studies of the abundance, life history and interactive behaviour of the component species will inform us about the ecology of the natural fauna of rock walls and caves.

SPONGES

These simple, sessile animals occur in many different shapes and forms, from boring to encrusting to massive forms up to 1 m high. The sponge is an aggregation of similar cells capable of coordinated action and with great plasticity in form and function, adept at re-aggregation and regeneration after mutilation. In most forms, a mineralised skeletal structure, consisting of a network of fibrous spongin and spicules made of silica or calcium carbonate, supports the body frame. Sometimes sand grains are also incorporated into the sponge structure.

Whatever their shape, all sponges are active filter-feeders with inhalant pores called ostia, connected to internal chambers, where plankton particles and detritus are trapped on the surface of cells. Water currents, set up by flagella which beat in unison and line the internal canals, flow through the body of the sponge and maintain a supply of food. The filtering efficiency of sponges is extraordinarily high at 95–99% and they filter out primarily bacterial picoplankton (< 2 μm) but also ultraplankton (2–5 μm), comprising cyanobacteria (Pile *et al.* 1996).

Distribution and habitat

Numerous factors influence sponge distribution in the shallow (to ~30 m) subtidal. They include: aspect (i.e. vertical, sloping or horizontal surfaces) and, related to this, shading, water movement, sedimentation and grazing. Only a few studies, outlined below, have considered these factors in southern Australia.

With respect to aspect, some species, such as *Halisarca* (an encrusting form) prefer vertical surfaces, some, such as *Cymbastela* and *Chondropsis*, prefer horizontal surfaces, and others, (e.g. *Haliclona*) show no preference or have inconsistent patterns (Knott *et al.* 2004b). In a study of water movement and sediment effects, Avril Watson (1976) examined a population of the globose sponge *Ancorina corticata* at Crawfish Rock, Westernport and found that individuals in strong current grew faster and much larger than those in slight current, where sedimentation also occurred. The faster growth was apparently due to the augmentation of the sponge's feeding currents in faster flows, although the increased sedimentation at sites of low flow may have exaggerated the difference, by reducing feeding activity there. Other reviews have emphasised that sediments can block the inhalant ostia, causing the sponge to expend energy to clear out the sediment, or cause tissue necrosis (Wilkinson and Vacelet 1979; Roberts *et al.* (2006b).

In shallow water, encrusting forms tend to live on shaded vertical faces in caves or under boulders. In open habitats in the deeper part of the photic zone, and below it in the circa-littoral zone, species-rich sponge gardens are found, especially in places of strong current, as discussed below.

Notable among southern described sponge garden habitats are:

- sponge beds off the New South Wales coast below ~18 m, where Roberts and Davis (1996) and Roberts *et al.* (2006a) recorded 82 species
- Port Phillip Heads in Victoria at 30–80 m depth, where Weidenmayer (1989) recorded 271 species,

42% of which appear endemic to the region (see below)

- Crawfish Rock, in Western Port, where ~300 species occur (Watson 1974)
- Spencer Gulf, South Australia (mainly 10–40 m depth) where 105 species have been recorded (Sorokin and Currie 2009), despite the severe negative impact of prawn trawling in deeper water
- deep-water habitats, such as the Benthic Protection Zone at 100–200 m depth in the Great Australian Bight (GAB), where Sorokin *et al.* (2007) recorded 351 species. Other deep-water habitats are described in detail later in this chapter.

Reproduction and dispersal

Reproductive patterns are astonishingly diverse in sponges (Ayling 1980; Maldonado and Riesgo 2008). Some species have separate sexes or partially separate sexes with some hermaphroditic individuals. Some species are fully hermaphrodite, producing eggs and sperm at one time, or sequential hermaphrodites (i.e. female first and then male, or vice versa, or having alternating sexes). Yet others reproduce asexually by:

- fragmentation, when pieces of the parent sponge break off and develop into another individual
- budding, when buds form and are released by the sponge
- the formation of gemmules – small, dormant, resistant bodies.

In sexual reproduction, sperm are released into the water and taken in via the inhalant ostia of another sponge, where they fertilise eggs attached internally. Regarding development, sponges can be oviparous or viviparous. In the former mode, fertilisation is external in many species, and embryos develop into larvae, equipped with cilia, enabling them to swim and disperse for hours to weeks, according to species, before settling. In a few species, the embryos settle directly as juveniles, without a free-swimming stage. In contrast, viviparous sponges are sperm-casters, releasing sperm, but retaining eggs for internal fertilisation. The developing embryos are then incubated within the sponge matrix and eventually released as free-swimming larvae (see Maldonado 2006).

Avril Ayling (1980) suggested from her study that the reproductive strategies of sponges vary according to the level of disturbance to which they are subject. In places of high disturbance, as on exposed coasts, larvae are large and viviparous and tend to be crawling larvae or settle quickly, whereas in low disturbance regimes, they are small and far more numerous and disperse more widely.

The diverse modes of reproduction and their complexities are illustrated by studies of six species in SW Australia (Table 13.1); notable among them are an oviparous species *Chondrilla australiensis* (common, brown or maroon, and crustose), and a viviparous species *Haliclona* sp. (maroon,

Table 13.1: Month of reproductive activity, presence of oocytes (O) (with diameter), embryos (E), larvae (L), sperm (S) and reproductive mode for six species of sponge.

Species	Month	Sex products	Reproductive mode
Echinodictyum clathrioides	Feb.	O (45μm);	Likely oviparous[1]
Coelosphaera sp.	Apr.	O E S[1]	Viviparous[1]
Haliclona sp.	Aug., Feb. – Apr. (+ asexual reproduction)	O E L	Broods embryonic larvae [1, 2]
Mycale sp.	Nov.–Jan.	O E L S	Viviparous, hermaphroditic[1]
Chondrilla australiensis	Feb. (asexual reproduction Apr.)	O (30–40 μm)	Oviparous, separate sexes[1,3]
Tethya sp.	Dec.–Mar. (spawning Mar.)	O develop Dec.–Mar.	Oviparous, separate sexes[1]

[1]Fromont (1999); [2]Whalan *et al.* (2005); [3]Usher *et al.* (2004a, b);

branching, and common in shallow caves and under overhangs). Adults of *C. australiensis* are male or female, and eggs develop over 4 weeks within special chambers. A week before spawning, nurse cells, lacking nuclei and flagella, attach themselves to the eggs and each other by filaments, and fuse with the eggs. Sexual reproduction occurs twice yearly and spawning of eggs and sperm takes place 10–11 days after the full moon in February and again in March, and lasts ~4 days. Apparently the lunar cycle, via either illumination or tidal flow, triggers spawning. *Haliclona* sp. is a hermaphroditic species, whose eggs are retained in brood chambers where they are fertilised and then develop over a month or so into larvae of ~0.6 mm size. Genetic analyses suggest that the brooded larvae, upon release, disperse locally over only a few hundred metres at most. This species also happens to produce novel chemicals of pharmaceutical value, and so is of interest for aquaculture. All the species in Table 13.1 reproduce sexually in summer–autumn, except *Haliclona* sp. which reproduces in winter as well as in summer, and it is likely that for summer breeders rising sea temperatures synchronise embryonic growth and larval release (Maldonado and Riesgo 2008).

Another southern species *Halisarca laxa*, which grows over the stalked ascidian *Pyura spinifera* in an apparently obligate relation, appears to reproduce asexually and spreads vegetatively to nearby ascidians, but it also reproduces sexually and is able to disperse over tens of kilometres (Davis *et al.* 1996). In many species, larval release is restricted to calm conditions and dispersal is very localised. However, a few species have very extensive geographic ranges, suggesting that their larvae can disperse over great distances. So far as is known, all sponge larvae are lecithotrophic (feeding on the egg's maternal yolk), and some species have symbiotic cyanobacteria in their eggs, providing photosynthates to the drifting eggs and enabling them to survive longer in the water column (see below). When they are ready to settle, larvae of shade-loving species respond negatively to light, which they detect with some pigmented cilia, and then use their cilia to help them settle on a suitable substratum (Maldonado 2006).

Ecology

Growth and longevity

In general, the growth rate of sponges is affected by the amount of sediment, light intensity, water flow and grazing. Thinly encrusting sponges grow slowly by lateral spreading, and their growth rate can be measured either by measuring changes in size of colonies or by clearing patches near existing colonies and measuring the rate of spread into the clearings. Using the former method, Avril Ayling found that the toxic encrusting sponge *Aplysilla rosea* grew at an average maximum rate of only ~0.3 mm^2 cm^{-1} day^{-1} (i.e. 0.3 mm^2 per cm border per day) and that 10 other encrusting sponge species grew even more slowly. However, when she damaged the *Aplysilla* patch, it regrew at ~20 times its natural rate, suggesting that damaged colonies have a mechanism for rapid regeneration – an adaptation that has obvious advantages in communities where sea-urchin grazing is common (Ayling 1983).

Using the second method referred to above, Kay and Keough (1981) cleared different sized patches on the vertical faces of jetty piles. They found that *Mycale* sp. grew at the fastest rate in the largest square patches of 0.25 m^2, with a rate of ~14 mm^2 cm^{-1} day^{-1}, but only at a rate of ~0.7 mm^2 cm^{-1} day^{-1} on the smallest patches of 0.01 m^2, although the smallest patches were grown over in a quarter of the time (50 days). In similar experiments with another encrusting sponge, *Crella* sp. grew much more slowly and at a rate of ~6 mm^2 cm^{-1} day^{-1} in the largest clearings, achieving over 60% cover in ~300 days. In the light of Ayling's (1983) findings on the rapid regeneration of damaged sponges, it is possible that Kay and Keough's rapid expansion rates were influenced by this mechanism.

Both the above experiments were done on vertical faces, and it is likely that the growth of sponges on horizontal faces is much slower. To test this, Knott *et al.* (2006) compared the growth of a common encrusting sponge, *Tedania anhelans*, on horizontal and vertical faces in Botany Bay, New South Wales. The sponge naturally grew on rock-faces with both aspects, but occupied more space on vertical faces. The authors found that it both grew faster and had better survival on vertical faces, compared with horizontal ones.

The lifespan of sponges is poorly known. Kay and Keough (1981) and Kay and Butler (1983) between them found that seven common sponges on pilings lived for > 2 years, *Dendrocia* sp. for > 4 years and *Mycale* sp. for > 5 years. However, Ayling (1983), on the basis of measured growth rates on undisturbed walls, estimated that colonies could be much older. For example, large colonies of *Aplysilla rosea* could be 50 years old and colonies of the slow-growing sponge *Eurypon* sp. > 78 years.

Role in benthic systems

Benthic–pelagic coupling

Sponges are functionally important in temperate systems and play a key role in a host of ecological processes (Wulff 2006; Bell 2008). With their high filtration efficiency and large quantities filtered, sponge beds can strip plankton, nutrients and dissolved organic matter from the water column, so transferring the pelagic productivity to the benthic system. Sponge beds even have the potential to control phytoplankton blooms. The interaction between the pelagic and benthic systems is termed 'benthic–pelagic coupling', and it is a two-way process, because material is not only transferred from the water column to the benthos, but also from the benthos back to higher trophic levels via predation and the release of dissolved organic matter.

On much of the southern coast with low nutrients, sponges are recognised as providing an important standing stock of silica, and sponge gardens, present in places of high current flow, are habitats that need to be conserved (Maldonado *et al.* 2010). Silicate, a dissolved form of silicon, is a major nutrient because it fuels primary production by enhancing the growth of diatoms, which require silicate to construct their skeletons. Benthic temperate sponges have high levels of 75–95% of biogenic silica, which is used to construct their skeletons.

Predators

The main predators of sponges are sea-stars, opisthobranchs, molluscs and fish, and important predatory species in these groups are discussed in Chapters 8, 9, 11, 16 and 17. Sea-urchins also occasionally eat sponges, but species living in urchin barrens, such as *Chondrilla australiensis*, *Clathria* sp. and *Darwiniella* sp., are chemically defended against grazing by the black urchin in New South Wales (Chapter 12). The high proportion of sand that occurs in the tissue of other species of sponges may also deter predation either directly as a deterrent, or indirectly by reducing their nutritional value (Wright *et al.* 1997). Similarly, the spicules present in encrusting sponges deter grazing by the black urchin (Ferguson and Davis 2008); whether spicules alone or combined with other deterrents referred to above deter fish grazing remains to be established in temperate regions.

Sponge associations

Symbiotic microalgae

Associations between sponges and blue-green algae (cyanobacteria) or other microalgae are common in shallow reef habitats, and the algae provide important nutrition for the host sponge. In a screening of eight sponge species around Sydney, Roberts *et al.* (1999) found that five of them had chlorophyll-*a* in their tissue, and that this imparted a tan or olive-brown colour to the surface tissue. On the basis of the typical colouration of sponges with symbiotic microalgae, the authors predicted that > 60% of sponges in their region (out of a sample of 100 species) had symbiotic microalgae, indicating that the symbiosis is almost as common in temperate as in tropical waters.

In general, sponges with symbiotic blue-green algae grow best in the light, and species without them grow best in the shade, although it is not clear whether cave-dwelling species are actively avoiding light or live in caves because they have less competition there. *Cymbastela* is a common phototrophic Australian genus with several temperate species. The photophysiology of one ochre- to yellow-brown species *Cymbastela* sp. was examined by Cheshire *et al.* (1995) who found that it was a net producer of photosynthetic carbon to a depth of ~11 m, which probably accounted for its presence in shallow illuminated habitats only to that depth. This meant that the host sponge could survive wholly on photosynthetic products in the form of glycerol and sugar phosphate produced by the symbiotic algae. This conclusion was strengthened by the experiments of Roberts *et al.* (2006b) with the congeneric *C. concentrica*. When this phototrophic sponge was placed under shade cloth, it lost symbiotic algae and weight, and became less active reproductively compared with controls.

Another example of a sponge–algal symbiosis is that of *Chondrilla australiensis*, which has blue-green algae in the outer cortex and bacterial symbionts throughout the tissues (Usher *et al.* 2001). Interestingly, in this species, the developing eggs, distributed throughout the sponge, are nourished by nurse cells, which transfer symbiotic cyanobacteria to the eggs. In this way, the blue-green algae are transmitted to the next generation of sponge. This is thought to be especially important for small, recruiting sponges, because it enables them to grow rapidly in sun-lit places, and to compete better with benthic algae.

Symbiotic fauna

The many intricate canals within the sponge matrix favour the development of associations with many other organisms, and one study has found a correlation between sponge canal volume and the numbers of animals living within (the endofauna). The most common inhabitants of sponges are crustaceans – mainly copepods, isopods, amphipods and shrimp (see Chapters 8, 15); others are polychaetes, ophiuroids and cnidarians (e.g. anemones, hydroids and attached scyphozoans), all of which form associations that may be mutualistic, commensal or parasitic (Tardent 1963; Van-Praët 1985).

Several examples among many such associations are:

a. The siliquariid mollusc *Tenagodus australis*, which lives obligately in its coiled tubes inside sponges, and shares the incoming planktonic food (Ludbrook and Gowlett-Holmes 1989). This is a mutualism in which the mollusc acquires living space and protection against predators, and its host sponge benefits from the mollusc pumping (Pansini *et al.* 1999).
b. The relationship between the doughboy scallop and an encrusting crellid or myxillid sponge, which is discussed in detail in Chapter 9;
c. The hydroid *Pararella* sp. lives within the canal system of some sponges and pokes its hydranth out through the ostia (JE Watson *pers. comm.*).
d. the hydroid *Zyzzyzus spongicolus* grows exclusively on the surface of an orange sponge, giving the sponge a fuzzy appearance (Watson 1982). These last two examples provide no obvious benefit to the host sponge and may be cases of a harmless commensalism.
e. Sponge crabs, such as *Austrodromidia* and *Dromia*, snip off pieces of sponge and attach them to their carapaces, where they grow and camouflage the crab – this is another example of a mutualism in which the benefits to the sponge are transport among different habitats, and possibly access to fine food particles.

The great variety of the above associations suggests, in the terms of Wulff (2006), 'that the uniquely homogeneous and readily re-arranged structure of sponges pre-adapts them to development of intimate associations which synergistically improve life for both partners'.

ASCIDIANS

Ascidians are a very diverse group of filter-feeding animals that live on rocky bottoms or shells, and abound from shallow water to abyssal depths. Some species have root-like hairs that help fix them on sandy bottoms. Although the adult stage is immobile, the group is placed in the phylum Chordata (Class Ascidiacea) because the larval stage is tadpole-like with a nerve chord, called a notochord, and has other features that indicate an affinity with vertebrates (see Kott 1997; Edgar 2008).

Ascidians fall into two groups: solitary and colonial species. Solitary species range in size from a few millimetres to ~30 cm; some of them are attached by stalks to the bottom and move with the swell or water currents. Colonial species are aggregations of zooids with several different types of organisation. Many are embedded in a common test, or have a common base, but are not connected to each other, while others are inter-connected with common cloacal systems, but with separate inhalant siphons opening direct to the water. Colonial species range in size from small cushions a few centimetres across to large sheets several square metres in area. Species in the Family Didemnidae are especially prolific in southern waters, with many small zooids that replicate prolifically and occupy large areas. Their stellate, calcareous spicules, which are embedded in the test, give skeletal support for large, sheet-like colonies that superficially resemble encrusting sponges.

All ascidians are active filter-feeders that pump water in through the inhalant siphon and out through the exhalant siphon, and on the way trap suspended particles of bacteria, phytoplankton and detritus in a mucous mesh. They are hermaphroditic, with both male and female gonads, which mature at different times to avoid self-fertilisation. However, solitary and colonial species have dramatically different reproductive and larval features. Solitary species are broadcast spawners with external fertilisation and they produce small larvae, which are usually widely dispersed. Colonial species, on the other hand, have a brood pouch, in which larvae develop after internal fertilisation, and upon release they soon settle in the vicinity, often after a few minutes (Ayre *et al.* 1997).

In all ascidians, the larvae are extraordinary mobile 'tadpoles', equipped with an eye-spot, a primitive brain, a notochord (the precursor of a backbone) and papillae for attachment at the front end. After a short swim, they attach themselves by the head to the bottom, and then absorb the tail. In some colonial species, the larvae are photopositive and then progressively more photonegative before settlement – a behaviour that would favour settling on rock walls or in caves – but there are few studies on Australian species, as described below (Svane and Young 1989).

The longevity of ascidians is largely known from the observations of Kay and Butler (1983). Three common species that lived on pilings, *Botrylloides leachii* and *Didemnum* spp. A and B – see Figs 13.1, 13.2), were annuals, with their larvae settling in summer, peaking in abundance in winter–spring and then dying off.

A few ascidians have evolved symbioses with photosynthetic zooxanthellae embedded in the outer test (see Chapter 9), or unicellular green algae (Kott 1980) – a symbiosis which has been shown to enhance the ascidian's growth rate by transfer of photosynthetic products (Olson 1986). Possibly, the most interesting such association is that of the colonial *Didemnum* spp. with the prochlorophyte *Prochloron* sp. In this relationship, the algal cells are loosely attached to the surface of the ascidian colony, imparting to it a greenish or pinkish colour (e.g. Duclaux *et al.* 1988). However, this association has not as yet been recorded in ascidian species south of ~32°S (Kott 1981).

Life history and ecology

The life history given below of 'bluebells', a small colonial species *Clavelina moluccensis*, and a large, solitary species *Pyura praeputialis* (formerly called *Pyura stolonifera* – see Rius and Teske 2011), illustrate the ecology of the two groups of ascidian. The former was intensively studied by Davis (1987, 1988a, b, 1989a, b; Davis and Butler 1989; Davis *et al.* 1991; Davis and Campbell 1996), and the latter by Anderson *et al.* (1976), Dalby (1995, 1996, 1997a, b), Marshall (2002) and others.

'Bluebells' *Clavelina moluccensis*

Studies on this species (Plate 35) were done on opposite sides of Gulf St Vincent and it showed remarkable differences in population dynamics between the two sites. At one site the population was annual, with colonies passing through phases of growth and degeneration. After recruitment in the first 1–2 months of the year, zooids increased in size and then degenerated from about mid-year, recovered later in the year, and then degenerated again at the end of the year. During degeneration, the siphons closed and the zooids shrunk within the test, producing tiny nodules 2–3 mm high on the basal mat. Within a month, the nodules regenerated into functional zooids. Growth of zooids was fast, with zooids reaching full size within a month. At the other site, degeneration of colonies recurred year round, but regenerated was rapid, with colonies often living for 2–4 years. Hence, overall density remained fairly constant.

In nature, the major predators at the sites were the side-gilled slug *Pleurobranchus hilli*, which grows to > 13 cm long, the flatworm *Pseudoceros* sp. and the red rock crab *Plagusia chabrus*. If a predator damaged a zooid, it rapidly regenerated, but if a juvenile was damaged it disappeared. In caging experiments, crabs ate 60% of zooids, while foraging leatherjackets ate none. This *Clavelina* contains quite high levels of vanadium, which is thought to deter fish predators (Davis 1988b).

Clavelina reproduces sexually and asexually. In sexual reproduction, the gonads ripen from August, sperm are released in the following months and fertilise eggs located in brood pouches, which acquire a conspicuous blue colour. Brood pouches at the two sites contained an average of 9–15 tadpoles. Tadpole larvae are released from late spring over 5–6 weeks at a size of ~4 mm, with larval release correlated with temperature. At one site the colonies died after larval release, but at the other the colonies repeatedly regenerated and reproduced for years (Davis 1989b).

In asexual reproduction, colonies subdivide to produce independent daughter colonies, which grow and survive for a few months. These fission events were found to be low at one site, but much higher (~40% of colonies) at the other, and almost always seemed to have been preceded by a degeneration process (Davis 1989a).

In imaginative experiments, Andy Davis placed small traps over colonies when tadpole larvae were mature and, after gently shaking the zooids, caught the newly released larvae and then let them go into the water column. The diver then followed them closely to see where they settled – indicated by a vigorous beating of their tails. The tadpoles always swam into the current, but their net movement was downstream. They swam a few metres above the bottom for up to 13 m (average distance only 2.2 m in ~2 min), and tended to settle on vacant space. A few settled on the solitary coral *Culicia* sp., and these were quickly eaten. Larval mortality was 3–18% over this short distance, but overall only 6% of those released settled successfully (Davis 1988a; Davis and Butler 1989). Interestingly, the larvae tended to settle at regular distances apart, suggesting that they could detect each other, and avoided settling too close – an example of a likely intra-specific competitive interaction.

Andy Davis then went on to examine the survival of settlers on different substrates. Those that settled on bare substrate had a high survival of ~80% over 10 days, whereas those that settled on sponges (e.g. *Mycale* sp.) had the lowest survival of ~5% over 10 days. In fact, most larvae that 'tested' *Mycale* sp., as well as other sponges, were deterred from settling, indicating that deterrence acted on contact with the sponge surface and was induced by chemical compounds in the sponge (Davis *et al.* 1991). Overall, < 1% of released larvae survived to adulthood (Davis 1987).

In summary, the study at two sites showed rather different life history strategies. At the site where *Clavelina* only bred once and then died, juvenile recruitment and survival were high, whereas at the other site where adult survival was high and colonies bred for several years, recruitment of juveniles was low and less important for the population. Many environmental factors could have brought about these differences, but the most likely ones could not be tested (see Davis 1989a). However, the study does illustrate how colonial species, with two reproductive modes, are well equipped for survival. Their budding ability provides the opportunity for colonies to spread rapidly over vacant space and enjoy greater resilience to loss from predation or damage (see Kingsford and Battershill 1998). Such detailed studies as the above are remarkable, and extremely demanding of time, and Andy Davis spent > 600 h underwater doing them.

The cunjevoi *Pyura praeputialis* and *Pyura dalbyi*

These species, once lumped together under *Pyura stolonifera*, were thought to be widespread on Southern Hemisphere temperate coasts until Rius and Teske (2011) and Teske *et al.* (2011) found that at least five cryptic species were involved. *P. praeputialis* is the 'cunjevoi' (an aboriginal name) and well known on the east Australian coast, where it forms extensive beds in the lower inter- and shallow subtidal zones. Here, it is used as bait and rock fishers have had a severe impact on intertidal populations (Fairweather 1991). As one shore fisher colourfully wrote: 'When you walk over it gingerly to recover your hook, it looks up malignantly and spits seawater into your eye' (Dakin 1960). *Pyura praeputialis* has recently been introduced to Chile (Castilla *et al.* 2002).

Pyura dalbyi is similar in appearance to 'cunjevoi' and occurs subtidally in Victoria and Tasmania to ~10 m depth; it was later introduced to Albany, Western Australia. Both species occur in Victoria, where *P. dalbyi* is yellow coloured and lives in calmer waters and *P. praeputialis* is a brown colour, with a thicker tunic and lives on exposed coasts. Curiously, the taller siphon is the inhalant one in *P. dalbyi* and the exhalant siphon in *P. praeputialis*. This is likely related to the flow regime, because the taller siphon would favour higher water intake rates in calmer water (Dalby 1997a, b). The picture is confused by an undescribed subtidal *Pyura* sp. present in SE Australia.

In New South Wales, *P. praeputialis* has mature gametes in April–May, with a minor resurgence in September, and at low tide ejects sperm and eggs via the exhalant siphon as a viscous matrix on the top of the tunic. In Victoria, spawning of *P. dalbyi* appears to occur earlier in January–February. Because these hermaphroditic species spawn synchronously and live in aggregations, they are useful to examine the fertilisation success of eggs (see also Chapter 11 for similar studies on sea-urchins). In such a study, Marshall (2002) found that fertilisation success strongly depended on the distance of the spawner from its nearest neighbour and the pair's location in relation to water movement direction. Spawners aligned in the water movement direction had high success rates of 60–90% when close together, and 0–40% when 3 m apart, whereas spawners side by side had low and variable success rates, even when close, and near to zero when > 1 m apart. Moreover, in other tests using self-sperm and non-self-sperm, fertilisation rates were more than three times higher with the latter, indicating that self-fertilisation in nature is probably low. Another factor affecting fertilisation of eggs is egg size, because eggs of large individuals (> 0.3 mm) are twice the size of those of small ones. Marshall *et al.* (2000) showed that high fertilisation success rates of 50–60% could be achieved in large eggs at sperm densities of only a fraction of those needed for small eggs. Hence, more distant, large neighbours could achieve higher cross-fertilisation levels than small neighbours.

After fertilisation, tadpole larvae begin to hatch ~12 h later at a size of ~1 mm and then swim for up to ~2 h before settling. When a larva detects an adult chemically, it seeks to settle and, upon attachment to the bottom, resorbs the tail, develops siphons within 2.5 days, and starts feeding at 4 days (Anderson *et al.* 1976). Hence, the tendency for this ascidian to live in aggregations is a consequence of the tadpole's settlement behaviour.

But what are the benefits of congested living? This question was examined by Dalby (1995) who found that living in aggregations increased the competition for food between individuals and reduced the growth rate. Yet he concluded that the behaviour was adaptive, because it brought numerous benefits that outweighed the costs. The benefits included: less chance of dislodgement by wave action; enhancement of food supply by baffling the water flow; reduced predation levels; and, perhaps most importantly, increased fertilisation success.

In feeding, solitary ascidians process large volumes of water by pumping it through them by the use of internal cilia or passively by orienting the inhalant siphon into the flow, and the exhalant siphon away from the flow. Knott *et al.* (2004a) showed that in the field *P. praeputialis* is consistently oriented with its inhalant siphon into the direction of the swell, and that the passive flow of water could contribute about half the total flow through the ascidian (probably a common phenomenon because it applies also to another ascidian, *Phallusia obesa* – AJ Butler *pers. comm.*). From Klumpp's (1984) equation, an average-sized *Pyura* of 5 g dry weight could actively pump up to 100 mL of water per minute through it, and one placed at a site with oceanic swell could increase that rate by 50%. Filtering efficiency of the species is 95–99% and, with a mucus sheet, it captures non-selectively particles of phytoplankton and detritus from 20 μm size down to bacteria of 0.5 μm size equally efficiently (Klumpp 1984). However, high silt loads in the water reduce filtration rates to low levels (see Robbins 1983).

As described in Chapter 8, these species have numerous symbionts – nemertean worms, copepods and amphipods – that live in the siphons, but curiously nemerteans occur only in *P. dalbyi* and copepods or amphipods in *P. praeputialis* (Dalby 1996). The main predator of *P. praeputialis* in New South Wales is the triton *Cabestana spengleri*, with crabs, the sea-star *Uniophora granifera*, rays, eels and oyster-catchers also consumers (Dalby 1995).

Ascidian associations

Ascidians form associations with many other animals, as noted in Chapter 8 for the relations between amphipods, copepods and nemertean worms and species of *Pyura*. These associates have been variously considered to be commensals or parasites. A curious, three-way 'defence' association of a different kind occurs between two ascidians, *Pyura spinifera* and *Cnemidocarpa pedata*, and the yellow or purple sponge *Halisarca laxa* in New South Wales, where urchin grazing is intense. Apparently, *Cnemidocarpa* first colonises the substratum and is then colonised by the sponge, and together they become the focus for the *Pyura*, which selectively settles there also. The sponge then grows all over both ascidians, providing protection from other fouling organisms (Davis 1996; Davis *et al.* 1996). It is difficult to know what selective pressures led to the evolution of such a complex non-obligate set of relationships, but Davis thought that the presence of *Cnemidocarpa* at a site would indicate that sea-urchin grazing was low, so that the ascidian signalled a refuge from predation. Alternatively, settlement of *Pyura* on *Cnemidocarpa* could be a reliable cue for finding the sponge *Halisarca*, and so pre-empting other foulers.

The external test of some ascidians also provides a substrate for animals and algae. However, some species with high concentrations of vanadium and an acidic test with low pH (< 2) are remarkably free of fouling organisms (Stoecker 1980).

BRYOZOANS

Bryozoans are colonial animals, consisting of many tiny zooids up to ~1 mm across that filter phytoplankton and protozoans from the water. The majority of species grow on a wide range of firm substrata, such as the shaded surfaces of reefs, pebbles, mollusc shells, ascidian tests and crab carapaces, while others grow on algae and seagrasses. They take a wide variety of forms. Some are non-calcareous (the soft bryozoans) and others are strongly calcified to provide structural support. Some are encrusting, while others are erect, tufted, branching or bushy in form – a few growing very large, to > 50 cm in height.

In shallow habitats, bryozoans are largely restricted to shaded surfaces, and in high wave energy zones soft bryozoans are especially prominent. In deeper water, bryozoans live in more open habitats; in places of moderate to strong tidal current with low sedimentation rates, bryozoan beds or mounds dominated by *Celleporaria* spp. sometimes occur. Below the photic zone at depths of 100–130 m, bryozoan thickets dominate the sea bed (Hageman *et al.* 1995). Anomalously, similar thickets occur as shallow as 10 m deep in Port Davey, south-west (SW) Tasmania, where dark brackish water blocks light penetration (Chapter 20).

Zooids have a tentacular crown (called a lophophore) that protrudes through an orifice, and together they form a filtration network to capture bacteria, phytoplankton and protozoans from water currents generated by cilia. Zooids are inter-connected by pores in the side-walls, allowing a slow percolation of nutrient fluids through a colony.

Life history

Reproduction in bryozoans is both asexual and sexual. The former occurs by the budding of zooids at the edge or end of a colony, and these then fuse with the neighbouring zooids. In sexual reproduction, most bryozoans are hermaphroditic, producing eggs and sperm. Some species release eggs into the water, while others brood their eggs in brood chambers. Larvae from non-brooding species live for up to ~2 months in the water and feed on plankton, whereas brooding species produce non-feeding larvae with a very short free life of hours to days. When ready to settle, larvae hover briefly above the bottom, usually on shaded surfaces, evert their adhesive sac, settle and grow into another colony.

The growth of bryozoans has seldom been studied locally, and then mainly on panels. Wass and Vail (1978) showed that the increase in radius of more or less round colonies was linear, (i.e. a plot of $(\text{area})^{0.5}$ versus time is linear). At Port Hacking, New South Wales, Vail and Wass (1981) found extreme variability in the growth of bryozoans. For example, some species grew rapidly, ceased growing and then started growing again, or regressed; others followed different patterns. Examples of some averaged growth rates, in terms of increase in area, are: *Schizoporella unicornis* ~4.6 cm^2 $week^{-1}$ over 6 months; *Celleporaria fusca* ~3.1 cm^2 $week^{-1}$ over 5 weeks (both encrusting species); and *Zoobotryon verticillatum*, an arborescent species, ~48 cm^2 $week^{-1}$ over 6 weeks.

The longevity of bryozoans varies as much as growth. Some crustose species are seasonal, living for < 1 year, but others live much longer. Kay and Butler (1983) recorded life spans of 2.5 to > 4 years for three species of *Celleporaria*, and 1.5–2 years for *Smittina* and *Biflustra*.

Predators

The defences of bryozoans are calcification of the walls of zooids, the development of spines and the production of deterrent chemicals (see below). Nevertheless, several generalist predators feed on bryozoans. They include some leatherjackets (Keough 1984b), some sea-stars (see Table 11.1) and some chitons, notably in the genera *Ischnochiton* (two species) and *Chiton* (four species) (Kangas and Shepherd 1984; Chapter 9). In addition, some specialised predators feed exclusively or substantially on bryozoans, notably nudibranchs (see Table 9.2), pycnogonids (see Table 15.9), and some flatworms (Ryland 1977).

Ecology of three species

Bock (1982) described the common southern Australian species and Ryland (1977) gave an account of the ecology of the group. The ecology of three very different species from the southern region – two of them widespread – illustrates the ecological diversity in the phylum.

Bugula neritina

This cosmopolitan species forms flexible, bushy colonies up to 80 cm high (see Plate 36). It is common in harbours and bays on Australian coasts, having apparently been transported to Australia and around the Pacific and Atlantic during the 19th century by shipping (Mackie *et al.* 2006). It is likely that disturbances, whether physical or by the discharge of toxic substances, that commonly occur in harbours increase the invasive success of this species (Clark and Johnston 2005).

Colonies are composed of hermaphrodite zooids in which chemically synchronised cross-fertilisation is apparently achieved between colonies. After fertilisation, embryos are brooded in large ovicells and a single colony may release up to 100 larvae during daylight, starting at dawn (Keough 1984a). The larvae are non-feeding and rely on nutrient reserves received before their release. Swimming larvae have alternate photopositive and photonegative behaviour. When ready to settle, they explore the surface of the substratum and then settle on a suitably rough surface. Normally they settle within an hour or so of release; however, if this is delayed, their reserves become exhausted and they lose the ability to metamorphose and settle (Walters 1992; Wendt 1996). Larvae tend to settle gregariously next to their siblings – a behaviour that is likely to reduce predation (Keough 1984a).

After settlement, growth can be rapid, but is highly variable, depending on the environment. In one study, *B. neritina* grew ~600 zooids in a month and was sexually mature in 6 weeks, but in other studies growth was much slower and colonies matured in ~1 year. Fast-growing colonies tend to grow larger, reproduce sooner and live longer (Hyman 1959; Keough 1989a, b). However, the bryozoan's life history sometimes includes an annual period of senescence, dying back in summer in warm temperate regions and in winter in cool-temperate regions (Keough and Chernoff 1987). Interestingly, Marshall *et al.* (2003) recorded a 'grandfather' effect, in which larger colonies produced larger larvae, which in turn reproduced at a younger age, lived longer and had greater fecundity.

A predator of this bryozoan is the introduced nudibranch *Polycera hedgpethi*, which can consume whole branches of a colony at one time. Bone and Keough (2005) found that such losses to predation (up to 20% of a colony) did not affect either the colony's reproduction or its reproductive output. However, other damage, such as might be inflicted by a storm, could reduce fecundity and delay the onset of reproduction. A chemical defence is also available, because *B. neritina* has a symbiotic bacterium that produces bryostatins. This chemical apparently deters fish and crab predators in some parts of its range (McGovern and Hellberg 2003). Whether it is a deterrent in Australian populations is yet to be examined. As an aside, bryostatins have proved to be useful anti-cancer compounds in the control of leukaemia and other cancers.

Parmularia smeatoni

This curious, disk-shaped species (Plate 37) is anchored to the bottom by a hollow, flexible stalk, which enables it to sway to and fro in the current. Colonies live in places of moderate to strong current to ~30 m depth, and the disks are always oriented normal to the current flow (see Shepherd 1983 and Brown *et al.* 2002).

The larva of *Parmularia* searches for a suitable settlement site, which may be a shell fragment, rock, sponge, rhodolith or seagrass roots. In Kirsty Brown's study, colonies appeared in mid-winter and grew at a rate of 0.15–0.3 mm day^{-1} to reach a size of 30–50 mm diam. in ~4 months. Many colonies died off after 8–9 months, but a few survived for up to ~1 year. In Shepherd's (1983) study, *Parmularia* settled in high densities on shell fragments in a megaripple field in an area of strong currents, and individuals grew to a size of ~25 mm in ~8 months at a rate which declined with size. A von Bertalanffy growth curve fitted to the data (K = 2.0 year^{-1}, $L\infty$ = 28.9 mm) indicated an average early growth rate of 0.11 mm day^{-1} from the first to the third month, slowing to an average of 0.07 mm day^{-1} over the next 3 months. In this unstable habitat *Parmularia* grew very long stalks to 25 cm, enabling it to better survive in a mobile sand dune. Mortality was high (48% per month), so that few individuals would have survived a year.

Membranipora membranacea

This cosmopolitan species forms thin, encrusting colonies over the blades of large brown algae, notably the kelps *Ecklonia* and *Macrocystis*, and some species of *Sargassum*. It is a protandrous hermaphrodite, going from a male stage to a transitional stage, where both eggs and sperm are released, and later a female stage where only eggs are released. Eggs are fertilised internally in vestibules between feeding tentacles by sperm from another colony. The fertilised eggs are then released without brooding and the emerging larvae drift and feed on the plankton for 4 weeks till they settle. Settlement, however, is not random, because the larvae test the surface and choose the highest points of the substratum on which to settle, where they are least susceptible to being overgrown (Walters and Wethey (1991). Sites with high water flow and low sediments are favoured by larvae (Duggins *et al.* 1990). In the Northern Hemisphere, the larvae settle in spring (Harvell and Grosberg 1988), and presumably they settle in the austral spring in the Southern Hemisphere.

The growth of colonies seems indeterminate – that is, they keep growing from a growing edge while there is space for expansion. When unimpeded by other colonies, the growth of *Membranipora* has been recorded at a rate of ~10 cm^2 week^{-1} for 10 weeks, but if space is depleted by competing colonies, growth ceases or is reduced to a fraction of this (Ellison and Harvell (1989). Colonies first reproduce after 40–60 days, with the oldest zooids being reproductive first. After ~3 months at the end of summer, colonies begin to senesce and die in cool temperate waters, but in warm temperate regions populations persist year round, although individual colonies may live for only ~6 weeks (Harvell *et al.* 1990; Harvell and Helling 1993).

Possibly the most remarkable aspect of this bryozoan's natural history is its ability to defend itself against predators and to produce an aggressive armoury against competitors. These inducible defences are triggered by the predator or competitor and are known mainly among a very few colonial invertebrates such as bryozoans and octocorals, which cannot escape (Adler and Harvell 1990). When a bryozoan larva settles on a kelp frond, other colonies may have previously settled on the same frond. In this case, when one colony's edge comes in contact with a competitor, growth of both ceases. Thereupon the larger colony produces stolons at the edges within 2 days. The stolons, sometimes called tower cells, are elongate buds that grow quickly from the marginal zooids at 3–4 mm a day to a length of ~10 mm, and these stolons project over the edge of the colony and stop the advancing front of the competitor. The larger colony can now surround the smaller one, and halt its growth. The stolons persist for a day or so and then disappear. However, this defence comes at a cost in terms of reduced growth rate and survival, and the colony is promptly induced to enter its reproductive phase (Harvell and Padilla 1990; Padilla *et al.* 1996). Interestingly, the production of stolonic outgrowths for defence or aggression occurs in other Australian bryozoans, and may be a widespread phenomenon (Osborne 1984).

In the case of a nudibranch attacking the edge of a colony: within 2 days of detecting the predator's cues, the colony produces spines on the frontal membrane of peripheral zooids and these are a deterrent to grazing. The cost of producing spines is similar to that described for stolons above. Thus it can be seen that the overall strategy of this bryozoan is to grow as fast as possible, but in the presence of a predator or competitor impeding growth, to reproduce as soon as possible (Harvell 1990, 1992).

ASSEMBLAGES OF SESSILE FAUNA

Shallow-water habitats

Sessile faunal assemblages, mostly studied on jetty piles, typically contain > 60 species from the five phyla previously noted. Hence, given that so many species must share space and planktonic food suspended in the water, it is clear that competition between them becomes important. Over time, as organisms die or are removed by disturbances (e.g. water movement or predators), patches of vacant space are created and become available for other species. The question then is which species will invade or colonise the space? Factors found to influence the outcome include: the ability of species to recruit into them; the size of the newly bared patch; the proximity of other organisms to the patch; and lastly the time needed to progress through successional phases. These factors are amenable to experimental manipulation, and have been studied at two sites, as discussed below. For reviews of these and related studies see Tardent (1963), Van-Praët (1985) and Butler (1995, 2008).

Interactions between species

Two studies – one in South Australia and one in Victoria – have examined the interactions between sessile fauna on jetties. In the former, Kay and Keough (1981) cleared patches of 600 cm^2 on jetty piles and recorded the recruitment and growth of animals on them, and their interactions over a year. From this and later work (Butler 1991), a hierarchical model was developed to explain what would happen in a cleared patch. Sponges and ascidians were good competitors but poor recruiters, whereas bryozoans and tubeworms were poor competitors but good recruiters. Hence, over time with low disturbance regimes, the good competitors would dominate the community. Kay and Keough (1981) presented a contact matrix (Fig. 13.1) showing the outcome of interactions between the various sessile species on pier pilings. Sponge and ascidian species generally overgrew bryozoans and tubeworms (the poor competitors), but within those phyla some pairs of species were equivalent; that is, stand-offs sometimes occurred, where both species stopped growing at their common boundary, and at other times each species won or lost encounters with about equal frequency. In other cases, one species was a clear winner in most interactions. However, among some of the dominant species the interactions formed a network rather than a hierarchy (Fig. 13.2), with no clear winner.

The studies of Russ (1980, 1982) at Portsea Pier, Victoria gave similar results to those of Kay and Keough, but Russ noted several other factors that influenced the outcome of species interactions, as follows:

- Size gave a competitive advantage to a colony, so that occasionally a large inferior competitor, such as a bryozoan, won over a smaller ascidian.
- Colony thickness was important: as ascidians and sponges increased in size, they became thicker, and this of itself gave them an advantage.
- Fish predation, notably by leatherjackets, could prevent the take-over of space by ascidians, the dominant competitors, and allow poor competitors, such as encrusting and arborescent bryozoans (e.g. *Bugula*) to become established (see Keough 1984c). The presence of arborescent bryozoans then increased the structural complexity of the community and provided newly settled ascidians with a refuge from fishes, so enhancing their survival.

Like the South Australian studies, Russ found that reversals were sufficiently common to give rise to networks and back-loops. Yet the end result remained hierarchical in terms of dominance: that is:

ascidians ≥ sponges > bryozoans > barnacles, tubeworms, tubiferous amphipods.

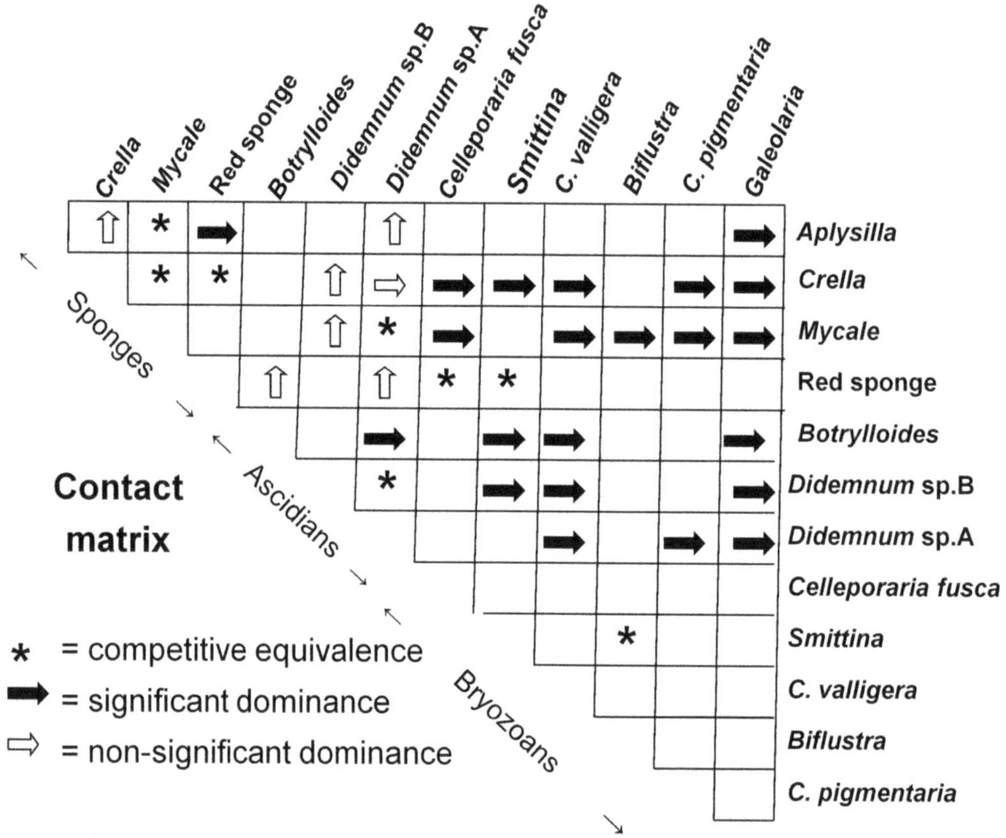

Figure 13.1: Contact matrix showing the net outcome of competitive interactions between pairs of species on jetty piles. Arrows point in the direction of the dominant species. Matrix redrawn from Kay and Keough (1981). Blank squares are present where the number of interactions was less than six.

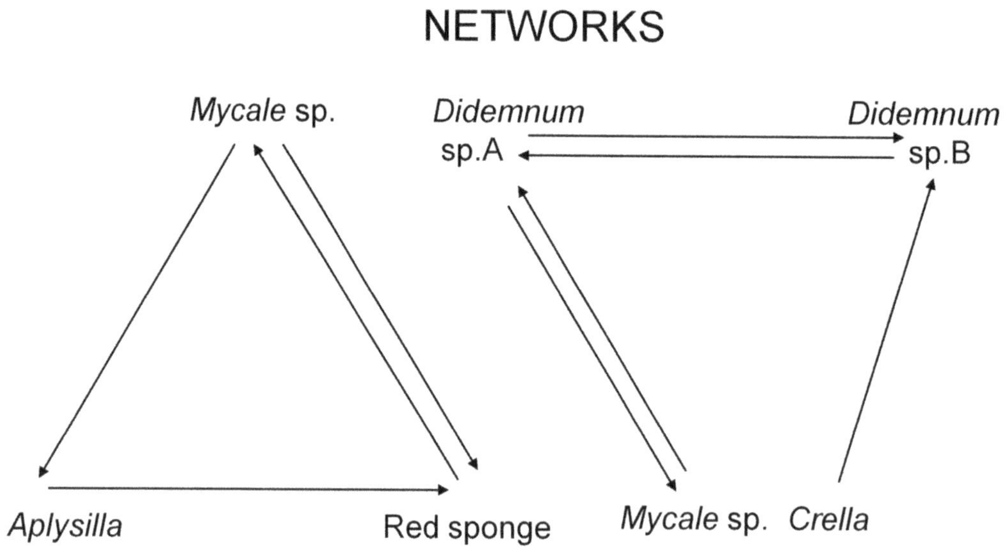

Figure 13.2: Two networks of overgrowth patterns from data in Fig. 13.1. Two parallel arrows pointing in opposite directions indicate that species are competitively equivalent. A single arrow points in the direction of dominance.

Succession in shallow reef habitats

Darren James and SA Shepherd

The faunal succession on natural and artificial reefs at ~5 m depth off Altona in upper Port Phillip Bay was studied by James (2005) over 8 years. The natural reef was patchy, smooth low basalt reefs with up to 0.5 m relief, which outcropped from a sandy bottom, and artificial concrete structures were placed there in an abalone ranching program. The concrete units were 1 × 0.3 m in area basally and 30 cm high, with channels running lengthwise above and below; the channels were 18 cm deep above and 6 cm deep below. The reefs were subject to periodic sedimentation, stirred up by local wind waves.

The succession on the artificial reef habitats is shown in Fig. 13.3. The first colonists were native oysters, barnacles, filamentous algae and tube diatoms (termed early colonists). After 1–2 years, the diatoms were replaced by red and green algae. Animals began to invade from surrounding reef early in the succession, and grazed the algae or fed on other colonising species. These invaders were:

- predatory species – i.e. the 11-armed sea-star *Coscinasterias muricata*, and the granular sea-star *Uniophora granifera*
- omnivorous species – i.e. the seagrass swimmer crab *Nectocarcinus integrifrons*, and the 6-armed sea-star *Meridiastra media*
- the herbivorous purple urchin *Heliocidaris erythrogramma*.

Blacklip abalone, *Haliotis rubra* also crawled onto the artificial habitats at low densities of up to 2 m^{-2}, and grazed algae in the vicinity, while their larvae also settled on the corallines. Thus, two processes were concurrent – the settlement and growth of sessile organisms, and their disappearance via predation or following disturbance.

After ~2 years sponges, bryozoans and ascidians (the late colonisers in Fig. 13.3) began to appear and gradually reached a cover of ~24% after 8 years. Sediment on the artificial reef varied over time according to wave intensity, and the high sediment load that appeared after ~20 months may have reduced the cover of early colonists and increased that of bare rock. The cover of organisms on the natural reef, mainly sponges and ascidians, also showed some instability at ~20 months, but then stabilised for the remaining 6 years (Fig. 13.4; Plate 38). A comparison of succession on the two reef habitats shows that the artificial reef was far from reaching the same degree of relative stability as the natural reef, even after 8 years.

Stability of communities

The above studies raise the question whether and when communities reach a stable end-point. Kay and

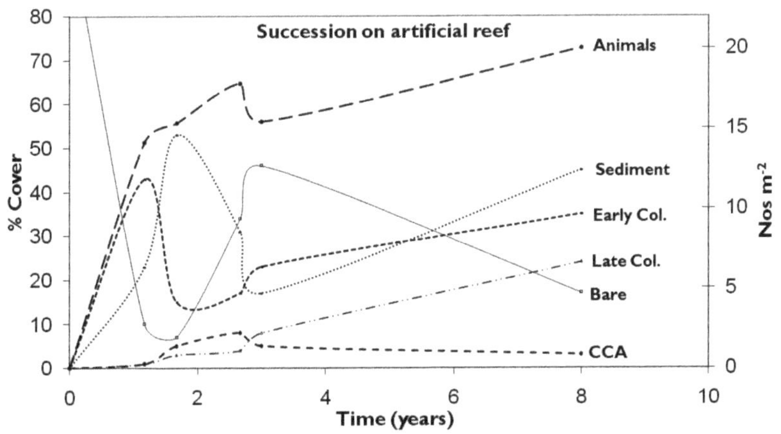

Figure 13.3: Changes in percentage cover of sessile organisms and uncolonised (= 'Bare') substratum on artificial reef of low relief in upper Port Phillip Bay over 8 years. Early colonisers (Early Col.) are barnacles, oysters and other bivalves (e.g. *Cleidothaerus albidus*). Late colonisers (Late Col.) are sponges, bryozoans and ascidians; CCA = crustose coralline algae; 'Sediment' refers to reef substratum covered with sediment. 'Animals' means the total abundance of animals (in numbers m^{-2} shown on vertical axis on right) recorded on the artificial reef (see text for species).

Butler (1983) considered this question for communities on two jetties that had been in place for > 70 years. First, they defined 'stability' in terms of the narrow boundedness of fluctuations in percentage cover of the various sessile species over time periods of several years. The communities were consistently dominated by long-lived species with lifespans of > 4 years (e.g. the sponge *Dendrocia* and the solitary coral *Culicia*) and their cover changed little over time. Hence, they were stable, despite a continual, slow turnover of space. However, neither of the above studies on jetty pilings had examined the process of the long-term development of a sessile community from a bare substratum.

Butler and Connolly (1996, 1999) followed changes in a sessile community over 14 years at a newly constructed jetty site at Point Lowly in upper Spencer Gulf and found that it took many years for the community to reach a stable point predicted by Butler's (1991) model. Even then, the authors cautioned that a stable end-point may not have been reached. Early in the succession, many unitary animals (hammer oysters, solitary ascidians, hydroids and the soft coral *Carijoa*) dominated the system. But over time these species disappeared or fell to small numbers and encrusting sponges came eventually to dominate the system, together with a few erect colonial species.

Darren James' (2005) study of artificial and natural reefs in Port Phillip Bay provide an interesting contrast to those on pilings. The periodic disturbances from sedimentation, predation and grazing ensured that, even on the long-established and relatively stable natural reef (Fig. 13.4), only about one third of the area was covered by late colonisers, and much of the bottom was relatively barren.

Studies of the fauna of razor-shells, discussed below, provide yet another picture.

Small, isolated substrata

The razor shell *Pinna bicolor* is a bivalve that roots on sandy bottoms and can live for up to 13 years. Above the sand-line, the shell presents the outer calcitic layer of its two faces, each with an area of up to ~150 cm^2 available for colonisation by sessile fauna. The fauna of razor shells differs markedly from that of jetty piles in that the encrusting fauna is mostly bryozoans and tubeworms. Sponges and ascidians are relatively uncommon. Keough (1984c) examined their interactions and found that tubeworms (*Galeolaria* and *Spirorbis*) were inferior to all other species, and were always overgrown by the rare sponges and ascidians, if they chanced to settle. No bryozoan ever overgrew another bryozoan, and colonies either fused if they were the same species, or had a stand-off if they were different species. The interaction matrix is given in Fig. 13.5. Because razor shells form small isolated patches on sandy bottoms, recruitment to them depends strongly on the recruiting ability of the sessile animals. Bryozoans and tubeworms were good recruiters and so prolifically colonised the shells.

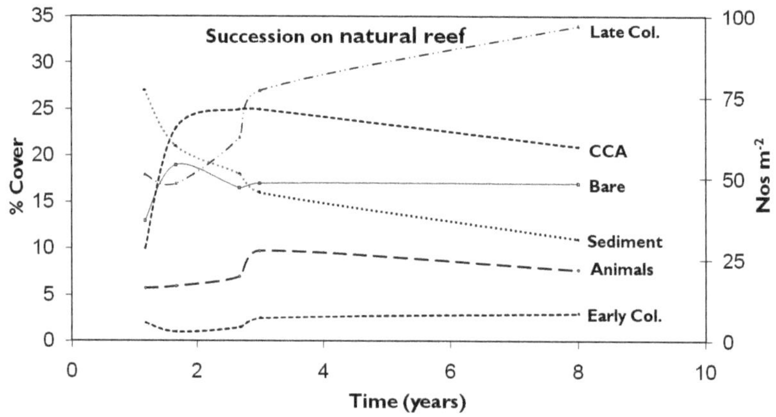

Figure 13.4: Changes in percentage cover of sessile organisms and uncolonised substratum on basalt reef of low relief in upper Port Phillip Bay over the same 8 year time period as in Fig. 13.3. For key to species composition see caption to Fig. 13.3.

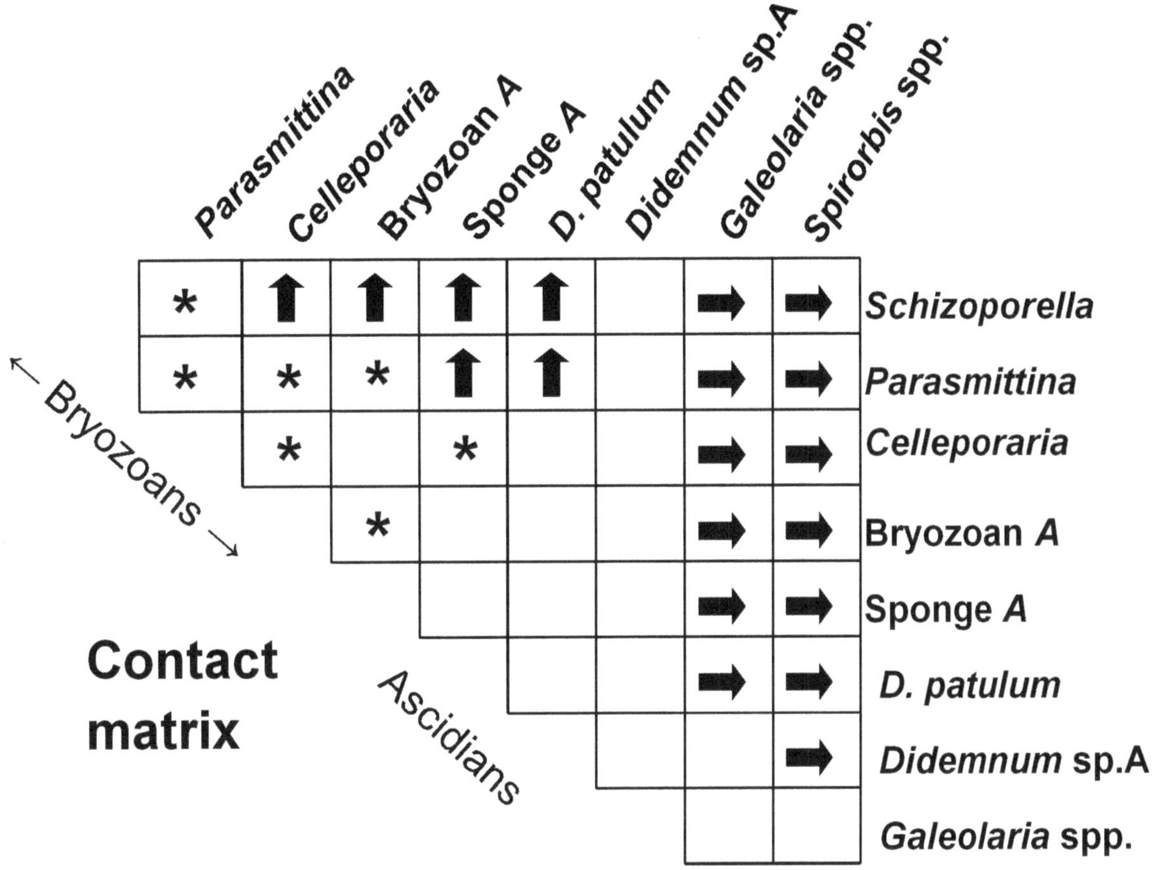

Figure 13.5: Contact matrix showing the net outcome of competitive interactions between pairs of species on the razor-shell *Pinna bicolor.* Arrows point in the direction of the dominant species. See Fig. 13.1 for the key to the meaning of symbols. Blank squares indicate that the number of interactions was less than three. Matrix redrawn from Keough (1984c).

Mike Keough also examined the effect of predators in the system. They were mainly leatherjackets, and these ate many of the two colonising didemnid ascidians, but only in the first month after they settled. Because these didemnids produce many calcareous spicules with growth, Mike Keough thought that they were only vulnerable or palatable to fish for a few weeks after settlement. Not surprisingly, where an assemblage depends so strongly on chance recruitment events of poor competitors, no equilibrium in the assemblage is ever reached (Keough and Butler (1983).

Patch size and location

The above studies on jetty piles and on razor shells suggest both the importance of patch size and their proximity to, or isolation from, similar habitats. These questions were examined experimentally by Keough (1984b) who found that non-isolated patches, as on jetties, were colonised by the rapid invasion of the surrounding species (ascidians and sponges); any larval recruits of other species were quickly overgrown. The size of the patch was unimportant. In contrast, for isolated patches size was important. Small patches were colonised by the good recruiters (bryozoans and tubeworms), but, with increasing size, patches were colonised more and more by ascidians and sponges. An exception to the model was the solitary coral *Culicia*. This species was a good competitor, because it grew slowly and could occupy space for a long time, but it was a poor recruiter. Yet it managed to colonise both pilings and razor shells probably because of its life history features (Butler 1991).

The constantly changing fauna on razor shells provides a good example of the importance of spatial and temporal scales to the concept of stability. At the large scale of long-established jetty pilings, mature assemblages were stable and the relative abundances of species did not change over time. But it could take many years for the fauna on a new jetty to reach this stable end-point. In level-bottom reef habitats, disturbances may prevent the community ever reaching a completely stable end-point dominated by sponges and ascidians. And, at the other extreme of scale, as in the case of a razor-shell, the picture is very different. Here, isolation of the substratum and the prolific recruiting ability of bryozoans and tubeworms together give them a marked advantage over ascidians and sponges, which can seldom recruit to such places and so cannot become dominant competitors.

Deep-water (> 30 m deep) reef habitats

These habitats to depths of > 100 m can now be described using multi-gas rebreathers and sophisticated video and benthic habitat mapping equipment. Such studies have been made in New South Wales, Western Australia and Victoria (Roberts 1996; Roberts and Davis 1996; Beaman *et al.* 2005; Edmunds *et al.* 2006, 2008a, b; Waddington *et al.* 2010). In such low-light habitats, reefs are dominated by sessile fauna – sponges, ascidians, cnidarians and bryozoans – with a macro-invertebrate fauna variously of polychaetes, decapods, amphipods, isopods, molluscs and echinoderms. Aspect, light, water movement and sedimentation are all critical factors determining the composition of species at any particular site, and the above studies all emphasise the small-scale patchiness of species and assemblages, which vary both over time and according to the various combinations of the above factors.

The studies at five sites are summarised in Table 13.2. On the lower west coast of Australia, the waters of the southward flowing Leeuwin Current are clear and nutrient poor, and the photic zone is deep. Here macroalgae are abundant and the dominant primary producers, while sponges are significant secondary producers, capturing pico- and microplankton (0.2–20 µm size). The western rock lobster is the main carnivore in this system (Chapter 15).

The two coastal Victorian sites studied by Matt Edmunds and colleagues show contrasting physical differences. The ocean site at Twelve Apostles is exposed to swell and clear water, and the site at Port Phillips Head is subject to strong tidal currents and low light penetration. The poor light reduces algal cover and the strong tidal currents greatly increase sponge and filter-feeding invertebrate cover.

After the channel deepening project at Port Phillip Heads had scoured the rock surfaces of their mainly sponge assemblages, subsequent monitoring of the scoured sites gave a crude idea of the successional process over time, as shown in Table 13.3 for different depths (Edmunds *et al.* 2008b). The succession had not reached a climax state over the time of the study and, given the slow growth of some sponge species, this might not be reached for many years. The most notable habitat shift occurred at the shallow 27 m site, where the sponge community was replaced by an *Ecklonia* forest. However, the authors cautioned that nearby control sites also showed large changes in cover over time, so that natural variation may have partly confounded the inferred succession.

New Zealand Star Bank lies on the outer shelf in eastern Bass Strait. The high relief granite outcrops occur over an area of ~100 km^2 and reach to within 10 m of the surface. *Macrocystis* forests are at shallow depths, and from 50–90 m depth dense gardens of sponges, octocorals, crustose coralline algae and many crinoids dominate the slopes. The black sea-urchin *Centrostephanus rodgersii* creates barrens to ~50 m depth (Table 13.2).

The New South Wales site differs from the others in having high crustose coralline, and low invertebrate, cover. Apart from the described effects of light penetration and water movement type, and likely biogeographic differences, the causes of other differences between sites are virtually unknown.

Table 13.2: Cover and biomass of sessile fauna on rocky substratum at deep (> 30 m) sites in southern Australia (after Roberts 1996; Roberts and Davis 1996; Beaman *et al.* (2005); Edmunds *et al.* 2006, 2008a, b; Waddington *et al.* 2010).

C = Percentage cover with standard error (s.e.); CCA = crustose coralline algae.

	W.A.	VICTORIA			NSW.
Site locations	Lancelin, Jurien Bay, Dongara	Twelve Apostles (SW of State)	Lonsdale Wall, The Rip, Port Phillip Heads	NZ Star Bank, E. Bass Strait	Sydney (North Head, Long Reef, Bungan Head)
Aspect	horizontal reef	low profile reef	vertical wall with ledges	high profile granite reef	low profile reef
Water movement	ocean swell	ocean swell	tidal currents (3–5 m s^{-1})	Ocean swell	ocean swell
Depth range (m)	35–75	35–45	40–60	46–90	30–60
Macroalgal cover (%)	5–50 (*Ecklonia*)	45.2 (Rhodophyta)	0.1	*Phyllospora* at 10–20 m	2–40 Only CCA
Biomass (kg m^{-2}, s.e.)	2.8 (0.8)				
Macroalgal lower depth limit (m)	60–80 (*Ecklonia*)	> 45	27 (*Ecklonia*), 50 m red algae	?	~30
Encrusting sponge (C)	~50	14.7 (2.8)	17.6	0–11	3.4 (0.9) at 30 m 2.5 (0.4) at 50 m
Biomass (kg m^{-2}, s.e.)	4.3 (0.8)	no data	no data	no data	no data
Erect sponge (C)	(includes encrusting sponge)	11.3 (4.0)	50.0	38–58	4.3 (0.7) at 30 m 7.1 (1.0) at 50 m
No. of sponge spp.	149	> 30 morphotypes	?	?	> 50
Hydrozoans (C)	–	1.8 (0.6)	0.1	5–25[1]	all cnidarians
Octocorals, anemones (C)	–	2.8 (1.2)	7.0	5–25[1]	1–2.5
Bryozoans (C)	–	1.1 (0.3)	0.2	5–25[1]	0–1.5
Ascidians (C)	–	0.4 (0.2)	2.1	–	0–1.3
Polychaetes, sipunculans (g m^{-2})	2–6	no data	no data	no data	no data
Decapods (g m^{-2})	2–4	no data	no data	no data	no data
Amphipods, isopods (g m^{-2})	1–2.5	no data	no data	no data	no data
Molluscs (g m^{-2})	0.5–1.5	no data	no data	no data	no data
Brittle-stars, sea-urchins (g m^{-2})	0.5–3	no data	no data	10–14 m^{-2} barrens[2]	no sea-urchins

[1] characterised as 'mixed gardens' containing sponges, octocorals and small erect and encrusting fauna;
[2] Barrens formed by sea-urchin *Centrostephanus rodgersii*.

Table 13.3: Generalised successional stages of deep-reef at 40–60 m depth, mainly sponge assemblages, 1–2 years after scouring at Port Phillip Heads (after Edmunds 2008; Edmunds *et al.* 2008a, b).

Depth (m)	Stage 1	Stage 2	Stage 3	Stage 4
27	Microalgae, invertebrates	Low, red and brown algal turf with overgrowing sponge	Thallose red algae and juvenile *Ecklonia*	*Ecklonia* canopy, red algal understorey, sponge
37–47	ditto. Overgrowing sponge from adjacent rock	Matrix of diverse, mostly crustose species	Layer of crustose and erect forms of hydroids, ascidians, and bryozoans	Increased spread of sponges, hydroids, and anemone, *Corynactis australis*
67	–	Sparse anemones and patches of sponge	Sponges and hydroid colonies	Isolated large, erect sponges, and overgrowing sponges

REFERENCES

Adler FR, Harvell CD (1990) Inducible defenses, phenotypic variability and biotic environments. *Trends in Ecology & Evolution* **5**, 407–410.

Anderson DT, White BM, Egan EA (1976) The larval development and metamorphosis of the ascidians *Pyura praeputialis* (Heller) and *Pyura pachydermatina* (Herdman) (Pleurogona, Family Pyuridae). *Proceedings of the Linnean Society of New South Wales* **100**, 205217.

Ayling AL (1980) Patterns of sexuality, asexual reproduction and recruitment in some subtidal marine demospongiae. *The Biological Bulletin* **158**, 271–282.

Ayling AL (1983) Growth and regeneration rates in thinly encrusting Demospongiae from temperate waters. *The Biological Bulletin* **165**, 343–352.

Ayre DJ, Davis AR, Billingham M, Llorens T *et al.* (1997) Genetic evidence for contrasting patterns of dispersal in solitary and colonial ascidians. *Marine Biology* **130**, 51–61.

Beaman RJ, Daniell JJ, Harris PT (2005) Geology-benthos relationships on a temperate rocky bank, eastern Bass Strait, Australia. *Marine and Freshwater Research* **56**, 943–958.

Bell JJ (2008) The functional role of marine sponges. *Estuarine, Coastal and Shelf Science* **79**, 341–353.

Bock PE (1982) Bryozoans (Phylum Bryozoa). In *Marine Invertebrates of Southern Australia. Part I.* (Eds SA Shepherd and IM Thomas) pp. 319–394. Flora and Fauna of South Australia Handbooks Committee, Adelaide.

Bone EK, Keough MJ (2005) Responses to damage in an arborescent bryozoan: effects of injury location. *Journal of Experimental Marine Biology and Ecology* **324**, 127–140.

Brown KM, Schmidt R, Bone Y (2002) Observations on ecological adaptations of *Lanceopora smeatoni* (MacGillivray), at West Island, South Australia. In *Bryozoan Studies 2001.* (Eds PN Wye-Jackson, CJ Buttler and ME Spencer-Jones) pp. 61–69. Balkema, Lisse, Netherlands.

Butler AJ (1991) Effect of patch size on communities of sessile invertebrates in Gulf St Vincent, South Australia. *Journal of Experimental Marine Biology and Ecology* **153**, 255–280.

Butler AJ (1995) Subtidal rocky reefs. In *Coastal Marine Ecology of Temperate Australia.* (Eds AJ Underwood and MG Chapman) pp. 88–105. University of New South Wales Press, Sydney.

Butler AJ (2008) Fauna of jetty piles, artificial reefs and biogenic surfaces. In *Natural History of Gulf St Vincent* (Eds SA Shepherd, S Bryars, IR Kirkegaard, P Harrison and JT Jennings) pp. 279–296. Royal Society of South Australia, Adelaide.

Butler AJ, Connolly RM (1996) Development and long term dynamics of a fouling assemblage of sessile marine invertebrates. *Biofouling* **9**, 187–209.

Butler AJ, Connolly RM (1999) Assemblages of sessile marine invertebrates: still changing after all these years? *Marine Ecology Progress Series* **182**, 109–118.

Castilla JC, Collins AG, Meyer CP, Guiñez R *et al.* (2002) Recent introduction of the dominant tunicate, *Pyura praeputialis*, to Antofagasta, Chile. *Molecular Ecology* **11**, 1579–1584.

Cheshire AC, Butler AJ, Westphalen G, Rowland B *et al.* (1995) Preliminary study of the distribution and photophysiology of the temperate phototrophic sponge *Cymbastela* sp. from South Australia. *Marine and Freshwater Research* **46**, 1211–1216.

Clark GF, Johnston EL (2005) Manipulating larval supply in the field: a controlled study of marine invasibility. *Marine Ecology Progress Series* **298**, 9–19.

Dakin WJ (1960) *Australian Seashores.* Angus and Robertson, Sydney.

Dalby JE (1995) Consequences of aggregated living in the ascidian *Pyura stolonifera*: evidence of non-contact intraspecific competition. *Marine and Freshwater Research* **46**, 1195–1199.

Dalby JE (1996) Nemertean, copepod, and amphipod symbionts of the dimorphic ascidian *Pyura stolonifera* near Melbourne, Australia: specificities to host morphs, and factors affecting prevalence. *Marine Biology* **126**, 231–243.

Dalby JE (1997a) Dimorphism in the ascidian *Pyura stolonifera* near Melbourne, Australia, and its evaluation through field transplant experiments. *P.S.Z.N. Marine Ecology (Berlin)* **18**, 253–271.

Dalby JE (1997b) Reproductive and electrophoretic evidence for genetic maintenance of dimorphism in the ascidian *Pyura stolonifera. Ophelia* **47**, 227–243.

Davis AR (1987) Variation in recruitment of the subtidal colonial ascidian *Podoclavella cylindrica* (Quoy & Gaimard): the role of substratum choice and early survival. *Journal of Experimental Marine Biology and Ecology* **106**, 57–71.

Davis AR (1988a) Effects of variation in initial settlement on distribution and abundance of *Podoclavella moluccensis* Sluiter. *Journal of Experimental Marine Biology and Ecology* **117**, 157–167.

Davis AR (1988b) Colony regeneration following damage and size-dependent mortality in the Australian ascidian *Podoclavella moluccensis* Sluiter. *Journal of Experimental Marine Biology and Ecology* **123**, 269–285.

Davis AR, Butler AJ (1989) Direct observations of larval dispersal in the colonial ascidian *Podoclavella moluccensis* Sluiter: evidence for closed populations. *Journal of Experimental Marine Biology and Ecology* **127**, 189–203.

Davis AR (1989a) Contrasting population dynamics and life histories in two populations of the colonial subtidal ascidian *Podoclavella moluccensis. Marine Ecology Progress Series* **51**, 107–119.

Davis AR (1989b) Temperature correlates with the daily release of larvae and their settlement in a temperate Australian ascidian. In *Reproduction, Genetics and Distributions of Marine Organisms: Proceedings of the 23rd European Marine Biology Symposium* (Eds JS Ryland and PA Tyler) pp. 61–65. Olsen & Olsen, Fredensborg.

Davis AP (1996) Association among ascidians: facilitation of recruitment in *Pyura spinifera. Marine Biology* **126**, 35–41.

Davis AR, Campbell DJ (1996) Two levels of spacing and limits to local population density for settled larvae of the ascidian *Clavelina moluccensis*: a nearest-neighbour analysis. *Oecologia* **108**, 701–707.

Davis AR, Butler AJ, van Altena I (1991) Settlement behaviour of ascidian larvae: preliminary evidence for inhibition by sponge allelochemicals. *Marine Ecology Progress Series* **72**, 117–123.

Davis AR, Ayre DJ, Billingham MR, Styan CA *et al.* (1996) The encrusting sponge *Halisarca laxus*: population genetics and association with the ascidian *Pyura spinifera. Marine Biology* **126**, 27–33.

Duclaux G, Lafargue F, Wahl M (1988) First report of *Prochloron* in association with the genus *Polysyncraton* didemnid ascidian (Tunicata). *Vie et Milieu* **38**, 145–148.

Duggins DO, Eckman JE, Sewell AT (1990) Ecology of understory kelp environments. II. Effects of kelps on recruitment of benthic invertebrates. *Journal of Experimental Marine Biology and Ecology* **143**, 27–45.

Edgar GJ (2008) *Australian Marine Life: the Plants and Animals of Temperate Waters.* 2nd edn. New Holland Publishers, Sydney.

Edmunds M (2008) 'Port Phillip Bay Channel Deepening Project. Rip bank monitoring'. Report No. 397 to Port of Melbourne Corporation. Australian Marine Ecology, Melbourne.

Edmunds M, Gilmour P, Power B, Shimata J *et al.* (2006) 'Port Phillip Bay Channel Deepening Project supplementary environmental effects statement – Marine Ecology Specialist Studies. Vol. 9. Deep reef Biota'. Report No. 357 to Port of Melbourne Corporation. Australian Marine Ecology, Melbourne.

Edmunds M, Pickett P, Judd A (2008a) *Reef Surveys at Twelve Apostles Marine National Park and the Arches Marine Sanctuary.* Parks Victoria Technical Series No. 56. Parks Victoria, Melbourne.

Edmunds M, Judd A, Sheedy E, Ong J (2008b) 'Port Phillip Bay Channel Deepening Project, Trial dredging program. Deep reef monitoring, Jan. 2008'. Report 396 to Port of Melbourne Corporation. Australian Marine Ecology, Melbourne.

Ellison AM, Harvell CD (1989) Size hierarchies in *Membranipora membranacea*: do colonial animals follow the same rules as plants? *Oikos* **55**, 349–355.

Fairweather PG (1991) A conceptual framework for ecological studies of coastal resources: an example of a tunicate collected for bait on Australian shores. *Ocean and Shoreline Management* **15**, 125–142.

Ferguson AM, Davis AR (2008) Heart of glass: spicule armament and physical defence in temperate reef sponges. *Marine Ecology Progress Series* **372**, 77–86.

Fromont J (1999) Reproduction of some demosponges in a temperate Australian shallow water habitat. *Memoirs of the Queensland Museum* **44**, 185–192.

Hageman S, Bone Y, McGowran B, James NP (1995) Modern bryozoan assemblages and distribution on the cool-water Lacepede Shelf, southern Australian margin. *Australian Journal of Earth Sciences* **42**, 571–580.

Harvell CD (1990) The ecology and evolution of inducible defenses. *The Quarterly Review of Biology* **65**, 323–340.

Harvell CD, Padilla DK (1990) Inducible morphology, heterochrony, and size hierarchies in a colonial invertebrate monoculture. *Proceedings of the National Academy of Sciences of the United States of America* **87**, 508–512.

Harvell CD (1992) Inducible defenses and allocation shifts in a marine bryozoan. *Ecology* **73**, 1567–1576.

Harvell CD, Grosberg RK (1988) The timing of sexual maturity in clonal animals. *Ecology* **69**, 1855–1864.

Harvell CD, Helling R (1993) Experimental induction of localized reproduction in a marine bryozoan. *The Biological Bulletin* **184**, 286–295.

Harvell CD, Caswell H, Simpson P (1990) Density effects in a colonial monoculture: experimental studies with a marine bryozoan (*Membranipora membranacea* L.). *Oecologia* **82**, 227–237.

Hyman LH (1959) *The Invertebrates: Smaller Coelomate Groups Chaetognatha, Hemichordata, Pogonophora, Phoronida, Ectoprocta, Brachiopoda, Sipuncula, the Coelomate Bilateria. Vol. 5.* McGraw Hill, New York.

James D (2005) Environmental site assessment for abalone ranching on artificial reef. PhD thesis, Deakin University, Geelong, Victoria.

Kangas M, Shepherd SA (1984) Distribution and feeding of chitons in a boulder habitat at West Island, South Australia. *Journal of the Malacological Society of Australia* **6**, 101–111.

Kay AM, Butler AJ (1983) 'Stability' of the fouling communities on the pilings of two piers in South Australia. *Oecologia* **56**, 70–78.

Kay AM, Keough MJ (1981) Occupation of patches in the epifaunal communities on pier pilings and the bivalve *Pinna bicolor* at Edithburgh, South Australia. *Oecologia* **48**, 123–130.

Keough MJ (1984a) Kin-recognition and the spatial distribution of larvae of the bryozoan *Bugula neritina*. *Evolution* **38**, 142–147.

Keough MJ (1984b) Effects of patch size on the abundance of sessile marine invertebrates. *Ecology* **65**, 423–437.

Keough MJ (1984c) Dynamics of the epifauna of the bivalve *Pinna bicolor*: interactions among recruitment, predation, and competition. *Ecology* **65**, 677–688.

Keough MJ (1989a) Variation in growth rate and reproduction of the bryozoan *Bugula neritina*. *The Biological Bulletin* **177**, 277–286.

Keough MJ (1989b) Dispersal of the bryozoan *Bugula neritina* and effects of adults on newly metamorphosed juveniles. *Marine Ecology Progress Series* **57**, 163–171.

Keough MJ, Butler AJ (1983) Temporal changes in species number in an assemblage of sessile marine invertebrates. *Journal of Biogeography* **10**, 317–330.

Keough MJ, Chernoff H (1987) Dispersal and population variation in the bryozoan *Bugula neritina*. *Ecology* **68**, 199–210.

Kingsford MJ, Battershill CN (1998) Subtidal habitats and benthic organisms of rocky reefs. In *Studying Temperate Marine Environments: a Handbook for Ecologists.* (Eds M Kingsford and C Battershill) pp. 84–114. Canterbury University Press, Christchurch.

Klumpp DW (1984) Nutritional ecology of the ascidian *Pyura stolonifera*: influence of body size, food quantity and quality on filter-feeding, respiration, assimilation efficiency and energy balance. *Marine Ecology Progress Series* **19**, 269–284.

Knott NA, Davis AR, Buttemer WA (2004a) Passive flow through an unstalked intertidal ascidian: orientation and morphology enhance suspension feeding in *Pyura stolonifera*. *The Biological Bulletin* **207**, 217–224.

Knott NA, Underwood AJ, Chapman MG, Glasby TM (2004b) Epibiota on vertical and on horizontal surfaces on natural reefs and on artificial structures. *Journal of the Marine Biological Association of the United Kingdom* **84**, 1117–1130.

Knott NA, Underwood AJ, Chapman MG, Glasby TM (2006) Growth of the encrusting sponge *Tedania anhelans* on vertical and on horizontal surfaces of temperate subtidal reefs. *Marine and Freshwater Research* **57**, 95–104.

Kott P (1980) Algal-bearing didemnid ascidians in the Indo-West Pacific. *Memoirs of the Queensland Museum* **20**, 1–47.

Kott P (1981) The ascidians of the reef flats of Fiji. *Proceedings of the Linnean Society of New South Wales* **105**, 147–212.

Kott P (1997) Tunicates (sub-Phylum Tunicata). In *Marine Invertebrates of Southern Australia. Part III.* (Eds SA Shepherd and M Davies) pp. 1092–1255. SARDI Aquatic Sciences and Flora and Fauna of South Australia Handbooks Committee, Adelaide.

Ludbrook NH, Gowlett-Holmes NH (1989) Chitons, gastropods and bivalves. In *Marine Invertebrates of Southern Australia. Part II.* (Eds SA Shepherd and IM Thomas) pp. 504–724. Flora and Fauna of South Australia Handbooks Committee, Adelaide.

Mackie JA, Keough MJ, Christidis L (2006) Invasion patterns inferred from cytochrome oxidase I sequences in three bryozoans, *Bugula neritina, Watersipora subtorquata*, and *Watersipora arcuata*. *Marine Biology* **149**, 285–295.

Maldonado M (2006) The ecology of sponge larvae. *Canadian Journal of Zoology* **84**, 175–194.

Maldonado M, Riesgo A (2008) Reproduction in the phylum Porifera: a synoptic overview. *Treballs de la Societat Catalana de Biologia* **59**, 29–49.

Maldonado M, Riesgo A, Bucci A, Rützler K (2010) Revisiting silicon budgets at a tropical continental shelf: silica standing stocks in sponges surpass those in diatoms. *Limnology and Oceanography* **55**, 2001–2010.

Marshall DJ (2002) *In situ* measures of spawning synchrony and fertilization success in an intertidal, free-spawning invertebrate. *Marine Ecology Progress Series* **236**, 113–119.

Marshall DJ, Styan CA, Keough MJ (2000) Intraspecific co-variation between egg and body size affects fertilisation kinetics of free-spawning marine invertebrates. *Marine Ecology Progress Series* **195**, 305–309.

Marshall DJ, Bolton TF, Keough MJ (2003) Offspring size affects the post-metamorphic performance of a colonial marine invertebrate. *Ecology* **84**, 3131–3137.

McGovern TM, Hellberg ME (2003) Cryptic species, cryptic endosymbionts, and geographical variation in chemical defences in the bryozoan *Bugula neritina*. *Molecular Ecology* **12**, 1207–1215.

Olson RR (1986) Light-enhanced growth of the ascidian *Didemnum molle/Prochloron* sp. symbiosis. *Marine Biology* **93**, 437–442.

Osborne S (1984) Bryozoan interactions: observations on stolonic outgrowths. *Australian Journal of Marine and Freshwater Research* **35**, 453–462.

Padilla DK, Harvell CD, Marks J, Helmuth B (1996) Inducible aggression and intraspecific aggression in a marine bryozoan, *Membranipora membranacea*. *Limnology and Oceanography* **41**, 505–512.

Pansini M, Cattanea Vietti R, Shiaporelli S (1999) Relationship between sponges and a taxon of obligatory inquilines: the siliquariid molluscs. *Memoirs of the Queensland Museum* **44**, 427–437.

Pile AJ, Patterson MR, Witman JD (1996) *In situ* grazing on plankton <10 µm by the boreal sponge *Mycale lingua*. *Marine Ecology Progress Series* **141**, 95–102.

Rius M, Teske PR (2011) A revision of the *Pyura stolonifera* species complex (Tunicata, Ascidiacea), with a description of a new species from Australia. *Zootaxa* **2754**, 27–40.

Robbins IJ (1983) The effects of body size, temperature and suspension density on the filtration and ingestion of inorganic particulate suspensions by ascidians. *Journal of Experimental Marine Biology and Ecology* **70**, 65–78.

Roberts DE (1996) Patterns in subtidal marine assemblages associated with a deep-water sewage outfall. *Marine and Freshwater Research* **47**, 1–9.

Roberts DE, Davis AR (1996) Patterns in sponge (Porifera) assemblages on temperate coastal reefs off Sydney, Australia. *Marine and Freshwater Research* **47**, 897–906.

Roberts DE, Cummins SP, Davis AR, Pangway C (1999) Evidence for symbiotic algae in sponges from temperate coastal reefs in New South Wales, Australia. *Memoirs of the Queensland Museum* **44**, 493–497.

Roberts DE, Cummins SP, Davis AR, Chapman MG (2006a) Structure and dynamics of sponge-dominated assemblages on exposed and sheltered temperate reefs. *Marine Ecology Progress Series* **321**, 19–30.

Roberts DE, Davis AR, Cummins SP (2006b) Experimental manipulation of shade, silt, nutrients and salinity on the temperate reef sponge *Cymbastela concentrica*. *Marine Ecology Progress Series* **307**, 143–154.

Russ GR (1980) Effects of predation by fishes, competition, and structural complexity of the substratum on the establishment of a marine epifaunal community. *Journal of Experimental Marine Biology and Ecology* **42**, 55–69.

Russ GR (1982) Overgrowth in a marine epifaunal community: competitive hierarchies and competitive networks. *Oecologia* **53**, 12–19.

Ryland JS (1977) Physiology and ecology of marine bryozoans. *Advances in Marine Biology* **14**, 285–443.

Shepherd SA (1983) The epifauna of megaripples: species' adaptations and population responses to disturbance. *Australian Journal of Ecology* **8**, 3–8.

Sorokin SJ, Currie DR (2009) *The Distribution and Diversity of Sponges in Spencer Gulf.* Publication F2008/001153-1. 61 pp. SARDI Aquatic Sciences, West Beach, Adelaide.

Sorokin SJ, Fromont J, Currie DR (2007) Demosponge biodiversity in the benthic protection zone of the Great Australian Bight. *Transactions of the Royal Society of South Australia.* **131**, 192–204.

Stoecker D (1980) Relationships between chemical defense and ecology in benthic ascidians. *Marine Ecology Progress Series* **3**, 257–265.

Svane I, Young CM (1989) The ecology and behaviour of ascidian larvae. *Oceanography and Marine Biology - an Annual Review* **27**, 45–90.

Tardent P (1963) Regeneration in the Hydrozoa. *Biological Reviews of the Cambridge Philosophical Society* **38**, 293–333.

Teske PR, Rius M, McQuaid CD, Styan CA *et al.* (2011) "Nested" cryptic diversity in a widespread marine ecosystem engineer: a challenge for detecting biological invasions. *BMC Evolutionary Biology* **11**, 176–189.

Usher KM, Kuo J, Fromont J, Sutton DC (2001) Vertical transmission of cyanobacterial symbionts in the marine sponge *Chondrilla australiensis* (Demospongiae). *Hydrobiologia* **481**, 15–23.

Usher KM, Sutton DC, Toze S, Kuo J *et al.* (2004a) Sexual reproduction in *Chondrilla australiensis* (Porifera: Demospongiae). *Marine and Freshwater Research* **55**, 123–134.

Usher KM, Sutton DC, Toze S, Kuo J *et al.* (2004b) Biogeography and phylogeny of *Chondrilla* species (Demospongiae) in Australia. *Marine Ecology Progress Series* **270**, 117–127.

Vail LL, Wass RE (1981) Experimental studies on the settlement and growth of Bryozoa in the natural environment. *Australian Journal of Marine and Freshwater Research* **32**, 639–656.

Van-Praët M (1985) Nutrition of sea anemones. *Advances in Marine Biology* **22**, 65–99.

Waddington KI, Meeuwig JJ, Evans SN (2010) Assessment of benthic biota of a deep coastal ecosystem by remote and *in situ* sampling. *Marine and Freshwater Research* **61**, 1164–1170.

Walters LJ (1992) Field settlement locations on subtidal marine hard substrata: is active larval exploration involved? *Limnology and Oceanography* **37**, 1101–1107.

Walters LJ, Wethey DS (1991) Settlement, refuges, and adult body form in colonial marine invertebrates: a field experiment. *The Biological Bulletin* **180**, 112–118.

Wass RE, Vail LL (1978) Encrusting Bryozoa exhibit linear growth. *Search* **9**, 42–44.

Watson AL (1974) The sponge and the environment: aspects of the ecology of *Ancorina corticata* (Choristida, Demospongiae). BSc (Hons) thesis. Monash University, Melbourne.

Watson AL (1976) Preliminary observations on the influence of water movement on population structure in *Ancorina corticata* (Carter) (Choristida: Demospongiae). *Proceedings of the New Zealand Ecological Society* **23**, 45–50.

Watson JE (1982) Hydroids (Class Hydrozoa) In *Marine Invertebrates of southern Australia. Part I.* (Eds SA Shepherd and IM Thomas) pp. 77–114. South Australian Government Handbooks Committee, Adelaide.

Weidenmayer F (1989) Demospongiae (Porifera) from northern Bass Strait. *Memoirs of the Museum of Victoria.* **50**, 1–342.

Wendt DE (1996) Effects of larval swimming duration on success of metamorphosis and size of the ancestrular lophophore in *Bugula neritina* (Bryozoa). *The Biological Bulletin* **191**, 224–233.

Whalan S, Johnson MS, Harvey E, Battershill C (2005) Mode of reproduction, recruitment and genetic subdivision in the brooding sponge *Haliclona* sp. *Marine Biology* **146**, 425–433.

Wilkinson CR, Vacelet J (1979) Transplantation of marine sponges to different conditions of light and current. *Journal of Experimental Marine Biology and Ecology* **37**, 91–104.

Wright JT, Benkendorff K, Davis AR (1997) Habitat associated differences in temperate sponge assemblages: the importance of chemical defence. *Journal of Experimental Marine Biology and Ecology* **213**, 199–213.

Wulff JL (2006) Ecological interactions of marine sponges. *Canadian Journal of Zoology* **84**, 146–166.

14 Cnidarians (hydroids, jellyfish, anemones and corals) and ctenophores

OVERVIEW

The ecology of the five classes of cnidarians varies widely within and between classes. Hydroids are sessile and commonly occur on shaded substrata and on macroalgae and seagrass. The hydroid is the dominant and most conspicuous form in the alternating generations of polyp and medusa stages. Here, the life history of three species is described. In contrast, in the jellyfishes the conspicuous medusoid form is pelagic, and the sessile, polyp form is minute and inconspicuous. The life histories of six common jellyfish species are described, and also their associations with fish and invertebrates. Jellyfish can play a keystone role and can bloom in certain conditions and have serious ecosystem effects. A small and poorly known group, the Stauromedusae, is then described, but their ecology is still largely unknown. Next, the diverse anthozoans, which have only a polyp phase, are considered. The ecology of two species of sea anemones and a zoanthid, *Zoanthus robustus* and its symbiosis with zooxanthellae, are then described. Temperate corals comprise stony and soft corals, and gorgonians. A common temperate representative of stony corals *Plesiastrea versipora* is long lived and notable because it lays down annual density bands with trace elements preserved in them. Hence, it provides insights into the palaeoecology of coastal waters. Soft corals are illustrated by the life history of *Drifa gaboensis* and general comments are added on gorgonian ecology. Finally, the benthic ctenophores, the platyctenes, are briefly mentioned.

INTRODUCTION

This diverse group of animals falls into five groups, variously termed Classes or Super-classes due to the higher classification being still in flux (Bouillon and Boero 2000). They are: Hydrozoa for hydroids; Scyphozoa, Cubozoa and Staurozoa for jellyfishes; and Anthozoa for a varied group that includes anemones, zoanthids, stony corals, soft corals and gorgonians. The first class is named after Hydra, the many-headed snake of Greek mythology, the second from *skuphos*, a cup, the third from *kubos*, a cube, the fourth from *stauros*, an upright paling, and the last after *anthos*, the Greek word for 'flower'. A distinguishing feature of cnidarians is that they have radial symmetry, and also two different body forms – polyps with a cylindrical body attached to the sea bed, and a free-swimming, hemispherical or trumpet-shaped medusoid form. In the Hydrozoa, these two forms have alternating generations, in which the polyp generation – the hydroid colony – gives rise to a medusa stage. In the Scyphozoa, Cubozoa and Staurozoa, the jellyfishes, the medusa stage is dominant and the polyp phase tiny and rarely noticed. In the Anthozoa –anemones, zoanthids and corals – only the polyp phase exists.

A unique feature of the whole group is the special stinging cells, called nematocysts, contained in the body wall or tentacles. Each microscopic stinging cell contains a coiled thread armed with a barb (sometimes poisonous) that is ejected on contact with prey or other stimulus.

HYDROIDS (SUPERCLASS HYDROZOA, CLASS HYDROIDOMEDUSA)

Hydroids have a complex life cycle comprising a sessile, colonial stage, which is the commonly seen hydroid, and a medusa stage, which is often only a few millimetres in diameter. This tiny medusa is shed from the hydroid and swims as a tiny jellyfish. The asexual colonial stage has many growth forms: some tree-like, some feather-shaped and others tiny and scarcely visible to the naked eye. They are anchored to the substratum by a root-like structure, the hydrorhiza, from which arises a flexuous stem, often bearing many polyps. The feeding polyps have a body cavity and mouth, which is surrounded by tentacles armed with nematocysts for the capture and ingestion of prey. Other polyps armed with nematocysts have a defensive function, and yet others, called gonophores, have a reproductive function.

Hydroids are most abundant in places of moderate to strong current, and they can live on a variety of firm substrata, such as reefs, jetty piles, mollusc shells and algal holdfasts. Most species have a preference for low light conditions and are commonly found in deeper water where light levels are low, or on shallower shaded surfaces – which may explain their greater cover on jetty piles than on natural reef, as noted by Carolina Zagal (2007). Many species are also early colonists of bare surfaces, often growing on old shells or pebbles (Watson 1982). The associations between hydroids and other species are described in detail below.

Some species, such as *Obelia*, *Sertularia*, *Plumularia* and *Aglaophenia*, are bioluminescent, but the benefits, possibly a deterrent to predators, are not fully known.

For further information, the reader is referred to the reviews of Watson (1982), Boero (1984), Gili and Hughes (1995) and Bouillon and Boero (2000).

Life cycles

Hydroid colonies are usually male or female, with the sex cells produced by a colony of the same sex. However, some species have both male and female reproductive polyps (the gonophores). In the asexual stage of some species, medusoid buds are formed in the gonophores and develop into medusae, which are shed and swim away as tiny jellyfish. In most species the medusae (1–3 mm size) are solitary, and bell- or umbrella-shaped, with tentacles around the bell for food capture. Medusae form the sexual stage. Within the bell lie the gonads, producing sperm or eggs according to their sex. Fertilisation can be external, with eggs and sperm shed into the water, or internal, with eggs fertilised and developing in the gonad. The resultant planula larva has a free-living, swimming or crawling stage lasting from a few hours to days; then, upon finding a suitable substratum, the planula develops into a hydroid colony. Thus, there is a regular alternation of asexual and sexual generations. The basic life cycle of a hydroid colony is illustrated in Fig. 14.1, together with the direction of reverse development (discussed below) known for a few species.

Many variations of the above basic life cycle exist. In about half of hydroid species, the free-swimming medusa stage is lost. In some species, the gonophore is a vestigial medusa, incubating eggs to become the larvae, and in others the female medusa is retained while the male is released. For example, in *Ectopleura crocea* (formerly called *Tubularia ralphi*), the hydroid's gonophores are male or female, and, after fertilisation, the eggs develop directly into crawling larvae. The larvae perform sluggish swimming and crawling movements, and then settle within 24 h by fixing themselves to the bottom. Within another 36 h, they develop stems and polyps with oral tentacles (Watson 1980).

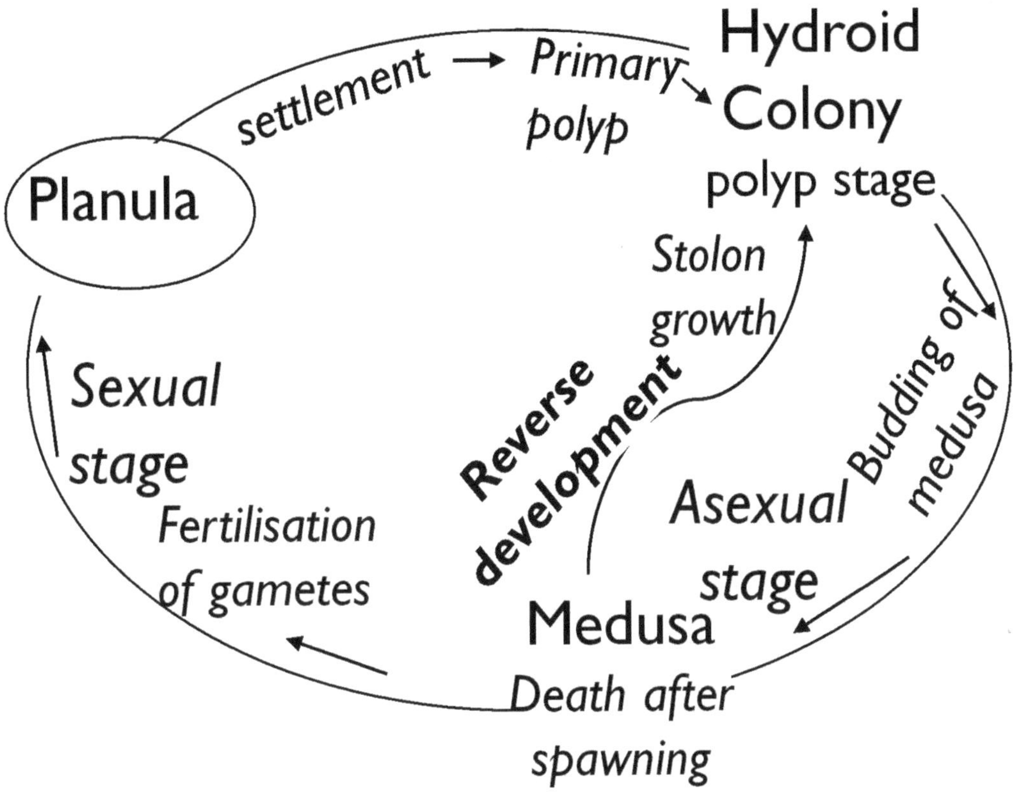

Figure 14.1: Basic life history of a hydroid, showing asexual stage from hydroid colony to medusa, and sexual stage from medusa to hydroid. The reverse development shown refers to *Turritopsis nutricula*.

In another southern hydroid *Staurocladia haswelli*, the medusa stage is not free-swimming, but crawls on the bottom (Watson and McInnes 1999). In this species, new small colonies, possessing spindle-shaped hydranths with seven capitate (i.e. knobbed) tentacles, arise from stolons attached to coralline algae. Medusa buds appear on the stolons and on the hydroid stem. The tiny medusae have four groups of divided tentacles and creep along the algal substrate, over time producing up to 18 tentacles that capture planktonic prey. New medusa buds later appear around the manubrium (i.e. behind the mouth). Eggs are also produced within the body of the medusa and upon fertilisation are liberated as planulae, usually in autumn. The planulae eventually settle on the bottom and develop into a new hydroid colony.

The lifespan of hydroids is indefinite or unknown. Unfavourable conditions can be avoided by dormant stages, polyps can regress and be replaced within a few days and in some species the hydrorhizae can persist indefinitely. Nevertheless, field studies show that some species are seasonal or annual, while others live much longer.

Gili and Hughes (1995) noted that, in low-light and calm-water habitats, hydroids with gonophores producing eggs and larvae tend to predominate compared with those producing free-living medusae. To illustrate the above plasticity in life histories, the life cycles of three common southern species are described in more detail.

A cosmopolitan species *Obelia geniculata*

This cosmopolitan species grows on several brown algae (see below). In Northern Hemisphere studies, *Obelia* is fertile at the end of spring and produces free-swimming medusae, which are released over a few days during the full moon in early summer (Elmhirst 1922). However, in warm conditions this *Obelia* can employ another method of reproduction. Berrill (1948) noted that at sea temperatures > 20°C massive outgrowths budded from the ends of stolons broke off and left the colony to settle elsewhere and establish new colonies. Like many other hydroids, *Obelia* has remarkable regenerative abilities. As short-lived polyps age they are resorbed and new ones regenerate, so continually rejuvenating the colony (Tardent 1963).

The magnificent hydroid *Ralpharia magnifica*

This hydroid, studied by Watson (1980), is easily the largest in southern Australia (Plate 39), growing to 5 cm across the extended tentacles. The species is always associated with the sheet-like soft coral *Erythropodium hicksoni*, in which it buries its hydrorhizae. It appears to prefer embayment conditions sheltered from the ocean swell, and also overhanging walls. It is a voracious predator, taking small crustaceans and even small fish up to 1 cm long.

Gonophores bud in November and within 2 months their cavities are filled with maturing, primitive medusae. Reproduction starts in January towards the peak in summer temperatures. Male and female medusae are liberated at night from the gonophores and they swim with jerking movements for several hours as they release sperm and eggs into the water, where fertilisation takes place. After ~20 h, stubby tentacles begin to form and over the next 2 days the rudimentary larva develops long tentacles and drifts along in the current until it comes into contact with an *Erythropodium* soft coral. It promptly settles on the coral and its stalk and hydrorhiza penetrate the coral's surface for anchorage, then grow rapidly upwards and produce a polyp and tentacles, all within a week or so. The protracted drifting larval life of 3–5 days is presumably an adaptation to provide ample time for the larva to find its specific soft coral substratum.

The red thimble jelly *Turritopsis nutricula*

Colonies of this species are common in bays and at sheltered sites on open coasts, where they grow in winter under ledges and in caverns and under jetties. They form small straggling bushes up to 4 cm high, with polyps terminating each stem. Colonies can live for several years, with growth phases periodically interrupted by shrinkage and regeneration. Medusa buds form in the polyps and are released in large numbers as tiny medusae of ~1 mm bell height in summer. The medusae reach sexual maturity in 25–30 days at a bell height of ~2 mm, and while adrift they feed and continually grow to reach 1–2 cm height. Swarms of these medusae, conspicuous with their bright orange gonads, are often seen swimming energetically near reefs on southern coasts (Plate 40).

A remarkable feature of *Turritopsis* is its ability to reverse its usual life cycle (see Fig. 14.1) (Piraino *et al.* 1996, 2004). In stressful conditions, such as starvation, changes in temperature or salinity, or senescence, the medusae react by rejuvenation – a reverse transformation to a polyp stage, which attaches to the bottom and grows into a new colony. The above authors stated that: 'the process would be hardly more remarkable if a butterfly were able to revert to its caterpillar stage'. Since its first discovery in *Turritopsis*, a reversal of the life cycle has been found in other hydrozoans and is now thought to be a widespread phenomenon. Although this hydroid has been popularly and extravagantly labelled as 'immortal', the reality is more prosaic, and it is likely that 'the potential for ontogeny reversal has limited meaning and is just an extreme example of a more general strategy for withstanding unfavourable periods …' (Piraino *et al.* 2004).

Feeding and predators

Hydroids are carnivores, capturing prey with their tentacles and nematocysts. Being attached to the bottom, they are most abundant in places of moderate to strong water flow, where food is brought to the tentacles. Their food is mainly small crustaceans, such as copepods and amphipods (1 μm–5 mm size), but they may also ingest nematodes, rotifers, polychaetes and other larvae. Some species can feed on benthic prey, and take in particulate matter and phytoplankton.

Two studies have shown that the feeding process in hydroids can be quite complex. A common cosmopolitan species *Pennaria disticha*, and a southern Australian species *Pennaria wilsoni*, have two kinds of tentacles – a lower whorl of fine ones and an upper whorl of knobbed ones around the mouth, with different types of nematocysts in each. Prey are first immobilised by nematocysts in the finer tentacles and later killed by the larger nematocysts in the knobbed tentacles. Pardy and Lenhoff (1968) noted experimentally that when prey contacted the finer tentacles they struggled for up to 7 minutes before ceasing to move, whereas those contacting the larger tentacles were immobilised in 6 seconds. The contact of prey with the fine tentacles initiated the hydroid's feeding response, consisting of bending the tentacles over the body towards the mouth. The prey were then killed by the nematocysts in the knobbed tentacles and finally engulfed in the mouth. Further studies showed that the hydroid in fact controlled the discharge of their nematocysts. For example, when the colony was well fed, the prey-killing nematocysts were inactivated, and there was also a community effect, in which a well fed polyp inactivated its neighbours. As the whole colony shares the digested food, it appears that the control of nematocyst discharge has evolved to conserve these complex structures (Clark and Cook 1986).

The predators of hydroids include both generalists and specialists. Generalists include benthic feeding fishes (e.g. small leatherjackets and moonlighters), grazing chitons, gastropods, sea-stars and polyclad flatworms, all of which probably take hydroids incidentally to other prey. However, the main predators are specialists, such as nudibranchs (see Table 9.2 where six hydroid-feeding species are recorded) and pycnogonids (Chapter 15).

Inter-specific relations

Hydroids often escape the intense competition for space on rocky bottoms by growing on other invertebrates or on algae. Hydroid species may be non-selective in habitat choice or substratum specialists, with the benefits and restrictions that specialisation confers. Of 50 common southern species described by Watson (1982), the majority are either weakly or non-selective, occurring on sponges, ascidians, bryozoans and other invertebrate substrata, including other hydroids; the rest live only on algae. For example, one species *Zyzzyzus spongicola* has a preference for soft-textured pink and orange sponges and, as noted above, *Ralpharia magnifica* is only found on the creeping soft coral *Erythropodium*.

The hydroids that colonise algae are known from the studies of Shepherd and Watson (1970) and Zagal (2007), as well as Jan Watson's own observations. In the 1970 study at one site, red algae were the most favoured substratum, with a few species found only on brown algae and none on green algae (although elsewhere several hydroid species do grow on green algae, notably *Caulerpa*). Of the 36 epiphytic hydroid species considered in the study, 47% were moderately selective of their red algal host, being found on an average of about seven species (range three to 18), while 14% were non-selective. Another 39% selected a brown algal substratum, occurring on an average of about two species (range one to six). Three species had an apparent obligate relation with an alga or algal group. *Obelia geniculata* was only found on the kelp *Ecklonia*, although later found also on the giant kelp *Macrocystis* and *Sargassum* (Watson 1982). *Silicularia bilabiata* was only found on two similar fucoids, *Scytothalia* and *Seirococcus* and, lastly, the minute *Hydrodendron daidalum* was obligately associated with the brown alga *Zonaria crenata*. These associations have been confirmed by later observations. Both physical and chemical factors were considered important in the above symbioses, because:

- some algae, especially red algae, with rough, or flat, laminar surfaces were favoured, while filamentous species were avoided
- the older parts of an alga were colonised first
- mucus-coated algal species probably depend on a positive chemotaxis of the planula larva for the alga in question, as found by Nishihira (1968).

Zagal (2007) specifically examined the abundance of hydroids on *Ecklonia* and *Sargassum* at coastal sites of central New South Wales. The main species present were *Obelia geniculata*, *Obelia* spp., *Campanularia* spp. and *Clytia hemisphaerica*. Carolina Zagal found that habitat features and the level of disturbance influenced the abundance of colonising hydroids and noted that hydroid cover varied on different zones of the kelp *Ecklonia*. It was greater on the edges of blades than in the centre, on crinkled parts of blades than on smooth parts, and in troughs rather than on ridges. The preferred zones of the kelp blade were:

Middle zone > distal zone > basal zone.

Using artificial plants, she was able to show an 'edge effect' in which the edges were favoured and also that the bushiness of the kelp increased hydroid cover. Water movement, with its whiplash effect on blades, removed more hydroids from the distal than from the middle zones of *Ecklonia* blades.

CLASSES SCYPHOZOA AND CUBOZOA

The jellyfishes in these groups comprise much of the gelatinous fauna of coastal waters and open seas (Kramp 1965). In contrast to the Hydrozoa, the medusa is the dominant and most conspicuous individual in the life cycle, with a bell diameter of up to 40 cm, and the polypoid phase is a small benthic larva. The medusa is otherwise very similar in form to a hydromedusa.

The sexes are separate, with the gonad sacs located under the bell, where they may hang down under the umbrella. Eggs are released into the water or retained in sub-umbrella brood chambers. After fertilisation and a brief free-swimming period, the planula larva settles and becomes attached to the sea bed. Here it develops into a polypoid larva, called the scyphistoma, which produces many tiny, free-swimming medusae (called ephyrae) that grow into adult medusae. The benthic scyphistoma can live for several years and produce large numbers of medusae. The basic life cycle is illustrated in Fig. 14.2.

Jellyfish associations

Symbioses

Jellyfishes are well known for their associations with other species (see reviews of Mansueti 1963; Kingsford 1993; Purcell and Arai 2001). In general, symbioses imply different nuances (Leung and Poulin 2008):

- **mutualism**, in which both parties enjoy reciprocal benefits
- **commensalism**, which benefits one party in terms of nutrition or protection without seriously harming the other
- **parasitism**, in which the advantages are wholly on one side, usually at the cost of the host, but not to the extent of killing it.

Jellyfish symbioses with fish are a temporary relationship, in which the jellyfish is a passive host, and the fish profit from a chance encounter, and live under the protection of its bell and tentacles. Records in southern waters are given in Table 14.1, but little else is known about these associations in this region.

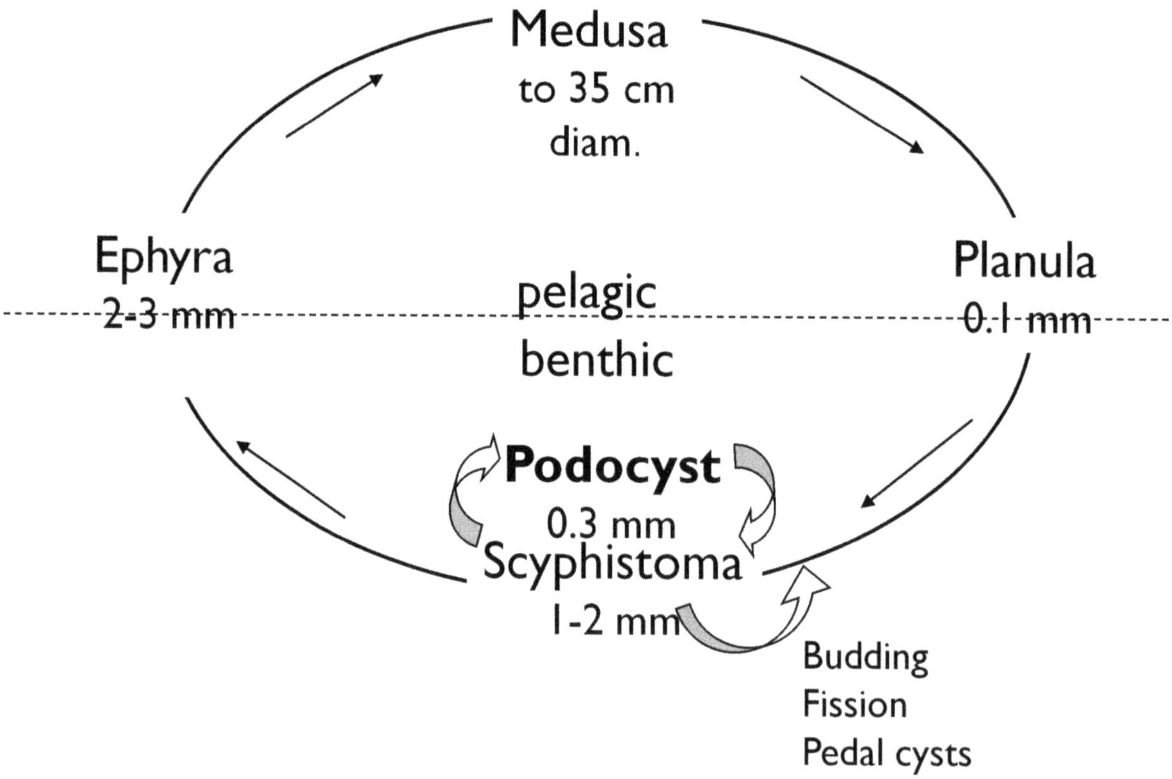

Figure 14.2: Basic life history of a medusa with typical sizes of the different stages. The podocyst stage and reversal by budding, fission and pedal cysts refer to *Aurelia coerulea*.

Table 14.1: Associations between jellyfishes as hosts and symbiotic species.

GSV = Gulf St Vincent. GAB = Great Australian Bight.

Symbiotic species/family	Jellyfish host	Location
Hardyheads *Atherinosoma microstoma*[1]	*Desmonema gaudichaudi*	SW Vic. SE SA
Silver trevally *Pseudocaranx georgianus*[1]		Encounter Bay, SA
Silver trevally *Pseudocaranx georgianus*[1]; Hydro-medusa *Eirene menoni*; amphipods[1]	*Pseudorhiza haeckeli*	Gulf St Vincent, Encounter Bay, SA
Yellow-tail scad *Trachurus novaehollandiae*[1] Horse mackerel *Trachurus declivis*[2] Mosaic leatherjacket *Eubalichthys mosaicus*[1] Copepod, *Paramacrochiron maximum*, also amphipods, spider crabs, shrimp[3]	*Catostylus mosaicus*	Eastern Australia; Bass Strait Eastern Australia
Silver trevally *Pseudocaranx georgianus*[1,5] Mosaic leatherjacket *Eubalichthys mosaicus*[1]	*Chrysaora southcotti*	GSV, Coffin Bay, SA, Bass Strait, Vic.
Bluebottle-fish *Nomeus gronovii*[2,6]	*Physalia aurita*	Eastern Australia
Juvenile leatherjackets (Monacanthidae)[1]	*Cyanea rosella*	Not given
Chinaman leatherjacket *Nelusetta ayraudi*[5]	Unspecified jellyfishes	Eastern GAB
Sparidae (black bream and snapper)[4]	Unspecified jellyfishes	Eastern Australia

[1]Southcott and Glover (1987); [2]Mansueti (1963); [3]Browne and Kingsford (2005); [4]Kingsford (1993); [5]B. Saunders *pers. comm.*; [6]Southcott (1967)

The fish species that associate with jellyfishes, mainly trevallies, trevallas and leatherjackets in southern Australia, are juveniles of species that are pelagic in their juvenile stage. These juveniles, up to a few centimetres long, and sometimes in large numbers, live among the jellyfishes' tentacles or under the bell for up to a year. The lines between the three categories of association listed above are in practice blurred, but it seems that the fish mainly derive shelter from the host, although some authors have observed fish nibbling on the tentacles, gonads or lobes of the jellyfish, or feeding on other organisms, either commensal with the jellyfish or caught in its tentacles (Mansueti 1963). A likely benefit to the fish is their enhanced survival under the jellyfish umbrella. In these symbioses, the fish lie somewhere along the continuum between commensalism and parasitism.

An exception to the above examples is the small pelagic fish *Nomeus gronovii* that, together with other species, lives as an adult among the tentacles of the bluebottle in eastern Australia. In this case *Nomeus* does feed on the tentacles of the bluebottle, which in turn can fatally sting *Nomeus* and digest it (Southcott 1967). However, the fish has an antibody to the toxins and can survive high doses. Besides, it has a complex skin structure that reduces nematocyst penetration. This association appears to be an example of a commensalism approaching mutualism.

Although jellyfishes often predate on small fish, there are few reports of them feeding on their commensal partners; hence, it is likely that the fish enjoy some sort of immunity from being stung, although there is no direct evidence. The existence of a mutualism is yet to be shown for southern species, although there are occasional reports of jellyfish benefiting from the removal of amphipod parasites.

The intriguing commensalism between a jellyfish and copepods was explored by Browne and Kingsford

(2005) in Botany Bay, New South Wales. Over 40 copepod species lived on the oral arms of *Catostylus mosaicus*. The most abundant of these was a copepod *Paramacrochiron maximum*, with > 5000 individuals sometimes found on a single jellyfish. Yet this copepod was scarce in the water column, strongly suggesting a commensalism between them. Species of *Paramacrochiron* are known as mucus feeders and it is likely that this species consumes mucus produced profusely by *Catostylus*. Possibly the copepod also feeds on zooplankton stunned by the jelly's nematocysts. Another suggested advantage for the copepod in such crowded living is the concentration of potential mates!

Predators

The predators of jellyfish include leatherjackets, scorpion-fishes, mackerels, tunas, sardines and anchovies, as well as many generalist pelagic species (Arai 2005; Pauly *et al.* 2009). The few observations in southern waters include leatherjackets and blue swimmer crabs – the latter often observed to swim up into the water column, seize a jellyfish, and take it down to the bottom for feeding (personal observations).

Life histories of some southern jellyfish

The dominant stage of jellyfish is pelagic and species are often widely distributed in the world's oceans. Here, we consider the ecology of seven common species: a box jelly *Carybdea rastoni*; the moon jelly *Aurelia* spp.; two related lion's mane jellies *Cyanea rosella* and *C. annaskala*; the mosaic jelly *Catostylus mosaicus*; Haeckel's jelly *Pseudorhiza haeckeli*; and, lastly, the white-spotted jelly *Phyllorhiza punctata*. For more information see Pitt and Kingsford (1999). The role of jellyfish in ecosystem processes is discussed in Chapter 19.

Cubomedusae

The cubomedusans, or box jellies, are distinguished from scyphomedusans by their box-like shape and the long tentacles that hang down from each corner of the box. The tentacles are notorious for the severity of their stings and for this reason box jellies are also called 'sea wasps'. Cleland and Southcott (1965) give a detailed account of the symptoms presented by such stings and their treatment.

The box jelly Carybdea rastoni

This box jelly (Plate 41) is common on reefs and in sandy depressions on the southern coasts of Australia and throughout the Indo-Pacific region, especially during summer months. Box jellies have long been known for the sophisticated 'image-forming' structure of their eyes, which places them at a high point in the evolution of cnidarians. Their complex eyes give them a wide view of their surroundings and enable them to distinguish objects. So, they tend to swim away from dark objects, such as an approaching diver, and towards light objects (Hartwick 1991).

In summer, box jellies often aggregate on the sandy sea floor by day. Towards dusk, they start swimming energetically by regular contractions of the bell at 80–100 beats min^{-1} diagonally towards the surface and down again. On the way they extend their tentacles to capture crustaceans, such as mysids, copepods, isopods and amphipods, and even small fishes. They complete the capture of prey by swimming upwards for a few seconds, turning upside down and resting, while the tentacles and captured prey fall down into the bell (Matsumoto 1995).

In box jellies, males release from the gonads strands of sperm, which are picked up by the females for fertilisation of their eggs. The resulting embryos develop into typical planulae, which after a short free-swimming stage attach themselves to the bottom and develop into scyphistoma polyps. Over time, the polyps release the next generation of juvenile box jellies.

Scyphomedusae

Eight species of jellyfish are common in southern Australian coastal waters, and we discuss some which bloom from time to time, notably in bays, gulfs and sheltered waters (see Chapter 19).

The moon jelly Aurelia *spp.*

The cosmopolitan genus *Aurelia* is now thought to be a complex of nine species, with two southern species: *Aurelia* sp. 7 (Plate 42) along the southern coast and the cryptogenic *Aurelia* sp. 1 on the east coast and elsewhere in the Pacific (Dawson *et al.* 2005). This species-complex is remarkable for its wide tolerance of fluctuating temperatures, salinities and nutrients, and the variability of its life history and abundance in different populations. Occasionally *Aurelia* sp. 7 blooms in semi-enclosed southern bays and estuaries, with densities of > 100 per m^3 – a behaviour that is thought to enhance fertilisation success and access to higher densities of zooplankton. Here, we summarise its most common life history pattern in temperate waters (reviewed by Lucas 2001).

Juvenile medusae, called ephyrae, are liberated from the benthic scyphistomae in late winter and develop into medusae in the spring. As temperatures increase, the medusae reach their maximum size of 20–30 cm bell diameter, with typical growth rates of 1–4 cm $week^{-1}$, depending on the food supply and the density of medusae. During summer, the medusae start to shrink as they release eggs and sperm. Their lifespan is mostly 4–8 months, although some may live up to 2 years in cold waters. After fertilisation of the eggs, the non-feeding planulae spend from 12 h to a week in the water column before settling on any firm substrate – whether rock, shells, polychaete tubes, ascidians or algae – to form the polyps of scyphistomae. These polyps can live for several years and can bud vegetatively to produce more polyps. The increase in temperature (or light) after the winter minimum is thought to initiate release of ephyrae, although food supply may also play a role.

In the Huon Estuary, SE Tasmania, vast blooms of > 170 million moon jellies weighing > 28 000 t fresh weight appear sporadically in summer and disappear by the end of January, causing a high mortality of farmed salmon. From the appearance of ephyrae (after strobilation), say on 1 October, medusae would reach sexual maturity in ~110 days at 10–15 cm bell diameter, and females would be brooding eggs within another week (Crawford *et al.* 2011). The colonial, sessile scyphistomae stage was also studied and seemed to be regulated by a combination of several density-dependent factors – water temperature, salinity, food availability and water quality. When conditions were good, scyphistomae colonies grew by asexual budding – likely a reproductive strategy to maximise production (Willcox *et al.* 2007, 2008).

The behaviour of the moon jelly is poorly understood. In one study (Costello and Colin (1994) it swam primarily upwards and downwards, pulsating for almost 100% of the time, indicating that it is a cruising predator, feeding non-selectively on zooplankton and spending almost all of its time in prey capture. The jet propulsive swimming of the medusa results in unsteady water flow around it, and this entrains slow-swimming zooplankton, but not fast-swimming copepods, which are likely to escape the net of tentacles. The daily food ration of the jelly has been estimated at 1.3% of the carbon content of an individual 30 cm bell diameter, and 10% of one 3 cm size (Dawson and Martin 2001).

The applied significance of moon jelly studies was that the many millions of salmon kept in sea cages in the Huon Estuary suffered a high mortality when medusa blooms entered the cages and discharged their nematocyst toxins into the fish. The development of blooms was apparently favoured by the settlement of jellyfish scyphistomae on the sea cage structures. Ultimately, the problem was resolved by towing the cages away from the blooms and allowing the toxins to be flushed out of the fish.

Lion's mane jellies Cyanea rosella *and* C. annaskala

C. rosella (Plate 43) has a southern Australian distribution and *C. annaskala* an eastern Australian one. These species have the same basic life history pattern as outlined for jellyfish above (see Fig. 14.1). In Port Phillip Bay, *C. rosella* spawns from late winter to early spring, followed by degeneration of its tentacles and death. The planulae settle and develop into benthic polyps, the scyphistomae, which can release ephyrae within 3 weeks. The scyphistomae may survive over summer and continue to release large numbers of ephyrae into the water column. Ephyrae are present throughout the year, but reach a maximum abundance from May to July, whereas adults are present only in June to September, when they reach a density of ~0.1 per m^3 (Fancett 1986).

C. rosella is a large jelly, with a bell diameter up to 30 cm. It swims slowly, with an extended net of tentacles up to 10 times the bell diameter in length. Hence, it can have a severe effect on the abundance of zooplankton, on which it feeds. Fancett (1988) found that the composition of prey items consumed was, in order of decreasing abundance:

larvaceans > cladocerans > fish eggs and larvae > copepods

Yet, copepods were the most abundant group present in the region, averaging 57% of all zooplankters in the Bay. Fancett explained the discrepancy by noting that copepods were generally able to escape unharmed, because they had a strong escape response and some apparent immunity to the jellyfish's nematocysts. Fish eggs and larvae were a preferred prey of the jellyfish, whose predation had a substantial impact on the abundance of fish larvae. For example, Fancett and Jenkins (1988) calculated that up to 20% of flounder eggs and larvae were removed during their development period by medusa predation.

Haeckel's jelly Pseudorhiza haeckeli

This species is a warm-temperate to tropical species, occurring in both coastal and oceanic waters. In Port Phillip Bay, it is common only in summer reaching a density of 0.1 per m^3 (Fancett 1986). Like the lion's mane jellyfish, this species shows a preference for fish eggs and larvae, and a selection for some, but not all, copepods. The composition of its diet was found by Fancett (1988) to be:

fish eggs/larvae (41%) > copepods (33%) > larvaceans > cladocera > crab zoeae

However, the feeding mode of this jelly was very different from *Cyanea*. It pulsated its bell energetically and swam rapidly, pumping water downwards through the arms, so that it filtered more water than *Cyanea*, and had a higher feeding rate. But, overall, its predatory impact on fish eggs and larvae was similar to that of *Cyanea* (Fancett and Jenkins 1988).

The mosaic or blue jelly Catostylus mosaicus

This jellyfish is common in lagoons and estuaries along the SE Australian coast, as far as Port Phillip Bay. The species forms summer blooms in surface waters (< 1 m depth) at densities of up to 2 per m^3. It is a large species with a bell diameter of 35 cm, and is commercially harvested in New South Wales and Victoria, where it is dried and exported to Asian countries for food (Kingsford *et al.* 2000). The following summary is taken from the many studies of Kylie Pitt and her associates in estuaries of New South Wales (Pitt 2000; Pitt and Kingsford 2000 a, b; 2003; Peach and Pitt 2005; Pitt *et al.* 2008).

This jelly has the same basic life history pattern as other jellies (Fig. 14.2), with an alternation of a sexual, medusoid stage and an asexual benthic stage. The medusa is sexually mature at 11–15 cm bell diameter, when it is ~2 months old; it reaches 20 cm bell diameter at 3 months. The jelly spawns serially from spring to autumn. Eggs are retained and, after fertilisation, larvae are apparently brooded by the female and then released as non-feeding planulae for 4–5 days, after which they settle in aggregations on a variety of firm substrata as polyploid scyphistomae. They quickly grow tentacles and ingest food, and start to release ephyrae after ~2 weeks by budding them off from the end of the scyphistomae and forming another underneath. The ephyrae, which usually appear in autumn and spring, grow into medusae, with a longevity of up to 13 months, and then the life cycle is repeated.

The mosaic jelly pulsates strongly and can swim at speeds of 2–6 m $minute^{-1}$. It occurs in aggregations well within estuaries, in salinities exceeding ~10 ppm, and can probably maintain its location in the upper reaches of estuaries by strong swimming. Its diet varies between night and day. Pitt *et al.* (2008) found that at night the nocturnally active shrimp *Lucifer* and emergent copepods comprised ~90% of the food eaten, whereas by day the diet was molluscan larvae (70%) and copepods (25%).

The white-spotted jelly Phyllorhiza punctata

This Indo-Pacific species extends down the west and east coasts of Australia, where it blooms seasonally in estuaries. Blooms in the Swan–Canning Estuary were described very early by Backhouse (1843) as 'multitudes of jelly-fish of great beauty ... with about ten brown, spongy masses...' In this estuary, small medusae (2 cm across) first appear in early summer, apparently from scyphistomae that have over-wintered from the previous season. These medusae become reproductive from 6 weeks of age at a size of 14–18 cm across, and start to produce planulae (Rippingale and Kelly 1995). The medusae continue growing to a maximum size of 50 cm and produce planulae until they die in the following winter.

The tiny planulae, < 0.5 mm long, settle on the bottom, grow tentacles, start feeding and develop into scyphistomae within 2–3 days. Each individual scyphistoma can then produce: (a) large numbers of buds asexually at rates of up to one per day, and these buds settle and develop into yet more scyphistomae; and (b) some ephyrae. From their study Rippingale and Kelly suggested that ephyrae production occurs annually after an increase in day length and rise in sea temperature, and their survival depends on a sufficiently high salinity, which is reached after winter runoff into the estuary has dissipated. As described below, these reproductive features enable this jellyfish to bloom spectacularly.

Blooming jellyfish

The elegantly simple structure and composition (95% sea water) of jellyfish and their complex life history enable them to grow very fast and to be highly efficient breeding machines. Their ability to produce sudden blooms is linked to their complex life histories (Boero *et al.* 2008). The asexual propagation of benthic polyps, and the release of many young medusae from each polyp, can quickly induce teeming thousands of medusae. Furthermore, rapid reproduction is ensured by their remaining in aggregates, which they achieve as they drift along in the current and respond uniformly to cues such as light and wave action. Other factors that give them a distinct advantage over many organisms are:

- Eutrophic conditions can stimulate jellyfish blooms by increasing the density of phytoplankton, and thus zooplankton, and thereby provide an abundant food supply for them.
- Reduced oxygen levels, resulting from the breakdown of plant material, and so on, are not particularly harmful, because many jellyfish can survive in hypoxic conditions, likely due to their ability to store oxygen within the jelly matrix. As commented by Graham (2008), 'in a sense jellyfish have their own SCUBA tank to use under short-term hypoxia'.

The occurrence of jellyfish blooms in southern coastal waters and their relationship with overfishing and high nutrient levels in ecosystems are discussed in Chapter 19.

STAUROMEDUSAE (CLASS STAUROZOA)

Carolina J Zagal

Stauromedusae are small, goblet-shaped jellyfish that spend their entire life attached to the substratum, as distinct from most classes of Cnidaria, which have free-swimming medusae in parts of their life cycle. They live attached to algae or other substrata by means of an adhesive disk on the base of a short stalk-like peduncle, with their mouth facing upwards. They can move, however, by detaching and re-attaching at will by somersaulting, using their adhesive organs found between their arms or tentacles for adhesion. Their main body part, the gelatinous umbrella, typically draws out into eight simple arms, or four arms each divided into two, and tipped with a bunch of short tentacles. They are found worldwide, from intertidal and subtidal zones to deep hydrothermal vents, but are seldom seen because of their cryptic colouration, which matches the substrata on which they live. Adult stauromedusae are typically small, ~1–4 cm in umbrella height.

Until recently, the Stauromedusae were placed as an order within the Scyphozoa, but genetic and other studies show that they are quite distinct, and require their own Class. Of the 52 species known globally (Mills and Hirano 2007), 12 are in the Southern Hemisphere, but more are certain to be discovered (Zagal *et al.* 2011). Four stauromedusan species are known from southern Australia: *Depastromorpha africana* (Plate 44) (from South Australia, Tasmania, South Africa and New Zealand); *Stenoscyphus inabai* (from Victoria, Tasmania, Japan and Korea); the recently described *Lucernariopsis tasmaniensis* (Plate 45) (Victoria, South Australia and Tasmania); and *Lipkea* sp. (New South Wales).

Most Stauromedusae are probably male or female, and their life cycle may be similar to that of jellyfish, except for the absence of a free-swimming medusa stage. Instead they develop creeping larvae that metamorphose into new adults.

Very little is known about their biology and ecology. The species that have been studied are carnivorous, and feed on small crustaceans, polychaetes and molluscs. They are preyed on by fish and nudibranchs. Some species seem to be quite rare, while others can be locally abundant, especially during the summer. Most southern species have been found in sheltered intertidal and subtidal habitats, except one species *Lucernaria australis* found at 385 m depth off Antarctica.

SEA ANEMONES (CLASS ANTHOZOA, ORDER ACTINIARIA)

In this group, the body consists of a column with a pedal disc at the base for attachment to the substratum and an oral disc at the other end with a circlet of tentacles. The tentacles are armed with nematocysts, and some species also have long threads, called acontia, used for defence, that are similarly armed, and protrude from the body or mouth of the anemone. One local species *Actinia tenebrosa* has acrorhagi – small, inflatable spherules surrounding the tentacles, also armed with nematocysts, and used for aggression. The epidermis is well supplied with mucus glands. The mouth leads to a large central cavity where food is digested and the undigested remains are passed out through the mouth.

Anemones are carnivorous and feed on plankton, worms and even small fish, which are paralysed by their stinging cells. Planktonic food sweeping past them may be caught by the tentacles and deposited in the mouth (Van-Praët 1985). They can also bend the column over and gather detritus and micro-benthos from the substratum with their tentacles. Another source of food arises from the symbiosis of anemones with *Symbiodinium* (zooxanthellae). These algae are located within animal cells in the tentacles and oral disc, and enable the anemone to benefit from exposure to the light. As in other similar symbioses, the zooxanthellae take up CO_2 and release much of it as glycerol, glucose and alanine to the host anemone.

Anemones are normally sedentary on rocky bottoms, but they can move about in several ways: by slowly gliding on their pedal disc at rates of up to 10 cm h^{-1}; by floating off and rolling over the bottom; or by walking on their tentacle tips. Several species, mostly in the genera *Calliactis*, *Paracalliactis* or *Adamsia*, allow themselves to become attached to shells occupied by *Paguristes brevirostris* and other hermit crabs. The crab strokes the anemone's body with its claws, whereupon it detaches and the crab then attaches the anemone to its protective shell. Presumably the relationship is a mutualism, because the anemone benefits from being carried about, the crab benefits from stinging defences, and each benefits from food caught by the other.

Reproduction can be both asexual and sexual, and in some species both are used. In asexual reproduction, parts of the disc split off by pedal laceration, and are left behind as the animal moves; these regenerate into small anemones. In other species, budding of young ones occurs when a tentacle breaks off and grows into another individual. In sexual reproduction, anemones may be hermaphroditic, with separate release of eggs and sperm, or with separate male and female sexes. Eggs may be fertilised in the body cavity or externally in the water. The resulting larvae are long lived in some species, but settle at an early stage in others.

The predators of sea anemones are poorly known in southern waters (Ottaway 1977). The snapper *Pagrus auratus* is a known predator, although anemones may be incidental to its diet. Another predator is the warm temperate to tropical toadfish *Arothron hispidus* on the east and west coasts of Australia. The pebble crab *Bellidilia laevis* has been observed feeding on the tentacles of the speckled anemone *Oulactis* sp. The crab approaches the anemone, which defensively extends its tentacles onto the claws of the crab, whereupon the crab backs off and eats the adhered tentacles (B. Saunders *pers. comm.*). Aeolid nudibranchs and some pycnogonids also predate on anemones (see Tables 9.2 and 15.10). Of all predators, the side-gilled slug *Pleurobranchaea maculata* is the most voracious, eating species in the genera *Actinia*, *Isactinia* and *Anthothoe*. In one study, one of these nudibranchs ate 100 *Actinia* in 3 days (Ottaway 1977). In these cases, the nudibranchs can sequester the anemones' nematocysts for their own defence (see Chapter 9).

The ecology of two common southern species, summarised below, illustrates the complexity of anemone life history and behaviour.

The white-striped anemone *Anthothoe albocincta*

This common anemone (Plate 46) has many colour forms and lives on firm substrate to > 25 m depth throughout SE Australia. It reproduces asexually and sexually. In asexual reproduction, individual polyps split into two to four equal parts by vertical fission to form identical, same-coloured polyps. This form of reproduction is common and maintains large, local populations. However, sexual reproduction is also important. In this reproductive mode, male and female anemones have a peak spawning season in March–April each year. The resultant larvae are plankton-feeders and disperse widely, contributing recruits to new populations, and maintaining genetic diversity within established populations (Billingham and Ayre 1996, 1997).

The diet of *Anthothoe* was examined by Shepherd and Gray (1985) who found that 65% of the food consumed consisted of small crustaceans (amphipods, ostracods and copepods), with invertebrate eggs comprising 19%, and the rest molluscan larvae and fragments of algae, polychaetes and bryozoans.

The waratah anemone *Actinia tenebrosa*

This anemone is widely distributed throughout southern Australia from the intertidal zone to ~5 m depth and has been intensively studied over much of its range mainly by Ottaway (1974, 1978, 1979a, b, 1980), Black and Johnson (1979), Ayre (1982, 1983, 1984a, b, c, 1987, 1988) and Ayre *et al.* (1991, 1997). The waratah anemone has separate sexes, as well as individuals without gonads, and reproduces both asexually and sexually. In asexual reproduction, juveniles are produced internally by budding in male, female and non-sexual anemones, and released after several weeks, or even longer – up to a year according to Ottaway (1979a) – generally at a size of ~7 mm across. The average number of brooded juveniles peaks in summer at four to 12 per adult, with occasionally many more; ~80% of anemones carrying brooded young. Upon their release, the young settle on a reef, usually within a few centimetres of their parent. The size of settlement of juveniles affects their survival, but, even when they settle at a relatively large size, mortality is high. Yet, it is still much lower than the mortality of juveniles produced by sexual reproduction and explains the huge advantage that internal brooding provides (Perrin *et al.* 1999).

A curious enquirer might well wonder how brooded juveniles escape being digested by the adult anemone while in the space where food is taken in. However, as food is captured, filaments surround it in a mucous envelope, and then secrete digestive enzymes. The juveniles develop in inter-septal spaces away from the immediate region of food digestion, so are not subjected to high enzyme concentrations. Nevertheless, it seems that brooded juveniles can resist low enzyme levels and are immune to nematocysts, because they can ingest food captured by the adult during the process of digestion (Ottaway 1974).

The significance of sexual reproduction to the species had long been hotly debated, but was finally resolved by Ayre *et al.* (1991) in their extensive genetic studies covering ~60 local populations in SE Australia. The gonads develop to a maximum size in summer and decline over the succeeding months, indicating an autumn–winter spawning season. However, the genetic studies showed that local populations were clones, produced asexually, and that sexual reproduction was a rare or highly variable event. Nevertheless, sexual reproduction was still important for the recruitment of widely dispersed colonists at a geographic scale.

The widespread occurrence of local populations of single clones raises interesting questions as to how the observed clonal territories are maintained at a local scale. First, Ottaway (1978) showed that intra-specific aggressive behaviour was common between individuals. In encounters between two anemones, the aggressor inflates its acrorhagi and sweeps them down against the victim as the nematocysts are fired. In such encounters, the defeated anemone (often the aggressor) moves away, and John Ottaway suggested that such behaviour effectively maintained space between individuals and reduced competition for food. Second, Ayre (1984d, 1987) showed that adults were able to distinguish between clone-mates and genetically distinct individuals, and that aggression was only directed against the latter group. Hence, by inter-clonal aggression, clonal territories were established and maintained, and the immigration of dissimilar adults prevented. David Ayre also showed that clones were highly adapted to the conditions in the immediate locality – implying that aggression was a component of clonal fitness. The close association of mutually tolerant clones can also reduce the individual cost of predation – called the 'Three Musketeer Effect' – due to the sheltering effect of close cooperative neighbours (Francis 1988). These conclusions supported Ottaway's (1978) findings that movement of adults was extremely localised, with about half a studied population moving < 5 cm, and the remainder 5–50 cm, over 2 years.

The growth rate of anemones is highly variable, as they can grow for a time, and then contract, especially if injured. Individuals can reach sexual maturity at ~15 mm diameter in 2 years, but average growth rates are low, at 0.3–1.6 mm year^{-1}, so that it would take 25–50 years to reach the average maximum size of 40 mm diam. (Ottaway 1980). These variable growth rates may well have been due in part to the anemone's opportunistic feeding on seldom available food in the intertidal zone, because Ayre (1984a) found little evidence of ingested food, but when present it was often insects, or occasionally, stranded salps.

The longevity of *Actinia tenebrosa* can be extraordinary, because it has few predators, and Ottaway (1980) estimated a longevity of ~50 years, with a maximum of > 210 years. However, Ayre (1984c) recorded much higher mortality rates in Western Australia, and proposed more modest estimates of longevity of 'at least a few years'.

ZOANTHIDS (ORDER ZOANTHINIARIA)

Adrienne Grant and SA Shepherd

The zoanthids are a colonial group, resembling anemones, but with the polyps joined by basal stolons or a thin membrane. Some species are symbiotic with invertebrates, chiefly sponges, such as the yellow zoanthid of SE Australia *Epizoanthus kareni*, which grows on shaded surfaces or on sponges. Other species, such as *Zoanthus robustus* (Plate 47), live in open habitats symbiotically with dinoflagellate algae (*Symbiodinium* sp., sometimes called zooxanthellae), and are found in shallow, well-lit habitats, usually at less than 10 m depth.

In *Zoanthus robustus* the algae live inside the host endoderm cells, surrounded by host-derived symbiosomes (membrane-bound compartments) under the polyp surface. To maintain the stability of the symbiosis, as in other symbiotic cnidarians, the zoanthid uses cell signals such as a 'host release factor' that stimulates the transfer of photosynthate from its symbiotic zooxanthellae to the host (Sutton and Hoegh-Guldberg 1990). The photosynthate is used by the zoanthid for growth, and in return the host supplies the zooxanthellae with nutrients.

Little is known about the ecology of *Z. robustus*, which occurs on low-lying reefs in colonies of interconnected polyps on moderately to fully exposed coasts, with tentacles retracted by day and extended at night to catch plankton. Hence, zoanthids can also derive nutrients from dissolved organic matter and by feeding on plankton. The dependence of zoanthids on these food sources is uncertain. Some species feed selectively on different prey items, while others seem rarely to capture zooplankton (Reimer 1971). In some species, if the zooxanthellae are lost, the zoanthid polyps rapidly decrease in size and eventually die (Trench 1974). In contrast, *Z. robustus* maintained in aquaria may produce offspring that lack algae, and these can survive for several years (A Grant unpublished data).

From a food web perspective, the remarkable nutritional plasticity of zoanthids suggests that they can occupy several trophic levels at the same time – as primary producers and as feeders on detritus and plankton (Trench 1974).

Z. robustus may also have practical use as a biomarker for copper toxicity levels in marinas and harbours, where copper in antifouling paint may be released. The zoanthid suffers subcellular damage to specific proteins in the polyps, so it can record long-term damage from fluctuating levels of copper (Grant *et al.* 2010).

STONY CORALS (ORDER SCLERACTINIA)

SA Shepherd and Adrienne Grant

Stony corals are closely related to anemones, except that they secrete a calcified external skeleton, which protects the polyp within. Together they form a corallite. Some species are solitary and others are colonial with lateral interconnections. The latter include several reef-building species widespread throughout southern Australia, and a few tropical species that extend south around SW Australia and east to the head of the GAB (Shepherd and Veron 1982).

The presence of reef-building species is remarkable, because they are rare examples of tropical corals that manage to survive in cool temperate regions, beyond the limits of most corals. We discuss the only species that has been studied in temperate waters.

The green and blue coral *Plesiastrea versipora*

This species is the most widespread coral in the Indo-Pacific region and occurs in southern Australia as far south as northern Tasmania, although in recent years its range is thought to have extended southwards with the extension of the East Australia Current (Johnson *et al.* 2011). The species has two basic forms: a flat, encrusting form (Plate 48) common on open coasts exposed to wave action; and a massive form growing to a height of ~2 m that lives in places of moderate to strong current in the deeper waters of the South Australian gulfs. Genetic studies have shown that, although the coral larvae are planktonic for an unknown period, genetic connectivity between populations is quite limited (Rodriguez-Lanetty and Hoegh-Guldberg 2002).

The geochemical ecology of the massive form was studied in detail by Samantha Burgess (2007) and Burgess *et al.* (2009). Colonies laid down annual density bands and lived for hundreds of years, so they became a valuable archive of past environments. The ages of numerous samples from the South Australian gulfs ranged from 100 to 381 years, with variable growth rates of 1–8 mm year^{-1}, measured as upward increment. Corallite density varied between individuals, and the high-density colonies grew more slowly than the low-density ones. The fastest growing colonies were those in optimal conditions of high light intensities in shallow water, a good nutrient supply and strong tidal flows. Yet, overall, there was no relation between growth and measured environmental factors.

Past sea temperatures and periods of coastal contamination were effectively preserved from trace elements present in the density bands, and Samantha Burgess used the oldest colonies to give a glimpse of the past environment of Gulf St Vincent since 1845. Sea surface temperatures, derived from estimates of the stable isotope of $\delta^{18}O$ (Fig. 14.3), showed a gradual warming trend of ~2°C in eastern Gulf St Vincent. In another coral collected off metropolitan Adelaide, lead, zinc, copper, tin and manganese levels greatly increased after European settlement, notably during the time of urban and industrial development. High lead levels from the 1960s–1990s were due to the proliferation of cars using lead petrol during that period. High zinc levels from the 1940s were due to stormwater inflow and the discharge of treated sewage < 10 km

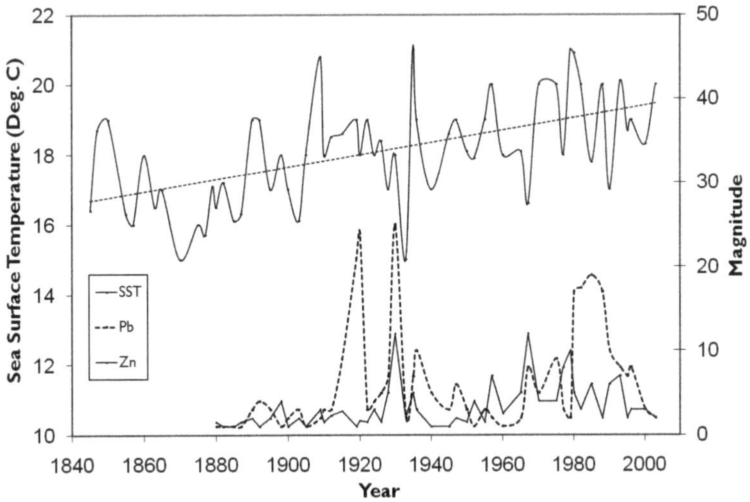

Figure 14.3: (Top curve) Sea temperatures, derived from low resolution ^{18}O analyses of density bands of a *Plesiastrea versipora* colony at Edithburgh, eastern Gulf St Vincent, and (lower curves) the magnitude of enrichment of trace metals, lead (Pb) and zinc (Zn), normalised to a pre-European baseline concentration, in a *Plesiastrea versipora* colony at Seacliff, western Gulf St Vincent (redrawn from Burgess (2007).

distant from the location of the coral. Heavy metals in treated effluent were reduced by 75–95% in the early 1990s, as sewage treatment improved and sludge effluent was diverted onto land. All these changes were faithfully reflected in the palaeo-signatures in the coral.

The unusual ability of *Plesiastrea* to adapt to cold-water climates has stimulated many studies of the symbiosis with its brown pigmented zooxanthellae, which are present in these green and blue corals. Zooxanthellae enhance the calcification rates of the coral by about four times during photosynthesis, compared with corals kept in the dark, and, even at low light intensities on a cloudy day, more energy is fixed by photosynthesis than is required by the host coral (Kevin and Hudson 1979). Nevertheless, apo-symbiotic corals kept in the dark for 2 years survived, and still produced the cell signals that would normally be used for the control of their symbiotic algae (Grant *et al.* 2004). In a 7-year study by Adrienne Grant of a colony shaded under an overhang, and receiving no direct sunlight in southern New South Wales, the average lateral extension rates measured in two directions was 6 mm year^{-1} and 9 mm year^{-1}, respectively.

The photosynthetic rates of *Plesiastrea* in the cold waters of 10–21°C in Port Phillip Bay have been found to be as high as those of tropical corals, due apparently to the increased photosynthetic pigment in zooxanthellae and near perfect temperature compensation in the coral (Howe and Marshall 2001). In the warmer half of the year the coral could subsist entirely on photosynthetic products, and in the cooler half it survived well on zooplankton.

Molecular studies have found that the regulation of the symbiosis is quite complex, because the coral contains two cell signalling compounds – a 'host release factor' (Grant *et al.* 1998) and a 'photosynthesis inhibiting factor' (Grant *et al.* 2001, 2003). The former stimulates the release of up to 42% of the carbon fixed by zooxanthellae to the host, mainly in the form of glycerol (Sutton and Hoegh-Guldberg 1990), thus providing the coral host with food; the latter partially inhibits algal photosynthesis (Grant *et al.* 2001). Together they control algal growth and avoid algal overpopulation (Grant *et al.* 2006). Thus, host cell signal regulation of their symbiotic zooxanthellae may contribute towards the adaptation of this coral to cold water.

SOFT CORALS AND GORGONIANS (ORDER ALCYONACEA)

Soft corals

These colonial animals, called octocorals, have many polyps, each with eight feather-like tentacles, and an internal skeletal structure of calcareous elements deposited in the body walls and, in the gorgonians, a rigid axis. They vary in form from encrusting species with few polyps to spectacular sea-fans a metre or more high. Colonies grow by budding and branching of polyps from the original settled larva.

The sexes are usually separate, and fertilisation of eggs takes place in the water, or in some cases on the female colony where the eggs are retained. A few species are hermaphroditic, liberating eggs and sperms at different times. Some species have evolved symbioses with zooxanthellae and these live in shallow, open habitats. On the other hand, species without zooxanthellae live in shaded or deep habitats, often in places of moderate water currents, where they extrude their polyps and feed on suspended material. This comprises particulate organic matter, phytoplankton and bacteria. Some species feed by day and others at night.

Octocorals lack stinging nematocysts and soft corals have no physical defence against predators. Yet they have relatively few predators, and this has been variously attributed to the presence in their tissues of high levels of toxic chemicals (terpenes) that deter fishes, or to calcitic needles, termed sclerites, that are also a deterrent. However, two groups, ovulid molluscs and some nudibranchs (Chapter 9), have specialised in feeding on octocorals, due to their ability to assimilate the toxic compounds and, in the case of nudibranchs, to use them for their own defence (Sammarco and Coll 1992). In eastern Australia, Coleman (2002) recorded two ovulids, *Ovula costellata* and *Prionovolva brevis*, on their respective host soft corals, *Cladiella* sp. and *Dendronephthys* spp.

The ecology of only one southern species of soft coral is known, and that was studied in a series of papers by Farrant (1985, 1986, 1987) and Farrant *et al.* (1987a, b) as described below.

Gabo I. soft coral *Drifa gaboensis* (formerly in the genus *Capnella*)

This soft, grey-blue coral (Plate 49) has a branching form, and grows to 40 cm high on vertical and horizontal rock surfaces from ~5 to 30 m depth in SE Australian waters west to the eastern GAB; it contains symbiotic zooxanthellae.

Reproduction is both asexual and sexual. Asexual reproduction is by fission of a colony. This is both a rare and protracted event, being recorded in only 4% of the colonies at Penny Farrant's Sydney site, and requiring several years to complete. The predominant reproductive mode is sexual, with separate sexes, although ~0.5% are hermaphrodites with male and female gonads in separate polyps. Sexual maturity occurs when colonies reach ~4 cm high at an age of 2–3 years. The gonads mature from November to May, followed by spawning over 3–4 weeks in early winter. Farrant (1985) estimated that an average-sized (200 g) colony with ~60 000 polyps released ~360 000 eggs annually. Released eggs are sticky and adhere externally to the female colony, where they are fertilised and where embryos develop into planulae larvae. The planulae then sink to the bottom and crawl slowly over the rocks, searching for a suitable settlement site, usually within ~1 m of the parent colony. They metamorphose in 2–3 days and grow into another colony. The larvae carry in their tissues zooxanthellae, acquired when embryos were being brooded, and in this way the symbiosis is transmitted to the next generation.

The mortality of settled larvae was high in Penny Farrant's study, with only ~0.25% surviving to 1 year. Mortality was due to algal abrasion, overgrowth by algae or sponges, or urchin grazing. From ~3 years of age, mortality declined until they reached ~13 years; longevity was ~20 years. The main predators of adults were the sea-urchin *Phyllacanthus parvispinus*, an ovulid mollusc *Globovula cavanaghi*, and a nudibranch *Ceratosoma amoenum*.

The soft coral's symbiosis with zooxanthellae appears to be an obligate one, but the contribution of photosynthate to the colony is < 50% of its total needs, so the coral must also rely on the capture of plankton, particulate matter and dissolved organic matter from the water.

Gorgonians

> *'Lower down there were to be seen swaying in the water under the reefs dark gorgonians or sea-fans.'*
>
> John Steinbeck, *The Log from the Sea of Cortez*

The horny-axis of gorgonians provides them with a firm structure and shape. Some are simple, unbranched whips, some have candelabra-like branches, some are feather-like, some are bushy and others have slender interlacing branches, fused to form broad fans. The ecology of southern species is virtually unknown.

Gorgonians require consistent water movement either from wave-induced surge or from moderate to strong tidal currents – conditions in which they can best feed (Sponaugle and LaBarbara (1991). Fan-shaped species, such as *Mopsella* spp., grow in one plane with their fans oriented normal to the current or water movement direction, but in turbulent or eddying water, bushy species such as the bramble coral *Acabaria* sp. are found. On exposed coasts, *Mopsella* is common in caves, vertical faces and under overhangs, while in deeper water, plumed species, such as *Pteronisis* and *Mopsea*, and sea-whips, such as *Primnoella*, become common. In places of strong current flow, species such as *Euplexaura* and *Echinogorgia* occur on reefs or attached to pieces of shell.

The preference of gorgonians for places of vigorous water movement means that they are a favoured feeding platform for the feather-star *Ptilometra macronema* and the basket star *Conocladus australis*, which are both often seen on their branches.

Gorgonians contain chemicals that are unpalatable to fish or cause them to vomit, and so they only have a few specialist predators. These are some nudibranchs (see Table 9.2) and six ovulid molluscs, which have apparent obligate associations with particular host gorgonians and mimic their background colour. These species and their respective hosts are: *Primovula cruenta* on *Echinogorgia*, *Primovula heleneae* and *P. tigris* on *Euplexaura*, *Globovula cavanaghi* on *Mopsella ellisi*, and *Prosimnia semperi* and *Phenacovolva angasi* on other species of *Mopsella* (Wilson 1993; Coleman 2002). In the case of the gorgonian *Euplexaura*, two to five ovulids live on each colony, and lay their egg capsules on the gorgonian fronds (Gowlett-Holmes and Holmes 1989; Gowlett-Holmes 2008).

Many species of shrimp (subfamily Pontoniinae) live commensally with other organisms, and two species, *Hamodactylus boschmai* and *Manipontonia psamathe*, have been recorded on the gorgonian *Mopsella ellisi*, which extends down the east Australian coast (Bruce 1970). These shrimps are only found on host species that produce abundant mucus, which contains waxes, esters and other substances of nutritional value – a very probable food source (AJ Bruce *pers. comm.*).

CTENOPHORES (PHYLUM CTENOPHORA)

Ctenophores or comb jellies are a small phylum of mainly pelagic jellies, similar to medusae in that they are carnivorous. They use tentacles to catch their prey, but otherwise they have no close evolutionary ties to cnidarians. They are oceanic, although they come into coastal waters, and fall outside the scope of this book – except for a very small and unusual group of benthic comb jellies, called platyctenes, which are discussed here.

In this small group, the body is flattened so that they creep along on a foot, and look like flatworms. They reproduce asexually by fragmentation, as well as sexually, and can sometimes be found in large numbers on a variety of hosts, including algae, ascidians and some echinoderms. Only four species are known from southern waters and three of them have been recorded on algae. These include *Coeloplana scaberiae* (Plate 50), which lives in an apparently obligate relationship with the brown alga *Scaberia* (Matsumoto and Gowlett-Holmes 1996). It is red or orange in colour, matching its algal substratum, and extends its feathery tentacles streaming from each end of the body to feed at night. Planktonic items captured on the tentacles are drawn down to the oral surface and then into the oral groove. A white sponge is always found under the ctenophore, suggesting that the sponge is the true host in a sort of innocent ménage-à-trois! A second species *Coeloplana thomsoni* is found at Rottnest I., Western Australia, living on the coralline alga *Jania* in the shallow subtidal zone (Matsumoto 1999), while a third species *C. willeyi* is found at 15 m depth on *Caulerpa* and red algae in southern Port Phillip Bay (Smith and Plant 1976). An undescribed fourth species lives under rocks and on *Zoanthus* in SE Australia (Gowlett-Holmes 2008).

REFERENCES

Arai MN (2005) Predation on pelagic coelenterates: a review. *Journal of the Marine Biological Association of the United Kingdom* **85**, 523–536.

Ayre DJ (1982) Inter-genotypic aggression in the solitary sea anemone *Actinia tenebrosa*. *Marine Biology* **68**, 199–205.

Ayre DJ (1983) The effects of asexual reproduction and inter-genotypic aggression on the genotypic structure of populations of the sea anemone *Actinia tenebrosa*. *Oecologia* **57**, 158–165.

Ayre DJ (1984a) The sea anemone *Actinia tenebrosa*: an opportunistic insectivore. *Ophelia* **23**, 149–153.

Ayre DJ (1984b) The effects of sexual and asexual reproduction on geographic variation in the sea anemone *Actinia tenebrosa*. *Oecologia* **62**, 222–229.

Ayre DJ (1984c) The effects of environment and population density on the sea anemone *Actinia tenebrosa*. *Australian Journal of Marine and Freshwater Research* **35**, 735–746.

Ayre DJ (1987) The formation of clonal territories in experimental populations of the sea anemone *Actinia tenebrosa*. *The Biological Bulletin* **172**, 178–186.

Ayre DJ (1988) Evidence of genetic determination of sex in *Actinia tenebrosa*. *Journal of Experimental Marine Biology and Ecology* **116**, 23–34.

Ayre DJ, Read J, Wishart J (1991) Genetic subdivision within the eastern Australian population of the sea anemone *Actinia tenebrosa*. *Marine Biology* **109**, 379–390.

Ayre DJ, Davis AR, Billingham M, Llorens T *et al.* (1997) Genetic evidence for contrasting patterns of dispersal in solitary and colonial ascidians. *Marine Biology* **130**, 51–61.

Backhouse J (1843) *A Narrative of a Visit to the Australian Colonies*. Hamilton, Adams, London.

Berrill NJ (1948) A new method of reproduction in *Obelia*. *The Biological Bulletin* **95**, 94–99.

Billingham M, Ayre DJ (1996) Genetic subdivision in the subtidal, clonal sea anemone *Anthothoe albocincta*. *Marine Biology* **125**, 153–163.

Billingham M, Ayre DJ (1997) Asexual reproduction and genetic determination of colour patterns within populations of the subtidal sea anemone *Anthothoe albocincta*. *Marine Ecology Progress Series* **156**, 121–130.

Black R, Johnson MS (1979) Asexual viviparity and population genetics of *Actinia tenebrosa*. *Marine Biology* **53**, 27–31.

Boero F (1984) The ecology of marine hydroids and effects of environmental factors: a review. *P.S.Z.N.I. Marine Ecology (Berlin)* **5**, 93–118.

Boero F, Bouillon J, Gravili C, Miglietta MP *et al.* (2008) Gelatinous plankton rule the world (sometimes). *Marine Ecology Progress Series* **356**, 299–310.

Bouillon J, Boero F (2000) The Hydrozoa: a new classification in the light of old knowledge. *Thalassia Salentina* **24**, 3–45.

Browne JG, Kingsford MJ (2005) A commensal relationship between the scyphozoan medusa *Catostylus mosaicus* and the copepod *Paramacrochiron maximum*. *Marine Biology* **146**, 1157–1168.

Bruce AJ (1970) Report on some commensal pontoniinid shrimps (Crustacea: Palemonidae) associated with an Indo-Pacific host (Coelenterata: Gorgonacea). *Journal of Zoology* **160**, 537–544.

Burgess SN (2007) Geochemical ecology of temperate reefs. PhD thesis. Australian National University, Canberra.

Burgess SN, McCulloch MT, Mortimer GE (2009) Structure and growth rates of the high-latitude coral: *Plesiastrea versipora*. *Coral Reefs* **28**, 1005–1015.

Clark SD, Cook CB (1986) Inhibition of nematocyst discharge during feeding in the colonial hydroid *Halocordyle disticha* (= *Pennaria tiarella*): the role of previous prey-killing. *The Biological Bulletin* **171**, 405–416.

Cleland JB, Southcott RV (1965) *Injuries to Man from Marine Invertebrates in the Australian Region*. National Health and Medical Research Series No. 12. Commonwealth of Australia, Canberra.

Coleman N (2002) *2002 Sea shells: Catalogue of Indo-Pacific Molluscs*. NC Underwater Geographic, Brisbane.

Costello JH, Colin SP (1994) Morphology, fluid motion and predation by the scyphomedusa *Aurelia aurita*. *Marine Biology* **121**, 327–334.

Crawford CM, Moltschaniwskyi NA, Willcox S (2011) Size and characteristics of aggregations of moon jellyfish (*Aurelia* sp.) in Tasmania, Australia. *Papers and Proceedings of the Royal Society of Tasmania*. **145**, 9–15.

Dawson MN, Martin LE (2001) Geographic variation and ecological adaptation in *Aurelia* (Scyphozoa, Semaostomeae): some implications from molecular phylogenetics. *Hydrobiologia* **451**, 259–273.

Dawson MN, Gupta AS, England MH (2005) Coupled biophysical global ocean model and molecular genetic analyses identify multiple introductions of cryptogenic species. *Proceedings of the National Academy of Sciences of the United States of America* **102**, 11968–11973.

Elmhirst R (1922) Cyclic conditions and rejuvenation in hydroids. *Nature* **109**, 208–209.

Fancett MS (1986) Species composition and abundance of Scyphomedusae in Port Phillip Bay, Victoria. *Australian Journal of Marine and Freshwater Research* **37**, 379–384.

Fancett MS (1988) Diet and prey selectivity of Scyphomedusae from Port Phillip Bay, Victoria. *Marine Biology* **98**, 503–509.

Fancett MS, Jenkins GP (1988) Predatory impact of Scyphomedusae on ichthyoplankton and other zooplankton in Port Phillip Bay. *Journal of Experimental Marine Biology and Ecology* **116**, 63–77.

Farrant P (1985) Reproduction in the temperate Australian soft coral *Capnella gaboensis*. *Proceedings of the Fifth International Coral Reef Congress. Vol. 4*. 27 May–1 June 1985, Tahiti (Eds C Gabrie and B Salval) pp. 319–324. Ecoles Practique des Haute etudes, Moorea, French Polynesia.

Farrant P (1986) Gonad development and the planulae of the temperate Australian soft coral *Capnella gaboensis*. *Marine Biology* **92**, 381–392.

Farrant P (1987) Population dynamics of the temperate Australian soft coral *Capnella gaboensis*. *Marine Biology* **96**, 401–407.

Farrant P, Borowitzka MA, Hinde R, King RJ (1987a) Nutrition of the temperate Australian soft coral *Capnella gaboensis* I. Photosynthesis and carbon fixation. *Marine Biology* **95**, 565–574.

Farrant P, Borowitzka MA, Hinde R, King RJ (1987b) Nutrition of the temperate Australian soft coral *Capnella gaboensis* II. The role of zooxanthellae and feeding. *Marine Biology* **95**, 575–581.

Francis L (1988) Cloning and aggression among sea anemones (Coelenterata: Actiniaria) of the rocky shore. *The Biological Bulletin* **174**, 241–253.

Gili J-M, Hughes RG (1995) The ecology of marine benthic hydroids. *Oceanography and Marine Biology - an Annual Review* **33**, 351–426.

Gowlett-Holmes KL (2008) *A Field Guide to the Marine Invertebrates of South Australia*. Notomares, Hobart.

Gowlett-Holmes KL, Holmes NJC (1989) Rediscovery of *Primovula (P.) heleneae* Cate and description of a new species of *Crenavolva* from South Australia (Mollusca: Gastropoda: Ovulidae). *Transactions of the Royal Society of South Australia* **113**, 205–214.

Graham W (2008) Jellyfish blooms. *Marine Scientist* **23**, 24–27.

Grant AJ, Rémond M, Hinde R (1998) A low molecular weight fraction from the symbiotic coral *Plesiastrea versipora* modifies the metabolism of symbiotic algae. *Marine Biology* **130**, 553–557.

Grant AJ, Rémond W, Withers KJT, Hinde R (2001) Inhibition of photosynthesis by a symbiotic coral. *Hydrobiologia* **461**, 63–69.

Grant AJ, Trautman DA, Frankland S, Hinde R (2003) A symbiosome membrane is not required for the actions of two host signalling compounds regulating photosynthesis in symbiotic algae isolated from cnidarians. *Comparative Biochemistry and Physiology Part A* **135**, 337–345.

Grant AJ, Starke-Peterkovic T, Withers KJT, Hinde R (2004) Aposymbiotic *Plesiastrea versipora* continues to produce cell-signalling molecules that regulate the carbon metabolism of symbiotic algae. *Comparative Biochemistry and Physiology Part A* **135**, 253–259.

Grant AJ, Rémond M, Starke-Peterkovic T, Hinde R (2006) A cell signal from the coral *Plesiastrea versipora* reduces starch synthesis in its symbiotic alga, *Symbiodinium* sp. *Comparative Biochemistry and Physiology Part A* **144**, 458–463.

Grant AJ, Trompf K, Seung D, Nivison-Smith L *et al.* (2010) Sub-cellular damage by copper in the cnidarian *Zoanthus robustus. Comparative Biochemistry and Physiology* **152**, Part C, 256–262.

Hartwick RF (1991) Observations on the anatomy, behaviour, reproduction and life cycle of the cubozoan *Carybdea rastoni. Hydrobiologia* **216/217**, 171–179.

Howe SA, Marshall AT (2001) Thermal compensation of metabolism in the temperate coral, *Plesiastrea versipora* (Lamarck, 1816). *Journal of Experimental Marine Biology and Ecology* **259**, 231–248.

Johnson CR, Banks SC, Barrett NS, Cazassus F *et al.* (2011) Climate change cascades: shifts in oceanography, species' ranges and subtidal marine community dynamics in eastern Tasmania. *Journal of Experimental Marine Biology and Ecology* **400**, 17–32.

Kevin KM, Hudson RCL (1979) The role of zooxanthellae in the hermatypic coral *Plesiastrea urvillei* (Milne Edwards and Halme) from cold waters. *Journal of Experimental Marine Biology and Ecology* **36**, 157–170.

Kingsford MJ (1993) Biotic and abiotic structure in the pelagic environment: importance to small fishes. *Bulletin of Marine Science* **53**, 393–415.

Kingsford MJ, Pitt KA, Gillanders BM (2000) Management of jellyfish fisheries, with special reference to the Order Rhizostomeae. *Oceanography and Marine Biology - an Annual Review* **38**, 85–156.

Kramp PL (1965) Some medusae (mainly scyphomedusae) from Australian coastal waters. *Transactions of the Royal Society of South Australia* **89**, 257–278.

Leung TLF, Poulin R (2008) Parasitism, commensalism, and mutualism: exploring the many shades of symbioses. *Vie et Milieu* **58**, 107–115.

Lucas CH (2001) Reproduction and life history strategies of the common jellyfish, *Aurelia aurita*, in relation to the ambient environment. *Hydrobiologia* **451**, 229–246.

Mansueti R (1963) Symbiotic behaviour between small fishes and jellyfishes, with new data on that between the Stromateid *Peprilus alepidotus*, and the Scyphomedusa, *Chrysaora quinquecirrha. Copeia* **1963**, 40–80.

Matsumoto GI (1995) Observations on the anatomy and behaviour of the cubozoan *Carybdea rastoni* Haacke. *Marine and Freshwater Behaviour and Physiology* **26**, 139–148.

Matsumoto GI (1999) *Coeloplana thomsoni* sp. nov., a new benthic ctenophore (Ctenophora: Platyctenida: Coeloplanidae) from Western Australia. In *The Seagrass Flora and Fauna of Rottnest Island, Western Australia* (Eds DI Walker and FE Wells) pp. 385–393. Western Australian Museum, Perth.

Matsumoto GI, Gowlett-Holmes KL (1996) *Coeloplana scaberiae* sp. nov., a new benthic ctenophore (Ctenophora: Platyctenida: Coeloplanidae) from South Australia. *Records of the South Australian Museum* **29**, 33–40.

Mills CE, Hirano YM (2007) Stauromedusae In *Encyclopedia of Tidepools and Rocky Shores.* (Eds Denny MW and Gaines SD) pp. 539–541. University of California Press, Berkley.

Nishihira M (1968) Brief experiments on the effect of algal extracts in promoting the settlement of the larvae of *Coryne uchidai* Stechow (Hydrozoa). *Bulletin of the Marine Biological Station of Asamushi* **13**, 91–101.

Ottaway JR (1974) Resistance of juvenile *Actinia tenebrosa* (Cnidaria: Anthozoa) to digestive enzymes. *Nauri Ora* **2**, 73–83.

Ottaway JR (1977) Predators of sea anemones. *Tuatara* **22**, 213–221.

Ottaway JR (1978) Population ecology of the intertidal anemone *Actinia tenebrosa* I. Pedal locomotion and intraspecific aggression. *Australian Journal of Marine and Freshwater Research* **29**, 787–802.

Ottaway JR (1979a) Population ecology of the intertidal anemone *Actinia tenebrosa* II. Geographical distribution, synonymy, reproductive cycle and fecundity. *Australian Journal of Zoology* **27**, 273–290.

Ottaway JR (1979b) Population ecology of the intertidal anemone *Actinia tenebrosa* III. Dynamics and environmental factors. *Australian Journal of Marine and Freshwater Research* **30**, 41–62.

Ottaway JR (1980) Population ecology of the intertidal anemone *Actinia tenebrosa* IV. Growth rates and longevity. *Australian Journal of Marine and Freshwater Research* **31**, 385–395.

Pardy RL, Lenhoff HM (1968) The feeding biology of the gymnoblastic hydroid, *Pennaria tiarella. The Journal of Experimental Zoology* **168**, 197–202.

Pauly D, Graham W, Libralato S, Morissette L *et al.* (2009) Jellyfish in ecosystems, online databases, and ecosystem models. *Hydrobiologia* **616**, 67–85.

Peach MB, Pitt KA (2005) Morphology of the nematocysts of the medusae of two scyphozoans *Catostylus mosaicus* and *Phyllorhiza punctata* (Rhizostomeae): implications for capture of prey. *Invertebrate Biology* **124**, 98–108.

Perrin MC, Thorpe JP, Solé-Cava AM (1999) Population structuring, gene dispersal and reproduction in the *Actinia equina* species group. *Oceanography and Marine Biology - an Annual Review* **37**, 129–152.

Piraino S, Boero F, Aeschbach B, Schmid V (1996) Reversing the life cycle: medusae transforming into polyps and cell transdifferentiation in *Turritopsis nutricula* (Cnidaria, Hydrozoa). *The Biological Bulletin* **190**, 302–312.

Piraino S, De Vito D, Schmich J, Bouillon J *et al.* (2004) Reverse development in Cnidaria. *Canadian Journal of Zoology* **82**, 1748–1754.

Pitt KA (2000) Life history and settlement preferences of the edible jellyfish *Catostylus mosaicus* (Scyphozoa: Rhizostomeae). *Marine Biology* **136**, 269–279.

Pitt KA, Kingsford MJ (1999) Jellyfish. In *Under Southern Seas: the Ecology of Australia's rocky reefs.* (Ed. N Andrew) University of New South Wales Press, Sydney.

Pitt KA, Kingsford MJ (2000a) Geographic separation of stocks of the edible jellyfish *Catostylus mosaicus* (Rhizostomeae) in New South Wales, Australia. *Marine Ecology Progress Series* **196**, 143–155.

Pitt KA, Kingsford MJ (2000b) Reproductive biology of the edible jellyfish *Catostylus mosaicus* (Rhizostomeae). *Marine Biology* **137**, 791–799.

Pitt KA, Kingsford MJ (2003) Temporal variation in the virgin biomass of the edible jellyfish *Catostylus mosaicus* (Rhizostomeae). *Fisheries Research* **63**, 303–313.

Pitt KA, Clement A-L, Connolly RM, Thibault-Botha D (2008) Predation by jellyfish on large and emergent zooplankton: implications for benthic-pelagic coupling. *Estuarine, Coastal and Shelf Science* **76**, 827–833.

Purcell JE, Arai MN (2001) Interactions of pelagic cnidarians and ctenophores with fish. *Hydrobiologia* **451**, 27–44.

Reimer AA (1971) Feeding behaviour in the Hawaiian zoanthids *Polythoa* and *Zoanthus*. *Pacific Science* **25**, 512–520.

Rippingale RJ, Kelly SJ (1995) Reproduction and survival of *Phyllorhiza punctata* (Cnidaria: Rhizostomeae) in a seasonally fluctuating salinity regime in Western Australia. *Marine and Freshwater Research* **46**, 1145–1151.

Rodriguez-Lanetty M, Hoegh-Guldberg O (2002) The phylogeography and connectivity of the latitudinally widespread scleractinian coral *Plesiastrea versipora* in the Western Pacific. *Molecular Biology* **11**, 1177–1189.

Sammarco PW, Coll JC (1992) Chemical adaptations in the Octocorallia: evolutionary considerations. *Marine Ecology Progress Series* **88**, 93–104.

Shepherd SA, Gray JD (1985) Food of the anemone *Anthothoe albocincta* at West Island, South Australia. *Transactions of the Royal Society of South Australia* **109**, 191–192.

Shepherd SA, Veron JEN (1982) Stony corals (Order Scleractinia or Madreporaria). In *Marine Invertebrates of Southern Australia. Part I.* (Eds SA Shepherd and IM Thomas) pp. 169–178. Flora and Fauna of South Australia Handbooks Committee, Adelaide.

Shepherd SA, Watson JE (1970) The sublittoral ecology of West Island, South Australia. 2. The association between hydroids and algal substrate. *Transactions of the Royal Society of South Australia* **94**, 139–146.

Smith BJ, Plant RJ (1976) A creeping ctenophoran (Platyctenea: Ctenophora) from Victoria, Australia. *Memoirs of the National Museum of Victoria* **37**, 43–46.

Southcott RV (1967) The Portuguese man-of-war or bluebottle. *Australian Natural History* **15**, 337–342.

Southcott RV, Glover CJM (1987) The occurrence of *Desmonema gaudichaudi* (Lesson) (Scyphozoa, Semaostomeae) in South Australian waters with records of fish-jellyfish symbioses. *Transactions of the Royal Society of South Australia* **111**, 131–132.

Sponaugle S, LaBarbara M (1991) Drag-induced deformation: a functional feeding strategy in two species of gorgonian. *Journal of Experimental Marine Biology and Ecology* **148**, 121–134.

Sutton DC, Hoegh-Guldberg O (1990) Host-zooxanthella interactions in four temperate marine invertebrate symbioses: assessment of effect of host extracts on symbionts. *The Biological Bulletin* **178**, 175–186.

Tardent P (1963) Regeneration in the Hydrozoa. *Biological Reviews of the Cambridge Philosophical Society* **38**, 293–333.

Trench RK (1974) Nutritional potentials in *Zoanthus sociatus* (Coelenterata, Anthozoa). *Helgoländer Wissenschaftliche Meeresuntersuchungen* **26**, 174–216.

Van-Praët M (1985) Nutrition of sea anemones. *Advances in Marine Biology* **22**, 65–99.

Watson JE (1980) The identity of two tubularian hydroids from Australia with a description and observations on the reproduction of *Ralpharia magnifica* gen. et sp. nov. *Memoirs of the Museum of Victoria.* **41**, 53–63.

Watson JE (1982) Hydroids (Class Hydrozoa). In *Marine Invertebrates of Southern Australia. Part I.* (Eds SA Shepherd and IM Thomas) pp. 77–115. Flora and Fauna of South Australia Handbooks Committee, Adelaide.

Watson JE, McInnes DE (1999) Hydroids from Ricketts Point and Black Rocks, Victoria. *Victorian Naturalist* **116**, 102–111.

Willcox S, Moltschaniwskyi NA, Crawford CM (2007) Asexual reproduction in scyphistomae of *Aurelia* sp.: effects of temperature and salinity in an experimental study. *Journal of Experimental Marine Biology and Ecology* **353**, 107–114.

Willcox S, Moltschaniwskyi NA, Crawford CM (2008) Population dynamics of *Aurelia* sp. scyphistomae in Tasmania, Australia. *Marine Biology* **154**, 661–670.

Wilson BR (1993) *Australian Marine Shells: Prosobranch Molluscs Part I.* Odyssey Publishing, Perth.

Zagal CJ (2007) Ecology of hydroids in natural and artificial habitats, PhD thesis. University of Sydney.

Zagal CJ, Hirano YM, Mills CE, Edgar GJ *et al.* (2011) New records of Staurozoa from Australian coastal waters, with a description of a new species of *Lucernariopsis* Uchida 1929 (Cnidaria, Staurozoa, Stauromedusae) and a key to Australian Stauromedusae. *Marine Biology Research* **7**, 651–666.

15 Larger crustaceans and sea spiders

OVERVIEW

The larger arthropods are mainly mobile species ranging in size from sea spiders and small mobile crabs to lobsters and giant crabs, but they also include sessile barnacles. The mobile species are omnivores or carnivores and play a significant function in transferring energy from herbivores to higher levels of the food web. This chapter first considers the ecology and life history of a subtidal barnacle *Balanus amphitrite*, a common shrimp *Palaemon intermedius*, and the behaviour of two cleaner shrimps *Stenopus hispidus* and *Periclimenes aesopius*. The ecology of three southern species of lobster is then described, and their larval and adult behaviour, growth rates and fishery management are compared. The lobster fisheries are the largest invertebrate fisheries in Australia, and the western rock lobster the most valuable species in terms of commercial catch. The ecology and shell exchange behaviour of several common hermit crabs are then discussed. Among the true crabs, the ecology of the spider crab *Halicarcinus ovatus*, symbiotic pea crabs (*Pinnotheres* spp.), the introduced European green crab *Carcinus maenas* and the giant crab *Pseudocarcinus gigas* are considered. Lastly, the ecology and life history of sea spiders are described. A common feature of marine arthropods is the complex series of larval stages, lasting from days to years, before the post-larvae settle onto reefs. The longevity of these species ranges from a year for smaller species to > 40 years for the giant crab.

INTRODUCTION

The arthropods are by far the most diverse phylum of free-living animals on Earth, and include the terrestrial insects and related groups as well as many marine representatives. They are distinguished by their hard exoskeleton and their segmented body. Most species in this group are mobile benthic predators or scavengers, with some exceptions, such as the barnacles, half crabs and some anomurans, which use a filtration system to capture plankton, and the mysids, krill and other smaller species groups, which feed on plankton while swimming in the water column. The smaller crustaceans (the copepods, ostracods, shrimps, mysids, amphipods, isopods and euphausids) associated with reefs are considered in Chapter 8. In this chapter, we have chosen some of the larger arthropods to illustrate the differing ecologies of a diversity of forms that live in reef habitats. These are, first, one of a few barnacles in the Class Cirripedia that extend into the subtidal zone, then some larger crustaceans, mainly crabs and lobsters in the Order Decapoda and, lastly, (an evolutionarily remote group within the arthropods) the sea spiders, in the Class Pycnogonida.

BARNACLES (CLASS CIRRIPEDIA)

Barnacles have been aptly described as crustaceans that have become attached by their heads to the bottom, and then have built around them cases of armour-like plates. Many species have evolved features that make them highly successful in occupying mid-intertidal shores, which they often dominate, but a few species extend into the subtidal zone and some into deep water, and so come within the ambit of this book. Barnacles have long been the bane of boat-owners, because they produce impervious glue that sticks to almost any hard surface and holds them fast in all conditions.

Within the shell-plates of the barnacle is the body, surrounded by the mantle cavity, with six pairs of feathery feet, the cirri, which unroll and protrude through the operculum (the lid) to feed. The cirri comb through the water and catch particles and plankton in the surrounding water. The ecology of the most advanced group of barnacles is illustrated below by a common cosmopolitan species *Balanus amphitrite* (controversially renamed *Amphibalanus amphitrite* by some). Further information can be found in the study of Egan and Anderson (1986) and the reviews of Crisp and Bourget (1985) and Barnes (1989).

The striped barnacle *Balanus amphitrite*

This species probably originated in the Indo-West Pacific, spreading from there around the world. It is common throughout southern Australia on any firm substratum to a depth of ~2 m. It is hermaphroditic and produces both sperm and eggs, but out-crossing is the general rule, by which sperm are deposited in the mantle cavity of a neighbouring individual through a long flexible penis several times the size of the barnacle. Fertilised eggs are brooded in the mantle cavity for several months before release as larvae. In the Sydney region, the breeding peak is in spring to autumn, when embryos are released as first stage nauplii, but some breeding persists at a low level throughout the year. Individuals can release up to 10 000 eggs in each brood and produce > 20 broods a year. Later, during summer, large numbers of recruits < 3 mm across (= basal length) settle on the bottom. The size of sexual maturity is ~7 mm diameter at an age of several months and the maximum size is ~20 mm diameter.

The wall plates grow at their edges and at the base at rates depending on temperature and water flow. The maximum growth rate recorded for this species is ~7 mm month^{-1}, a rate that slows markedly with increasing size. A maximum lifespan of 1.4 years in the Mediterranean, and 5–6 years in South Africa, has been recorded for this species.

After release, the larvae pass through six naupliar stages with progressively increasing complexity and reach a final cyprid stage in ~17 days, when they seek to settle in aggregations near adults of the same species; in Bayliss' (1993) South Australian study this was on the pneumatophores of mangroves.

SHRIMPS, LOBSTERS AND CRABS (ORDER DECAPODA)

The basic body plan of this group of crustaceans is a hard carapace enfolding the gills and body, an abdomen, and a series of appendages comprising antennae in front, mouth parts, the maxillae, and walking or swimming legs. A great diversity of species with equally diverse ecologies live on reefs in cryptic microhabitats, where they are protected from fishes (their main predators) and emerge at night to feed. Some filter plankton and detritus from the water, some are scavengers, herbivores or carnivores, and a few are parasitic or commensal. We now consider a range of species, from small to large, in several different families.

Shrimps (Infra-order Caridea)

This is a diverse group of small crustaceans, rarely larger than ~6 cm, whose five pairs of legs are adapted for swimming rather than walking. Some are pelagic, while others live under rocks, among algae or in seagrasses, and some live closely associated with benthic animals, either as a symbiont or a parasite (Poore 2004).

The food of shrimps includes epiphytes on algae or seagrass, detritus, and small benthic protists or invertebrates, such as forams, polychaetes, crustaceans and molluscs. Their predators are crabs and various fishes – mainly scorpion fishes, weedfishes and leatherjackets. Howard (1984) argued that shrimps play a key role in the food web by strongly influencing the structure and dynamics of lower trophic levels and by transferring their productivity to higher levels, particular predatory fishes.

Some shrimps live in association with other benthic animals (Table 15.1), but the benefits (if any) to the host are unclear. Shrimps living under the spines of sea-urchins feed on surface tissues covering the spines, but may also be omnivorous. Hence, the sea-urchins provide both food and protection for the shrimp. The symbiosis of the shrimp *Gnathophylloides mineri* was studied by Vaitilingon *et al.* (2007) in New South Wales waters. The shrimp was cryptically coloured to match its host sea-urchin, and was observed to move from one host to another – indicating it could visually recognise its preferred host urchin from other species.

Table 15.1: Shrimp species associated with benthic animals in southern Australia (after Poore 2004; Gowlett-Holmes 2008).

Shrimp species	Host
	Molluscs
Anchistus custos	Scallops, razor shells (*Pinna bicolor*)
Alpheus astrinx	In burrows of bivalve *Limatula strangei*
	Ascidians
Notopontonia platycheles	Branchial cavity of *Herdmania grandis*
Pseudopontonia minuta	In *Polycarpa clavata* (formerly *P. clava*)
	Sea-urchins
Arete indicus	Under spines of *Echinometra mathaei*
Athanas granti	Under spines of *Heliocidaris erythrogramma*
Gnathophyllum taylori	Under spines of *Centrostephanus rodgersii*
Gnathophylloides mineri	Under spines of *Tripneustes gratilla*
Periclimenes carinidactylus	**Crinoid** *Cenolia trichoptera*
Synalpheus fossor *Synalpheus streptodactylus*	**Sponges**

Shrimps associated with bivalves, crinoids or ascidians likely feed on mucus streams from their host. The association between the shrimp *Notopontonia platycheles* with the red-mouthed ascidian *Herdmania grandis* is among the better known examples (Berggren 1999). This shrimp lives in the branchial cavity of its host, sometimes as a male–female pair and sometimes with other inhabitants, such as pea crabs or even the rare, large-tailed cardinal fish *Vincentia macrocauda*. At one site, 75% of the ascidians had shrimp occupants.

The shrimp genus *Synalpheus* has especial interest. Nine species recorded in southern waters are associated with sponges (two are listed in Table 15.1) with depth ranges variously going to 250 m (Davie 2002). Some have been found in substantial numbers in the canals of individual sponges, which provide both habitat and food. Further, this genus of shrimp is the only known marine organism demonstrating eusociality, in which members groom or guard closely related nest-mates within kin-structured colonies and show other behaviours more typical of social insects (Duffy 2002).

Ecological studies on southern species have focused on seagrass beds, where shrimp tend to be far more abundant than in other marine habitats. Here, we discuss the ecology of *Palaemon intermedius* (formerly in the genus *Macrobrachium*), one of the commonest southern species in reef and seagrass habitats (Howard 1984; Gray 1991; Walsh and Mitchell 1998; Paul 2001), and add notes on an introduced Japanese shrimp, *Palaemon macrodactylus*, which is present in South Australia and on New South Wales coasts, and on cleaner shrimps – an intriguing but poorly known group of shrimps.

The red-spotted shrimp *Palaemon intermedius*

At Gray's (1991) sites around Sydney, juveniles < 3 mm carapace length (CL), appeared in highest numbers in autumn and spring, although present in other seasons. Females reached sexual maturity in about 6 months at ~5 mm CL and maximum size in about a year at ~8 mm CL, when males are then ~5.5 mm CL. Egg-bearing females were present in the population all year round. The largest cohorts disappeared during winter, suggesting a longevity of about a year for the species. In Victorian waters, the reproductive season is much shorter, from spring to early summer, and juveniles appear soon after, peaking in summer (Howard 1984). Here, half the food eaten by *P. intermedius* was detritus and the rest was subdivided between polychaetes, amphipods, mysids and rissoid gastropods.

Cleaner shrimps

Fish-cleaning behaviour is a special kind of symbiosis involving an interaction between a cleaner species and a client fish, in which the cleaner benefits by removing and eating external parasites and necrotic tissue from the client, which in turn benefits by their removal. The mutualism has led to the evolution of reliable signals by the cleaner and host to reduce misunderstandings and conflict – especially for the shrimp to avoid being eaten by the client! These behaviours in the banded cleaner shrimp have been studied by Johnson (1969, 1977) and in a species of *Periclimenes* by Chapuis and Bshary (2010), with a review by Jonasson (1987).

Two species of cleaner shrimp are known in the southern region: the tropical banded cleaner shrimp *Stenopus hispidus*, whose range extends down the east and west Australian coasts, and the southern species *Periclimenes aesopius.*

The banded cleaner shrimp Stenopus hispidus

This species is distinguishable by the red and white banding extending along the body and on its third legs – a colouration that typically advertises that the species is willing to clean clients of parasites (see Chapter 16). Initially, a male shrimp tends to be solitary, engaging in threat or appeasement behaviours when approached by others. Aquarium studies have shown that at this stage they are able to recognise other shrimp individually – an ability rarely known among marine invertebrates. After initial hostility, a male becomes attached to a female following a courtship involving mutual antennal interplay. Together they then form a strong pair bond and set up a cleaning station during daylight hours near the entrance to a cave or crevice, where fish come to be cleaned (Zhang *et al.* 1998; Edgar 2008). As with cleaner fish, it is likely that a pair of cooperating cleaner shrimp provide a better cleaning service than one cleaning alone (Bshary *et al.* 2008).

In *Periclimenes longicarpus*, cleaning is done early and late in the day. When approached by a fish, the shrimp claps its claws, producing an audible signal to indicate its identity as a cleaner and to help avoid predation. In this case, the clapping can be seen as an advertisement of a service to improve the mutual exchange between cleaner and customer. Claw clapping has not, however, been reported for the banded cleaner shrimp.

The temperate cleaner shrimp Periclimenes aesopius

This cleaner shrimp (Plate 51), though seldom seen, is usually associated with large sponges and sometimes the swimming anemone *Phlyctenactis tuberculata*; it also occurs on jetty piles. The structural features of sponge habitats and jetty piles serve well as prominent cleaner stations, where the shrimp engages in a dance, rocking to and fro – a behaviour that apparently serves as an advertisement of services offered. Leafy seadragons and toadfishes have been observed being cleaned at a station, and the shrimp also feeds on detritus and possibly sponge mucus (Shepherd *et al.* 2005; Gowlett-Holmes 2008).

Spiny or rock lobsters (Infra-order Achelata)

Spiny lobsters are the largest and among the most abundant of decapods on rocky substrates, and hence are of great ecological importance, as well as providing the basis of the most valuable fisheries of southern Australia. For the latter reason, they have also been the most researched group of marine species in Australia. Fortunately, much of the research has been brought together in texts edited by Bruce Phillips and colleagues, and in conference proceedings. Relevant references are: Cobb and Phillips (1980), Phillips *et al.* (1994), George (2005) and Phillips (2006).

Three species of spiny lobster, all in the family Palinuridae, occur in the southern half of Australia: the western lobster *Panulirus cygnus* on the west Australian coast as far north as North-west Cape (22°S); the southern rock lobster *Jasus edwardsii* along the south coast and up the west and east coasts to ~30°S; and the warm temperate eastern or packhorse lobster *Sagmariasus* (formerly *Jasus*) *verreauxi* up the east Australian coast from Bass Strait to the Queensland border (~28°S). *Sagmariasus verreauxi* is the most primitive and the largest of all spiny lobsters. Some slight overlap in distribution occurs between them at the edges of the southern rock lobster's range in SW and SE Australia, where they share the same habitat, but the abundance of the southern lobster in the region of overlap is generally low.

Life history

The basic ecology and life history of the three species are very similar, so we shall first give a general picture and then discuss the individual differences between them. After the female moults, mating takes place with the males, described by Gray (1992) as 'a tangle of legs in a head to head and belly to belly affair' and the male deposits a sperm package on the female. As the package breaks open, ripe eggs are extruded and fertilised, and then placed by the female in a brood chamber under her tail fan, where she incubates and grooms them till hatching. The emerging larva, initially a naupliosoma then a phyllosoma (plural phyllosomata), begins a long life of 1–2 years in the plankton. Phyllosomata drift far offshore, passing through many sequential development stages, the last of which is called the puerulus (meaning 'a little child') and they eventually return to the coast to settle on reef in shallow water. Over the next few years, the growing lobsters may gradually move offshore to depths of up to 300 m or more. During their inshore phase, rock lobsters rest communally in dens by day and emerge at night to forage. This life cycle, shown schematically for the western rock lobster (Fig. 15.1), is broadly applicable to the other two species. Comparative life history attributes of the three species are listed in Table 15.2.

Ecology of lobsters – larvae

The behaviour of phyllosoma and puerulus larvae and how they manage to drift offshore and return to the coast a year or more later, were long a mystery, but has been linked to transport by currents for the western and southern lobsters (Booth and Phillips 1994).

Western rock lobster larvae

The western lobster first stage phyllosoma larvae emerge from the egg mass mainly at 30–80 m depth, and first rise to the surface, but later sink by day to a

Table 15.2: Comparative life history features of three species of spiny lobster in southern Australia, with references. CL = carapace length.

Feature	*Panulirus cygnus*	*Jasus edwardsii*	*Sagmariasus verreauxi*
Maximum size (mm CL). Male ■ female	180 ■ ~160[2,3]	235 ■ 180[1,2]	264 ■ 242[1,2]
Adult moulting: m. ■ f.	Spring ■ winter[2,4]	spring ■ autumn[2]	Winter (?)m,f[2]
Female sexual maturity L (mm) ■ Age (years)	90–100 ■ 6–7[6]	80–110 ■ 4–5[5,6,7]	166 ■ 4[6,8]
Broods per year	1–2[6]	1[6]	1[6]
Incubation period (weeks)	3–10[6]	16–20[6]	12[6]
No. of larval stages Larval life (months)	15[6] 9–11[6]	15–17[6,9] 9–24[6,9]	17[6] 8–12[6]
Puerulus life (days)	15[10]	19[10]	20–30[9,10]
Peak settlement season	Spring–summer[11]	(SA) winter[9,11]. (Tas) winter–spring[11]	(NSW)spring–summer[8]
Temperature range (°C)	16–26[2]	8–18[2]	16–22[2]
Depth range (m)	0–220	0–100	0–200

[1]Morgan (1980); [2]Booth (2006); [3]Phillips *et al.* (1980); [4]Gray (1992); [5]Frusher *et al.* (1999); [6]George (2005); [7]Lewis (1981); [8]Montgomery (1999); [9]Olsen and Shepherd (2006); [10]Booth and Phillips (1994); [11]Bruce *et al.* (2000).

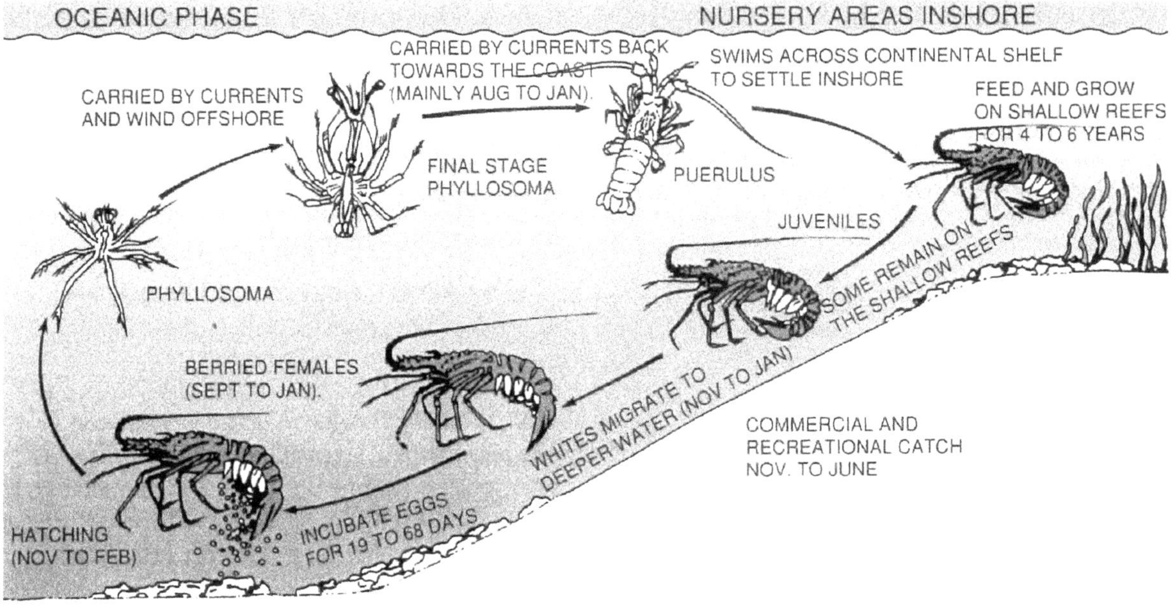

Figure 15.1: Life history of the western rock lobster *Panulirus cygnus* (original by BF Phillips and redrawn from Kailola *et al.* 1993).

depth of 30–60 m, and rise to the surface at night. Later phyllosoma stages sink even further, up to 100 m depth, during daylight. This daily vertical movement of larvae is critical for their rapid transport offshore, and movement rates of > 5 km per day have been recorded. In offshore larval surveys, Phillips (1981) found the highest larval densities 375–1030 km from shore in the SE Indian Ocean and some larvae were 1500 km from land. As phyllosomata develop, they live deeper in the water column, where water currents transport them back towards the shore. By the time they have metamorphosed to pueruli, they are < 120 km from the coast, and the pueruli, which by now are up to 12 mm CL, alternate between swimming near the surface shorewards by day, and resting on the bottom at night.

Finally, the pueruli, which can firmly grasp any firm object, settle on the bottom and metamorphose into juvenile lobster. Knowledge of their grasping ability enabled biologists to develop collectors, made of artificial seaweed, which they used to quantify the strength of puerulus settlement from year to year on west Australian coasts. Puerulus settlement strength has now been monitored at several sites annually for some decades, and both environmental factors and the size of the parent stock have been found to influence settlement strength (Caputi *et al.* 1993, 2003, 2010). For example, strong El Niño events cause a reduction in the strength of flow of the Leeuwin Current, leading to a poor puerulus settlement inshore in southern populations. The poor puerulus settlement in turn results in reduced population size of the lobster 3 years later. Hence, lobster catches by fishers can be routinely predicted from annual indices of puerulus settlement (Fig. 15.2). In more northern populations at the Abrolhos Is, some 60 km offshore, the size of the parent spawning stock, and also westerly wind strength, are significant factors affecting settlement strength. When spawning adults are abundant and winds are weak, puerulus settlement is high at the islands, and vice versa with strong westerly winds and heavily fished stocks (Fig. 15.3). During strong westerly winds, easterly flowing surface currents increase in strength, and transport pueruli inshore and away from the islands; this results in greater settlement on inshore reefs, but less at the Abrolhos Is.

Since 2006, poor puerulus settlements have consistently happened, and confounded the above conclusions (Caputi 2008; Brown 2009; Caputi *et al.* 2010). The causes are unclear, but climate change (sea

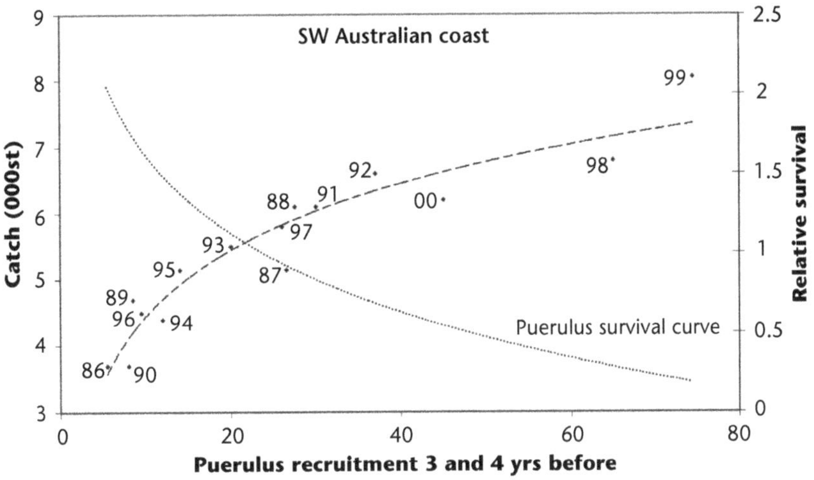

Figure 15.2: Relation between the western rock lobster catch in the SW Western Australian Management Zone and puerulus abundance in coastal collectors 3 and 4 years before, for 1986–2000. The decline in relative survival with increasing abundance of pueruli is shown by a dotted line. Redrawn from Caputi *et al.* (2003).

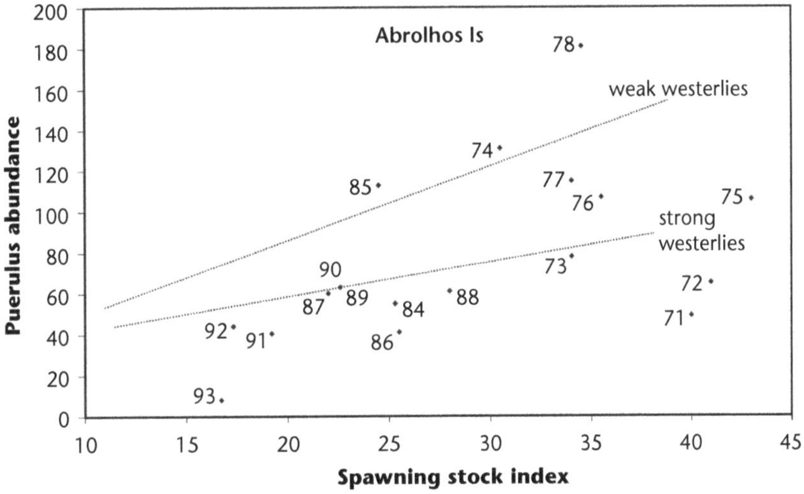

Figure 15.3: A stock-recruitment relation for the western rock lobster at the Abrolhos Is for 1971–1993, incorporating environmental effects of westerly winds on puerulus abundance in collectors (redrawn from Caputi *et al.* 1993).

temperature increases of ~1°C since ~1960) and more frequent, positive Indian Ocean Dipole effects weakening onshore winds – see Chapter 2) have been implicated. Climate change is predicted to keep increasing sea temperatures, but with less severe storm events, and the many other species that settle on the collectors are now being used to provide indicators of such changes (de Lestang *et al.* 2011). Another complexity is that the size of sexual maturity is declining, partly due to the warmer waters and possibly also to an evolutionary effect of fishing causing younger maturity and slower growth.

Southern rock lobster larvae

The ecology and movement of phyllosomata of the southern rock lobster are not as well known as those of the western lobster and existing information is summarised by Bruce *et al.* (2000). With these data and knowledge of ocean currents, Bruce *et al.* (2007) later modelled the movement of phyllosomata throughout southern Australia, and estimated the percentage larval retention and dispersal for different regions. After hatching in spring to early summer, the larvae drift with surface currents offshore for several months as they pass through successive development stages. Early stage phyllosomata are found mainly in southern shelf waters, and mid- to late phyllosomata further (sometimes > 500 km) offshore; they are rare on the west coast of Australia or north of Flinders I., Bass Strait.

Finally, the late stage phyllosomata develop into pueruli and swim shorewards. Larval surveys have found little evidence of a complex pattern of vertical movement of larvae in the water column by day or night. Larvae were mainly in the upper 100 m of the water column, with some down to 200 m, but were commonly near the surface, especially at night. How the non-feeding pueruli can orient themselves and swim shorewards has long been a mystery. Recent studies (Jeffs *et al.* 2005) suggest that underwater sounds, such as waves breaking on the coast that are detectable up to 40 km from shore, provide a directional cue to pueruli. Another possible cue is surface waves reflected back from the coast, which may be detectable by the pueruli swimming near the surface at night.

The development rate of phyllosomata in southern waters varies according to sea temperature, and their movement according to regional ocean currents. The extent of dispersal from the natal region was estimated roughly by Barry Bruce and colleagues' modelling study and is summarised in Table 15.3 in terms of the sources of larvae settling in six regions spanning the southern coast. Larvae drift up to 1000 km from shore and their net dispersion is in an easterly direction. Hence, larvae from populations in SW Australia spread right across the southern region, but an important component is retained within the region and is almost the only source for local lobster populations. Similarly, larvae from Regions 2 and 3 are a larval source for Tasmania, as well as being a local source, while Regions 4–6 are essentially larval sinks, dependent on larvae being transported from the west.

The studies of Rob Lewis (1977) and Mick Olsen's drift card experiments (Olsen and Shepherd 2006), later confirmed by Bruce *et al.* (2007), showed that in SE South Australia and western Bass Strait, larvae upon hatching in October are first transported by westerly flowing currents up to 1300 km over ~5 months to the NW, and return by reverse currents up to 1500 km to the SE in ~4 months. Phyllosoma larvae develop much faster in the warmer South Australian waters (up to 5°C warmer) than in Tasmanian waters, and have a correspondingly shorter larval life. These larvae have a peak settlement in July–August in SE South Australia and Linnane *et al.* (2010b) showed that this peak coincided with the eastward coastal current driven by strong NW winds prevailing at that time. The results give some hope for the development of a puerulus settlement index to predict lobster abundance in the region, as is emerging from similar, long-term east Tasmanian studies (Gardner

Table 15.3: Predicted modal age (months) of southern rock lobster larvae, and range of ages at settlement, with larval source regions, numbered 1–6 for six southern Australian regions (from Bruce *et al.* 2007).

GAB = Great Australian Bight.

Region	Age (months) (range)	Regional larval sources (%)					
		1	2	3	4	5	6
1. SW West Australia (125°–130° E)	10 (7–23)	94	5	0	0	0	0
2. Eastern GAB and Gulfs (130°–139° E)	7–10 (7–23)	53	41	5	< 1	0	1.3
3. SE of SA (139°–140° 30′ E)	9 (7–23)	44	39	15	0	0	2
4. Portland, Vic. region (141°–143°E)	10 (7–23)	14	60	20	1	0	4
5. Eastern Vic. (142° 15′–151° E)	11 (8–23)	20	42	26	6	0	6
6. Tasmania (143°–151° E)	9–15 (8–27)	20	27	25	13	< 1	6

et al. 2001), where commercial catches are correlated with puerulus settlement strength 5 years before.

Eastern rock lobster larvae

The development and ecology of larvae of this species are thought to be similar to the southern lobster (Montgomery 1999; MacDiarmid and Booth 2005). After a summer hatching in deep water at the main spawning grounds off northern New South Wales (30–33°S), the developing larvae are carried south by the East Australia Current (EAC) and eventually metamorphose to pueruli near the edge of the shelf. The pueruli then swim shorewards and settle on coastal shallow reefs. Monitoring of collectors along the coast (Montgomery and Craig 2005) showed strong puerulus settlement in spring to summer, with highest abundances off Sydney and south to Ulladulla (36°S). The authors also found high inter-annual variation, which they suggested resulted from environmental factors (e.g. continental shelf width) that affected the strength of the EAC.

The phyllosoma of the eastern lobster is the only larva whose feeding has been examined (Nelson *et al.* 2002). When prey was offered, the phyllosoma used its mandibles to manipulate and take it to the mouth. Brine shrimp were rapaciously shredded and ingested, and when a jellyfish was presented the phyllosoma quickly shredded it, but became entangled in the gelatinous material. Hence, hard prey, such as copepods, are the most likely type of prey to be taken in nature. After ingestion, particles took only 4–5 min to pass from the mouth through the gut.

Ecology of lobsters – from juvenile to adult

Western rock lobster Panulirus cygnus

Juvenile studies have been carried out in great detail on the shallow inshore reefs off the west Australian coast where post-pueruli settle in large numbers. Juveniles remain here till they are ~5 years old, when they migrate to deeper water (for ecological studies see Berrill (1976), Cobb (1981), Joll and Phillips (1984), Jernakoff (1987, 1990) and Jernakoff *et al.* (1987, 1993). Newly settled post-pueruli (< 25 mm CL) first find a cryptic habitat in small holes and fissures on coastal reefs down to ~30 m depth, where algae, sponges or seagrass offer additional protective cover. They are often solitary, but must sometimes share an available crevice with others, where they compete for space and food. In these situations, they aggressively wave their long antennae and emit a rasping noise as a threat display to warn their neighbours.

Older juveniles (2+ years) become more gregarious and live increasingly by day in dens with many other young lobsters (often > 100 per den), emerging at night to forage for food. At the most intensively studied site, Seven Mile Beach near Dongara, they foraged mainly in the seagrasses, *Amphibolis*, *Heterozostera* and *Halophila*, growing on reefs or sandy patches. Their nocturnal movements, recorded with electromagnetic tags over 460 'lobster-nights', showed that they did one or more of the following:

- started foraging around sunset, and generally returned to the same den or one nearby at around sunrise
- went on long circuitous excursions for a median distance of 155 m (maximum 400 m) away from the den
- travelled on specific pathways to and from an algal bed, sometimes 200–300 m away, indicating a knowledge of the local topography
- made short excursions in the vicinity of their dens, with a few (~10%) not leaving the den.

The diet of post-pueruli, juveniles (2+ years old) and adults is summarised in Table 15.4. The presence of corallines in the gut indicated that they foraged in *Amphibolis* beds on the limestone reefs, which offered more protection from predators than the other seagrasses. Moreover, calcium was needed by post-moult lobsters for a new exoskeleton and Jernakoff's (1987) study showed that recently moulted individuals ate twice the amount of corallines as pre- and inter-moult juveniles. Edgar (1990a, b) took the studies a stage further by examining the effect of lobster predation on the bottom community. He showed that:

- The juvenile diet reflected the abundance and availability of benthic fauna.
- Juveniles preferred sedentary prey such as small molluscs, particularly the trochid snail *Cantharidus lepidus (now Strigosella lepida)* and the chiton *Stenochiton cymodocealis*, that were seasonally abundant in *Amphibolis* beds; juveniles only switched to coralline algae when the above prey became scarce.

Table 15.4: Diet (percentage composition by volume) of western rock lobster post-pueruli, older juveniles and adults in near-shore waters and deep offshore waters (after Joll and Phillips 1984, Jernakoff *et al.* 1993 and Waddington *et al.* 2008).

Traces indicated by +.

Food type	Post-pueruli	Juveniles[1] (2+years)	Adults[2] (deep)
Corallines (e.g. *Corallina, Metagoniolithon*)	35	30	–
Other algae (e.g. *Caulerpa cactoides*)	+	7	2
Seagrass	+	10	–
Molluscs (e.g. mussels, gastropods)	29	16	3
Crabs (including *Panulirus*[3])	–	~5	62
Other crustaceans (amphipods, isopods)	6	~4	10
Worms (polychaetes, enteropneusts)	1	1	–
Other (detritus, sponge, bait, unidentified)	29	27	23

[1] average of two sites, Seven Mile Beach and Cliff Head, Western Australia.
[2] average for three sites, Lancelin, Jurien Bay, and Dongara, Western Australia.
[3] cannibalism and consumption of the shed exoskeleton are indistinguishable.

- Depending on mollusc density, juveniles had a nightly consumption rate of ~200 molluscs, which provided them with a very rich diet.
- The lobsters on this molluscan diet grew at twice the rate of those living at sites with low molluscan abundance.
- A major competitor of the juvenile lobsters was the blue swimmer crab *Portunus armatus*, which ate twice as many molluscs eaten by the lobsters.

In deeper water (35–60 m), the western lobster is mainly carnivorous, as it switches increasingly to crabs, reflecting the differences in prey availability and the choice of prey in deep water (Waddington *et al.* 2008, 2010).

From the studies of Cobb (1981) and Howard (1988), the major predators of juveniles appear to be the octopus *Octopus maorum*, the giant cuttlefish *Sepia apama* and several fish species, particularly the sand bass *Psammaperca waigiensis*. Of these three, the octopus is likely the most serious predator. In a comparison by Phillips *et al.* (1980), 91% of predated rock lobsters were taken by octopus, compared with 6% by cuttlefish and 3% by fish. Cobb considered that the communal sheltering behaviour of the lobsters in dens served as a defence against octopus predators, as the dense phalanx of stout spiny antennae presented a formidable barrier to attack.

Migration from shallow inshore reefs to deeper water is an ongoing process from ~4 years of age. Immature lobsters moult during November of each year and their new exoskeleton is a light colour. These 'whites', as they are known, move seawards to deeper reefs from about November and older lobsters soon follow, so that by age 6, when animals are all mature, they have moved to deep water of 30–150 m depth. Movement is generally < 50 km from the nursery grounds to the breeding areas offshore in a westerly to NW direction, whereupon at the edge of the shelf movement becomes northward. Some undertake remarkable long-distance migrations of up to 438 km at rates of ~1 km day^{-1} and at depths to 180 m (Chubb *et al.* 1999).

Up to age 6, lobsters moult about every 100 days, maturing sexually 1–2 years after the 'white' stage. Inshore growth rates vary with food supply and the growth rate at Cliff Head, a site of relatively fast growth, is shown in Fig. 15.4. Growth rates increase as they migrate offshore, and approach the generalised curve as shown in the figure. The lifespan likely exceeds 30 years and a lobster kept in an aquarium lived for over 20 years.

Southern rock lobster

The ecology of the southern lobster is similar in many features to that of the western lobster, although specific studies are few (see Fielder 1965a; Lewis 1981; MacDiarmid *et al.* 1991; Linnane *et al.* 2005; reviews of Frusher *et al.* 1999; Booth 2006; and Hoare 2008). In their early solitary, post-puerulus stage, juveniles forage close to their home crevice, while larger animals forage in the near vicinity of shelter, the distance depending on the abundance of food. Using electromagnetic tags, Ramm

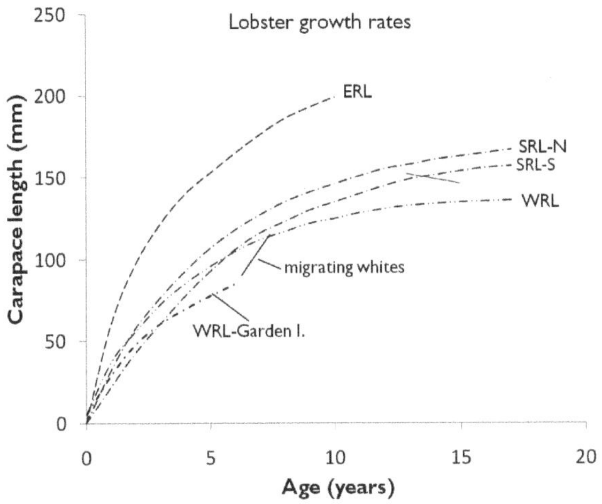

Figure 15.4: Generalised growth curves from the time of puerulus settlement for three rock lobster species. The western rock lobster (WRL) growth curve for Cliff head is redrawn from Morgan (1980), with a curve for an inshore, slower growing sub-population at Garden I. The added continuous line shows the increased growth rate of 'whites' migrating offshore (redrawn from Phillips *et al.* 1980). The southern rock lobster curves are for the northern (SRL-N) and southern (SRL-S) zones, respectively, in South Australia (derived from McGarvey *et al.* 1999). The eastern rock lobster (ERL) growth curve is derived from Montgomery (1992, 1999).

(1980) found that over several months sub-adults foraged at night within ~24 m of their home lair, with no net movement away from the reef. Hoare (2008) used an index of topographic complexity of fissures to show that lobster abundance reflected the complexity of the habitat. Hence, we might infer that in areas of high habitat complexity the higher densities of lobster would reduce the available food and increase lobster foraging distances. Similarly, differences in topographic complexity and food abundance would explain the wide differences in foraging distances recorded by various authors.

Larger juveniles live gregariously in dens and forage at night, with peak feeding activity 1–2 h after sunset, while large males tend to live alone or with cohabiting females in a shelter. Available shelter is occupied by lobsters in accordance with a dominance pattern, determined by size. When shelter is limited, the approach of a larger lobster to an occupied crevice may be sufficient to evict the smaller occupant. If the latter resists the larger one may aggressively grasp it by its carapace and, by flapping its tail fan rapidly several times, haul it out of the shelter with a series of jerking pulls. However, an occupant may vigorously defend its shelter. In this case, it emerges partly out of its shelter, stands high on its legs, and waves the first pair of walking legs in a threat display at the challenger, sometimes also striking at him with its antennae; frequently this is enough to cause an immediate retreat of the challenger (Fielder 1965b).

Lobsters also move about locally during the female moulting and mating season, either to improve the quality of their shelter or, in the case of a female, to choose a larger male guarding a shelter. Many larger males will tend to be sedentary, maintaining guard over shelters with numerous females, while others will go searching in nearby shelters for unguarded receptive females (MacDiarmid 1989). The ability of this, and other, lobsters to visit a series of shelters, or to remain within a 'home range,' implies that they can form a mental map of their area, so that at any time they know their position in relation to home (Vannini and Cannicci 1995).

The southern lobster is a generalist predator. The small post-pueruli take cryptic animals, such as brittle stars, small crustaceans and bivalves, while larger juveniles and adults take molluscs, crabs, sea-urchins, bryozoans and algae, with some preference for calcareous food to harden the exoskeleton. In the SE of South Australia, sea-urchins and red algae over a year together averaged more than half the diet (Table 15.5; Hoare 2008). In a novel study in SE Tasmania using chemical tracers, Guest *et al.* (2009) compared lobster diets inside and outside reserves. The results showed that: availability was important, with lobsters switching to molluscs when they became more abundant; and, surprisingly, ascidians were a significant part of the diet, something not previously known for lobsters (Table 15.5).

Lobsters have many predators (SEFC 1982). Juveniles are eaten by many fishes, such as flatheads, wrasses and morwongs, and larger ones are eaten by octopuses and most large predatory reef fishes, notably bearded rock cod *Pseudophycis barbata*, conger eels and rock ling. The octopus *Octopus maorum* is a serious predator of lobsters trapped in pots. Despite strong evidence that escape gaps in pots reduce predation by lobsters (see Chapter 10), these have not yet been adopted in all States (see Table 15.6). In a study of the fate of undersize lobsters returned to the water by lobster fishers, the small lobsters no sooner reached

Table 15.5: Diet of the southern rock lobster in (a) at four sites, Maria I. and Governor I., SE Tasmania (after Guest *et al.* 2009), and (b) at four sites, Ringwood Reef, Southend, Cape Douglas and Middle Point, in SE of South Australia (after Hoare 2008).

At Tasmanian sites, the range in percentage contributions of six food types are shown, and at sites in SE of South Australia, the range in percentage frequency of occurrence of the seven groups are shown. R = reserve site; F = fished site.

Site	Blacklip abalone	Ascidians[1]	Gastropods	Urchins[3]	Crabs	Other	Brown algae	Red algae
Maria I. (R)	0–19	0–10	57–91[2]	4–22	–	–	0–9[7]	0
Maria I. (F)	0–8	80–93	0–7[2]	0–13	–	–	0–9[7]	0–1
Governor I. (R)	0–32	24–53	0–11[2]	42–49	–	–	0–6[7]	0
Governor I. (F)	0–25	10–37	0–16[2]	58–71	–	–	0–10[7]	0–1
SE of SA	0–2	–	0–13[4]	15–33	6–33[5]	6–14[6]	0–5[8]	13–38

[1]*Herdmania*, *Pyura* and *Cnemidocarpa*; [2] *Turbo undulatus*; [3] *Heliocidaris* and *Centrostephanus* (Tasmania only); [4] Trochidae (*Phasianella*, *Thalotia* and *Phasianotrochus*); [5] *Plagusia*, lobster (exuviae?) and unidentified prey; [6] hydroids, bryozoans and bivalves; [7]*Ecklonia* and *Phyllospora*; [8] unidentified prey.

the bottom than they were voraciously attacked by wrasses and leatherjackets, which first crippled them by biting at the soft tail-fan muscles from below, and then devoured the immobilised lobster more leisurely (unpublished data). Other predators are gummy sharks, wobbegongs and seals.

Extensive tagging studies have been done in three States to determine lobster migration and growth (Gardner *et al.* 2003; Linnane *et al.* 2005, 2012). Out of > 14 000 tag-recaptures off South Australia, 68% moved < 1 km, 28% moved 1–20 km and the rest greater distances (maximum 127 km) over 1–2 years at liberty; the longer movements were directed offshore. In Tasmania, lobster movement was even more localised than the above, with > 90% of ~40 000 tag-recaptures moving < 5 km over 1–2 years at liberty. However, a few off NW Tasmania moved > 8 km into deeper water and some multiple recaptures off SE and east Tasmania showed seasonal movement between inshore and offshore areas. In Sanctuary Zones in SE Tasmania, Barrett *et al.* (2009) found that 95% of tagged individuals remained within 400 m of their original tagging site and only 3% moved further, often over sandy bottom between reefs. Hence, over time, the abundance and biomass of lobsters increased enormously in the reserve, in spite of the increasing density. For example, in the Maria I. reserve, the biomass of legal sized lobsters increased 20–28 times according to sex, and egg production some four times. In contrast, in another sanctuary in South Australia, McGarvey (2004) found that 68% of tagged lobsters migrated out of the sanctuary, with some crawling > 100 km offshore. In Victoria, lobster movement is also localised, except east of Cape Otway where longer offshore migrations have been recorded. The benefits of Sanctuary Zones for research and lobster conservation are discussed by Childress (1997) and Booth (2006), and in a broader context in Chapter 20.

Overall, the above studies suggest that the southern lobster tends to have fidelity to its home reef and that migration is most likely in specific circumstances such as:

- where suitable rocky bottom habitat is limited, as in locations with extensive sandy plains
- from some inshore waters or in areas with high lobster densities (but notably not in SE Tasmania)
- among a few nomadic individuals, and among immature females, again with the exception of some Tasmanian sites.

The growth of southern lobsters is highly variable over its range, and varies according to sex, temperature, food availability and other factors. Growth rates are highest towards the western edge of their range and lowest in southern Tasmania. For example, females mature sexually at 4–5 years of age; at this age they are ~11.4 cm CL in western South Australia, 7.5–9.3 cm CL in SE South Australia and 6.5 cm CL in southern Tasmania. Moulting occurs two to three times a year in small lobsters, and annually upon reaching maturity, and for 10–12 days after moulting the lobster is soft-shelled, as it absorbs water to expand its volume by ~20%. Once the shell hardens, the animal cannot grow further during that inter-moult period. Generalised growth curves in the warmer waters of western South Australia, and the cooler southern waters, are shown in Fig. 15.4, after converting weight-by-age information to length (see Table 15.5 in McGarvey *et al.* 1999). These curves represent average growth rates for males and females, because males grow faster and to a larger size than females.

The annual natural mortality rate is thought to be 10–20%, giving a longevity of > 20 years. Mating takes place annually in autumn and the male deposits a sperm package on the female's underside. As with other species, the sperm package quickly breaks down and the female then extrudes her eggs from the gonopores and deposits them under the tail, where she carries them till they hatch some 3–5 months later (Table 15.2). Ovary development and preparation for the next moulting starts the following December–January.

Eastern rock lobster

The ecology of the eastern lobster differs little from that of the western and southern lobsters (see Montgomery 1995, 1999; Montgomery and Craig 2005; MacDiarmid and Booth 2005). After settlement, post-pueruli are solitary within holes and crevices, but become more gregarious with age, when they live in dens with many others and use their spiny antennae for defence. They spend ~3 years on inshore reefs and are commonest in southern New South Wales, although in recent years they have been increasingly captured in NE Tasmania. During this inshore phase, they forage at night for up to 25 m from their den, searching for worms, small bivalves, crabs, sea-stars, sea-urchins and fish. The lobster uses its mandibles to crush most prey, but for large cockles it chips at the edges to gain entry, and for a sea-urchins the lobster 'inserts its front legs through the mouth and neatly cuts around the entire rim of the oral opening, rather like a can opener, before scooping out the roe and gut' (MacDiarmid and Booth 2005). Tasmanian studies on lobster feeding in urchin barrens suggest the eastern lobster may have played a key role controlling urchin barrens before they were over-exploited in the 1960s (see Chapter 12). Their main predators are carnivorous fish, such as wrasses, octopuses, sharks and rays.

At a size of 12–13 cm CL and ~3 years of age (see growth curve in Fig. 15.4), the lobsters move offshore in autumn–winter to deeper water, where they aggregate for moulting and mating. Then, when females are carrying eggs, they return inshore over spring and summer to waters < 100 m deep. In addition, mature lobsters are thought to undertake a northward migration of up to several hundred kilometres against the current to the main spawning grounds off northern New South Wales, where major spawning takes place. Larvae from these spawning areas then drift south with the EAC, and are transported inshore in eddies that break off from the current to populate coastal waters south to NE Tasmania. In recent years, the range of lobster has extended southwards 80–100 km with the further penetration of the EAC south (Johnson *et al.* 2011).

Management of lobster fisheries

Because humans are by far the most serious predator of lobsters, some mention must be made of management of the fisheries in the five southern States. Management aims to sustain populations and catches, and ecosystem stability. These goals are achieved by regulating the catch by a variety of means in order to maintain an adequate spawning stock size. Principal controls currently used by the States are outlined in Table 15.6. Recruitment indices based on abundance of settling pueruli, and estimated juvenile abundance, are now used in all States to estimate settlement strength and, where possible, to project future harvest levels, assisting management objectives (Lipcius and Cobb 1994; reviews in Phillips 2006).

In Western Australia, catches had been maintained at fairly consistent levels at an estimated egg production level of 22% of the unfished stock, and in a stable management environment until 2009. However, subsequent catastrophic declines in puerulus settlement from 2007–2011 (de Lestang *et al.* 2011), and predicted flow-on effects to the fishery 4 years later, have created concern about the adequacy of breeding stocks. This concern was reinforced by the findings of Babcock *et al.* (2007) that adult lobster densities were 50 times higher in a Sanctuary Zone at Rottnest I. than in adjoining fished areas. Other possible causes of the recruitment decline are changes in the Indian Ocean Dipole and warming sea temperatures (see Chapter 2; Brown 2009).

In SE Australia, the fisheries have been generally sustained, but, starting from 1999 in western South Australia and progressively to the east, catches have

Table 15.6: Management measures and some fishery parameters in commercial lobster fisheries in southern Australia.

M and F are the instantaneous natural and fishing mortality rates, and percentage egg production is present percentage of an unfished population conserved, with target percentage in bold and in brackets. f = female; m = male. In S.A. S = Southern Zone, *N* = Northern Zone. Data sources are Brown and Phillips (1994), Lipcius and Cobb (1994), Hobday and Punt (2001), Phillips and Melville-Smith (2006), Montgomery and Craig (2005), Buxton *et al.* (2006) and Linnane *et al.* (2013).

	Palinurus	*Jasus*				*Sagmariasus*
Measures	**WA**	**WA**	**SA**	**Vic.**	**Tas.**	**NSW**
Minimum size CL (mm) (approximate age in years)	76, 77 (3–4)	98.5 (?)	98.5(S), 102(N)	105(f), 110(m)	105(f), 110(m)	104 (~4)
Maximum size (mm)	115 (f only)	None	None	None	None	200
Closed season	July–Nov	July–Oct	May–Oct.	Jun.–Nov	Sept.–Oct	None
Escape gaps per pot	3	None	None	1	1–2	None
Licence limitation	Yes	Yes	Yes	Yes	Yes	Share fishery
Annual commercial catch (t)	~11 000	37	1250 (S); 310(N)	~400	~1500	~150
Quotas	No	No	Yes	Yes	Yes	Yes
Declines in catch in 2010 (%)	–	43	37(S) 57(N)	~45	~17	Slow increase from 1995
Egg production (%) **[target]**	15–20 **[20]**	?	< 10(S); 9–14(N)**[12]**	**[20]**	12 **[20]**	?

declined in each State, most severely in the west and least in the east (Table 15.6). Quotas have been progressively reduced (but not reached) in several States and puerulus settlement is declining in some regions (Linnane *et al.* 2010 a,b,c, 2013). The causes are unclear, but climate change (increased upwelling strength) seems implicated and breeding stock decline through overfishing is certain. The > 90% depletion of natural lobster populations in eastern Tasmania through fishing (Barrett *et al.* 2009) provides the clearest evidence of overfishing, and the percentage spawning biomass of 12% of the unfished population is dangerously low (Table 15.6). As the southern lobster comprises a single stock, in which larvae are transported towards the east (Table 15.3), the decline or collapse of 'larval source' populations to the west are predicted to progressively affect recruitment to the more easterly 'sink' populations in a domino effect. The recorded declines lend support to this hypothesis.

Management of the eastern lobster has a different history. In New South Wales disastrous open-access policies were fervently espoused in all fisheries (e.g. Francois and Gorman 1980), leading to severe depletion of stocks and a decline of ~80% in the catch. Subsequently, in 1994, controls were belatedly introduced and stock rebuilding undertaken (Montgomery and Craig 2005), with as yet only a modest success.

Hermit crabs (Infra-order Anomura)

Hermit crabs lack a calcified exoskeleton on their abdomen and so for protection they have adapted to live in empty gastropod shells and, in rare cases, cavities in stones. The shells are vital for their survival, reproduction and protection from predators, and do not restrict mobility. In all, shell acquisition is essentially a unique form of habitat selection, and this ability has contributed to the evolutionary success of hermit crabs. Studies of the > 30 southern hermit crabs are few and those that exist have focused on their feeding and shell-selection behaviour. Global reviews are given by Conover (1978), Hazlett (1981), Barnes (2003) and Williams and McDermott (2004).

Associations with other species

Shell choice

Empty gastropod shells are often in short supply and most hermit crabs seize the chance to occupy whatever becomes available of appropriate size. However, they do have preferences in terms of internal shell volume, aperture size and shape, while damaged or worn shells tend to be avoided. Too small a shell can restrict growth and reduce the number of eggs a female can carry. With each moult and increase in size, the hermit crab must seek a larger shell. Besides, even the best shells tend to wear out within a year from physical abrasion. For these reasons, the hermit crab is often in search of a bigger and better shell and, if it cannot find a recent empty one – a favoured searching site is an octopus lair – it must seek a shell-exchange with another hermit crab. The choice of shells that a crab makes is quite complex in nature, depending on shell availability and the crab's propensity for a change, as well its preference. This was tested experimentally by Liszka and Underwood (1990).

In almost all species, a set of ritualised shell-exchange behaviours has evolved, as described by Abrams (1982a, b) for an Australian species *Clibanarius virescens*. An exchange of shells can arise in two situations between two hermit crabs: (1) a competitive one in which only one crab benefits, and (2) a mutualistic one in which both crabs benefit. In the latter case, the exchange rate is high (~70%). In competitive situations, Peter Abrams derived the following principles, from laboratory and field experiments:

- Crabs living in adequate shells retain their shells (~90% of the time) in the face of crabs in poor shells (i.e. shells that are too big, too small or broken).
- Instigators of exchange are more persistent and more willing to escalate a fight when their own shell quality is lower.
- Defenders seldom leave their shell, unless it is of poor quality.
- On average, the victory in a fight goes to the crab that is larger and heavier.

These principles in general accord with the theory of asymmetric contests developed by Smith and Parker (1976). That theory predicts that, because no two individuals are exactly equal (here, it would be in terms of shell suitability, and crab size and strength) there is an asymmetry in the resource-holding ability of the two parties to a contest. Hence, contests are common among hermit crabs because they live in a perpetual state of house hunting!

Symbionts and epibionts

Amphipods and copepods commonly live under the shells of hermit crabs. Hamond (1973) found that several species of copepod lived under the shell of the hermit crab *Diogenes*, and probably fed on its faeces and shared its captured food. Vader (1995) and Vader and Myers (1996) recorded several amphipod species living under the shells of the hermit crab genera *Strigopagurus*, *Dardanus* and *Paguristes*. The symbioses were all facultative (i.e. a matter of choice) except for an amphipod *Liljeborgia*, which apparently was an obligate symbiont with hermit crabs. In this case, the amphipod was probably a predator of the hermit crab eggs – indicating a parasitic rather than a symbiotic relationship.

Species from many other phyla (e.g. bryozoans, polychaetes, molluscs, flatworms and ascidians) also live on or under the shells of southern hermit crabs. But the commonest symbiont of all is anemones – often to be seen on *Paguristes brevirostris*. The association with anemones is especially interesting, because elsewhere symbiotic anemones are known to protect their hermit crab host from predation by octopuses (Ross and von Boletzky 1979). Because hermit crabs resurrect old, sometimes buried, shells to create a new habitat, they have been called 'ecosystem engineers', albeit on a small scale (Williams and McDermott 2004).

Ecology of some southern hermit crabs

The hermit crab Clibanarius virescens

This Indo-West Pacific species extends down the east and west coasts of Australia (Poore 2004), and its copulatory behaviour has been examined by Hazlett (1996). In the Sydney region, it occupies shells of the gastropods *Austrocochlea*, *Subninella*, and *Morula*, among others. As in other species, males, stimulated by female pheromones, become more active during the breeding season and search for a mate. When one finds a female, he first grasps her shell with his walking legs and, in this position, may lightly tap the top of her shell aperture with his claw. The female then starts to shed her moult, sometimes with the assistance of the male, who pulls the exuviae (the remains of her exoskeleton) out of her shell. After her moult, the male turns the female round and alternately taps and rubs the rim of her shell with a claw, at the same time rocking the shell lightly to and fro. Meanwhile, the female has bouts of feeling the male's mouth appendages with her claws. After several minutes of this foreplay, the pair partly emerge from their shells and copulate, during which the male transfers spermatophores to the female. Unlike some other species of *Clibanarius*, the male does not fight when another male approaches, but merely offers a visual display.

The above courtship ritual was observed when the males occupied a *Thais* shell, and Hazlett then evicted the males from their shells and put them in *Polinices* shells, which have rounded apertures. In their new 'homes', none of the males attempted copulation, but rather tried to initiate a shell exchange with males in *Thais* shells. This and other studies confirmed that the shell-type has an undoubted effect on reproductive success (Contreras-Garduño and Córdoba-Aguilar 2006).

After fertilisation of her eggs, the female releases them into the water column within a few hours and the resulting larvae are planktonic for some weeks. Brooding of eggs is unknown in other species in the group, except for the southern hermit crab *Paguristes frontalis*, which may be unique among Australian hermit crabs in brooding eggs to a post-larval stage (Morgan 1987).

Another curious feature of this *Clibanarius* is the occurrence of sex change from female to male under circumstances, which are as yet unknown (Hazlett 1981).

The main food ingested by *C. virescens* is organic detritus, macroalgae and small invertebrates (Kunze and Anderson 1979). Its two claws are equal in size and it uses either of them to scoop up detritus or both of them to tear apart larger food items and pass them to the mouth, where they are shredded into particles by the mandibles.

The hermits Paguristes *spp. and* Dardanus setifer

The diets of three of the 10 southern species of *Paguristes* and one *Dardanus* have been examined and are summarised in Table 15.7. It is evident that the three species of *Paguristes* are mainly herbivorous, but also take small crustaceans and gastropods. Both *P. frontalis* and *P. sulcatus* are active by day and are able to prise small abalone up to ~4 cm size off rocks (Mower and Shepherd 1988). *Paguristes squamosus* lives mainly intertidally and collects detritus on the mat of fine hairs that covers its limbs and carapace, so that brushing of their surfaces provides additional food. Another source of food comes from flocculent detritus that the crab can filter from the water using the dense set of brushes on its fore-limbs (Kunze and Anderson 1979).

True crabs (Infra-order Brachyura)

Unlike some of the earlier groups, the abdomen in true crabs is folded underneath and most species have a pair of well-developed claws and four pairs of walking legs. Periodic moulting is needed for the crab to grow and females can only mate just after moulting, when the shell is soft. Males are sometimes seen carrying around the female – indicating she has recently moulted – and the male is ensuring that no other male will cuckold him. Eggs are borne by the female internally or externally. In some species, eggs are released at an early stage of development and in others at a much later stage. In the former case, the hatched larvae pass through many larval stages (variously called nauplii or zoeae) until they settle as miniature adults. The life history of a few species, mainly exploited or introduced, as well as the diet of several others, are summarised here.

Diet and feeding activity

The wide range of habitats in which crabs live is matched by their diversity in diet and mode of feeding, and several studies have examined the relation between diet and the morphology of the mouthparts and the digestive enzymes of the crab. The diet of 12 reef-associated crab species from seven different families, and from habitats ranging from the intertidal to the outer continental slope, is given in Table 15.8.

Table 15.7: Diet of four species of hermit crab (after Kunze and Anderson 1979 and Mower and Shepherd 1988).

Hermit crab	Diet
Paguristes squamosus	Filamentous, coralline and macroalgae, crustaceans, gastropods
Paguristes frontalis	Filamentous (13%), coralline (31%), other algae (36%), animal (15%)
Paguristes sulcatus	Filamentous (14%), coralline (24%), other algae (57%), animal (5%)
Dardanus setifer	Macroalgae, dead animals scavenged in detritus

Two species are filter-feeders (in part), three are herbivores, seven are omnivores taking algae and invertebrates, and two are carnivores. Some of them hunt for food by day and some by night (Table 15.8), and a few are crepuscular (i.e. they emerge to feed only around sunrise and sunset). The behaviour of one crepuscular crab, the red rock crab *Plagusia chabrus*, at West I., South Australia was apparently determined by its predator, as described by Shepherd (1973).

This rock crab lives by day in narrow crevices to avoid its main predator, the blue-throated wrasse, and emerges in dim light around dawn and dusk to browse on algae and search for animal prey, when the wrasse is roosting in caves. Abalone, the prey of both crab and wrasse, only emerge at night after both predators have retired. The successive activity periods of the three species around sunset are shown in Fig. 15.5; a similar, but reverse, sequence of emergent behaviour occurs around sunrise.

Several of the crabs listed in Table 15.8 predate on limpets or gastropods, both of which are equipped with strong protective shells. This raises the question as to how the predator gains access to the flesh. Skilleter and Anderson (1986) examined this for a pebble crab *Ozius truncatus* and, in the case of its whorled gastropod prey, found that the crab holds the shell close to its body with the apex of the shell near the mouth. The large, crushing claw of the crab

Table 15.8: Diet of some brachyuran, anomuran and other crabs, with information on their depth range, daily period of activity, and feeding type.

Species are classified according to feeding type: C = carnivore; F = filter feeder; H = herbivore; O = omnivore, and time of activity, where known, is indicated: D = daytime; N = night time. (br) indicates the crab burrows in sand under rocks by day.

Crab species	Depth (m)	Type, Active	Diet
New Zealand porcelain crab *Petrolistes elongatus*	0–12	F, H	Diatoms, algal fragments, detritus[1]
Hairy stone crab *Lomis hirta*	0–5	F, O	Brown algae (50%), red algae (23%), animal (20%)[5]
Giant spider crab *Leptomithrax gaimardii*	2–800	O	Algae, carrion[6]
European green crab *Carcinus maenas*	0–5	O N	Algae; gastropods, bivalves, crustaceans, fish[1]
Red rock crab *Nectocarcinus integrifrons*	0–20	H N	Seagrass, plant material[1,2]
Velvet crab *Nectocarcinus tuberculosus*	0–50	O N	Algae, seagrass; molluscs, other animals[1, 7]
Shore crab *Ozius truncatus*	Intertidal	C N	Limpets, gastropods, barnacles, carrion[4]
Crested forehead crab *Megametope carinata*	1–5	O N (br)	Brown algae red algae (44%); molluscs crustaceans, echinoderms (49%)[5]
Red-spotted hairy crab *Pilumnus rufopunctatus*	1–5	H N	Green algae (6%), brown algae (49%), red algae (25%), seagrass (7%); mollusc (5%)[5]
Swift-footed crab *Leptograpsus variegatus*	Intertidal	O D,N	Green, red algae; limpets, barnacles, carrion[4]
Red rock crab *Plagusia chabrus*	0–50	O N	Algae; sponge, bryozoans, hydroids, molluscs[3,6]

[1]Johnston and Freeman (2005); [2]Klumpp and Nicholls (1983); [3]Edgar (2008); [4]Skilleter and Anderson (1986); [5]Mower and Shepherd (1988); [6]Saunders (2009); [7]Salindeho and Johnston (2003).

then chips away at the edge of the shell until it can crush or break open the remainder of the shell. For shelled molluscs that attach themselves by suction to the rock (e.g. limpets and abalone) the crab inserts its claws under the shell edge and levers the animal from the rock, sometimes chipping the edge in the process. Mower and Shepherd (1988) found that several crab species were able to lever small abalone up to a size of 3–4 cm from the rock, and in the process chipped the shell edges and, after consuming the meat, left the empty shells – thereby enabling biologists to estimate the mortality of small abalone due to crab predation.

Life histories of five crabs

The life history of the following five well studied species or groups of species provides interesting contrasts between crabs of very different sizes and habitats. These are: the tiny spider crab *Halicarcinus ovatus* and its close relative *Amarinus laevis*; symbiotic pea crabs in several genera; the introduced European green crab *Carcinus maenas*; and the very large, deep-water king crab *Pseudocarcinus gigas*.

Three-pronged spider crabs Halicarcinus ovatus *and* Amarinus laevis

This common *Halicarcinus* ranges widely in the shallow subtidal throughout southern Australia on open coasts and in sheltered bays and inlets. Many colour forms occur, from red, orange, cream and brown to black, with mottled carapace and legs. Lucas (1975, 1980) suggested that it may be a species-complex, because he found slight larval differences in different populations. John Lucas studied the life history of the species in Cockburn Sound, Western Australia, where the crab lives on shallow reefs among coralline algae, sponges and ascidians, and under rocks. He also studied the life history of a closely related spider crab *Amarinus laevis* – an estuarine species (then called *Halicarcinus australis*) in the Swan Estuary (Lucas and Hodgkin 1970 a, b).

These two species are among the smallest of all crabs, with maximum carapace widths of ~15 mm. They have especial interest because their life histories are adapted to a small size, which severely limits the energy available for reproduction. Both species have a 1-year life span, although a few may survive to the second year. After several moults as a juvenile, *H. ovatus* undergoes a final pubertal moult on attaining sexual maturity and then stops growing. Females extrude eggs within a few days of the pubertal moult and continue to carry and release eggs for almost all their reproductive life, totalling up to ~10 000 eggs per female.

Copulation also begins at the pubertal moult and two patterns of copulation can occur. In one, female copulation takes place after a brief courtship when the female is still hard-shelled. Alternatively, a lengthy pre-moult courtship goes on until the female moults – an event that precipitates copulation. The female also has the ability to store spermatozoa, so that one copulation is enough to fertilise several

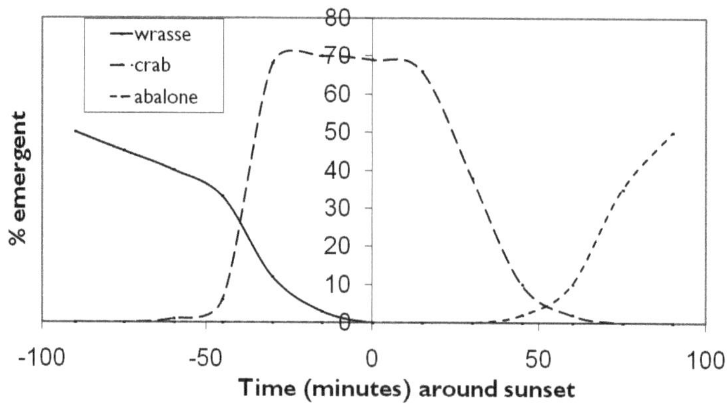

Figure 15.5: The activity periods around sunset of the predatory wrasse *Notolabrus tetricus*, the crab *Plagusia chabrus* and the abalone *Haliotis roei* at West I., South Australia (redrawn from Shepherd 1973, with later data from Shepherd and Clarkson (2001).

successive egg masses. The average-sized female carries ~850 eggs at a time in a brood cavity formed by four pairs of swimmerets (i.e. limbs) on the abdominal segments. During the ensuing incubation, the female maintains a current flow of water over the egg mass to irrigate them and ensure that they survive the development period of ~29 days.

After hatching, the larvae are pelagic and drift in the water column for ~23 days before settling on the bottom as instars – the immature juvenile stage. Larvae are most abundant in summer and least abundant in winter. After settling, the instars grow rapidly to maturity within a few months. The relative abundances of the different life history stages were followed in more detail by Lucas for the estuarine species *Amarinus laevis* (Fig. 15.6). Here, mature males and egg-bearing females were abundant from August to February, while instars were rare outside the summer months.

In a review of the life history of spider crabs (Family Hymenosomatidae), Lucas (1980) considered that they compensated for their small size and limited available energy by:

- producing continuous batches of a few, relatively large eggs, without interruption from moulting cycles
- retaining eggs till they hatch in order to maximise their survival
- having an abbreviated planktonic larval development stage to reduce mortality, restrict dispersal and ensure a high larval retention near the natal habitat.

For all these reasons, *H. ovatus* is most abundant in enclosed bays and inlets, which are, in effect, nurseries for local populations.

The ecology of the estuarine *Amarinus laevis* is similar to that of *H. ovatus*, except that the larval development period is less (14 days), and the number of eggs per egg mass held by females about four times greater. In this species, an abbreviated larval development period helps to retain larvae within an estuary and to avoid adverse conditions that are likely with increasing distance from the natal habitat.

Common predators of *H. ovatus* are benthic foraging fish, such as juvenile Australian salmon and flathead (Saunders 2009), while predators of *Amarinus laevis* are black bream, tailor (*Pomatomus saltatrix*) and mulloway.

Pea crabs

Four species of tiny pea crabs occur in southern Australia and all four live in an obligate association with either molluscs or, more rarely, with other sedentary invertebrate species (Poore 2004). One species, *Orthotheres haliotidis*, lives in the mantle cavity of two

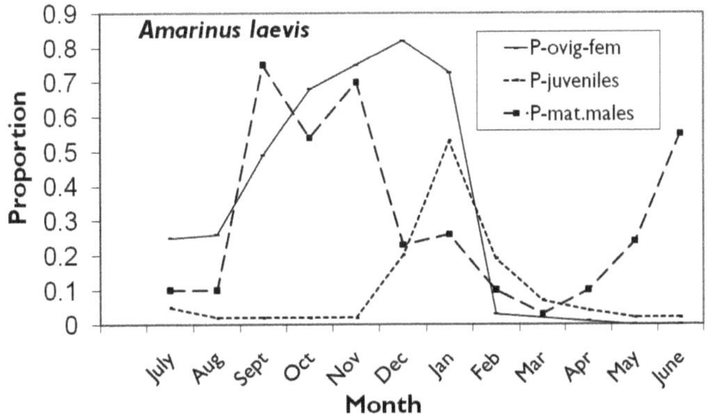

Figure 15.6: The estuarine spider crab *Amarinus laevis* showing seasonal changes in abundances of (a) proportions of ovigerous females to total females (P-ovig-fem); (b) proportions of juveniles with carapace width < 6 mm to all juveniles (P-juveniles); and (c) proportions of sexually mature males to all males (P-mat.males), in a population in the lower Swan Estuary, Western Australia in 1964–5 (redrawn from Lucas and Hodgkin 1970a).

abalone species, *Haliotis squamata* and *H. coccoradiata*, off SW and SE Australia, respectively; another, *Ostracotheres holothurensis*, lives in the anal cavity of sea cucumbers and in the inhalant siphon of the red-mouthed ascidian *Herdmania grandis*; and two species, *Ostracotheres subglobosus* and *Pinnotheres hickmani*, live in mussels, cockles and other species. The pea crab is not 'the guardian' of its host, as Aristotle thought (Bierbaum and Ferson 1986), but is rather a parasite, because the crab steals food from its host and is a cause of chronic stress to it. *Pinnotheres hickmani* (Plate 52) is the most abundant pea crab in southern waters and its association with the blue mussel, *Mytilus galloprovincialis*, was studied by Pregenzer (1978, 1979).

The pre-symbiotic larvae of *Pinnotheres* are planktonic and include four to five zoeal stages, followed by a final megalope, taking in all ~25 days. Thereupon, in mid-summer, the first crab stage is reached, having a megalopal appearance, a hard carapace < 1 mm wide and long swimming hairs. This is the invasive stage when the crab swims about in search of a host mussel and, on finding one, settles permanently on its gills. After settling, the pea crab grows to a maximum size of ~11 mm carapace width (CW) in the next 12 months, and passes through five more development phases. During this time, the pea crab feeds on the food-laden strings of mucus destined for ingestion by the mussel.

The immediate effect of the invader on the mussel is a reduction in pumping rate of ~25% and, in the longer term, the mussel's gills may be damaged and tissue production reduced. The incidence of infestation can be high; in both of Pregenzer's study sites (Port Phillip Bay and Botany Bay) 50–80% of mussels contained pea crab symbionts.

At a size of 2–5 mm CW, the male pea crab starts to travel from mussel to mussel in search of a female with whom to copulate. After mating, males do not guard the female, but leave at once to search for other females. Males are generally much smaller than females, suggesting, according to Soong (1997), an adaptation to facilitate easy entry and departure from host mussels. Females become egg-bearing at a size of 9–10 mm CW and have a life span of up to 3 years. They produce batches of up to 9000 eggs per batch, which hatch some 3–5 weeks later (Cheng 1967).

The European green crab Carcinus maenas

The green crab (Plate 53) was transported by shipping to Port Phillip Bay, Victoria, in the late 1800s and since then has become established in all southern States, most recently in the early 1990s in Tasmania, where it has become especially common. In South Australia, the crab became mainly established in degraded habitats, but has declined since 2001, possibly due to predation by a native competitor crab (Tanner 2007; Wiltshire *et al.* 2010). Excellent reviews of the biology and ecology of the species include those of Crothers (1967, 1968), Thresher (1997), Thresher *et al.* (2003) and Klassen and Locke (2007). *Carcinus maenas* has a broad habitat in shallow water to ~5 m depth on rocky, sandy, seagrass or muddy bottoms in sheltered waters, but is absent from exposed coasts.

The green crab has separate sexes and puberty is reached in ~1 year at a CW of 2–3 cm. The female produces one or two clutches of eggs a year, with up to 185 000 eggs in a clutch. The male selects his mate several days before she is due to moult and carries her around clasped under his body until she moults. They then mate and the male deposits spermatophores into pouches located near the oviduct openings, where they remain viable for many months. After fertilisation, the female attaches the eggs to the hairs on her swimmerets and carries them for several months, all the time ventilating them by fanning her abdomen. After hatching, the larvae are planktonic for 5–7 weeks (depending on sea temperature) and pass through six stages – five zoeae and a final megalope (pl. megalopa). Larval life is long enough for larvae to disperse long distances and field studies have recorded megalopa up to 45 km offshore. The zoeae actively swim up or down in the water column to assist movement out of estuaries or offshore and the megalopa use tidal currents to return inshore for metamorphosis and settlement on the sea bed.

The time of reproduction varies throughout the green crab's global range from winter to spring in some cool-temperate regions to all year round in other regions. The reproductive season is not yet known for southern Australia, although young-of-the-year have been recorded all year round in Tasmania (Thresher *et al.* 2003). Juveniles live and feed in inshore waters and later move into deeper water of 5–10 m.

The growth rate in Australia is yet to be determined, but the green crab is known to moult up to 18 times before puberty and annually after that. It has a life span of ~6 years and a maximum size of ~9 cm CW (Klassen and Locke 2007).

In its native habitats, the green crab, like *Plagusia* (Fig. 15.5), actively feeds in the crepuscular period around dawn and dusk (Svane, in Thresher *et al.* 2003). The food of the green crab includes bivalves, notably mussels and oysters, gastropods, polychaetes, sea-urchins, crustaceans and small fish. In Tasmanian waters, the crab has had an impact on the cockle *Katelysia* and mussels, and has completely displaced the grapsid crab *Paragrapsus gaimardii* (Walton *et al.* 2002; Johnson 2008). As yet, no impact on the Tasmanian oyster and mussel culture industries has occurred (Thresher 1997). The rate of spread of the green crab in SE Australia has been slow – at ~1.7 km a year along the coast – a speed which is about the same as the crab's crawling pace. Hence, despite its long larval life, its rate of invasive spread is slow, implying that highly localised recruitment is the norm. The sudden appearance of the green crab in South and Western Australia apparently occurred with human assistance.

Predators of the green crab include other crabs, gulls, fishes such as flathead and flounder, rays and seals, and it is possible that the high abundance of rays and the blue swimmer crab *Portunus armatus* in southern waters of mainland Australia control densities of the green crab there.

The giant or king crab Pseudocarcinus gigas

This is the largest brachyuran crab species in the world, and reaches a size of 240 mm carapace length (CL) and a maximum weight of ~17 kg. The king crab (Plate 54) ranges from SW Australia to central New South Wales and is found on deep reef and coarse to fine bryomol carbonate sediment habitats on the shelf edge and slope, primarily between 80 and 400 m depth. It has been fished commercially since 1992 over much of the southern part of its range, with peak catches taken at ~150 m depth. The need to manage the fishery has stimulated extensive research on the life history and biology of the crab (see Levings *et al.* 2001; McGarvey *et al.* 2002; Levings 2008; Gardner *et al.* 2009; Levings and Gill 2010).

The giant crab prefers a temperature range of 12–14°C and is well adapted to live in deep water on the steep terrain of the continental margin, where its ecology reflects the exigencies and physical character of this habitat. The males moult in summer and the females during winter, when mating with recently moulted females takes place. However, moulting only occurs every few years as discussed below. After a moult and mating, the females store the sperm for at least 4 years, and use them to fertilise successive broods. The females brood annually and produce up to 2.5 million eggs in a batch, attaching them to their swimmerets, where they are carried till hatching some 4 months later. Caleb Gardner (1997) noted that nemertean worms and amphipods sometimes infested the egg masses, causing some decline in brood size.

The eggs hatch into larvae progressively from September in SW Western Australia to November in SE Tasmania, due to the warmer waters in the west. During the hatching process, the female vigorously opens and closes her abdominal flap to wash the hatchlings out of the egg mass. Upon hatching, the larvae are planktonic for ~2 months, while they go through five zoeal stages and are transported in an easterly direction by the strong surface currents then flowing across southern Australia (Gardner and Quintana 1998). After settlement, the juvenile crabs reach sexual maturity at 11–14 cm CL, with slight differences between the sexes (Gardner and Williams 2002).

The giant crab has a nomadic lifestyle within a thermal niche, moving with cooler upwelled water in summer onto the shallower outer and mid-continental shelf, where there is more food, and leaving for deeper water in autumn as the shelf fills with the relatively warmer waters of the surface layer forced shorewards by rapidly cycling cold fronts typical of the austral winter.

The crab is carnivorous and uses its very large right claw, the largest of any known decapod, to crush gastropod and hermit crab shells. Dietary studies (Levings *et al.* 2001) gave the following average food intake by the king crab: sea-stars, 22%; juvenile *P. gigas* 8%; carrion 26%; and gastropods 38%. There were minor differences between the sexes, with males eating more sea-stars and females more gastropods.

Males grow larger than females, and so have the advantage of a much larger right pincer, enabling them to crush larger prey as they move inshore to feed.

Tagging studies have shown that the king crab crawls southwards on the west Australian coast in a general westerly direction along the southern coast, but northwards on the east Tasmanian coast. Along this last N–S coast, the interface between the EAC and Antarctic waters is at 200–300 m depth, and this is where the majority of the area's catch comes from. In SW Australia (long. 117–121°), the average annual distance crawled was 35 km, on the SE mainland coast it was ~6.6 km, and off Tasmania ~11 km, except for a few tagged crabs which, remarkably, crawled > 400 km. Along all coasts, the king crab crawled against the current, following the scent of its prey.

From returns of tagged crabs after periods at liberty of up to 10 years, Levings (2008) determined the intermoult period and moult increment for a range of size classes for each sex, for each of the southern States. West Australian crabs were the slowest growing, with longer moulting periods and smaller moult increments, and Tasmanian crabs the fastest, in a gradation of slower to faster growth rates west to east along Australia's southern margin (Table 15.9). There was no strong evidence of a terminal moult, as in some crabs. The giant crab cannot be aged in absolute terms, but it is evident from the slow growth rates recorded that the life span must exceed 40 years for females and 25 years for males.

Genetic studies found that the giant crabs comprise a single stock across its southern range. Along-shelf movement against the current by adults, and the extensive long-distance transport of larvae together ensure a well mixed stock, and replenishment of eastern populations. The possibility exists that eastern populations are maintained by recruitment from west Australian populations.

The evolution and adaptation of the giant crab are well summarised by Levings (2008). As the crab cannot control its internal temperature, it must occupy a narrow thermal niche. So, it lives at a depth that gives easy access to warmer and cooler water in an upwelling environment. Downslope movement into cold water conserves energy when it cannot feed and upslope movement gives access to a richer food supply for reproduction and growth. In autumn when upwelling ceases and winter downwelling starts, the water becomes warmer and males crawl into deeper, colder shelf waters and mate with moulting females.

The fishery for the giant crab is managed by size limits (15 cm CL for most States) and quotas, so as to conserve enough egg production (~30–34% of the virgin population) to ensure future recruitments (Gardner *et al.* 2009). Although the States manage their own king crab fisheries, they can be detrimentally affected by Commonwealth demersal trawl fisheries, which destroy the bryozoan habitats where the crabs live.

SEA SPIDERS (CLASS PYCNOGONIDA)

SA Shepherd and DA Staples

This unusual group of small and inconspicuous arthropods is distantly related to the Crustacea and more closely to the Chelicerata, which includes spiders, scorpions and mites, as its evolutionary relations are gradually becoming clearer (e.g. Damen 2010). Southern Australian pycnogonids, or sea spiders, as the group is commonly known, have four pairs of short to long walking legs. The long-legged species are

Table 15.9: Frequency and amount of moult increment of adult male and female king crab *Pseudocarcinus gigas* in Western Australia and Tasmania.

	Female 141–150 mm CL		Male 141–150 mm CL	
	Moult increment (mm)	Intermoult period (years)	Moult increment (mm)	Intermoult period (years)
Western Australia	13.21 ± 0.36	9	25.33 ± 1.20	6
Tasmania	19.60 ± 0.33	5	28.24 ± 0.63	3

well adapted for clambering among algal and animal clumps, while the short, stubby legged ones are usually found on firm substrata, preying on sedentary organisms. They are protected by an external cuticle, which they must moult regularly in order to grow. The body size of temperate species ranges from < 1 mm with a leg span of a few millimetres to larger animals with a leg span of up to ~5 cm. In the deep sea and Antarctic waters, much larger species occur, with the largest having a leg span of 75 cm. Having such a compact body and long, thin legs, they do not need a respiratory system or excretory organs. Their general ecology is reviewed by Arnaud and Bamber (1987) and Australian species by Staples (1977, 1997, 2005).

Food, feeding and dispersal

Sea spiders are almost all carnivorous and feed by inserting their proboscis into hydroids, soft corals, anemones, polychaetes or bryozoans, and sucking out the fluids within. For example, Staples (1997) observed that a feeding sea spider *Pycnogonum aurilineatum* tore at the body wall of the anemones, *Actinia tenebrosa* and *Corynactis australis*, by using the claws of its first pair of legs and then inserted its proboscis through the exposed tissue. The diet of > 30 species of southern sea spiders, listed in Table 15.10, shows that hydroids and bryozoans are the prey of most pycnogonid species. Further, although many of them feed on a range of hydroid species, indicating a

Table 15.10: Diet of pycnogonids (after Staples 1977, 1997; Staples and Watson 1987).

Prob. = probably (i.e. inferred from association).

Pycnogonid species	Prey
Nymphonidae	
Nymphon novaehollandiae	Prob. polychaetes
N. aequidigitatum	Hydroids, *Eudendrium* spp., bryozoans, sponges.
N. molleri	Prob. bryozoans, hydroids, spionid worms
N. dubitabile, N. conirostrum	Hydroids, bryozoans
Callipallenidae	
Callipallene spp.	Hydroids (12 species)
Pseudopallene ambigua	Bryozoans, notably *Orthoscuticella* spp.
P. pachychiera	Bryozoans
Stylopallene tubirostris	Bryozoans, notably *Amathia* spp.
S. cheilorhynchus	Bryozoans *Amathia* spp., *Orthoscuticella* spp.
S. longicauda	Bryozoan *Amathia biseriata*
S. dorsospinum	Bryozoans
Pallenoides stylirostrum	Hydroid *Symplectoscyphus subdichotomus*
Pallenoides spp.	Hydroids *Campanularia integra, Sertularia tenuis*
Parapallene avida	Hydroids and bryozoans
P. famelica	Hydroid *Gymnangium longirostrum*
P. australiensis	Hydroid *Halopteris glutinosa*
Propallene cyathus	Bryozoan *Vittaticella fusca*
Oropallene minor	Hydroids *Amphisbetia operculata, Symplectoscyphus subdichotomus*
Ammotheidae	
Achelia assimilis	Athecate hydroids

(Continued)

Table 15.10: (Continued)

Pycnogonid species	Prey
A. transfugoides	Hydroids *Stereotheca elongata, Amphisbetia minima*
Ammothea australiensis	Polychaete *Galeolaria caespitosa*
A. ovatoides	Bryozoan *Orthoscuticella ventricosa*
Ammothea spp.	Hydroids *Ectopleura crocea, Bimeria, Halecium, Sertularia, Stereotheca,Plumularia, Halicornopsis, Aglaophenia*
Tanystylum sp.	Hydroids *Ectopleura crocea, Campanularia, Sertularia, Sertularella, Halocordyle wilsoni*
Cilunculus sp.	Hydroid *Synthecium elegans*
Austrodecidae	
Austrodecus staplesi	Hydroids *Symplectoscyphus, Dictyocladium, Halicornopsis*
Pycnogonum aurilineatum	Coral *Culicia hoffmeisteri*; anemones *Actinia tenebrosa, Corynactis australis*
Phoxichilidiidae	
Anoplodactylus evansi	Hydroids *Eudendrium* spp., polychaetes, bryozoans, opisthobranchs
Anoplodactylus spp.	Hydroids (10 species)
Endeidae *Endeis* sp.	Hydroids *Campanularia integra*

broad diet, some (e.g. *Parapallene australiensis*) are found exclusively on only one species, implying an obligate relationship.

Predation by the sea spider *Anoplodactylus evansi* (Plate 55) on opisthobranchs was examined by Rogers *et al.* (2000). In experimental trials, 13 species were eaten and the sea spider used the claws on its front four legs to hold the slimy prey and the back four legs for stability on the substratum. It then tore off pieces of flesh with its pincers or alternatively inserted its proboscis directly into the prey. In the field, both the sea spider and its main prey, the sea-hare *Aplysia parvula*, were cryptic by day and emerged at night to feed, so increasing the encounter rate between predator and prey. However, the sea-hare fed on the red alga *Delisea*, which contains powerful chemical deterrents, and sequestered them into its tissues (Chapter 9). So, Rogers *et al.* (2000) then tested whether the presence of this chemical taken up by the sea-hare had any deterrent effect on the predatory spider. They found that the spider could tolerate only low levels of the chemical, but that it could minimise its own exposure by predating on small sea-hares and avoiding their digestive glands, which stored the chemical.

Sea spiders can crawl or swim, and Staples (1997) described their swimming in these terms: 'By vigorously treading water, slender forms can . . . lift themselves into the water column, where with little further effort, they can drift with the current. Descent is regulated by folding the legs dorsally to adopt a 'plummeting' position.' This behaviour clearly enables dispersal of a group of animals that brood their eggs.

Life history

Sea spiders are male or female and, in a remarkable adaptation, the male carries the eggs on 'ovigers' – special appendages found on the ventral surface of the first trunk segment. Ovigers are often absent in the female, and in some species absent in both sexes, in which case eggs are cemented directly to the ventral surface of the male.

Mating behaviour is known for a few species. The male first grasps the female who then releases eggs from the ovary. After fertilisation, she transfers them to the male who cements them individually in a ball, or in mass as a bracelet, around its egg-bearing legs where they remain till hatching. Frequently a male will have multiple matings and carry several groups of eggs in different stages of development.

After hatching, the larva, called a protonymphon, passes through a series of developmental stages over the next 1–2 months and finally leaves the male at a stage of development that varies with species. In some species, the protonymphon stays on the male's body until it can live independently, and in others it leaves the male soon after hatching and attaches itself parasitically to an invertebrate host, often favouring a soft fleshy one (e.g. a nudibranch, polychaete, sea cucumber, anemone or hydroid).

Several examples of such parasitism are now known (Staples and Watson 1987; Staples 1997). One sea spider in the genus *Anoplodactylus* encysts the hydroid *Sarsia radiata* (now possibly *S. exima*). Another species in the genus *Tanystylum* becomes attached to the hydroid *Pennaria wilsoni* (previously *Halocordyle wilsoni*). Both use their pincers (called chelae) for attachment. In the former case, the male releases the tiny newly hatched larvae 0.05 mm long onto the hydroid where they become encysted. After metamorphosis, the larvae, now ~0.7 mm long, emerge from their capsules within the hydroid by tearing away the cap with their claws, and commence a free life.

Many sea spiders closely match the colour of their host, due to the transparent nature of the cuticle, which reveals the colour of their gut content (Staples (1997). A common example is *Parapallene australiensis*, which has the same crimson colour as its host, the hydroid *Halopteris glutinosa* (formerly *H. buski*), on which it feeds. Hence, although predation on the spider by foraging fish (e.g. morwongs) has been recorded, few are likely to be taken.

Life history of *Parapallene avida*

This is the only known Australian life history study of a southern sea spider (Hooper 1980). *Parapallene avida* lives on algae, mainly *Sargassum* spp., in association with the hydroids *Campanularia* and *Clytia*. Adults are most abundant in summer, when breeding occurs, and at this time the upper legs (femurs) of the females, where the ovaries are located, become distended as eggs grow within them. The males and females then mate in the following manner. During a brief courtship of ~ 30 minutes, the male climbs on the back of the female and remains in that copulatory position for the next 3.5 h. He aligns the genital pores on his legs with those on the female's legs and then fertilises her eggs. The male then collects the eggs from each of the female's femurs, up to a total of ~60 eggs, and places them onto his ovigers, helped initially by the spines on the ovigers and the sticky coating of the eggs. The female then moves off and the male proceeds to cement the eggs in place using the cement glands on his legs.

A month or so after fertilisation, the embryos develop rudimentary appendages and they then go through several moults, eventually assuming a free-living stage between the second and third month. After leaving the male, they become attached to hydroids and go through further moulting stages. Juveniles apparently reach maturity at ~1 year of age in the following summer and finally disappear the next winter, suggesting a longevity of ~18 months, although survival of some to the next year cannot be ruled out. The life history is illustrated by Hooper's data on the numbers in each stage at his study site (Fig. 15.7).

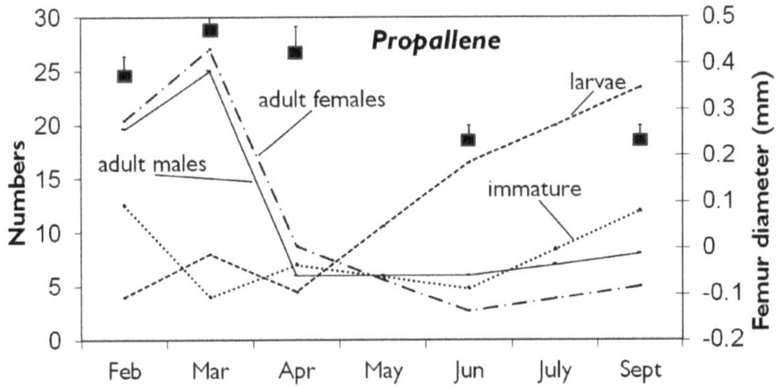

Figure 15.7. Plot of change in numbers per standardised sample of adult males and females, larval protonymphon and immature stages of *Propallene avida* on *Sargassum* over a year. Solid black squares show changes in mean diameter of female femurs containing eggs, with standard error bars, over the same period (redrawn from Hooper 1980).

REFERENCES

Abrams P (1982a) Intraspecific shell exchange in the hermit crab *Clibanarius virescens* (Krauss). *Journal of Experimental Marine Biology and Ecology* **59**, 89–101.

Abrams P (1982b) Frequencies of interspecific shell exchanges between hermit crabs. *Journal of Experimental Marine Biology and Ecology* **61**, 99–109.

Arnaud F, Bamber RN (1987) The biology of Pycnogonida. *Advances in Marine Biology* **24**, 2–97.

Babcock RC, Phillips JC, Lourey M, Clapin G (2007) Increased density, biomass and egg production in an unfished population of western rock lobster (*Panulirus cygnus*) at Rottnest Island, Western Australia. *Marine and Freshwater Research* **58**, 286–292.

Barnes M (1989) Egg production in Cirripedes. *Oceanography and Marine Biology - an Annual Review* **27**, 91–166.

Barnes DKA (2003) Local, regional and global patterns of resource use in ecology: hermit crabs and gastropod shells as an example. *Marine Ecology Progress Series* **246**, 211–223.

Barrett N, Buxton C, Gardner C (2009) Rock lobster movement patterns and population structure within a Tasmanian Marine Protected Area inform fishery and conservation management. *Marine and Freshwater Research* **60**, 417–425.

Bayliss DE (1993) Spatial distribution of *Balanus amphitrite* and *Elminius adelaidae* on mangrove pneumatophores. *Marine Biology* **116**, 251–256.

Berggren M (1999) *Notopontonia platycheles*, a little-known shallow-water shrimp from western Australia (Decapoda: Pontoniinae). *Journal of Crustacean Biology* **19**, 180–187.

Berrill M (1976) Aggressive behaviour of post-puerulus larvae of the western rock lobster *Panulirus longipes* (Milne-Edwards). *Australian Journal of Marine and Freshwater Research* **27**, 83–88.

Bierbaum RM, Ferson S (1986) Do symbiotic pea crabs decrease growth rate in mussels? *The Biological Bulletin* **170**, 51–61.

Booth JD (2006) *Jasus* species. In *Lobsters: Biology, Management, Aquaculture and Fisheries*. (Ed. BF Phillips) pp. 340–358. Blackwell, Oxford.

Booth JD, Phillips BF (1994) Early life history of spiny lobster. *Crustaceana* **66**, 271–294.

Brown R (2009) Western rock lobster risk assessment workshop. Western Australian Department of Fisheries, Perth. <http://www.rocklobsterwa.com> (Accessed September 2010)

Brown RS, Phillips BF (1994) The current status of Australia's lobster fisheries. In *Spiny Lobster Management*. (Eds BF Phillips, JS Cobb and J Kittaka) pp. 33–63. Blackwell, Oxford.

Bruce BD, Bradford RW, Griffin DA, Gardner C, Young JW (2000) 'A synthesis of existing data on larval rock lobster distribution in Southern Australia'. FRDC Report 96/107. CSIRO Marine and Atmospheric Research, Hobart.

Bruce B, Griffin D, Bradford R (2007) 'Larval transport and recruitment processes of southern rock lobster'. FRDC Report 2002/007. CSIRO Marine and Atmospheric Research, Hobart.

Bshary R, Grutter AS, Willener AST, Leimar O (2008) Pairs of cooperating cleaner fish provide better service quality than singletons. *Nature* **455**, 964–966.

Buxton C, Barrett N, Haddon M Gardner C, Edgar G (2006) 'Evaluating the effectiveness of marine protected areas as a fisheries management tool'. FRDC Final Report. Project No. 1999/162. Tasmanian Aquaculture and Fisheries Institute, University of Tasmania, Hobart.

Caputi N (2008) Impact of the Leeuwin Current on the spatial distribution of the western rock lobster (*Panulirus cygnus*) and implications for the fishery of Western Australia. *Fisheries Oceanography* **17**, 147–152.

Caputi N, Chubb CF, Brown RS (1993) Relationships between spawning stock, environment, recruitment and fishing effort for the western rock lobster, *Panulirus cygnus*, fishery in Western Australia. *Crustaceana* **68**, 213–226.

Caputi N, Chubb CF, Melville-Smith R, Pearce A *et al.* (2003) Review of relationships between life history stages of the western rock lobster, *Panulirus cygnus*, in Western Australia. *Fisheries Research* **65**, 47–61.

Caputi N, Melville-Smith R, de Lestang S, Pearce A *et al.* (2010) The effect of climate change on the western rock lobster (*Panulirus cygnus*) fishery of Western Australia. *Canadian Journal of Fisheries and Aquatic Sciences* **67**, 85–96.

Chapuis L, Bshary R (2010) Signalling by the cleaner shrimp *Periclimenes longicarpus*. *Animal Behaviour* **79**, 645–647.

Cheng TC (1967) Marine molluscs as hosts for symbioses, with a review of known parasites of commercially important species. *Advances in Marine Biology* **5**, vii–390.

Childress MJ (1997) Marine reserves and their effects on lobster populations: report from a workshop. *Marine and Freshwater Research* **48**, 1111–1115.

Chubb CF, Rossbach M, Melville-Smith R, Cheng YW (1999) 'Mortality, growth and movement of the western rock lobster (*Panulirus cygnus*)'. FRDC Report 1995/020. Fisheries Western Australia, Perth.

Cobb JS (1981) Behavior of the Western Australian spiny lobster, *Panulirus cygnus* George, in the field and laboratory. *Australian Journal of Marine and Freshwater Research* **32**, 399–409.

Cobb JS, Phillips BF (Eds) (1980) *The Biology and Management of Lobsters*. 2 volumes. Academic Press, New York.

Conover MR (1978) The importance of various shell characteristics to the shell-selection behaviour of hermit crabs. *Journal of Experimental Marine Biology and Ecology* **32**, 131–142.

Contreras-Garduño J, Córdoba-Aguilar A (2006) Sexual selection in hermit crabs: a review and outlines of future research. *Journal of Zoology* **270**, 595–605.

Crisp DJ, Bourget E (1985) Growth in barnacles. *Advances in Marine Biology* **22**, 199–244.

Crothers JH (1967) The biology of the shore crab *Carcinus maenas* (L.). 1. The background—anatomy, growth and life history. *Field Studies* **2**, 407–434.

Crothers JH (1968) The biology of the shore crab *Carcinus maenas* (L.). 2. The life of the adult crab. *Field Studies* **2**, 579–614.

Damen WGM (2010) Hox genes and the body plans of chelicerates and pycnogonids. In *Hox Genes: Studies from the 20th to the 21st Century. Advances in Experimental Medicine and Biology Vol. 689.* (Ed. JS Deutsch) pp. 125–132. Springer, New York.

Davie PJF (2002) Crustacea: Malacostraca: Phyllocarida: Hoplocarida: Eucarida (Part 1) In *Zoological Catalogue of Australia. Vol. 19.3.* (Eds A Wells and WWK Houston). CSIRO PUBLISHING, Melbourne.

de Lestang S, How J, Foster S (2011) 'Assessing possible environmental causes behind the reduced colonisation of Western Rock Lobster puerulus collectors by a wide suite of species.' FRDC Report 2008/085. Fisheries Research Report No. 218. Dept of Fisheries, Perth.

Duffy JE (2002) The ecology and evolution of eusociality in sponge-dwelling shrimp. In *Genes, Behavior, and Evolution in Social Insects.* (Ed. T Kikuchi) pp. 1–37. University of Hokkaido Press, Sapporo, Japan.

Edgar GJ (1990a) Predator-prey interactions in seagrass beds. I. The influence of macrofaunal abundance and size-structure on the diet and growth of the western rock lobster *Panulirus cygnus* George. *Journal of Experimental Marine Biology and Ecology* **139**, 1–22.

Edgar GJ (1990b) Predator-prey interactions in seagrass beds. II. Impacts of the western rock lobster *Panulirus cygnus* George on epifaunal gastropod populations. *Journal of Experimental Marine Biology and Ecology* **139**, 33–42.

Edgar GJ (2008) *Australian Marine Life: the Plants and Animals of temperate waters.* 2nd edn. Reed New Holland, Sydney.

Egan EA, Anderson DT (1986) Larval development of *Balanus amphitrite* Darwin and *Balanus variegatus* Darwin (Cirripedia, Balanidae) from New South Wales, Australia. *Crustaceana* **51**, 188–207.

Fielder DR (1965a) The spiny lobster, *Jasus lalandei* (H. Milne-Edwards) in South Australia: food, feeding and locomotor activity. *Australian Journal of Marine and Freshwater Research* **16**, 351–367.

Fielder DR (1965b) A dominance order for shelter in the spiny lobster *Jasus lalandei* (H. Milne-Edwards). *Behaviour* **24**, 236–245.

Francois DD, Gorman T (1980) Unique management measures in NSW abalone fishery. *Australian Fisheries* **39**, (9), 26–28.

Frusher S, Prescott J, Edmunds M (1999) Southern rock lobsters. In *Under Southern Seas: the Ecology of Australia's Rocky Reefs.* (Ed. N Andrew) pp. 107–113. University of New South Wales Press, Sydney.

Gardner C (1997) Effect of size on reproductive output of giant crabs *Pseudocarcinus gigas* (Lamarck): Oziidae. *Marine and Freshwater Research* **48**, 581–587.

Gardner C, Quintana R (1998) Larval development of the Australian giant crab *Pseudocarcinus gigas* (Lamarck 1818) (Decapoda: Oziidae) reared in the laboratory. *Journal of Plankton Research* **20**, 1169–1188.

Gardner C, Williams H (2002) Maturation in the male giant crab *Pseudocarcinus gigas*, and the potential for sperm limitation in the Tasmanian fishery. *Marine and Freshwater Research* **53**, 661–667.

Gardner C, Frusher SD, Kennedy RB, Cawthorn A (2001) Relationship between settlement of southern rock lobster pueruli, *Jasus edwardsii*, and recruitment to the fishery in Tasmania, Australia. *Marine and Freshwater Research* **52**, 1271–1275.

Gardner C, Frusher SD, Haddon M, Buxton C (2003) Movements of the southern rock lobster *Jasus edwardsii* in Tasmania, Australia. *Bulletin of Marine Science* **73**, 653–671.

Gardner C, Ziegler P, Jennings S (2009) 'Reassessing giant crab (*Pseudocarcinus gigas*) size limits to optimise value and sustainability of the fishery'. FRDC Project 2006/022. Tasmanian Aquaculture and Fisheries Institute, University of Tasmania, Hobart.

George RW (2005) Evolution of life cycles, including migration, in spiny lobsters. *New Zealand Journal of Marine and Freshwater Research* **39**, 503–514.

Gowlett-Holmes K (2008) *A Field Guide to the Marine Invertebrates of South Australia.* Notomares, Hobart.

Gray CA (1991) Demographic patterns of the palaemonid prawn *Macrobrachium intermedium* in southeastern Australia: spatial heterogeneity and the effects of species of seagrass. *Marine Ecology Progress Series* **75**, 239–249.

Gray H (1992) *The Western Rock Lobster,* Panulirus cygnus. *Book 1. A Natural History.* Westralian Books, Perth.

Guest MA, Frusher SD, Nichols PD, Johnson CR *et al.* (2009) Trophic effects of fishing rock lobster *Jasus edwardsii* shown by combined fatty acid and stable isotope analyses. *Marine Ecology Progress Series* **388**, 169–184.

Hamond R (1973) Four new copepods (Crustacea: Harpacticoida, Canuellidae) simultaneously occurring with *Diogenes senex* (Crustacea: Paguridea) near Sydney.

Proceedings of the Linnean Society of New South Wales **97**, 165–201.

Hazlett BA (1981) The behavioural ecology of hermit crabs. *Annual Review of Ecology and Systematics* **12**, 1–22.

Hazlett BA (1996) Precopulatory behaviour of the hermit crabs *Diogenes avarus* and *Clibanarius virescens. Australian Journal of Zoology* **44**, 487–492.

Hoare M (2008) The effects of southern rock lobster (*Jasus edwardsii*) translocation on inshore rocky reef communities in South Australia. BSc (Hons) thesis. Flinders University, Adelaide.

Hobday D, Punt AE (2001) Size-structured population modelling and risk assessment of the Victorian southern rock lobster, *Jasus edwardsii*, fishery. *Marine and Freshwater Research* **52**, 1495–1507.

Hooper JNA (1980) Some aspects of the reproductive biology of *Parapallene avida* Stock (Pycnogonida: Callipallenidae) from northern New South Wales. *Australian Zoologist* **20**, 473–481.

Howard RK (1984) The trophic ecology of caridean shrimps in an eelgrass community. *Aquatic Botany* **18**, 155–174.

Howard RK (1988) Fish predators of the western rock lobster (*Panulirus cygnus* George) in a nearshore nursery habitat. *Australian Journal of Marine and Freshwater Research* **39**, 307–316.

Jeffs AG, Montgomery JC, Tindle CT (2005) How do spiny lobster post-larvae find the coast? *New Zealand Journal of Marine and Freshwater Research* **39**, 605–617.

Jernakoff P (1987) Foraging patterns of juvenile western rock lobsters, *Panulirus cygnus* George. *Journal of Experimental Marine Biology and Ecology* **113**, 125–144.

Jernakoff P (1990) Distribution of newly settled western rock lobsters, *Panulirus cygnus. Marine Ecology Progress Series* **66**, 63–74.

Jernakoff P, Phillips BF, Maller RA (1987) A quantitative study of nocturnal foraging distances of the western rock lobster, *Panulirus cygnus* George. *Journal of Experimental Marine Biology and Ecology* **113**, 9–21.

Jernakoff P, Phillips BF, Fitzpatrick JJ (1993) The diet of post-puerulus western rock lobster, *Panulirus cygnus* George, at Seven Mile Beach, Western Australia. *Australian Journal of Marine and Freshwater Research* **44**, 649–654.

Johnson VR (1969) Behavior associated with pair formation in the banded shrimp *Stenopus hispidus* (Olivier). *Pacific Science* **23**, 40–50.

Johnson VR (1977) Individual recognition in the banded shrimp *Stenopus hispidus* (Olivier). *Animal Behaviour* **25**, 418–428.

Johnson CR (2008) Impacts of introduced marine species: Tasmania as a laboratory. In *Marine Ecology.* (Eds SD Connell and BM Gillanders) pp. 578–579. Oxford University Press, Melbourne.

Johnson CR, Banks SC, Barrett NS, Cazassus F *et al.* (2011) Climate change cascades: shifts in oceanography, species' ranges and subtidal marine community dynamics in eastern Tasmania. *Journal of Experimental Marine Biology and Ecology* **400**, 17–32.

Johnston D, Freeman J (2005) Dietary preference and digestive enzyme activities as indicators of trophic resource utilization by six species of crab. *The Biological Bulletin* **208**, 36–46.

Joll LM, Phillips BF (1984) Natural diet and growth of juvenile western rock lobsters *Panulirus cygnus* George. *Journal of Experimental Marine Biology and Ecology* **75**, 145–169.

Jonasson M (1987) Fish cleaning behaviour in shrimp. *Journal of Zoology* **213**, 117–131.

Kailola PJ, Williams MJ, Stewart PC, Reichelt RE *et al.* (1993) *Australian Fisheries Resources.* Bureau of Resource Sciences and FRDC, Canberra.

Klassen G, Locke A (2007) 'A biological synopsis of the European green crab, *Carcinus maenas*'. Canadian Manuscript Report Fisheries and Aquatic Science No. 2818. Fisheries and Oceans Canada, Moncton.

Klumpp DW, Nicholls PD (1983) Utilisation of the seagrass *Posidonia australis* as food by the rock crab *Nectocarcinus integrifrons* (Latreille) (Crustacea: Decapoda: Portunidae). *Marine Biology Letters* **4**, 331–339.

Kunze J, Anderson DT (1979) Functional morphology of the mouthparts and gastric mill in the hermit crabs *Clibanarius taeniatus* (Milne-Edwards), *Clibanarius virescens* (Krauss), *Paguristes squamosus* McCulloch and *Dardanus setifer* (Milne-Edwards) (Anomura: Paguridae). *Australian Journal of Marine and Freshwater Research* **30**, 683–722.

Levings AH (2008) A Life history model for the giant crab *Pseudocarcinus gigas*. PhD thesis, Deakin University, Warrnambool, Victoria.

Levings AH, Gill PG (2010) Seasonal winds drive water temperature cycle and migration patterns of southern Australian giant crab *Pseudocarcinus gigas*. In *Biology and Management of Exploited Crab Populations under Climate Change.* (Eds GH Kruse, GL Eckert, RJ Foy, RN Lipcius, *et al.*) pp. 461–477. Alaska Sea Grant, University of Alaska Fairbanks.

Levings A, Mitchell BD, McGarvey R, Matthews J *et al.* (2001) 'Fisheries biology of the giant crab, *Pseudocarcinus gigas.*' FRDC Projects 93/220 and 97/132. School of Ecology and Environment, Deakin University, Geelong.

Lewis RK (1977) Rock lobster puerulus settlement in the South East. *South Australian Fishing Industry Council* **14**, 9–11.

Lewis RK (1981) 'Southern rock lobster (*Jasus novaehollandiae*) Zone S review'. Internal report no. 88. South Australian Dept Fisheries, Adelaide.

Linnane A, Dimmlich W, Ward T (2005) Movement patterns of the southern rock lobster, *Jasus edwardsii*, off South Australia. *New Zealand Journal of Marine and Freshwater Research* **39**, 335–346.

Linnane A, Gardner C, Hobday D, Punt A *et al.* (2010a) Evidence of large-scale spatial declines in recruitment patterns of southern rock lobster *Jasus edwardsii* across south-eastern Australia. *Fisheries Research* **105**, 163–171.

Linnane A, James C, Middleton J, Hawthorne P *et al.* (2010b) Impact of wind stress anomalies on the seasonal pattern of southern rock lobster (*Jasus edwardsii*) settlement in South Australia. *Fisheries Oceanography* **19**, 290–300.

Linnane A, Sloan S, McGarvey R, Ward T (2010c) Impacts of unconstrained effort: lessons from a rock lobster (*Jasus edwardsii*) fishery decline in the northern zone management region of South Australia. *Marine Policy* **34**, 844–850.

Linnane A, Hobday D, Frusher S, Gardner C (2012) Growth rates of juvenile southern rock lobster (*Jasus edwardsii*) estimated through a diver-based tag-recapture program. *Marine and Freshwater Research* **63**, 110–118.

Linnane A, McGarvey R, Hoare M, Hawthorne P (2013) The importance of conserving recruitment pulses in rock lobster (*Jasus edwardsii*) fisheries where puerulus settlement is low or highly sporadic. *Marine Biology Research* **9**, 97–103.

Lipcius RN, Cobb JS (1994) Introduction: ecology and fishery biology of spiny lobsters. In *Spiny Lobster Management.* (Eds BF Phillips, JS Cobb and J Kittaka) pp. 1–30. Blackwell, Oxford.

Liszka D, Underwood AJ (1990) An experimental design to determine preferences for gastropod shells by a hermit crab. *Journal of Experimental Marine Biology and Ecology* **137**, 47–62.

Lucas JS (1975) The larval stages of some Australian species of *Halicarcinus* (Crustacea, Brachyura, Hymenosomatidae). III. Dispersal. *Bulletin of Marine Science* **25**, 94–100.

Lucas JS (1980) Spider crabs of the family Hymenosomatidae (Crustacea, Brachyura) with particular reference to Australian species: systematics and biology. *Records of the Australian Museum* **33**, 148–247.

Lucas JS, Hodgkin EP (1970a) Growth and reproduction of *Halicarcinus australis* (Haswell) (Crustacea, Brachyura) in the Swan Estuary, Western Australia. 1. Crab instars. *Australian Journal of Marine and Freshwater Research* **21**, 149–162.

Lucas JS, Hodgkin EP (1970b) Growth and reproduction of *Halicarcinus australis* (Haswell) (Crustacea, Brachyura) in the Swan Estuary, Western Australia. II. Larval stages. *Australian Journal of Marine and Freshwater Research* **21**, 163–173.

MacDiarmid A (1989) Size at onset of maturity and size-dependent reproductive output of female and male spiny lobsters *Jasus edwardsii* (Hutton) (Decapoda, Palinuridae) in northern New Zealand. *Journal of Experimental Marine Biology and Ecology* **127**, 229–243.

MacDiarmid A, Booth J (2005) Crayfish. In *The Living Reef: the Ecology of New Zealand's Rocky Reefs.* (Eds N Andrew and M Francis) pp. 120–126. Craig Potton Publishing, Nelson.

MacDiarmid A, Hickey B, Maller RA (1991) Daily movement patterns of the spiny lobster *Jasus edwardsii* (Hutton) on a shallow reef in northern New Zealand. *Journal of Experimental Marine Biology and Ecology* **147**, 185–205.

McGarvey R (2004) Estimating the emigration rate of fish stocks from marine sanctuaries using tag-recovery data. *Fishery Bulletin* **102**, 464–472.

McGarvey R, Ferguson GJ, Prescott JH (1999) Spatial variation in mean growth rates at size of southern rock lobster, *Jasus edwardsii*, in South Australian waters. *Marine and Freshwater Research* **50**, 333–342.

McGarvey R, Levings AH, Matthews JM (2002) Moulting growth of the Australian giant crab *Pseudocarcinus gigas*. *Marine and Freshwater Research* **53**, 869–881.

Montgomery S (1992) Size at first maturity and at onset of breeding in female *Jasus verreauxi* (Decapoda: Palinuridae) from New South Wales waters, Australia. *Australian Journal of Marine and Freshwater Research* **43**, 1373–1379.

Montgomery S (1995) Patterns in landings and size composition of *Jasus verreauxi* (H. Milne-Edwards, 1851) (Decapoda, Palinuridae) in waters off New South Wales, Australia. Proceedings of the Fourth International Workshop on Lobster Biology and Management, 1993. *Crustaceana* **68**, 257–266.

Montgomery S (1999) Eastern rock lobster. In *Under Southern Seas: the Ecology of Australia's Rocky Reefs.* (Ed. N Andrew) pp. 115–117. University of New South Wales Press, Sydney.

Montgomery S, Craig JR (2005) Distribution and abundance of recruits of the eastern rock lobster (*Jasus verreauxi*) along the coast of New South Wales, Australia. *New Zealand Journal of Marine and Freshwater Research* **39**, 619–628.

Morgan GR (1980) Population dynamics of spiny lobsters. In *The Biology and Management of Lobsters. Vol. II.* (Eds JS Cobb JS and BS Phillips) pp. 189–217. Academic Press, New York.

Morgan GJ (1987) Abbreviated development in *Paguristes frontalis* (Milne Edwards, 1836) (Anomura: Diogenidae) from southern Australia. *Journal of Crustacean Biology* 7, 536–540.

Mower AGJ, Shepherd SA (1988) The crab fauna of West Island, South Australia: their abundance, diet and role as predators of abalone. *Transactions of the Royal Society of South Australia* **112**, 83–86.

Nelson MM, Cox SL, Ritz DA (2002) Function of mouthparts in feeding behaviour of phyllosoma larvae of the packhorse lobster, *Jasus verreauxi* (Decapoda: Palinuridae). *Journal of Crustacean Biology* **22**, 595–600.

Olsen AM, Shepherd SA (2006) Historic drift bottle experiments show reversing surface water masses in western Bass Strait waters: implications for lobster larval dispersal. *Transaction of the Royal Society of South Australia* **130**, 113–122.

Paul LK (2001) Factors associated with habitat selection, competition and coexistence in the palaemonid shrimps of the Barker Inlet-Port River Estuary. BSc (Hons) thesis. University of Adelaide, Adelaide.

Phillips BF (1981) The circulation of the southeastern Indian Ocean and the planktonic life of the western rock lobster. *Oceanography and Marine Biology - an Annual Review* **19**, 11–39.

Phillips BF (Ed.) (2006) *Lobsters: Biology, Management, Aquaculture and Fisheries.* Blackwell, Oxford.

Phillips BF, Melville-Smith R (2005) Sustainability of the western rock lobster fishery: a review of past progress and future challenges. *Bulletin of Marine Science* **76**, 485–500.

Phillips BF, Melville-Smith R (2006) *Panulirus* species. In *Lobsters: Biology, Management, Aquaculture and Fisheries.* (Ed. BF Phillips) pp. 359–384. Blackwell, Oxford.

Phillips BF, Morgan GR, Austin CM (1980) Synopsis of biological data on the western rock lobster, *Panulirus cygnus* George, 1962. FAO fisheries Synopsis No. 128. Rome.

Phillips BF, Cobb JS, Kittaka J (Eds) (1994) *Spiny Lobster Management.* Blackwell, Oxford.

Poore GCB (2004) *Marine Decapod Crustacea of Southern Australia.* CSIRO PUBLISHING, Melbourne.

Pregenzer C Jr (1978) *Pinnotheres hickmani* in wild and cultured *Mytilus edulis* in Port Phillip Bay, Victoria. *Australian Journal of Marine and Freshwater Research* **29**, 127–139.

Pregenzer C Jr (1979) Effect of *Pinnotheres hickmani* on neutral red clearance of *Mytilus edulis. Australian Journal of Marine and Freshwater Research* **30**, 547–550.

Ramm DC (1980) Electromagnetic tracking of rock lobsters (*Jasus novaehollandiae*). *Australian Journal of Marine and Freshwater Research* **31**, 263–269.

Rogers CN, de Nys R, Steinberg PD (2000) Predation on juvenile *Aplysia parvula* and other small Anaspidean, Ascoglossan, and Nudibranch gastropods by pycnogonids. *The Veliger* **43**, 331–337.

Ross DM, von Boletzky S (1979) The association between the pagurid *Dardanus arrosor* and the actinian *Calliactis parasitica.* Recovery of activity in 'inactive' *D. arrosor* in the presence of cephalopods. *Marine Behaviour and Physiology* **6**, 175–184.

Salindeho IR, Johnston DJ (2003) Functional morphology of the mouthparts and proventriculus of the rock crab *Nectocarcinus tuberculosus* (Decapoda: Portunidae). *Journal of the Marine Biological Association of the United Kingdom* **83**, 821–834.

Saunders B (2009) *Shores and Shallows of Coffin Bay: an Identification Guide.* South Australian Government, Adelaide.

SEFC (South Eastern Fisheries Committee) (1982) 'Southern rock lobster'. Fishery Situation Report 7. CSIRO, Cronulla.

Shepherd SA (1973) Studies on southern Australian abalone (genus *Haliotis*). 1. Ecology of five sympatric species. *Australian Journal of Marine and Freshwater Research* **24**, 217–257.

Shepherd SA, Clarkson PS (2001) Diet, feeding behaviour, activity and predation of the temperate blue-throated wrasse, *Notolabrus tetricus. Marine and Freshwater Research* **52**, 311–322.

Shepherd SA, Teale J, Muirhead D (2005) Cleaning symbiosis among inshore fishes at Althorpe Island, South Australia and elsewhere. *Transactions of the Royal Society of South Australia* **129**, 193–201.

Skilleter GA, Anderson DT (1986) Functional morphology of the chelipeds, mouthparts and gastric mill of *Ozius truncatus* (Milne Edwards) (Xanthidae) and *Leptograpsus variegatus* (Fabricius) (Grapsidae) (Brachyura). *Australian Journal of Marine and Freshwater Research* **37**, 67–79.

Smith JM, Parker GA (1976) The logic of asymmetric contests. *Animal Behaviour* **24**, 159–175.

Soong K (1997) Some life history observations on the pea crab, *Pinnotheres tsingtaoensis*, symbiotic with the bivalve mollusc, *Sanguinolaria acuta. Crustaceana* **70**, 855–866.

Staples D (1977) Pycnogonida . . . the 'nobodies'. *Oceans* **1**, 122–125.

Staples D (1997) Sea spiders or pycnogonids (Phylum Arthropoda). In *Marine Invertebrates of Southern Australia. Part III.* (Eds SA Shepherd and M Davies) pp. 1040–1072. SARDI Aquatic Sciences and SA Government Printer, Adelaide.

Staples D (2005) Pycnogonida. In *Marine Parasitology.* (Ed. K Rohde) pp. 222–226. CSIRO PUBLISHING, Melbourne.

Staples DA, Watson JE (1987) Associations between pycnogonids and hydroids. In *Modern Trends in the*

Systematics, Ecology and Evolution of Hydroids and Hydromedusae. (Ed. J Bouillon) pp. 215–225.Oxford University Press, Oxford.

Tanner JE (2007) The influence of the introduced green crabs (*Carcinus maenas*) on habitat selection by juvenile native blue crabs (*Portunus pelagicus*). *Estuaries and Coasts* **30**, 601–606.

Thresher RE (Ed.) (1997) 'Proceedings of the First International Workshop on the Demography, Impacts and Management of Introduced Populations of the European Crab, *Carcinus maenas*'. Centre for Research on Introduced Marine Pests. Technical Report No. 1. CSIRO Marine Laboratories, Hobart.

Thresher RE, Proctor C, Ruiz G, Gurney R *et al.* (2003) Invasion dynamics of the European shore crab, *Carcinus maenas*, in Australia. *Marine Biology* **142**, 867–876.

Vader W (1995) *Liljeborgia* species (Amphipoda, Liljeborgiidae) as associates of hermit crabs. *Polski Archiwun Hydrobiogogii* **42**, 517–525. [Polish Archives of Hydrobiology]

Vader W, Myers AA (1996) Amphipods living in association with hermit crabs in S.E. Australia. I. Five new Ischynoceridae. *Bolletino del Museo Civico di Storia Naturale de Verona* **20**, 263–292.

Vaitilingon D, Gleeson C, Williamson J (2007) Host selection of sea urchins by the ectosymbiotic shrimp *Gnathophylloides mineri* (Schmitt 1933): a behavioural assessment. Abstract. In *Program and Abstract Book. Marine Science in a Changing World. AMSA 45th Annual Conference.* 9–13 July 2007, (Eds T. O'Hara, N Hall) p.115. Australian Marine Sciences Association, Melbourne.

Vannini M, Cannicci S (1995) Homing behaviour and possible cognitive maps in crustacean decapods. *Journal of Experimental Marine Biology and Ecology* **193**, 67–91.

Waddington KI, Bellchambers LM, Vanderklift MA, Walker DI (2008) Western rock lobsters (*Panulirus cygnus*) in western Australian deep coastal ecosystems (35–60 m) are more carnivorous than those in shallow coastal ecosystems. *Estuarine, Coastal and Shelf Science* **79**, 114–120.

Waddington KI, Meeuwig JJ, Evans SN, Bellchambers LM. (2010). Assessment of the benthic biota of a deep coastal ecosystem by remote and *in situ* sampling. *Marine and Freshwater Research* **61**, 1164–1170.

Walsh CJ, Mitchell BD (1998) Factors associated with variations in abundance of epifaunal caridian shrimps between and within estuarine seagrass meadows. *Marine and Freshwater Research* **49**, 769–777.

Walton WC, MacKinnon C, Rodriguez LE, Proctor C *et al.* (2002) Effect of an invasive crab upon a marine fishery: green crab *Carcinus maenas* predation on a venerid clam, *Katelysia scalarina*, in Tasmania (Australia). *Journal of Experimental Marine Biology and Ecology* **272**, 171–189.

Williams JD, McDermott JJ (2004) Hermit crab biocoenoses: a world-wide review of the diversity and natural history of hermit crab associates. *Journal of Experimental Marine Biology and Ecology* **305**, 1–128.

Wiltshire K, Rowling K, Deveney M (2010) 'Introduced marine pests in South Australia: a review of records and distribution mapping'. SARDI Research Report Series No. 468, Adelaide

Zhang D, Lin J, LeRoy Cresswell R (1998) Mating behavior and spawning of the banded coral shrimp *Stenopus hispidus* in the laboratory. *Journal of Crustacean Biology* **18**, 511–518.

PART 4

MARINE VERTEBRATES

The next three chapters focus on the ecology of some species representative of the upper levels of the food web. Species have been selected from the bony fishes, the cartilaginous fishes (i.e. the sharks and rays) and the marine mammals. Reef fishes have traditionally been broadly divided into five feeding types – herbivores, omnivores, carnivores, planktivores and piscivores. Chapter 16 describes the ecology of examples of mainly bottom-feeders from the first three groups and Chapter 17 considers planktivores, which are often associated with reefs, and some larger predators, up to the sharks, some of which can range trans-oceanically. Some of these larger predators have become depleted through overfishing in recent decades. Lastly, Chapter 18 describes the ecology of some pinnipeds (seals and sea lions) and whales at the highest levels of the food web. These species are still recovering from their decimation by hunting in the 19th century.

16 Bottom-feeding fishes

OVERVIEW

Three bottom-feeding types of fish – herbivores, omnivores and carnivores – are treated in this chapter. Herbivorous species are relatively few and their abundance declines with increasing latitude. Their feeding ecology is well known, but knowledge of their life history is still sketchy. The feeding ecology of the eight most common southern species is described. This varies widely from territorial species –an odacid (herring cale) and a pomacentrid (scalyfin) – that browse within localised areas, to wide-ranging kyphosids and girellids (black fish, drummer, buff bream and zebra fish) that feed in schools, to a solitary aplodactylid grazer (sea carp). Omnivorous species include two species of sweep, three 'pickers', moonlighter, western talma and mado, and various species of leatherjacket; their diverse life history strategies and ecology are summarised. The benthic carnivores considered range from small microcarnivores (i.e. weedfishes and two seadragons) to four species of morwong – magpie perch, red morwong, banded morwong and blue morwong. Then we consider the ecology and life history of 11 macrocarnivores – the labrid fishes, including blue groper, foxfish, rainbow cale and eight other species – and discuss briefly their ecological significance and the effect of fishing on them. Finally, the chapter concludes with a review of the symbiosis between cleaner fishes and their fish clients in southern waters and their respective behaviours.

INTRODUCTION

Fishes are a conspicuous part of the reef fauna, and are also highly important because they influence many other reef organisms by their herbivory, predation and role as prey for other species higher in the food web. Some species are territorial, some have limited home ranges and others range widely in gulfs or on continental shelves. Because many fish species are also harvested for food, their population biology is of interest to fisheries biologists. In this chapter, we discuss the better-known reef-associated species according to their feeding type – herbivores, omnivores and micro- and macrocarnivores. The division is approximate, because many herbivores are omnivorous or carnivorous as juveniles, and some bottom-feeders are also opportunistic planktivores. However, the classification is useful, because it aids in understanding the trophic links on reefs and the impact of higher predators on species lower in the web.

HERBIVORES

Fish herbivory is important ecologically because it provides a link between the primary producers at the base of the food web and large consumers. But feeding on algae poses formidable problems for fishes, such as accessing sufficient nutrients to enable them to meet their energy needs and overcoming the chemical defences that plants have evolved to deter herbivory (Horn 1989, Choat and Clements 1998). A striking pattern of herbivore distribution is that their numbers decline with increasing latitude – thought to be due to the decreasing digestive efficiency of herbivorous fishes in colder waters (Floeter *et al.* 2005; but see Clements *et al.* 2009). As sea temperatures have increased in SE Australia in the last century, however, many herbivorous species have expanded their range southwards to SE Tasmania (Last *et al.* 2011).

On temperate Australian reefs, herbivores usually comprise 8–15% of total fish numbers; however, due to their large mean size, herbivorous fishes on SW Australian reefs, notably *Kyphosus* spp., can contribute most of the local fish biomass (Edgar *et al.* 2007). Most temperate herbivorous species, other than blennies, are browsers; that is, they feed selectively on single algal plants rather than grazing indiscriminately on mixtures of turfing algae, as is common among tropical fishes. Juveniles, however, invariably feed either in whole or part on animal matter, usually small crustaceans. The commonest species on southern reefs are in the genera *Olisthops, Parma, Kyphosus, Girella* and *Aplodactylus*, with several species in most of the genera.

Herring cale *Olisthops cyanomelas* (Family Odacidae)

Food and foraging

Once placed in the genus *Odax*, this species has now been found to be only distantly related to the New Zealand species of *Odax* (Clements *et al.* 2004). The herring cale is common on southern Australian coasts and is often seen by divers as it moves sinuously among kelp, skulling its pectorals and undulating its dorsal fin. Survey densities of up to 8 per 500 m^2 have been recorded in Victoria and about half that in South Australia (Jones 1992; unpublished data). However the true densities may be double these figures because the herring cale rests hidden in caves and crevices for half the day. The foraging behaviour of herring cale has been studied in New South Wales and South Australia (Andrew and Jones 1990; Shepherd and Baker 2008a) and it varies remarkably over its range.

In the Sydney region, herring cale feeds only on *Ecklonia* most of the year by taking feeding bites from the laminae (see Chapter 4), but in August to October, during the breeding season, females gather in groups and change their feeding behaviour. They eat the meristem (at the end of the stipe) and primary laminae of the kelp, in this way totally clearing patches of 1–100 m^2 in *Ecklonia* forests. Females then aggregate in these clearings for mating with males, who live in the surrounding forest, each aggressively territorial in his individual patch. Hence, as interpreted by Jones (1999a), kelp clearing and patch creation at this time of year are a by-product of the aggregation of females for mating. A rapid regrowth of young *Ecklonia* plants then occurs in the clearings and females apparently return each year to the same traditional clearings for spawning.

The above behaviour may be unique to the east coast because it has not been recorded on southern coasts. In Port Phillip Bay, Victoria, herring cale feeds extensively on crayweed *Phyllospora comosa*, and the herbivore's abundance was highly correlated with *Phyllospora* cover (Wheatley 2000). In South Australia, herring cale browses selectively on the secondary laminae of *Ecklonia*, where the sori (sites of spore release) are located. It also feeds on the receptacles and tips of fucoid algae (*Cystophora* and *Sargassum*) and sometimes on the green alga *Caulerpa* and the brown alga *Pachydictyon*. The daily foraging activity also varies between sites. At Althorpe I., near Spencer Gulf entrance, it emerged from its sleeping crevices at dawn and fed on *Ecklonia*, then retired to its crevices, only to re-emerge towards dusk to feed on fucoids. The overall diet comprised *Ecklonia* 72%, *Cystophora* 14% and articulated corallines 14%. At Flinders I., in the eastern Great Australian Bight (GAB) it gradually emerged during the morning and fed most intensely towards dusk (Fig. 16.1). There, the overall diet was *Cystophora* 48%, *Sargassum* 31%, *Zonaria* 11%, filamentous algae 5% and corallines 3%, consumed in accordance with the following preference order:

Sargassum spp. /*Cystophora monilifera* >
C. subfarcinata > *C. brownii*

But how does the herring cale cope with the high loads of polyphenolic chemicals (tannins) present in kelps and fucoid algae? Clements (1991) and Seeto *et al.* (1996) showed that this species has a unique digestive system. It has a short gut with an alkaline pH, a feeding apparatus that finely macerates the food algae and a complex microflora in the lower intestines, resembling that in termites. Together, these features enable the fermentation and digestion of mannitol and sugars, as well as tannins – all present in fucoids and kelps – even in cold Tasmanian waters.

Another interesting facet of herbivore feeding is the tendency for some species to feed maximally in the late afternoon. This has been explained by the diel feeding hypothesis, derived from the fact that algae have higher nutritional value in the afternoon as they accumulate photosynthates (Zemke-White *et al.* 2002). Optimal foraging theory predicts that herbivorous fish should take advantage of this and increase their feeding during the afternoon. The data so far for herring cale give some support for the theory, because an afternoon increase in foraging was seen at both sites, in addition to being high after dawn at one of the sites (Fig. 16.1). Yet, it is likely that the picture is more complex, with other factors (e.g. assimilation rates, and gut retention time) playing a role in feeding behaviour.

Growth and reproduction

Odacids are protogynous hermaphrodites, maturing first as females at ~25–30 cm length at a probable age of 3–4 years according to temperature, and changing sex 1–2 years later (Clements 2003). Longevity is 10–12 years. Herring cale has an August–October spawning season and eggs are released into the water where fertilisation takes place. Larvae settle in summer in kelp or fucoid forest.

Scalyfin *Parma* spp. (Family Pomacentridae)

Six closely related species of scalyfin occur in temperate Australian waters and here we shall discuss two of

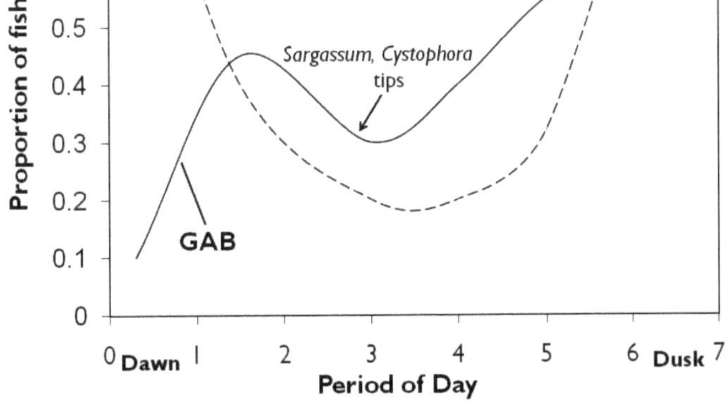

Figure 16.1: Smoothed curves showing the proportion of local populations of herring cale *Olisthops cyanomelas* emergent from shelter at Althorpe I. and Flinders I. in the eastern Great Australian Bight (GAB) from dawn till dusk. The algae selected at different times of day are shown for each site. Daylight hours were divided into seven equal periods and the proportion emergent censused 16 times during daylight hours (redrawn from Shepherd and Baker 2008a).

them: white-ear *Parma microlepis* from the SE Australian coast (Moran and Sale 1977) and Victorian scalyfin *Parma victoriae* from the southern coast (Norman and Jones 1984; Jones and Norman 1986; Shepherd *et al.* 2008). Scalyfin are all strictly territorial species that aggressively defend an area of reef containing a shelter hole, a roof area for nesting and a patch of turfing algae on which to browse. Females lay adhesive eggs, which are deposited in a nest and tended by the male until they hatch.

Scalyfin farm their 'garden' patch by weeding out unwanted brown algae that might otherwise dominate the patch and reduce the growth of preferred understorey red algae. This interesting behaviour has given rise to questions about their food preferences and territory sizes, as discussed below.

White-ear *Parma microlepis*

This species has a varied diet, feeding on benthic algae (66%) and invertebrates, mainly polychaetes (29%). The life history of white-ear has been well studied by Tzioumis and Kingsford (1995, 1999). The two sexes and juveniles are easily distinguishable by colour. Juveniles grow rapidly and by 2 years of age reach half their maximum size ($L\infty$) of 13.6 cm (Fig. 16.2). They mature sexually at an age of ~5 years and a length of 11–12 cm and may live for up to 37 years, although few exceed 22 years.

The breeding season extends from spring to early summer and spawning often occurs after a storm event, but is not synchronised within a population. During the breeding season, the male prepares a nesting site and the female then skims over the nest and deposits her eggs in a dense layer on the rocky bottom. The eggs have adhesive filaments that attach them firmly to the rock. The male then fertilises the eggs and vigorously guards them from the many possible predators and bottom feeders, such as crimson-band wrasse, senator wrasse, red morwong and rock cale. The vigorous aggression of white ears against predatory intruders was found by Moran and Sale (1977) to be related more to competition for space, especially during the breeding season, than to competition for food. Hence, they suggested that the purpose of territories was rather to defend their shelter holes, especially against egg-predators, than to protect a food supply.

The eggs hatch into pelagic larvae some 9–12 days after laying and the emerging larvae are pelagic for 2–4 weeks before settling in shallow reef habitats. After settlement, the small juveniles live in sheltered reef habitats until they can establish their own territories. Hence, the only dispersive stage for white-ear is

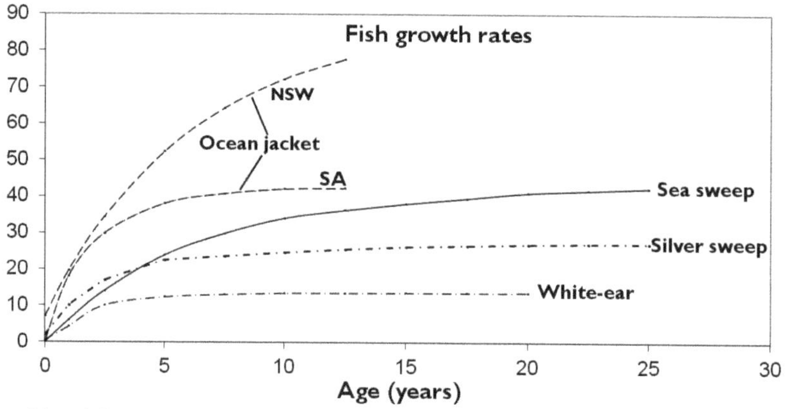

Figure 16.2: Growth rates of four fish species: a herbivore *Parma microlepis* (after Tzioumis and Kingsford 1999); two omnivores, silver sweep *Scorpis lineolatus* (after Stewart and Hughes 2005) and sea sweep *Scorpis aequipinnis* (after Saunders *et al.* 2010 and Coulson *et al.* 2012); and a carnivore, ocean jacket *Nelusetta ayraudi* at sites in the eastern Great Australian Bight, and off New South Wales (after Grove-Jones and Burnell 1991 for female fish and Miller and Stewart 2009 for both sexes).

during the brief pelagic larval stage. This life history prompted Curley and Gillings (2009) to examine the connectivity of white-ear populations along the New South Wales coast. They found an absence of genetic structure, suggesting that gene flow via dispersal of the larvae by the East Australia Current, coupled with inshore reverse flows was enough to prevent genetic differentiation.

Victorian scalyfin *Parma victoriae*

The distribution of this scalyfin is strongly influenced by reef habitat characteristics. Jones (1992) in a study along the Victorian coast found that average densities at 10 moderately exposed sites ranged from < 1 to 17 (average 6.9) per 500 m^2, compared with only 1.3 (standard error 0.1) per 500 m^2 at 148 similarly exposed sites throughout South Australia (Shepherd *et al.* 2008; unpublished data). Geoff Jones also found that scalyfin density:

- was inversely related to kelp cover, due to the fishes' preference for feeding below the kelp zone, where food algae were more abundant
- was directly related to the heterogeneity (i.e. the complexity) of the reef surface, which increased from near zero on smooth reefs to > 20 per 500 m^2 on highly complex reefs (see also Lincoln Smith and Jones 1995).

The low densities in South Australian waters are also likely related inversely to the high cover of fucoid and *Ecklonia* forests often to depths of 30 m or more, well beyond survey depths.

This scalyfin feeds only on red algae and Jones and Norman (1986) found in 'cafeteria' experiments, in which many different red algae were laid out on trays for the fish to choose from, that they recognised different species and were highly selective in their choice. However, at Pope's Eye at the entrance to Port Phillip Bay, where their studies were done, they found that territories were small (mostly only 3–6 m^2) and that in these small areas they had little choice but to eat the available algae in about the proportions in which they occurred. On the other hand, in the occasional larger territories, where the fish had more choice, they showed distinct preferences for particular red algae, such as *Champia* and *Rhodoglossum*.

Scalyfin densities were very high at Pope's Eye, due to the highly creviced habitat, and their territories were often contiguous. To test the factors determining territory size, the algal standing crop was experimentally reduced in some territories and scalyfin densities reduced in other areas. Reducing the food supply had no effect on territory size, but removal of neighbours resulted in the remaining scalyfins at once expanding their abutting borders into the newly vacant territory and so increasing their own territories sometimes by up to 300%. Norman and Jones (1984) concluded from these experiments that territory size was determined, not by the food supply, but by population density.

At two South Australian sites situated in *Ecklonia* forests, similar to that at Pope's Eye, but with far lower scalyfin densities, Shepherd *et al.* (2008) found that territory sizes averaged 11–13 m^2 and red algal food was twice as abundant as at Pope's Eye. At a third site in the eastern GAB, scalyfin territories were in a fucoid forest, where red algae were few and almost entirely epiphytic on the fucoids. Here, territories were > 80 m^2. How can these large differences in territory size be reconciled?

Numerous explanations for territory sizes of fish have been proposed (Grant 1997). Food supply and defence costs are critical factors (the resource defence theory) and food abundance will often determine territory size. The different findings of the above studies can be reconciled by distinguishing between two situations. At high scalyfin densities, fish are constrained within compressed territories due to pressure from their neighbours, so that territory size does not reflect food abundance. This was the case at Pope's Eye. But at lower densities the fish can expand their territories to match the available food supply. Under this model, scalyfin would select a territory size that reflects the available food, provided that the fitness benefits of defending the resource against intruders exceed the costs of defence.

The foraging tactics of scalyfin also vary according to territory size and food abundance (see Shepherd *et al.* 2008). In small territories, the species adopts 'saltatory' searching in which the fish darts here and there with many changes of direction, whereas in large territories it prefers 'cruise searching', in which the fish cruises in a straight line picking food as it goes. The former tactic is more efficient in small territories and the latter in large territories.

The Victorian scalyfin has a life history very similar to that of white-ear, with a similar growth rate and age of sexual maturity. However, the recorded longevity of 15 years was based on scale readings, which may under-estimate the true longevity (Brown 1982, cited in Tzioumis and Kingsford 1999). This scalyfin reproduces in early to mid-summer and when the female has laid her sticky eggs in the nest, carefully prepared by the male, he devotes his time to their paternal care. One monitored male fish at West I. spent 63% of the time fanning the eggs, either with a pectoral or caudal fin or by expelling water through the mouth, and 23% of the time driving off other fishes that approached too closely (Shepherd *et al.* 2008).

Drummer and blackfish (Families Kyphosidae and Girellidae)

These two closely related families have many species and are the most abundant herbivores on southern reefs, often foraging in schools of a hundred or more. All species have rows of closely set teeth that can be used for scraping filamentous algae from the reef surface. Yet their feeding activity is so spatially dispersed that they have no obvious effects on algal communities. The diet and feeding behaviour of the two families are better known than their life history, which has been studied only in a few species. One group, the kyphosids, has evolved a fermentation process to digest chemically defended brown algae. This process requires a constant body temperature and food supply, long food retention time, anaerobic conditions and control of gut pH, as discussed below.

Silver drummer or buff bream *Kyphosus sydneyanus*

This is the largest (to 85 cm) temperate Australian herbivore and is often seen in schools of 20–30 fish, which are attracted by the presence of a diver. Adult silver drummer, or buff bream as they are called in Western Australia, feed intensely at dawn and dusk on a diet primarily of brown and green algae (Table 16.1), whereas juveniles feed by day on more delicate green and red algae (Clements and Choat 1997; Moran and Clements 2002; Clements 2003). Silver drummer has a long gut ending in the hindgut chamber where symbiotic microbes ferment the plant material, enabling digestion of complex carbohydrates. Fermentation is a slow process, lasting > 24 h, hence the unusual dawn and dusk feeding behaviour that allows food algae to be broken down.

Silver drummer has a spring–summer spawning season and a long planktonic larval stage. Juveniles settle on inshore reefs and mature sexually at a size of ~35 cm at an age of 6 years. Longevity is > 60 years.

Table 16.1: Mean percentage gut contents of adult (> 30 cm) and juvenile (< 10 cm) silver drummer, *Kyphosus sydneyanus*.

Adult data from the coast of New South Wales (after Clements and Choat 1997) and juvenile data from New Zealand (after Moran and Clements 2002).

	K. sydneyanus	
Food	**Adult**	**Juvenile**
Brown algae – sundry	–	8
Dictyopteris sp.	0.8	–
Phyllospora comosa	31.6	–
Sargassum spp. (thalli)	26.6	–
Sargassum spp. (receptacles)	4.1	–
Green algae – *Ulva* sp.	0.9	–
Other green algae	35.3	18
Red algae	0.8	69
Crustaceans	–	2

Adults of a related species, western buffalo bream *Kyphosus cornelii* from the west coast of Australia, have a diet almost entirely of red algae. However, juveniles eat some animal matter, mainly crustaceans, which comprises up to 10% of their diet, before they are large enough to develop fermentative microbial digestion (Rimmer 1986, Rimmer and Wiebe 1987). A curious social behaviour of this species is the maintenance of cleared territories on the shallow rock platforms at Rottnest I., Western Australia and further north to the Abrolhos Is (Berry and Playford 1992). Adults were found to occur individually in many adjoining polygonal territories averaging ~12 m^2 in area, each distinctively bounded by a 'hedgerow' of *Sargassum* or other fucoid algae. Adults browsed red algae within their territories and aggressively defended them against intruders, whether neighbours or vagrants. At the same time, schools of several hundred vagrant buffalo bream lived independently in adjoining waters, and often trespassed into the defended territories taking a few bites of the turf, before being chased away by the 'owners'. In deeper water, buffalo bream congregated in nearby caves and crevices, and seemed to move and feed within defined home ranges.

The causes of the above social behaviour are still unclear. It is not related to spawning, because the species spawns year-round. More likely, the schooling behaviour of some provides a mechanism to circumvent the disadvantages of territoriality – which include increased mortality from ospreys (*Pandion haliaetis*) and competition from neighbours. Hexagonal territories will tend to arise in homogeneous environments with equally spaced neighbours.

Girellids

Of the five southern species in this group, the diet of three has been examined. These are: rock blackfish *Girella elevata* from the east coast; zebra fish *Girella zebra* from the south coast; and luderick *Girella tricuspidata* from the SE coast. All three species live on shallow, coastal reefs and feed on a range of algae (Table 16.2), but only the life history of luderick is known.

Table 16.2: Mean percentage gut contents by volume of adults of three species of *Girella* – two species from New South Wales (after Clements and Choat 1997) and one species from South Australia. B Saunders (*pers. comm.*)

Food	***G. elevata*** **rock blackfish**	***G. tricuspidata*** **luderick**	***G. zebra*** **zebra fish**
Green algae			
Caulerpa sp.	–	10.3	
Ulva sp.	20.2	–	3
Filamentous spp.	19.2	17.3	+*Struvea* sp.
Brown algae			
Ecklonia radiata	9.2	–	2[1]
Phyllospora comosa	20.8	–	
Sargassum sp.	7.2	2.5	
Red algae			
Filamentous species	21.5	18.6	92[2]
Foliose species	1.2	–	
Corallines	–	0.4	–
Detritus	+	35.0	< 1
Crustaceans	+	15.8	1[3]
Other animal matter	0.7	0.1	1[4]

[1]*Cystophora* receptacles; [2] 80% of the red algae were six species of delicate, branching species, including *Polysiphonia* sp., *Callithamnion* sp. and *Champia* sp.; [3]amphipods and caprellids; [4]ophiuroids and polychaetes.

Although microbial fermentation is important to silver drummer in digesting algae, it is not so for girellids, which function quite differently. The latter have shorter gut lengths and rely on their own digestive enzymes to hydrolyse algal polysaccharides, mainly from green and red algae (Clements and Choat 1997). Both plant and animal material are nutritionally complementary, with the latter selected to supplement the nitrogen-poor diet that would otherwise result from an algal diet (Raubenheimer *et al.* 2005).

Rock blackfish Girella elevata *and luderick* Girella tricuspidata

A seasonal study of the rock blackfish by Bell *et al.* (1980) gave a similar picture of its diet to that given in Table 16.2, but with more detail about the particular species eaten. In addition to algae, adult blackfish ate salps and ascidians (9%), hydroids (4%), bryozoans (3%) and sponges (2%), while juveniles (< 13 cm) ate crustaceans (8%), bryozoans (7%), ascidians (6%) and hydroids (2%). Overall, from juvenile to adult, animal material eaten declined from 58% to 12%, while plant material increased from 42% to 90%.

Luderick, described by Doak (1991) as 'a wary, furtive herbivore, rather like wild sheep', has a similar diet to rock blackfish (Bell and Pollard 1989). Like other herbivorous species, juveniles are largely carnivorous. Luderick spawn in late spring–summer and juveniles settle out of the plankton into estuarine seagrass beds at a size of ~2 cm in late summer. They mature at a size of ~25 cm at an age of 3 years, and live for up to 24 years in the heavily exploited estuaries of New South Wales and longer elsewhere; natural mortality rate estimates range from 0.16–0.3 (Clements 2003; Edgar 2008; Gray *et al.* 2010).

Zebra fish Girella zebra

B Saunders and SA Shepherd

Zebra fish forage in schools, and they typically swim about as a coordinated group throughout the water column. When one fish dives to the bottom and starts to forage, they all follow suit and then just as quickly resume swimming. About half an hour before sunset, the school breaks up and individuals head for caves and crevices where they spend the night, emerging at dawn to re-school. Their diet (Table 16.2) is largely red algae, browsed from the reef or from the seagrass *Amphibolis*. But they also feed opportunistically on planktonic species, such as salps and polychaete epitokes when they are prolific inshore. Juveniles (to 2 cm) feed on small crustaceans, polychaetes and filamentous green algae.

At West I., South Australia, during observations (by SAS) over many years, they seemed to follow the same course day after day over their extensive home range. They tended to arrive at the Island at about the same time every day and, after foraging briefly on the reef or on seagrass epiphytes, left to continue their course. In summer they were often infested with ectoparasites, and the presence of a cleaner station at the Island may have induced their regular daily attendance there around midday for cleaning by cleaner fish (Shepherd *et al.* 2005).

Tide-pools or the shallow subtidal zone provide nursery habitat for juveniles, and with increasing size they migrate into deeper water. Juveniles have been observed being cleaned by the shrimp *Palaemon serenus*.

Sea carp or marblefish *Aplodactylus* spp. (Family Aplodactylidae)

Three species occur in temperate Australian waters: the western sea carp *Aplodactylus westralis* on the SW coast; southern sea carp *A. arctidens* on the southern coast; and rock cale *A. lophodon* on the SE coast. Here, we discuss the southern sea carp, whose ecology is known from Clements' (2003) study in New Zealand.

The sea carp is a sluggish and solitary fish that rests by day at sheltered sites, but becomes active at dawn and dusk when it grazes non-selectively (like a lawn mower) on turfing algae. In one study of its diet, Russell (1983) found that its food comprised 15% brown algae (e.g. *Glossophora* and *Zonaria*), 80% red algae (mainly low, turfing species) and the remainder green algae (*Codium*). A later study by Choat and Clements (1992) found even more (~97%) red algae in the gut. The sea carp has both a gizzard-like stomach to break down the algae and a long, capacious gut that employs microbial fermentation to digest the algae (Mountfort *et al.* 2002).

Sea carp is a slow-growing species and spawns from winter to spring. Juveniles settle at a size of 6–7 cm after a prolonged planktonic phase and mature at ~35 cm length. The maximum size is ~60 cm and longevity is 30–40 years.

OMNIVORES

The commonest omnivorous reef fishes in southern Australia are sweep *Scorpis* spp., leatherjackets, and dusky morwong *Dactyophora nigricans*. Here, we discuss the ecology of two species of sweep, three omnivorous 'pickers' and 15 species of leatherjacket. The ecology of dusky morwong is virtually unknown.

Sweep *Scorpis* spp. (Family Kyphosidae)

Sea sweep *Scorpis aequipinnis* ranges throughout southern Australia on rocky reefs to depths of > 40 m, while silver sweep *Scorpis lineolatus* ranges along the east Australian coast south to Tasmania to ~30 m depth. Sea sweep reaches a length of ~47 cm (rarely to 50 cm) and has a longevity of 68 years, and silver sweep reaches a length of ~37 cm and lives to > 50 years (Stewart and Hughes 2005; Saunders *et al.* 2010; Coulson *et al.* 2012). Sea sweep has often been recorded as feeding on zooplankton, but Richard Saunders' study and that of Newman *et al.* (2005) recorded a mixed algal/animal diet or one of red algae (Table 16.3). The latter authors suggested that sea sweep feed throughout the day, with about a third of them feeding on planktonic organisms. The variability of prey items indicates high feeding flexibility, in which sweep variously feed both in open water and on the benthos. Such flexibility in response to the availability of different foods in different parts of the species' range is clearly adaptive, and an important aspect of foraging theory.

Both the habitat and diet of sea sweep change with increasing size. The smallest fish are found in small schools in sheltered caves on shallow reefs, where they feed substantially on small crustaceans. Older fish move to more exposed habitats, and algae comprise a greater proportion of the diet. A growth curve for both species of sweep is given in Fig. 16.2.

Sea sweep reaches sexual maturity at ~14–16 years at ~36–38 cm size according to sex, whereas silver sweep reaches sexual maturity at 2–3 years at ~17 cm size; this startling difference is thought to be attributable to selection pressures on silver sweep at a young age. Both species have a late autumn–winter spawning season. Larval life is short and larvae settle at a size of ~13 mm in rock pools or shallow water.

Both species of sweep have a schooling behaviour and are believed to be site-attached. Their long life renders them susceptible to serial depletion of local populations (as already recorded for silver sweep) and eventually overfishing.

The moonlighter *Tilodon sexfasciatus*, mado *Atypichthys strigatus* (Family Scorpididae), and western talma *Chelmonops curiosus* (Family Chaetodontidae)

The moonlighter or 6-banded coral fish ranges from SW Australia to the central Victorian coast and the western talma to the SE of South Australia. The mado ranges from central Victoria and Tasmania to SE Queensland. The first two species live on reefs to a depth of ~30–40 m and the third species to 150 m depth. They are all 'pickers', foraging mainly on the bottom or on vertical faces, but their diet can be quite variable. The moonlighter and talma are mainly microcarnivores, but they also ingest algae, while the mado is a planktivore that feeds extensively on the bottom.

The moonlighter *Tilodon sexfasciatus*

The time budget of this species was examined at two South Australian sites – Cape Jervis, SE Gulf St Vincent, and Althorpe I. off SW Yorke Peninsula at 3–5 m depth (Shepherd and Baker unpublished). The moonlighter sleeps at night in a shelter hole or cave and emerges ~30 min after sunrise to forage until ~50 min before sunset, when it retreats again to its shelter. The home range was ~1200 m^2 at Cape Jervis and ~400 m^2 at Althorpe I.

During the day, the fish spends its emergent time either foraging, often in groups of two or three, swimming about or resting on the bottom (Fig. 16.3). It

Table 16.3: Mean percentage diet of sea sweep *Scorpis aequipinnis* (6–38 cm length) in western Victoria (after Newman *et al.* 2005), moonlighter *Tilodon sexfasciatus* (unpublished data), western talma *Chelmonops curiosus* (after Currie and Sorokin 2010, and unpublished data), and mado *Atypichthys strigatus* (modified after Glasby and Kingsford 1994).

Glasby and Kingsford distinguished the diet of two size classes < 8 cm, but, being similar, they are combined here. Data are dry weight for sweep, percentage volume for moonlighter, western talma and mado.

Food components	Sea sweep	Moonlighter		W. talma	Mado	
		8–12 cm	15–20 cm		< 8 cm	8–13 cm
Seagrass	–	–	3.75	–	–	–
Green algae	8.3	–	2.5	–	–	–
Brown algae	0.2	18.2	10.1	–	–	–
Red algae						
(filamentous/ foliose species)	42.6	25.0	7.5	29.1	–	–
Algae (unspecified)	–	–	–	–	4.1	10.4
Sponges	–	15.0	20.0	–	–	0.2
Ascidians	–	–	–	50.0	–	–
Benthic animals						
Isopods, amphipods	3.6	26.2	13.3	–	4.5	2.3
Shrimps	–	10.0	6.3	–	–	–
Crabs	–	1.2	12.6	–	–	–
Molluscs	0.5	–	–	–	–	0.4
Echinoderms	–	–	3.8	–	–	–
Polychaetes	–	–	5.0	19.0	0.2	0.1
Other	1.8	–	–	1.8	1.0[1]	0.6[1]
Planktonic species						
Jellyfish	7.5	–	–	–	–	–
Egg casings (cephalopod?)	12.6	–	–	–	–	–
Crustaceans	–	–	–	–	86.4	79.0
Meroplankton[2]	–	–	–	–	3.2	7.0
Residue (unidentified/ detritus)	18.8	4.6	15.1	–	–	–

[1]parasitic species; [2]defined as eggs or larvae of benthic and nektonic species.

periodically retreats into a cave: the total daily retreat time was ~37% at Cape Jervis and 30% at Althorpe I. Foraging activity was highest during the morning and gradually declined towards dusk, when resting time reached a peak. During foraging excursions, the moonlighter divided its time mainly between algal turfs (43%) and fucoid forests (51%), and time foraging varied only slightly between the three size classes observed, 5–10 cm, 10–15 cm, and 15–20 cm. The total number of daily feeding bites varied little between the three size classes of fish (Fig. 16.4), Moonlighters feed on a wide range of invertebrates found in algae or among cave fauna (Table 16.3).

Moonlighters at each site were also part-time cleaners and, during summer months, regularly cleaned up to ~20 species of fish of their ectoparasites (Shepherd *et al.* 2005). Cleaning only took place after midday and was done mainly by juveniles (Fig. 16.4). Cleaning behaviour is discussed in more detail below.

Western talma *Chelmonops curiosus*

Like the moonlighter, western talma sleeps in a shelter hole and forages in a small home range of up to ~500 m^2 by day. The talma emerges from shelter soon after sunrise and forages till an hour or so before sunset. Feeding peaks during the morning and gradually declines during the afternoon, as resting time increases (Fig. 16.3). At three sites in eastern Gulf St Vincent, where its behaviour was observed for ~5 h per site from dawn to dusk, 42% of feeding bites were made among the sessile fauna on vertical faces or in caves, 52% in turf algae and the rest in sand or in the canopy of fucoid algae. This talma usually forages alone or with a partner, especially during the breeding season. The diet is probably broader than shown in Table 16.3, because casual observations show that they may also forage on small crustaceans and sponges.

Mado *Atypichthys strigatus*

This small (usually < 20 cm) species is often the most abundant fish on reefs in coastal New South Wales, where it usually occurs in large schools. Its diet (Table 16.3) is flexible because it feeds both on plankton (mainly crustaceans) and benthic invertebrates. Cleaning parasites from other species of reef fish provides a small dietary supplement. The mado is often seen in disturbed habitats and, to discover the reason for this, Glasby and Kingsford (1994) experimentally cleared patches of kelp, and created 'disturbed' barrens habitats. They found that mado had a marked response to disturbances, which temporarily flushed out benthic invertebrates, on which they fed voraciously. Not surprisingly, mado are also secondary feeders, following eastern blue groper and red morwong, and feeding on invertebrates flushed out by them or on left-over urchin and molluscan remains – food they could not otherwise access with their small mouths.

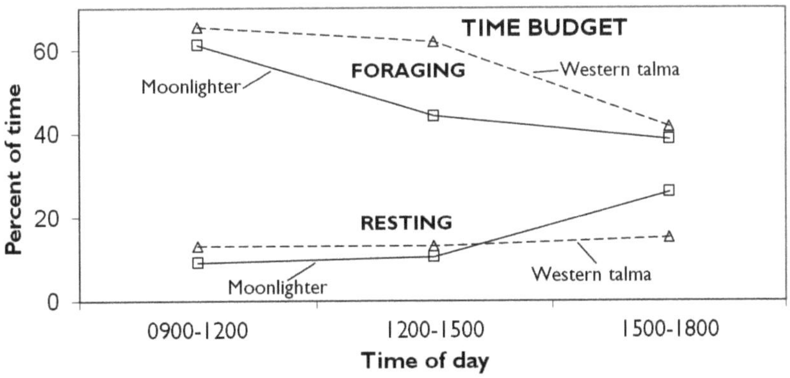

Figure 16.3: Averaged percentage of time spent foraging and resting by moonlighters *Tilodon sexfasciatus* at Althorpe I. and Cape Jervis, South Australia (□), and western talma *Chelmonops curiosus* (Δ) at three sites in eastern Gulf St Vincent for three time periods of the day. Swimming time for both species (not shown) was ~25–30%. Each point is the mean of 18 replicates for moonlighters and 36 replicates for western talma (after Shepherd and Baker unpublished data).

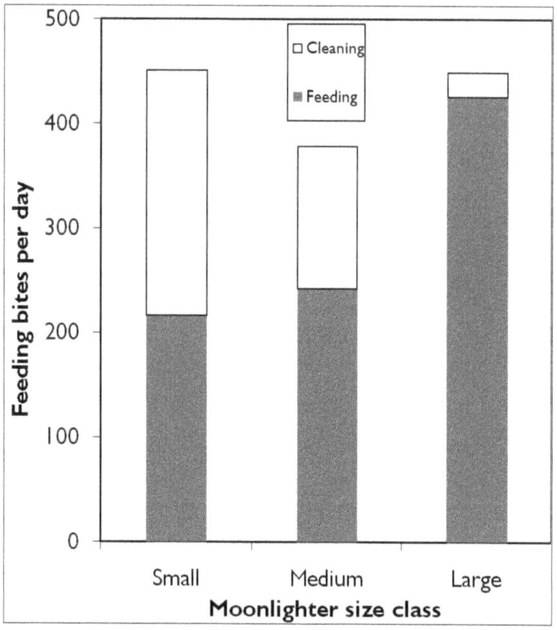

Figure 16.4: Mean total number of feeding bites during foraging, and cleaning bites during cleaning operations, of three size classes (5–10 cm, 10–15 cm and 15–20 cm) of the moonlighter *Tilodon sexfasciatus* at Cape Jervis (Shepherd and Baker unpublished data).

Leatherjackets (Family Monacanthidae)

Leatherjackets are a highly diverse group with 22 species in the southern half of Australia, 21 of which are endemic. They live on coastal reefs, in seagrass or on sandy bottoms, often to 100 m depth. One species, the ocean jacket *Nelusetta ayraudi*, as its common name indicates, is oceanic and lives its adult life in deep continental shelf and slope waters to 350 m deep. Another species, the mosaic leatherjacket *Eubalichthys mosaicus*, has an extraordinary symbiotic association as a juvenile. As larvae approach the settling stage, they seek out particular species of jellyfish, notably *Catostylus mosaicus* and *Chrysaora southcotti*, and shelter among the tentacles under the jelly's bell. Here they feed in part on tiny crustaceans caught by the jellyfish, as well as on some of its tentacles, until they are large enough to find suitable bottom habitat (Southcott 1987; Hutchins 1999).

Leatherjackets have separate sexes, often with different colour patterns and sometimes with intermediate colours – which can easily cause confusion in identification. The life history of three species is reasonably well known, as described below.

Food and feeding behaviour

The broad diet of leatherjackets may be the secret of their success in southern waters. The diet of 14 species (Table 16.4) shows that all except the ocean jacket are omnivorous. In the one known example of a juvenile diet (*Monacanthus chinensis*), the animal component was much higher than in adults, as we have noted above for herbivorous fishes. For two species (horseshoe and toothbrush leatherjackets), diets have been examined in different regions within a State, and these were markedly different, reinforcing the view that leatherjackets are opportunistic omnivores and scavengers, ingesting whatever is available. Leatherjacket foraging can have a profound effect on sessile communities. Russ (1980) found that juvenile toothbrush leatherjackets ate large amounts of colonial ascidians (40% of the gut contents), so preventing their monopolisation of the habitat.

The foraging behaviour of three species is known for sites in New South Wales, South Australia and Tasmania – the fanbellied leatherjacket in seagrass beds in Botany Bay (Conacher *et al.* 1979), the horseshoe leatherjacket at Althorpe I. (unpublished data), and juvenile bridled leatherjacket in Tasmania (Robertson and White 1986). The first two studies found that fish foraged during daylight hours and retired at night to rest in seagrass (Botany Bay) or in reef crevices (Althorpe I.). In Botany Bay, feeding peaked during the morning and gradually declined during the afternoon, whereas at Althorpe I. there were two feeding peaks, one in the morning and another during the mid afternoon (Fig. 16.5). The third study on bridled leatherjacket found a quite different pattern. Juveniles were omnivorous, feeding mainly on red algae and small crustaceans immediately after sunrise and towards dusk.

A curious behaviour of the horseshoe leatherjacket at Althorpe I. was its propensity for secondary feeding. Fish frequently followed western blue groper and fed on material stirred up from the bottom by the groper's feeding or on material under boulders which the groper overturned (see Shepherd and Brook 2005). When threatened by larger conspecifics or

Table 16.4: Diet (percentage volume in guts) of 14 leatherjacket (LJ) species.

Gut sample sizes were mainly in the range 10–40. Ad. = adult; Juv = juveniles. The 'Worm' group includes polychaetes, sipunculans and echiurans. Where the main food eaten is a single group, this is indicated below, e.g. Corall = coralline algae; Urchin = echinoids; Bivalv = bivalve molluscs. Percentage of diet unaccounted for is sediment, shell fragments, forams or unidentifiable material. In multiple studies, dietary data are given in the order of the references.

Species	Sea-grass	Algae	Sponges	Worm	Bryozoans	Cnidarians	Crustaceans	Echinoderms	Molluscs	Ascidians
Bridled LJ[1,6]	3.6	32	–	24	–	20	12	0.7	9.4	–
Acanthaluteres spilomelanurus	–	46.3	–	–	1.7	2.4	7.5	–	7.5	–
Toothbrush LJ[1,5,6]	16.1	57.1	4.0	1.6	0.4	16.3	0.6	6.1	0.7	–
Acanthaluteres	–	10.1	53.0	–	1.0	–	–	–	–	36.7
vittiger		42.9	6.3	0.8	3.5	11.8	23.8	–	4.0	–
Spinytail LJ[2]	–	98	1	–	–	–	+	–	–	1
Acanthaluteres brownii		Corall								
Pygmy LJ[1]	27.0	–	–	13.5	–	–	54.1	5.4	–	–
Brachaluteres jacksonianus								Urchin		
Degen's LJ[1,9]	2.2	26.9	0.3	11.1	16.2	5.5	28.8	–	7.1	–
Thamnaconus degeni	20.1	6.0	–	–	6.2	–	6.4	7.3	25.4	–
Fanbellied LJ3,4 Ad	30.5	29.5	–	10.9	–	1.3	7.8	–	9.8	0.5
Monacanthus chinensis Juv	13.8	18.2	–	–	12.7	–	55.4	–	–	0.1
Rough LJ[1,7,9]	3.1	20	6.4	3.4	6.3	10.3	9.9	15.6	8.3	5.8
Scobinichthys	43	5	–	38	–	–	14	–	–	–
granulatus	5.5	8.4	–	3.1	3.8	4.8	2.3	7.1	21.9	–
Brown-striped LJ[6]	–	11.9	25.5	0.4	1.1	9.6	6.5	12.2	19.9	–
Meuschenia australis						Hydroid		Urchin		

(*Continued*)

Table 16.4: (Continued)

Species	Sea-grass	Algae	Sponges	Worm	Bryozoans	Cnidarians	Crustaceans	Echinoderms	Molluscs	Ascidians
Six-spine LJ[2,3,6]	–	5	40	–	–	10	–	30	–	–
Meuschenia freycineti	64.8	11.4	–	0.4	–	11.8	1.7	–	6.1	0.4
	–	23.7	10.2	0.5	0.9	0.9	7.8	27.8	24.5	0.9
Horseshoe LJ[2,5,10]	–	3	75	–	–	3	+	10	~3	2
Meuschenia hippocrepis	–	45.0	–	3.8	–	2.5	10.3	–	15.4	10.0
		11.5	74.3	7.5	1.0	–	0.1	–	3.5	1.7
Velvet LJ[1]	3.7	–	10.7	13.9	65	–	2.4	0.5	4.2	–
Meuschenia **scabra**									Bivalv	
Yellow-fin LJ[3]	30.2	55.0	–	0.4	–	–	3.2	–	3.2	–
Meuschenia trachylepis										
Ocean Jacket[1,8]	–	–	–	–	–	–	1.4	–	98.6	–
Nelusetta ayraudi		0.3	–	–	–	–	12.7	13.5	Cephal	–
Gunn's LJ[1]	–	16.7	33.3	33.3	–	16.7	–	–	–	–
Eubalichthys gunnii						Hydroid				
Mosaic LJ[1]	0.3	23.5	47.8	–	7.2	8.5	–	–	0.5	12.1
Eubalichthys mosaicus						Hydroid				

[1]Currie and Sorokin (2010); [2]Coffin Bay data – B Saunders (*pers. comm.*); [3]Bell *et al.* (1978); [4]Conacher *et al.* (1979); [5]Althorpe I. data – S.A. Shepherd unpublished; [6]Last (1983); [7]Burchmore *et al.* (1984); [8]Lindholm (1984, but not included in Table are salps (Class Thaliacea) 19%, and fish 23%; [9]I Svane: data from two sites in Spencer Gulf (*pers. comm.*); [10]data from > 20 m depth – Rodgers *et al.* (2013).

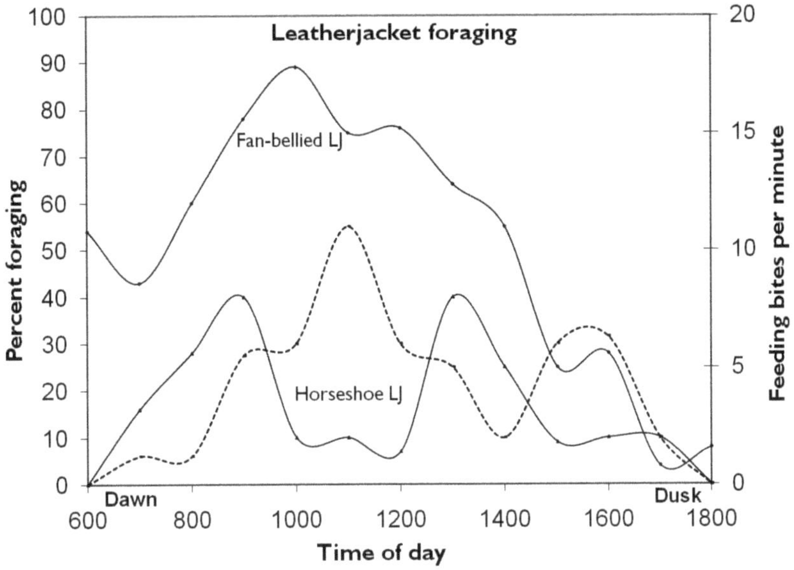

Figure 16.5: Daily foraging behaviour of fan-bellied leatherjacket *Monacanthus chinensis* (upper), and horseshoe leatherjacket *Meuschenia hippocrepis* (lower two graphs). Data for fan-bellied leatherjacket is the percentage of fish feeding at different times of day (after Conacher *et al.* 1979). Data for horseshoe leatherjacket are the mean percentage of the time of individual fish engaged in foraging (as opposed to resting or swimming). Each data point is the mean of three 5-minute records of the activity–time budget of different individuals. Feeding bite rates during foraging is shown with a dotted line (unpublished data).

competitors, it erected its dorsal spine as a threat or defensive display.

Movement

The movement of leatherjackets has been studied by several authors, who have examined whether they have territories (which are defended against all intruders) or home ranges (which are shared with conspecifics) (Thomson 1959; Barrett 1995a; Wheatley 2000; Edgar *et al.* 2004). The last three studies on the toothbrush (*Acanthaluteres vittiger*), six-spine (*Meuschenia freycineti*), horseshoe (*Meuschenia hippocrepis*) and brown-striped (*Meuschenia australis*) leatherjackets showed that they were site-attached to a reef. The home ranges were < 2500 m^2 for toothbrush and ~1 ha for brown-striped, but in general they stayed within a radius of ~100–200 m of the home reef. A gap of sandy bottom > 50–60 m wide between reefs proved sufficient to deter movement from one reef to another.

Similarly, Jim Thomson's study in Lake Macquarie, New South Wales on five species (fanbelly, yellow-fin, rough, mosaic and pygmy leatherjackets) found that, in the short term, movement was mainly localised. However, a few individuals moved distances of up to 10 km within the Lake over several months and during the spring–summer spawning season some undertook extensive migrations out of the Lake and northwards for tens of kilometres along the coast.

Studies of the fish fauna in seagrass habitats in New South Wales (Bell *et al.* 1978; Bell and Pollard 1989) noted that the larvae of three leatherjacket species examined by them (see Table 16.4) settled in seagrass beds in estuaries, and subsequently migrated into reef habitats on more exposed coasts. The ocean jacket (see below) follows a similar, but far greater, migration pattern.

Leatherjacket life histories

The ocean jacket Nelusetta ayraudi

Unlike most other leatherjackets, which live in coastal waters, the adult ocean jacket lives in shelf slope waters to > 350 m depth in southern and eastern Australia to northern New South Wales (Grove-Jones and Burnell 1991; Miller and Stewart 2009). Its habitat is near the seabed on reefs, rhodolith substrates, sandy bottoms or in sparse seagrass, where fishes of about the same size-classes often form schools. The female:male sex ratio is ~2:1 and, upon reaching sexual maturity at ~3 years of age, they form spawning aggregations, with multiple spawnings annually. These occur in late April–May in the eastern GAB, in August in northern New South Wales, and ~2 months later in southern New South Wales. Females produce ~1.1 million eggs at ~50 cm length and age 5 years, and ~2 million at ~80 cm length and at age ~10 years.

In the eastern GAB, the eggs and hatching larvae are thought to float and be carried eastwards to the coast by the Leeuwin Current during winter, so that juveniles eventually settle in inshore waters some 6 months later. Rod Grove-Jones and Andy Burnell monitored their growth in the bays of southern Eyre Peninsula, and found that they remained there and grew to ~18 cm size in the first year. From an age of ~1 year, they migrate seawards to deeper inner shelf waters, and by 2 years of age live further offshore at a depth of 40–80 m, where they sometimes form mid-water schools. A westerly pre-spawning migration starts in late summer, peaking in April, when they swim for distances of > 70 km, at speeds of ~6 km day^{-1}. A return migration then starts in May, but they swim only about half the above distance at a much slower speed. With increasing age, they move further down the shelf slope > 200 km offshore, while others have been recorded swimming for up to 1000 km south-easterly along the shelf edge to the SE of South Australia.

According to Grove-Jones and Burnell (1991), the ocean jacket grows rapidly to a size of 40–42 cm, and lives for ~10 years (Fig. 16.2). However, their studies were in the shallower part of the ocean jacket's depth range and, in view of the records of fish up to 1 m long (May and Maxwell 1986), it is likely that they live considerably longer and grow to a larger size. In the eastern GAB and New South Wales, the estimated annual mortality rate was 65–69%, of which fishing accounted for ~25–50%.

The ocean jacket is unique in that it is the only carnivorous species of leatherjacket (Table 16.4). Yet Lindholm (1984) noted that it had a long, coiled intestinal tract, usually associated with herbivorous species, and suggested that it was an adaptation to ingesting a large amount of food – favouring a fast growth rate.

A large fishery for the ocean jacket existed in New South Wales in the early 1970s, and in the eastern GAB in the late 1980s, each with annual catches of 800–1200 t. Both fisheries came close to collapse after a few years, but since 2001 the New South Wales fishery has recovered somewhat, with annual catches of ~400 t (Miller and Stewart 2009). In South Australia, the fishery never recovered, but a trawl fishery later developed with annual ocean jacket catches of ~200 t and discards of another 200 t (Moore *et al.* 2009). No size or catch limits exist anywhere in Australia, although both the above studies warned that the ocean jacket is vulnerable to overfishing.

The toothbrush leatherjacket Acanthaluteres vittiger

This species lives in a range of habitats – reef, seagrass and sandy bottoms – and has a diet that varies accordingly (Table 16.4). It has separate sexes, and a fast growth rate (Fig. 16.6), but is short lived, with males living for 2–3 years and females for ~5 years (Barrett 1995b). In Tasmania, juveniles first appear in autumn at ~6 cm size in near-shore *Phyllospora* forests and reach sexual maturity at an age of 2 years at a size of

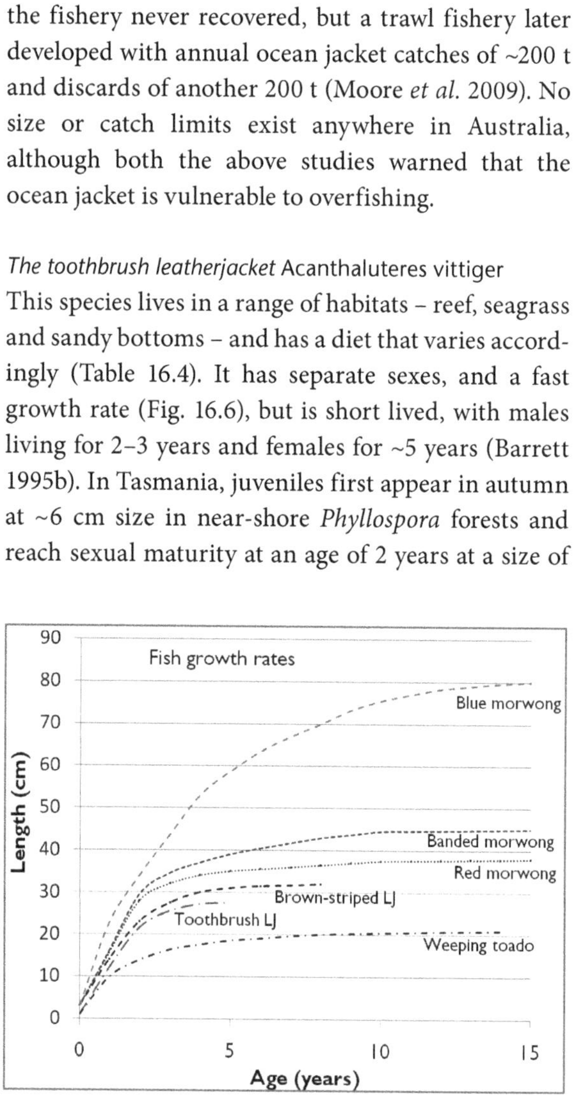

Figure 16.6: Growth rates of blue morwong (male) *Nemadactylus valenciennesi* (after Coulson *et al.* 2010), banded morwong *Cheilodactylus spectabilis* (after Ewing *et al.* (2007), red morwong (both sexes) *Cheilodactylus fuscus* (after Lowry 2003), male brown-striped leatherjacket *Meuschenia australis* and female toothbrush leatherjacket *Acanthaluteres vittiger* (both after Barrett 1995b), and weeping toadfish *Torquigener pleurogramma* in Gulf St Vincent (after Humphries 2008).

~22 cm. Males court females assiduously during the spawning season from September to January, often nudging or nipping them and displaying 'fluttering' behaviour (i.e. contractions of the body in vigorous 5-second spasms).

The brown-striped leatherjacket Meuschenia australis

Juveniles of this species settle in September, and it has a similar growth rate to the toothbrush leatherjacket (Fig. 16.6), with males growing ~3 cm larger than females. Longevity is 7–8 years for males and ~9 years for females. The spawning season also occurs from spring to mid-summer (Barrett 1995b).

The six-spine leatherjacket Meuschenia freycineti

The spawning season of this species is from October to March and post-larvae settle in shallow waters at 1–2 m depth within a few weeks and then swim into adjoining seagrass habitat. They remain in this habitat for about a year and then move into reef habitat at a size of ~20 cm. The growth rate (not shown) is very similar to that of the brown-striped leatherjacket (Jenkins and Wheatley 1998; Wheatley 2000).

BENTHIC CARNIVORES

In reef communities, carnivorous fishes are prominent, comprising > 80% of all fish species. These can be divided into open-water feeders and benthic feeders. Open-water feeders include mid-water species that prey on many smaller fish and zooplankton (see Chapter 17). Benthic feeders consume a wide range of sessile and mobile invertebrates associated with reefs and may also range into adjoining sandy or seagrass habitats to feed. They can be roughly subdivided into microcarnivores and macrocarnivores.

Benthic feeding microcarnivores

Blennies, triplefins and weedfishes

This is a heterogeneous group of common, but small and inconspicuous, fishes that live in cryptic habitats among algae on reefs or in sand. Most species rarely exceed ~12 cm in size, many are undescribed and their ecology is largely unknown. Here, we briefly discuss the life histories of nine species of *Heteroclinus* from Tasmania (Gunn and Thresher 1991). These species all feed on a crustacean diet of amphipods and small decapods. The sexes are separate, but they are unusual among fishes in that they are all viviparous. Males have testes and a strong muscular intromittent organ with which to fertilise females internally, apparently during late winter. During the reproductive season (spring–mid-summer), females brood batches of eggs in follicles within ovisacs. Females give birth to batches of larvae almost fortnightly throughout the long reproductive season. After birth, the larvae swim actively near the bottom, feeding on minute planktonic crustaceans and develop into juveniles over the next 3–7 weeks, according to species. They settle in algae or seagrass at a size of 1–2 cm and then start to eat small mysids and amphipods. They grow rapidly to reach ~5 cm length in 3 months, and 8–10 cm in a year. Reproductive fish are all 1 year old, indicating that they die at the end of one reproductive season. Viviparity (giving birth to live young) is rare among fishes due to the high cost of low fecundity. These clinid fishes have evolved to minimise this cost by simultaneously developing multiple batches of eggs and embryos over a prolonged spawning season.

Pufferfishes or toadfishes (Family Tetraodontidae)

Ten species of puffers are seen on reefs and sandy bottoms in southern Australia. They all have teeth fused into a solid beak, enabling them to consume a wide range of bottom fauna. The diet of one southern species, weeping toadfish or toado *Torquigener pleurogramma* (Plate 56; Table 16.5), suggests it is opportunistic, feeding indiscriminately on whatever is available – mainly small molluscs and crustaceans. Potter *et al.* (1988) also studied the diet of juveniles and adults, and found that both age groups fed mainly on polychaetes and amphipods.

Pufferfishes have two peculiarities. First, if disturbed, they can inflate their swim bladders with water, so increasing their body size, and second they accumulate tetrodotoxin (TTX) in their liver, ovary and skin. Both features are a likely defence mechanism to deter predators, one of which is the gummy shark.

The way in which puffers accumulate TTX is curious. Like the blue-ringed octopus (see Chapter 15), they cannot produce this toxin themselves, but take

Table 16.5: Diet (percentage volume) of a puffer, the weeping toadfish *Torquigener pleurogramma* (after Humphries 2008), and three morwongs – magpie perch *Cheilodactylus nigripes* (after Cappo 1995), red morwong *Cheilodactylus fuscus* (juveniles < 20 cm after Lockett and Suthers 1998; adults after Bell 1979), and banded morwong *Cheilodactylus spectabilis* (adults) (after Leum and Choat 1980).

Food type	Toadfish	Magpie perch	Red morwong		Banded morwong
			Juvenile	Adult	
Amphipods, isopods	24.5	26.6	54.2	15.6	68
Ostracods, tanaids, cladocerans	–	7.5	12.8	–	–
Mysids	–	2.7	0.1	–	–
Decapods (crabs, shrimp)	37.3	3.2	1.2	25.3	11
Polychaetes	1.5	10.0	12.3	24.0	4
Molluscs	33.9	6.6	4.1	21.3	2
Sea-urchins	–	–	0.1	4.6	2
Sea-stars, brittle-stars	–	–	1.2	1.0	9
Sipunculans	–	–	–	4.2	–
Ascidians	–	–	0.4	1.5	–
Cnidarians (hydroids, anemones)	–	–	0.1	0.7	–
Other	–	4.5	1.6	1.2	4
Sediment; unidentified	2.8	38.9	11.9	0.6	–

up TTX-producing bacteria from their prey. Noguchi *et al.* (2006) suggested that puffers even select prey which contains the toxic bacteria. Even more curious is the fact that pufferfishes with their TTX are a favoured delicacy – called fugu – in Japan.

The population biology of the weeping toadfish is known from the work of Potter *et al.* (1988) and Humphries (2008). The toadfish has a December–January spawning season in Western Australia, and juveniles enter estuaries as juveniles, frequenting the shallows for ~7 years before returning to fully marine conditions. They mature sexually at ~2 years of age. This toadfish is also common in the South Australian gulfs where they grow to a size of ~21 cm and have a longevity of ~14 years (Fig. 16.6) – twice that of Western Australian weeping toadfishes. Joshua Humphries found that the natural mortality rate was 30–36% a year, but the cause of death is a mystery, because the only known predator in the region, yellow-tail kingfish, is depleted through overfishing.

Seahorses and seadragons (Family Syngnathidae)

This group of sluggish fishes is extraordinary in several ways. They have:

- an upright posture and a long, tube-like snout used to catch sessile and other invertebrates
- a reproductive system that is unique among fishes.

They live among dense stands of algae or in seagrass and, being weak swimmers, rely on camouflage and an outer covering of bony armour for survival. The group has been a life-long interest of Rudie Kuiter, whose impressive works (Kuiter 1987, 2003) have been drawn on for this account. The family is fully protected in most southern States, due to potential over-exploitation in the Chinese medicine and aquarium trades. We discuss here two charismatic species: the leafy seadragon *Phycodurus eques* and the weedy seadragon *Phyllopteryx taeniolatus*. With few exceptions, notably the cave-dwelling sawtail pipefish *Maroubra perserrata* and upside-down pipefish *Heraldia nocturna*, temperate pipefish live in seagrass habitats and are not considered here (see Howard and Koehn 1985).

Leafy seadragon Phycodurus eques

This iconic species ranges along the southern Australian coast from the Abrolhos Is south and east to Wilsons Promontory. Its ecology has been studied by Rod Connolly and associates in Encounter Bay, South Australia (Connolly *et al.* 2002a, b). The leafy seadragon inhabits kelp and fucoid forests and seagrass (or at the boundary between them), usually at depths of 5–25 m. Here, the strap-like fucoids *Seirococcus* or *Scytothalia* often occur, and the leafy frequently rests under these algae, where its appendages closely mimic the algal straps – a camouflage first noted by Tenison-Woods (1883): '. . . their tattered cerements are like shape and colour to the seaweed they frequent, so that they hide and feed with safety. The long ends of ribs which seem to poke through the skin to excite our compassion are really protective resemblances, and . . . allure the prey within reach of these awful ghouls'.

Rod Connolly and colleagues found that individual leafies were recognisable by their facial markings, so enabling the first and only estimation of total abundance at a site. At the West I. Reserve, this was estimated to be ~57 fish ha^{-1}. Individuals marked with ultrasonic tags moved within well defined home ranges, mainly < 5 ha. Most remained sedentary for long periods of time (up to almost 3 days), and then suddenly took off on a long-distance swim – usually < 70 m, but occasionally > 200 m. This could happen at any time of day or night. During tracking over 10 days at West I., nine leafies spent ~4% of the time among brown-algal covered boulders, 15% among fucoids/kelp, 10% on sand and 62% in seagrass – showing some preference for the seagrass *Posidonia*. But why should they prefer to be in seagrass?

The answer lies both in the behaviour of a major prey, mysids, and the mode of attack by leafies. Mysids prefer to swarm on sandy bottoms, which at Connolly's study site is fringed by seagrass or macroalgae and, if attacked, they break into small groups and escape into seagrass (see Chapter 8). The leafies are ambush predators (i.e. they remain motionless) and to capture a mysid (or a shrimp) move only their snout with a rapid head-flick. So they prefer to remain at the edge of a seagrass bed or algal forest, where they are well hidden and can strike at mysids at the edge of a swarm. Flynn and Ritz (1999) showed that seahorses had 80% success rate in attacks from a vegetation vantage point, compared with 20% success in open habitats. Very large individuals, however, may capture large shrimp, prawns and even squat lobsters (Kuiter 2003; Currie and Sorokin 2010).

The mating season of leafies is primarily November–December and individuals are thought to be monogamous at least for one season (Wilson *et al.* 2003). In anticipation of mating, males and gravid females often congregate in October. The male tail swells and becomes a bright yellow and he pairs with a female in a prolonged courtship ritual that may last many days. When they are both ready, the female swims up against the male and manoeuvres her body to deposit 200–300 eggs on his tail, from the tail-end forwards, pushing them into place with her body. The soft skin of the male then folds around the eggs and holds them secure and the male fertilises them, possibly via ducts in the pouch. After an incubation of ~8 weeks, the eggs hatch over about a week and hatchlings emerge tail-first. The male distributes them over the bottom and the young swim off near the bottom into shallower depths of 5–7 m, where they often form small groups. After ~2 days from hatching, they start to feed and grow rapidly, reaching 8.5 cm length in 3 weeks and 30 cm in a year (Fig. 16.7). By this time, they become sexually mature and some move offshore into deeper water. They reach full size (~43 cm) in 2 years, and have a lifespan probably exceeding 10 years.

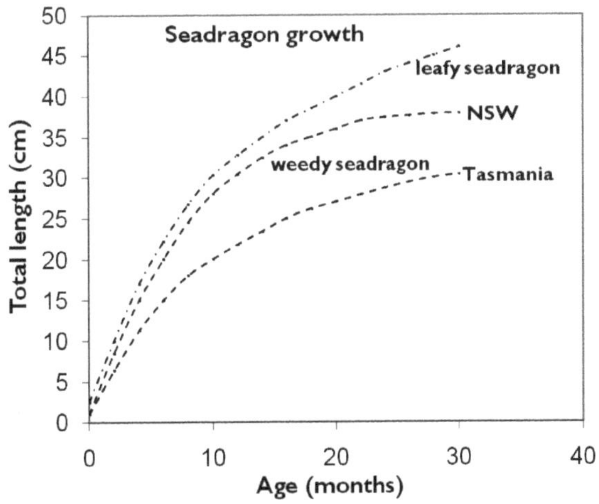

Figure 16.7: Average growth rates of the leafy seadragon *Phycodurus eques* at Victor Harbor, South Australia (after Kuiter 1987) and the weedy seadragon *Phyllopterix taeniolatus* in New South Wales and Tasmania (after Sanchez-Camara *et al.* 2011).

Weedy seadragon Phyllopteryx taeniolatus

This species has a southern range from Port Macquarie, New South Wales to Geraldton, Western Australia. Its habitat is in macroalgal beds of *Ecklonia* or fucoids, or in seagrasses, and has a home range of ~5700 m^2 that overlaps between males and females. Martin-Smith (2011) showed that individual weedies could be photo-identified by abdominal markings and from this he estimated densities of 20–60 ha^{-1} in Tasmania, similar to those in New South Wales. In the only dietary study of the species, mysids comprised 89% of prey consumed and other small crustaceans the remainder (Currie and Sorokin 2010).

In New South Wales, the weedy has a 6-month breeding season, from June to December, with a peak (55%) in male pregnancy in the last 2 months (see Sanchez-Camara and Booth 2004, Sanchez-Camara *et al.* 2005 and Forsgren and Lowe 2006).

In Tasmania, males incubate eggs from October to March and at least two clutches could occur in the breeding season (Martin-Smith 2011).

Before breeding starts, a male courts a female for 2–4 weeks with an elaborate and prolonged ritual. Courtship involves tail curling behaviour – in which the male approaches a female and curls his tail away from her and, if agreeable, she responds with the same tail-gesture. In this posture, the pair then swim parallel, both nodding their heads to and fro towards each other. This behaviour is repeated by the pair many times during the day until dusk, but sometimes the female may switch to courting another male. A few days before mating, female behaviour changes and she starts swimming up into the water column whenever a male approaches. Eventually he 'takes the hint' and follows her upwards, and then they progressively become closer to one another. Finally, the male tries aligning his brood patch with her cloaca. He practises this for 3–4 days, till the eggs can be seen within her enlarged cloaca. At last they make contact and the female transfers the 100 or so eggs to his brood patch, and they separate. After receiving the eggs, the male shakes his tail, swims in circles and fertilises them with sperm from the vas deferens duct, which is very close to the cloaca.

This species does not form pair bonds, and a female may court two males and transfer eggs to both of them within 2 weeks. The gestation period lasts 30–38 days and most males have a single pregnancy, although two pregnancies ~2 months apart can also occur. In both New South Wales and Victoria, males tended to move into shallow water in their last stage of pregnancy, presumably to provide a nursery habitat for the young in macroalgal beds.

Kuiter noted that toadfishes and leatherjackets are predators of eggs attached to the male weedy and he observed parts of the weedy's tail bitten off together with part of the egg-mass. To avoid this, some pregnant males carry algae on and around the eggs to help conceal them from such predators. Embryos are retained to an advanced stage of development and they hatch as juveniles and promptly start feeding. The hatching and growth of juveniles are coordinated with the life cycle of their main food – mysids.

In New South Wales, the weedy has a maximum size of ~42 cm, but in Victoria and Tasmania it reaches a larger size of ~46 cm at a slower growth rate (Fig. 16.7), and the reproductive season is ~3 months later. Here, annual survival is 82–85% and longevity is > 10 years – both almost double those in New South Wales (Sanchez-Camara *et al.* 2011; Martin-Smith 2011); these differences are in agreement with the metabolic theory of ecology discussed below.

Morwongs (Family Cheilodactylidae)

Morwongs are a prominent part of the temperate Australian fish fauna, living on reefs and sometimes in seagrass. The family is characterised by its large pectoral fins and thick fleshy lips, which are adapted for foraging among algal turfs. The fish employ the 'suck and sort' method of feeding by which the mouth is rapidly opened to produce a powerful suctorial force (also a loud crack) as it ingests small invertebrates and sediment, the latter being ejected through the gill chambers or the mouth.

Members of the family are notable for their long pelagic larval stages, often lasting 7–10 months. The late larval stage has a very thin, laterally compressed body and is known as 'paper-fish', and sometimes found on shallow reefs (McCormick and Francis 2003).

Here, we consider four of the eight southern species, aspects of whose ecology are known (Lowry and Cappo 1999). A fifth species, jackass morwong *Nemadactylus macropterus*, is a deep-water species, usually living at 40–300 m depth, but occasionally visiting inshore waters, particularly in the southern part of its range. This species has a similar life history and longevity to the red morwong (Jordan 2001) and is an important component of the SE Australian trawl fishery, producing ~1000 t annually.

Magpie perch Cheilodactylus nigripes

This common, site-attached, southern species has a small home range, often of only 30–50 m^2, where it hides at night more or less permanently in a shelter hole and forages by day among turfing algae for small invertebrates (Cappo 1995). At Mike Cappo's study site in western Gulf St Vincent, the diet comprised a variety of prey species (Table 16.5), whereas other studies found a dominance (> 86%) of amphipods.

Two later studies (Connell 2002, Wellenreuther and Connell 2002) examined the causes of different feeding rates of magpie perch among turf algae. First, they found that fish abundances were higher in open habitats with high cover of turfing algae than in those with a kelp canopy and little turf. Next, they manipulated prey densities within the turf and found that fish could detect differences in prey densities, and so they concentrated their feeding (often with twice to three times the intensity) in turf patches with more abundant prey. The third factor examined was the effect that fur seals (the main predator of the morwong) had on their feeding. In the absence of seals, bite rates averaged ~6.4 bites $min.^{-1}$ on large boulders covered with algal turf, whereas when seals were cavorting about, as they often did near their haul-out sites, with periodic sorties to the bottom, bite rates fell to < 2 $min.^{-1}$. Sean Connell thought this was due either to (a) the increased vigilance of magpie perch towards their predators or (b) their tendency to avoid movement that would attract predator attention. It is likely that both factors play a part, because, from the author's own observations at the same and other sites (e.g. Shepherd *et al.* 2010), when seals appear, all benthic fish cease foraging. They do not necessarily hide, but lie still with heads just above the kelp, intently observing the seals' movements. When a seal approaches within ~3 m, the fish rapidly disperse and hide – termed 'submergent behaviour'. Other predators of magpie perch listed by Mike Cappo are cormorants, rock ling, giant cuttlefish, snook and flathead, while the large isopod ectoparasite *Creniola laticauda* was often seen attached to the head.

Cappo (1980) examined the growth rate of magpie perch by ageing them from the scales. These suggested a longevity of ~4 years, likely an under-estimate given the limitations of scale-ring counts. The growth rate was almost identical to that of the red morwong in its first 4 years (Fig. 16.6).

Red morwong Cheilodactylus fuscus

This species is conspicuous on eastern Australian reefs and forages by sucking or wrenching animals from the rocky bottom. The species is carnivorous, with a broad diet that changes with increasing size (Table 16.5). The feeding behaviour also changes with size. Juveniles feed in algal turfs, mostly on small crustaceans continuously through the day, with bite rates of up to 18 $min.^{-1}$ at depths to ~5 m (Lockett and Suthers 1998). From a size of ~20 cm, fish move to deeper water (5–20 m) where larger prey becomes available. Here, adults aggregate during the day in boulder habitat and spend much of the time resting on the bottom within a home range of ~2000 m^2. At and after dusk, they make extensive forays, ranging over ~4000 m^2 searching for food (Lowry and Suthers 1998).

Males and females can be visually sexed in the field and it appears that, around Sydney, female groups mainly lived within the Harbour, and male groups outside (Schroeder *et al.* 1994). Sexual maturity is reached at ~30 cm length, with an autumn–early winter spawning season. The larvae have a long pelagic stage of 8–12 months, and finally settle in inshore waters, after which juveniles grow rapidly to their maximum size at ~8 years of age (Fig. 16.6)

The red morwong is seldom taken by line fishers, but is popular with spear-fishers. It is long lived, with a lifespan of > 40 years (Lowry 2003) and is used by Sydney water authorities as an indicator of heavy metal bio-accumulation. Body tissues typically have high levels of mercury, arsenic and zinc around sewage outfalls (McLean *et al.* 1991).

Banded morwong Cheilodactylus spectabilis

This species has a similar reef habitat and behaviour to the red morwong and replaces that species in Tasmania and Bass Strait, where it is moderately common to 40 m deep. Like the other morwongs discussed, it is usually a site-attached species with a diet of benthic prey (Table 16.5). However, it has a very large home range of up to 7 ha, in which it forages in loose groups. Females and juveniles tend to live in shallow inshore areas, while males dominate deeper waters (McCormick 1989; Jones 1999a).

Growth is rapid (Fig. 16.6) and sexual maturity is reached at an age of 4 years, with a maximum size of ~70 cm at ~15 years for males and 58 cm for females, which grow more slowly. Remarkably, it has a longevity of > 95 years (Ewing *et al.* 2007). Females spawn serially in late summer–autumn and, according to Wade Doak (1991), the female swims into deeper water to find a male and, when successful, she lies prostrate 'while the male vibrates over her, belly to belly' and fertilises her eggs. The larval stage is ~180 days and larvae are carried up to 200 km from the coast before being returned to inshore waters.

The banded morwong does not readily take a bait, but has been seriously impacted by gill-net fishing in near-shore waters for the 'live-fish market. The fishery is managed by various input controls, including a keyhole size limit of 36–46 cm and a seasonal spawning closure. Catches since the early 1990s have declined by > 60%, together with the median age of fish, and models predict that the population is likely to decline further. Curiously, this decline has been accompanied by an increase in growth rate, a reduction in age at maturity (from 4 to < 3 years), and some increase in recruitment (Ziegler *et al.* 2007) – thought to be a density-dependent response to a severe reduction in population size and size structure.

Blue morwong (or queen snapper) Nemadactylus valenciennesi

This morwong (Plate 57) occurs on exposed reefs of the lower west coast of Western Australia and eastwards to Wilsons Promontory, Victoria to a depth of 240 m. It is one of the fastest growing morwongs (Fig. 16.6), reaching a maximum size of ~84 cm (males) and 70 cm (females). The age of sexual maturity is ~3 years at ~50 cm length on the lower west coast of western Australia, but higher at ~7 years on the south coast. Available evidence suggests that major spawning occurs in February–June on the lower west coast and for a shorter period (summer to early autumn) on the south coast, where fewer fish reached sexual maturity. Larvae are then transported by the Leeuwin Current along the south coast where settlement occurs. Juveniles on the south coast then grow and migrate back towards the lower west coast to the major spawning grounds (Coulson *et al.* 2007, 2010). Longevity is ~21 years, with a low annual natural mortality of ~20% and a similar fishing mortality rate from the substantial fishing pressure by commercial gill-netting.

Macrocarnivores

Wrasses (Family Labridae)

This is one of the most diverse fish families, both globally and in Australia. In southern temperate waters, there are > 90 species, which are diverse in colour, size and shape, and most with two sex-related colour forms, as well as different juvenile colouration. Many species have complex sex lives. Some mature as females and later switch to males (protogyny), others do the reverse (protandry), while yet others have separate sexes (gonochorism). The family is distinctive because they use their pectoral fins for normal swimming (libriform locomotion) and their tail only for rapid escape from predators. Labrids are generally carnivorous, with teeth and lips adapted for benthic foraging or picking, and powerful jaws with formidable crushing ability. Many of the larger southern species are fished, and we discuss the three largest species in the Hypsigenyini Tribe, one unusual odacid (Tribe Odacini) and then the ecology of seven well-studied labrids in other wrasse tribes. Aspects of their life history and ecology are summarised in Table 16.6.

Blue groper genus Achoerodus

The two southern blue gropers, the eastern species *A. viridis*, and the western species *A. gouldii*, have broadly similar life histories and ecology (except that the western is much larger and longer-lived). The two species probably evolved when the Bassian Isthmus emerged and isolated populations from their common ancestor. The eastern groper has been well studied by Gillanders and Kingsford (1993, 1998) and Gillanders (1995a, b, 1996, 1997a, b, 1999) and the western groper by Shepherd (2006), Shepherd and Brook (2005, 2007), Coulson *et al.* (2007, 2009) and Bryars *et al.* (2012).

After a planktonic life of 1–2 months (Leis and Hay 2004), larvae settle at a size of 7–8 mm and metamorphose into juveniles at ~1 cm size; in eastern groper this is either in seagrass in estuaries or on sheltered coastal reefs, and in western groper in sheltered bays,

Table 16.6: Life history summaries for 11 labrid species, giving sexual characteristics and average length (cm) at which sex change (SC) of protogynous and diandrous species occurs, size and age of sexual maturity, reproductive season, and geographic and depth range, with references (Ref.).

P = protogynous hermaphrodite; G = gonochorist; D = diandrous hermaphrodite.

Species	SC	Maturity: size (cm) age (years)	Max. size (cm) age (years)	Spawning season	Geographic range	Depth (m)
Eastern blue groper[1]	P	25–35	100	July–Oct.	NSW, E. Vic., N. Tas.	0–40
Achoerodus viridis	48–58	2–4	35			
Western blue groper[2]	P	~65	175	July–Oct.	WA–W. Vic.	0–60
Achoerodus gouldii	80–92	17	70			
Western foxfish[3]	P	24	48	Oct.–Feb.	WA–SA, Qld–E. Tas.	15–80
Bodianus frenchii	37[e] 32[f]	5	78			
Purple wrasse[4]	G	~12	40	July–Jan.	Mid-SA–Vic.–Tas.	0–90
Notolabrus fucicola	–	2	25	Peak Oct.		
Blue-throat wrasse[4]	P	13–25	45	July–Jan.	W. SA–SE Aust., NSW	0–30 (–150)
Notolabrus tetricus	32[a,b]	2	14[a] 23[b]	Peak S-O		
Crimsonband wrasse[5]	P	18	45	Apr.–Oct.	Bass St.–S. Qld.	4–40
Notolabrus gymnogenis	27	1.8	10			
Brown-spotted wrasse[6]	P	20[e]	35	July–Nov.	WA–	0–20
Notolabrus parilus	16–28	2.9	12		mid SA	
Maori wrasse[5,6]	P	18[c] 15[e]	45[c] 32[e]	Jan.–March	WA, SA, Vic., NSW	(0–)
Ophthalmolepis lineolata	29	2[c] 2.6[e]	14			10–60
Senator wrasse[4,5]	D	9	18[c] 30[d]	Aug.–Jan.	WA–Tas.–NSW	0–30
Pictilabrus laticlavius	14–25	1	5[c] 11[d]	Peak Oct.		
Rosy wrasse[4]	P	10–15	25	Aug.–Jan.	S. NSW– Tas.–S.WA.	0–220
Pseudolabrus mortonii	18–20	2–3	11			
Western king wrasse[6]	P	18[e]	4[e]	Apr.–June	WA	0–45
Coris auricularis	20–32	2.8[e]	12[e]			

[a]in South Australia; [b]in Victoria and Tasmania; [c]in New South Wales; [d]in Tasmania; [e]off Perth, WA; [f]south coast, WA.
[1]Gillanders (1999); [2]Coulson *et al.* (2009); [3]Cossington *et al.* (2010); [4]Barrett (1995a, b), Saunders *et al.* (2010); [5]Morton *et al.* (2008b); [6]Lek (2011).

back-reefs or lagoons, but never in seagrass. In the latter species, recruitment was found to be highly variable from year to year and this variability was related to the strength of the Leeuwin Current. When the Current was strong, recruitment was poor on inshore reefs, and the reverse occurred when the Current was weak. Because the larvae may have to swim some distance into near-shore waters, it seems that return to the coast is achieved more easily when the Current weakens.

With increasing size, juveniles of both species gradually leave their juvenile habitat and move offshore to deeper water on more exposed reefs, where spawning takes place. Sex change from female to male occurs at an age of 35–39 years in western groper, and 14–18 years in eastern groper, and in both species this is accompanied by a change in body colour from green to blue in western groper and to reddish-brown to blue in the eastern groper (see Table 16.6 for other details). The growth rate of gropers is slow (Fig. 16.8) and diverges after sex change, as males continue growing to a larger size than females.

Western groper live in social groups of one male and two to three females and some juveniles. The males are zealously territorial and vigorously defend their boundaries, as indicated by mouth-holding conflicts that sometimes occur at boundaries between neighbouring groups. The social grouping of eastern groper is uncertain.

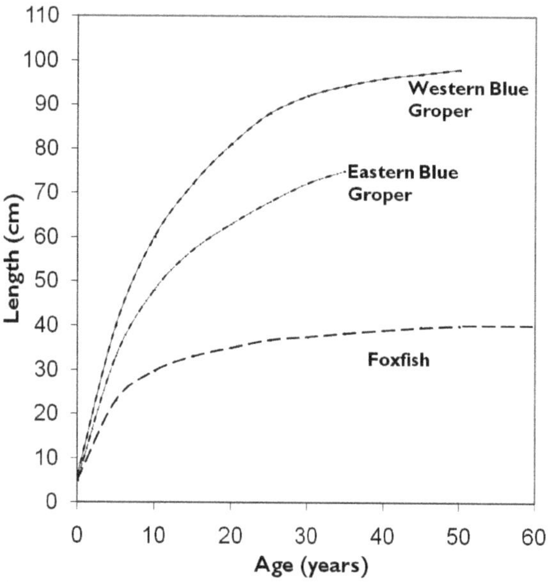

Figure 16.8: Growth curves of male western blue groper *Achoerodus gouldii* (WBG) (after Coulson *et al.* (2009), eastern blue groper *Achoerodus viridis* (EBG) (after Gillanders 1999) and foxfish *Bodianus frenchii* in southern Western Australia (after Cossington *et al.* 2010).

Gut content analyses for the two species (Table 16.7) show that juveniles feed largely on amphipods, other small crustaceans and tiny bivalves (e.g. mussels), while adults feed on a wide range of crustaceans, sea-urchins and molluscs. Adults of western groper use several modes of feeding. As they swim over their large home ranges, they pause now and then and use a ram-and-bite mode to capture chitons, limpets and abalone. Like their eastern relative, they also use their powerful crushing jaws to eat other gastropods and urchins. These large items are supplemented by 'popcorn snacks' of tiny crustaceans in turf algae, such as *Caulerpa* mats. Here they use a bite–suction feeding mode, in which they bite off ~50 cm^2 chunks of turf and sift out the small crustaceans living in them. When groper were offered a choice of different prey types in 'cafeteria' field experiments (Shepherd 2006), their preference order was:

Crabs > abalone > sea-urchins > algal mats

Western groper rest in crevices or caves at night and forage or swim by day over home ranges that increase with size of fish. Juveniles < 10 cm have small ranges of ~20 m^2, which increase with size. While appearing to lazily skull with their pectoral fins for much of the day over their territories, they are actually vigilant in defending their borders against intruding neighbours. In a longer term study of their movement using acoustic tags, Bryars *et al.* (2012) estimated that adult home ranges (defined as the area normally used 95% of the time) and their core ranges (defined as the area used 50% of the time) averaged ~45 000 m^2 and 18 000 m^2, respectively, at coastal sites. Occasionally an adult left its home range, wandered up to ~1 km along the coast and returned.

Time spent foraging, compared with resting or swimming, varies with site. At islands in the eastern GAB, western groper spent 20–25% of the time foraging for food, whereas at a sheltered island in Investigator Strait it spent half the time foraging (Shepherd and Brook 2005, 2007; Shepherd 2006). In contrast, eastern groper spent all its time foraging at sites in New South Wales over ranges of unknown size. Here, the much

Table 16.7: Diet (percentage volume) of eastern blue groper *Achoerodus viridis* (EBG) (after Gillanders 1995b), western blue groper *Achoerodus gouldii* (WBG) (after Shepherd (2006), foxfish *Bodianus frenchii* (after Platell *et al.* 2010) and adult rainbow cale *Heteroscarus acroptilus* (after Choat and Clements 1992).

EBG data were averaged over autumn and spring, and show only the main items. WBG data were taken in summer, and the average given for two sites for juveniles and three sites for adults. Foxfish data were averaged for all seasons.

	EBG		WBG		Foxfish	
	Juv.	Adults	Juv.	Adults	Adults	Rainbow cale
Food type	(< 15 cm)	(> 30 cm)	(< 10 cm)	(> 60 cm)	~32 cm	
Algae	+	+	–	5.5	2.3	12.3
Amphipods, isopods	12.5	1	57.5	28.5	2.5	–
Decapods (crabs, shrimp)	–	+	4.5	23.0	28.1	–
Crustacean fragments	26	4	–	–	–	4.0
Polychaetes	+	+	+	5.0	4.7	0.01
Gastropods, chitons	+	+	6.0	11.6	32.4	24.6
Bivalves	8.7	51.2	30.5	3.5	5.5	–
Sea-urchins	–	13	–	11.0	10.9	–
Other echinoderms	+	+	+	3.7	8.8	–
Ascidians	–	+	–	+	1.1	–
Other	+	+	1.0	3.5	1.2	7.8
Sediment; detritus; unidentified	–	–	–	–	2.4	50.1

greater time spent foraging was possibly because it foraged much more in algal turfs and urchin barrens compared with the western groper (Gillanders 1995b).

Both species are targeted by fishers and, because they are slow-growing, late-maturing and long-lived, they are especially susceptible to overfishing. Other factors putting them at risk are their variable recruitment and their protogynous life history, because fishing will deplete the males first, and lead to sperm limitation and reduced fertilisation success. Hence, they have been protected from spear-fishing and commercial fishing in New South Wales. In Western Australia, where size and bag limits offer some protection, the western groper is the second most important scalefish in the demersal and long-line fishery along the south coast, and at risk from depletion through fishing (McAuley and Simpfendorfer 2003). In South Australia, this species is protected in the Gulfs and partially conserved through size and bag limits. However, this protection is weak, because a fishing mortality of ~62% a year still occurs from recreational fishing and likely contributes to the decline in abundance of the species over the eastern half of its range (Shepherd and Brook 2007).

Foxfish Bodianus frenchii

The foxfish is the longest living labrid globally, with a lifespan of 78 years. Juveniles live on shallow, inshore reefs, often swimming upside down in caves. With increasing size, they gradually move offshore to depths of 15–40 m. Life history, diet and growth rate information is given in Tables 16.6, 16.7 and Fig. 16.7, respectively, and shows that this species shares many ecological features with other species in the Tribe Hypsigenyini discussed above. During the spring–summer reproductive season, spawning takes place daily. Females change sex to males at an age of ~29 years. Like groper, the foxfish is targeted by fishers and its life history features render it equally vulnerable. In Western Australia, it is classified as a 'high-risk' species and is subject to seasonal fishing closures along the lower west coast, and other measures.

Rainbow cale Heteroscarus acroptilus

This species has its teeth fused together into a beak and has features similar to tropical scarid parrotfishes – hence its place in the Tribe Odacini, within the

Family Labridae. It has apparently evolved from a herbivorous ancestor to a carnivore (Clements *et al.* 2004). Rainbow cale (Plate 58) ranges throughout southern Australian coasts to SW Western Australia and mid-New South Wales, and lives in cryptic habitats in algal forests or seagrass to a depth of ~20 m. Its life history and ecology are largely unknown, except that, like many labrids, it is a protogynous hermaphrodite. From a dietary study by Choat and Clements (1992), the species is thought to forage on reefs in turfs and on macroalgal surfaces, targeting small invertebrates, and taking some algae incidentally (Table 16.7).

Other wrasse species

Diet

Wrasses are probably the commonest fishes on temperate reefs in southern Australia, with often two to three species sharing the same reef habitat. Some species have ranges restricted to the west, east or SE Australian coasts, while others range more widely throughout temperate waters. All of them are carnivorous and with increasing size develop powerful jaws with great crushing ability. Larger wrasses may specialise on particular kinds of food, but most are generalist feeders, consuming a wide range of larger prey, including species protected in shells or with a hard exoskeleton, such as trochids, abalone, sea-urchins and crabs. Their success is no doubt due largely to their flexibility in foraging behaviour and diet, which vary according to available food, predator preference and risk of predation.

Of the above factors, food preference is influenced by prey size, morphology and palatability. For example, the prey choice of the bluethroat wrasse was determined by field experiments using multiple and two species choice, and no choice, at one site (Shepherd and Clarkson 2001). The results in general terms were this:

Chitons > abalone > gastropods > ophiuroids > sea-stars

In other multiple choice experiments, the preferences were in this order:

Polynoid worms > tethered crabs > shrimp > gastropods, ophiuroids, terebellids, sea-urchins

Another interesting example of this flexibility in diet is the opportunistic feeding of some wrasses on faeces and vomit of fur seals in southern waters. As seals gather in large numbers at their haul-out sites, they defecate wastes after feeding excursions. Their faeces and vomit then provide an abundant source of nutrients for the wrasses living on nearby reefs (Krajewski and Sazima 2010).

By day, wrasses forage over reefs searching for prey and at dusk they retreat to crevices and caves for the night. Some species (e.g. the purple wrasse) exude a mucous envelope for nocturnal protection. However, they can be easily aroused and, if woken by torchlight, they immediately respond by searching around them for food.

The diet of seven species is summarised in Table 16.8, and it is evident that the proportions of each prey group vary widely between species and between regions. However, juveniles of all species eat small prey, mainly amphipods, shrimp and tiny molluscs, as illustrated for bluethroat wrasse (Table 16.8). Similarly, the smallest wrasse of those discussed here, the senator, with an unusually small mouth (even when an adult) has a diet more similar to that of juveniles. Curiously, Morton *et al.* (2008a) noted that the senator wrasse used ambush hunting to capture tiny, cryptic amphipods. Fish would lie on their sides, orienting their mouth towards a particular microhabitat and use a combination of ram and suction to capture moving prey.

On the east and west Australian coasts, where several wrasse species co-occur in reef habitats, interesting questions arise as to resource partitioning and regional dietary trends. In New South Wales, Morton *et al.* (2008a) examined the diet of three abundant species – Maori wrasse, crimsonband wrasse and the senator wrasse, and in Western Australia Lek *et al.*(2011) similarly examined the diets of three common wrasses – western king, brown-spotted and Maori (Table 16.8). In both studies, the wrasses consumed a wide range of food types, but showed substantial inter-specific differences in feeding micro-habitats and feeding behaviour, as well as within-species differences in diet with increasing size. Together these differences reduced competition between them and facilitated their co-existence in reef habitats.

Table 16.8. Diet (percentage volume) of seven species of wrasse: adult purple wrasse *N. fucicola* (after Edmunds 1990); blue-throat wrasse *N. tetricus* (juveniles 10–20 cm, adults 30–40 cm, after Shepherd and Clarkson 2001); crimson-band wrasse *Notolabrus gymnogenis* and Maori wrasse *Ophthalmolepis lineolata* (NSW) (both after Morton *et al.* 2008a); and senator wrasse *Pictilabrus laticlavius*, (NSW data after Morton *et al.* 2008a; Tasmanian data after Edmunds 1990); western king wrasse *Coris auricularis* and brown-spotted wrasse *Notolabrus parilus* (both after Lek *et al.* 2011).

		Blue-throat			Maori		Senator		Western	Brown-
Food type	**Purple**	**Juv.**	**Adult**	**Crimson**	**NSW**	**WA**	**NSW**	**Tas.**	**king**	**spotted**
Algae	–	–	–	2.2	1.0	1.0	1.1	–	1.5	0.2
Bryozoans	6.1	–	–	–	< 0.1		–	–	–	–
Cnidaria	–	–	–	0.4	0.2		0.1	–	–	–
Polychaetes[1]	–	5.5	–	3.9	8.2	10.8	5.0	–	8.7	4.4
Amphipods, isopods etc.	17.9	64.0	1.5	18.2	11.7	12.1	25.7	63.9	36.2	12.2
Crabs, shrimps	41.1	2.0	35.0	28.6	10.3	21.3	17.5	7.3	19.6	29.3
Gastropods	20.9	18.0	39.5	26.0	28.0	11.8	36.2	10.1	14.9	22.3
Chitons		2.0	8.0	3.1	7.4	6.6	3.2	12.4	0.4	3.4
Bivalves	–	5.0	7.0	9.0	12.2	5.5	5.3	–	3.0	12.3
Sea-urchins	–	–	7.0	1.4	4.9	14.9	0.6	–	3.5	10.5
Sea-stars	–	–	–	–	–	0.8	–	–	0.2	–
Brittle-stars		5.5	–	2.0	2.1	4.4	0.6	–	3.6	2.7
Ascidians	–	–	–	–	< 0.1	–	0.7	–	–	–
Other (e.g. fish)	14.0	–	2.0	0.4	2.5	8.5	0.5	6.2	5.3	2.7
Sediment, or unidentifiable	–	–	–	2.7	8.6	2.3	1.2	–	2.7	–

[1]includes a small proportion of sipunculans, nemerteans, echiurans and nematodes.

Social lives

In most southern species, individual fish develop first as females and some years later a few change sex and become males (termed sequential protogynous hermaphroditism). The ecological and social circumstances under which the switch occurs were studied for the bluethroat wrasse at West I. by Shepherd *et al.* (2010).

This wrasse lives in haremic social groups, usually of 20–30 females and one or two dominant males, giving a female:male (F:M) sex ratio of 15–20:1. If numbers of females increase or a male is captured or dies, another large female will change sex to male, so maintaining a high F:M sex ratio. Hence, sex change is thought to be socially controlled in accordance with an induction–inhibition model, combining both stimulatory and inhibitory processes (Ross 1990). Under this model, sex change is inhibited socially by the presence of a dominant male, but induced when he disappears. At West I., after the male bluethroat disappeared, a large female changed first its colour and then its shape to those of a male over ~8 weeks.

Another model of sex change, known for purple wrasse *Notolabrus fucicola*, is that of a secondary gonochorist (Barrett 1995b; Denny and Schiell 2002). In this species, fish start to develop as females, but before they are mature some change sex and become males. The F:M sex ratio averaged 3:2 (quite close to equality) so the social structure seems to be that males compete for females, rather than one in which a male maintains many haremic females, as in the protogynous model.

Yet a third model, thought to apply to the senator wrasse *Pictilabrus laticlavius*, is diandry, in which some juveniles develop directly into males at ~1 year of age, and then other males develop from females much later in life – an example of protogyny (Morton *et al.* 2008b). Thus, some males are primary and others secondary. In this species, the F:M ratio is 5–10:1, as in many protogynous species.

Habitat and range

Interesting questions arise as to which factors most affect wrasse abundance and intra- and inter-specific differences in home-range size. Some species (e.g. senator wrasse) are restricted to sheltered habitats, due to their small size and limited swimming ability (Fulton and Bellwood 2002). Reef complexity – the number of crevices, caves and holes in a reef – and algal cover also have strong effects on the abundance of species and, by implication, on home-range size. For example, Tuya *et al.* (2009) found that the density of wrasses in SW Australia increased up to three times with an increase in reef complexity. This increased complexity provided fish with both protection against larger predators, and also many more micro-habitats for potential prey, such as shrimp, crabs and urchins. Another factor was the presence in kelp habitats of medium to high cover of red algal patches, which were favoured by many gastropod grazers and other invertebrates, and so gave wrasses access to more prey.

Home-range sizes and wrasse movement, based on observations of tagged fish, are known for several species (Table 16.9). Sex-changing species investigated were all found to be site-attached, with home ranges usually estimated in the short term by observation. These vary in size according to species. Males are territorial and vigorously defend their territories, whereas the females in a haremic group share a common home range. However, in the long term some individuals wander far beyond their home range, as exemplified by several acoustically tagged large bluethroats, which over a year ranged for > 1 km along a site on the north coast of Kangaroo I. (Bryars *et al.* 2011). In contrast, the purple wrasse differs from all the above species, not only in social life and mode of sex-change, but also in behaviour. It is not site-attached and individuals range over large areas.

Life history

Spawning in temperate wrasses has rarely been observed and possibly occurs in many species in the late afternoon, at sunset or offshore. Morton *et al.* (2008b) observed courtship and spawning by the senator wrasse. During courtship, the male flared his dorsal and anal fins, and repeatedly flexed his

Table 16.9: Home ranges in terms of area (m^2) and maximum measured displacement (D) in metres of three species of wrasse over the short and long term (after Barrett 1995a; Shepherd and Clarkson 2001; Edgar *et al.* 2004; Bryars *et al.* 2011). 'Short term' means the home range of species, measured over days to weeks, and 'long term' refers to studies for a year or more. Ad. = adults; Juv. = juveniles.

	Range	
Species	**Short term Long term**	**Long term**
Purple wrasse	> 1750 m^2 ; D = 80 (Tas.)	Up to 1 ha; D uncertain (Tas.)
Notolabrus fucicola		
Blue-throat wrasse	Ad. ~800 m^2; D = 30 (Tas.) ~2300 m^2; Juv. 135 m^2 (SA)	< 2500 m^2; D = 60 m (Tas.); Max. 4.8 ha; D = 1.2 km (SA)
Notolabrus tetricus		
Senator wrasse	175 m^2; D = 10 m (Tas.)	< 2500 m^2; D = 20 m (Tas.)
Pictilabrus laticlavius		
Rosy wrasse	330–375 m^2 (Tas.)	< 2500 m^2 (Tas.)
Pseudolabrus mortonii[1].		

[1]previously called *Pictilabrus psittaculus*

body from side to side as he swam beside the female. He did this most of the year, suggesting that courtship was a way of ensuring female subordination and inhibited them from changing sex to male. In this way, he could fertilise more eggs and so ensure more progeny for himself. Jason Morton (2007) described a rarely witnessed spawning episode of the senator wrasse around midday at the beginning of the spawning season in these words: 'Pre-spawning behaviour involved the male performing repeated courting passes above the female who was resting within algal shelter. A rapid, vertical swim was initiated by the female and followed by the male. At the apex of the ascent, eggs and sperm were shed in a cloud of gametes.'

After fertilisation, resultant wrasse larvae are planktonic for up to 3 months before settling on inshore reefs. The growth rates and longevity of wrasses vary widely among and within species (Fig. 16.9, Table 16.6). For two species – bluethroat and senator – growth rates are faster in warmer waters, but maximum size and longevity are greater in the cooler waters of western Victoria and Tasmania (see below).

On the west coast of Australia, Elaine Lek (2011) compared the growth rates of three wrasse species – western king, brown-spotted and Maori – at different latitudes (i.e. at Jurien Bay and off Perth, 250 km to the south). The western king wrasse *Coris auricularis* grew to ~40 cm by 12 years of age off Perth (growth trajectory, not shown, being similar to the ocean jacket – see Fig. 16.2), and ~5 cm smaller in Jurien Bay. The brown-spotted wrasse *Notolabrus parilus* and the Maori wrasse *Ophthalmolepis lineolata* off Perth both had similar growth rates to the Maori wrasse in New South Wales (Fig. 16.9), with all three reaching a maximum size of 31–35 cm at 12 years of age. In Tasmania, the rosy wrasse has a very similar growth rate to the senator wrasse (Barrett 1995b).

In the several studies of wrasse life histories at different latitudes, a trend of often slower growth, larger maximum size and greater longevity at cooler and higher latitudes has been noted for several species (Table 1966). A few species (e.g. the rosy wrasse) grow considerably larger and likely live longer in deep, cold

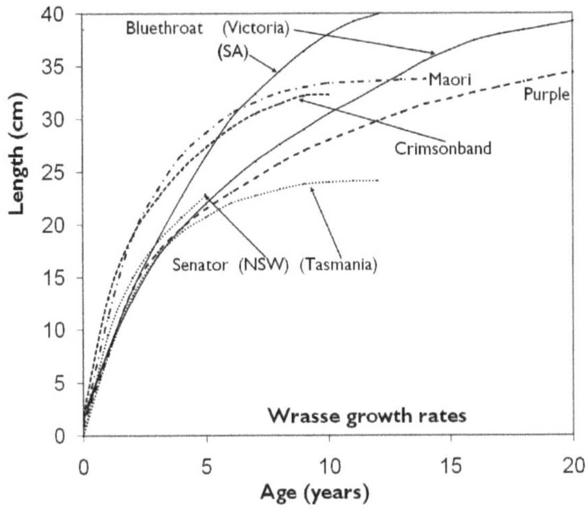

Figure 16.9: Growth curves for five species of wrasse: bluethroat *Notolabrus tetricus* in South Australia (after Saunders *et al.* 2010), and in western Victoria (after Smith *et al.* 2003); Maori wrasse *Ophthalmolepis lineolata*, crimsonband wrasse *Notolabrus gymnogenis* and senator wrasse *Pictilabrus laticlavius* (all in New South Wales after Morton *et al.* 2008b); purple wrasse, *Notolabrus fucicola* in Tasmania (after Ewing *et al.* 2007); and senator wrasse in Tasmania (after Barrett 1995a).

water than the maximum sizes and ages shown in Table 16.6. Such trends are in accordance with the metabolic theory of ecology, which predicts that demands for energy are less in cooler waters, hence more energy is available for growth (Charnov and Gillooly 2004).

Ecological significance and effect of fishing

It is widely acknowledged that wrasses have a significant effect on invertebrates in reef systems and it is sometimes suggested that they may be keystone predators (i.e. they have an ecosystem effect disproportionately greater than other species). This is still an open question, although undoubtedly the larger species, such as groper, may reduce the abundance of urchins, which can create barrens (Chapter 12) and smaller species may control recruitment of some molluscs (e.g. Shepherd and Clarkson 2001).

Wrasses are extensively fished and small commercial fisheries take them in all southern States, with few controls imposed. In addition, recreational fishing on wrasses has generally increased as other, more preferred, species have declined in abundance. Sex-changing species cannot withstand the same levels of fishing as other fish species, because, as fishing intensity increases, males decrease disproportionately in numbers, being more susceptible to capture. Sex ratios then become severely biased and eventually males disappear altogether, because females are caught before sex change can occur. Female mean sizes also decrease and this measure, as well as sex ratios, can be used to estimate fishing intensity in local populations (Shepherd and Baker 2008b; Shepherd *et al.* 2010). Studies of the effects of marine protected areas on wrasses confirm the effect of fishing on wrasse size and sex ratio (Smith *et al.* 2003; Kleczkowski *et al.* 2008; Barrett *et al.* 2007), and show that larger fish may increase in average size by ~10%, and abundance by up to 10 times. Modelling of these species (Heppell *et al.* 2006) confirmed that spatial closure of adult habitats is the optimal management measure to maintain wrasse populations

Fish-cleaning symbiosis (Plates 57,58)

This is undoubtedly one of the most remarkable of ecological adaptations among fish. All fish suffer from parasites that attach themselves to the scales or gills and feed on mucus or tissue from the host fish. Eventually the host fish is weakened and may become an easier prey for predators. In this situation, a few fish species and shrimp species (see Chapter 15) have evolved to offer a parasite-cleaning service to the local fish community. The phenomenon is little known in temperate regions (but see Hutchins 1991; Shepherd *et al.* 2005; Morton 2007; Shepherd and Baker 2008b), because it is uncommon and less predictable than in tropical waters, where parasite infection is more widespread.

This behaviour has many of the features of a business in a free market economy. A person offering a service has an advertised place of business, where clients go to obtain the service. Fish cleaning has evolved in the same way. Cleaner fish set up business at a prominent site, usually a large rock, cave or even a large sponge, and client fishes go there to be cleaned of their parasites. The cleaner advertises itself in several ways. Some cleaners, such as moonlighters or clingfish, have bold transverse stripes, while others, such as weedy whiting, have prominent eye-spots near the tail. To indicate its desire to be cleaned, the client adopts a near-vertical pose, tail down and fins extended. The cleaner fish then emerges and swims over the client's body, nipping off the attached parasites and taking mucus and dead tissue. The angle of the client's pose may also be important, and in the Shepherd *et al.* (2005) study, the clients that adopted a near-vertical pose received more than twice the attention of cleaners compared with those resting horizontally. There are also secondary stimuli and, in the case of the cleaner clingfish, Hutchins (1991) wrote that the client 'also receives some tactile stimulation from the small cleaner fish as it brushes its ventral fins across the skin, not unlike the stroking of a cat . . . the fish at times appearing to be almost totally oblivious to its surroundings.' But Grutter (2001) reminds us that parasite infection is still the main stimulus for a fish seeking cleaning.

Cleaning often takes place at a fixed time of day, when many fish of several species gather to be cleaned. And, as in a market economy, there may be problems of queue-jumping clients and cheating cleaners that take bites out of the client's scales or flesh, causing it to twitch and shake noticeably, and swim off in irritation.

The main recorded cleaners in South Australia are moonlighter *Tilodon sexfasciatus* (see above), pencil weed whiting *Siphonognathus beddomei*, western cleaner clingfish *Cochleoceps bicolor*, black-spotted wrasse *Austrolabrus maculatus* and old wife *Enoplosus armatus*. In New South Wales, juvenile rosy wrasse *Pseudolabrus mortonii* and the clingfishes *C. orientalis* and *Aspasmogaster costatus* are cleaners (Jones 1999b, Morton 2007). In Western Australia, juvenile western king wrasses are also cleaners (Lek 2011). The western cleaner clingfish sets up its cleaning stations near prominent sponges, stalked ascidians and jetty piles (Hutchins 1991). Clingfish are unusual in that they are equipped with a suction disk – a modified ventral fin – and this is a helpful aid for sliding over a fish's body and clinging to a substrate.

Much yet remains to be learnt of this astonishing symbiosis and the complex social behaviours that have evolved. But what is known demonstrates the high intelligence of fish, in particular the

Machiavellian or 'cunning' behaviour of cheats, the self control of well-behaved cleaners, the trust they build up with local clients and their avoidance of aggressive clients (reviewed by Bshary 2006).

REFERENCES

Andrew NL, Jones GP (1990) Patch formation by herbivorous fish in a temperate Australian kelp forest. *Oecologia* **85**, 57–68.

Barrett NS (1995a) Short- and long-term movement patterns of six temperate reef fishes (Families Labridae and Monacanthidae). *Marine and Freshwater Research* **46**, 853–860.

Barrett NS (1995b) Aspects of the biology of six temperate reef fishes (Families Labridae and Monacanthidae). PhD thesis. University of Tasmania, Hobart.

Barrett NS, Edgar GJ, Buxton CD, Haddon M (2007) Changes in fish assemblages following 10 years of protection in Tasmanian marine protected areas. *Journal of Experimental Marine Biology and Ecology* **345**, 141–157.

Bell JD (1979) Observations on the diet of red morwong, Cheilodactylus fuscus Castelnau (Pisces: Cheilodactylidae). Australian Journal of Marine and Freshwater Research 30, 129–133.

Bell JD, Pollard DA (1989) Ecology of fish assemblages and fisheries associated with seagrasses. In *Biology of Seagrasses: a Treatise on the Biology of Seagrasses with Special Reference to the Australian Region* (Eds AWD Larkum, AJ McComb and SA Shepherd) pp. 565–609. Elsevier, Amsterdam.

Bell JH, Burchmore JJ, Pollard D (1978) The feeding ecology of three sympatric species of leatherjackets (Pisces: Monacanthidaea) from a *Posidonia* seagrass habitat in New South Wales. *Australian Journal of Marine and Freshwater Research* **29**, 631–643.

Bell JH, Burchmore JJ, Pollard DA (1980) The food and feeding habits of the rock blackfish, *Girella elevata* Macleay (Pisces: Girellidae), from the Sydney region, New South Wales. *Australian Zoologist* **20**, 391–405.

Berry PP, Playford PE (1992) Territoriality in a subtropical kyphosid fish associated with macroalgal polygons on reef platforms at Rottnest Island, Western Australia. *Journal of the Royal Society of Western Australia* **75**, 67–73.

Brown GD (1982) The population structure of *Parma victoriae* (Pisces: Pomacentridae) BSc (Hons) thesis. Monash University, Melbourne.

Bryars S, Rogers P, Huveneers C, Payne N *et al.* (2011) 'What size do marine park sanctuary zones need to be to protect the western blue groper *Achoerodus gouldii*, harlequin fish *Othis dentex*, and bluethroat wrasse *Notolabrus tetricus*?' Report to Department of Environment and Natural Resources. Wildlife Conservation Project No. 0638, Adelaide.

Bryars S, Rogers P, Huveneers C Payne N *et al.* (2012) Small home range in southern Australia's largest resident reef fish, the western blue groper (*Achoerodus gouldii*): implications for adequacy of no-take marine protected areas. *Marine and Freshwater Research* **63**, 552–563.

Bshary R (2006) Machiavellian intelligence in fishes. In *Fish Cognition and Behavior.* (Eds C Brown, K Laland and J Krause) pp. 223–244. Blackwell Publishing, London.

Burchmore JJ, Pollard DA, Bell JD (1984) Community structure and trophic relationships of the fish fauna of an estuarine *Posidonia australis* seagrass habitat in Port Hacking, New South Wales. *Aquatic Botany* **18**, 71–87.

Cappo MG (1980) Some aspects of the ecology and population biology of magpie perch (*Cheilodactylus nigripes*, Richardson 1850), (Teleosteii; Cheilodactylidae) living beneath a South Australian pier. BSc (Hons) thesis. University of Adelaide, Adelaide.

Cappo MG (1995) The population biology of the temperate reef fish *Cheilodactylus nigripes* in an artificial reef environment. *Transactions of the Royal Society of South Australia* **119**, 113–122.

Charnov EL, Gillooly JF (2004) Size and temperature in the evolution of fish life histories. *Integrative and Comparative Biology* **44**, 494–497.

Choat JH, Clements KD (1992) Diet in Odacid and Aplodactylid fishes from Australia and New Zealand. *Australian Journal of Marine and Freshwater Research* **43**, 1451–1459.

Choat JH, and Clements KD (1998) Vertebrate herbivores in marine and terrestrial environments: a nutritional perspective. *Annual Review of Ecology and Systematics* **29**, 375–403.

Clements KD (1991) Endosymbiotic communities of two herbivorous labroid fishes, *Odax cyanomelas* and *O. pullus*. *Marine Biology* **109**, 223–229.

Clements KD (2003) Herbivorous fishes. In *The Living Reef: the Ecology of New Zealand's Rocky Reefs.* (Eds N Andrew and M Francis) pp. 128–135. Craig Potton Publishing, Nelson, New Zealand.

Clements KD, Choat JH (1997) Comparison of herbivory in the closely related marine fish genera *Girella* and *Kyphosus*. *Marine Biology* **127**, 579–586.

Clements KD, Alfaro ME, Fessler JL, Westneat MW (2004) Relationships of the temperate Australasian labrid fish tribe Odacini (Perciformes: Teleostei). *Molecular Phylogenetics and Evolution* **32**, 575–587.

Clements KD, Raubenheimer D, Choat JH (2009) Nutritional ecology of marine herbivorous fishes: ten years on. *Functional Ecology* **23**, 79–92.

Conacher MJ, Lanzing WJR, Larkum AWD (1979) Ecology of Botany Bay. II. Aspects of the feeding ecology of the fanbellied leatherjacket, *Monacanthus chinensis* (Pisces:

Monacanthidae), in *Posidonia australis* seagrass beds in Quibray Bay, Botany Bay, New South Wales. *Australian Journal of Marine and Freshwater Research* **30**, 387–400.

Connell SD (2002) Effects of a predator and prey on a foraging reef fish: implications for understanding density-dependent growth. *Journal of Fish Biology* **60**, 1551–1561.

Connolly RM, Melville AJ, Keesing JK (2002a) Abundance, movement and individual identification of leafy seadragons, *Phycodurus eques* (Pisces: Syngnathidae). *Marine and Freshwater Research* **53**, 777–780.

Connolly RM, Melville AJ, Preston KM (2002b) Patterns of movement and habitat use by leafy seadragons tracked ultrasonically. *Journal of Fish Biology* **61**, 684–695.

Cossington S, Hesp SA, Hall NG, Potter IC (2010) Growth and reproductive biology of the foxfish *Bodianus frenchii*, a very long-lived and monandric protogynous hermaphroditic labrid. *Journal of Fish Biology* **77**, 600–626.

Coulson PG, Potter IC, Hesp SA, Hall NG (2007) 'Biological parameters required for managing western blue groper, blue morwong and yellowtail flathead'. FRDC Report. Project 2004/057. Centre for Fish and fisheries research, Murdoch University, Perth.

Coulson PG, Hesp SA, Hall NG, Potter IC (2009) The western blue groper (*Achoerodus gouldii*), a protogynous hermaphroditic labrid with exceptional longevity, late maturity, slow growth, and both late maturation and sex change. *Fishery Bulletin* **107**, 57–75.

Coulson PG, Hesp SA, Potter IC, Hall NG (2010) Life cycle characteristics of the blue morwong *Nemadactylus valenciennesi* compared with those of other species of Cheilodactylidae. *Marine and Freshwater Research* **61**, 104–118.

Coulson PG, Potter IC, Hall NG (2012) The biological characteristics of *Scorpis aequipinnis* (Kyphosidae), including relevant comparisons with those of other species and particularly of a heavily exploited congener. *Fisheries Research* **125–126**, 272–282.

Curley BG, Gillings MR (2009) Population connectivity in the temperate damselfish *Parma microlepis*: analyses of genetic structure across multiple spatial scales. *Marine Biology* **156**, 381–393.

Currie DR, Sorokin SJ (2010) 'A preliminary evaluation of the distribution and trophodynamics of demersal fish from Spencer Gulf'. SARDI Research Report Series No. 424. Publication No. F2010/000088/1, Adelaide.

Denny CM, Schiell DR (2002) Reproductive biology and population structure of the banded wrasse, *Notolabrus fucicola* (Labridae) around Kaikoura, New Zealand. *New Zealand Journal of Marine and Freshwater Research* **36**, 555–563.

Doak W (1991) *World of New Zealand Fishes*. Hodder & Stoughton, Auckland.

Edgar GJ (2008) *Australian Marine Life: the Plants and Animals of Temperate Waters*. 2nd edn. Reed New Holland, Sydney.

Edgar GJ, Barrett NS, Morton AJ (2004) Patterns of fish movement on eastern Tasmanian rocky reefs. *Environmental Biology of Fishes* **70**, 273–284.

Edgar GJ, Barrett NS, Crane K, Bancroft K (2007) 'Ecological monitoring of subtidal reefs in different management zones of the Jurien Bay Marine Park, 1999–2007'. Tasmanian Aquaculture and Fisheries Institute. Internal Report 1–28, Hobart.

Edmunds M (1990) Ecology of fishes on Tasmanian rocky reefs. BSc (Hons) thesis. University of Tasmania, Hobart.

Ewing GP, Lyle JM, Murphy RJ, Kalish JM *et al.* (2007) Validation of age and growth in a long-lived temperate fish using otolith structure, oxytetracycline and bomb radiocarbon methods. *Marine and Freshwater Research* **58**, 944–955.

Floeter SR, Behrens MD, Ferreira CEL, Paddack MJ *et al.* (2005). Geographical gradients of marine herbivorous fishes: patterns and processes. *Marine Biology* **147**, 1435–1447.

Flynn AJ, Ritz DA (1999) Effect of habitat complexity and predatory style on the capture success of fish feeding on aggregated prey. *Journal of the Marine Biological Association of the United Kingdom* **79**, 487–494.

Forsgren KL, Lowe CG (2006) The life history of weedy seadragons, *Phyllopteryx taeniolatus* (Teleostei: Syngnathidae). *Marine and Freshwater Research* **57**, 313–322.

Fulton CJ, Bellwood DR (2002) Patterns of foraging in labrid fishes. *Marine Ecology Progress Series* **226**, 135–142.

Gillanders BM (1995a) Reproductive biology of the protogynous hermaphrodite *Achoerodus viridis* (Labridae) from south-eastern Australia. *Marine and Freshwater Research* **46**, 999–1008.

Gillanders BM (1995b) Feeding ecology of the temperate marine fish *Achoerodus viridis* (Labridae): size, seasonal and site-specific differences. *Marine and Freshwater Research* **46**, 1009–1020.

Gillanders BM (1996) Elements in otoliths may elucidate the contribution of estuarine recruitment to sustaining coastal reef populations of a temperate reef fish. *Marine Ecology Progress Series* **141**, 13–20.

Gillanders BM (1997a) Comparison of growth rates between estuarine and coastal reef populations of *Achoerodus viridis* (Pisces: Labridae). *Marine Ecology Progress Series* **146**, 283–287.

Gillanders BM (1997b) Patterns of abundance and size structure in the blue groper, *Achoerodus viridis* (Pisces: Labridae): evidence of links between estuaries and coastal reefs. *Environmental Biology of Fishes* **49**, 153–173.

Gillanders BM (1999) Blue groper. In *Under Southern Seas: the Ecology of Australia's Rocky Reefs*. (Ed. N Andrew) pp. 188–193. University of New South Wales Press, Sydney.

Gillanders BM, Kingsford MJ (1993) Abundance patterns of *Achoerodus viridis* (Pisces: Labridae) on estuarine and exposed rocky reefs: possible linkages. In *Proceedings of the Second International Temperate Reef Symposium. Vol. 1*. 7–10 January 1992, Auckland, New Zealand. (Eds CN Battershill, DR Schiell, GP Jones, RG Creese *et al.*) pp. 93–98. NIWA Marine, Wellington.

Gillanders BM, Kingsford MJ (1998) Influence of habitat on abundance and size structure of a large temperate fish, *Achoerodus viridis* (Pisces: Labridae). *Marine Biology* **132**, 503–514.

Glasby TM, Kingsford MJ (1994) *Atypichthys strigatus* (Pisces: Scorpididae): an opportunistic planktivore that responds to benthic disturbances and cleans other fishes. *Australian Journal of Ecology* **19**, 385–394.

Grant JWA (1997) Territoriality. In *Behavioural Ecology of Teleost Fishes*. (Ed. J-GJ Godin) pp. 83–103. Oxford University Press, Oxford.

Gray CA, Ives MC, Macbeth WG, Kendall BW (2010) Variation in growth, mortality, length and age compositions of harvested populations of the herbivorous fish *Girella tricuspidata*. *Journal of Fish Biology* **76**, 880–899.

Grove-Jones R, Burnell AF (1991) 'Fisheries biology of the ocean jacket (Moncanthidae: *Nelusetta ayraudi*) in the eastern waters of the Great Australian Bight, South Australia. FRDC Report'. Grant DFSO1Z. South Australian Department of Fisheries, Adelaide.

Grutter AS (2001) Parasite infection rather than tactile stimulation is the proximate cause of cleaning behaviour in reef fish. *Proceedings. Biological Sciences* **268**, 1361–1365.

Gunn JS, Thresher RE (1991) Viviparity and the reproductive ecology of clinid fishes (Clinidae) from temperate Australian waters. *Environmental Biology of Fishes* **31**, 323–344.

Heppell SS, Heppell SA, Coleman FC, Koenig CC (2006) Models to compare management options for a protogynous fish. *Ecological Applications* **16**, 238–249.

Horn MH (1989) Biology of marine herbivorous fishes. *Oceanography and Marine Biology - an Annual Review* **27**, 167–272.

Howard RK, Koehn JD (1985) Population dynamics and feeding ecology of pipefish (Syngnathidae) associated with eelgrass beds of Western Port, Victoria. *Australian Journal of Marine and Freshwater Research* **36**, 361–370.

Humphries J (2008) Population biology of weeping toado (*Torquigener pleurogramma*) in the northern parts of the gulfs of South Australia. BSc (Hons) thesis. Flinders University, Adelaide.

Hutchins B (1991) Southern Australia's enigmatic clingfishes. *Australian Natural History* **23**, 627–631.

Hutchins B (1999) Leatherjackets. In *Under Southern Seas: the Ecology of Australia's Rocky Reefs*. (Ed. N Andrew) pp. 194–201. University of New South Wales Press, Sydney.

Jenkins GP, Wheatley MJ (1998) The influence of habitat structure on nearshore fish assemblages in a southern Australian embayment: comparison of shallow seagrass, reef–algal and unvegetated sand habitats, with emphasis on their importance to recruitment. *Journal of Experimental Marine Biology and Ecology* **221**, 147–172.

Jones GP (1992) Interactions between herbivorous fishes and macro-algae on a temperate rocky reef. *Journal of Experimental Marine Biology and Ecology* **159**, 217–235.

Jones GP (1999a) Herbivorous fishes. In *Under Southern Seas: the Ecology of Australian Rocky Reefs*. (Ed. N Andrew) pp. 202–209. University of New South Wales Press, Sydney.

Jones GP (1999b) The wrasses. In *Under Southern Seas: the Ecology of Australian Rocky Reefs*. (Ed. N Andrew) pp. 181–188. University of New South Wales Press, Sydney.

Jones GP, Norman MD (1986) Feeding selectivity in relation to territory size in a herbivorous reef fish. *Oecologia* **68**, 549–556.

Jordan AR (2001) Age, growth and spatial and interannual trends in age composition of jackass morwong, *Nemadactylus macropterus*, in Tasmania. *Marine and Freshwater Research* **52**, 651–660.

Kleczkowski M, Babcock RC, Clapin G (2008) Density and size of reef fishes in and around a temperate marine reserve. *Marine and Freshwater Research* **59**, 165–176.

Krajewski JP, Sazima I (2010) Fish feed on faeces and vomits of New Zealand fur seals: marine mammals bring nutrients from open to inshore waters. *Marine Biodiversity Records* **3**, e81.

Kuiter RH (1987) Note sur les soins parentaux, l'eclosion et l'elevage des Dragons de mer (Syngnathidae). *Revue Francais d'Aquariologie* **14**, 113–122.

Kuiter RH (2003) *Seahorses, Pipefishes and their Relatives*. Revised edn. TMC Publishing, Chorleywood, UK.

Last PR (1983) Notes on the feeding ecology of four sympatric leatherjackets (Pisces: Monacanthidae) from Tasmania. *Tasmanian Fisheries Research* **25**, 17–25.

Last PR, White WT, Gledhill DC, Hobday AJ *et al.* (2011) Long-term shifts in abundance and distribution of a temperate fish fauna: a response to climate change and fishing practices. *Global Ecology and Biogeography* **20**, 58–72.

Leis JM, Hay AC (2004) Larval development of *Achoerodus viridis* (Pisces: Labridae), the Australian eastern blue groper. *Ichthylogical Research* **51**, 46–51.

Lek E (2011) Comparisons between three co-occurring and reef-dwelling labrid species at two different latitudes. PhD thesis. Murdoch University, Perth.

Lek E, Fairclough DV, Platell ME, Clarke KR *et al.* (2011) To what extent are the dietary compositions of three abundant, co-occurring labrid species different and related to latitude, habitat, body size and season? *Journal of Fish Biology* **78**, 1913–1943.

Leum LL, Choat JH (1980) Density and distribution patterns of the temperate marine fish *Cheilodactylus spectabilis* in a reef environment. *Marine Biology* **57**, 327–337.

Lincoln-Smith MR, Jones GP (1995) Fishes of shallow coastal habitats. In *Coastal Marine Ecology of Temperate Australia* (Eds AJ Underwood and MG Chapman) pp. 240–254. University of New South Wales Press, Sydney.

Lindholm R (1984) Observations on the Chinaman leather-jacket *Nelusetta ayraudi* (Quoy & Gaimard) in the Great Australian Bight. *Australian Journal of Marine and Freshwater Research* **35**, 597–599.

Lockett MM, Suthers IM (1998) Ontogenetic shift and feeding activity in the temperate reef fish *Cheilodactylus fuscus. Proceedings of the Linnean Society of New South Wales* **120**, 105–116.

Lowry MB (2003) Age and growth of *Cheilodactylus fuscus*, a temperate rocky reef fish. *New Zealand Journal of Marine and Freshwater Research* **37**, 159–170.

Lowry MB, Cappo M (1999) Morwongs. In *Under Southern Seas: the Ecology of Australian Rocky Reefs.* (Ed. N Andrew) pp. 173–180. University of New South Wales Press, Sydney.

Lowry MB, Suthers IM (1998) Home range, activity and distribution patterns of a temperate rocky-reef fish, *Cheilodactylus fuscus. Marine Biology* **132**, 569–578.

Martin-Smith K (2011) Photo-identification of individual weedy seadragons *Phyllopteryx taeniolatus* and its application in estimating population dynamics. *Journal of Fish Biology* **78**, 1757–1768.

May JL, Maxwell JGH (1986) *Field Guide to Trawl Fish from Temperate Waters of Australia.* CSIRO PUBLISHING, Melbourne.

McAuley R, Simpfendorfer C (2003) 'Catch composition of the Western Australian temperate demersal gillnet and longline fisheries, 1994–1999'. Fisheries Research Report No 146. Department of Fisheries, Perth.

McCormick MI (1989) Spatio-temporal patterns in the abundance and population structure of a large temperate reef fish. *Marine Ecology Progress Series* **53**, 215–225.

McCormick M, Francis M (2003) Red moki and other foragers. In *The Living Reef: the Ecology of New Zealand's Rocky Reefs.* (Eds N Andrew and M Francis) pp. 168–181. Craig Potton Publishing, Nelson, New Zealand.

McLean AG, Miskiewicz AG, Roberts EA (1991) Effect of three primary treatment sewage outfalls on metal concentrations in the fish *Cheilodactylus fuscus* collected along the coast of Sydney, Australia. *Marine Pollution Bulletin* **22**, 134–140.

Miller ME, Stewart J (2009) The commercial fishery for ocean leatherjackets (*Nelusetta ayraudi*, Monocanthidae) in New South Wales, Australia. *Asian Fisheries Science* **22**, 257–264.

Moore A, Patterson H, Pham T (2009) 11. Great Australian Bight trawl sector. In *Fishery Status Reports 104.* (Eds DT Wilson, R Curtotti and GA Begg) pp. 232–249. Bureau of Rural Sciences, Canberra.

Moran MJ, Clements KD (2002) Diet and endogenous carbohydrases in the temperate marine herbivorous fish *Kyphosus sydneyanus. Journal of Fish Biology* **60**, 1190–1203.

Moran MJ, and Sale PF (1977) Seasonal variation in territorial response, and other aspects of the ecology of the Australian temperate pomacentrid fish *Parma microlepis. Marine Biology* **39**, 121–128.

Morton JK (2007) The ecology of three species of wrasse (Pisces: Labridae) on temperate rocky reefs of new South Wales, Australia. PhD thesis. University of Newcastle.

Morton JK, Platell ME, Gladstone W (2008a). Differences in feeding ecology among three co-occurring species of wrasse (Teleostii: Labridae) on rocky reefs of temperate Australia. *Marine Biology* **154**, 577–592.

Morton JK, Gladstone W, Hughes JM, Stewart J (2008b) Comparison of the life history of three co-occurring wrasses (Teleosti: Labridae) in coastal waters of south-eastern Australia. *Marine and Freshwater Research* **59**, 560–574.

Mountfort DO, Campbell J, Clements KD (2002) Hindgut fermentation in three species of marine herbivorous fish. *Applied and Environmental Microbiology* **68**, 1374–1380.

Newman DM, Jones PL, Laurenson LJ (2005) A preliminary examination of the diet and feeding of sea sweep *Scorpis aequipinnis* (Pisces: Scorpididae) from coastal waters of south-western Victoria, Australia. *Hydrobiologia* **533**, 175–186.

Noguchi T, Arakawa O, Takatani T (2006) TTX accumulation in pufferfish. *Comparative Biochemistry and Physiology* **1**, Part D. 145–152.

Norman MD, Jones GP (1984) Determinants of territory size in the pomacentrid reef fish, *Parma victoriae. Oecologia* **61**, 60–69.

Platell ME, Hesp SA, Cossingham SM, Lek E *et al.* (2010) Influence of selected factors on the dietary composition of three targeted and co-occurring temperate species of reef fishes: implications for food partitioning. *Journal of Fish Biology* **76**, 1255–1276.

Potter IC, Cheal AJ, Loneragan NR (1988) Protracted estuarine phase in the life cycle of the marine pufferfish *Torquigener pleurogramma. Marine Biology* **98**, 317–329.

Raubenheimer D, Zemke-White WL, Phillips RJ, Clements KD (2005) Algal macronutrients and food selection by the omnivorous marine fish *Girella tricuspidata. Ecology* **86**, 2601–2610.

Rimmer DW (1986) Changes in diet and the development of microbial digestion in juvenile buffalo bream, *Kyphosus cornelii. Marine Biology* **92**, 443–448.

Rimmer DW, Wiebe WJ (1987) Fermentative microbial digestion in herbivorous fishes. *Journal of Fish Biology* **31**, 229–236.

Robertson CH, White RWG (1986) Feeding patterns of *Nesogobius* spp., *Gymnapistes marmoratus, Neoodax balteatus*, and *Acanthaluteres spilomelanurus* from a Tasmanian seagrass meadow. *Australian Journal of Marine and Freshwater Research* **37**, 481–489.

Rodgers GG, Linnane AJ and Huveneers C (2013) Contrasting diet of two temperate reef fish species (*Notolabrus tetricus* and *Meuschenia hippocrepis*) as determined from commercial rock lobster bycatch samples. *Transactions of the Royal Society of South Australia* **137**, 80–89.

Ross DM (1990) The evolution of sex change mechanisms in fishes. *Environmental Biology of Fishes* **29**, 81–93.

Russ GR (1980) Effects of predation by fishes, competition, and structural complexity of the substratum on the establishment of a marine epifaunal community. *Journal of Experimental Marine Biology and Ecology* **42**, 55–69.

Russell BC (1983) The food and feeding habits of rocky reef fish of north-eastern New Zealand. *New Zealand Journal of Marine and Freshwater Research* **17**, 121–145.

Sanchez-Camara J, Booth DJ (2004) Movement, home range and site fidelity of the weedy seadragon *Phyllopteryx taeniolatus* (Teleostei: Syngnathidae). *Environmental Biology of Fishes* **70**, 31–41.

Sanchez-Camara J, Booth DJ, Turon X (2005) Reproductive cycle and growth of *Phyllopteryx taeniolatus. Journal of Fish Biology* **67**, 133–148.

Sanchez-Camara J, Martin-Smith K, Booth DJ, Fritschi J *et al.* (2011) Demographics and vulnerability of a unique Australian fish, the weedy seadragon, *Phyllopteryx taeniolatus. Marine Ecology Progress Series* **422**, 253–264.

Saunders RJ, Bryars SR, Fowler AJ (2010) 'Preliminary consideration of the biology of several of South Australia's marine fish species of conservation or management interest'. SARDI Research Report Series No. 426, Adelaide.

Schroeder A, Lowry M, Suthers I (1994). Sexual dimorphism in the red morwong. *Australian Journal of Marine and Freshwater Research* **45**, 1173–1180.

Seeto GS, Veivers PC, Clements KD (1996) Carbohydrate utilisation by microbial symbionts in the marine herbivorous fishes *Odax cyanomelas* and *Crinodon lophodon. Journal of Comparative Physiology. B, Biochemical, Systemic, and Environmental Physiology* **165**, 571–579.

Shepherd SA (2006) Ontogenetic changes in diet, feeding behaviour and activity of the western blue groper, *Achoerodus gouldii*. In *The Marine Fauna and Flora of Esperance, Western Australia Volume 2 (Proceedings of the 12th International Marine Biology Workshop)* (Eds FE Wells, DI Walker and GA Kendrick) pp. 477–494. Western Australian Museum, Perth.

Shepherd SA, Baker JL (2008a) Investigator Group Expedition 2006: flexible foraging ecology of a temperate herbivore, the herring cale, *Olisthops cyanomelas*, in South Australia. *Transactions of the Royal Society of South Australia* **132**, 147–162.

Shepherd SA, Baker JL (2008b) Reef fishes of lower Gulf St Vincent. In *Natural History of Gulf St Vincent.* (Eds SA Shepherd, S Bryars, I Kirkegaard, P. Harbison and JT Jennings) pp. 297–316. Royal Society of South Australia, Adelaide.

Shepherd SA, Brook JB (2005) Foraging ecology of the western blue groper, *Achoerodus gouldii*, at Althorpe Islands, South Australia. *Transactions of the Royal Society of South Australia* **129**, 202–208.

Shepherd SA, Brook JB (2007) Distribution and ontogenetic shifts in habitat and abundance of the temperate western blue groper, *Achoerodus gouldii. Journal of Fish Biology* **71**, 1457–1478.

Shepherd SA, Clarkson PS (2001) Diet, feeding behaviour, activity and predation of the blue-throated wrasse, *Notolabrus tetricus. Marine and Freshwater Research* **52**, 311–322.

Shepherd SA, Teale J, Muirhead D (2005) Cleaning symbiosis among inshore fishes at Althorpe Islands, South Australia. *Transactions of the Royal Society of South Australia* **129**, 193–201.

Shepherd SA, Clark M, Ferguson G (2008) Investigator Group Expedition 2006: habitat-dependent foraging behaviour and diet of the scalyfin, *Parma victoriae*, in South Australia. *Transactions of the Royal Society of South Australia* **132**, 134–146.

Shepherd SA, Brook JB, Xiao Y (2010) Environmental and fishing effects on the abundance, size and sex ratio of the blue-throated wrasse, *Notolabrus tetricus*, on South Australian coastal reefs. *Fisheries Management and Ecology* **17**, 209–220.

Smith DC, Montgomery I, Sivakumaran KP, Krusie-Golub K *et al.* (2003) 'The fisheries biology of bluethroat wrasse (*Notolabrus tetricus*) in Victorian waters'. FRDC report No. 97/128. Marine and Freshwater Resources Institute, Queenscliff.

Southcott RV (1987) The occurrence of *Desmonema gaudichaudi* (Lesson) (Scyphozoa, Semaeostomeae) in South Australian waters with records of fish–jellyfish symbioses. *Transactions of the Royal Society of South Australia* **111**, 131–132.

Stewart J, Hughes JM (2005) Longevity, growth, reproduction and a description of the fishery for silver sweep *Scorpis lineolatus* off New South Wales *New Zealand Journal of Marine and Freshwater Research* **39**, 827–838.

Tenison-Woods JE (1883) *Fish and Fisheries of Australia.* T Richards, Government Printer, Sydney.

Thomson JM (1959) Some aspects of the ecology of Lake Macquarie, N.S.W., with regard to an alleged depletion of fish. X. The movements of fish. *Australian Journal of Marine and Freshwater Research* **10**, 375–387.

Tuya F, Wernberg T, Thomsen MS (2009) Habitat structure affects abundances of labrid fishes across temperate reefs in south-western Australia. *Environmental Biology of Fishes* **86**, 311–319.

Tzioumis V, Kingsford MJ (1995) Periodicity of spawning of two temperate damselfishes: *Parma microlepis* and *Chromis dispulus. Bulletin of Marine Science* **57**, 596–609.

Tzioumis V, Kingsford MJ (1999) Reproductive biology and growth of the temperate damselfish *Parma microlepis. Copeia* (2), 348–361.

Wellenreuther M, Connell SD (2002) Response of predators to prey abundance: separating the effects of prey density and patch size. *Journal of Experimental Marine Biology and Ecology* **273**, 61–71.

Wheatley MJ (2000) Ecology of populations and assemblages of temperate reef fish in Port Phillip Bay, Australia. PhD thesis. Monash University, Melbourne.

Wilson AB, Ahnesjö I, Vincent ACJ, Meyer A (2003) The dynamics of male brooding, mating patterns, and sex roles in pipefishes and seahorses (Family Syngnathidae). *Evolution* **57**, 1374–1386.

Ziegler PE, Lyle JM, Haddon M, Ewing GP (2007) Rapid changes in life-history characteristics of a long-lived temperate reef fish. *Marine and Freshwater Research* **58**, 1096–1107.

Zemke-White WE, Choat JH, Clements RD (2002) A re-evaluation of the diel feeding hypothesis for marine herbivorous fishes. *Marine Biology* **141**, 571–579.

17 Planktivores and large reef-associated carnivores

OVERVIEW

This chapter discusses first three planktivorous species – hulafishes, bullseyes and trevally – all associated with reefs, which provide them shelter and sometimes food. Then we discuss some of the larger common carnivorous species with differing ecologies associated with reefs in some way. These include two poorly known site-attached species: the harlequin fish and the blue devil, which are both ambush predators. Each of them is spectacular in appearance, and iconic, and also long-lived. Next, some heavily exploited, roving carnivorous fish – breaksea cod, snapper, Westralian dhufish, yellowtail kingfish and barracouta – are discussed. Finally, the largest top predators, the sharks and rays, are discussed. These are illustrated by six sharks – the Port Jackson, grey nurse, two bronze whalers, wobbegongs and white shark – and the smooth stingray, which predates on and around reefs.

INTRODUCTION

The planktivores and larger carnivores considered here are either open-water species or species associated in some way with reefs. The plankton-feeding species typically live in schools and forage along reef edges or further offshore, while the benthic carnivores both shelter and feed in them. Some of these carnivores are ambush predators that lay hidden in crevices and caves for most of the time, and occasionally emerge to hunt for prey. Others are roving species that prey on other mid-water species and return to reefs at night, where they nestle in crevices or among sponges or other attached fauna. The white shark is pelagic, but often cruises over reefs.

PLANKTIVORES

Planktivores play an important role in reef communities. As plankton is transported by currents over a reef, planktivores typically gather in shoals on the upstream side and consume large quantities of plankton and detritus, at recorded consumption rates of up to one plankter per second. This feeding activity imports a large amount of energy to the reef, and the faecal matter provides a substantial input of nutrients to the reef (e.g. Bray *et al.* 1981).

Shoaling is a common behaviour in planktivorous fishes and benefits them in many ways. A fish shoal is safer from predation than an individual due to increased corporate awareness, and the fishes have greater foraging success because of the greater chances of finding a patch of plankton. Here, we shall discuss three common species groups on southern reefs – hulafishes, bullseyes and trevally.

Hulafishes *Trachinops* spp. (Family Plesiopidae)

Hulafishes, so named because of their distinctive 'dancing' behaviour, are common inhabitants of caves and crevices, where they form small to very large shoals. They are also common on vertical walls, which give access to planktonic food over a wide depth range, as well as protection from predators (Clynick 2008). Of the four southern species of *Trachinops*, the eastern hulafish, *T. taeniatus* ranges down the east coast to eastern Victoria, the southern hulafish *T. caudimaculatus* from the eastern Great Australian Bight (GAB) to Wilsons Promontory, and the yellowhead hulafish from Gulf St Vincent to the Abrolhos Is; only the two species below have been studied.

Eastern hulafish *Trachinops taeniatus*

The eastern hulafish swims in shoals of up to 800 individuals on shallow reefs and forages high in the water column on plankton, as well as on the bottom. Gregson and Booth (2005) studied their shoaling behaviour in relation to patches of planktonic food. They found that hulafish shoals tended to target large food patches containing high densities of copepods, with the result that copepod densities within patches were positively related to the size of fish shoals. The most likely explanation was that large shoals were better at locating food patches, and they were quickly joined by other fish observing their foraging success. The hulafish diet comprised 60% copepods and 40% bivalve larvae, plus a few polychaetes and euphausids.

The eastern hula is also useful as a negative indicator of sewage pollution in eastern Australia. In the region of outfalls, it suffers a higher mortality, slower growth and reduced reproduction success, compared with unpolluted sites, resulting in a sharp reduction in abundance (Smith and Suthers 1999; Gaston *et al.* 2004). Hence, monitoring of the species provides a valuable bio-indicator of the intensity and extent of contamination around outfalls.

Southern hulafish *Trachinops caudimaculatus*

The southern hulafish has been recently studied in Port Phillip Bay (Hunt *et al.* 2011; Ford and Swearer 2013). It spawns its eggs in nests on reefs in spring and, after hatching the larvae drift for up to ~8 km in local tidal currents and settle after 30–40 days on nearby reefs. The settling larvae grow to ~3 cm in the first month and reach a maximum size of 10–15 cm a few months later. In surveys of ~20 reefs around the margins of the Bay, recruitment between reefs varied greatly, due to three significant factors:

- The number of adults on a reef was a strong positive influence on the number of settlers.
- The abundance of plankton had a strong positive influence on settler abundance.
- The number of shelter holes (i.e. reef complexity), providing refuges from predation also influenced settlement.

By day, the southern hulafish shoals feed on plankton close to a reef, and at dusk fish retreat into small reef crevices until dawn. In this habitat, they face many predators, both benthic and pelagic and, in a unique study, Ford and Swearer (2013) examined the interactive effect of both predator types on them on artificial reef systems. Benthic predators are cryptic fishes, such as threefins, weedfishes and rock ling, and pelagic ones included squid, Australian salmon, snapper and yellowtail scad. At high hulafish densities, pelagic predators were the most significant, but as densities declined, shoaling behaviour was reduced and hulafishes foraged near the bottom in the domain of benthic predators, which then took their toll. The study also noted the benefits of prey shoaling, in enhancing food finding and corporate vigilance, reducing attacks and confusing predators.

Bullseyes *Pempheris* spp. (Family Pempheridae)

Bullseyes, noted for their deep bodies and large eyes half the length of the head, are common on southern coasts, and especially so under jetties where they prefer the low light conditions. They are fast swimmers with upward-pointing mouths, so are well equipped to catch mobile and elusive prey. They mainly range down to 30–40 m depth and are site-attached, returning to the same daytime refuge under a ledge or in the gloom of a cave after nocturnal feeding excursions over adjoining reef, seagrass or sand. These gregarious fish are far more abundant on reefs of high relief with abundant caves and crevices than on reefs of low relief (Harman *et al.* 2003). Six species occur in southern waters, but little is known about them except for dietary studies of three species.

In SW Australia two species, the slender bullseye *Parapriacanthus elongatus*, and the rough bullseye *Pempheris klunzingeri*, are common between 5 and 35 m depth. They migrate at dusk from their reef shelter sites onto sandy bottoms up to a kilometre away, and feed on the bottom on the infauna and epifauna, and in the water column on planktonic species (Table 17.1). The slender bullseye feeds at the deeper end of the depth range and the rough bullseye at the shallower end. Both species fed maximally straight after sunset. In addition, the diet of the rough bullseye changes markedly with increasing size. The smallest size group (< 7 cm) feeds > 70% on mysids, whereas adults (15–17 cm) feed < 15% on mysids, and much more on amphipods and errant polychaetes. Platell and Potter (1999, 2001) in a comparative study of the diets of many species noted that, although many species shared the same depth region for feeding, their diets differed in the relative composition of prey taxa, and also with size, so reducing competition between and within species.

The common bullseye *Pempheris multiradiata* is another common species on southern reefs and its diet is known from the preliminary studies of Currie and Sorokin (2010). Like other bullseyes, it feeds almost entirely on small, planktonic crustaceans (Table 17.1). Predators of bullseyes in southern waters are the Australian sea lions and little penguins (McIntosh *et al.* 2006).

Table 17.1: Diet of three species of bullseyes and silver trevally.

Data for slender bullseye *Parapriacanthus elongatus*, rough bullseye *Pempheris klunzingeri* (after Platell and Potter 1999), for common bullseye *Pempheris multiradiata* (after Currie and Sorokin 2010), and for silver trevally *Pseudocaranx georgianus* at two sites (after French *et al.* 2013).

Food type	Slender bullseye	Rough bullseye	Common bullseye	Silver trevally	
				Perth	Esperance
Algae, seagrass	–		–	–	–
Amphipods, isopods, ostracods, etc.	60.8	38.0	77.8	11.0	45.6
Decapods (crabs, shrimp)	7.7	12.1	–	26.4	15.9
Mysids	18.1	22.3	8.7	–	–
Crustacean fragments	–	–	–	0.4	
Polychaetes	12.8[1]	21.1	13.5	2.0	5.1
Gastropods, chitons	–	0.1	–	19.5	5.0
Bivalves	–	4.2	–	7.1	5.8
Cephalopods	–	0.3	–	5.6	2.4
Echinoids	–	–	–	6.0	–
Ophiuroids	–	–	–	2.2	1.7
Fish	0.6	1.6	–	19.9	18.3
Other	–	–	–	–	0.2
Sediment, detritus	–	0.3	–	–	–

[1] Errantia (i.e. pelagic, crawling species) or Sedentaria (i.e. tube-dwelling species).

Silver trevally *Pseudocaranx georgianus* (Family Carangidae)

This is a schooling fish that lives in coastal waters around reefs, and above sandy bottoms or in seagrass. It ranges from northern New South Wales to SW Australia. It is a vigorous plankton feeder and is sometimes seen 'humping above the water as krill and other plankton are gulped down' (Doak 1991). But this trevally is versatile in its feeding and, if plankton are not available, it will forage on the bottom, taking a variety of benthic invertebrates. The species has been studied in SW Australia and in New South Wales (Rowling and Raines 2000; Farmer *et al.* 2005; French *et al.* 2012).

Silver trevally is a long-lived, slow-growing species, reaching a maximum size of ~70 cm at an age of > 25 years. Growth rates for eastern and western Australia (Fig. 17.1), both based on fishes taken in deeper water, are dramatically different. However, growth of trevally from western shallower waters, where individuals only up to 12 years of age are found, is similar to that in New South Wales. Juveniles live in shallow waters around reefs and seagrass, moving about in small groups, and, with increasing size, gradually move into deeper water. The oldest fish live at depths of 60–120 m. Sexual maturity is reached at 3–4 years of age at a size of 28–31 cm, ~8 cm larger than in New South Wales. Silver trevally is a broadcast spawner and is thought to form spawning aggregations during the spawning season lasting from August to December in the west and extending to April in the east and south.

In New South Wales, silver trevally is taken by the trawl fishery, but catches have declined by 70–80% since the early 1980s. An effect of intense fishing is that the fastest growing fish are preferentially captured, leaving slower growing individuals to reproduce. Over time, the growth rate, size of sexual maturity, maximum size and productivity gradually decline. This is an evolutionary effect and thought to be the explanation for the growth differences between the east and west coasts of Australia.

In Western Australia, this trevally shares the same habitats with snapper from juvenile to adult stages. Although both species feed on reefs and sandy bottoms, their diets differ due to their different dentition, jaw morphology and behaviour, with silver trevally feeding on smaller and more mobile prey than snapper, and so reducing competition for food resources (Table 17.1).

A closely related southern species, sand trevally *Pseudocaranx wrightii*, lives and feeds on sandy bottoms (Platell *et al.* 1997), and so is outside the scope of this book.

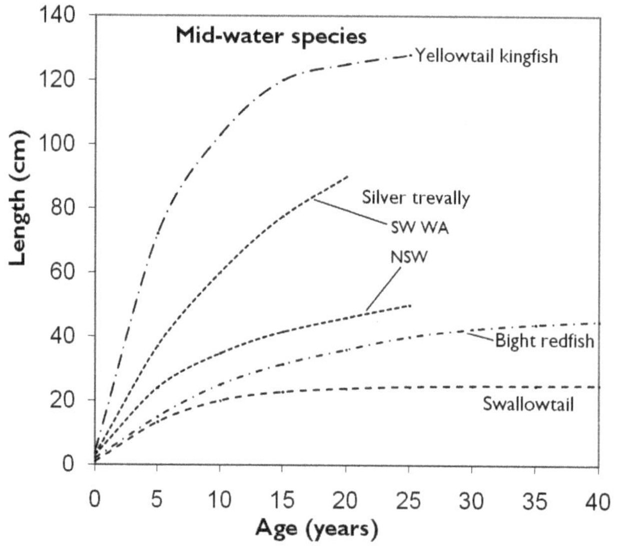

Figure 17.1: Growth of silver trevally *Pseudocaranx georgianus* from deep water in SW Western Australia (after Farmer *et al.* 2005) and in New South Wales (after Rowling and Raines 2000), yellowtail kingfish *Seriola lalandi* (after Stewart *et al.* 2004), bight redfish *Centroberyx gerrardi* and swallowtail *Centroberyx lineatus* (both after Saunders *et al.* 2010).

A ROVING CARNIVORE

Breaksea cod *Epinephelides armatus* (Family Serranidae)

This cod is endemic to shelf waters of SW Australia, ranging eastwards to the Recherche Archipelago in the GAB, and lives in reef habitat. It can grow to 50 cm size and is important for recreational fishers. The diet of this carnivore was studied by Platell *et al.* (2010) who found that it was a ram-feeder, eating mainly small fishes, most of them bullseyes (Table 17.2). They also noted a gradual shift in the diet of the fish with increasing size. The smallest size class (< 10 cm) ate 67% amphipods and 33% small fish, whereas adults had a diet comprised almost entirely of fish.

Breaksea cod has separate sexes, and females mature sexually at 7 years of age and ~30 cm size. The spawning season is prolonged over 8 months from October to May, with multiple spawnings occurring over this period. The cod is slow-growing with a longevity of ~20 years (Fig. 17.2).

Like other slow-growing, long-lived species, the cod is of some concern for management, because the minimum size at capture (30 cm) is the same as that of sexual maturity. Hence, males will be more susceptible to size-selective fishing, potentially causing a biased sex ratio and severely limiting sperm production.

AMBUSH PREDATORS AND NIGHT STALKERS

S Bryars and SA Shepherd

This group of fishes includes many cryptic and seldom seen fishes, such as moray and conger eels, scorpionfishes, gurnard perches, harlequin fish, rock cods and devil fishes. This section considers two species of conservation interest, the harlequin fish and the blue devil.

Harlequin fish *Othos dentex* (Family Serranidae)

This species (Plate 59) is distributed across southern Australia from Exmouth Gulf to Port Phillip Bay, but is possibly now extinct in Victorian waters. It reaches a maximum size of ~76 cm, and lives cryptically in caves and crevices or rests in the open on reefs to depths of > 30 m. It is a demersal piscivore and hunts during the daytime by sitting and waiting to ambush prey. The eyes are situated atop the head and able to

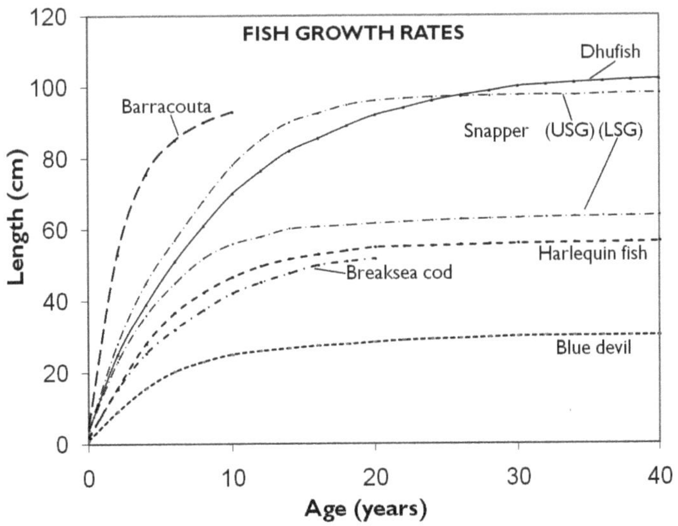

Figure 17.2: Growth rates of breaksea cod *Epinephelides armatus* (after Moore *et al.* 2007), barracouta from Tasmania (after Grant *et al.* 1978), male dhufish *Glaucosoma hebraicum* (after Hesp *et al.* 2002), snapper *Pagrus auratus* in upper and lower Spencer Gulf (after Fowler *et al.* 2004), harlequin fish *Othos dentex* and southern blue devil *Paraplesiops meleagris* (both after Saunders *et al.* 2010).

move independently of one another, and so are particularly well adapted for hunting. The teeth are long, sharp and thin, and also well suited to capturing fish. Acoustic tracking data indicate that harlequin fish activity is confined mainly to the daytime (SB unpublished data). Tracking data at a coastal reef site showed distinct movements of individuals up and down the reef slope during daylight – presumably to different ambush locations in search of prey. Prey includes leatherjackets, wrasses (*Saunders* et al. *2010*) and bullseyes, but preferred prey species are unknown.

The harlequin fish is site-attached with a home range of no more than a few hundred metres of coastline. However, at times a fish may go on longer excursions beyond its normal home range, and then return to its home base several days later (Bryars *et al.* 2011). The reason for such movements is likely due to storm activity. In addition to tracking work, photographic studies of individuals, using pattern recognition of the spots on the head, have recorded multiple re-sightings of the same individuals at the same places across many years (Bryars 2011).

The harlequin fish is a slow-growing species (Fig. 17.2) with a longevity of at least 42 years. Both features render it vulnerable to depletion from fishing.

Southern blue devil *Paraplesiops meleagris* (Family Plesiopidae)

This devilfish (Plate 60) has a similar geographic distribution to that of the harlequin fish above, while a sister species, the eastern blue devil *Paraplesiops bleekeri*, ranges along the New South Wales coast. Both species live to depths of ~40 m and are cryptic in caves by day, although emergent fish are sometimes seen by divers in the daytime. The diet of the southern blue devil is crabs, gastropods and fishes (Saunders *et al.* 2010). This species is also site-attached and, like harlequin fish, individuals can be identified by facial recognition. Repeated surveys of reefs using this facial recognition technique have shown that they have small territories of a few tens of metres in reef length, with some fish resident on the same reef for many years (Bryars 2011). Their territoriality is verified by observations of conflicts between an invader and a resident, as well as by their aggressive behaviour towards divers.

The growth rate of the southern blue devil is even slower than that of the harlequin fish, and the species may be longer lived, with a maximum known age of 59 years (Fig. 17.2). During the summer reproductive period, males and females form mating pairs and lay clutches of many thousands of small (~1 mm diameter) adhesive eggs in crevices or on the ceiling of caves. Males then guard the eggs until they hatch about a week later. Egg guarding by the same male in the same cave has been observed in consecutive summers. The newly hatched larvae are small (~4–5 mm length) and planktonic (SB unpublished data), but larval duration is unknown. Recruitment does not appear to be common because the distinctively coloured, small juveniles of ~5–10 cm are rarely seen.

PELAGIC HUNTERS

This section discusses the life history and ecology of seven pelagic species, the three largest of which have been overfished to varying degrees.

Snapper *Pagrus auratus* (sometimes placed in *Chrysophrys*) (Family Sparidae)

Snapper ranges throughout subtropical and temperate Australian coastal waters in reef and sand habitats to a depth of ~200 m, and is particularly abundant in bays and gulfs (e.g. the South Australian gulfs and Shark Bay), where they are intensely fished (Kailola *et al.* 1993; Henry and Gillanders 1999). They occupy a broad daytime feeding niche (Table 17.2), eating benthic invertebrates and both pelagic and bottom-living fishes opportunistically. Consumed fishes include herring (clupeids), leatherjackets, weedfishes, pipefishes, sardines and toadfishes. Juveniles in Spencer Gulf feed mainly on worms and shrimps, gradually shifting to crabs and fishes with increasing size (see also Jones 1981), while in SW Australia juveniles feed much more on brittle-stars and small crustaceans, but their diet is otherwise similar with increasing size (French *et al.* 2012). In deeper shelf waters (40–200 m), snapper eat almost entirely bottom-living fishes (96%), with 4% invertebrates (Bulman *et al.* 2001).

Table 17.2: Diet of four species: breaksea cod *Epinephelides armatus* (percentage volume) (after Platell *et al.* 2010); small and large snapper *Pagrus auratus* (percentage volume) in northern Spencer Gulf (after Lloyd 2010; Saunders *et al.* 2012a) and in SW Australia (after French *et al.* 2012); yellowtail kingfish *Seriola lalandi* (percentage weight) in northern and southern Spencer Gulf (SG) (after Fowler *et al.* 2003); and West Australian dhufish *Glaucosoma hebraicum* (percentage volume) (after Platell *et al.* 2010).

		Snapper				Yellowtail kingfish		
		Spencer Gulf		SW Aust.				
Food type	Breaksea cod	< 10 cm	> 20 cm	< 40 cm	> 40 cm	North	South	Dhufish
Algae, seagrass	0.1	–	4.0	–	4.1	4.3	–	1.5
Amphipods, isopods, mysids	0.6	1.1	2.7	3.4	2.5	4.6	1.2	34.4
Decapods (crabs, shrimps)	17.2	7.1	57.8	14.0	16.6	23.2	–	10.8
Crustacean fragments	–	8.3	–	1.3	–	–	–	–
Polychaetes	–	57	0.2	3.3	1.5	–	–	1.9
Gastropods, chitons	–	–	0.1	3.4	6.4	1.9	–	–
Bivalves	–	–	4.3	5.6	4.3	6.2	–	–
Cephalopods	0.6	1.1	1.3	–	3.7	–	6.7	6.4
Sea-urchins	–	–	1.1	13.4	14.1	–	–	0.6
Other echinoderms	–	–	0.7	45.8[1]	16.8	–	–	–
Ascidians, salps	–	–	0.6	–	–	–	16.9	–
Fish	81.1	16.5	15.4	9.8	29.3	48.3	75.0	43.8
Other; unidentified	–	–	10.9	–	0.4	11.5[2]	0.2	–

[1] 96% were ophiuroids; [2] Not shown is the presence of some grain and fish pellets.

Life history

In the northern part of its range, snapper spawn in winter, but in the southern part in summer, when sea temperatures rise above ~19.5°C. Snapper is a group-spawning species and Doak (1991) described a mass spawning event in these words: 'Males rush around among the females, both sexes rising in groups to release eggs and sperm'. Mass spawning takes place in bays and gulfs, with the best known areas in Shark Bay, Western Australia, the upper South Australian gulfs, Port Phillip Bay, Victoria and the estuaries of New South Wales (Fowler and Jennings 2003; Hamer and Jenkins 2004; Saunders 2009; Saunders *et al.* 2012b). The spawning season lasts for several months and at its peak spawning activity is almost daily.

Spawned eggs hatch after 36 h into larvae ~2 mm size, and the plankton-feeding hatchlings drift with the currents for 3–4 weeks before settling on soft bottoms in sheltered bays, estuaries and gulfs at a size of ~1.2 cm. Soft sediments have many small-scale features that favour juvenile survival. These are holes, burrows, mounds and epifauna, such as ascidians and sponges, all of which help provide refuge from predation for young fish (Thrush *et al.* 2002; Ross *et al.* 2007; Saunders *et al.* 2012a). Reefs and the sand–rock interface also provide food and cover.

Snapper grows to a maximum size of 1.3 m at rates which vary widely over its range. The two extreme growth rates from Spencer Gulf (Fig. 17.2) illustrate this variability, which Lloyd (2010) attributed to differences in the available food. In the southern Gulf, gut fullness of fish was much lower and, notably, crab consumption was 40% lower than in the northern Gulf. Earlier studies of snapper in Victoria (Sanders and Powell 1979) gave growth rates that were intermediate between the two extremes in Spencer Gulf. Longevity of the species is ~60 years, but recent studies have rarely recorded fish > 30 years old, due to intense fishing throughout its range.

Snapper reaches sexual maturity at an age that varies over its distribution (i.e. at ~2 years in South Australia, at 3 years in New South Wales and at 4–6 years in Victoria). In New Zealand, some maturing juveniles are initially female before changing sex to become male – a behaviour not yet known for Australian populations.

Movement

On settlement in sheltered habitats, juvenile snapper establish feeding territories and for the next 3–4 years remain in their nurseries. Juveniles then leave the nursery areas and disperse throughout adjoining coastal waters, sometimes moving distances of several hundred kilometres. For example Jones (1990) recorded movement of tagged fish from upper Spencer Gulf into the eastern GAB and into Gulf St Vincent, and Sanders (1974) noted long-distance movements from eastern Victoria to southern Queensland and from Port Phillip Bay to eastern South Australia. At an age of ~6 years, adults settle down in the region where they have moved, swimming from reef to reef, with local average movements of ~0.25 km day^{-1} (Jones 1981; Fowler *et al.* 2005; Fowler and McGlennon 2011). The record for time at liberty of tagged fish is that of an adult snapper tagged in mid-Spencer Gulf in the early 1960s by CSIRO and recaptured in the same area > 20 years later (K Jones *pers. comm.*).

Recruitment to the fisheries

Recruitment has proved to be extremely variable throughout southern waters, with many years of poor recruitment interspersed by an occasional strong recruitment year up to 200 times greater than other years. Continuous studies in Spencer Gulf since the early 1970s have shown strong recruitment years only in 1973, 1979, 1991 and 1997. The drivers of such variability seem to be sea temperature, larval food supply and variable mortality, but their interactions are ill understood (Hamer and Jenkins 2004; Fowler and McGlennon 2011). In Port Phillip Bay, juvenile recruitment strength depends on planktonic prey abundance and its effect on larval diet (Murphy *et al.* (2012).

Snapper is intensely fished in subtropical and temperate Australian waters, with a current total commercial catch of ~1750 t year^{-1}, down from 2500 t year^{-1} in the early 1980s; more than a third of the Australian catch is taken in the South Australian gulfs. The State fisheries are maintained by the occasional strong recruitment years and catches fluctuate between years accordingly. Management is dependent on fishery-based statistics, and is complex because:

- The dense schools that form during the spawning season are heavily fished by fishers who, over the years, have become increasingly efficient in locating them.
- The principal measure of fish abundance, catch per unit effort, is hyperstable due to the aggregation of fish during spawning; hence, it is of limited use as an index of abundance, and needs to be interpreted cautiously.

Fisheries are managed by input controls (minimum sizes, spawning closures, longline-hook limits and other gear limitations). McGarvey *et al.* (2011) examined the performance of the various management strategies and found that, although all measures are important, controlling exploitation levels by spawning closures was the most efficient measure, because it increased recruitment levels.

Stock structure

Genetic studies have shown several distinct stocks throughout southern Australia. One stock occurs from northern New South Wales to eastern Victoria, with others in western Victoria, South Australia and Western Australia. Shark Bay differs notably because separate stocks exist in the east and west parts of the Bay, indicating a lack of movement of adult fish between them (Johnson *et al.* 1986).

West Australian dhufish *Glaucosoma hebraicum* (Family Glaucosomatidae)

Dhufish is a fishing icon, endemic to Western Australia. It is a site-attached species living in reef habitats to a depth of 50 m and ranges from Shark Bay to the Recherche Archipelago. This species has a large, rounded mouth, without canines, but with slender, finely pointed, small teeth, adapted for suction feeding on small fishes.

The diet of dhufish changes with size. Juveniles < 10 cm eat > 80% small crustaceans (e.g. amphipods) and gradually change their diet with size, so that large fish > 80 cm size feed > 80% on fish. Table 17.1 summarises the diet of small to large fish combined. This species and two other benthic species described in Chapter 16 – western foxfish and breaksea cod – share the same habitat. However, their different dentitions and feeding modes together reduce competition for food between them and facilitate their co-existence (Platell *et al.* 2010).

Peak larval settlement occurs in January–February and juveniles are first seen in April–May at 6–8 cm size within narrow crevices on low-relief limestone reefs and drop-offs. Later they move to reefs of higher relief. Growth is rapid and very similar to that of snapper in upper Spencer Gulf (see Fig. 17.2) and in SW Australia, and both species have a similar longevity of > 40 years. Fish become sexually mature at 3–4 years of age at a size of 30–35 cm, and grow to a maximum size of ~1 m for males and ~93 cm for females. Dhufish form spawning aggregations and have a complex social behaviour, in which a few dominant males monopolise some females, while other males and females spawn in pairs.

Dhufish is intensely fished in SW Australia and mortality is ~20% per annum, raising serious concerns about the sustainability of the fishery (Hesp *et al.* 2002).

Yellow-tail kingfish *Seriola lalandi* (Family Carangidae)

This pelagic species ranges widely throughout subtropical and temperate waters of Australia and the SW Pacific, and schools of them often aggregate around reefs and islands. They are sensitive to sound and, attracted by a diver's bubbles, will circle around inquisitively and then just as suddenly disappear. They also have very small swim bladders, enabling them to dive rapidly to > 70 m depth. An extraordinary altruistic behaviour of the kingfish reported by Doak (1991) is that it will hover close by an injured companion to give it support.

Although fished commercially and recreationally over their range, and cultured extensively in Spencer Gulf until recently, there are few studies of their ecology (see review of Henry and Gillanders 1999). Curiously, the escape of cultured fish from fish cages prompted studies on ways of discriminating between wild and cultured fish, due to fears of the genetic and other effects of escapees on wild populations (Fowler *et al.* 2003; Gillanders and Joyce 2005).

The kingfish is a schooling fish, feeding mainly on fishes, but it also feeds on a range of other species, such as cephalopods (Table 17.2). The common fish and cephalopods consumed in Spencer Gulf were: red bait, blue sprat, blue mackerel, tommy ruff, weeping toado, squid and cuttlefish. However, some differences in the food eaten by escapees and wild fish in Spencer Gulf are also evident. These may be partly due to food availability at the sites sampled (e.g. the greater abundance of crabs and prawns in the upper Gulf and salps in the lower Gulf, and partly to the previous feeding history of the escapees). When confined in cages, they were fed pellets from the surface and this explains their occasional feeding on floating plant material, and also their unusual behaviour of milling around small craft expecting to be fed!

Their spawning behaviour is well described by Moran *et al.* (2007). Spawning occurs in summer when temperatures rise above ~17°C. Courtship between a pair lasts for about an hour, with the male taking up a position under the female, his snout under her gonoduct. They swim erratically, with occasional bursts of speed and intermittent stalling mid-water with bodies touching. About 15 minutes before spawning, the male starts to nip her oviduct – a behaviour thought to provide tactile stimulation that triggers spawning. Spawning then occurs in a circular swimming frenzy and lasts for ~20 s. Sometimes a second male joins the frenzy and adds his sperm for good measure. After fertilisation, eggs hatch in 2–3 days, according to temperature, and thereafter growth is rapid (Fig. 17.1). Juveniles feed at sunrise and sunset on zooplankton and small fish, and females reach sexual maturity at an age of 3 years and a size of 60–70 cm; males can mature as young as 1 year of age.

Tagging studies show that individual fish range over ~50 km, but occasionally they have been recorded swimming > 3000 km across the Tasman Sea to New Zealand and to offshore islands of the SW Pacific (Gillanders *et al.* 2001). Yellowtail kingfish range throughout the Pacific, but the Australian stocks are considered a separate subspecies from those in the north Pacific. The main commercial fishery for kingfish is now in New South Wales, with an annual production of ~130 t, down from ~500 t during the 1980s. Annual catches are ~50 t in Western Australia and < 0.5 t in South Australia, where a large commercial fishery once existed in the 1950s–1960s. The species is overfished in all three States, and schools, once abundant in coastal waters, are now very rarely seen (Stewart *et al.* 2004; K Jones *pers. comm.*).

Nannygai, bight redfish and swallowtail, *Centroberyx* spp. (Family Berycidae)

Nannygai *Centroberyx affinis* ranges along the east Australian coast from southern Queensland to Bass Strait, bight redfish *Centroberyx gerrardi* ranges from SW Australia to Bass Strait, and swallowtail *Centroberyx lineatus* from Shark Bay to SE New South Wales. The three species are all similar in appearance and have some similarities in their ecology. They are schooling species that often congregate around reefs and above soft bottoms from 5 to > 300 m deep. There is some evidence that smaller fish inhabit inshore waters and gradually move offshore with increasing size (Kailola *et al.* 1993; Saunders *et al.* 2010). Nannygai feed on small fish, crustaceans and molluscs, and also are reported to swim high above the reef in the upper water column at night to feed on plankton.

All three species are long lived. Bight redfish lives for > 70 years, swallowtail for > 30 years, and nannygai for at least 16 years. The growth rate of two of the species (Fig. 17.1) shows that are they are relatively slow growing, with little further growth after 20 years. Bight redfish reach sexual maturity at an age of 9 years, and are multiple spawners during a 4-month spawning season from February to May (Brown and Sivakumaran 2007; Stokie and Krusic-Golub 2005). Nannygai are sexually mature at 4 years of age and have a similar spawning season to bight redfish.

Nannygai and bight redfish are taken by commercial trawlers off New South Wales and in the GAB, respectively.

Barracouta *Thyrsites atun* (Family Gempylidae)

This large (1.4 m) predatory fish ranges throughout the southern half of Australia and once comprised one of the largest southern fisheries, mainly in SE Australia, with an annual production of > 4000 t. The fishery apparently collapsed in the 1970s, but then the fish fell out of favour due to the prevalence of parasites in the flesh; the present status of the stocks is uncertain. It is taken occasionally by recreational fishers and is a minor by-catch in commercial trawl fisheries in SE Australia.

Barracouta feeds in surface waters on krill and other small crustaceans, pilchards, anchovy, jack mackerel and squid. Its biology is known largely from studies of Maurice Blackburn and colleagues (see Blackburn 1957 and references). The size of sexual maturity is ~50 cm at an age of 2–3 years, but the spawning season varies with different stocks. The species has a preferred temperature range of 12–18°C and the population off eastern Victoria makes a spawning migration northwards to New South Wales in winter–spring, returning south after the spawning season. The central Bass Strait population migrates westwards for hundreds of kilometres in spring to summer for spawning, returning east after spawning. Two other stocks in SW Australia and the GAB spawn in late autumn–winter. Growth of barracouta is very rapid (Fig. 17.2) and longevity is ~10 years.

SHARKS AND RAYS (CLASS CHONDRICHTHYES)

If success is measured by the survival of ancient forms, sharks and rays must be the most successful of all fishes, with an ancestry going back > 400 million years. They are distinguished from teleost fishes by having a skeleton of calcified cartilage, special organs (claspers) for transferring sperm to females to fertilise her developing eggs, and back rows of replacement teeth that eventually replace the functioning ones from behind, as the latter wear out. The class has two reproductive strategies – viviparity and oviparity. In the former, the shark or ray retains the developing eggs and embryos within the uterus until they emerge as juveniles; in the latter strategy, the shark or ray lays an egg case containing a fertilised egg in a reef crevice where it develops and eventually hatches into a juvenile.

Another remarkable feature of this group of fishes is that they can detect electric fields with their electrosensory system, termed the ampullae of Lorenzini. These are clustered in the head and enable the shark or ray to navigate its way and detect prey (Tricas 2001). In this section, we consider a few reef-associated and roving species whose ecology has been studied.

Port Jackson shark (*Heterodontus portusjacksoni*) (Family Heterodontidae)

This small shark ranges in coastal waters along the southern half of Australia, and the pioneering observations by Ken O'Gower (1995) have been followed by the detailed work of Powter and Gladstone (2008 a,b,c,d, 2009) and Powter *et al.* (2010). Adults come into shallow reef areas annually during the reproductive season from July to November, and David Powter observed their mating behaviour on several occasions. After following a female about, the male grasps the end of her pectoral fin and wraps his tail around her tail, and then unrolls and inserts his left clasper into her vent. They both roll over and lie upside down on the bottom for several minutes, and sometimes another male joins in the mating frenzy, before they finally separate. The female later lays an egg in a casing surrounded by a spiral flange, and she either puts it in her mouth and screws it deep into a narrow crevice or else leaves it at a crevice entrance to be washed in by the swell (Whitley 1983). During the reproductive season, the female will lay an average of 16 eggs.

When not mating, these sharks rest on the bottom by day, often in clustered groups. It has been suggested that social aggregation is an important behaviour of this shark, but Powter and Gladstone (2009) noted that females sometimes sheltered in gutters to avoid males, and at other times aggregated for mating with one or more males. In South Australia and Victoria, large groups of a hundred or more commonly come into gulfs and bays in spring to lay their eggs on reefs.

Port Jackson sharks go foraging at night, feeding on a wide range of demersal and pelagic species (Table 17.3). In New South Wales, they do not feed on

Table 17.3: Diet of juvenile (< 50 cm) and adult (> 75 cm for males and > 90 cm for females) Port Jackson sharks *Heterodontus portusjacksoni* (after Powter *et al.* 2010), bronze whaler *Carcharias brachyurus* (after Rogers and Huveneers 2009), juvenile (< 1 m) and adult (> 2 m) dusky whaler *Carcharias obscurus* (percentage occurrence data) (after Simpfendorfer *et al.* 2001), ornate wobbegong *Orectolobus ornatus* (after Huveneers *et al.* 2007b), western wobbegong *Orectolobus hutchinsi* (after Chidlow 2001) and white shark *Carcharodon carcharias* (percentage occurrence) (after Bruce 1992).

Data are percentage weight unless otherwise stated.

	Port Jackson shark		Bronze whaler	Dusky whaler		Ornate wobbe-gong	Western wobbe-gong	White shark
Food type	< 50 cm	Adults	Adults	< 1 m	> 2 m			Adults
Decapods (crabs, shrimp, prawns)	31.9	6.5	0.1	0.4	–	–	–	22
Worms	31.3[1]	0.6	–	–	–	–	–	–
Cephalopods	0.8	19.0	14.2	52.7	44.2	13.5	33.4	33
Other molluscs	5.5	0.8	–	–	–	–	–	–
Fish	28.9	73.0[2]	78.8[3]	58.3	61.1[4]	86.5[5]	66.6[6]	22
Sharks, rays	–	–	6.9	0.8	27.8	–	–	22
Seals	–	–	–	–	–	–	–	11
Dolphins	–	–	–	–	–	–	–	44
Sea birds	–	–	–	–	–	–	–	11
Other	1.6	0.1	–	–	–	–	–	–

[1] nemerteans, polychaetes and echiurans; [2] mainly flounder, sole and flathead; [3]94% sardines, and rest anchovy, snapper, wrasse and leatherjacket; [4] mainly tuna, sardines, wrasses, morwongs and leatherjackets; [5] mainly moray, snapper and luderick; [6] labrids, mullet, flathead, leatherjackets and eels.

sea-urchins, but elsewhere urchins, molluscs and crustaceans are variously common in the diet (Currie and Sorokin 2010); in the vicinity of reefs, they are often seen with purple-stained lips from feeding on the purple urchin *Heliocidaris*. At the end of the reproductive season, the sharks leave inshore waters in New South Wales and swim south as far as southern Tasmania. Lincoln Smith (1999) recorded a swimming speed of 6.5 km day^{-1} and a distance of 250 km swum in 31 days by one shark. In the next winter, the sharks return to the same reef and site as they had occupied the previous year, and do so year after year, showing strong fidelity to the same area – an indication of a well-developed spatial memory (O'Gower 1995).

The eggs remain in crevices for 10–11 months until hatching. Initially, the egg capsule is sealed by mucous plugs at each end, but after ~4 months oxygen within the capsule becomes exhausted. The plugs then dissolve and open the capsule to sea water, so that movement of the embryo within draws in sea water, enabling it to breathe (Rodda and Seymour 2008). The egg capsule is susceptible to predation by numerous predators, such as the crested horn shark, eastern blue groper and the black stingray, and Powter and Gladstone (2008b) found that only 11% of the eggs survived to hatch into young sharks. The hatchlings then swim into nearby seagrass beds, where they live for several years and forage mainly on the infauna and epifauna there (Table 17.3). In this habitat, they have a home range of up to 40 000 m^2, centred around the sand–seagrass interface.

The dorsal fin spines of this shark lay down annual growth rings and these have been used to estimate its growth and longevity (Tovar-Ávila *et al.* 2009). Females were found to grow more slowly than males at first, but grew ~20 cm larger than males to a maximum size of ~1.3 m (Fig. 17.3); growth rates differ very little between the States (Izzo and Rodda 2012). Other life history features of this shark are summarised in Table 17.4. The annual mortality rate is ~20% for juveniles and 7% for adults.

At least two stocks of the Port Jackson shark exist in SE Australia: one from New South Wales southwards to eastern Victoria and a second one west of Wilsons Promontory. It is commonly taken as by-catch in demersal otter trawl and shark fisheries, but is usually returned to the water. Although susceptible to overfishing, this is unlikely to occur under present fishing practices (Tovar-Ávila *et al.* 2007, 2010).

Grey nurse shark *Carcharias taurus* (Family Odontaspidae)

This cosmopolitan species lives in temperate waters and ranges along the southern half of Australia to depths of 200 m at the shelf edge. However, it mainly frequents inshore waters < 40 m deep, where it rests in gutters and caves in reefs by day and forages at night near the bottom. It has the unusual behaviour of gulping down air to maintain buoyancy, so enabling it to hover just above the bottom. The grey nurse feeds voraciously on many species of fish, such as Australian salmon, kingfish and other sharks, and groups of them have been observed herding schooling fishes by 'tail thumping' on the water surface before attacking.

Using satellite tags, Bansemer and Bennett (2011) recorded the annual mating migration of mature males and non-gravid females northwards along the east Australian coast in late winter to early spring. Pregnant females stayed in the north until the next

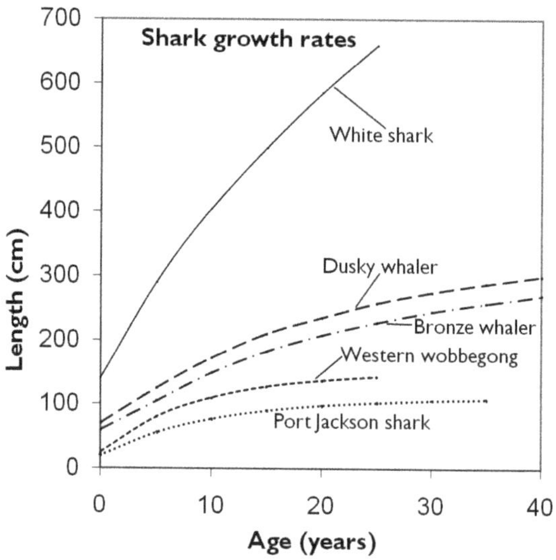

Figure 17.3: Growth rate of Port Jackson shark averaged for males and females (after Tovar-Ávila *et al.* 2009), western wobbegong *Orectolobus hutchinsi* (after Chidlow *et al.* 2007), bronze whaler *Carcharinus brachyurus* in the western Indian Ocean (after Jones 2008), dusky whaler *Carcharinus obscurus* (after Simpfendorfer *et al.* 2002), and white shark *Carcharodon carcharias* (after Cailliet *et al.* 1985). Length data for sharks are given as fork length for the two whalers, and as total length for the Port Jackson shark, western wobbegong and white shark.

winter–spring and then returned south to pup. In all, this shark spent 95% of its time at temperatures of 17–24°C and 74% of its time at depths of < 40 m. Once common on the New South Wales coast, the grey nurse numbers declined so seriously (> 99%, judging from catch rates) through fishing and beach meshing that it was declared a protected species there in 1984 (Pollard *et al.* 1996; Reid *et al.* 2011).

Males and females mature at a size of ~2.2 m at 6–12 years of age and can live for > 50 years (Table 17.4). It is an ovoviviparous species, producing one to two pups every 2 years. Courtship takes place in October–November and is a lengthy process, with the male following a female and nipping her flanks or fins until mating occurs. Eggs are fertilised in the oviduct before a variety of egg capsules are produced – some with fertilised eggs and others unfertilised. The egg capsules pass into the uteri where they stay for 5–6 months. Several embryos ~10 cm size hatch in each uterus during this period and feed on the egg yolk as they develop teeth. The largest in each uterus then feeds on other embryos and unfertilised eggs – a behaviour called uterine cannibalism – and, after a total gestation period of 9–12 months, one or two pups are born at 1–1.2 m size. Growth of juveniles is ~2 cm a month (Gilmore *et al.* 1983).

Dusky sharks *Carcharinus obscurus* and bronze whalers *Carcharinus brachyurus* (Family Carcharhinidae)

Dusky sharks range throughout Australian waters to > 500 m depth and bronze whalers range over southern temperate waters, south to Bass Strait in inshore waters on reefs and over seagrass, and offshore in open waters to ~100 m depth. The two species feed on pelagic fishes, often schooling species, but more often on benthic fishes on reefs and in seagrass; juveniles appear to feed more on cuttlefish, octopus and squid

Table 17.4: Life history features of common sharks and a ray; the Port Jackson shark *Heterodontus portusjacksoni* (after Tovar-Ávila *et al.* 2007, 2009, 2010); western wobbegong *Orectolobus hutchinsi* (after Chidlow *et al.* 2007); gulf wobbegong *Orectolobus halei* (after Chidlow 2001; Huveneers *et al.* 2007c, 2011; Corrigan *et al.* 2008); grey nurse shark *Carcharias taurus* (after Pollard *et al.* 1996); dusky whaler *Carcharhinus obscurus* (after Simpfendorfer *et al.* 2002); bronze whaler *Carcharhinus brachyurus* (after Jones 2008); white shark (after Cailliet *et al.* 1985; Francis 1996); and smooth stingray *Dasyatis brevicaudata* (after Last and Stevens 2009).

BF = breeding frequency in years.

Species	Length at maturity (cm): Age at maturity (years) Female	Male	Max. length (m): Longevity (years)	Litter size	Gestation period (months) BF	Birthing season
Port Jackson shark	90 15	77 12	1.3 35+	10–16 Eggs	– Annual	July–Nov. (eggs laid)
Western wobbegong	110 ~10	110 ~10	1.5–2 25+	18–29	9–11 2–3 years	July–Sept.
Gulf wobbegong	159	178	3	17–47	10–11 3 years	Sept.–Dec.
Grey nurse shark	220 9–10	220 6–7	4.3 55	2	9–12	Spring
Dusky whaler	~280 ~20	~210 15–20	3.7 > 45	5–14	11 2–3 years	Summer
Bronze whaler	245 19	216 13–19	3.7 ~45	7–22	15–21 2–3 years	Summer
White shark	~490 9–10	~430 9–10	6–8 25–30	6–13	12–18? 2–3 years?	Spring–summer
Smooth stingray	~100[1] width	~100 width	3 17 (?)	6–10	?	Spring–summer

[1]The smooth stingray is measured by width rather than length.

than on fish, as shown by the dietary study of small to large duskies (Table 17.3). In South Australia, the diet of duskies is similar to that in Table 17.3, and the fish mainly eaten are barracouta, mackerel and snapper (Rogers 2011; Rogers *et al.* 2012).

The life history of the two species is similar (Table 17.4). They are both long lived, and slow growing, producing young once every 2–3 years. However, duskies grow slightly larger (Fig. 17.3) and produce fewer young than bronzies. These features make them very susceptible to overfishing, which can lead to depletion of the breeding stock and recruitment failure. The two whalers have among the poorest rebound potential of any known sharks, largely due to their late age at maturity (Smith *et al.* 1998) and catches in Australian States have either declined or are unsustainable.

Dusky sharks migrate extensively in the southern Australian region. On the lower west Australian coast, they migrate to the NW shelf in winter to mate, returning southwards to give birth the next autumn. In New South Wales, duskies migrate northwards in winter, to later return to the south, and on the southern coast seasonal offshore–inshore migrations from shelf waters have been recorded (Kailola *et al.* 1993). A small group of sub-adults, marked with pop-up satellite tags in lower Spencer Gulf in summer, foraged there for 1–3 months, and then migrated westwards to the SW Australian outer shelf waters (one even to the Abrolhos Is) – an average distance of 2225 km at ~12 km day^{-1} over 6 months (Rogers 2011). These results suggest extensive movement of dusky whalers between the west and southern coasts.

Bronzies are the most common whalers in the eastern GAB and South Australian gulfs region, and comprise ~80% of all whalers caught by fishers. Extensive tagging of juvenile to sub-adult whalers in this region showed that their movement was generally localised during 1–2 years at liberty. About 25% were caught < 5 km, and 33% between 5 and 50 km, from their tagging sites, while the remainder were recaptured within

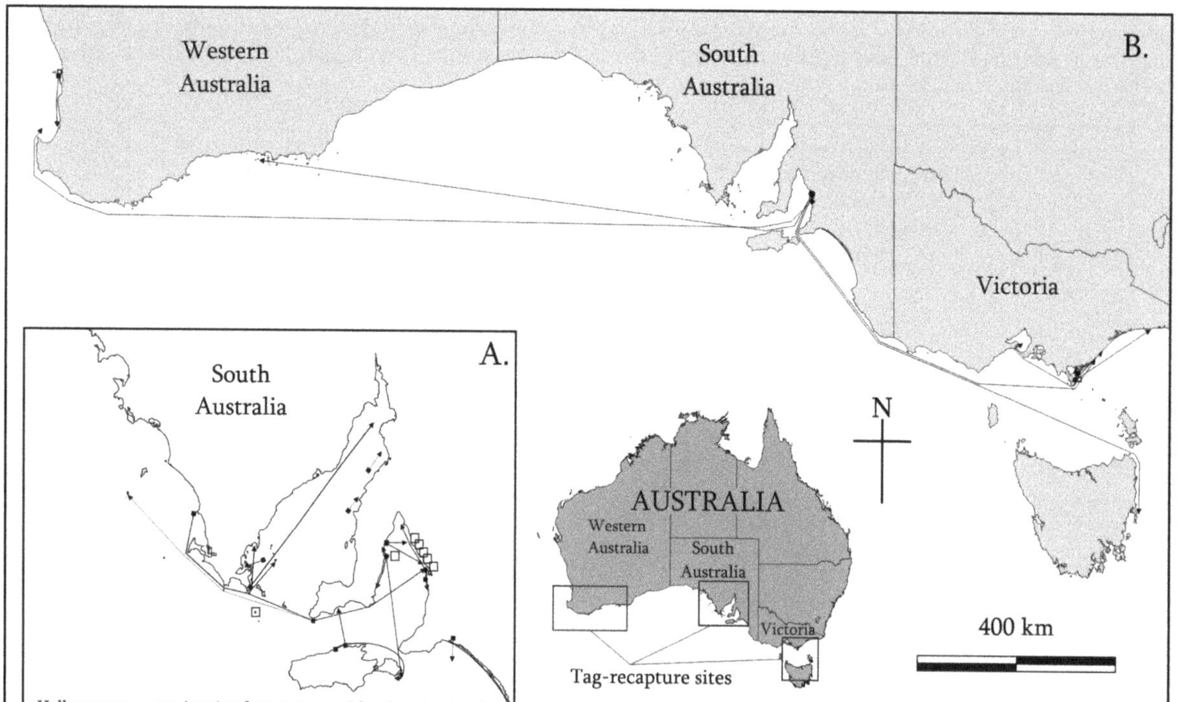

Figure 17.4: Longest recorded movements of tagged bronze whalers *Carcharinus brachyurus* and (inset) the common locally recorded movements of bronze whalers in the Gulfs region (after Rogers 2011).

370 km distance (Fig. 17.4). From his analysis of the results, Paul Rogers concluded that the gulfs are nurseries for juveniles and that they often return to the mid- and upper gulfs in spring and summer year after year, after localised movements of a few hundred kilometres during the rest of the year. In this sense, they show fidelity to a region.

Wobbegongs *Orectolobus* spp. (Family Orectolobidae)

Three species occur in Australian temperate waters. They are notable for their cryptic colouration that matches reef habitats, where they frequently rest by day and hunt by night. The three species are: the western wobbegong *O. hutchinsi* on the SW Australian coast; the gulf wobbegong *O. halei*, which ranges widely in southern Australia; and the spotted wobbegong *O. maculatus* on southern and eastern coasts; the last two species are the most common.

Wobbegongs are masterful ambush predators. They may be seen lying motionless, but alert, on the bottom in a reef crevice or crack, where they can swiftly seize a passing fish. In studies of the New South Wales species, fishes comprised the bulk of the diet, with cephalopods (mainly octopus) a minor component. The fishes eaten by the ornate wobbegong were mainly demersal species, such as morays, snapper and luderick (Table 17.3), whereas the other species of wobbegong ate mainly groper, mackerel, drummer and sharks (Huveneers *et al.* 2007b). The gulf wobbegong tended to eat more pelagic species than the others. Overall, the trophic level of wobbegongs is high (~4.3), and at a similar level to marine mammals.

The habitat of the gulf wobbegong was studied by Carraro and Gladstone (2006). Out of six reef habitat types in New South Wales, this wobbegong had a clear preference for resting by day in structurally complex reef (i.e. sponge gardens, boulder habitats in urchin barrens, and on artificial reefs made of metal frames). It showed no preference for fringe habitat with kelps, and avoided sandy bottoms and seagrasses. Many identified individuals showed site-fidelity to particular resting sites, always at depths of < 20 m for up to 3 months – the period of frequent surveys. Cryptic sites were probably selected to avoid larger shark predators, such as tiger sharks, bull sharks and white sharks. Later, Huveneers *et al.* (2007a) studied movement of the gulf wobbegong using acoustic tags. This species also showed site-fidelity for up to 2 years. Typically, a wobbegong rests in a cave by day and makes excursions at night for hunting prey, and then returns to the same cave. It occupies this cave for weeks to months, and then goes on a more distant excursion for several weeks.

Wobbegongs are all oviparous and, at birth, a juvenile shark still has internal yolk-sac reserves, which are thought to provide it with food reserves until it can feed independently (Huveneers *et al.* 2011). The above three wobbegong species have the same basic life histories (Table 17.4): that is, a reproductive cycle every 3 years, a long gestation period of ~1 year, and a few live young produced triennially in early summer. The growth of two species has been described. The western wobbegong has a slow growth rate, reaching its maximum size of ~1.5 m in 24 years (Fig. 17.3) and the ornate wobbegong has a similar growth trajectory, but reaches a larger size of ~2.2 m length in 30 years (Chidlow 2001).

These life history features render wobbegongs highly susceptible to overfishing, as shown by the serious decline (> 60%) in catch in the New South Wales fishery within 10 years (Corrigan *et al.* 2008) and by the decline in Western Australia.

The white shark *Carcharodon carcharias* (Family Lamnidae)

The white shark is near the apex of marine food webs and, in keeping with that position, has a reputation of mythical proportions, which has inspired much research and many books. Yet much of its ecology and biology remains unclear.

Diet and behaviour

Young white sharks feed mainly on crabs, benthic fish and small elasmobranchs (e.g. rays and skates). The diet of large white sharks (Table 17.3) is opportunistic and varies according to the availability of large prey in different seasons. Dolphins, sharks, rays and seals are common prey. In April–June at seal colonies, when fur seal and sea lion pups are learning to swim and hunt in the shallows, the sharks will stay in the vicinity to take advantage of easy prey. Abalone divers have reported seeing blue gropers, leatherjackets and wrasses following a white shark fearlessly, and feeding secondarily on the left-overs from predated seals. At other times of the year, white sharks follow and feed on schooling fishes. For example, they will follow snapper to their spawning sites, as in the upper South Australian gulfs or eastern Victoria (Palmer 1981; Bruce 1992).

The shark's attack mode on adult seals is described by McCosker (1985) and Klimley et al. (1996). On sighting a seal, the shark rushes up from below and takes a massive bite at the mid-body or haunches, often releasing the seal and returning later to consume it. Seals are most vulnerable when at the surface and this is where most attacks take place. Similarly, 80–90% of attacks on humans take place at the surface.

Injuries inflicted on seals at colonies were examined by Shaughnessy (2006) and Shaughnessy *et al.* (2007) on Kangaroo I., South Australia. Australian sea lion pups and adult females were attacked most often in summer and autumn, with a peak of 17% in numbers injured in January. Most attacks were at the head and mid-body (42%) and the rear (39%), and females were more vulnerable than males because they often had to return to shore to feed their pups, whereas males foraged well out to sea.

Evidence, some of it anecdotal, is accumulating on the social organisation among white sharks, notably when hunting or feeding. This includes a dominance hierarchy where some are dominant and others subordinate, threat displays and swimming rituals that a shark will use to establish its social priority (Sperone *et al.* 2010). Sharks have a high brain mass to body mass ratio, and this provides them with the potential to develop complex social behaviours and bonds (reviewed by Jacoby *et al.* 2012).

Movement

The movements of white sharks have been followed extensively with satellite tags (Strong *et al.* 1992; Bruce *et al.* 2006; Robbins 2007). Four patterns of adult movement have been observed:

- They may spend days to months resident in one area, and 'island-patrol' as long as food remains available. Then they will suddenly leave and engage in one of the following.
- They may undertake 'inter-island cruising' where a shark swims from one seal colony to another, often many kilometres apart, guided possibly by 'an odour corridor' or by its electrosensory system, which can detect the Earth's magnetic field. By these means, the shark can travel a long way in a straight line in the absence of visual bottom landmarks. At other times, a shark will follow schooling snapper, mullet, tailor or Australian salmon, feeding on them as opportunity arises.
- A shark may leave an area and cruise much longer distances along a 'white shark highway' around the Australian coast (Fig. 17.5). The longest recorded distance swum by one tagged shark was ~6000 km in 220 days. In addition to these long-distance cruises, white sharks on the east Australian coast migrate seasonally northwards to warmer water towards winter and return southwards towards summer. Such movements all suggest some fidelity to temperate Australian waters. Most of the time on the shark highways is spent at depths < 10 m, with occasional off-shelf excursions to depths of 500 m or more.
- Lastly, trans-oceanic cruising has been recorded between Australia and New Zealand, and between South Africa and Australia. In the latter case, a female white shark swam to the Australia NW shelf and, after a short stay, returned home – a total distance of > 11 000 km in 99 days at an average speed of 4.7 km h^{-1} (Bonfil *et al.* 2005). On this cruise, she spent two thirds of her time in surface waters, probably using celestial clues and/or the Earth's magnetic field for guidance. She migrated to Australia during the mating season and the authors suggested that if she returned pregnant to South Africa to give birth that would be consistent with the genetic evidence!

In summary, southern white sharks show site fidelity at multiple scales – oceanic, continental and local – persistently returning to the their familiar home network of coastal hotspots. However, at one site, Neptune Is, where an ecotourism cage diving industry has developed, the presence of boats for > 200 days a

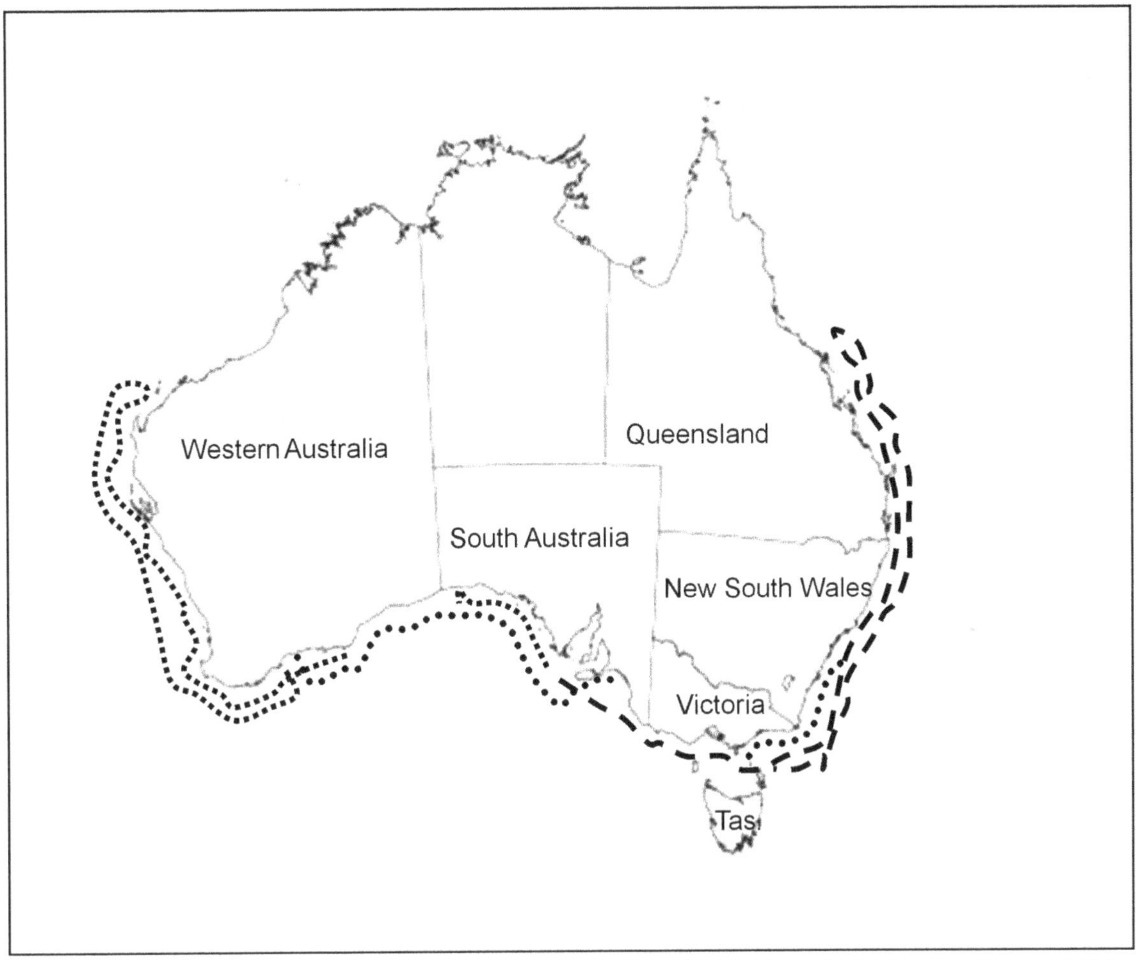

Figure 17.5: Map of Australia showing movements of white sharks along shark 'highways' (after Bruce and Stevens 2004; Bruce *et al.* 2006). All sharks, except one, were tagged in lower Spencer Gulf. The exception, tagged at Corner Inlet, central Victoria, moved up the New South Wales coast.

year, releasing berley to attract the white sharks, has markedly changed their behaviour. They now remain at the Island for 4–7 days at a time, compared with 1–2 days a decade before, and daily sightings have increased to ~7.

On its many cruises, the white shark generally swims at speeds of 3–4 km h^{-1}, staying near the surface or near the bottom, but seldom mid-water. Sexual segregation is also common, with females often remaining around inshore islands and males around offshore islands. Sometimes this may be due to the absence of females offshore when they are mating and giving birth in spring and summer (Robbins 2007). A size-based feeding hierarchy, in which younger male sharks defer to larger females, has also been observed, and this may also contribute to the segregation of the sexes.

The movement of juveniles is known only from a study of one white shark just a few weeks old, fitted with a satellite tag (Dewar *et al.* 2004). The behaviour of this young shark was quite different from that of adults. At night it remained in the top 50 m of the water column, with frequent excursions to the surface, and during the day it made deep dives of up to 165–240 m to the bottom, where it probably foraged. Generally, juveniles prefer warmer waters than adults – which helps to maximise their growth – but this individual was surprisingly tolerant of colder waters as low as 12°C.

Life history

The life history of the white shark is summarised in Table 17.4. After a spring–summer birth, the young shark grows rapidly (Fig. 17.3), reaching sexual maturity at an age of 9–10 years. Mating was once observed by Francis (1996), who wrote: 'At the beginning (they seemed to be) fighting, as one animal appeared to be attempting to grasp the other with its great mouth … however they eventually became motionless, one under the other, turning over from time to time belly to belly. This copulation lasted some forty minutes before the animals finally parted and glided off in opposite directions'.

After mating, some six to 13 pups develop within the uterus, but uterine cannibalism is unlikely, because advanced embryos are nourished by unfertilised eggs within the uterus (termed oophagy). Pups are born at a size of 1.2–1.5 m.

Although protected in Australia, the white shark is in low numbers and is susceptible to incidental capture in beach meshing and commercial fishing. It was listed in 2000 as 'vulnerable' on IUCN's Red List of threatened species (IUCN 2003).

Rays and skates

The body form of these groups of fishes is essentially a large disc formed by the fusion of the head and pectoral fins, with a short to long tail. Most of the 150 Australian species live on sandy bottoms or in estuaries, but two large species – the smooth stingray *Dasyatis brevicaudata* discussed below and the southern eagle ray *Myliobatis australis* – often rest in caves and forage on reefs. These rays have a long tail that carries a venomous barb, used for defence.

Smooth stingray *Dasyatis brevicaudata* (Family Dasyatidae)

This is the largest stingray in the world, reaching a length of > 4 m and a wing span of > 3 m. The species occurs widely in temperate waters of the Southern Hemisphere and has recently extended its range down the east Tasmanian coast (Last *et al.* 2011).

A ray tagged at a New Zealand site with satellite tags spent most of its time on the sea bed, and had a home range of < 25 km (Le Port *et al.* 2008). It overwintered in deeper water of 100–200 m and spent the rest of the year in shallow water. However, it is not clear whether this is the case in southern Australian waters. At West I., South Australia, where their numbers were monitored for 5 years, two recognisable individuals were resident in the locality for several years, resting in caves by day, and periodically emerging to forage on abalone and other molluscs on nearby reefs (Shepherd 1990). They also foraged on crustaceans and worms on sandy bottoms and in seagrass both by day and at night.

The life history of the smooth stingray is only sketchily known (Table 17.4). Courtship and mating occur mid-water. A receptive female is followed by a male who grips her disc, flips upside-down under her and then inserts a clasper into her vent, all the time waving his tail from side to side like a dog. This stingray is viviparous and developing embryos are nourished by the mother's uterine secretions.

Sharks and dolphins are the main predators of rays. A dolphin was once observed at the surface with a tight grasp on the forewing of a large smooth stingray. After shaking the ray violently, the dolphin took a large bite out of the wing, despite the ray desperately arching its tail to inflict a sting, and then released and pursued it for a second bite (personal observations). This stingray, as well as other rays and the Port Jackson shark, use their venomous spines for defence. The venom contains many protein fractions, such as serotonin and others. In humans, a sting provokes intense local pain, nausea and muscle cramps for up to 48 h, and is treatable by irrigation of the wound in hot water (Slaughter *et al.* 2009).

Conservation issues

Overfishing of apex predators, such as the sharks and rays, is a serious global issue. Firstly, they are especially vulnerable because of their life history features – slow growth, late sexual maturity, long lifespans, low fecundity and reproductive rate, and low natural mortality. In consequence, very few shark fisheries have been sustainably managed in Australia or elsewhere (reviewed by Stevens *et al.* 2000). Secondly, as top predators, their removal can seriously affect the structure and function of coastal food webs. Trophic cascades will likely occur as their prey, the mesopredators, increase and deplete prey lower in the food web (see Chapters 19 and 20). Such depletions go largely unrecognised because of the paucity of monitoring and research on the many non-commercial species affected (Myers *et al.* 2007).

ECOSIM models, which predict changes in fish and invertebrate biomass over time, can be used to predict ecosystem effects, but southern studies have not specifically examined the effects of shark collapses. Stevens *et al.* (2000) suggested that, due to factors such as prey-switching, changes in diet with size and cannibalism on sharks, the ecosystem effects of shark declines were likely to be diffuse and weak. Yet, some anecdotal evidence is available. Fishing for sharks in Tasmania caused an increase in abundance of one of its prey, the octopus, which, in turn, depleted populations of its prey, the southern rock lobster (Stevens *et al.* 2000). A similar cascade was reported in 1980 by AM Olsen (*pers. comm.*) in the SE of South Australia after overfishing of school shark there. In an ECOSIM study of another temperate system, Okey *et al.* (2004) predicted that collapse of shark populations would increase the abundance of dolphins, sea lions and many smaller fish predators, and declines of larger fish predators and octopods. Specific studies are needed to predict more precisely the effects of overfishing sharks in southern waters.

REFERENCES

Bansemer CS, Bennett MB (2011) Sex- and maturity-based differences in movement and migration patterns of grey nurse shark, *Carcharias taurus*, along the eastern coast of Australia. *Marine and Freshwater Research* **62**, 596–606.

Blackburn M (1957) The relation between the food of the Australian barracouta, *Thyrsites atun* (Euphrasen), and recent fluctuations in the fishery. *Australian Journal of Marine and Freshwater Research* **8**, 29–54.

Bonfil R, Meyer M, Scholl MC, Johnson R *et al.* (2005) Transoceanic migration, spatial dynamics and population linkages of white sharks. *Science* **310**, 100–103.

Bray RN, Miller AC, Geesey GG (1981) The fish connection: a trophic link between planktonic and rocky reef communities. *Science* **214**, 204–205.

Brown LP, Sivakumaran KP (2007) 'Spawning and reproductive characteristics of Bight redfish and deepwater flathead in the Great Australian Bight trawl fishery'. FRDC Report No. 2003/003. Primary Industries Research Victoria, Queenscliff.

Bruce BD (1992) Preliminary observations on the biology of the white shark, *Carcharodon carcharias*, in South Australian waters. *Australian Journal of Marine and Freshwater Research* **43**, 1–11.

Bruce BD, Stevens J (2004) 'Tracking the movement patterns of large white sharks in Australian waters'. Report to the Aquarium Institute of Western Australia. CSIRO, Hobart.

Bruce BD, Stevens JD, Malcolm H (2006) Movement and swimming behaviour of white sharks (*Carcharodon carcharias*) in Australian waters. *Marine Biology* **150**, 161–172.

Bryars S (2011) 'Monitoring marine fishes of conservation concern on Adelaide's coastal reefs: results of 2009/2010 surveys for the southern blue devil and harlequin fish'. Report to the Adelaide and Mount Lofty Ranges Natural Resources Management Board. Department of Environment and Natural Resources, Adelaide.

Bryars S, Rogers P, Huveneers C, Payne N *et al.* (2011) 'What size do marine park sanctuary zones need to be to protect the western blue groper *Achoerodus gouldii*, harlequin fish *Othis dentex*, and bluethroat wrasse *Notolabrus tetricus*?' Report to Department of Environment and Natural Resources. Wildlife Conservation Project No. 0638, Adelaide.

Bulman C, Althaus F, He X, Bax NJ *et al.* (2001) Diets and trophic guilds of demersal fishes of the south-eastern Australian shelf. *Marine and Freshwater Research* **52**, 537–548.

Cailliet GM, Natanson LJ, Welden BA, Ebert DA (1985) Preliminary studies on the age and growth of the white shark, *Carcharodon carcharias*, using vertebral bands. In *Biology of the White Shark: a Symposium Memoirs of the Southern California Academy of Sciences. Vol. 9* (Eds G Sibley, JA Seigel and CC Swift) pp. 49–60. California Academy of Sciences, Los Angeles.

Carraro R, Gladstone W (2006) Habitat preferences and site fidelity of the ornate wobbegong shark (*Orectolobus ornatus*) on rocky reefs of New South Wales. *Pacific Science* **60**, (2), 207–223.

Chidlow JA (2001) The biology of wobbegong sharks (Family Orectolobidae) from south-western Australian waters. MSc thesis. James Cook University, Townsville.

Chidlow JA, Simpfendorfer CA, Russ GR (2007) Variable growth band deposition leads to age and growth uncertainty in the western wobbegong shark, *Orectolobus hutchinsi*. *Marine and Freshwater Research* **58**, 856–865.

Clynick BG (2008) Characteristics of an urban fish assemblage: distribution of fish associated with coastal marinas. *Marine Environmental Research* **65**, 18–33.

Corrigan S, Huveneers C, Schwartz TS, Harcourt RG *et al.* (2008) Genetic and reproductive evidence for two species of ornate wobbegong shark *Orectolobus* spp. on the Australian east coast. *Journal of Fish Biology* **73**, 1662–1675.

Currie DR, Sorokin SJ (2010) The distribution and trophodynamics of demersal fish from Spencer Gulf. *Transactions of the Royal Society of South Australia* **134**, 198–227.

Dewar H, Domeier M, Nasby-Lucas N (2004) Insights into young of the year white shark, *Carcharodon carcharias*, behavior in the southern California Bight. *Environmental Biology of Fishes* **70**, 133–143.

Doak W (1991) *World of New Zealand Fishes*. Hodder and Stoughton, Auckland.

Farmer BM, French DJW, Potter IC, Hesp SA *et al.* (2005) 'Determination of biological parameters for managing the fisheries for mulloway and silver trevally in Western Australia'. FRDC Project 2002/004. Centre for Fish and Fisheries Research, Murdoch University, Perth.

Ford JR, Swearer SE (2013) Shoaling behaviour enhances risk of predation from multiple predator guilds in a marine fish. *Oecologia* **172**, 387–397.

Fowler AJ, Jennings PR (2003) Dynamics in 0+ recruitment and early life history for snapper (*Pagrus auratus*, Sparidae) in South Australia. *Marine and Freshwater Research* **54**, 941–956.

Fowler AJ, McGlennon D (2011) Variation in productivity of a key snapper, *Chrysophrys auratus*, fishery related to recruitment and fleet dynamics. *Fisheries Management and Ecology* **18**, 411–423.

Fowler AJ, Ham JM, Jennings PR (2003) 'Discriminating between cultured and wild yellowtail kingfish (*Seriola lalandi*) in South Australia'. SARDI Aquatic Sciences Publication No. RD03/0159. SARDI Aquatic Sciences, Adelaide.

Fowler AJ, Gillanders BM, Hall KC (2004) 'Adult migration, population replenishment and geographic structure for snapper in SA'. FRDC Final Report. No 2002/001. SARDI Aquatic Sciences Publication No. RD04/0163, Adelaide.

Fowler AJ, Gillanders BM, Hall KC (2005) Relationship between elemental concentration and age from otoliths of adult snapper (*Pagrus auratus*, Sparidae): implications for movement and stock structure. *Marine and Freshwater Research* **56**, 661–676.

Francis MP (1996) Observations on a pregnant white shark with a review of reproductive biology In *Great White Sharks: the Biology of* Carcharodon carcharias. (Eds AP Klimley and DG Ainley) pp. 157–174. Academic Press, San Diego.

French B, Platell ME, Clarke K, Potter IC (2012) Ranking of length-class, seasonal and regional effects on dietary compositions of the co-occurring *Pagrus auratus* (Sparidae) and *Pseudocaranx georgianus* (Carangidae). *Estuarine, Coastal and Shelf Science* **115**, 309–325.

Gaston TF, Kostoglidis A, Suthers IM (2004) The ^{13}C, ^{15}N and ^{34}S signatures of a rocky reef planktivorous fish indicate different coastal discharges of sewage. *Marine and Freshwater Research* **55**, 689–699.

Gillanders BM, Joyce TC (2005) Distinguishing aquaculture and wild yellowtail kingfish via natural elemental signatures in otoliths. *Marine and Freshwater Research* **56**, 693–704.

Gillanders BM, Ferrell DJ, Andrew NL (2001) Estimates of movement and life history parameters of yellowtail kingfish (*Seriola lalandi*): how useful are data from a cooperative tagging program? *Marine and Freshwater Research* **52**, 179–192.

Gilmore RB, Dodrill JW, Linley PA (1983) Reproduction and embryonic development of the sand tiger shark, *Odontaspis taurus* (Rafinesque). *Fishery Bulletin* **81**, 201–225.

Grant CJ, Cowper TR, Reid DD (1978) Age and growth of snoek, *Leionura atun* (Euphrasen), in south-eastern Australian waters. *Australian Journal of Marine and Freshwater Research* **29**, 435–444.

Gregson MA, Booth DJ (2005) Zooplankton patchiness and the associated shoaling response of the temperate reef fish *Trachinops taeniatus*. *Marine Ecology Progress Series* **299**, 269–275.

Hamer PA, Jenkins GP (2004) High levels of spatial and temporal recruitment variability in the temperate sparid *Pagrus auratus*. *Marine and Freshwater Research* **55**, 663–673.

Harman N, Harvey ES, Kendrick GA (2003) Differences in fish assemblages from different reef habitats at Hamelin Bay, south-western Australia. *Marine and Freshwater Research* **54**, 1–8.

Henry G, Gillanders B (1999) Snapper and yellowtail kingfish. In *Under Southern Seas: the Ecology of Australia's Rocky Reefs*. (Ed. N Andrew) pp. 159–163. University of New South Wales Press, Sydney.

Hesp SA, Potter IC, Hall NG (2002) Age and size composition, growth rate, reproductive biology, and habitats of the West Australian dhufish (*Glaucosoma hebraicum*) and their relevance to the management of this species. *Fishery Bulletin* **100**, 214–227.

Hunt TL, Ford JR, Swearer SE (2011) Ecological determinants of recruitment to populations of a temperate reef fish, *Trachinops caudimaculatus* (Plesiopidae). *Marine and Freshwater Research* **62**, 502–509.

Huveneers C, Harcourt R, Otway N (2007a) Observation of localised movements and residence time of the wobbegong shark *Orectolobus haleu* at Fish Rock, NSW, Australia. *Cybium* **30**, 103–111.

Huveneers C, Otway NM, Gibbs S, Harcourt RG (2007b) Quantitative diet assessment of wobbegong sharks (genus *Orectolobus*) in New South Wales. *ICES Journal of Marine Science* **64**, 1272–1281.

Huveneers C, Walker TI, Otway NM, Harcourt RG (2007c) Reproductive synchrony of three sympatric species of wobbegong shark (genus *Orectolobus*) in New South Wales, Australia: reproductive parameter estimates necessary for population modelling. *Marine and Freshwater Research* **58**, 765–777.

Huveneers C, Otway NM, Harcourt RG, Ellis NM (2011) Quantification of the maternal–embryonal nutritional relationship of elasmobranchs: case study of wobbegong sharks (genus *Orectolobus*). *Journal of Fish Biology* **78**, 1375–1389.

IUCN (2003) IUCN *Red List of Threatened Species*. Cambridge, UK. <www.redlist.org> (Accessed February 2013).

Izzo C, Rodda KR (2012) Comparative rates of growth of the Port Jackson shark throughout its southern range. *Marine and Freshwater Research* **63**, 687–694.

Jacoby DMP, Croft DP, Sims DW (2012) Social behaviour in sharks and rays: analysis, patterns and implications for conservation. *Fish and Fisheries* **13**, 399–417.

Johnson MS, Creagh S, Moran M (1986) Genetic subdivision of stocks of snapper, *Chrysophrys elongatus*, in Shark Bay, Western Australia. *Australian Journal of Marine and Freshwater Research* **37**, 337–345.

Jones GK (1981) Biological research on snapper (*Chrysophrys auratus* syn. *elongatus*) and an analysis of the fishery in northern Spencer Gulf. *SAFIC* **5**, (6), 5–8.

Jones GK (1990) 'Growth and size related movements of snapper in South Australian waters'. South Australian Department of Fisheries Internal Report, Adelaide.

Jones GK (2008) 'Review of the fishery status for whaler sharks (*Carcharinus* spp.) in South Australian and adjacent waters'. FRDC Report. Project No. F2004/067. SARDI Aquatic Sciences Publication No. F2007/000721-1. SARDI Aquatic Sciences, Adelaide.

Kailola PJ, Williams MJ, Stewart PC, Reichelt RE *et al.* (1993) *Australian Fisheries Resources*. Bureau of Resource Sciences and FRDC, Canberra.

Klimley P, Pyle P, Anderson SD (1996) The behaviour of white sharks and their pinniped prey during predatory attacks. In *Great White Sharks: the Biology of* Carcharodon carcharias (Eds P Klimley and DG Ainley) pp. 175–192. Academic press, San Diego.

Last PR, Stevens JD (2009) *Sharks and Rays of Australia*. 2nd edn. Harvard University Press, Cambridge.

Last PR, White WT, Gledhill DC, Hobday AJ *et al.* (2011) Long-term shifts in abundance and distribution of a temperate fish fauna: a response to climate change and fishing practices. *Global Ecology and Biogeography* **20**, 58–72.

Le Port A, Sippel T, Montgomery JC (2008) Observations of mesoscale movements in the short-tailed stingray, *Dasyatis brevicaudata* from New Zealand using a novel PSAT tag attachment method. *Journal of Experimental Marine Biology and Ecology* **359**, 110–117.

Lincoln-Smith M (1999) Reef sharks and rays. In *Under Southern Seas: the Ecology of Australia's Rocky Reefs*. (Ed. N Andrew) pp. 210–217. University of New South Wales Press, Sydney.

Lloyd M (2010) Spatial variation in the growth of snapper (*Chrysophrys auratus*) in South Australian gulf waters: an investigation of diet as a contributing factor. BSc (Hons) thesis. University of Adelaide, Adelaide.

McCosker JE (1985) White shark attack behaviour: observations of and speculations about predator and prey strategies. In *Biology of the White Shark: A Symposium. Memoirs of the Southern California Academy of Sciences. Vol. 9* (Eds G Sibley, JA Seigel and CC Swift) pp. 123–135. California Academy of Sciences, Los Angeles.

McGarvey R, Burch P, Feenstra JE (2011) Use of management strategy comparison scatterplots to identify clusters of high-performing strategies. *Fisheries Management and Ecology* **18**, 349–359.

McIntosh RR, Page B, Goldsworthy SD (2006) Dietary analysis of regurgitates and stomach samples from free-living Australian sea lions. *Wildlife Research* **33**, 661–667.

Moore SE, Hesp SA, Hall NG, Potter IC (2007) Age and size compositions, growth and reproductive biology of the breaksea cod *Epinephelides armatus*, a gonochoristic serranid. *Journal of Fish Biology* **71**, 1407–1429.

Moran D, Smith CK, Gara B, Poortenaar CW (2007) Reproductive behaviour and early development in yellowtail kingfish (*Seriola lalandi* Valenciennes 1833). *Aquaculture* **262**, 95–104.

Murphy HM, Jenkins GP, Hamer PA, Swearer SE (2012) Interannual variation in larval survival of snapper (*Chrysophrys auratus*, Sparidae) is linked to diet breadth and prey availability. *Canadian Journal of Fisheries and Aquatic Sciences* **69**, 1340–1351.

Myers RA, Baum JK, Shepherd TD, Powers SP *et al.* (2007) Cascading effects of the loss of apex predatory sharks from a coastal ocean. *Science* **315**, 1846–1850.

O'Gower AK (1995) Speculations of a spatial memory for the Port Jackson shark *Heterodontus portusjacksoni* (Meyer) (Heterodontidae). *Marine and Freshwater Research* **46**, 861–871.

Okey TA, Banks S, Born AF, Bustamante RH *et al.* (2004) A trophic model of a Galápagos subtidal rocky reef for evaluating fisheries and conservation strategies. *Ecological Modelling* **172**, 383–401.

Palmer E (1981) The white shark. *SAFIC Magazine* **5**, (2), 33–34.

Platell ME, Potter IC (1999) Partitioning of habitat and prey by abundant and smaller-sized species of the Triglidae and Pempheridae (Teleostei) in coastal waters. *Estuarine, Coastal and Shelf Science* **48**, 235–252.

Platell ME, Potter IC (2001) Partitioning of food resources amongst 18 abundant benthic carnivorous fish species in marine waters of the lower west coast of Australia. *Journal of Experimental Marine Biology and Ecology* **261**, 31–54.

Platell ME, Sarre GA, Potter IC (1997) The diets of two co-occurring marine teleosts, *Parequula melbournensis* and *Pseudocaranx wrightii*, and their relationships to body size and mouth morphology, and the season and location of capture. *Environmental Biology of Fishes* **49**, 361–376.

Platell ME, Hesp SA, Cossington SM, Lek E *et al.* (2010) Influence of selected factors on the dietary compositions of three targeted and co-occurring temperate species of reef fishes: implications for food partitioning. *Journal of Fish Biology* **76**, 1255–1276.

Pollard DA, Lincoln-Smith MP, Smith AK (1996) The biology and conservation status of the grey nurse shark (*Carcharias taurus* Rafinesque 1810) in New South Wales, Australia. *Aquatic Conservation: Marine and Freshwater Ecosystems* **6**, 1–20.

Powter DM, Gladstone W (2008a) The reproductive biology and ecology of the Port Jackson shark, *Heterodontus portusjacksoni* in the coastal waters of eastern Australia. *Journal of Fish Biology* **72**, 2615–2633.

Powter DM, Gladstone W (2008b) Embryonic mortality and predation on capsules of the Port Jackson shark *Heterodontus portusjacksoni* (Meyer). *Journal of Fish Biology* **72**, 573–584.

Powter DM, Gladstone W (2008c) Demographic analysis of the Port Jackson shark *Heterodontus portusjacksoni* in the coastal waters of eastern Australia. *Marine and Freshwater Research* **59**, 444–455.

Powter DM, Gladstone W (2008d) Habitat preferences of Port Jackson sharks, *Heterodontus portusjacksoni*, in the coastal waters of eastern Australia. *Proceedings of the Linnean Society of New South Wales* **129**, 151–165.

Powter DM, Gladstone W (2009) Habitat-mediated use of space by juvenile and mating adult Port Jackson sharks, *Heterodontus portusjacksoni*, in eastern Australia. *Pacific Science* **63**, 1–14.

Powter DM, Gladstone W, Platell M (2010) The influence of sex and maturity on the diet, mouth morphology and dentition of the Port Jackson shark, *Heterodontus portusjacksoni*. *Marine and Freshwater Research* **61**, 74–85.

Reid DD, Robbins WD, Peddemors VM (2011) Decadal trends in shark catches and effort from the New South Wales, Australia, shark meshing program 1950–2010. *Marine and Freshwater Research* **62**, 676–692.

Robbins RL (2007) Environmental variables affecting the sexual segregation of great white sharks *Carcharodon carcharias* at the Neptune Islands, South Australia. *Journal of Fish Biology* **70**, 1350–1364.

Rodda KR, Seymour RS (2008) Functional morphology of embryonic development in the Port Jackson shark *Heterodontus portusjacksoni* (Meyer). *Journal of Fish Biology* **73**, 573–584.

Rogers PJ (2011) Habitat use and dietary dynamics of pelagic sharks in coastal and shelf ecosystems off southern Australia. PhD thesis. Flinders University, Adelaide.

Rogers PJ, Huveneers C (2009) 'Diet and reproductive biology of pelagic sharks in southern Australia: understanding their ecology to enhance policy development and conservation management'. SARDI Aquatic Sciences Publication No. F2009/000162-1. Research Report Series No.362. SARDI Aquatic Sciences, Adelaide.

Rogers PJ, Huveneers C, Page B, Hamer DJ *et al.* (2012) A quantitative comparison of the diets of sympatric pelagic sharks in gulf and shelf ecosystems off southern Australia. *ICES Journal of Marine Science.* In press.

Ross PM, Thrush SF, Montgomery JC, Walker JW *et al.* (2007) Habitat complexity and predation risk determine juvenile snapper (*Pagrus auratus*) and goatfish (*Upeneichthys lineatus*) behaviour and distribution. *Marine and Freshwater Research* **58**, 1144–1151.

Rowling KR, Raines LP (2000) 'Description of the biology and an assessment of the fishery for silver trevally *Pseudocaranx dentex* off New South Wales'. FRDC Project 97/125. NSW Fisheries Research Institute, Cronulla, Sydney.

Sanders MJ (1974) Tagging indicates at least two stocks of snapper *Chrysophrys auratus* in south-east Australian waters. *New Zealand Journal of Marine and Freshwater Research* **8**, 371–374.

Sanders MJ, Powell DGM (1979) Comparison of the growth rates of two stocks of snapper (*Chrysophrys auratus*) in south-east Australian waters using capture–recapture data. *New Zealand Journal of Marine and Freshwater Research* **12**, 343–350.

Saunders RJ (2009) The reproductive biology and recruitment dynamics of snapper, *Chrysophrys auratus*. PhD thesis. University of Adelaide.

Saunders RJ, Bryars SR, Fowler AJ (2010) 'Preliminary consideration of the biology of several of South Australia's marine fish species that have conservation or management interest'. SARDI Research Report Series No. 143. No. F2009/000693-1. SARDI Aquatic Sciences, Adelaide.

Saunders RJ, Fowler AJ, Gillanders BM (2012a) The use of food resources by 0+ snapper, *Chrysophrys auratus*, from northern Spencer Gulf, South Australia. *Marine and Freshwater Research* **63**, 680–686.

Saunders RJ, Fowler AJ, Gillanders BM (2012b) The spawning dynamics of snapper, *Chrysophrys auratus*, in northern Spencer Gulf, South Australia. *New Zealand Journal of Marine and Freshwater Research*. In press.

Shaughnessy PD (2006) Instances of predation on fur seals by white sharks in South Australia. *Australian Mammalogy* **28**, 107–110.

Shaughnessy PD, Dennis TE, Berris M (2007) Predation on Australian sea lions *Neophoca cinerea* by white sharks *Carcharodon carcharias* in South Australia. *Australian Mammalogy* **29**, 69–76.

Shepherd SA (1990) Studies on southern Australian abalone (genus *Haliotis*). XII. Long-term recruitment and mortality dynamics of an unfished population. *Australian Journal of Marine and Freshwater Research* **41**, 475–492.

Simpfendorfer CA, Goodreid A, McAuley RB (2001) Diet of three commercially important shark species from Western Australian waters. *Marine and Freshwater Research* **52**, 975–985.

Simpfendorfer CA, McAuley RB, Chidlow J, Unsworth P (2002) Validated age and growth of the dusky shark, *Carcharinus obscurus*, from Western Australian waters. *Marine and Freshwater Research* **53**, 567–573.

Slaughter RJ, Michael D, Beasley G, Lambie BS *et al.* (2009) New Zealand's venomous creatures. *Journal of the New Zealand Medical Association* **122** published online <http:nzma.org.nz/122-1290/3494/> (Accessed July 2011).

Smith AK, Suthers IM (1999) Effects of sewage effluent discharges on the abundance, condition and mortality of hulafish, *Trachinops taeniatus* (Plesiopidae). *Environmental Pollution* **106**, 97–106.

Smith SR, Au DW, Show C (1998) Intrinsic rebound potentials of 26 species of Pacific sharks. *Marine and Freshwater Research* **49**, 663–678.

Sperone E, Micarelli P, Andreotti S, Spinetti S *et al.* (2010) Social interactions among bait-attracted white sharks at Dyer Island (South Africa). *Marine Biology Research* **6**, 408–414.

Stevens JD, Bonfil R, Dulvy NK, Walker PA (2000) The effects of fishing on sharks, rays and chimaeras (chondrichthyans), and the implications for marine ecosystems. *ICES Journal of Marine Science* **57**, 476–494.

Stewart J, Ferrell DJ, van der Walt B (2004) Sizes and ages in commercial landings with estimates of growth, mortality and yield per recruit of yellowtail kingfish (*Seriola lalandi*) from New South Wales, Australia. *Marine and Freshwater Research* **55**, 489–495.

Stokie TK, Krusic-Golub K (2005) 'Age estimation of Bight redfish and deepwater flathead in the Great Australian Bight trawl fishery, 2004/05'. AFMA Project No. RO4/1098. Primary Industries Research Victoria, Queenscliff.

Strong WR, Murphy RC, Bruce BD (1992) Movements and associated observations of bait-attracted white sharks, *Carcharodon carcharias*: a preliminary report. *Australian Journal of Marine and Freshwater Research* **43**, 13–20.

Thrush S, Shultz D, Hewitt JE, Talley D (2002) Habitat structure in soft sediment environments and abundance of juvenile snapper *Pagrus auratus*. *Marine Ecology Progress Series* **245**, 273–280.

Tovar-Ávila J, Day RW, Walker TI (2010) Using rapid assessment and demographic methods to evaluate the effects of fishing on *Heterodontus portusjacksoni* off far-eastern Victoria. *Journal of Fish Biology* **77**, 1564–1578.

Tovar-Ávila J, Walker TI, Day RW (2007) Reproduction of *Heterodontus portusjacksoni* in Victoria, Australia: evidence of two populations and reproductive parameters for the eastern population. *Marine and Freshwater Research* **58**, 958–965.

Tovar-Ávila J, Troynikov VS, Walker TI, Day RW (2009) Use of stochastic models to estimate the growth of the Port Jackson shark, *Heterodontus portusjacksoni*, off eastern Victoria. *Fisheries Research* **95**, 230–235.

Tricas TC (2001) The neuroecology of the elasmobranch electrosensory world: why peripheral morphology shapes behavior. *Environmental Biology of Fishes* **60**, 77–92.

Whitley GP (1983) *Australian Sharks*. 2nd edn. Lloyd O'Neil, Melbourne.

18 Marine mammals

OVERVIEW

This chapter describes the feeding, ecology and life history of three pinnipeds: the New Zealand fur seal *Arctocephalus forsteri*, the Australian fur seal *Arctocephalus pusillus* and the Australian sea lion *Neophoca cinerea*. They all have a moderate longevity of 20–25 years and feed at a high trophic level on cephalopods, seabirds and fish. Juvenile growth rates are similar, but there are substantial differences in maximum size between the species and the sexes. Populations of the two fur seal species are increasing, but the Australian sea lion has suffered a high mortality from gill-net fishing, and declines still occur in some important breeding colonies. We then describe the ecology and life history of three species of whale seen in southern waters. The pygmy blue whale *Balaenoptera musculus brevicauda* visits several coastal upwelling regions to feed on krill on its way to and from tropical calving grounds. The southern right whale *Eubalaena australis* migrates into coastal calving grounds annually in winter at many sites along the southern Australian coast to calve and mate and then returns to Southern Ocean feeding grounds. The orca or killer whale *Orcinus orca* is most often seen in SE Australian waters near seal colonies, where it feeds on fish, dolphins, seals and whales. The orca has a high level of intelligence, as shown by a suite of complex cultural and social behaviours. Because marine mammals feed at moderate to high trophic levels, they play a significant role in marine ecosystems.

INTRODUCTION

Marine mammals evolved ~50–60 Ma in the shallow western Tethys Sea, roughly where the Mediterranean–Persian Gulf now lies, and over time underwent major adaptations as they adopted an aquatic life. The Pinnipedia diverged and evolved probably ~10 Ma in the Northern Hemisphere and later crossed the equator into the Southern Hemisphere. Southern waters are now a diversity hotspot for marine mammals, with 14 species present, and constitute a key conservation region (Pompa *et al.* 2011). Two orders, the Pinnipedia (seals) and Cetacea (dolphins and whales), live in southern temperate waters and are notable because they suckle their young and have a regulated body temperature. Two seal species and whales were hunted almost to extinction from the 19th to 20th century and populations have only started to recover in recent decades. In addition, elephant seals once numbered > 10 000 on King I., Bass strait, but were completely wiped out early in the 19th century. We describe below the ecology of three seal species, and three whale species, omitting the smaller, open sea species – the dolphins.

SEALS AND SEA LIONS (ORDER PINNIPEDIA)

The two species of fur seal, New Zealand fur seal *Arctocephalus forsteri* and Australian fur seal *Arctocephalus pusillus*, have broadly similar life histories, which differ in key features from that of the Australian sea lion *Neophoca cinerea* (Table 18.1). The species also differ in their distribution and foraging ecology. The two fur seals have overlapping ranges: the New Zealand species from SW Western Australia to SE South Australia and Tasmania; the Australian fur seal from Kangaroo I., central South Australia to SE Australia, and the sea lion from the Abrolhos Is, Western Australia to The Pages – small islands ~15 km east of Kangaroo I.

The pelt of a sea lion is a layer of hair underlain by a layer of blubber that provides insulation and enables it to dive deep and stay for a protracted period on the bottom. In contrast, fur seals have two layers of hair in their pelt – a sleek, coarse and hairy outer layer and a fine, dense, lower layer of fur that traps air and insulates them. However, they cannot dive as deep or as long as sea lions, because, as the air is compressed, water starts to penetrate the layer of fur, reducing its insulation value (Shaughnessy 1999). To achieve long hours of diving, sea lions and seals have evolved four mechanisms:

- They exhale before diving to minimise the amount of air in their lungs.
- Their haemoglobin concentrations in the blood are very high and they have a large blood volume, which together provide a large storage capacity for oxygen.

Table 18.1: Life history and population characteristics of: Australian sea lions *Neophoca cinerea* (after Gales *et al.* 1997; Shaughnessy 1999; Goldsworthy *et al.* 2007, 2009; Lowther 2011); New Zealand fur seal *Arctocephalus forsteri* (after Shaughnessy *et al.* 1994, 2010; Goldsworthy and Shaughnessy 1994; McKenzie *et al.* 2005, 2007a, b; Shaughnessy 2011); and Australian fur seal *Arctocephalus pusillus doriferus* (after Shaughnessy 1999; Kirkwood *et al.* 2005; Gibbens and Arnould 2009).

BS = Bass Strait; KI = Kangaroo I.

Life history and population characteristics	Australian sea lion	NZ fur seal	Australian fur seal
Breeding cycle	17–18 months	Annual	Annual
Breeding season	Variable	Nov.–Jan.	Late Oct.–Dec.
Length of breeding season	5–7 months	3 months	2 months
Age at maturity (years)	4.5–6	4–5 (female) 4–5 (male)[1]	3–4 (female) 6 (male)[1]
Reproductive success of females	71% at Seal Bay	25–67% at Cape Gantheaume	53–55%
Diapause period (months)	3.5–5 months	~4 months	Probably
Post-implantation period	14–15 months	~8 months	1 year in all
Pup weaning age (months)	15–18 months	8–12 months (average 285 days)	8–11 months (rarely to 2 years)
Pup mortality (pre-weaning)	40–50%	3–15%[2]	15–28%
Longevity	~25 years	~24 years	21 years
Aust. population size	~13 000 (2011)	~85 000 (1999)	92 000 (2002)
Number of breeding sites	~80	26(WA); 27(SA); 1(Tas.)[3]	13 (BS);1 (KI)
Number of haul-out sites	~150	Many	Many

[1] able to hold breeding territories from 9–10 years of age; [2] in first 2 months; [3] breeding colonies defined here as those with juveniles or pups (after Shaughnessy *et al.*(1994).

- Their heart rate slows to < 10 beats a minute (called brachycardia) during a dive and oxygenated blood flow is limited to two essential organs – the brain and heart.
- They can effectively store CO_2 and lactic acid produced and expel it at the surface.

Australian seals and sea lions are all colonial breeders. They come ashore at communal breeding colonies to give birth to their young and then mate. Soon after, they leave to go foraging, but the females must return frequently to their colony to suckle their young until they are weaned. The two species of fur seal breed annually, whereas the Australian sea lion breeds non-annually and asynchronously between populations.

Australian sea lion *Neophoca cinerea* (Family Otariidae)

Breeding cycle

After hunting decimated sea lion populations in the 19th century, their distribution contracted westwards, as Victorian and southern New South Wales' colonies became extinct. The most easterly breeding colony (and the largest) is at The Pages and many smaller colonies are on the brink of extinction. The largest surviving colonies are on islands off the central South Australian coast, with just three sites, Dangerous Reef, Seal Bay and The Pages, together producing 40% of pups for the species.

This sea lion has a 17–18 month breeding cycle, and prolonged pupping and lactation periods compared with seals (Table 18.1). After mating, an embryonic diapause of several months – a period of suspended development – occurs, followed by a post-implantation period of ~14 months; together the two periods fit with a total gestation period of 17–18 months.

Each breeding colony has its unique breeding season, which is asynchronous with neighbouring populations, and gradually changes over time, so that over the life time of a female she will pup at all times of the year. In consequence, females of a colony retain high fidelity to their natal site and colonies only 60 km apart differ genetically (Campbell *et al.* 2008). This uniquely different breeding system has led to highly subdivided populations at a fine scale, with a consequent high risk of local extinction of many small colonies (see below). But how could such a system have evolved?

It has been suggested that the breeding cycle is a response to the nutrient-poor waters of southern Australia – due to the warm Leeuwin Current of low productivity flowing eastwards for much of the year. On this view, a prolonged gestation and lactation period would help the young survive a poor or scarce food supply and subdivided populations would spread the limited resources. Alternatively, Goldsworthy *et al.* (2009) proposed another scenario – the 'family–farm hypothesis'. They pointed out that upwellings occurred seasonally along the central South Australian coast and substantially increased the productivity of shelf waters (see Chapter 2) – which incidentally may explain why 75% of sea lion populations are in this region, with their colonies located near suitable foraging grounds. The pups would need to acquire a repertoire of foraging skills, and they learn these not only from their mothers and siblings (as in the family–farm hypothesis), but also from unrelated members of the colony (see Lowther *et al.* 2012). So perhaps the 'shopping mall' hypothesis is a more appropriate description! In any event, the prolonged learning period would favour strong site-fidelity of females, and a prolonged lactation period, during which pups can supplement their mother's milk with bottom prey.

Diet and foraging

For an animal that has evolved to hunt in bottom habitats and feed its pup frequently until weaning, it would need to hunt in the vicinity of the colony to minimise travelling distance and also to spend the maximum time hunting on the bottom for prey. Costa and Gales (2003) confirmed these predictions for the Seal Bay colony on southern Kangaroo I. by placing data loggers on lactating mothers to determine their diving practices. Mothers had to work hard in their hunt for food. They travelled an average of 25 km to dive sites and dived there night and day, with ~11 dives an hour and surface intervals between dives of only 1–2 minutes. Average diving depth was 61 m and the maximum was 105 m. As their pregnancies advanced during the late lactation period, they dived even deeper and longer – probably to meet the greater needs of a growing foetus and a suckling pup back at the colony. Costa and Gales also noted that maximising their bottom time had the added benefit of minimising predation on them by sharks, which tend to attack a seal at the surface, where they are most vulnerable. In contrast to females, males foraged far more widely – up to ~200 km from their colonies.

The Seal Bay study was later repeated for the Dangerous Reef colony in lower Spencer Gulf (Lowther and Goldsworthy 2011a). Here, time spent foraging by mothers was similar, but their milk had lower lipid fat content, pup birth size was larger and growth slower – all indicating the local adaptation of maternal investment to a particular site.

Diving behaviour based on tracking data and stable isotope analysis has been recorded at numerous sites over the sea lion's range (Goldsworthy *et al.* 2009; Lowther and Goldsworthy 2011a; Lowther *et al.* 2011). Foraging ability improves with age and size. Juveniles start to forage at ~4 months old and for the next few months dive mostly to ~10 m depth for short dives of < 1 minute duration. By the time they are weaned at 15–18 months old, individual dives last for 2–3 minutes and go to maximum depths of 50–70 m, and in another few months to 90 m depth (Fowler *et al.* 2006).

A curious specialisation in sea lion foraging behaviour occurs at every site so far studied. Two groups of foraging types exist, with few intermediates:

- an inshore type, whose females forage within 2–9 km of the colony at shallow depths of 15–30 m, mainly in seagrass or on shallow reefs
- an offshore type, whose females forage 20–50 km from the colony at 45–65 m deep.

The proportions of the two types vary among the colonies, according to the distance of the colony from deep water and the foraging possibilities. Foraging trips last 1–2 days. A switch by a female in foraging type appears to be rare, as does migration from a colony. The food eaten also differs, with the offshore group feeding at higher trophic levels.

The picture now emerging seems to be that females are generalist bottom foragers that develop specialist strategies for particular habitats. They acquire a spatial memory for their favoured feeding sites, enabling them to maximise their foraging success. Social learning is an important element in foraging, because young sea lions learn the behaviour and habitat of prey species, and culturally acquire the necessary hunting skills from parents and others. In the light of such an investment, long-term fidelity to a site is clearly adaptive.

Another consequence of differences between colonies in the foraging conditions is that significant differences in milk fat levels, offspring growth rates and time spent by mothers nursing pups ashore occur in different colonies, so further emphasising localised adaptation and the flexibility of maternal investment (Lowther and Goldsworthy 2011b).

Bottom foraging is probably a more dependable place for food than the pelagic environment in waters of low or seasonal productivity. However, the precise diet is difficult to determine because the scats contain very few identifiable hard parts, apparently because the stomach contents are ground to paste by stomach stones (called gastroliths). So, McIntosh *et al.* (2006) used digestive tract contents from dead sea lions and regurgitates to estimate the percentage presence of two categories of prey species: cephalopods; and fish, invertebrates and seabirds (Table 18.2). Clearly, bottom-living species (i.e. octopus, cuttlefish, lobster and reef-dwelling sharks and fish) are the main prey groups. This was confirmed by video cameras attached to sea lions, showing them resting on projecting rocks above the reef bottom, watching and waiting for cryptic prey to appear from crevices (S Goldsworthy *pers. comm.*).

Growth and age

A method of ageing the sea lion was first developed by Rebecca McIntosh, who counted growth layers in teeth. From these, she was able to estimate the growth rates of males and females and their longevity (Table 18.1; Fig. 18.1). Growth is rapid in the first 5 years, and at 10 years of age they are almost fully grown, with males reaching a much larger size than females. The oldest individuals have been estimated to be 24 years of age.

Population status

The Australian sea lion is distributed over ~80 breeding colonies in its range. Long-term census data since the 1980s (> 20 years for some large sea lion colonies) give a mixed picture of their status. Some larger colonies appear stable, others are in decline and a few have

Table 18.2: Diet of Australian sea lion *Neophoca cinerea*, New Zealand fur seal *Arctocephalus forsteri* and Australian fur seal *Arctocephalus pusillus doriferus*.

Data for: Australian sea lion after McIntosh *et al.* (2006); New Zealand fur seal (percentage biomass) after Page *et al.* (2005a); and Australian fur seal after Kirkwood *et al.*(2008) (percentage numbers in terms of reconstructed biomass). McIntosh *et al.* (2006) showed cephalopod component (percentage numbers) independently of other prey (percentage identified prey items), but data are here combined.

Food type	Australian sea lion	NZ fur seal		Australian fur seal
		Female	Male	
Cephalopods				
Squid	7	6.8	13.4	10.3
Cuttlefish	13	–	1.0	–
Octopus	7	–	0.9	0.4
Unidentified	1	39.1	20.2	–
Crustaceans, fish, birds				
Seabirds (nearly all little penguin)	4	27.3	37.1	–
Fish	26[1]	29.8[4]	26.0[6]	89.3[7]
Crustaceans	19[2]	8.9[5]	1.4[5]	–
Other	23[3]	–	–	–

[1]mainly swallowtail (54%), leatherjackets (12%), flathead, bullseye, whiting and mackerel (each 4%); [2]lobster 16% and crab 3%; [3]mainly cat shark; [4]redbait 11%, barracouta 3%, jack mackerel 2% and 20 other species 13.8%; [5]mainly prawns; [6]redbait 7%, leatherjackets 5%, lanternfish 1%, swallowtail 2% and 13 other species 19%; [7]barracouta 30%, redbait 22%, jack mackerel 8%, gurnard 7%, bearded/red cod 7%, flathead 7%, and 10 other species 8%.

become extinct. Many existing small ones have a high probability of extinction, should only a slight increase in pup mortality occur (Goldsworthy *et al.* 2009), and two serious trends are:

- Colonies towards the northern edge of the sea lions' range in Western Australia, e.g. on islands on the mid-west coast, are currently declining at a rate of ~1% per breeding season in pup production.
- The large Seal Bay colony on Kangaroo I. has declined since 1982 at a rate of ~4% per breeding season. At this rate, the colony there will be halved within another 30 years.

Mortality and predators

The mortality of pups is a serious issue for the survival of colonies. In surveys at Seal Bay, pup mortality typically fell into three categories:

- **trauma**, as a result of crushing, or aggression by large males – accounting for up to 60% of known deaths
- **still births and starvation/emaciation**, the latter due mainly to female rejection or mother–pup separation, – accounting for 35% of deaths
- **shark attacks**, accounting for another 3%.

The main predators of sea lions are sharks, but the incidence of their predation is not known. However, shark bite scars on living sea lions, indicating failed attempts, do provide evidence of the pattern of predation. Over 5 years of records from Seal Bay, an average of 36 sea lions per annum (i.e. ~24% of the population) received shark injuries, mainly to the forepart or rear end of the body. A third of those attacked were adult females and a quarter were juveniles, with the highest incidence of injuries in summer during the pupping season, and the lowest in winter. Killer whales, occasionally seen in the region, are another predator.

A more serious cause of mortality to sea lions has been entanglement in nets, as by-catch in the shark gill-net fishery, causing a mortality of > 60 sea lions per year (Page *et al.* 2004). Fishing effort in this fishery is high (~20 000 km of nets set per year on the bottom), and the fishery completely overlapped the foraging range of the sea lions. When gill-net fishing was banned in lower Spencer Gulf near the Dangerous Reef colony, the population there recovered spectacularly, indicating a strong link between fishing and sea lion mortality. Eventually in 2011, gill-net fishing was banned by the Commonwealth within 20 km distance of 48 sea lion colonies, but it remains to be seen whether this will halt the population decline.

New Zealand fur seal *Arctocephalus forsteri* (Family Otariidae)

Similar to the sea lions above, the great majority in numbers (77%) of this fur seal breed on the central South Australian coast (Plate 61). The summer breeding season is short, and pregnant females come ashore to their colony, usually on smooth, granitic islands, 1–2 days before pupping. Here they form haremic groups of five to eight females, defended by a dominant male, give birth, mate with the male, and stay with their pups ~10 days, before leaving to go foraging to replenish milk and energy reserves. Embryonic diapause lasts ~4 months and active gestation another 8 months, so giving an annual cycle.

Early in the lactation period, a female goes foraging for 3–5 days at a time and then returns to feed her pup for 2 days before leaving again. This pattern persists till late in her lactation period of ~10 months (Table 18.1), after which she stays at sea for longer – up to ~16 days at a time. By the next breeding season, her pup is fully weaned and she very likely will return to the same territory for the next one or two seasons to mate with the same dominant male. Throughout the breeding season, the male does not feed, but stays on shore, defending his territory and his females, by barking at intruders, using threatening postures and, if necessary, engaging in chest-to-chest conflicts.

For the first 2 months of life, pups depend entirely on their mothers for nutrition from their milk. They then start to explore shallow pools close by, and within another 2 months they are foraging in the shallows. At 6–8 months, they spend ~6 h a day at sea diving to depths of ~5 m, rarely up to 20 m, for an average duration of ~ 30 s each dive. By ~8 months, their foraging efficiency has greatly improved and they gradually switch to diving at night to feed on fish prey that migrate towards the surface at night (e.g. anchovies, lanternfish, redbait and pilchards). At this age, their diet comprises fish (43%), crustaceans (36%) and squid (20%), in addition to their mothers' milk (Baylis *et al.* 2005).

From a weaning age of ~1 year until they reach breeding age, young fur seals spend most of their time at sea, occasionally coming ashore to areas far from the territories in which they were reared. In the first 3–5 years, they forage in oceanic waters up to 1000 km from land, where they are thought to target lantern fishes in surface waters. Growth is rapid in the first 10 years, with males growing faster than females and putting on a distinct spurt of growth soon after first sexual maturity at ~5 years until an age of 9 years (Fig. 18.1). Longevity is similar to that of the sea lion (Table 18.1).

Female fur seals are restricted in the distance they can swim from their colony because they always need to return to shore to suckle their pups. They feed both in the water column and on the bottom, and their prey is mostly fish, squid and little penguins in shelf waters (Table 18.2). They dive only at night to maximum depths of ~70 m, remaining on the surface during the day. Adult males, being larger and unconstrained by dependent pups, can forage further offshore along the shelf break > 100 km offshore for periods of up to 26 days and to 380 m depth (Page *et al.* 2005a, b). These areas support high densities of fish and squid. In addition to these prey, males target seabirds, often

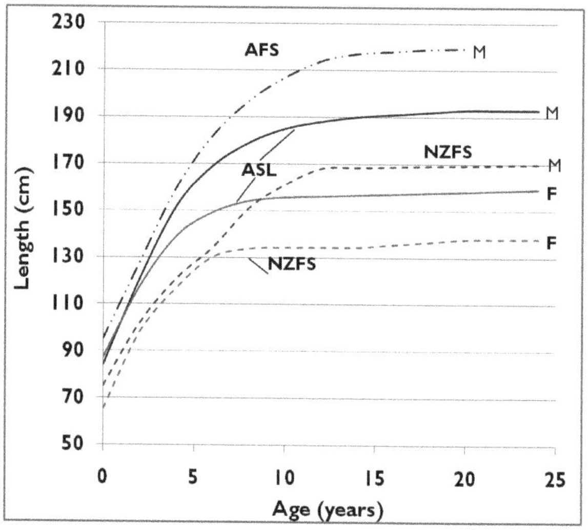

Figure 18.1: Growth rates of: male (M) Australian fur seal *Arctocephalus pusillus doriferus* (AFS) (after Arnould and Warneke 2002); male (M) and female (F) Australian sea lions *Neophoca cinerea* (ASL) (continuous lines) (after McIntosh 2007); and male and female New Zealand fur seal *Arctocephalus forsteri* (NZFS) (dashed lines) (after McKenzie *et al.* 2007a). Growth rate of female Australian fur seal (not shown) is very similar to that of the female Australian sea lion.

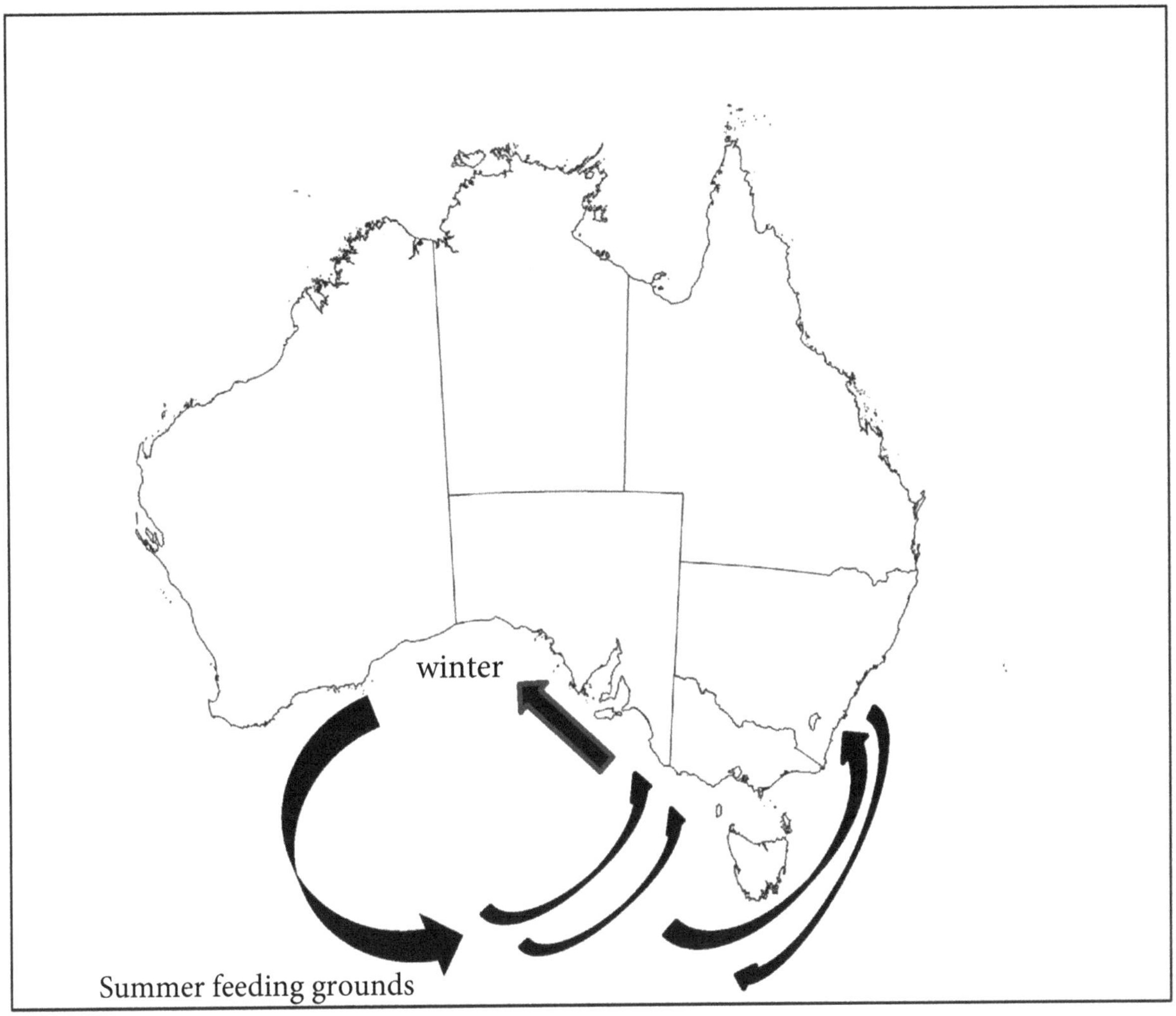

Figure 18.2: Annual migration route of southern right whales (redrawn from Burnell 2001).

penguins, with small numbers of shearwaters, gannets and albatross.

Male diving behaviour on these trips is curious. Most dives (80–90%) are foraging dives, in which he actively swims and hunts for prey. These dives are interspersed by drift dives in which he swims down to a depth of 50–150 m, and then drifts in the water column for up to 10 minutes at a time. Page *et al.* (2006) thought that drift dives likely occurred between bouts of 'gorge' feeding, when the seal rested to process the food eaten. And the drift dives occur at night, because the seal can rest in complete darkness without the risk of being detected by white sharks, which are known to locate seals visually from their silhouettes against the water surface.

An interesting feature, noted by Bradshaw *et al.* (2000), is that the fur seal's growth rate and pup condition are dependent on the presence of upwelling systems near fur seal colonies. These upwelling systems occur off the central South Australian coast where the highest densities of this fur seal occur, and the high seasonal primary productivity of these waters (Chapter 2) is without doubt critically important in supporting prey fish populations throughout the year. Baylis *et al.* (2008) investigated this aspect further by comparing the foraging activity of lactating mothers at a site on southern Kangaroo I. close to an upwelling region during and after the upwelling season. During the upwelling (March–April), the mothers foraged in the upwelling zone, averaging 7-day trips 122 km from home at average depths of 38 m, with each dive lasting ~2 minutes, and bottom time of 1 minute. During winter they averaged 17-day trips to the far off subtropical oceanic front 350 km offshore, at average depths of 47 m, with similarly long dives. The long winter trips accord with optimal foraging theory, which predicts that they should choose the best feeding sites in each season.

In contrast to the Australian sea lion, New Zealand fur seal populations and the number of breeding colonies have generally been increasing throughout the range of the species: a trend that appears to have accelerated in recent decades. For example, along the mid-South Australian coast where 77% of the Australian population occurs, the abundance of pups has been increasing; for example, at ~10% a year on Kangaroo I, since 1989, equivalent to a 10-fold increase up to 2011 (Shaughnessy *et al.* 1994; Shaughnessy 2011).

Australian fur seal *Arctocephalus pusillus doriferus* (Family Otariidae)

This is the largest known fur seal and is believed to have migrated to southern Australia from South Africa ~12 000 years ago in the late Pleistocene. Before sealing commenced, numbers are thought to have exceeded 200 000 seals, but by 1923 at the end of seal harvesting only nine of the original 26 colonies persisted on offshore islands in Bass Strait; however, culling continued till 1971. Currently, more colonies have become established (Table 18.1), and the fur seal has recently extended its range to Kangaroo I., where a small breeding colony now exists (Shaughnessy *et al.* 2010). This fur seal has a similar life history to the New Zealand fur seal (Table 18.1), but with a slightly earlier breeding season and faster growth rates that are intermediate between those of the New Zealand fur seal and the Australian sea lion (Fig. 18.1).

Satellite tagging studies in Bass Strait show that this fur seal forages at 40–100 m depth on the continental shelf (Arnould and Hindell 2001; Kirkwood *et al.* 2006; Arnould and Kirkwood 2008). Lactating females tend to stay closer to home than males and in the summer following birth, forage for ~4-day trips and then return to suckle their young for ~3 days; in the following winter, foraging trips extend to a week. On these trips, they keep mainly within ~100 km of their home colony and dive both day and night about eight times an hour to average depths of 58 m.

Male foraging trips last ~7 days and they often forage to distances of 200 km or more from their home colony in similar habitats to females. A few go as far as the GAB and southern Tasmania > 500 km distant. They spent 65–97% of their time at sea travelling on average 77 km a day. Gibbens and Arnould (2009) thought that such a demanding travel and diving schedule could only be accomplished by a very large seal, such as this species.

On most dives they swim to the bottom, feeding on mid-water and benthic species, and both males and females have their favoured foraging 'hotspots' where they frequently return to forage. One such site, described by Arnould and Kirkwood (2008), is a high-relief granite outcrop at 60–80 m depth in NE Bass Strait (Beaman *et al.* 2005, and see Table 13.2). Common schooling fish around this reef are butterfly perch and eastern orange perch, both of which are hunted here by this fur seal on its foraging excursions from Gabo I.

The diet of this seal has been studied at numerous sites and has been shown to vary widely with location, season and sampling mode (whether scat, regurgitates or stomach contents), and even inter-annually and between individuals (e.g. Gales *et al.* 1993; Hume *et al.* 2004; Arnould *et al.* 2011). For example, a 9-year study at Seal Rocks in northern Bass Strait found that subtle ecosystem changes occurred between years with changing oceanographic conditions, and these affected the abundances of the seal's main prey – barracouta, redbait and squid (Table 18.2).

The recovery of populations of the Australian fur seal has been slow since sealing and culling ended, and only started to increase in the late 1990s (Kirkwood *et al.* 2005; Gibbens and Arnould 2009; Lancaster *et al.* 2010). Since then, populations have grown slowly in the more northerly breeding colonies, declined slowly in the southern ones, and been stable at others. Overall the average population growth rate is slow (~2.2% year^{-1}), and thought to be due to a poor feeding environment in the nutrient-poor Bass Strait waters.

WHALES

'Leviathan,
Hugest of the living creatures, on the Deep
Stretcht like a promontory sleeps or swims
And seems a moving land, and at his gills
Draws in, and at his trunk spouts out a sea.'

John Milton, *Paradise Lost*, 1677

Whales have a reputation of mythical proportions – for their size, their ecological importance and their cultural significance. In this section, we consider three species: the pygmy blue whale *Balaenoptera musculus brevicauda*, the southern right whale *Eubalaena australis*, and the orca *Orcina orca*. The first two are broadly distributed throughout the Southern Hemisphere and are listed as 'endangered' under federal legislation, while the orca lives in all oceans of the world.

Pygmy blue whale *Balaenoptera musculus brevicauda* (Family Balaenopteridae)

The blue whale is the largest living animal, growing to > 30 m length and weighing up to 180 t. Two subspecies exist: the 'true blue' whale *Balaenoptera musculus musculus* that lives south of ~60°S in summer, but occasionally migrates north to temperate waters; and the slightly smaller pygmy blue whale *Balaenoptera musculus brevicauda* that lives in temperate waters north of ~55°S and is known mainly from the southern Indian, Pacific and Atlantic Oceans. Knowledge of the pygmy blue in southern coastal waters is largely due to the studies of Peter Gill (2004) – and see reviews of Branch *et al.* (2007), McCauley and Jenner (2010), and Gill *et al.* (2011). After an estimated 360 000 were captured between 1904 and 1979, some 6000 now survive in southern waters.

Given its size and insatiable demand for food (3–4 t day^{-1}), this filter-feeding whale searches for the densest patches of prey and its behaviour is strongly influenced by the feeding opportunities. It feeds almost exclusively on swarms of krill *Nyctiphanes australis*, occasionally on salps and rarely on copepods, which are too small to be filtered out by its baleen. These prey species are all abundant in upwelling regions and the whales accordingly aggregate there in the upwelling season. In southern Australia these areas are:

- in the Perth Canyon, off Rottnest I., Western Australia from October to December each year, when the population is migrating south after breeding/calving off NW Australia, and later in April to August, on the whales' return to the tropics
- in the Bonney upwelling annually from November to April, and in adjacent upwelling areas extending from the west coast of Tasmania and Cape Otway, Victoria, to the eastern GAB, a distance of > 1000 km, at depths to ~500 m
- in feeding grounds off Eden, New South Wales in October–November.

The whale feeds on surface krill swarms using two techniques – lunge feeding and skimming (Gill 2004). In the former, the whale accelerates as it approaches a swarm, rolls on its side with mouth agape and takes in huge volumes of water and prey. The whale then closes its mouth and lies stationary as it ejects water through its baleen and retains the krill. In the less common skimming behaviour, it skims through surface swarms continuously filter-feeding. When the krill swarms sink into deep water, the whale employs continuous diving to feed and in Peter Gill's study the whales did this at 30–90 m. Because swimming is energetically costly for large whales, how do they manage this deep foraging?

In one study, a video camera was attached to a whale and its behaviour recorded (Williams *et al.* 2000). In an average 10-minute dive, the blue whale swam slowly by continuously stroking to ~20 m depth and then glided to the bottom at 80 m, where it fed for several minutes before returning to the surface by a mixture of stroking and gliding. Passive gliding is greatly helped by the changes in buoyancy and by the gradual collapse of the lungs as the whale is compressed by hydrostatic pressure during descent, and by the reverse process during ascent. Were it not for this energy-saving strategy, the whale would not be able to dive for such long periods to such depths.

In the eastern Indian and west Pacific Oceans, the pygmy blue migrates between its wintering grounds in Indonesian–New Guinea waters and southern 'greener pastures' (i.e. the summer feeding grounds south of Australia at the subtropical convergence). The west Australian coast is a whale highway for this whale on these migrations and the Perth Canyon is a 'way station for opportunistic feeding' by the migrating whales, as they gather to feed on krill within the canyon (Rennie *et al.* 2009). This study showed that as the Leeuwin Current flows strongly to the south, deep water from the nutrient-rich Leeuwin Undercurrent upwells at 200–400 m depth, supporting high phytoplankton production and the growth of krill swarms within the Canyon. From records of the whales' long, low-frequency sounds, McCauley and Jenner (2010) calculated that each year ~1100 pygmy blues steadily migrate from their wintering grounds south along the west Australian coast at an average speed of 9 km h^{-1}.

The whales then move eastwards from the SW corner of Australia along the south coast to the eastern GAB, where they forage in upper slope waters to depths of 1500 m. They continue SE along the 200 m isobath to the Bonney coast, SE South Australia, where the shelf is narrow, the upwelling strongest and krill densities the highest (Gill *et al.* 2011). These are the principal southern feeding grounds and the whales arrive and stay there from January to April, when krill swarms start to decline. Here, they forage on the shelf almost entirely at depths of < 200 m. Over the same period, other pygmy blues may come to the same feeding grounds from Bass Strait. In Peter Gill's surveys, up to 30 pygmy blues, mostly mature ones, but also some females with their young calves 6–7 m long, gathered in the same region.

From April, the whales retrace their path, returning west and north to their wintering grounds in Indonesian waters. However, the total pool of pygmy blue whales is much larger than just this western–southern group, and an eastern group, about which little is known, is thought to traverse the east Australian coast in a similar migratory pattern. Other groups exist south of other Southern Hemisphere continents.

After a 12-month gestation period, the pygmy blues calve in July–September in their wintering grounds and have a lactation period of 7 or more months, when the juveniles are ~16 m long. The pygmy blue has a maximum size of ~24 m and a lifespan of up to 50 years, compared with the true blue, which has a life span of ~90 years.

The Antarctic true blues have their own distinctive call and long-term acoustic records of whale calls in the Perth Canyon indicate that this sub-species occasionally makes northward winter migrations into the eastern Indian Ocean. However, the northern extent of their migration is still unknown (Stafford *et al.* 2004).

The southern right whale *Eubalaena australis* (Family Balaenopteridae)

The southern right is similar to the pygmy blue in that it lives and feeds for much of the year south of Australia near the subtropical convergence. During winter, it migrates north to warmer coastal calving grounds off southern Australia (Plate 62), while other populations migrate to New Zealand, South Africa and South America. Researchers have studied their behaviour, reproduction and feeding off each of the four continents. The best known Australian calving grounds are in SW Australia and at the head of the GAB, with other small calving grounds off the central South Australian and Victorian coasts. The calving grounds are gradually increasing as the populations grow. Two separate stocks of the southern right are recognised – a western stock that calves at Western and South Australian sites, and a small eastern stock that calves off Warrnambool, Victoria, and off New South Wales and New Zealand (Carroll *et al.* 2011). It is noteworthy that off Patagonia southern right whales, in response to repeated attack by orcas, changed their calving grounds to inshore embayments apparently to avoid predation (Sironi *et al.* 2008). Possibly near-shore calving areas in Australia have similarly been selected by southern rights to minimise predation.

The western stock is slowly recovering at a rate of ~6–7% a year, after commercial extinction in the 1830s–1850s, when 98% of the stock was taken, and subsequent Soviet illegal fishing from 1951–1971 (Tormosov *et al.* 1998). The western stock is thought to number ~3000 whales, whereas the eastern stock is still critically low and numbers < 80.

Researchers can identify individuals from their marking patterns and, so far, > 1200 individuals have been recorded and their movements followed. Consistent with the two stocks, two winter migration patterns are known: one northwards up the east Tasmanian coast and the other up the west Tasmanian coast to calving grounds close to shore in Western and South Australia during the winter fasting period (Fig. 18.2). Each year, the females return to the same calving grounds, showing a form of cultural memory. In Steve Burnell's (2001) long-term study, 92% of the whales showed site fidelity, although some males appeared to visit several sites over winter, where they likely mated with several females. Genetic studies indicate that calves learn site fidelity to summer feeding grounds from their mothers, a learned behaviour that persists for many generations (Valenzuela *et al.* (2009).

Upon arrival at the breeding grounds early in the southern winter, pregnant females give birth and then mate. The peak of the breeding season is mid-August. The average interval between calving by females is 3.6 years, and, if ~9% of calves are stillborn or miscarry, as Burnell (2001) believed, then about one in four females at a wintering ground will produce a viable calf each year. Further, the Australian populations have lower calf production during El Niño years, possibly due to greater pregnancy failure and increased juvenile mortality. This was attributed to climate-driven reduction in krill abundance by Leaper *et al.* (2006).

Small to large numbers of whales variously aggregate at the southern breeding grounds and, after a brief courtship, in which a females vocalises to attract mates, she eventually rolls over, takes a deep breath and copulates with a chosen male. Best *et al.* (2003) thought that the female initiated social groupings on these occasions to incite competition among potential partners. Competition between males occurs in two ways: first they compete for access to a position next to a female, to enable insemination and, subsequently, sperm competition will take place after she has been inseminated several times. By inciting numerous competing males, she increases her chance to mate with the best quality male and secure conception.

The gestation period is 1 year, and on her next visit to the calving site a year later, the female gives birth to a ~6 m sized calf. After another 1–2 months, she leaves on her southerly, counter-clockwise migration to feeding grounds at 40–50°S (Fig. 18.2), and continues to suckle her young for a year (Burnell and Bryden 1997). Young southern right whales grow rapidly in the first year of life, but then growth slows markedly as they gradually approach their maximum size of ~15 m for males and 17 m for females (Fig. 18.3).

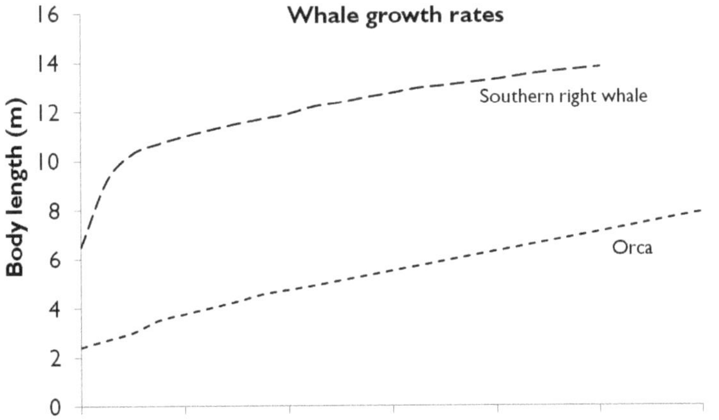

Figure 18.3: Growth rate of the southern right whale *Eubalaena australis* (after Best and Schell 1996) and the orca (after NMFS 2008).

Females calve for the first time at ~8 years of age at a length of ~12.4 m.

Feeding studies (Tormosov *et al.* 1998) showed that, when north of ~40°S, 92% of the food is copepods, but south of 50°S it is 99% krill (*Euphausia superba*). The right whale's feeding behaviour has some similarity to that of the blue whale (Hamner *et al.* 1988). When krill swarms are near the surface, the whale skims at the surface at high speed, with its upper jaw out of the water, taking in vast quantities of krill and water. When the krill swarm at depth, the whale hyperventilates for ~90 s before diving for periods lasting initially for ~4 minutes. This is energetically costly and they very soon start to dive for shorter and shorter periods, and then move on. Curiously, the southern right has developed tail-sailing skills. In moderate winds, it lifts its tail high out of the water, puts its head down and sails downwind, sometimes feeding on krill at the same time.

Killer whale or orca *Orcinus orca* (Family Delphinidae)

The killer whale is the largest member of the dolphin family, growing up to ~10 m size and weighing ~5 t. It is found in all seas of the world and in southern Australia is seen most often in pods of five to 15 in cooler SE Australian waters near seal colonies (reviewed by Morrice 2004). On genetic evidence, the Southern Hemisphere ecotype is now considered to be a distinct species from the north Pacific and Atlantic ecotypes, with distinctive diet and behaviour (Morin *et al.* 2011).

This whale uses echo-location to produce a sonic picture of its environment and the shape and size of prey. It is known to prey on other whales, sharks, seals, dolphins, stingrays, fish and squid, but its diet in southern waters is not known, except for the numerous reports of attacks on some of the above prey. Of these reported attacks, 31% were on whales, 19% on fur seals, 12% on dolphins, 19% on sharks and 19% on fish (Australian salmon and trumpeter), with one attack on a dugong in Shark Bay, Western Australia (Morrice 2004). An attack by a pod of 11 orcas on a blue whale and her calf on the Bonney coast was recorded by Cotton (1944). Other attacks on sperm whales, southern right whales and humpback whales, some of them related to whaling activities, were reviewed by Jefferson *et al.* (1991). In one attack astonishingly observed underwater by Bill Gladstone (1988), the orca attacked a sunfish ~1 m long, and, over 2 hours, tore one piece after another off the fish, until all was gone. During this period, it kept diving to the bottom, remaining there for 3–5 minutes at a time.

The cooperation between pods of orcas in their attacks is legendary. When they attack large whales, some bite the flukes and flippers while others attack the lips and tongue (a favoured morsel). Pods may also encircle or herd dolphins or seals to stop them escaping, and in the Antarctic they collaborate to capture seals. To accomplish this, they fan out near ice floes and first begin 'spy-hopping, (i.e. lift their heads out of the water to see if seals are hauled out on the floe). When an orca detects a seal, it summons the rest of the pod and together they line up in formation and carry out a wave-wash attack, in which they charge underwater to the floe, produce a breaking wave up to 1 m high. The floe breaks up and the seal is tipped into the water and captured (Pitman and Durban 2012).

Of great fame is the human–orca cooperation that existed for ~80 years at the whaling station in Twofold Bay, near Eden, New South Wales during the 19th century. The orcas attracted the attention of shore look-outs when a whale was detected, and the whalers and orcas then cooperated in an attack on the whale. The orcas herded the large whale into positions for convenient attack and the whalers moved in for the kill. The whalers then allowed the orcas to feed on the whale's tongue and lips, and claimed the rest of the carcass (Wellings 1944; Mitchell and Baker 1980).

The life history and cooperative behaviour of the orca are well reviewed by NMFS (2008). Mating occurs over a several month period. During mating, the male turns belly up with penis erect under a willing female. After a gestation of ~15 months, females calve in summer at intervals of 3–8 years – an event that has been observed on several occasions. It takes place underwater and is quickly followed by several pod members repeatedly lifting the newborn out of the water for several seconds at a time over the next few hours. The mother suckles her young for a year, occasionally for up to 2 years, with the young supplementing her milk with fish from a young age. Juveniles become sexually mature at ~5–6 m size, but reproductive maturity is some years later. Females have their first calf on average at ~15 years of age, usually producing around five calves over the next 25 years. Annual pup mortality is ~43%, but by adulthood mortality has fallen to ~5% a year for females, and somewhat more for males. Females live for up to 90 years, and have the longest post-reproductive life span of any non-human mammal. Mothers maintain close social relations with their sons for life – thought to be beneficial in foraging and providing support during conflicts (Foster *et al.* 2012).

For the first year or two, calves remain close to their mothers, but this maternal association slowly weakens as juveniles reach adolescence and spend much of their time with siblings, sharing in chasing, leaping and fast porpoising. Remarkably, altruistic behaviour is well known among orcas. Those without offspring sometimes provide parental-like care for juveniles that are not their own, as, for example, when a mother gives birth to a second calf. Adult males are known to 'baby-sit' groups of these young ones. Care-giving behaviour is also known and on one recorded occasion two adults cradled a calf, injured from a ferry's propeller, to prevent it from rolling over and drowning. Culturally learned social behaviour and traditions are well known among adolescents and adults. Orcas communicate vocally, and vocal traditions, called dialects, are inherited maternally, and thought to help young orcas survive by transmission of information on food availability (Yurk *et al.* 2002). Social behaviours include: aerial displays, breaching, surface splashing and sexual behaviour, the last providing courtship skills to young males. Even a greeting ceremony has been documented when two pods, separated for a time, meet again. The pods merge excitedly, engage in physical contact and vocalise loudly.

Growth, as known for the orca in the NE Pacific (Fig. 18.3), is almost linear for the first 14 years and then slows. The growth of southern orcas is unknown but probably follows a similar trajectory. Longevity is 80–90 years for females and ~20 years less for males.

There is some evidence that orcas migrate seasonally between Antarctic and Australia, spending their winters in the Antarctic, where they forage on seals and penguins, and their summers in Australian and subtropical waters. An intriguing possibility is that these migrations may be for bodily maintenance (Durban and Pitman 2012). In the Antarctic, their skin accumulates a thick covering of diatoms and they migrate to warmer waters where they shed their outer skin and regenerate the skin tissues without serious loss of body warmth.

Concluding remarks

The marine mammals discussed in this chapter are high in the food web. As reviewed by Pauly *et al.* (1998), the trophic levels of the species discussed here are: southern right and pygmy blue whales both 3.2; the orca 4.5; Australian sea lion 4.2; and the two fur seals 4.1 (where Levels 1–4 are, respectively, primary producers, herbivores, third level predators and top predators – see Chapter 19). These values for marine mammals are derived from multiple feeding studies and they define the ecological role of the species in the ecosystem. Hence marine mammals, notably those at Level 4 or above, can have a significant impact on fish species at a lower level, many of which are also subject to commercial and recreational fishing.

The iconic status of marine mammals and the listing of some species as threatened on the IUCN Red List (e.g. the pygmy blue whale) render their habitats worthy of protection, especially in view of the many anthropogenic threats to them. These include: collisions with vessels, noise pollution, gear entanglement, harassment and entanglement in marine debris. Hence, Butler *et al.* (2002) argued persuasively for the protection of blue whale feeding grounds.

The by-catch of Australian sea lions in the gill-net fishery and in lobster pots is an ongoing issue whose outcome is uncertain until protective measures are fully assessed (Shaughnessy *et al.* 2003). Another issue likely to increase in importance arises from the steadily increasing fur seal populations and their negative impact on fisheries.

REFERENCES

Arnould JPY, Hindell MA (2001) Dive behaviour, foraging locations, and maternal-attendance patterns of Australian fur seals (*Arctocephalus pusillus doriferus*). *Canadian Journal of Zoology* **79**, 35–48.

Arnould JPY, Kirkwood R (2008) Habitat selection by female Australian fur seals (Arctocephalus pusillus doriferus). Aquatic Conservation: Marine and freshwater Systems 17, S53–S67.

Arnould JPY, Warneke RM (2002) Growth and condition in Australian fur seals (*Arctocephalus pusillus doriferus*) (Carnivora: Pinnipedia). *Australian Journal of Zoology* **50**, 53–66.

Arnould JPY, Cherel Y, Gibbens J, White JG *et al.* (2011) Stable isotopes reveal inter-annual and inter-individual variation in the diet of female Australian fur seals. *Marine Ecology Progress Series* **422**, 291–302.

Baylis AMM, Page B, Peters K, McIntosh R *et al.* (2005) The ontogeny of diving behaviour in New Zealand fur seal pups (*Arctocephalus forsteri*). *Canadian Journal of Zoology* **83**, 1149–1161.

Baylis AMM, Page B, Goldsworthy SD (2008) Effect of seasonal changes in upwelling activity on the foraging locations of a wide-ranging central-place forager, the New Zealand fur seal. *Canadian Journal of Zoology* **86**, 774–789.

Beaman RJ, Daniell JJ, Harris PT (2005) Geology–benthos relationships on a temperate rocky bank, eastern Bass Strait, Australia. *Marine and Freshwater Research* **56**, 943–958.

Best PB, Schell DM (1996) Stable isotopes in southern right whales (*Eubalaena australis*) baleen as indicators of seasonal movements and growth. *Marine Biology* **124**, 483–494.

Best PB, Schaeff CM, Reeb D, Palsboll PJ (2003) Composition and possible function of social groupings of southern right whales in South African waters. *Behaviour* **140**, 1469–1494.

Bradshaw CJA, Davis LS, Lalas C, Harcourt RG (2000) Geographical and temporal variation in the condition of pups of the New Zealand fur seal (*Arctocephalus forsteri*): evidence for density dependence and differences in the marine environment. *Journal of Zoology* **252**, 41–51.

Branch TA, Stafford KM, Palacios DM, Allison C *et al.* (2007) Past and present distribution, densities and movements of blue whales *Balaenoptera musculus* in the Southern Hemisphere and northern Indian Ocean. *Mammal Review* **37**, 116–175.

Burnell SR (2001) Aspects of the reproductive biology, movements and site fidelity of right whales off Australia. *The Journal of Cetacean Research and Management* **2**, 89–102.

Burnell SR, Bryden MM (1997) Coastal residence periods and reproductive timing in southern right whales, *Eubalaena australis*. *Journal of Zoology* **241**, 613–621.

Butler A, Althaus F, Furlani D, Ridgway K (2002) 'Assessment of the conservation values of the Bonney upwelling area'. Report to Environment Australia. CSIRO Marine Research, Hobart.

Campbell RA, Gales NJ, Lento GM, Baker CS (2008) Islands in the sea: extreme natal site fidelity in the Australian sea lion, *Neophoca cinerea*. *Biology Letters* **4**, 139–142.

Carroll E, Patenaude N, Alexander A, Steel D *et al.* (2011) Population structure and individual movement of southern right whales around New Zealand and Australia. *Marine Ecology Progress Series* **432**, 257–268.

Costa DP, Gales NJ (2003) Energetics of a benthic diver: seasonal foraging ecology of the Australian sea lion, *Neophoca cinerea*. *Ecological Monographs* **73**, 27–43.

Cotton BC (1944) Killer whales in South Australia. *Australian Zoologist* **10**, 293–294.

Durban JW, Pitman RL (2012) Antarctic killer whales make rapid, round-trip movements to subtropical waters: evidence for physiological maintenance migrations? *Biology Letters*

Foster EA, Franks DW, Mazzi S, Darden SK *et al.* (2012) Adaptive prolonged postreproductive life span in killer whales. *Science* **337**, 1313.

Fowler SL, Costa DP, Arnould JPY, Gales NJ *et al.* (2006) Ontogeny of diving behaviour in the Australian sea lion: trials of adolescence in a late bloomer. *Journal of Animal Ecology* **75**, 358–367.

Gales R, Pemberton D, Lu CC, Clarke MR (1993) Cephalopod diet of the Australian fur seal: variation due to location, season and sample type. *Australian Journal of Marine and Freshwater Research* **44**, 657–671.

Gales NJ, Williamson P, Higgins LV, Blackberry MA *et al.* (1997) Evidence for a prolonged post-implantation period in the Australian sea lion (*Neophoca cinerea*). *Journal of Reproduction and Fertility* **111**, 159–163.

Gibbens J, Arnould JPY (2009) Age-specific growth, survival, and population dynamics of female Australian fur seals. *Canadian Journal of Zoology* **87**, 902–911.

Gill PC (2004) Ecological linkages in the Bonney upwelling blue whale feeding area. PhD thesis. Deakin University, Geelong.

Gill PC, Morrice MG, Page B, Pirzl R *et al.* (2011) Blue whale habitat selection and within-season distribution in a regional upwelling system off southern Australia. *Marine Ecology Progress Series* **421**, 243–263.

Gladstone W (1988) Killer whale feeding observed underwater. *Journal of Mammalogy* **69**, 629–630.

Goldsworthy SD, Shaughnessy PD (1994) Breeding biology and haul-out pattern of the New Zealand fur seal, *Arctocephalus forsteri*, at Cape Gantheaume, South Australia. *Wildlife Research* **21**, 365–376.

Goldsworthy SD, Peters KJ, Page B (2007) 'Foraging ecology and diet analysis of Australian sea lions'. SARDI Report No. 251. Publication F2007/001024. SARDI Aquatic Sciences, Adelaide.

Goldsworthy SD, McKenzie J, Shaughnessy PD, McIntosh RR *et al.* (2009) 'An update of the Report: understanding the impediments to the growth of Australian sea lion populations'. SARDI Report No. 356. Publication F2008/000847-1. SARDI Aquatic Sciences, Adelaide.

Hamner WM, Stone GS, Obst BS (1988) Behavior of southern right whales, *Eubalaena australis*, feeding on the antarctic krill, *Euphausia superba*. *Fishery Bulletin* **86**, 143–150.

Hume F, Hindell MA, Pemberton D, Gales R (2004) Spatial and temporal variation in the diet of a high trophic level predator, the Australian fur seal (*Arctocephalus pusillus doriferus*). *Marine Biology* **144**, 407–415.

Jefferson TA, Stacey PJ, Baird RW (1991) A review of killer whale interactions with other marine mammals: predation to co-existence. *Mammal Review* **21**, 151–180.

Kirkwood R, Gales R, Terauds A, Arnould JPY *et al.* (2005) Pup production and population trends of the Australian fur seal (*Arctocephalus pusillus doriferus*). *Marine Mammal Science* **21**, 260–282.

Kirkwood R, Lynch M, Gales N, Dann P *et al.* (2006) At-sea movements and habitat use of adult male Australian fur seals (*Arctocephalus pusillus doriferus*). *Canadian Journal of Zoology* **84**, 1781–1788.

Kirkwood R, Hume F, Mindell M (2008) Sea temperature changes mediate annual changes in diet of Australian fur seals in Bass Strait. *Marine Ecology Progress Series* **369**, 297–309.

Lancaster ML, Arnould JPY, Kirkwood R (2010) Genetic status of an endemic marine mammal, the Australian fur seal, following historical harvesting. *Animal Conservation* **13**, 247–255.

Leaper R, Cooke J, Trathan P, Reid K *et al.* (2006) Global climate drives southern right whale (*Eubalaena australis*) population dynamics. *Biology Letters* **2**, 289–292.

Lowther AD (2011) Alternate foraging strategies and population structure of adult female Australian sea lions. PhD thesis. University of Adelaide, Adelaide.

Lowther AD, Goldsworthy SD (2011a) Detecting alternate foraging ecotypes in Australian sea lion (*Neophoca cinerea*) colonies using stable isotope analysis. *Marine Mammal Science* **27**, 567–586.

Lowther AD, Goldsworthy SD (2011b) Maternal strategies of the Australian sea lion (*Neophoca cinerea*) at Dangerous Reef, South Australia. *Australian Journal of Zoology* **59**, 54–62.

Lowther AD, Harcourt RG, Hamer D, Goldsworthy SD (2011) Creatures of habit: foraging habitat fidelity of adult female Australian sea lions. *Marine Ecology Progress Series* **369**, 297–309.

Lowther AD, Harcourt RG, Goldsworthy SD, Stow A (2012) Population structure of adult female Australian sea lions is driven by fine-scale foraging site fidelity. *Animal Behaviour* **83**, 691–701.

McCauley RD, Jenner C (2010) 'Migratory patterns and estimated population size of Pygmy Blue Whales (*Balaenoptera musculus brevicauda*) traversing the Western Australian coast based on passive acoustics'. IWC Scientific Committee, Agadir, Morocco.

McIntosh RR (2007) Life history and population demographics of the Australian sea lion. PhD thesis. La Trobe University, Bundoora, Victoria.

McIntosh RR, Page B, Goldsworthy SD (2006) Dietary analysis of regurgitates and stomach samples from free-living Australian sea lions. *Wildlife Research* **33**, 661–669.

McKenzie J, Parry LJ, Page B, Goldsworthy SD (2005) Estimation of pregnancy rates and reproductive failure in New Zealand fur seals (*Arctocephalus forsteri*). *Journal of Mammalogy* **86**, 1237–1246.

McKenzie J, Page B, Goldsworthy SD (2007a) Growth strategies of New Zealand fur seals in southern Australia. *Journal of Zoology* **272**, 377–389.

McKenzie J, Page B, Shaughnessy PD, Hindell M (2007b) Age and reproductive maturity of New Zealand fur seals (*Arctocephalus forsteri*) in southern Australia. *Journal of Mammalogy* **88**, 639648.

Mitchell E, Baker AN (1980) Age of reputedly old killer whale, *Orcinus orca*, 'Old Tom' from Eden, Twofold Bay, Australia. *Report of the International Whaling Commission* **Special Issue 3**, 143154.

Morin PA, Archer FI, Foote AD, Vilstrup J *et al.* (2010) Complete mitochondrial genome phylogeographic analysis of killer whales (*Orcinus orca*) indicates multiple species. *Genome Research* **20**, 908–916.

Morrice MG (2004) 'Killer whales in Australian territorial waters'. Report to Whale & Dolphin Society. Deakin University, Warrnambool, Victoria.

NMFS (2008) *Recovery Plan for Southern Resident Killer Whales (*Orcinus orca*).* National Marine Fisheries Service, NW region, Seattle, Washington. <www.nmfs.noaa.gov/pr/recovery/plans.htm> (Accessed August 2011).

Page B, McKenzie J, Goldsworthy SD (2005) a Inter-sexual differences in New Zealand fur seal diving behaviour. *Marine Ecology Progress Series* **304**, 249–264.

Page B, McKenzie J, McIntosh R, Baylis A *et al.* (2004) Entanglement of Australian sea lions and New Zealand fur seals in lost fishing gear and other marine debris before and after Government and industry attempts to reduce the problem. *Marine Pollution Bulletin* **49**, 33–42.

Page B, McKenzie J, Sumner MD, Coyne M *et al.* (2006) Spatial separation of foraging habitats among New Zealand fur seals. *Marine Ecology Progress Series* **323**, 263–279.

Page B, McKenzie J, Hindell M, Goldsworthy SD (2005b) Drift dives by male New Zealand fur seals (*Arctocephalus forsteri*). *Canadian Journal of Zoology* **83**, 293–300.

Pauly D, Trites AW, Capuli E, Christensen V (1998) Diet composition and trophic levels of marine mammals. *ICES Journal of Marine Science* **55**, 467–481.

Pitman RL, Durban JW (2012) Cooperative hunting behavior, prey selectivity and prey handling by pack ice killer whales (*Orcina orca*), Type B, in Antarctic Peninsula waters. *Marine Mammal Science* **28**, 16–36.

Pompa S, Ehrlich PR, Ceballos G (2011) Global distribution and conservation of marine mammals. *Proceedings of the National Academy of Sciences of the U.S.* **108**, 1360–1365.

Rennie S, Hanson CE, McCauley RD, Pattiaratchi C *et al.* (2009) Physical properties and processes in the Perth Canyon, Western Australia: links to water column production and blue whale abundance. *Journal of Marine Systems* **77**, 21–44.

Shaughnessy PD (1999) Seals and sea lions. In *Under Southern Seas: the Ecology of Australia's Rocky Reefs* (Ed. N Andrew) pp. 219–227. University of New South Wales Press, Sydney.

Shaughnessy PD (2011) 'Abundance of New Zealand fur seal pups on Kangaroo Island in 2010–11'. Report to Department of Environment and Natural Resources, South Australia.

Shaughnessy PD, Gales NJ, Dennis TE, Goldsworthy SD (1994) Distribution and abundance of New Zealand fur seals, *Arctocephalus forsteri*, in South Australia and Western Australia. *Wildlife Research* **21**, 667–694.

Shaughnessy PD, Kirkwood R, Cawthorn M, Kemper C *et al.* (2003) Pinnipeds, cetaceans and fisheries in Australia: a review of operational interactions. In *Marine Mammals: Fisheries, Tourism and Management Issues* (Eds N Gales, M Hindell and R Kirkwood) pp. 136–152. CSIRO PUBLISHING, Melbourne.

Shaughnessy P, McKenzie J, Lancaster ML, Goldsworthy SD *et al.* (2010) Australian fur seals establish haulout sites and a breeding colony in South Australia. *Australian Journal of Zoology* **58**, 94–103.

Sironi M, Lopez JC, Bubas R, Carribero A *et al.* (2008) 'Predation by killer whales (*Orcinus orca*) on southern right whales (*Eubalaena australis*) off Patagonia, Argentina: effects on behaviour and habitat choice'. Report SC/60/BRG29 to International Whaling Commission.

Stafford KM, Delwayne R, Tolstoy M, Chapp E *et al.* (2004) Antarctic-type blue whale calls recorded at low latitudes in the Indian and eastern Pacific Oceans. *Deep-sea Research. Part I, Oceanographic Research Papers* **51**, 1337–1346.

Tormosov DD, Mikhaliev YA, Best PB, Zemsky VA *et al.* (1998) Soviet catches of southern right whales *Eubalaena australis*, 1951–1971. *Biological Conservation* **86**, 185–197.

Valenzuela LG, Sironi M, Rowntree VJ, Seger J (2009) Isotopic and genetic evidence for culturally inherited site fidelity to feeding grounds in southern right whales (*Eubalaena australis*). *Molecular Ecology* **18**, 782–791.

Wellings CE (1944) The killer whales of Twofold Bay, N.S.W., Australia, *Grampus orca*. *Australian Zoologist* **10**, 291–293.

Williams TM, Davis RW, Fulman LA, Francis J *et al.* (2000) Sink or swim: strategies for cost-efficient diving by marine mammals. *Science* **288**, 133–136.

Yurk H, Barrett-Lennardt L, Ford JKB, Matkins CO (2002) Cultural transmission within maternal lineages: vocal clans in resident killer whales in southern Alaska. *Animal Behaviour* **63**, 1103–1119.

PART 5

MARINE ECOSYSTEMS AND THEIR CONSERVATION

The previous 15 chapters have described in detail the ecology of many species from algae to whales on reef and in coastal regions of southern Australia. In this Part, we transcend the detail and examine the way in which natural marine systems (i.e. ecosystems) function as a whole. Chapter 19 discusses food webs, and Chapter 20 their conservation in the face of the many threats that affect their integrity and function.

An ecosystem can be defined broadly as a functional unit that includes all the species interacting with each other and the environment within its boundaries. At one scale, southern Australia is one large ecosystem, but this is far too big to treat or manage as a single unit. So, ecologists prefer to think of ecosystems as much smaller units that interact and can be combined for particular purposes to form larger, yet still manage-able, units, such as the SW region, the eastern Great Australian Bight, the Gulfs and Bass Strait.

At local levels within marine systems, two other terms are used – community and assemblage. A community may be defined as a collection of interacting species occurring together in the same place and an assemblage as a related group of organisms within a community (Fauth *et al.* 1996). However, these definitions are by no means universally accepted and they are often used simply to mean the species co-occurring in one place.

Every ecosystem shows organisation, which is the outcome of the interactions between species and their consequent adjustments, and Darwin described them as 'bound together by a web of complex relations'. The concept of food webs arose in this manner and it provides a way of organising a little of this complexity. Food webs can be defined as 'a network of consumer–resource interactions among a group of organisms, populations or aggregate trophic units' (Polis and Winemiller 1996), and they are essentially maps that describe the organisms, what they eat, and by what they are eaten. The component species or species groups are generally organised in a hierarchic way into trophic levels, reflecting their distance from the original source of energy or food production, as shown in the figure in the Introduction to this book.

Finally, in Chapter 20, we consider the conservation of reef ecosystems. The major threats are pollution of various kinds, mainly near population centres and, more pervasively, overfishing. The former is managed with varying success under the environmental legislation of the States, and the latter by fishing controls and marine protected areas (MPAs). The southern States and the Commonwealth all have ongoing programs of development of MPAs, and their current status and the many challenges are reviewed.

REFERENCES

Fauth JE, Bernardo J, Camara M, Resetarits WJ *et al.* (1996) Simplifying the jargon of community ecology: a conceptual approach. *American Naturalist* **147**, 282–286.

Polis GA, Winemiller KO (1996) *Food Webs: Integration of Patterns and Dynamics*. Chapman and Hall, New York.

19 Food webs

'Every theory of the course of events in nature is necessarily based on some process of simplification, and is to some extent therefore a fairytale.'

Sir Napier Shaw

OVERVIEW

Food webs can be conceptualised as nodes connected by links, and this chapter describes five food webs, each at widely different scales, and three of them with functional insights drawn from modelling studies. We then describe concepts of key species in web links according to their influence in the system (e.g. keystones, bio-engineers and measures of interaction strength). The influence of biodiversity on food webs and concepts of productivity and stability are considered. Fishing may have complex effects on ecosystems, which may include: the loss of top predators, termed 'fishing down the food web'; indirect effects such as destruction of the benthic habitat, discarded by-catch and its consumption by scavengers; and, lastly, switches to alternative states via algal or jellyfish blooms.

INTRODUCTION

An understanding of ecosystem dynamics (i.e. the way in which a system will respond to fishing, climate change, pollution or other perturbation) is critically important for management of the system. Of these perturbations, fishing is arguably the most pervasive, because the most serious destabilisations or disturbances to ecosystems have been caused by overfishing.

The dynamics of systems can vary widely, depending on the mode of control. The three types of control are bottom-up, top-down and wasp-waist. Bottom-up control is when the abundance of large predators is controlled by the lower trophic groups; conversely, top-down control occurs when top predators control the abundance of their prey at lower trophic levels. Wasp-waist systems typically are those where small pelagic fishes, such as sardines, exert both top-down control of lower trophic levels (zooplankton), and also bottom-up control of higher levels (large fishes, seabirds and seals). As a generalisation, bottom-up control is considered more prevalent in cooler regions and top-down control in warm temperate to tropical regions. Southern systems appear to be transitional because elements of all three types of control may co-occur in one system, as discussed below.

Another concept important for understanding the dynamics of ecosystems is their productivity. Three different types of system, elaborated by Sommer *et al.* (2002), occur in southern Australia. These are:

- **Upwelling systems**, with high nutrient and silicon levels carried up from deep water (see Chapter 2). This nutrient regime favours a diatom-rich phytoplankton community that is passed on to higher levels of the web. It has been well described as the 'muscle food chain'.
- **Coastal regions with low nutrient levels.** These regions are widespread in southern temperate waters and the low nutrient system is accentuated after prolonged stratification of the water column in summer. The waters are dominated by picoplankton and primary production is low and channelled through a microbial loop (picoplankton – flagellates – ciliates). This is a low energy food chain and supports fewer top predators than upwelling systems.
- **Eutrophic coastal systems**, with high nutrients produced from urban run-off and sewage discharge. In these systems, summer blooms of macroalgae, toxic dinoflagellates and jellyfish blooms may prevail.

Food webs are readily constructed from observations and data on diets of different species, and even by inference. When presented as a single diagram, they may give the illusion of comprehensiveness. However, as pointed out by Paine (1988), they can never be complete, because some species are cryptic or poorly known, some are transient and others, notably some molluscs and fish, change their diet with increasing size, making their allocation to one trophic level rather arbitrary. In reality, all food webs omit an enormous amount of detail and have been criticised because the systems they describe are open-ended and the chosen scales are arbitrary, depending on data of variable quality and sometimes 'best guesses'; other subjective factors, such as time available to the researcher, also influence the outcome (Peters 1988).

In this chapter, we first describe some common features of webs and then illustrate them with southern Australian examples. We then examine the strength of interactions between species, and the relations between biodiversity and ecosystems, and lastly we discuss the effect of fishing and nutrient enrichment as major stresses that can precipitate regime shifts to an alternative state.

FOOD WEB PATTERNS WITH EXAMPLES

> *'A food web is a map that describes which kinds of organisms in a community eat which other kinds.'*
>
> Pimm *et al.*, 1991

After noting that webs are designed to present the general principles underlying complex communities, and necessarily differ in many details, Pimm and colleagues in the above review emphasise the following common features of webs:

- Cycles are rare. A cycle occurs where, for example, A eats B eats C eats A.
- The average proportions of top predators, intermediate species and basal species are roughly constant between webs.
- Species at higher trophic levels have fewer prey and predators than species at lower levels;
- The length of many webs is three to four trophic levels, but in marine webs five levels are very common.
- The average number of links between species is roughly twice the number of species in any given web.

These generalities of course submerge particular features, and most are contentious. For example, many fish and invertebrates feed at higher levels as they grow larger (i.e. they eat their way up the food web). Others do the reverse; for example, herbivorous fish may start life as planktivores or carnivores and switch to algae as they mature. Furthermore, these generalities are abstractions that still need to be explored and verified in many southern marine habitats. The four examples below illustrate the widely different scales and conditions for which webs can be described. The first is in a coastal island system, seasonally enriched with nutrient run-off from an adjoining seabird colony; the second is also in a coastal system but with low nutrient levels; the third covers a vast area with extensive, seasonal, coastal upwellings; and the fourth is a bay with high nutrient levels from urban input.

West I., South Australia

This example is a descriptive account of a benthic reef web for a semi-sheltered boulder slope at 4–5 m on the lee of West I., resulting from > 40 years of studies there (Shepherd and Womersley 1970; Kangas and Shepherd 1984; Clarkson and Shepherd 1985; Shepherd and Gray 1985; Shepherd and Turner 1985; Shepherd and Daume 1996; Shepherd and Clarkson 2001; Shepherd *et al.* 2008, 2010). The web describes an area of only ~800 m^2 and is at about the smallest scale realistically possible (Fig. 19.1). The focus is largely on the interactions between commercial species of abalone and other species, so it shows only a tiny fraction of all the interactions present. Species with a similar diet and feeding mode are often called guilds, and in Fig. 19.1 guilds are shown in boxes, with sometimes > 1 guild per box for simplicity, although this submerges subtle differences that may exist between the component species. The web has some interesting features:

- Omnivory, here used in its general sense to mean the consumption of both plant and animal material, is low, with only two species of leatherjacket and two species of sweep feeding on both plants and animals. Note, however, that in food web studies the term omnivory is sometimes used to mean feeding at two or more trophic levels and on this definition most fish species would be omnivorous (Pimm *et al.* 1991).
- Vast supplies of drift algae are transported by wave action from more exposed areas to this site; the algal drift supports a large abalone population and many gastropods, and in consequence attached macroalgae are little grazed.
- Top predators, such as sharks and seals, are rare. Although the Island is a haul-out site for New Zealand fur seals, they forage offshore and seldom in the Island's vicinity;
- Important species interactions and control mechanisms for abalone are indicated with a heavy line in Fig. 19.1. They are:
  1. Small gastropods and chitons graze on algal turf and crustose coralline surfaces, maintaining a high cover of these corallines, which are a preferred settlement micro-habitat for settling abalone larvae. Otherwise, the corallines would quickly become smothered by an algal turf and die, leaving no habitat for the larvae to settle on.
  2. Greenlip abalone recruitment varied four-fold between years according to sea temperature (Chapter 9), indicating bottom-up forcing of recruitment to the population.
  3. Stingrays exerted a strong top-down, density-dependent control on adult abalone abundance.
  4. The establishment of a seal haul-out site in 1991 fortuitously revealed a cascade effect down the web. Seals sharply reduced wrasse numbers and this allowed the wrasses' prey, juvenile abalone, to quickly increase in abundance, so revealing a strong interaction between wrasses and a preferred prey species (Shepherd *et al.* 2010).

Jurien Bay Marine Park

Jurien Bay, situated on the warm temperate central west coast of Australia and covering ~780 km^2, has a unique combination of offshore reefs, islands and lagoons. The reefs are dominated by *Ecklonia* forests, and the lagoons by seagrass and sand. The diverse topography provides a range of habitats for a very high diversity of tropical and temperate species, and the marine park has management goals aimed at conserving this richness. Major fisheries in the area are western rock lobster, dhufish, baldchin groper and Roe's abalone. A study by Lozano-Montes *et al.* (2011) identified 80 functional groups of plants or animals and used an ecopath-ecosim model to understand the web processes. It is the most detailed model yet constructed in Australia.

Such models have been developed among other things to predict the impact of various management strategies on marine systems and their application in Australia has been spurred by the Australian adoption of

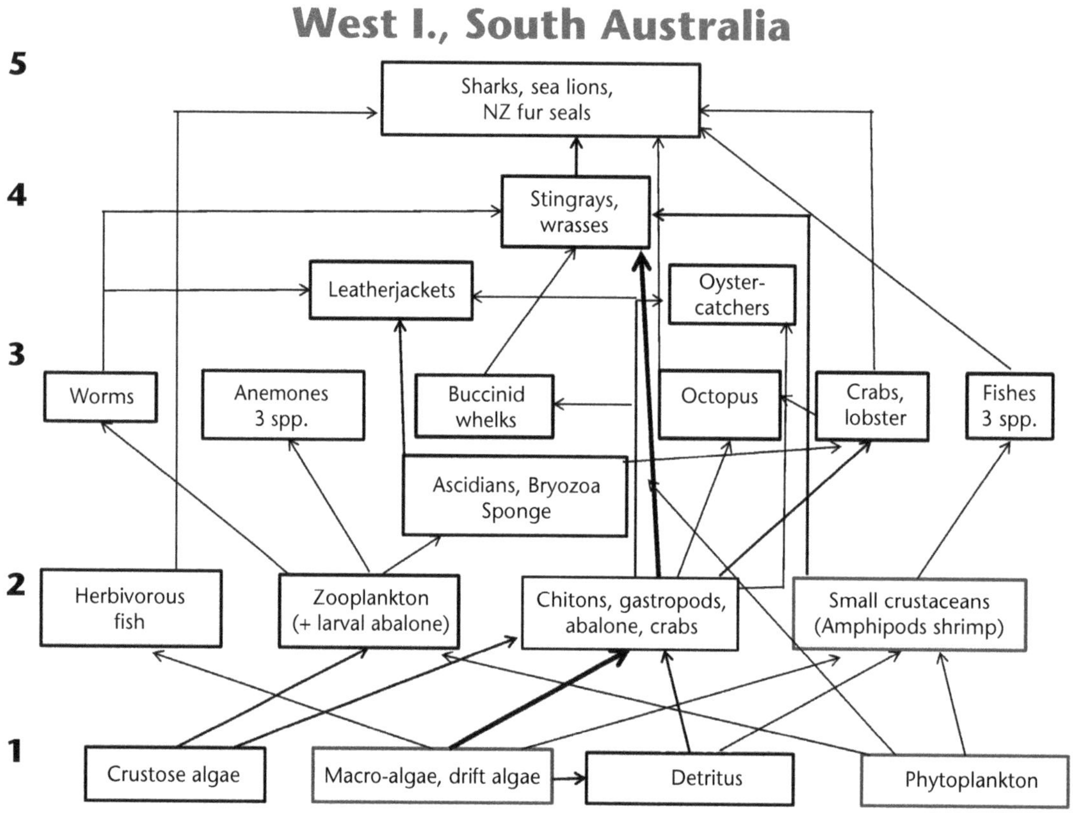

Figure 19.1: Food web constructed for a boulder slope at 4–5 m depth at West I., South Australia, with emphasis on those species relevant to abalone species. Arrows show the direction of flow of food and energy, and the heavy arrows indicate the major interactions. Approximate trophic levels are indicated at left. The crustose algae are encrusting corallines and red and brown algae (*Cruoria* sp., *Archaeolithoderma* sp., etc.); the herbivorous fish are sea carp, herring cale, dusky morwong, silver drummer, scaly fin and zebra fish; the worms include nemerteans, flatworms and polychaetes; the buccinid is the carnivorous gastropod *Pleuroploca australasiae*; the anemones are *Anthothoe albocincta, Isanemonia australis* and *Epiactis* sp.; the octopuses are the blue-ring octopus *Hapalochlaena maculosa* and the southern octopus *Octopus maorum*; the three carnivorous fishes (Level 3) are magpie perch, ornate cowfish and leafy sea dragon; the omnivorous leatherjackets are the horseshoe leatherjacket and the spiny-tailed leatherjacket; the wrasses are the blue-throated, senator and snakeskin, as well as the western blue groper; the stingray is the smooth stingray *Dasyatis brevicaudata*; the shark is the white shark (a rare visitor), and the sea lion is the Australian sea lion. See Chapters 8–10, 15–18 and papers listed in the text for more detail about the various species in the web.

a national oceans policy based on ecosystem management. The ecopath-ecosim model is among the best of such models, and is based on (a) biomass estimates of primary production and of all the principal species in the system, and (b) the feeding, growth and mortality of these species, and their losses to fishing. The model is then tuned to balance the biomass flowing from the primary producers through the intermediate to top trophic levels. However, as a cautionary note, these models are based on a simplified view of the system (i.e. the thermodynamic laws of physics), and have similar limitations to food web theory in neglecting many complexities.

In Jurien Bay, the model estimated that the total throughput of biomass was 15 300 t km^{-2} year^{-1}, with ~35% of this energy being converted to detritus; the transfer efficiency between trophic levels was ~9%. Fishing catches were ~0.5 t km^{-2} year^{-1}, of which 71% was western rock lobster. The main components of the system, appropriately grouped, and their trophic links are shown in Fig. 19.2.

The model was then manipulated to predict the changes that would take place if the Bay were completely closed to fishing. The predictions for this scenario were: (a) the biomass of many exploited species

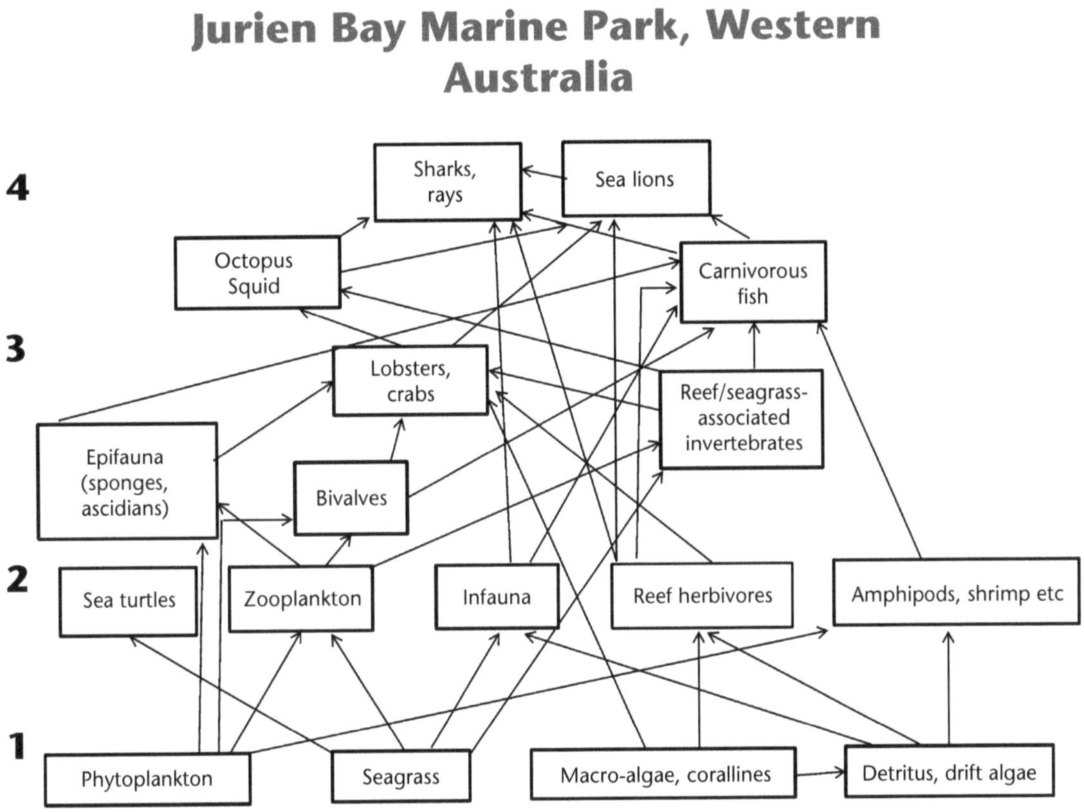

Figure 19.2: Food web for Jurien Bay, Western Australia showing selected functional groups and their trophic links, using the same conventions as in Fig. 19.1. The sharks are whalers, grey nurse, long-nose, wobbegongs and white sharks, and the sea lion is the Australian sea lion. The carnivorous fishes are dhufish, baldchin groper, breaksea cod, wrasses, blennies and leatherjackets. The reef-associated herbivores are mainly herbivorous fish, sea-urchins, turbinid snails, abalone and small gastropods.

(snapper, sharks and dhufish) would increase by two to five times and (b) lobster biomass would increase by ~70%. These increases would cascade down the web and cause up to 30% declines in biomass of their prey (cephalopods, pelagic and reef fishes, foxfish and groupers). Next, the model was used to test the significance of primary production to the system, by increasing the biomass of *Ecklonia* by 10%. The results substantially increased the biomass of many invertebrate groups, such as juvenile lobster, crabs and infauna, as well as some commercial fishes. The simulation demonstrated that the Jurien Bay system was dominated by benthic functional groups, and driven by bottom-up forces (i.e. the primary producers – macroalgae and seagrasses) that provide habitat and food for marine invertebrates and fish.

Eastern Great Australian Bight (eGAB)

This region extends from the head of the GAB to Cape Catastrophe at the southern tip of Eyre Peninsula, South Australia, but the models below extended the eastern boundary to Cape Jaffa, in SE South Australia, and included the lower South Australian Gulfs and Investigator Strait. The area covers 154 000 km^2, and includes shelf waters to 200 m deep between 132° and 139.7°E longitude. Food web models were prepared by Bulman *et al.* (2011) and Goldsworthy *et al.* (2011) partly to predict the impact of sardine fishing on higher level predators, especially seals and sea-birds, and partly to elicit the underlying mechanisms of web control, particularly by mid-trophic species.

First, the model was manipulated to simulate various levels of fishing. Under a 50% increase in the sardine quota, tuna, sea-birds, squid and many piscivorous fish would seriously decline. Tuna stocks were already depleted, having declined by 95% from their pristine size, and they would decline by another 20% over the next two decades under existing quotas. The model also suggested that the 'predator gaps' resulting from the historical reduction in populations of tuna, shark and fur seal populations were filled by cephalopods.

In this system, the small pelagic species with short life histories (sardines, redbait, mackerel and anchovy) typically dominate Trophic Level 3, and are the main channel funnelling energy from plankton to large fish. However, krill or cephalopods could fill the same role, because they arguably have the same function (Bulman et al. 2011).

In their eGAB modelling, Goldsworthy *et al.* (2011) highlighted the importance of sardines to the apex predators, but conceded a possibly pivotal role for squid. When the squid filled the 'predator gap' they

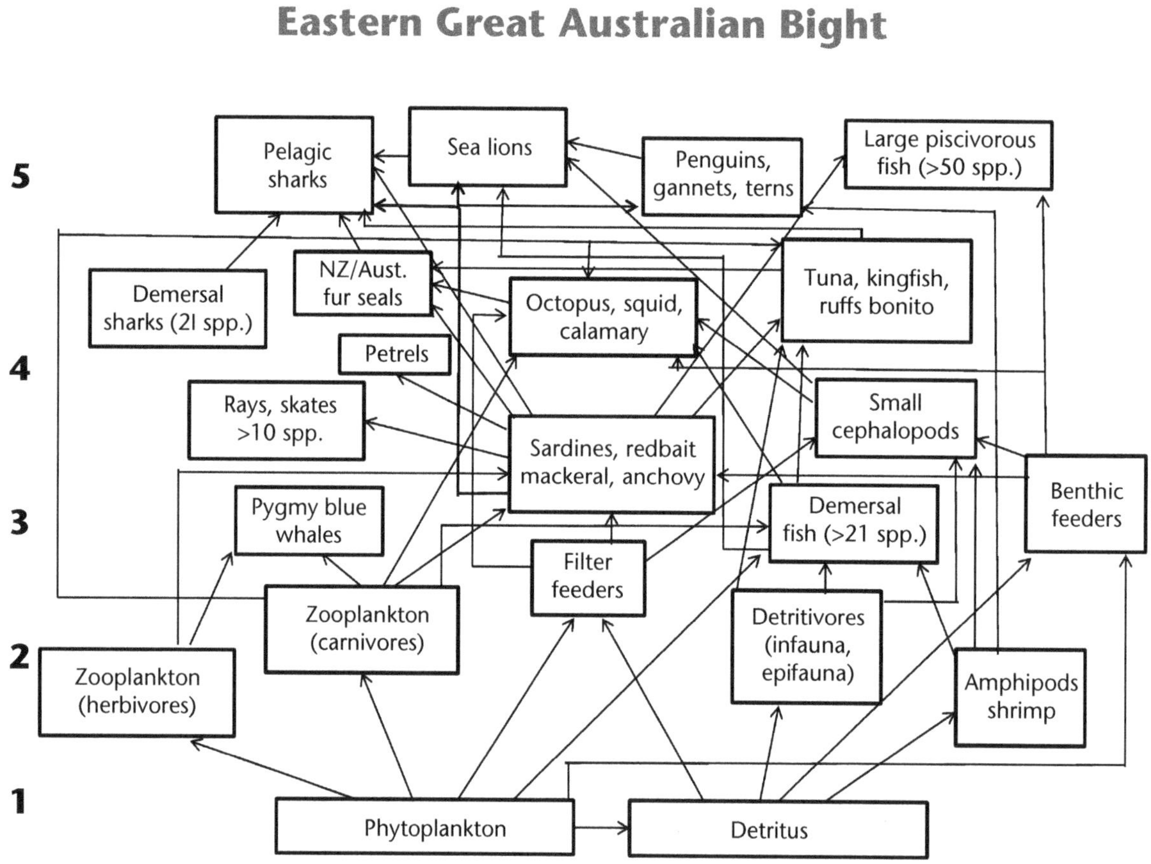

Figure 19.3: Simplified pelagic food web for the eastern Great Australian Bight showing selected functional groups and their trophic links, according to the conventions in Fig. 19.1. The sharks are the blue, mako, hammerhead, thresher, bronze whaler, dusky whaler and white shark. The sea-birds also include the shy albatross, silver gull, little tern, silver gull and shearwaters. The sea lion is the Australian sea lion. The large piscivorous fish include pink snapper, trevally, wrasses, many reef species, flathead, old wives and leatherjackets. The benthic feeders are crabs, prawns and lobster, the small cephalopods include *Euprymna* and *Histioteuthis*, and the many small demersal fishes include barber and butterfly perch, moonlighters, bullseyes and whiting. Filter feeders include bivalves and worms, mainly polychaetes.

would have reduced the food available to other higher predators and, being short lived, their own biomass would have gone back to detritus. Overall, the sardines (the small pelagics) exert a bottom-up control of higher trophic levels, although a small part (that of squid) may exert some waist-wasp control by restricting energy flow to top competing predators! Bulman et al. (2011) compared numerous predator–prey interactions in the system, and they also concluded that the interactions of fur seal and pelagic sharks with squid were dominating elements in the system. They also noted that different branches of the eGAB food web could display different forcing, with switches between them possible depending on fishing pressures. In a comparison between the eGAB system and that in eastern Bass Strait and eastern Tasmania, Cathy Bullman and colleagues found that the latter system, though still largely bottom-up forced, had more top-down controlling elements. Hence, under stresses induced by climate change, the mid-trophic levels (small pelagic fish) may well be at the centre of a regime shift.

Port Phillip Bay

This almost enclosed bay of ~1930 km^2 is an example of an embayment subject to nutrient inputs of ~7500 t N year^{-1} from sewage treatment works for a surrounding population of > 3.5 million, and river/stormwater inputs (Murray and Parslow 1999). The Bay has a similar range of habitats to Jurien Bay, albeit more sheltered, and is subject to intense recreational and commercial fishing on scalefish (annual catch 1200–2500 t), abalone (50 t year^{-1}), and cultured mussels (600 t year^{-1}). Three different models were used by Fulton and Smith (2004) to compare their performance under different scenarios:

- the status quo
- a five-fold increase and decrease in nutrient loads
- a fisheries optimisation module, so that fishing mortality met economic, social and ecological objectives. The economic and social components related to employment value of fisheries, and the ecological component maintained an optimal ecosystem structure, with a mandatory stock rebuilding policy.

The purpose of the study was to identify the best management strategy, given the intense fishing pressures and existing nutrient loads. Without going into the detail of the models, the overall conclusions were that:

- Changes in nutrient levels had a much larger and more widespread impact than changes in fishing pressure. If productivity increased, so did the fisheries' take. Similarly, if it fell, the take did also.
- The economic strategy gave increased catches and fishing mortality levels in some fisheries, notably in the high value fishery for lobster, but produced a decline in many fish stocks and a collapse in shark and flatfish.
- Under the ecological strategy, when stock rebuilding is given little weight, the abalone catch increases and then the abalone stocks collapse. If charismatic or preferred recreational species are favoured, other components of the system continue to suffer. But when conservation of all groups is emphasised, a more balanced system results and shark populations recover. However, this balance comes at the cost of a catch reduced by about one third.
- Under a compromise strategy, the depletions suffered under an economic strategy are avoided and, as the economic weightings are further reduced, the optimal strategy becomes the ecological one. Thus balanced objectives are required both across the sectors of human interest and the affected trophic groups in the system.

Bass Strait and Tasmanian shelf waters

This region covers 137 000 km^2 of Tasmania coastal shelf waters north to the Victorian coast and includes ~2300 km^2 of reef habitat. An ecopath-ecosim model was developed to test impending climate-change shifts in primary production (Watson *et al.* 2013). With climate change, warm, low-nutrient northern waters of the East Australia Current will flow more strongly down the east Tasmanian coast and reduce the primary productivity of waters there (Chapter 2).

Like those described above, the model covered the food web from Level 1 (primary producers) to Level 4 (dolphins, seals, sea-birds, sharks, rays and large carnivorous fish) and showed changes under modelled

10–30% increases and declines in primary productivity. In the 10% decline scenario, a substantial decline in fish catches occurred as shown in Table 19.1, and similar declines of 20–30% biomass occurred in other benthic animals, with lesser declines of up to ~5% in species of conservation interest (e.g. dolphins, seals and sea-birds). Conversely, modelled increases in primary productivity showed similar increases in biomass or catches of these species groups.

It may be concluded from this study that primary productivity is the main driver of the production of marine resources; hence, with increasing temperatures under global warming, the food web will change substantially as marine productivity declines, energy transfer through the system is reduced and fisheries are impacted. However, the authors caution that the modelled scenarios grossly over-simplify the many uncertainties of climate change, and that increased seasonal upwellings in some sub-regions are likely with reverse positive effects.

INTERACTIONS BETWEEN SPECIES

Many advances have been made since the construction of descriptive food webs developed in the 1960s. Ecologists began to quantify the flow of energy and matter through the web, and this led to experimental and other studies on the strength of interactions between species. From this emerged the recognition of species that were dynamically important to the system, and such species were variously called 'keystones', foundation species, bio-engineers or simply 'key species', according to the perception of their role. By identifying the most influential links, and excluding all others, the community 'is distilled into an interaction web, a subset including only species that dominate biomass and/or regulate the flow of energy and matter' (Polis 1994). In this way, the important processes that most influence the community are identified.

Table 19.1: Approximate projected decline in landings of commercially fished species under a 10% reduction in primary productivity (extracted from Watson *et al.* 2013).

Commercial fishery	Trophic level	Projected decline (%)
Abalone	2.0	30
Arrow squid	3.7	33
Giant crab	2.4	45
Octopus	3.4	44
Reef fishes	2.4	34
S. rock lobster	2.4	60
Shark	~4.0	26
Southern calamari	3.4	33

Keystone species and bio-engineers

All species are not equal, and it is well known that some play a critically important role in structuring the system in which they are found. For example, canopy algal species are a major source of energy on temperate coastal reefs and they also provide structure that supports many other species. Those species that have a strong influence in communities – an influence that is disproportionately large relative to their abundance – first came to be called **keystones**, because their presence was crucial in maintaining the organisation and diversity of their community. Such species may be quite rare in a community and very few have been so designated in southern Australia; however, this may be partly due to the failure to recognise or name them as such. To illustrate the concept, we shall give several southern examples of what may be keystones:

- The spiny lobster on eastern Tasmanian reefs has been shown to play a unique predatory role by controlling sea-urchin abundance. When their densities are decimated by fishing, the black sea-urchin, released from predation pressure, increases and gives rise to urchin barrens, from which > 500 species of algae and invertebrates are lost (see Chapter 12).
- Conversely, in New South Wales the black sea-urchin was thought to be a keystone, because it controlled algal growth and enabled the persistence of limpets which would otherwise disappear (Chapter 12).
- In a pelagic upwelling system, the blue whale has been called a keystone, because, by consuming huge quantities of krill daily, it controls the

abundance of many other species in the system (May *et al.* 1979; Power *et al.* 1996; Chapter 18). Similarly, jellyfish have been called a keystone in some systems, although this role can be intermittent, given the sporadic nature of their outbreaks (Chapter 14; Piraino *et al.* 2002; and see below).
- Wrasses are keystones if they control the abundances of other key species (e.g. molluscs or urchins). Shepherd and Clarkson (2001) discussed the possible keystone role of the blue-throated wrasse in the above West I. system, because it controlled the abundance of abalone and likely other grazing molluscs. However, they concluded on balance that the wrasse did not quite meet the definition, because it lacked a structuring role in that system. There the prey abalone fed passively on drift algae. However, in other systems where abalone or urchins actively graze the algae from the reef surface and modify the habitat, the wrasse might well fulfil this role.

The above examples show that a keystone function may be context-dependent (i.e. it may be specific to a particular environment) (Mills *et al.* 1993). Indeed, the last authors discussed how the term, originally applied to a predator, was later extended to prey species, host species, habitat modifiers and mutualism-promoters! Eventually, the acquired vagueness of the term led Hurlbert (1997) to state that it meant nothing more than being 'important for something' – perhaps 'appealing and harmless, but . . . stultifying'. Yet, others argue for retention of the term, because keystone species are 'small elements sustaining complex ecological buildings' and 'sustain the community by keeping its diversity high' (Piraino *et al.* 2002). Overall, the term has been partly superseded by efforts to measure the functional importance and interaction strength of species in a system.

A recent study (Libralato *et al.* 2006) used ecosystem models to quantify the impact matrix of jellyfish in an ecosystem. This approach has much promise, given the extreme difficulty in assessing the role of a species experimentally. Pauly *et al.* (2009) then showed that the keystone character of jellyfish varied greatly between systems. For example, *Cyanea rosella* in Port Phillip Bay would rate as a keystone if its blooms had a strong modifying effect on the Bay's communities (Chapter 14).

Another term becoming fashionable in ecosystem studies is '**ecosystem engineers**' or '**bio-engineers**', used to describe the key role that some species play in the creation, maintenance or modification of habitats (Jones *et al.* 1994, Wright and Jones 2006). Whereas keystones exert their effects largely through predation, and so emphasise the outcome, engineers physically modify the environment, and so emphasise a process. Temperate southern examples are:

- Macroalgal canopy species, such as kelps on temperate reefs, reduce light, retard water flow, reduce sedimentation (by sweeping the bottom), and create a micro-habitat under their canopy or within their root system for many other species (Chapters 4, 5; Schiel 2006; Connell *et al.* 2008). The seagrass *Amphibolis* growing on reef would have a similar bio-engineering function.
- Sea-urchins, by forming barrens under certain conditions, would also qualify. In the Tasmanian example, a keystone (lobster) has a major effect because it controls the impact of one engineer (urchin) on another (kelp), with flow-on effects on many other community organisms (Jones *et al.* 1994).
- Molluscs, such as mussels, oysters, cockles and mud-snails, are clear examples of species that modify their physical habitats in various ways and create a hard substratum or structure with many interstices or cavities for a host of other species (Thomsen *et al.* 2010). For example, the blue mussel provides a micro-habitat for high densities of polychaete worms, isopods and amphipods (Chapman *et al.* 2005; Chapter 9), and hammer oysters form low to high (> 3 m) relief reefs in the South Australian gulfs, with high densities of ophiuroids and small crustaceans.
- Calcified macroalgae, mainly coralline turfs, by their inter-meshed branches resist wave action, trap detritus and provide a complex habitat and shelter for a large number of organisms, such as polychaetes and small crustaceans (Chapman *et al.* 2005). Rhodoliths similarly provide cryptic habitat for many species (Chapter 7; Nelson 2009).

- Even in pelagic systems bio-engineers are present. The chlorophyll *a* in phytoplankton absorbs and scatters light, so reducing light penetration and raising the maximum depth to which primary production goes (Breitburg *et al.* 2010). Gelatinous zooplankton (jellyfish and ctenophores) can also control abundances of many species (e.g. some bio-engineering oysters) by consuming their larvae.

Interaction strengths

As community ecology developed with the recognition of strongly and weakly interacting species, it became important to find ways of estimating their strength. Interactions are complex, usually referring to predation or herbivory, but they also include mutualisms and parasitism. They were early defined as the average direct per capita effect of one species on another species, and various experimental, empirical and modelling ways of estimating them are reviewed by Laska and Wootton (1998) and Montoya *et al.* (2006). A crude, but easily understood, predictive measure of strength is predator-to-prey body size ratio – the larger the ratio, the stronger the interaction. Of the many more sophisticated measures developed, an observational index – based on predator consumption rates, predator behaviour and predator and prey densities – gave a per capita index of interaction strength that matched independent experimental manipulations (e.g. Wootton 1997). A simplified example of interaction strengths between oystercatchers and their intertidal prey is given in Appendix 1.

From theoretical studies on interaction strengths in food web dynamics, several predictions have been made (McCann *et al.* 1998). First, large numbers of weak interactions, which characterise food webs, are important in promoting stability and persistence, because they tend to dampen oscillations between consumers and their resources. Hence, they 'may be the glue that binds natural communities together' (McCann *et al.* 1998). Second, generalist-dominated food webs should show less variable dynamics than those dominated by specialist species.

BIODIVERSITY AND ECOSYSTEMS

The effect of biodiversity on the productivity and stability of ecosystems has long been debated and ecologists have puzzled over why so many species evolved when many of them had similar roles in ecosystems. In one famous analogy, species were likened to rivets in a plane. The loss of a few would weaken the plane, but it would still fly. Yet, the loss of very few more would precipitate a catastrophic crash. Others thought that species were more like the crew and passengers on the plane. As long as the pilots remained, the plane would fly safely, but lose the pilots and disaster would ensue no matter how many passengers were aboard. Schiel (2006) carried the analogy further and suggested that some key species, such as bio-engineers, were more like bolts that held communities together, so that their loss caused a catastrophic failure in the ecosystem. Slowly, evidence has accumulated that some species are critically important in ecosystems (i.e. they are bolts), while a few others may be redundant (i.e. they are like rivets or passengers). However, there is general agreement that retaining as many species as possible is critically important.

Gradually, a unifying biodiversity theory has started to emerge, with increasing evidence that most species are important if we consider multiple years, places, functions and environmental fluctuations (Steiner *et al.* 2005; Isbell *et al.* 2011; Hooper *et al.* 2012). Here, we summarise current insights about the mechanisms underlying the functional role of species in maximising the productivity and stability of ecosystems (Duffy *et al.* 2007; Loreau 2010). Many of the ideas suggested below are derived from modelling or enclosed experimental systems. They need validation for marine systems and so should still be regarded as tentative propositions. Besides, their application to open systems, where populations are connected and energy transported across habitats, is often unclear.

Productivity

It is well established for land plant communities that more species means greater productivity. In formal terms, the mechanisms driving the increase in primary production and standing stock are functional complementarity and selection of more productive species. Put more simply, the reasons seem obvious. For example, in an algal community, a greater diversity of algal shapes with differing life history features allows the community as a whole to take better advantage of available light and space. An illustration of this is the late succession in an algal community (Chapter 7; Shepherd 1981). As dominant species became established and increased their cover, they created structural complexity and allowed low-light species, and later stress-tolerant species, to establish under the canopies (Chapter 7). Selection occurred as more productive species increased in biomass and some rare species found a niche. The end result was a gradual increase in productivity of the community. Thus, for any set of, say, 100 species, one will be more productive than the others and, by increasing the size of the subset of 100 species in the community, the chances of including the most productive species, and hence total productivity, increases.

At higher trophic levels, the balance of evidence for many systems suggests that 'increasing prey diversity often leads to lower total . . . impact by higher trophic levels' (Duffy *et al.* 2007). Evidence supporting this proposition comes from two ideas – variance-in-edibility and the dilution hypothesis. The former means that predators will have a weaker impact on more diverse prey and the latter refers to the tendency for populations of individual prey species to be smaller as prey diversity increases. There is also evidence that herbivore impact on algal biomass will decrease with greater algal diversity (Hillebrand and Cardinale 2004).

Stability

This concept is multi-faceted, but generally has three aspects: resistance to disturbance; resilience (i.e. the rate of recovery after a disturbance); and constancy over time (Tilman 1999; Schindler *et al.* 2010). Traditionally, ever since Darwin, ecologists have believed that complex species-rich communities are more stable than simple ones, but the causes have been largely unexplored. This has changed in recent years and two related causes have been proposed – the **portfolio effect** and the **insurance hypothesis**.

The portfolio effect states that the average of many community variables, such as biomass of component species, is less variable than that based on a few species. The effect is so-called because of the long-standing economics principle that a more diversified portfolio is less volatile than one with few holdings. The insurance hypothesis involves the buffering effect and the performance-enhancing effect. Because different species respond differently to a fluctuating environment, over time there will be a slight selection favouring species that are better adapted to the current conditions and have a higher than average productivity (Yachi and Loreau 1999). Furthermore, different kinds of perturbations may occur (e.g. pulse disturbance, press disturbance, invasions of species and over-exploitation), with multiple or interacting effects, but these have rarely, if ever, been investigated (Ives and Carpenter 2007).

Another aspect of stability is a community's resistance to invasion by exotic species (Sakai *et al.* (2001). Greater saturation of niches in species-rich communities reduces the availability of limiting resources. Hence, greater diversity reduces the chance that an invading species can establish a viable population. An example of the converse – the loss of native species – is Crawfish Rock, Western Port (Shepherd *et al.* 2009). Here, the loss of 66% of the algal species from a pristine species-rich community through increased sedimentation, allowed invasive species to establish, possibly by the opening up of free space (see Chapter 7). Another reason, independent of the community's resistance, is the fact that an invader has no native predators, so giving it a 'head start' compared with native species (Tilman 1999). Given the complexities, uncertainty and our ignorance in the whole field, Tilman's advice that 'the safest policy is to preserve as much diversity as possible' is self-evident (and see review of Hooper *et al.* 2005).

ECOSYSTEM EFFECTS OF FISHING

Fishing changes not only the abundance of target species at different levels of the food web, but also the structure and function of the ecosystem. We first consider fishing effects at the top of the food web and then effects lower down.

Fishing down the food web

The decline of top predators is a global phenomenon and was first clearly enunciated by Pauly *et al.* (1998) who coined the term 'fishing down the food web'. This refers to the sequence of events in heavily exploited systems. As the top predators are fished out or become depleted, fishers turn to other species lower in the web. These species then collapse and the collapses of other species follow sequentially through the system (Fig. 19.4). Two other consequences can then follow: the removal of key species can lead to restructure of the system as discussed above, and the system may become unstable and switch to an alternative state, as discussed below.

In southern systems, the top predators – sharks and large piscivorous fishes – are the most susceptible to overfishing or being taken as by-catch (Figs 19.1–3), and almost all species of exploited sharks have either collapsed or declined seriously (e.g. grey nurse sharks, wobbegongs in New South Wales, dusky and bronze whalers in southern states). Notably, the school shark and whiskery shark stocks are depleted, having declined by 84% and 77%, respectively (Stevens *et al.* 2000). Other indicators of population decline come from the shark-meshing program in New South Wales, where coastal shark populations fell by 94% in the 40 years until 1990 (Reid and Krogh 1992; Chapter 17). As large whaler sharks declined, they were temporarily replaced by the more fecund and wide-ranging tiger and hammerhead sharks. The Jurien Bay modelling study described above also suggested that sharks and large fishes had extensively declined under fishing.

Large piscivorous fishes have fared no better than sharks. Barracouta populations collapsed in the 1970s, southern bluefin tuna, yellow-tail kingfish and mulloway are depleted, and snapper catches have declined by 30% since the early 1980s (Chapter 17). The consequences of top predator declines have been reviewed by Heithaus *et al.* (2008) and Ferretti *et al.* (2010).

Figure 19.4: Schematic diagram of fishing down the food web (redrawn from Pauly 2003).

Besides the direct impact on prey species, the loss of sharks often exerts a top-down impact on their prey, either directly or indirectly, by affecting their behaviour (the risk effect). The various effects are listed below with southern examples:

- Mesopredators (e.g. small sharks, rays and skates) may increase in abundance due to predator release. For example, the piked dogfish *Squalus megalops* increased five-fold on SE Australian trawl grounds from 1977 to 1997 (Graham *et al.* 2001).
- Foraging and other prey behaviour may be affected (i.e. risk effects), as, for example, when prey give up resources in order to reduce their predation risk. Two studies suffice to illustrate the effect. In Shark Bay, tiger sharks invade inshore feeding grounds annually in summer, causing dolphins and dugongs to abandon foraging there and switch to deeper, less productive habitats with lower predation risk. Similarly, at West I. the arrival of fur seals caused magpie perch to reduce their foraging time in exchange for increased vigilance. This reduced fish growth rates and also cascaded down the web to increase abundance of their prey – crustaceans. The arrival of seals also caused a steep decline in blue-throated wrasse numbers, because many fled to deeper water habitats for a less stressful life, with cascade effects on prey abalone (Chapter 16; Connell 2002; Shepherd *et al.* 2010).
- Top predator declines can produce trophic cascades – an effect more pronounced in coastal and demersal systems. For example, sea-birds, seals and large fishes may increase (predator release), with cascading effects at lower levels (see previous examples and the Jurien Bay model). These effects are, however, context-dependent and may vary between ecosystems.

The generalised picture in Fig. 19.5 summarises some of the known fishing effects on ecosystems described above.

Fishing lower levels and wasp-waist control

A common biodiversity pattern in upwelling systems is a rich benthic flora and fauna at the base of the food chain and many species at the top, with relatively few species in high abundance at intermediate levels (Cury

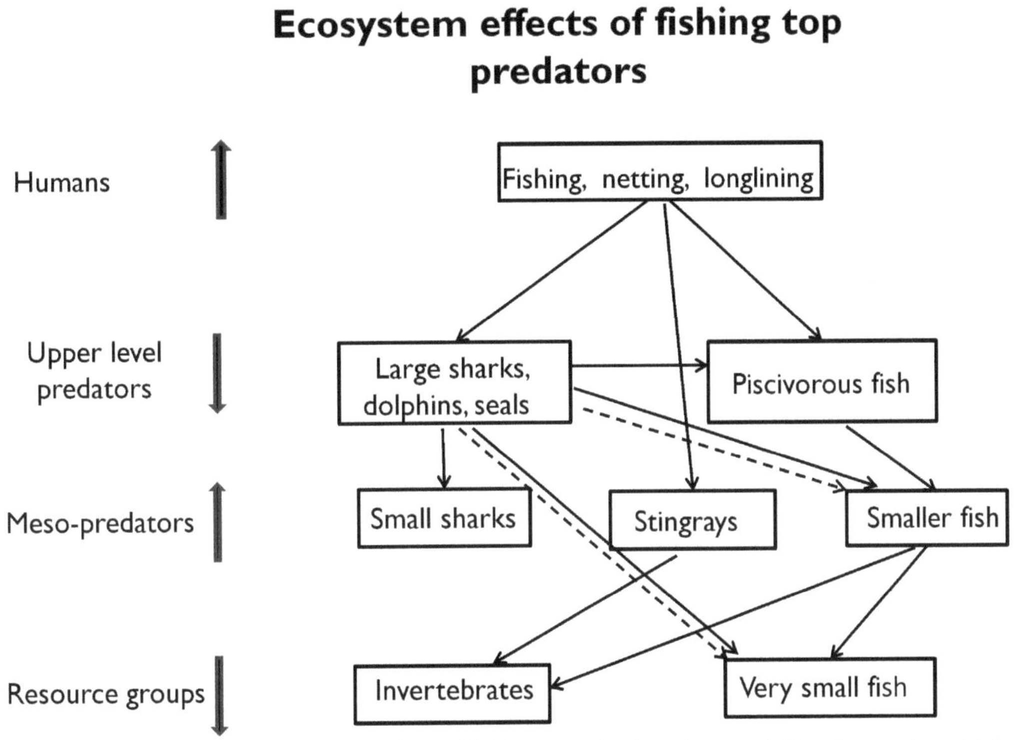

Figure 19.5: Summary of some ecosystem effects of fishing upper level predators or taking them as by-catch in netting and longlining operations. Block arrows at left indicate overall population trends, arrows indicate trophic interactions and dashed arrow indicates behavioural effects (modified from Ferretti *et al.* 2010).

et al. 2000; Smith *et al.* 2011). Small pelagic fish – pilchards, anchovy, sardines and jack mackerel – dominate these intermediate levels and provide food for higher levels. In such systems, wasp-waist control describes the trophic linkages when there is top-down control of zooplankton by the small pelagic fishes and bottom-up control by them of higher level predators (e.g. large pelagic fish, sea-birds and whales). Under wasp-waist control, zooplankton (typically krill) is consumed by schools of small pelagic fish (e.g. sardines, anchovy) and competition for these prey is greatly reduced as fishing intensity increases and reduces the abundance of the small pelagics. This allows other fish species and squid to increase, and they then delay recovery of the small pelagics. At the same time, the small pelagics affect the food supply of higher trophic levels that depend on them for prey (e.g. seals, little penguins and tuna). So, control through this trophic link may exert itself upwards in the web as food becomes limiting for higher predators.

Wasp-waist control is well known in some upwelling systems, but the extent to which it applies to southern systems is still a moot point. As discussed above in the eGAB ecosystem example, control at various trophic levels in this region, and beyond in SE Australia, is complex, with control varying between species, and possibly some wasp-waist control by squid. But, the important implication of all controlling mechanisms, whether wasp-waist, bottom-up or top-down, is that fluctuations in the abundance of intermediate trophic levels, whether due to overfishing or climate change, may have serious ramifications up and down the food web. For the above reasons, the management of southern sardine stocks is conservative, with quotas varying according to environmental conditions and abundances predicted from annual egg surveys (Smith *et al.* 2011).

Indirect effects of fishing

Although this book is primarily concerned with the ecology of firm substrata, the distinction between firm and soft substrata is sometimes vague. Soft substrata may have an epifauna of shellfish, such as oysters and razor shells, or an accumulation of dead shells or rhodoliths, which variously provide a firm substrate for a rich epifauna of erect sponges or bryozoans, and many interstices for other fauna. In particular, beds of the hammer oyster *Malleus meridianus* can accumulate shells to create reef structures often ~20 cm high, and rarely to ~2 m high. Furthermore, at times, a thin veneer of sand over a firm substratum is swept away to reveal the underlying reef, which then acquires an epifauna.

Trawling seriously impacts benthic assemblages with structure, as described above, and has been likened to the clear felling of rain forests. Although the impacts are occasionally described, the ecological implications are rarely known. For example, Tanner (2003) found that 20 experimental trawl passes over a seabed dominated by bryozoans were enough to destroy 36% of them. Later, in a 35-year, long-term comparison of sea bed communities in Gulf St Vincent, prawn trawling was found to have destroyed ~975 km^2 of a hammer oyster/razor shell assemblage in the SE Gulf, ~80% of bryozoans in a bryozoan assemblage over ~900 km^2 in the central Gulf (leaving bare sand) and ~1500 km^2 of eel-grass in Investigator Strait (Shepherd and Sprigg 1976; Tanner 2005). Similarly, in Spencer Gulf, Svane *et al.* (2009) found that prawn trawling had strong effects on the structure of the benthic habitats – effects that were proportional to the intensity of trawling. The abundance, percentage cover and biomass of epifaunal species all declined with increasing trawling time. Affected habitats included sponge and hammer oyster beds in the mid to upper Gulf, and rhodolith beds further south. The ecosystem spillovers of these changes are significant. The macrofauna, through continuous bioturbation recycles nutrients and silica, and their loss, can stimulate dinoflagellate growth, including harmful algal blooms (Olsgard *et al.* 2008). However, the overall trawling impacts on the Spencer Gulf infauna remain unclear, but likely accumulate over time. One short-term study found only slight effects, mainly on some polychaetes (Drabsch *et al.* 2001).

Another significant ecosystem impact arises from the discarded by-catch of trawling. The main by-catch species in the Spencer Gulf study (Svane *et al.* 2008) were leatherjackets, silverbelly, sand trevally and squid. The small proportion that floated was rapidly eaten by dolphins and sharks, while the rest sank to the sea bed and was consumed by a range of scavengers – mainly leatherjackets and blue crabs during the day and isopods (sea-lice) at night. Other studies on trawling by-catch in SW Australia and on New South Wales coasts, reviewed by Kennelly (1995), showed a substantial by-catch of many species, including juvenile snapper, bream, mulloway, sharks, rays and cephalopods.

The gill-net and longline shark fisheries caused the long-term decline of sea lions and captured large numbers of dolphins, until closures were introduced in 2011 (Chapter 18). Gill-nets and longlines each have a by-catch that includes over 125 fish species, and comprises 5% of the catch in Bass Strait and 18% in South Australia. Discards account for ~25–30% of the total catch (Walker *et al.* 2005).

The above impacts of fishing all change the function of an ecosystem by affecting ecologically dependent species, genetic diversity and trophic level balance. Historically, almost all fishing has been selective as fishers target specific preferred species, or select for size, or for season, or spatially. To minimise changes in ecosystem function, the concept of balanced exploitation has slowly emerged (i.e. to proportionally take all usable species from an ecosystem, in proportion to their productivity). This would 'achieve sustainable overall yields while maintaining ecosystem services and function' (Zhou *et al.* (2010), and requires a major shift in the philosophy of fishing, as aptly illustrated in Fig. 19.6.

The long-term ecosystem consequences of these impacts are virtually unknown in southern temperate waters, but the development of ecopath, ecosim and ecospace modelling methods should aid their

Figure 19.6. Discriminated groups cry out for balanced exploitation of the sea, in accordance with Ecosystem Based Fishery Management (EBFM) (courtesy of Zhou *et al.* 2010).

evaluation, and help ensure maintenance of biodiversity and the integrity of local ecosystems (Pauly et *al.* 2000).

ALTERNATIVE STATES AND REGIME SHIFTS

Ecosystems are complex and characteristically have non-linear dynamics and threshold effects. If the system is destabilised by one or more severe disturbances, it may cross a threshold and 'flip' sharply and dramatically to another state; in other cases, the change may be more gradual. In either event, the ability of the system to return to its pristine state depends on the scale and intensity of the disturbance and the system's resilience. Principal disturbances causing regime shifts in southern waters are excessive nutrients that produce eutrophic conditions, overfishing and climate change (reviewed by Folke *et al.* 2004).

One dramatic example in eastern Tasmania described in Chapter 14 was the regime shift from kelp forest to urchin barrens, precipitated by overfishing and climate change. Here, we discuss three other potentially serious regime shifts, evident as blooms of harmful micro- and macroalgae and jellyfish. They commonly occur in disturbed ecosystems and in recent decades have increased globally in frequency and intensity.

Microalgal blooms

In southern Australia, microalgal blooms (red- and brown-tides) have occurred in bays near all southern capital cities due to the presence of excess nutrients. Some species of blue-green algae and dinoflagellates are toxic, and a few of them have high levels of neurotoxins and/or cause oxygen depletion in the water. Dense blooms can devastate shellfish and fish farms, and kill virtually all the fish in an enclosed body of water. The most serious recorded bloom was of the dinoflagellate *Chattonella marina*, which killed 7000 t of caged tuna in Port Lincoln harbour in 1996 (Marshall 2002). Other blooms of toxic dinoflagellates (genera *Gymnodinium* and *Alexandrium*) cause paralytic shellfish poisoning in the ports of Hobart, Melbourne and Adelaide (Hallegraeff 1993). For example, in the Derwent and Huon Estuaries in SE Tasmania, blooms occur annually in late summer. They are triggered by a rainfall event, maintained by calm conditions and subside with increasing wind stress and declining winter temperatures (Hallegraeff *et al.* 1995; Clementson *et al.* 2004) Outbreaks of the blue-green alga *Nodularia* have been a common occurrence in the Peel-Harvey estuary (Hillman *et al.* 1990), and red- and brown-tides of dinoflagellates and diatoms occur seasonally in the Swan–Canning estuary (Anon. 2005). Such outbreaks are usually short lived and occur mainly in estuaries or ports, and hence are beyond the scope of this book (but see the excellent reviews of Turner and Tester (1997) and Waite and Suthers (2007).

Macroalgal blooms

This type of bloom differs from microalgal blooms in three ways: they are not chemically toxic, they have a wide range of ecological effects and they can last much longer (see Valiela *et al.*1997). In Chapter 7, we discussed the many species of turfing and filamentous epiphytic algae – *Ulva, Enteromorpha, Chaetomorpha, Ralfsia, Hincksia, Feldmannia* and *Ceramium* – that often bloom on reefs in coastal waters in response to elevated nutrients, but are seldom reported. One serious case in eastern Gulf St Vincent, reported by Connell *et al.* (2008) is described in Chapter 7. Here, many near- to offshore reefs once dominated by large canopy algae underwent a regime shift under the influence of high nutrient levels and switched to a turf community. Whether the former fucoid communities will re-establish once nutrient inputs are reduced to low levels will depend on whether their functioning and diversity are irreversibly degraded.

The worst and most persistent recorded blooms were those in the Peel-Harvey estuary (131 km^2), where *Cladophora, Chaetomorpha* and *Ulva* successively bloomed in response to elevated nutrients. Lavery *et al.* (1991) described the periodicity of the intense blooms and the factors causing them (availability of phosphorus, nitrogen, seasonal differences in light attenuation, etc.). Eventually, the problem was solved by construction of a channel to Cockburn Sound that allowed the nutrients to be flushed out to sea. The effects of such blooms can be serious, causing anoxia and degradation of the whole ecosystem, as occurred in the above estuary when ~250 000 tonnes wet weight of drift *Cladophora* rotted around the shoreline.

Jellyfish blooms

Blooms of large gelatinous plankton, such as jellyfish (cnidarians medusa, ctenophores and salps) are increasing in frequency and severity throughout the world. Species can be rare for years and then suddenly bloom dramatically and just as suddenly disappear again. Sometimes blooms can persist and become a permanent feature, with devastating consequences. Hence, blooms can be more than a mere nuisance by blocking fishing nets or killing farmed fish, and may signify a regime shift in some ecosystems. Although some blooms are seasonal events, many appear to be human-caused and related to overfishing, eutrophication, aquaculture operations, habitat modification, acidification and translocation. Several mechanisms are suggested below, but these remain unproven because they occur on a scale that makes their testing almost impossible (reviewed by Mills 2001; Boero *et al.* 2008; Richardson *et al.* 2009; Pauly *et al.* 2009; Brotz *et al.* 2012):

- An estimated 90% decline of predatory fish has occurred in coastal regions worldwide (Myers and Worm 2003) and this can lead to an alternative state in which a coastal system becomes dominated by short-lived, rapidly reproducing species, such as squid and jellyfish. Southern waters are no exception, and have witnessed the decline of many large, long-lived species, such as sharks, snapper and kingfish.
- Jellyfish may be favoured in systems affected by increasing temperatures due to climate change (see Chapter 3), which stratify water bodies with low nutrient levels.

- Intense trawling of the sea bed can decimate benthic fishes, which ingest the polyp stage of jellyfish as well as the jellyfish themselves, so their removal can lead to the dominance of jellyfish.
- The overfishing of sardine stocks, which predate on the benthic stages (ephyrae) of jellyfish and small jellyfish, can lead to a cascade in which jellyfish come to dominate the system. This has occurred in two major sardine fisheries globally.

The consequences of jellyfish blooms can be disastrous for fisheries, as diverse fish communities are replaced by a jellyfish monoculture. A positive feedback loop may then come into play, because the jellyfish, by their intense predation on fish eggs and larvae, reinforce the 'alternative state' of the ecosystem. Yet, some systems seem to be far less vulnerable to jellyfish outbreaks than others and Pauly *et al.* (2009) explored the likely responsible factors. First, they examined the 'keystone' character of jellyfish to see when they are important for the structure and function of a marine system. They concluded that jellyfish are most likely to become a keystone species and induce an ecosystem switch when:

- The ecosystem is small or enclosed. In temperate Australian waters, all recorded jellyfish blooms have been in estuaries, lagoons, bays or gulf waters, as described below.
- The jellyfish have a great impact on the populations of zooplankton prey. This is highly probable in partly enclosed waters, because jellyfish have extraordinarily high clearance rates (see Acuña *et al.* 2011)
- Jellyfish are an important prey of higher predators, such as sardines in the above example.

In southern Australia, jellyfish blooms have been recorded throughout the region in recent decades, but few have studied their causes or consequences and few baseline data have been collected. For example, *Catostylus mosaicus* blooms spectacularly in estuaries and coastal lagoons of eastern Australia (Pitt *et al.* 2008), *Cyanea rosella* and *Pseudorhiza haeckeli* bloom annually in Port Phillip Bay (Fancett 1986), and *Aurelia aurita coerulea* and *C. rosella* can be common in Tasmanian estuaries (Kingsford *et al.* 2000; R Stuart-Smith *pers. comm.*). *Cyanea rosella*, *Aurelia aurita coerulea* and *Chrysaora southcotti* bloom seasonally in Coffin Bay, South Australia (Saunders 2009) and also in the South Australian gulfs and estuaries (Southcott 1982). *Phyllorhiza punctata* blooms in the Swan-Canning Estuary in Western Australia and also in eastern Australia (Rippingale and Kelly (1995). The effect of these blooms on fisheries is unknown and modelling their impact is clearly needed, as stressed by Pauly *et al.* (2009).

In conclusion, understanding the many factors that control marine ecosystems is a monumental task, and still in the early stages of development. The use of energy flow models is a start, but the variability of marine systems and random processes in space and time increase the complexity of the systems, and incorporating them is a future challenge.

APPENDIX 1

Here, we provide an empirical example of interaction strength, using an observational approach, with the

Interaction strength per metre of coast per 6 months = no. prey consumed × (total no. prey)$^{-1}$. Total no. prey = prey density × coastline length, i.e. 500 m. Results are given below.

Prey species	Simulated density (no./m of coast)	Prey consumed (percentage of total)	Interaction strength
Limpet *Cellana tramoserica*	6	345 (74.4)	0.115
Chiton *Plaxiphora albida*	4	40 (8.8)	0.020
Turbinid *Turbo undulatus*	3	35 (7.5)	0.023
Abalone *Haliotis roei*	1	25 (5.4)	0.050
Muricid *Dicathais orbita*	2	18 (3.9)	0.018

author's unpublished data from West I., South Australia, on predation by the sooty oystercatcher *Haematopus fuliginosus* intertidally. This species forages daily for several hours around low water over ~500 m of coast, much of it in boulder habitat. Two parent oystercatchers occupy a nest among boulders on the sheltered side of the Island annually and the pair rear a chick in spring and summer, and feed it daily for 6 months by transporting intertidal molluscs to the nest for feeding the chick (Aplin and Cockburn (2012). The birds then fly off to the mainland for the rest of the year. In all, 439 shells were collected from a midden next to the nest over the nesting period and the species identified and measured. Other observations were made on the foraging behaviour of the oystercatcher. In order to calculate interaction strength (IS) for simulated densities, it was assumed that the birds foraged over 500 m of the Island's north shore for 6 months and the shells in the midden represented the total prey captured and devoted to the chick.

To calculate the interaction strength for the parent oystercatchers, their food intake would also need to be included.

REFERENCES

Acuña JL, López-Urrutia AL, Colin S (2011) Faking giants: the evolution of high prey clearance rates in jellyfishes. *Science* **333**, 1627–1629.

Anon (2005) Algal blooms in the Swan–Canning estuary: patterns, causes and history. *River Science* **3**, 1–12. <www.swanrivertrust.wa.gov.au> (Accessed September 2011).

Aplin LM, Cockburn A (2012) Ecological selection and sexual dimorphism in the sooty oystercatcher, *Haematopus fuliginosus*. *Austral Ecology* **37**, 248–257.

Boero F, Bouillon J, Gravili C, Miglietta MP *et al.* (2008) Gelatinous plankton: irregularities rule the world (sometimes). *Marine Ecology Progress Series* **356**, 299–310.

Breitburg DL, Crump BC, Dabiri JO, Gallegos CL (2010) Ecosystem engineers in the pelagic realm: alteration of habitat by species ranging from microbes to jellyfish. *Integrative and Comparative Biology* **50**, 188–200.

Brotz L, Cheung WWL, Kleisner K, Pakhomov E *et al.* (2012) Increasing jellyfish populations: trends in large marine ecosystems. *Hydrobiologia* **690**, 3–20.

Bulman CM, Condie SA, Neira FJ, Goldsworthy SD *et al.* (2011) 'The trophodynamics of small pelagic fishes in the southern Australian ecosystem and the implications for modelling of southern temperate fisheries'. FRDC Report 2008/023. CSIRO Marine and Atmospheric Research, Hobart.

Chapman MG, People J, Blockley D (2005) Intertidal assemblages associated with natural *Corallina* turf and invasive mussel beds. *Biodiversity and Conservation* **14**, 1761–1776.

Clarkson PS, Shepherd SA (1985) Distribution and feeding of archaeogastropods in a boulder habitat at West Island, South Australia. *Journal of the Malacological Society of Australia* **7**, 35–44.

Clementson LA, Parslow JS, Turnbull AR, Bonham PI (2004) Properties of light absorption in a highly coloured estuarine system in south-east Australia which is prone to blooms of the toxic dinoflagellate *Gymnodinium catenatum*. *Estuarine, Coastal and Shelf Science* **60**, 101–112.

Connell SD (2002) Effects of a predator and prey on a foraging reef fish: implications for understanding density-dependent growth. *Journal of Fish Biology* **60**, 1551–1561.

Connell SD, Russell BD, Turner DJ, Shepherd SA *et al.* (2008) Recovering a lost baseline: missing kelp forests from a metropolitan coast. *Marine Ecology Progress Series* **360**, 63–72.

Cury P, Bakun A, Crawford RJM, Jarre A *et al.* (2000) Small pelagics in upwelling systems: patterns of interaction and structural changes in "wasp-waist" ecosystems. *ICES Journal of Marine Science* **57**, 603–618.

Drabsch SL, Tanner JT, Connell SD (2001) Limited infaunal response to experimental trawling in previously untrawled areas. *ICES Journal of Marine Science* **58**, 1261–1271.

Duffy JE, Cardinale BJ, France E, McIntyre PB *et al.* (2007) The functional role of biodiversity in ecosystems: incorporating trophic complexity. *Ecology Letters* **10**, 522–538.

Fancett MS (1986) Species composition and abundance of Scyphomedusae in Port Phillip Bay, Victoria. *Australian Journal of Marine and Freshwater Research* **37**, 379–384.

Ferretti F, Worm B, Britton GL, Heithaus MR *et al.* (2010) Patterns and ecosystem consequences of shark declines in the ocean. *Ecology Letters* **13**, 1055–1071.

Folke C, Carpenter S, Walker B, Scheffer M *et al.* (2004) Regime shifts, resilience, and biodiversity in ecosystem management. *Annual Review of Ecology Evolution and Systematics* **35**, 557–581.

Fulton EA, Smith ADM (2004) Lessons learnt from a comparison of three ecosystem models for Port Phillip Bay, Australia. *African Journal of Marine Science* **26**, 219–243.

Goldsworthy SD, Page B, Rogers P, Ward T (2011) 'Establishing ecosystem-based management for the South Australian sardine fishery: developing ecological performance indicators and reference points to assess the need for ecological allocations'. SARDI Research report Series No. 529. Publication No. F2010/000863-1. SARDI Aquatic Sciences, Adelaide.

Graham KJ, Andrew NL, Hodgson KE (2001) Changes in relative abundance of sharks and rays on Australian southeast fishery trawl grounds after twenty years of fishing. *Marine and Freshwater Research* **52**, 549–561.

Hallegraeff GM (1993) On the global increase of harmful algal blooms. *Wetlands (Australia)* **12**, 2–15.

Hallegraeff GM, McCausland MA, Brown RK (1995) Early warning of toxic dinoflagellate blooms of *Gymnodinium catenatum* in southern Tasmanian waters. *Journal of Plankton Research* **17**, 1163–1176.

Heithaus MR, Frid A, Wirsing A, Worm B (2008) Predicting ecological consequences of marine top predator declines. *Trends in Ecology and Evolution* **23**, 202–210.

Hillebrand H, Cardinale BJ (2004) Consumer effects decline with prey diversity. *Ecology Letters* **7**, 192–201.

Hillman K, Lukatelich RJ, McComb AJ (1990) The impact of nutrient enrichment on nearshore and estuarine ecosystems in Western Australia. *Proceedings of the Ecological Society of Australia* **16**, 39–53.

Hooper DU, Chapin FS, Ewel JJ, Hector A *et al.* (2005) Effects of biodiversity on ecosystem functioning: a consensus of current knowledge. *Ecological Monographs* **75**, 3–35.

Hooper DU, Adair EC, Cardinale BJ, Byrnes JEK *et al.* (2012) A global synthesis reveals biodiversity loss as a major driver of ecosystem change. *Nature* **486**, 105–109.

Hurlbert SH (1997) Functional importance *vs* keystoneness: reformulating some questions in theoretical biocenology. *Australian Journal of Ecology* **22**, 369–382.

Isbell F, Calcagno V, Hector A, Connolly J *et al.* (2011) High plant diversity is needed to maintain ecosystem services. *Nature* **477**, 199–202.

Ives AR, Carpenter SR (2007) Stability and diversity of ecosystems. *Science* **317**, 58–62.

Jones CG, Lawton JH, Shachak M (1994) Organisms as ecosystem engineers. *Oikos* **69**, 373–386.

Kangas M, Shepherd SA (1984) Distribution and feeding of chitons in a boulder habitat at West Island, South Australia. *Journal of the Malacological Society of Australia* **6**, 101–111.

Kennelly SJ (1995) The issue of bycatch in Australia's demersal trawl fisheries. *Reviews in Fish Biology and Fisheries* **5**, 213–234.

Kingsford MJ, Pitt KA, Gillanders BM (2000) Management of jellyfish fisheries, with special reference to the Order Rhizostomeae. *Oceanography and Marine Biology - an Annual Review* **38**, 85–156.

Laska MS, Wootton JT (1998) Theoretical concepts and empirical approaches to measuring interaction strength. *Ecology* **79**, 461–476.

Lavery PS, Lukatelich RJ, McComb AJ (1991) Changes in the biomass and species composition of macroalgae in a eutrophic estuary. *Estuarine, Coastal and Shelf Science* **33**, 1–22.

Libralato SV, Christensen V, Pauly D (2006) A method for identifying keystone species in food web models. *Ecological Modelling* **195**, 153–171.

Loreau M (2010) Linking biodiversity and ecosystems: towards a unifying ecological theory. *Philosophical Transactions of the Royal Society of London. Series B, Biological Sciences* **365**, 49–60.

Lozano-Montes HM, Loneragan NR, Babcock RC, Jackson K (2011) Using trophic flows and ecosystem structure to model the effects of fishing in the Jurien Bay Marine Park. *Marine and Freshwater Research* **62**, 421–431.

Marshall J-A (2002) Comparative ecophysiology, chemotaxonomy and ichthyotoxicity of *Chattonella marina* (Rhaphidophyceae) from Australia and Japan. PhD thesis. University of Tasmania, Hobart.

May RM, Beddington JR, Clark CW, Holt SJ *et al.* (1979) Management of multispecies fisheries. *Science* **205**, 267–277. 10.1126/science.205.4403.267.

McCann K, Hastings A, Huxel GR (1998) Weak trophic interactions and the balance of nature. *Nature* **395**, 794–798.

Mills CE (2001) Jellyfish blooms: are populations increasing globally in response to changing ocean conditions? *Hydrobiologia* **451**, 55–68.

Mills LS, Soulé ME, Doak DF (1993) The keystone-species concept in ecology and conservation. *Bioscience* **43**, 219–224.

Montoya JM, Pimm SL, Solé RV (2006) Ecological networks and their fragility. *Nature* **442**, 259–264.

Murray AG, Parslow JS (1999) Modelling of nutrient impacts in Port Phillip Bay – a semi-enclosed marine Australian ecosystem. *Marine and Freshwater Research* **50**, 597–611.

Myers RA, Worm B (2003) Rapid worldwide depletion of predatory fish communities. *Nature* **423**, 280–283.

Nelson WA (2009) Calcified macroalgae – critical to coastal ecosystems and vulnerable to change: a review. *Marine and Freshwater Research* **60**, 787–801.

Olsgard F, Schaaning MT, Widdicombe S, Kendall MA *et al.* (2008) Effects of bottom trawling on ecosystem functioning. *Journal of Experimental Marine Biology and Ecology* **366**, 123–133.

Paine RT (1988) Food webs: road maps of interactions or grist for theoretical development? *Ecology* **69**, 1648–1654.

Pauly D (2003) Ecosystem impacts of the world's marine fisheries. *Global Change Newsletter* **55**, 22–23.

Pauly D, Christensen V, Dalsgaard S, Froese R *et al.* (1998) Fishing down marine food webs. *Science* **279**, 860–863.

Pauly D, Christensen V, Walters C (2000) Ecopath, Ecosim and Ecospace as tools for evaluating ecosystem impacts of fisheries. *ICES Journal of Marine Science* **57**, 697–706.

Pauly D, Graham W, Libralato S, Morissette L *et al.* (2009) Jellyfish in ecosystems, online data bases, and ecosystem models. *Hydrobiologia* **616**, 67–85.

Peters RH (1988) Some general problems for ecology illustrated by food web theory. *Ecology* **69**, 1673–1678.

Pimm SL, Lawton JH, Cohen JE (1991) Food web patterns and their consequences. *Nature* **350**, 669–674.

Piraino S, Fanelli G, Boero F (2002) Variability of species' roles in marine communities: change of paradigms for conservation priorities. *Marine Biology* **140**, 1067–1074.

Pitt KA, Clement A-L, Connolly RM, Thibault-Botha D (2008) Predation by jellyfish on large and emergent zooplankton: implications for benthic-pelagic coupling. *Estuarine, Coastal and Shelf Science* **76**, 827–833.

Polis GA (1994) Food webs, trophic cascades and community structure. *Australian Journal of Ecology* **19**, 121–136.

Power M, Tilman JA, Menge BA, Bond WJ *et al.* (1996) Challenges in the quest for keystones. *Bioscience* **46**, 609–620.

Reid DD, Krogh M (1992) Assessment of catches from protective shark meshing of New South Wales between 1950 and 1990. *Australian Journal of Marine and Freshwater Research* **43**, 283–296.

Richardson AJ, Bakun A, Hays GC, Gibbons MJ (2009) The jellyfish joyride: causes, consequences and management responses to a more gelatinous future. *Trends in Evolution and Ecology* **24**, 312–322.

Rippingale RJ, Kelly SJ (1995) Reproduction and survival of *Phyllorhiza punctata* (Cnidaria: Rhizostomeae) in a seasonally fluctuating salinity regime. *Marine and Freshwater Research* **46**, 1145–1151.

Sakai AK, Allendorf FW, Holt JS, Lodge DM *et al.* (2001) The population biology of invasive species. *Annual Review of Ecology and Systematics* **32**, 305–332.

Saunders B (2009) *Shores and Shallows of Coffin Bay: an Identification Guide.* Eyre Peninsula NRM Board, Adelaide.

Schiel DR (2006) Rivets or bolts? When single species count in the function of temperate reef communities. *Journal of Experimental Marine Biology and Ecology* **338**, 233–252.

Schindler DE, Hilborn R, Chasco B, Boatright CP *et al.* (2010) Population diversity and the portfolio effect in an exploited species. *Nature* **465**, 609–612.

Shepherd SA (1981) Ecological strategies in a deep water red algal community. *Botanica Marina* **24**, 457–463.

Shepherd SA, Clarkson PS (2001) Diet, feeding behaviour, activity and predation of the temperate blue-throated wrasse, *Notolabrus tetricus. Marine and Freshwater Research* **52**, 311–322.

Shepherd SA, Daume S (1996) Ecology and survival of juvenile abalone in a crustose coralline habitat in South Australia. In *Survival Strategies in Early Life Stages of Marine Resources.* (Eds Y Watanabe, Y Yamashita and Y Oozeki) pp. 297–313. Balkema, Rotterdam.

Shepherd SA, Gray JD (1985) Food of the anemone *Anthothoe albocincta* at West Island, South Australia. *Transactions of the Royal Society of South Australia* **109**, 191–192.

Shepherd SA, Sprigg RC (1976) Substrate, sediments and subtidal ecology of Gulf St Vincent and Investigator Strait. In *Natural History of the Adelaide Region* (Ed. CR Twidale) pp. 161–174. Royal Society of South Australia, Adelaide.

Shepherd SA, Turner JA (1985) Studies on southern Australian abalone (genus *Haliotis*). VI. Habitat preference, abundance and predators of juveniles. *Journal of Experimental Marine Biology and Ecology* **93**, 285–298.

Shepherd SA, Womersley HBS (1970) The sublittoral ecology of West Island, South Australia. 1. Environmental features and the algal ecology. *Transactions of the Royal Society of South Australia* **94**, 105–138.

Shepherd SA, Clark M, Ferguson G (2008) Habitat-dependent foraging behaviour and diet of the scalyfin, *Parma victoriae*, in South Australia. *Transactions of the Royal Society of South Australia* **132**, 134–146.

Shepherd SA, Watson JE, Womersley HBS, Carey J (2009) Long-term changes in macroalgal assemblages after increased sedimentation and turbidity in Western Port, Victoria, Australia. *Botanica Marina* **52**, 195–206.

Shepherd SA, Brook JB, Xiao Y (2010) Environmental and fishing effects on the abundance, size and sex ratio of the blue-throated wrasse, *Notolabrus tetricus*, on South Australian coastal reefs. *Fisheries Management and Ecology* **17**, 209–220.

Smith ADM, Brown CJ, Bulman CM, Fulton EA *et al.* (2011) Impacts of fishing low-trophic level species on marine ecosystems. *Science* **333**, 1147–1150.

Sommer U, Stribor H, Katechakis A, Sommer F *et al.* (2002) Pelagic food web configurations at different levels of nutrient richness and their implications for the ratio fish production: primary production. *Hydrobiologia* **484**, 11–20.

Southcott RV (1982) Jellyfishes (Classes Scyphozoa and Hydrozoa). In *Marine Invertebrates of Southern Australia Part I* (Eds SA Shepherd and IM Thomas) pp. 115–159. Fauna and Flora of South Australia Handbooks Committee, Adelaide.

Steiner CF, Long ZT, Krumins JA, Morin PJ (2005) Temporal stability of aquatic food webs: partitioning the effects of species diversity, species composition and enrichment. *Ecology Letters* **8**, 819–828.

Stevens JD, Bonfil R, Dulvy NK, Walker PA (2000) The effects of fishing on sharks, rays, and chimaeras (chondrichthyans), and the implications for marine ecosystems. *ICES Journal of Marine Science* **57**, 476–494.

Svane I, Roberts S, Saunders T (2008) Fate and consumption of discarded by-catch in the Spencer Gulf prawn fishery, South Australia. *Fisheries Research* **90**, 158–169.

Svane I, Hammett Z, Lauer P (2009) Impacts of trawling on benthic macro-fauna and -flora of the Spencer Gulf prawn fishing grounds. *Estuarine, Coastal and Shelf Science* **82**, 621–631.

Tanner JE (2003) The influence of prawn trawling on sessile benthic assemblages in Gulf St Vincent, South Australia. *Canadian Journal of Fisheries and Aquatic Sciences* **60**, 517–526.

Tanner JE (2005) Three decades of habitat change in Gulf St Vincent, South Australia. *Transactions of the Royal Society of South Australia* **129**, 65–73.

Thomsen MS, Wernberg T, Altieri A, Tuya F *et al.* (2010) Habitat cascades: the conceptual context and global relevance of facilitation cascades via habitat formation and modification. *Integrative and Comparative Biology* **50**, 158–175.

Tilman D (1999) The ecological consequences of changes in biodiversity: a search for general principles. *Ecology* **80**, 1455–1474.

Turner JT, Tester PA (1997) Toxic marine phytoplankton, zooplankton grazers, and pelagic food webs. *Limnology and Oceanography* **42**, 1203–1214.

Valiela I, McClelland J, Hauxwell J, Behr PJ *et al.* (1997) Macroalgal blooms in shallow estuaries: controls and ecophysiological and ecosystem consequences. *Limnology and Oceanography* **42**, 1105–1118.

Waite AM, Suthers IM (2007) Open water: plankton ecology. In *Marine Ecology* (Eds SD Connell and BM Gillanders) pp. 281–315, Oxford University Press, Melbourne.

Walker TI, Hudson RJ, Gason AS (2005) Catch evaluation of target, by-product and by-catch species taken by gillnets and longlines in the shark fishery of south-eastern Australia. *Journal of Northwest Atlantic Fishery Science* **35**, 505–530.

Watson R, Nowara G, Tracey S, Fulton B *et al.* (2013) Ecosystem model of Tasmanian waters explores impacts of climate change induced changes in primary productivity. *Ecological Modelling* **56**, 115–129.

Wootton MS (1997) Estimates and tests of per capita interaction strength: diet, abundance, and impact of intertidally foraging birds. *Ecological Monographs* **67**, 45–64.

Wright JR, Jones CG (2006) The concept of organisms as ecosystem engineers ten years on: progress, limitations and challenges. *Bioscience* **56**, 203–209.

Yachi S, Loreau M (1999) Biodiversity and ecosystem productivity in a fluctuating environment: the insurance hypothesis. *Proceedings of the National Academy of Sciences of the United States of America* **96**, 1463–1468.

Zhou S, Smith ADM, Punt AE, Richardson AJ (2010) Ecosystem-based fisheries management requires a change to the selective fishing philosophy. *Proceedings of the National Academy of Sciences of the United States of America* **107**, 9485–9489.

20 Conservation and management

'In wildness is the preservation of the world.'
Henry David Thoreau, 1862

OVERVIEW

Marine Protected Areas (MPAs), which afford partial protection, and Sanctuary Zones (SZs), which exclude all extractive activities, are valuable tools to protect species, habitats and ecosystems from human-induced disturbances, notably fishing, but also various forms of pollution. We summarise the progressive establishment of MPAs in southern Australia and their principal purposes and benefits – for biodiversity conservation. We discuss selection criteria and their limitations, and then review the recorded effects of protection in reef communities in terms of increases in fish and invertebrate abundance and biomass over time, and trophic cascades, as the changes at higher levels of the food web affect lower levels. We consider, with examples, the benefits and limitations of SZs in fisheries management and argue that fishery management needs to be integrated with the establishment of MPA networks to better provide for ecosystem-based management (EBM) of marine systems.

INTRODUCTION

The relentless increase in human population pressure worldwide, including the southern coastal regions of Australia, has seen an increase in the severity of threats to marine systems. Broadly, there are five major threats: overfishing; habitat degradation; exotic organisms; pollution; and most recently increasing CO_2 (ocean warming, acidification and rising sea levels). The impacts of these threats, together with the variable failure of traditional management, have long generated concerns about the adequacy of current controls and, over time, Australia has committed to important management approaches. In 1982, it endorsed the UN World Charter for Nature by adopting the precautionary approach. In 1992, Australia signed the international Convention on Biological Diversity, and endorsed the Outcomes Statement from the UN Conference on Environment and Development, by which it agreed to the precautionary approach, ecosystem management and the creation of networks of Marine Protected Areas (MPAs) to conserve coastal ecosystems (Nevill 2010).

In essence, the precautionary approach rests on the idea that action should be taken to avoid possible harm in the absence of scientific certainty, and the ecosystem approach is based on the idea that harvesting within an ecosystem should be managed so as to preserve the essential structure and function of that system. The protection of 'ecosystem services' grew out of the latter approach.

Historically, three broad management tools have been used to conserve marine systems in relation to fishing impacts. One is to impose controls on fished species, such as size limits, closed seasons and quotas; another is to control specific activities, such as trawling and dredging; and the third approach is to create spatial closures. MPAs are spatial closures that can protect ecosystems against specific threats and they range from Sanctuary Zones (SZs), sometimes termed 'no-take areas', which provide complete protection from harvesting, to multi-purpose, Partial Protection Zones. Such zones may, for example, prohibit particular activities or they may be temporal controls, such as seasonal closures of spawning sites. Two other small classes of reserve are: Total Exclusion (or Restricted Access) Zones for special purposes; and Indigenous Protected Areas (IPAs) that are entrusted to Aboriginal people to manage and conserve. As yet, two small IPAs have been declared in temperate Australia – in NW Tasmania and in the far west of South Australia (Nursey-Bray 2011).

In southern States, threats from the many types of pollution have been controlled by State bodies (e.g. environment protection agencies) and these are outside the scope of this book.

Here, we shall discuss first the purpose and benefits of ecosystem protection in SZs, and the benefits of MPAs where some extractive activities are allowed. We then describe the pragmatics of site selection in southern Australia, review the ecosystem effects over time of SZs, and finally the ecological and social benefits of MPAs to the ecosystem-based management (EBM) goals embraced by State governments.

PURPOSES AND BENEFITS

The primary objectives of a system of SZs have been long debated by their advocates. Some see their primary role as a means of conserving biodiversity, with its ancillary goals of protecting ecosystem function, habitats, whole assemblages, and threatened or depleted species – a role that has been widely accepted in Australia. Others see their role as a means to restore depleted fish stocks – a view common in regions where overfishing has been severe (review of Vandeperre *et al.* 2011). Yet this role has gained little traction in Australia, despite the fact that management of its fisheries has been average by world standard (see 'Fisheries Management' section below). Complete protection in SZs is neither feasible nor necessary for all coastal waters, and MPAs that seek to protect habitats by restricting the most destructive activities, such as bottom trawling, are extensively used. Hence, both MPAs and SZs have an important part to play in the precautionary management of fisheries.

Networks of MPAs, and especially SZs, have many benefits. The immediate ones are: all habitats are freed as much as possible from human-induced disturbance; endemic and threatened species have an increased level of protection; and fish, relieved of harvesting pressure, can increase in size and abundance. The flow-on benefits are often unpredictable because the whole ecosystem slowly returns to a more pristine state (see below). Other important, if less obvious, benefits include: ensuring reservoirs of genetic diversity for many species; protection of fish spawning and nursery areas; and reducing the risk of collapses of exploited species. Human benefits are the provision of sites for the scientific study of natural processes and for tourism, recreational and educational purposes.

MPA ESTABLISHMENT

Experience in southern States

By 1991, a total of 51 MPAs, most of them SZs < 1 km^2 in area, had been established in southern Australia, largely on an ad hoc basis, to conserve particular ecological features or species (e.g. seal colonies) (McNeill 1994). With the impetus of congresses on the topic (Fairweather and McNeill 1992) and Australian ratification of international agreements, a national strategic plan was adopted by the Australian States and Commonwealth to establish a comprehensive, adequate and representative (CAR) system of MPAs and SZs in each bioregion of the States by 2012 (ANZECC 1998; Nevill and Ward 2009). Such a system was intended to:

- protect a full range of habitats and communities in each region (comprehensiveness)

- ensure that each SZ area was of sufficient size to maintain connectivity of the network and with some redundancy in protected habitats (adequacy)
- ensure that examples of all habitats and conservation features are represented including: iconic sites; the habitats of unique, migratory, rare, threatened, vulnerable or depleted species; centres of endemism; and natural refugia (representativeness).

The value of a network lies in maintaining connectivity between the flora and fauna within and between sanctuaries – which depends on the dispersal potential of the component species – and ensuring that all major elements of biodiversity are safeguarded. This is illustrated in Fig. 20.1, which shows the range of larval dispersal distances of algae, invertebrates and fish in coastal waters. Thus, while macroalgae and some invertebrates should be self-sustaining within an SZ, recruitment of some invertebrates and many benthic fishes may be partly or wholly decoupled from local populations within the SZ.

To achieve a CAR system, some have argued for a few, large reserves to be established progressively at iconic sites, strategically placed within the different bioregions – see Commonwealth of Australia (2006). The first and largest of these was the Great Australian Bight Marine Park, set up in 1995 and extending for 205 km from the Western Australian border (129°E) to 132°E (Fig. 20.2). Conservation and Sanctuary Zones were established in State waters to 3 nautical miles offshore, and abutting them a Mammal Protection Zone extended up to 20 nautical miles further offshore to protect the southern right whale calving region, with a narrow Benthic Protection Zone going for a further 226 nautical miles offshore (McLeay *et al.* 2003).

Other workers have favoured a coastal network of small, stepping-stone reserves that are more highly protected. This divergence of approach is reflected in the various MPA systems later set up in southern Australia. In New South Wales, six large, multiple-use MPAs were set up, each containing a network of small SZs, as at the Solitary Is, Port Stephens, Batemans Bay and Jervis Bay (Banks and Skilleter 2010; Beeton *et al.* 2012). In Western Australia also, large, multiple-use MPAs have been established since 1990, containing small SZs within them covering ~15% of the total MPA areas (DEC 2010). In Victoria, some 11 SZs and 13 larger no-fishing MPAs, termed Conservation Zones, were set up across the State in 2002, purportedly based on CAR principles (Wescott 2006). The largest of these covers 221 km^2 around Wilsons Promontory (i.e. an SZ of 155 km^2 over the southern part, a tiny Exclusion Zone around a

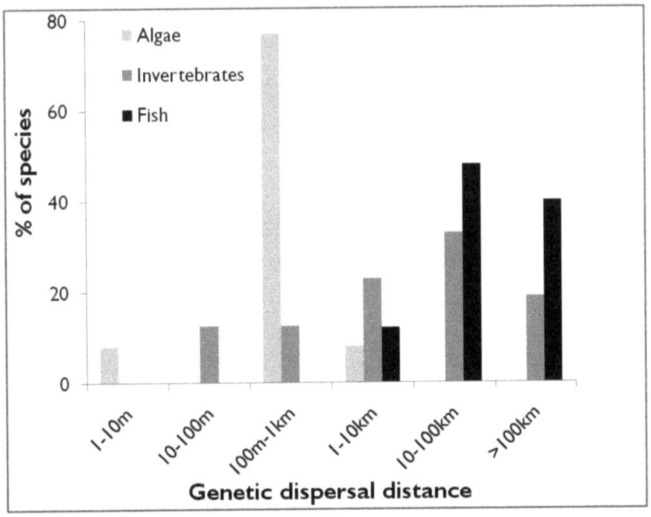

Figure 20.1: Mean dispersal distances of a range of benthic species of macroalgae (13 species), sessile and sedentary invertebrates (48 species) and demersal fish (25 species). Data extracted from literature and redrawn from Kinlan and Gaines (2003). Pelagic species and asexually reproducing species are excluded, but organisms that drift as adults (e.g. drifting macroalgae) are included.

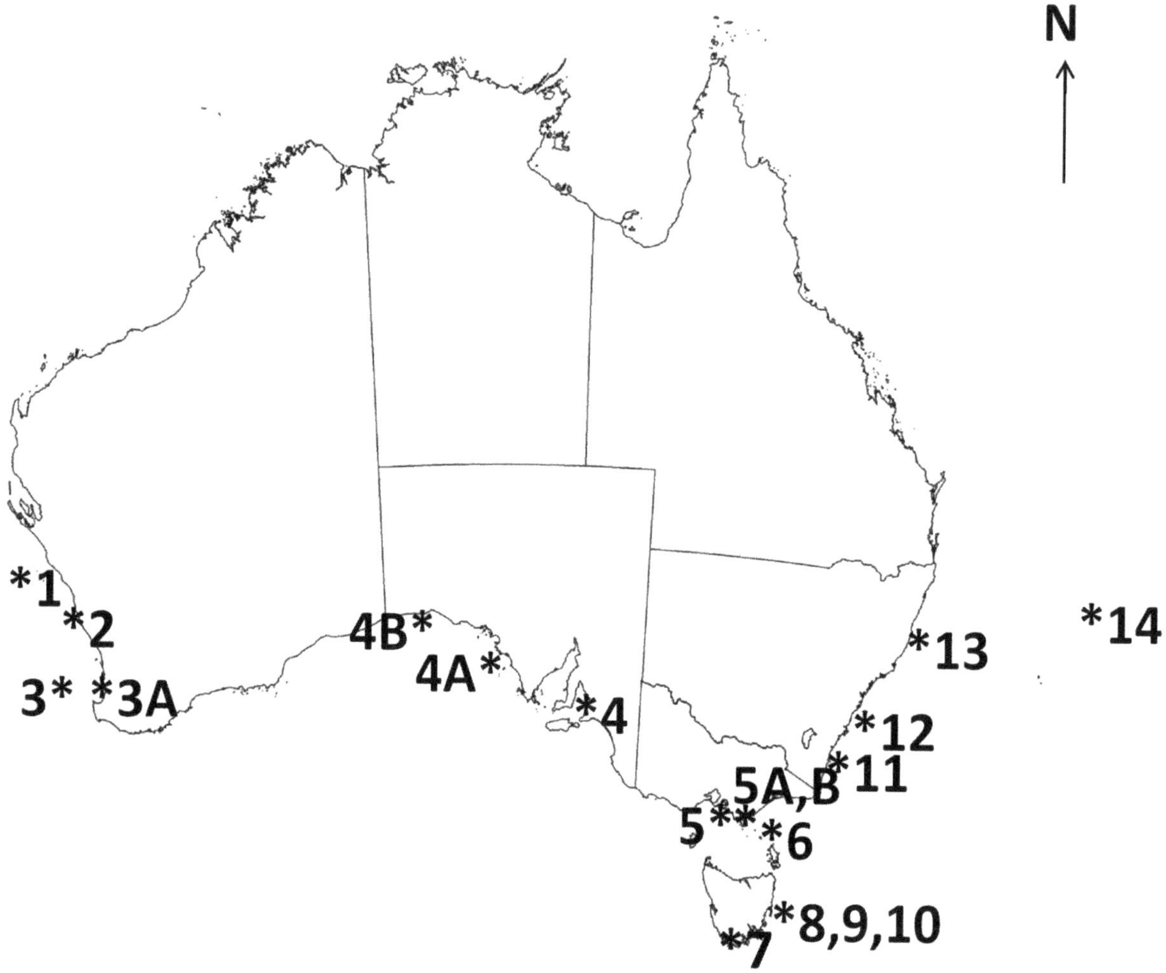

Figure 20.2: Map showing some southern SZs that have been studied: those numbered 1–14 by Edgar *et al.* (2009b), and those with letters as below. The names of the SZs are: 1– Abrolhos Is; 2 – Jurien Bay; 3 – Rottnest I.; 3A – Marmion Lagoon; 4 – Aldinga Reef; 4A – Point Labatt, studied by Currie and Sorokin (2009); 4B monitored by Goldsworthy *et al.* (2010); 5 – Port Phillip Heads; 5A – Cape Paterson; 5B – Wilsons Promontory; 6 – Kent Group; 7 – Port Davey; 8 – Tinderbox, Ninepin Point; 9 – Maria I., 10 – Governor I.; 11 – Batemans Bay; 12 – Jervis Bay; 13 – Fly Point Halifax Park; 14 – Lord Howe I.

breeding site of the Australian fur seal on Kanowna I. and the rest a Conservation Zone (Parks Victoria 2006). In South Australia, after a detailed review by Janine Baker (2004) and extensive consultation, the outer boundaries of MPAs incorporating Habitat Protection Zones allowing recreational fishing only, and Managed Use Zones allowing commercial fishing, were established in 2012, with SZs prohibiting all fishing nested within them declared in 2013 (Kirkman 2013).

In Tasmania, 20 MPAs with some (often minor or no) fishing restrictions were set up by 2009. Notable among them are those at: Port Davey, extending over an area of 20 × 30 km; Maria I., with SZ over 7 km of coast and covering 685 ha; Tinderbox, recently extended to 6 km of coast and covering ~140 ha; and Ninepin Point covering 6 km and 730 ha; and Governor I., extending for 1 km of coast and covering ~50 ha. The percentage of State waters within SZs and MPAs are summarised in Table 20.1.

In the light of these modest achievements, Nevill and Ward (2009) concluded in their review that State governments overall had so far shown 'only a token commitment to marine conservation' with inadequate areal coverage, inadequate protection levels and too few SZs.

Table 20.1: Percentage of jurisdictional waters of the Commonwealth and four southern States within Marine Protected Areas (MPAs) and Sanctuary Zones (SZs) by late 2011.

State waters include all waters in bay and gulfs, and elsewhere out to 3 nautical miles from shore, and Commonwealth waters to the edge of the Continental shelf (Government of South Australia 2011).

	MPAs	SZs
Commonwealth	9.8	< 2
New South Wales	36	6
Victoria	12	5.3
Tasmania	2.7	1.1
South Australia	44	~5[1]
Western Australia	12	2.5

[1] this provisional estimate includes > 350 km of coast in the Eucla Bioregion. If this is excluded, the figure is ~2% for State waters (Kirkman 2013).

Biodiversity conservation as a criterion

Here, biodiversity refers to the totality of representative habitats and communities, and their species (especially threatened ones) and ecosystem processes. Although biodiversity conservation has been the primary objective in the Australian MPA program, limited knowledge of the biodiversity of coastal waters has made it difficult in practice to select the best sites. Iconic sites with a wide range of habitats, depths and degree of exposure to swell, as are often found in offshore islands, are favoured – hence, the reserves in the Abrolhos Is, Western Australia, the Kent Group in Bass Strait and the Solitary Is, New South Wales. Within these reserves, smaller SZs have been selected for maximum protection. But such large MPAs need to be supplemented by a network of mainland coastal sites, where fishing and other pressures are highest and habitats may be distinctive. This raises the question how MPAs and the (usually) smaller SZs should be selected when the richness of local species is poorly known, if at all, as is often the case.

To resolve the problem, the use of surrogates – defined as any set of ecological features that can be used to guide conservation planning – has been proposed as a cost-effective means of measuring biodiversity. One intertidal study in the Solitary Is (Smith 2005) showed that the best local surrogate for invertebrate diversity was shelled molluscs, followed by crustaceans and then cnidarians. In a study of reefs at 13–20 m depth along the New South Wales coast, Bill Gladstone (2007) found that fish diversity from survey data was an efficient biodiversity surrogate and far better than the simpler habitat selection criteria often used. He concluded that, of 13 sites surveyed over 140 km of coast, 75% of the 106 fish species recorded would be included in a network of reserves that included just 30% of the surveyed sites. On this basis, he inferred that MPAs should cover at least 30% of the coast (an inference that assumes adequate coverage of the range of habitats). On the other hand, two related studies in Jervis Bay, southern New South Wales, provided alternative perspectives.

Vanderklift *et al.* (1998) recommended caution in the use of surrogates. The use of higher taxonomic groups (e.g. family level) as surrogates for species diversity was inefficient, although the use of generic levels was more satisfactory. In the sequel to the work of Matt Vanderklift and colleagues, Ward *et al.* (1999) concluded that simple habitat categories (reef, sand, etc.) determined from aerial photography provided the best surrogates for species richness. The strength of these two studies was that they drew on earlier, very detailed studies of Jervis Bay's biodiversity (CSIRO 1994) that justified the conclusions. In contrast to the above, the review of Keough *et al.* (2007) of two Victorian MPAs, and their SZs within, found only weak correlations between the diversity of fish, mobile invertebrates and algae, suggesting that the diversity of one group cannot be used as a surrogate for other groups. Instead, they suggested that detailed assessments of individual MPAs should be undertaken progressively to describe the biodiversity of significant groups, such as ascidians, crustaceans and bryozoans, across the MPA network.

Optimisation algorithms (e.g. Marxan – see Fernandes *et al.* 2005; Watts *et al.* 2009) have often been recommended for the selection of a combination of sites that best meets CAR principles. These procedures typically depend on the input of habitat mapping data, where available, size and spacing guidelines, and information on faunal groups; they can also include other stipulations, such as avoidance of fishing sites. Recent reviews of network design (Pérez-Ruzafa *et al.* 2008; McLeod *et al.* 2008) recommended that the optimal size of an SZ is: 16–144 km^2 (the former study) and 100–400 km^2 (the latter study, which also advocated spacing them 15–20 km apart). This system would ensure that species with short larval stages would be self-seeding in an SZ and that species with longer larval stages would have adequate connectivity between mutually replenishing SZs.

Two serious limitations of all methods of site selection have become evident:

- Rare, endangered or (near-)threatened species have a much higher risk of extinction than other species and the habitats in which they are found need to be given priority in conservation in accordance with vulnerability and irreplaceability criteria. However, many such species and their habitats are still unknown.
- Serious problems have also emerged in the political process of establishing SZ networks. Stakeholder demands tend to become dominant and can easily frustrate desirable outcomes. These can be especially strong in human-dominated seascapes, so that SZs may end up being located in places of little biodiversity significance – the 'worthless lands' scenario described by Edgar (2011). Political concerns have tended to be alleviated by minimising the size and number of SZs, and/or by granting concessions to stakeholders.

To tackle the 'at-risk species' problem, Edgar *et al.* (2008b) proposed that IUCN (2001) terrestrial criteria be used to identify high priority marine areas for conservation. These criteria involve the identification of key biodiversity areas (KBAs), determined from knowledge of the distribution and abundance of 'at-risk' species or of ecological sites of high significance. Many KBAs can be readily identified throughout southern Australia, and examples are: the blue whale feeding sites (Chapter 18); small endemic or relict species, as in upper Spencer Gulf; and the unique deep-water fauna in the Bathurst Channel, Port Davey (see below). The use of marine KBAs is still in its infancy in Australia, but Edgar *et al.* (2008a) provide examples of their application in the Galapagos Is, where 'at-risk' species abound.

To resolve the selection biases that creep into the establishment process through stakeholder demands and short-term interests, KBAs should be adopted by regulators as non-negotiable priority SZ areas in the public consultation process, while sites without the essential features of KBAs would be negotiable with stakeholders. As argued by Peterson *et al.* (2004), 'environmental decision making rooted in consensus theory leads to the dilution of … conservation … and legitimises unsustainable social constructions of reality'.

BENEFITS OF MPAS FOR REEF COMMUNITIES

Given the pervasive effect of fishing on marine ecosystems (Chapter 19), it is unsurprising that the cessation of fishing in SZs may have profound consequences, which will vary according to the region protected. Yet, despite a legion of studies, much uncertainty persists about specific effects and few confident predictions can be made about the ecological changes that will take place at any one site. Here, we review the factors involved and the effects recorded in 14 SZs established in temperate Australian waters in recent decades, as summarised in Figs 20.2–20.5.

In addition to the response of SZs to current threats, it is important to note that other benefits of SZs, and to some extent MPAs, are to prevent the realisation of future possible threats, provide sites in which to record climate change, and simply to conserve the biodiversity of the areas.

Factors influencing outcomes

The effectiveness of an SZ for the protection of species or group of species from current threats may be influenced by many factors (see Edgar *et al.* 2009a, b). The most relevant are:

- **Time since establishment.** The effects of protection are expected to accumulate over time. For example, the increase in biomass of fish in different SZs protected for different time periods (Fig. 20.3) shows that initially a rapid increase occurs, slowing over time, but that fish biomass may be still increasing even after 36 years,. The density of large (> 45 cm) fishes also increases in a more or less linear fashion over time, at least for some decades (Fig. 20.4).

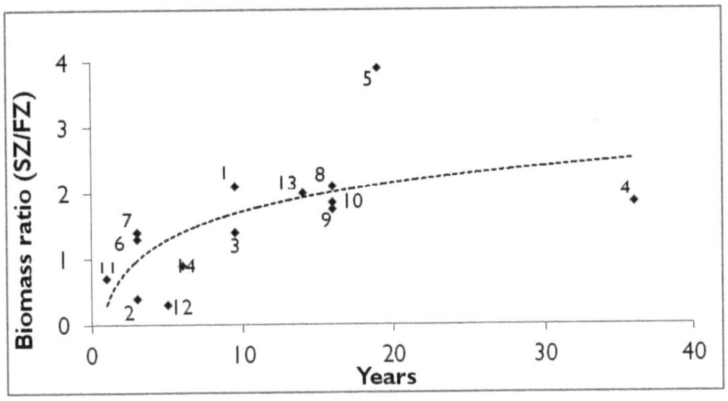

Figure 20.3: Increase in fish biomass, expressed as the biomass ratio of Sanctuary Zone to Fished Zone (SZ/FZ) plotted against years since establishment for 14 southern SZs (plotted and numbered 1–14 in Fig. 20.2, but excluding Sites 3A, 4A, 5A and 5B) with logarithmic curve of best fit (redrawn from Edgar *et al.* 2009b, but on a linear scale).

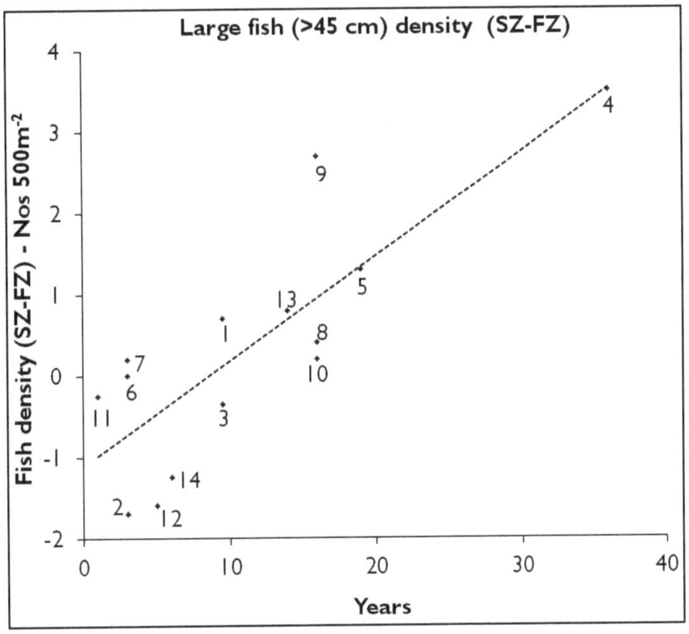

Figure 20.4: Differences in density of large fishes (> 45 cm) between Sanctuary and Fished Zones (= SZ–FZ) versus time period since establishment of SZ (redrawn from Edgar *et al.* 2009b, but on a linear scale).

- **Size, design and habitat features.** Small or narrow SZs will have fewer and less obvious effects than larger ones, because the boundaries are closer, and even slight benefits are likely to spill over to adjoining waters and dilute any observable effect. Some habitats (e.g. a flat reef with little relief, or one scoured by sand) may show no change, because there is no protective habitat for fish. Reef habitats extending beyond an SZ boundary may show migration of species from the SZ (Barrett *et al.* 2009b).
- **Species-specific life history features.** Species that range widely over large distances, such as sharks, are likely to benefit the least from SZs, whereas

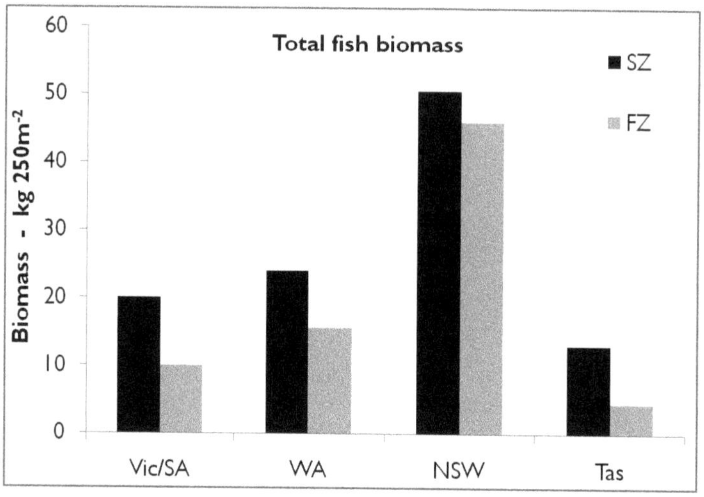

Figure 20.5: Total mean biomass of fishes along transects of standardised size in the 14 SZs (numbered 1–14 in Fig. 20.2, but excluding 3A, 4A, and 5A) and in control fished zones (FZ) in the five States (redrawn from Edgar and Stuart-Smith 2009).

site-attached species will benefit the most from protection. Conversely, the latter group suffer most by an impact, such as an oil spill. An example of an impact, albeit a natural one, occurred at West I., South Australia, when New Zealand fur seals established a haul-out site there and decimated the local blue-throated wrasse population; however, the population recovered after some years as juveniles recruited to the population and adopted a more vigilant and wary behaviour (Shepherd *et al.* 2010; Chapter 16).

- **Metapopulation dynamics.** Recruitment to a sub-population of a species in an SZ may depend on outside larval sources. For example, if an SZ protects only a sink sub-population, which is sustained by a larval source sub-population outside the SZ, then the sub-population in the SZ may collapse (Lipcius *et al.* 2005). This event occurred at West I., where the reserve was set up to protect the local abalone population, although a large part of the population was outside the reserve. The collapse of the unprotected population from fishing was quickly followed by recruitment failure and the collapse of the whole metapopulation (Shepherd and Brown 1993). Small SZs are more likely to suffer this effect than larger ones.
- **Previous fishing intensity.** Past fishing history can have significant and persistent effects. Species previously heavily fished in an SZ can take a long time to recover after protection, depending on recruitment strength and frequency, but the ultimate increase in fish biomass will be much more dramatic than in lightly fished areas (Barrett *et al.* 2007). In intensely fished areas surrounding an SZ, spillover from the SZ would increase, with possibly other changes at reference (control) sites. Unfished species would be expected to show little or no change, unless affected by an indirect cascade.
- **Proximity to external contaminant sources.** A factor seldom considered is the effect of the many contaminants on the fauna and flora in MPAs. Those located in poorly flushed bays and estuaries are more likely to be affected than those on open coasts, in deeper water or with moderate to strong currents (reviewed by McKinley and Johnston 2010; McKinley *et al.* 2011a). For example: sites downstream from fish farms or treated sewage outlets may well show increased abundance of some species; and those receiving industrial effluents or stormwater run-off are likely to be negatively affected.

EFFECTS OF PROTECTION

As shown in the previous section, the response of an MPA to protection depends on a variety of natural features and human-related disturbances in the surrounding region. In relatively undisturbed or remote sites, little change, other than natural fluctuations in populations, would be expected in subsequent monitoring studies (e.g. Edgar and Barrett 2012). In such cases, the benefits of a reserve are the provision of a scientific reference point and also to protect against future contingencies, threats or climatic events, as in example (c) below. Four examples, listed (a) to (d), of MPAs that have so far shown no overt effects of protection are given first, and followed by six examples, listed (e) to (j), where substantial benefits have followed.

a. **Port Davey and its contributory channels.** This is a unique region with habitats and marine life found nowhere else in Australia, but with strong affinities to the deep fjords of Chile and New Zealand. The surface layers of the channels have tannin-stained brackish water that precludes establishment of macroalgae below a few metres depth, allowing deep-water invertebrates to carpet shallow-to-deep reef surfaces. Mobile invertebrates are depauperate, and sponges, jewel anemones (*Corynactis* spp.), octocorals (gorgonians, sea whips and soft corals), bryozoans, ascidians and tube-worms dominate the unique communities. Many of the species are distinctive and endemic, with some rare deep-water ones known from depths of > 300 m (Edgar and Barrett 2010; Edgar *et al.* 2010). Little fishing has occurred and the region is near-pristine, but the region is threatened by boating, fishing, dive tourism, nutrient enrichment, introduced species and climate change (Barrett and Edgar 2010; Edgar *et al.* 2010).
b. **Point Labatt, South Australia.** A small SZ of 278 ha was created here in 1986 to protect a colony of the endangered Australian sea lions (Chapter 18). The SZ covers a granitic headland on an exposed coast in the eastern Great Australian Bight and extends to a depth of > 30 m ~2 km offshore (Fig. 20.2). After 19 years of protection, Currie and Sorokin (2005, 2009) found no significant differences between the fauna and flora within the SZ and those at adjoining external reference sites. Exploited species, such as greenlip abalone and rock lobster, were quite rare within and outside the SZ, and the only moderately common exploited mollusc was blacklip abalone that occurred at densities of 20–50 per 100 m^2 from 5–15 m depth both within and outside the SZ. The lack of any effect is likely due to a combination of the high faunal/floral variability between sites (acknowledged by the authors) and low fishing intensity in the area, given the inaccessibility of the site for most of the year. However, poaching in this remote reserve is distinctly possible.
c. **Cape Paterson, central Victoria.** The MPA was established in 2002 far from boating access, but with nearby parking. The MPA includes the Bunurong SZ, which extends along the coast for ~4 km. Reef habitats in the SZ contain a rich macroalgal flora and fauna, and monitoring surveys were conducted from its creation until 2010 (Pritchard *et al.* 2011). The surveys did not show any changes arising from protection. However, they did record substantial inter-annual fluctuations in the abundance of macroalgae (e.g. *Macrocystis*), reef fishes, blacklip abalone and urchins, both inside the SZ and at external reference sites. These fluctuations were thought to be attributable to the long-term sea temperature increases in SE Australia and El Niño events during the period – in itself a significant finding.
d. **Great Australian Bight Marine Park.** This park illustrates the fact that wide-ranging species may not benefit from an SZ when the threat is distant. This long-established and inaccessible SZ did not protect the Australian sea lion in its nine breeding colonies along the Bunda Cliffs from by-catch mortality in the shark gill-net and lobster fisheries. The seals forage up to 200 km from shore in deep shelf waters, where adults are caught in nets and traps and suffer a mortality of ~2.8% per breeding cycle – enough to cause negative or near zero population growth rates to the highly vulnerable populations (Goldsworthy *et al.* (2010; see Chapter 18).

In contrast to the above studies, most monitored MPAs after a few years have shown distinct effects of protection on exploited species as well as subtle

Table 20.2: Summary of effects of 16 years of protection in Sanctuary Zones (SZs) on reef communities established in 1991 in SE Tasmania.

Data presented are: biomass ratios of Sanctuary Zones in 2008 to Fished Zones (FZs) in 1992 (i.e. = SZ/FZ) for southern rock lobster and large fish (> 45 cm length) and density ratios (= SZ/FZ) for sea-urchins and gastropods at Maria I. and Tinderbox/Ninepin Point, SE Tasmania (after Edgar *et al.* 2009b);

Species' group	Maria I.	Tinderbox Ninepin Point
Large fish (> 30 cm) – biomass	4.7	20
Rock lobster – biomass	9.3	25
Sea-urchins[1] – density	0.56	0.61
Gastropods[2] – density	0.72	0.70

[1] Purple urchin *Heliocidaris erythrogramma* and black urchin *Centrostephanus rodgersii*
[2] Abalone (*Haliotis rubra*), whelks, tritons, and turbinids.

complexities, such as trophic cascades down the food web. Notable examples are given here for fishes and invertebrates, as well as evidence of trophic cascades.

e. **SE Tasmania.** These are the best studied SZs and they show a complex picture, with not all fish species increasing in size and biomass (Table 20.2; Barrett *et al.* 2007; Edgar and Stuart-Smith 2009). The seriously overfished bastard trumpeter *Latridopsis forsteri* showed the most dramatic increase. This long-lived schooling fish has a strong, but sporadic, recruitment every 7–10 years and lives for its first 4–5 years on near-shore reefs before migrating offshore at a size of ~50 cm. The species has a broad home range of ~5 km of coast, over which it makes daily excursions. At Maria I. and Tinderbox, one strong recruitment pulse was enough to drive an eight to 10 times increase in trumpeter abundance in the first 10 years after protection, whereas at the two very small SZs, Ninepin Point and Governor I., the trumpeter made no recovery. The reason for the latter failure was attributed to two factors – the small size of the reserves and the intense gill-net fishing and lobster pot by-catch that constantly occurred at their boundaries. Hence, any recruiting trumpeter would very soon be captured at the reserve boundary.

Other larger fishes that increased in abundance at Maria I. (compared with controls) were: banded morwong *Cheilodactylus spectabilis*; the unfished globefish *Diodon nichthemerus*; sea carp *Aplodactylus arctidens*; and blue-throated and senator wrasses *Notolabrus tetricus* and *Pictilabrus laticlavius*. Generally, un- or little exploited species – such as purple and rosy wrasses *N. fucicola* and *Pseudolabrus mortonii*, respectively, bullseyes *Pempheris* spp., hulafishes *Trachinops* spp., the leatherjackets *Acanthaluterus vittiger* and *Meuschenia australis*, herring cale *Olisthops cyanomelas*, and others – showed little change.

In the other small SZs, one of the few fish to show any benefit from protection was the blue-throated wrasse, likely due to its small home range. Other species with home ranges of several hundred metres showed no benefit; hence, losses or gains by a species in a small SZ apparently depended on the extent of their home ranges, and the high likelihood of 'wandering' individuals crossing a boundary.

f. **Western Australia.** Here, broadly similar results to those in Tasmania were recorded at several sites (Kleczkowski *et al.* 2008; Watson *et al.* 2007). At Rottnest I., a small SZ of 1.1 km^2 over a shallow reef 5–10 m depth was established in 1988. After 16 years, the biomass of some large fish species – dhufish *Glaucosoma hebraicum* and breaksea cod *Epinephelides armatus* – increased three to five times. In contrast, the biomass of other large species – yellowtail kingfish *Seriola lalandi*, snapper *Chrysophrys auratus* and western blue groper *Achoerodus gouldii* – which were rare outside the SZ, showed only a very slight increase within it. Other targeted fishes – silver trevally *Pseudocaranx georgianus*, King George whiting *Sillaginodes punctata* and tarwhine *Rhabdosargus sarba* – showed no difference in abundance between populations within and outside the SZ. This negative result is unsurprising given the large ranges of the three species.

g. **Abrolhos Is.** Here, four large MPAs, where only lobster fishing was allowed, were established in 1994 over areas of 13–27 km^2. In extensive surveys of shallow (10 m) and deeper (24 m) reefs, Watson *et al.* (2007) found that, after 11 years of protection from fishing, several targeted fish species showed substantial increases in abundance in the MPAs. For example, coral trout *Plectropomus leopardus* (up to three times), spangled emperor *Lethrinus nebulosus* (two times), red-throat emperor *L. miniatus* and dhufish *G. hebraicum* (three to eight times). Snapper *C. auratus* and baldchin groper *Chaerodon rubescens* were also more abundant in the MPAs, but this varied with time of year and depth; some non-targeted fishes, mainly labrids, were also more common there. Nardi *et al.* (2004) had previously found even larger increases in abundance of some of the above species and Watson *et al.* (2007) attributed the discrepancies between the two studies to poaching and/or differences in sampling techniques. Kim Nardi and colleagues used underwater visual census techniques, while Dianne Watson and colleagues used baited remote underwater video (BRUV) systems; hence, the discrepancies may have been due to differing fish behaviour towards divers and bait, respectively.

Large, piscivorous fish are significant consumers of small fishes, and in particular the coral trout (*Plectropomus leopardus*) largely feeds on pomacentrid, labrid and scarid fishes. In the Abrolhos Is quasi-SZs, the strong recovery of coral trout was accompanied by the decline of five species in these three fish families, thus providing persuasive evidence of a trophic cascade (Watson *et al.* 2007).

h. **Marmion Lagoon, Western Australia**. Here, an MPA of 95 km^2 was set up in 2000 with three tiny SZs within, totalling only 0.4 km^2. Recreational fishing was allowed within the MPA, but not in the SZs. The area comprises limestone reefs with numerous caves and crevices and a sandy bottom. In one of the few studies to examine the total macrofauna and -flora, Kylie Ryan (2008) compared the three SZs and nearby reference sites in the MPA after 4–6 years of protection. Densities of fauna and flora after protection and in adjoining reference areas are summarised in Table 20.3. Large fish predators, such as baldchin and western blue gropers and dhufish, although once common in the lagoon, remained rare, leaving in the SZs only 'weedy, fast-growing species'. Of these, large (20–25 cm) size classes of some fishes (e.g. western king wrasse, horseshoe leatherjacket and brown-spotted wrasse) were more abundant in the SZs. However, the scalyfin *Parma mccullochi* was commoner in fished areas, likely because it competes with the western lobster for shelter holes and availability of holes declined as lobster densities increased. Curiously, the western buffalo bream *Kyphosus cornelii* increased in abundant in the SZs for reasons unknown.

The strong dominance of western lobsters in SZs likely caused a trophic cascade, with a steep decline in abundance of its prey species – the purple urchin, the egg urchin, hermit crabs and the gastropod *Mitra chalybeia* (Table 20.3). These prey species are mainly herbivores and their decline in SZs likely induced the shift in the algal communities.

Kylie Ryan observed that in Marmion Lagoon 'traditional management measures, such as bag and size limits had not prevented overfishing', but she concluded that the SZs were much too small to provide long-term refuges for either wide-ranging fishes or the western lobster. Hence, they would need to be much larger to have long-term conservation benefits in terms of protecting sexually mature lobsters or most species of fish.

i. **Jervis Bay.** This MPA in southern New South Wales came into force in 2002 and covered 220 km^2, of which ~50 km^2 were protected in 14 SZs located in different parts within and outside the Bay. The Bay was noteworthy as one of overlap between subtropical and temperate flora and fauna, with a rich diversity of fishes (216 species recorded in surveys). After 4 years of protection (Barrett *et al.* 2008), the most notable changes in the SZs were: a 30% increase in abundance, and ~8 cm increase in mean size, of red morwong *Cheilodactylus fuscus*; an overall doubling in abundance of fish > 30 cm size; and upward trends in abundance of the wide-ranging snapper *Chrysophrys auratus* and yellowfin bream *Acanthopagrus australis*. Remarkably, the endangered grey nurse shark *Carcharias taurus* was increasingly encountered during the surveys, suggesting that

Table 20.3: Mean densities of fishes and invertebrates in number 200 m^{-2} (mean length in one fish species) and biomass of algae (fresh weight) in SZs and adjoining Fished Zones (FZs) in Marmion Lagoon, Western Australia after 6 years of protection (after Ryan 2008).

CL = carapace length (cm). All results listed are significant.

Species	SZs	FZs
Fishes – Western king wrasse *Coris auricularis* mean length (cm)	19.9	16.0
Brown-spotted wrasse *Notolabrus parilus*	1.1	1.5
Common scalyfin *Parma mccullochi*	3.1	5.9
Western buffalo bream *Kyphosus cornelii*	1.5	0.2
Invertebrates Western rock lobster *Panulirus cygnus* (all)	29.0	49.8
Adults (> 97 cm CL)	10.3	2.6
Sea cucumbers *Stichopus* spp.	9.2	3.6
Purple urchin *Heliocidaris erythrogramma*	11.6	72.8
Egg urchin *Holopneustes porosissimus*	0.6	3.2
Sea-stars (3 spp. *Plectaster, Patiriella* spp.)	0.3	14.5
Macroalgae – biomass fresh weight (g m^{-2}) *Sargassum* spp.	370	248
Hypnea spp.	38	15
Botryocladia	6.4	2.8
Dictyomenia	22.8	4.8

they had a recruitment site in caves within SZs near the Bay's northern headland.

j. **Comparing different protection levels in New South Wales.** A key and unique comparative study of the effects of protection in four markedly different types of zone is that of McKinley *et al.* (2011b). The different zones are: Habitat Protection Zones (HPZs) where fishing is allowed and is moderate; fully protected SZs; 'Modified' Zones (MZs) which are heavily fished, and also affected by pollution, nutrient-enrichment and urbanisation, and Unmodified Zones (UZs) in fished but near-pristine estuaries. The HPZs, SZs and UZs were in Batemans Bay and Jervis Bay, described above, and the MZs were in Port Jackson and Port Hacking in the Sydney region. In all, 53 fish species, of which 23 were targeted by recreational fishers, were recorded using the BRUV method and the study estimated the abundance of specific fish groups in each zone. The studies found that the abundance of three of the commonest targeted fish species – snapper *Chrysophrys auratus*, silver trevally *Pseudocaranx georgianus* and yellowfin bream *Acanthopagrus australis* – in the four zones was in this order:

MZs (10.2) > UZs (2.4) > SZs (2.2) > HPZs (0.3)

Numbers are the mean combined density of the three species per BRUV drop. The average maximum numbers per BRUV drop of the 23 fish species targeted by recreational fishers in the four zones were in this order:

MZs (11) > SZs (4) > UZs (3) > HPZs (1.5)

Despite the limitations of the study (e.g. lack of replication and no before-and-after comparisons), two conclusions are evident. First, the MZs had much greater abundances of targeted species, largely due to their high nutrient levels, which increased the productivity of the Port Jackson and

Port Hacking systems. Second, the SZs were effective compared with HPZs in Batemans Bay and Jervis Bay in providing some protection to targeted fish species. Hence, nutrients, among other things, strongly affect fish abundance, although such 'polluted' areas are rarely selected for protection. This policy may be worth reviewing if increased fish abundance per se is a critical element in selecting MPAs.

Key species in MPAs

Both western and southern rock lobster underpin major invertebrate fisheries in southern temperate waters, and both are also functionally important in reef systems, as we shall see in the next section on trophic cascades. The effect of lobster protection has been studied at two sites. At Maria I., after 16 years of protection, the southern rock lobster biomass increased 10 times – almost all of which was due to the increase in numbers and growth of large lobsters > 113 mm carapace length (Edgar *et al.* 2009b). Barrett *et al.* (2009b) examined the question whether the Maria I. SZ became 'saturated' with lobsters, causing them to crawl over the boundary, perhaps in search of food (a density-dependent effect). Out of > 4000 tagged lobsters, < 3% moved between reefs and even fewer into adjoining fished areas, indicating a very low 'spillover' effect (see Chapter 15). The increases in biomass and density were similar at Tinderbox, but no differences in lobster abundance occurred in SZs at Ninepin Point or Governor I. The negative results in the last two very small SZs were attributed by Barrett *et al.* (2009a) to poaching and an edge effect. The latter effect was inferred from the declining mean size of lobsters away from the centre to the SZ's edge (< 500 m), suggesting that over time the growing lobsters crawled out of the SZs and were captured. These results are similar to those of McGarvey (2004), where 68% of tagged lobsters moved out of a very small lobster sanctuary in SE Spencer Gulf.

At Rottnest I., the increase in abundance of the western rock lobster was even more dramatic (Babcock *et al.* 2007). After 16 years of protection, lobster densities increased ~34 times within the SZ, and densities of those above legal minimum size (77 mm carapace length) increased ~50 times. The study noted that the western lobster is a social animal, forming aggregations to mate or to share shelters, and that the carrying capacity of coastal reefs was apparently far higher than previously thought, as both lobster biomass and egg production increased by > 100 times inside the SZ. Here the spillover effect was strong, as lobsters emigrated from the SZ into the surrounding low-density fished areas.

Trophic cascades

Trophic cascades are secondary or indirect effects that usually occur at lower levels of the food web and often take many years to become manifest after cessation of fishing. If the top predators steeply decline under fishing, densities of prey fish and invertebrates may increase and algal communities may also change. In short, a series of cascades can ripple through the system following removal of top predators by fishing, with these population changes ultimately affecting the abundance of species two to three trophic levels down the food web. After fishing ceases, the reverse process back to full recovery can take many years (see Edgar *et al.* 2009b). Babcock *et al.* (2010) noted in their review that, on average, direct effects were evident after 5 years and indirect effects after 13 years. The best known example of cascading indirect effects is from Maria I.

At this site, the abundance and size of rock lobsters slowly increased, but it took some 7–15 years (depending on sex) for them to grow large enough to consume large purple urchins (*Heliocidaris*) as well as other urchins and grazing molluscs. The purple urchins emerge from the protection of crevices at a large size and they only started to decline significantly (as a consequence of lobster predation) after ~7 years. Blacklip abalone are also preyed upon by rock lobsters and they also showed a complex response to increasing predator abundance (Barrett *et al.* 2009a). Small blacklip (85–134 mm size) customarily emerge from crevices to forage at night, when they are susceptible to capture by foraging lobsters. After a decade of protection, this size group declined by > 90% in abundance, but larger animals (135–200 mm) remained stable in numbers but grew in size, suggesting that they might have a refuge in size from predating lobsters. Manipulative studies by Pederson *et al.* (2008)

indicated that complex mechanisms were operating here (see Chapter 12). The increased abundance of lobsters changed the behaviour of abalone and they remained cryptic and deep within crevices until they were on average some 4 cm larger than in fished areas, and less susceptible to predation. Yet, predation on the smallest size group still increased substantially.

The decline in abundance of the various grazing species (i.e. ~56% of urchins and ~72% of gastropods, including small abalone, whelks, tritons and turbinids) would be expected to result in an increase in the macroalgal canopy, as grazing declined. This began to occur after about 8 years at Maria I., and by 15 years macroalgal cover had increased substantially (Babcock *et al.* 2010). As yet, no change in the diversity of algal species has been recorded.

SANCTUARY ZONES AND FISHERIES MANAGEMENT

Traditional management controls (quotas, size limits, etc.) designed to prevent overfishing can still fail, even in well-managed fisheries, and many fisheries (e.g. barracouta, jack mackerel, striped trumpeter, school shark, gemfish, ocean leatherjacket, yellow-tail kingfish and warehou) have collapsed or declined seriously in southern Australia in recent decades. Australian fisheries management is no better than average in a global sense, as shown by Mondoux *et al.* (2008) where Australia ranked as number 31 in a comparison of 53 maritime countries (notwithstanding the assertions of Kearney *et al.* (2012) to the contrary). The most recent review (Australian Government 2011) listed > 60% of fish species and species groups assessed as in poor or very poor condition in terms of abundance in the SW, SE and east coasts of Australia.

In general, the States perform stock assessments and manage only the dozen or so of their most commercially valuable species, leaving the great majority of fished species without assessment and few or no controls, other than size or bag limits (see Chapter 19). The serial depletion of unmanaged species is a continuing problem. Hence, it is increasingly accepted that an urgent need exists to ramp up protection of coastal waters to protect biodiversity and to safeguard fisheries. The benefits of MPAs for targeted fish species are many, with detailed reviews by Baker *et al.* (1996), Ward *et al.* (2001), Roberts *et al.* (2005), Vandeperre *et al.* (2011) and many others. We summarise here six established benefits and give specific southern temperate applications to reef-dwelling and pelagic fishes, lobsters and abalone.

1. **Subject to the provisos discussed above, populations of previously fished species increase in abundance and biomass, and develop a natural age structure, with many more reproductively active age classes.**
   The effects of protection depend on the life history and ecology of the exploited species. Pelagic fish that range widely beyond the boundaries of a small SZ may gain little protection from one, although even here there may be surprises, as some pelagic species have both 'wanderers' that range widely, and 'residents' that tend to remain within a restricted area. However, sedentary species, and fish with small home ranges, receive maximum protection from an SZ. Rock lobsters are especially interesting in this regard, because their movement behaviour varies greatly over their range (Chapter 15 and see above) and the benefits vary accordingly.
   In Tasmania, Buxton *et al.* (2006) modelled the benefit of spatial closures for the southern lobster, whose movements were known to be slight (generally < 1 km) (see Chapter 15). Hence, biomass spillover would be low, with little enhancement of the fishery. The fishery was modelled on the bases: that a closure was established over a single fishing block, covering 55 km of coast; and fishing continued outside the closure **with no reduction in quota**. This modelling process was repeated for many sections of the coast from north to south. Within each modelled closure, biomass greatly increased, sometimes up to 33 times, and egg production likewise increased by up to 24%. Outside the modelled closures, fishing would be more intense, because fishers compensated for the loss of catch within the closures by catching more lobsters outside. Hence, biomass and egg production would decline over the rest of the region. The modelling showed that, overall, the gains made inside the

closures would be offset by the losses outside, so providing no conservation benefits to the fishery. These conclusions contrast sharply with other studies on lobsters (see Baker *et al.* 1996). For example, the Mediterranean lobster *Palinurus elephas* had an estimated annual spillover of only 5.2% from a reserve, but still produced an average annual net benefit of 10% to the catch (Goñi *et al.* 2010).

Two general conclusions can be drawn from this modelling study:

a. Where spillover of an exploited species from an SZ is negligible, there may be no management benefits of closures for that species. As a corollary, Colin Buxton and colleagues noted that if emigration of lobsters from these large closures increased to 30%, the spillover would be enough to compensate for the loss of yield and the increased egg production in the closure would be a positive fishery benefit. Note that the Tasmanian scenario does not apply to Western and South Australia, where lobster spillover is much higher (Chapter 15).
b. quota reduction (or a buy-back scheme) matching the biomass of the species 'locked-up' in the SZ is an essential element in SZs with little emigration of that species. Such management measures are themselves a positive initiative where egg production is admittedly perilously low, as is the case with the southern lobster along Tasmania's east coast (Chapter 15).

Buxton *et al.* (2006) also argued that the benefits of MPAs would be zero for the sedentary, self-recruiting blacklip abalone (Chapter 9), except where the fishery had collapsed. This is clearly wrong given: the genetic connectedness of blacklip, with a neighbourhood size of 500 km, and the complexity of factors influencing larval dispersal (reviewed by Morgan and Shepherd 2006); and the enhanced larval spillover, resulting from the synergistic effects, described under Point 2 below. In Roe's abalone *Haliotis roei*, with a neighbourhood distance of only 6–15 km, an exploited population was replenished by larval spillover from a neighbouring reef in Waterman Reserve 3 km away (Wells and Keesing 1990).

Greenlip abalone *H. laevigata* has a well-defined metapopulation structure, often extending over 20–30 km of coast (Baker *et al.* 1996; Morgan and Shepherd 2006; Chapter 9). In South Australia, some 29 metapopulations collapsed under intense fishing and many showed marked spatial contraction to apparent remnant larval source local populations (LPs). Closures established over such larval sources may well have prevented the collapse of downstream LPs (Shepherd and Brown 1993; Shepherd and Rodda 2001; Shepherd *et al.* 2001; Chapter 9). Yet, a cautionary note is necessary. MPAs are not a panacea for over-exploited species, due to environmental vagaries and the complexities of metapopulation dynamics. As emphasised by Lipcius *et al.* (2005), the only strategy ensuring a positive outcome is one based on an understanding of the larval dispersal and determinants of spatial recruitment of the species in question.

2. **Many more eggs/larvae are produced for export from the SZ.**

Four synergistic effects are present here. First, as described above, the number of age classes of an exploited species will greatly increase over time, depending on the emigration rate of adults. For example, in greenlip abalone individuals up to 30 years old can be found in a reserve, whereas in fished populations individuals > 8 years of age are rare (unpublished data; Shepherd and Breen 1992). Second, density increases greatly over time and induces a third effect – enhanced fertilisation success of eggs in free-spawning species. This is the reverse of the Allee effect, defined as fertilisation loss due to distance between free-spawners (see Chapter 9, where an example is given for greenlip abalone). Fourth, larger individuals have elevated reproductive capacity, because the fecundity of a species increases exponentially with size. These effects in combination produce a steep exponential increase in larval export from an SZ as population density increases.

Another published example is the value of an SZ in protecting the spawning site of a pelagic fish species. For some species (e.g. snapper), temporary spatial closures during the spawning season are valuable in facilitating spawning in conditions undisturbed by fishing. In the case of King George whiting, the main spawning sites for SE Australia are on specific deep reefs at the entrances to the two South Australian gulfs and possibly in western Bass Strait. This whiting has a metapopulation structure over its extensive range, and larger and more fecund adults aggregate at deeper spawning sites. Jenkins *et al.* (2003) argued the significant management benefits in having such spawning sites protected within SZs. The larval life of whiting is 3–4 months and the extent of larval dispersal varies over its range according to the regional hydrodynamics. Hence, a network of SZs located at upstream spawning sites would be of great value in maintaining adequate egg production of this intensely fished species.

3. **The genetic diversity of populations is protected, so reversing the intense selective pressures on the exploited species.**
The exploitation of marine populations can induce rapid evolutionary effects on the life history of the exploited species. Fishers targeting large individuals will tend to take the fastest growing ones, leaving behind stunted and slower growing ones. Hence, fishing will have a selective effect over time, leaving slower growing fish that mature at a smaller size and younger age and invest relatively more energy in greater egg production than in growth. The selection pressure is greater in small populations than in large ones and an example for greenlip abalone is given in Chapter 9, where population growth rates *inter alia* declined by 40% in only four generations. Overall, the productivity of a fishery will slowly decline as the life history of a species adapts to the higher mortality rates (reviewed by Kuparinen and Merila 2007). Small, isolated populations can evolve rapidly, as in the abalone example above, but for much larger populations, measurable life history changes may take 20–30 years. Establishment of SZs that protect small sub-populations of a species can help maintain genetic diversity and minimise these selection pressures.

4. **Where destructive fishing gear has been used, habitat quality will improve and by-catch of non-target species will cease.**
The destructive fishing gear referred to here is mainly trawling and dredging gear, which in reef systems will destroy all epifauna, such as sponge beds, rhodolith beds and oyster beds, which in turn harbour a very rich associated invertebrate fauna. For example, in Gulf St Vincent, prawn trawling destroyed ~975 km^2 of hammer oyster and razor shell beds (Shepherd *et al.* 2008). In Tasmania, oyster reefs and inshore scallop beds were similarly destroyed by dredging, and along with them a rich associated fauna (Edgar and Samson 2004). No attempt at restoration of any of them has yet been attempted although this is feasible because MPAs are declared in degraded areas (Beck *et al.* 2011). In principle, recovery of most reef systems should be possible provided the system has not changed to an alternative stable state and species are not functionally extirpated, as happened to the hammer oyster in lower Gulf St Vincent and oysters and scallops in Tasmania.

5. **Spillover of emigrating adults and the export of eggs/larvae of spawners from the SZ will occur for many species.**
The spillover of adult fish and invertebrates from an SZ into adjoining fished areas is notoriously difficult to measure and is usually inferred from the increased density of species within an SZ or from increased catches outside it, as shown in the examples given above. The evidence is strongest (a) where a gradient in density or size occurs from the centre to the edge of an SZ, as shown above for the western rock lobster at Rottnest I., the southern lobster in Tasmania or (b) from tagging experiments within reserves (see McGarvey 2004; Chapter 15).
Some have raised doubts, as in the lobster modelling study of Buxton *et al.* (2006), whether the spillover from a reserve could ever match the loss to a fishery of the production locked up within the reserve. In a review of this question, Gaylord *et al.* (2005) found that, in addition to enhanced larval export, the life history features of a species have important effects on yield. Their modelling indicated that 'reserve

networks may have the potential to enhance fishery yields under a surprisingly large number of circumstances'. Such enhancement was predicted to be greatest in species that are relatively long lived, and not migratory, but move within local regions and show post-dispersal density dependence. Where fishing intensity is high outside a reserve, the positive effects will be more dramatic, as shown by Stobart *et al.* (2009) in the Mediterranean, where spillover kept increasing after 16 years of protection. A local Tasmanian example is the movement of bastard trumpeter from the Maria I. MPA. After successful recruitment, trumpeter grew to reproductive size over 3 years within the MPA, and then spilled over to reproduce elsewhere. At that time, gill-net fishers had almost completely fished out trumpeter in inshore waters, leaving almost none of a reproductive size (Barrett *et al.* 2007).

6. **Sanctuary Zones as reference areas**
   SZs provide reference areas within which the effects of fishing in fished areas can be compared and so aid in understanding the ecosystem effects of exploitation and interactions between fishing and other anthropogenic stressors (Edgar *et al.* 2008). In particular, the role of climate change in population declines of fished species can only be differentiated from fishing effects when changes in fished and unfished areas are compared over time. Furthermore, experiments can be conducted within SZs without external interference, as shown by the many long-term studies in the West I., South Australia SZ (see Chapter 9). A specific New Zealand example of a reference benefit was provided in a study of a high local incidence of tail fan necrosis in the southern rock lobster. Comparisons between fished areas and SZs, where the incidence of necrosis was found to be almost zero, made it clear that the disease was due to handling and holding the lobsters in pots (Freeman and MacDiarmid 2009).
7. **SZs complement management by assisting a precautionary and ecosystem approach**
   Three issues are involved here: (a) the value of a precautionary approach as a buffer against uncertainty; (b) an ecosystem approach to management of marine ecosystems; and (c) an adequate monitoring system to assess the effectiveness of spatial management; as expanded on below.

Precautionary approach

Even well-managed fisheries can fail, due in part to the inherent uncertainties in stock assessment and also the risk of adverse or more extreme climate conditions that reduce the resilience of fished stocks. Many authors, such as Lauck *et al.* (1998), have suggested that fisheries management strategies should always incorporate precautionary SZs as a 'bet hedging' component against the risk of collapse (Fig. 20.5). Desirable features of such SZs will vary according to the managed species and should include spawning and/or aggregation grounds, larval source areas and nursery habitats. SZs are a more certain way of protecting such vulnerable life history stages because they embrace the entire habitat and all other species within it (Roberts *et al.* 2005).

Ecosystem-based management

The overarching goal of EBM is to sustain the capacity of marine systems to provide ecosystem services through strategies aimed at minimising the risk of irreversible change to assemblages of species and ecosystem processes (Halpern *et al.* 2010). The Australian national strategy enshrined EBM in the management of exploitable resources (COAG 1992). Here, SZs become immediately relevant to EBM goals, because they can spatially target the main stressors, whether they are fisheries, coastal development or coastal discharges. They can be deliberately located over areas that will prevent or minimise impacts to sensitive species, habitats or ecosystems, and so promote EBM goals.

Monitoring

Monitoring is crucial when evaluating spatial management within and outside MPAs and SZs, and the many human-generated threats to ecosystems (see reviews of Day 2008, Claudet and Guidetti 2010, Murphy and Jenkins 2010, Smith *et al.* 2011, White *et al.* 2011). The various threats to rocky reef systems, and their vulnerabilities to such threats, have been reviewed at international workshops where spatial scale, frequency and impact of threats, and the resistance and resilience of the ecosystems to those threats, were considered (Halpern *et al.* 2007). The resulting vulnerability values are listed in Table 20.4. Note that the most serious threats recognised globally are sediment and organic pollution inputs, coastal development, fishing and invasive species. A slightly less serious group of threats includes diseases, algal blooms, hypoxia and climate change. All of these threats are variously present in temperate Australian waters, especially near cities and adjoining estuaries. Rising sea temperatures associated with climate change require particular comment, as such effects are already prominent on the SW and SE Australian coasts (see Chapters 1, 2 and 12). To meet these challenges, Alastair Hobday (2011) advocated networks of MPAs in near-shore areas to track changes in the distributions of species and, interestingly, for pelagic systems he proposed mobile MPAs to track key species as their ranges shifted southwards.

Optimal requirements (modified from Day 2008) for monitoring of reef habitats require:

- establishment of a regular and systematic survey program in which the abundance of flora and fauna is measured quantitatively at selected sites over time
- estimation of levels and kinds of activities and impacts in the area

Table 20.4: Average scores of selected ecosystem vulnerabilities of three types of coastal reef systems to major threats (after Halpern *et al.* 2007).

Oyster reefs would include those formed by different species of oyster, mussels and razor shells. Values of 2–2.5 are shown in darker shading and values of 1.5–2.0 in light shading.

Threat	Kelp forest	Rocky reef	Oyster reef
Freshwater input	0	1.5	1.7
Sediment input	1.2	2.0	2.2
Nutrient input[1]	0	1.7	0
Pollutant input[2]	1.0	2.2	2.8
Coastal engineering	0	1.9	3.0
Coastal development	1.2	2.5	3.2
Aquaculture	0	1.9	1.5
Fishing			
Destructive demersal	1.5	2.7	3.1
Pelagic	0	2.6	3.1
Recreational	2.3	2.6	1.3
Climate change			
Sea level	1.6	1.5	1.8
Sea temperature	2.0	1.9	0.8
Acidification	0	1.1	0.7
Species invasion	1.3	2.5	2.0
Disease	0.7	1.8	2.1
Harmful algal blooms	0.4	1.7	2.5
Hypoxia	1.0	1.6	2.9
Commercial activity	0	1.4	0
Benthic structures	0	1.7	0
Ecotourism	0.8	1.7	0.3

[1] into oligotrophic water; [2] non-point, organic.

- the above survey program is conducted both within MPAs, and in outside control areas
- periodic evaluation of the monitoring program and management responses.

A critical gap in some monitoring studies is the lack of monitoring in control areas outside MPAs, where intense fishing pressure (displacement effects) can strongly influence their effectiveness (Claudet and Guidetti 2010).

After developing a model to evaluate performance indicators that have been used to assess whether MPAs were fulfilling their function, Smith *et al.* (2011) applied the model to many species under a range of ecosystem conditions (e.g. low to high productivity, nutrient pollution, climate change and different fishing pressures). They found that the biomass of a range of benthic fishes, including some sharks, was superior to all other indicators, including structural indicators relating to habitats, and recommended their use. Furthermore, standing biomass of reef fish assemblages provided a robust indicator of two important features of ecosystems and fishing pressures on them (Mora *et al.* 2011):

- energy flow and biodiversity
- regional human population density, which in fished areas varies inversely with fish biomass (cf. Chapter 16, where size and abundance of bluethroat wrasse reflected distance from human population centres).

Of the various monitoring methods available, Murphy and Jenkins (2010) concluded that underwater visual census (UVC) and baited remote underwater video stations (BRUVS) were the two most cost-effective and efficient means of obtaining habitat and fish assemblage data. In a comparison of the two techniques, Colton and Swearer (2010) found that the UVC method recorded more species and individuals, but BRUVS more of the mobile large predators. The UVC method (RLS 2011), which is now widely used around Australia and indeed globally for reef monitoring to ~20 m depth, gathers quantitative data on: (a) fish density and size structure (and hence biomass); (b) mobile invertebrate density; and (c) algal and other substrate cover. BRUVS are typically used for fish abundance and biomass estimates at greater depths (> 20 m) or at sites with low fish abundance.

Historically, the responsibility for fisheries management and for MPAs in southern Australia has been undertaken by different government bodies, with almost no attempt to integrate management of the ecosystem. As well described by Jones (2007), this divergence has been reinforced by the different ethical perspectives of fishery scientists and conservationists. Fishery scientists are influenced by utilitarian, resource management views and the sustainable harvesting of fish stocks, with limited concern for the sustaining ecosystem, whereas conservationists are preservationist, and see fish and other marine species and their habitats as part of a broad ecosystem with intrinsic ecological values. In Jones' view, fishery science is reductive, intra-disciplinary and applied, and conservation science holistic and trans-disciplinary. The objectives of each differ widely and their 'storylines' are in sharp contrast with each other. Although both approaches have merit and logic, middle ground is needed if management of coastal regions is to be holistically ecosystem-based. This will require fishery managers, conservation interests and fishers to collaborate closely or even to fall under one administrative body, as advocated by Nevill (2012) and Beeton *et al.* (2012).

Community benefits

Ultimately, the success of MPAs requires the weaving of ecology with sociological factors. The local communities and fishers must be motivated to care for the closed areas. Hence, management needs to develop educational programs about the many benefits, because reserve areas are valuable to increase public understanding and awareness of natural systems and the human impacts on them. As demonstrated by Bill Ballantine (1999) after 35 years of experience in New Zealand, MPAs promote a range of new activities, such as eco-tourism, underwater photography, educational programs and naturalist endeavours. Many of these goals are severely compromised without reserves. Underlying conservation concerns and the management of MPAs are our concerns for the future. Will our grandchildren lament the loss of species and destruction of marine habitats, in the plaintive words of Aldo

Leopold (1964)? 'Our grandparents who did the killing were our agents … in the sense they shared the conviction, which we have only now begun to doubt, that it is more important to multiply people and comforts than to cherish the beauty of the land in which they live.'

Conclusions

'We must place biodiversity conservation at the centre of ocean governance.'

Sylvia Earle and D Laffoley, 2006

As human pressures on coastal waters have increased, increasing attention has been directed towards conservation of species, habitats and ecosystems in temperate Australian waters, and the number and size of MPAs and SZs have slowly increased since the 1970s. Studies of the effects of protection have also expanded rapidly, so that broad conclusions on their benefits and limitations are beginning to emerge, although much remains to be learnt. The ecological changes following establishment of MPAs are hard to predict, because they vary geographically and according to: the size, shape, location and management of the MPA; the fish and invertebrates concerned; their life history features; the type and diversity of habitats; and many environmental factors such as depth, exposure, water currents and temperature. The potential is vast to improve understanding of the many aspects of the ecological processes involved and their variability over southern coasts. Monitoring of MPAs is developing and acquiring great importance as proof of MPA effectiveness is increasingly demanded by decision makers and the community. MPAs and SZs are to be viewed as complementary to the many measures needed to conserve marine ecosystems and manage the numerous fisheries under EPM principles, which themselves are only slowly emerging.

REFERENCES

ANZECC (1998) *Guidelines for establishing the National Representative System of Marine Protected Areas.* Australian New Zealand Environment and Conservation Council Task Force on Marine Protected Areas. Environment Australia, Canberra.

Australian Government (2011) *SOE (State of Environment) Report 2011.* Department of Sustainability, Environment, Water, Population and Communities, Canberra. <www.environment.gov.au/soe/2011/index.html> (Accessed November 2011).

Babcock RC, Phillips JC, Lourey M, Clapin G (2007) Increased density, biomass and egg production of an unfished population of Western Rock Lobster (*Panulirus cygnus*) at Rottnest Island, Western Australia. *Marine and Freshwater Research* **58**, 286–292.

Babcock RC, Shears NF, Alcala AC, Barrett NS *et al.* (2010) Decadal trends in marine reserves reveal differential rates of change in direct and indirect effects. *Proceedings of the National Academy of Sciences of the United States* **107**, 18256–18261.

Baker JL (2004) 'Towards a system of ecologically representative marine protected areas in South Australian marine bioregions'. Technical Report. Department for Environment and Heritage, Adelaide.

Baker JL, Shepherd SA, Edyvane K (1996) The use of marine fishery reserves to manage benthic fisheries, with emphasis on the South Australian abalone fishery. In *Developing Australia's Representative System of Marine Protected Areas: Criteria and Guidelines for Identification and Selection* (Ed R Thackway) pp. 103–113. Department for Environment, Canberra.

Ballantine B (1999) 'Marine reserves in New Zealand: the development of the concept and principles'. Paper for workshop on MPAs, KORDI, Korea.

Banks SA, Skilleter GA (2010) Implementing marine reserve networks: a comparison of approached in New South Wales (Australia) and New Zealand. *Marine Policy* **34**, 197–207.

Barrett NS, Edgar GJ (2010) Distribution of benthic communities in the fjord-like Bathurst Channel ecosystem, south-western Tasmania, a globally anomalous estuarine protected area. *Aquatic Conservation: Marine and Freshwater Ecosystems* **20**, 397–406.

Barrett NS, Edgar GJ, Buxton CD, Haddon M (2007) Changes in fish assemblages following 10 years of protection in Tasmanian marine protected areas. *Journal of Experimental Marine Biology and Ecology* **345**, 141–157.

Barrett NS, Edgar GJ, Polachek A, Lynch T *et al.* (2008) *Ecosystem Monitoring of Subtidal Reefs in the Jervis Bay Marine Park 1996–2007.* TAFI, University of Tasmania, Hobart.

Barrett NS, Buxton CD, Edgar GJ (2009a) Changes in invertebrate and macroalgal populations in Tasmanian marine reserves in the decade following protection. *Journal of Experimental Marine Biology and Ecology* **370**, 104–119.

Barrett NS, Buxton CD, Gardner C (2009b) Rock lobster movement patterns and population structure within a Tasmanian Marine Protected Area inform fishery and conservation management. *Marine and Freshwater Research* **60**, 417–425.

Beck MW, Brumbaugh RD, Airoldi L, Carranza A *et al.* (2011) Oyster reefs at risk and recommendations for conservation, restoration and management. *Bioscience* **61**, 107–116.

Beeton RJS, Buxton CD, Cutbush GC, Fairweather PG *et al.* (2012) 'Report of the independent scientific audit of marine parks in New South Wales'. NSW Dept of Primary Industries and Office of Environment and Heritage, Sydney.

Buxton C, Barrett N, Haddon M, Gardner C *et al.* (2006) 'Evaluating the effectiveness of marine protected areas as a fisheries management tool'. FRDC Final Report. Project No. 1999/162. Tasmanian Aquaculture and Fisheries Institute, University of Tasmania, Hobart.

Claudet J, Guidetti P (2010) Improving assessments of marine protected areas. *Aquatic Conservation: Marine and Freshwater Ecosystems* **20**, 239–242.

COAG (1992) *The National Strategy for Ecologically Sustainable Development.* Australian Government Publishing Service, Canberra.

Colton MA, Swearer SE (2010) A comparison of two survey methods: differences between underwater visual census and baited remote underwater video. *Marine Ecology Progress Series* **400**, 19–36.

Commonwealth of Australia (2006) *A Guide to the Integrated Marine and Coastal Regionalisation of Australia. Version 4.0.* <www.environment.gov.au/coasts/mbp/publications/imcra/pubs/imcra4.pdf> (Accessed January 2012).

CSIRO (1994) *Jervis Bay Baseline Studies: Final Report. Vols 1–3.* CSIRO Division of Fisheries, Perth.

Currie DB, Sorokin SJ (2005) 'Point Labatt marine reserve: evaluating the effects of reserve closure on the biota of subtidal reefs'. SARDI Research Report Series No. 103. Publication No. RD04/0123-2. SARDI Aquatic Sciences, West Beach.

Currie DB, Sorokin SJ (2009) Evaluating the effects of reserve closure on algae, invertebrate and fish assemblages at a temperate South Australian marine reserve. *Journal of the Marine Biological Station of the UK* **89**, 651–661.

Day J (2008) The need and practice of monitoring, evaluating and adapting marine planning and management – lessons from the Great Barrier Reef. *Marine Policy* **32**, 823–831.

DEC (Dept Environment and Conservation) (2010) Website. <www.dec.wa.gov.au/mpra> (Accessed November 2011).

Earle SA, Laffoley D (2006) Protecting the earth's last frontier. *International Herald Tribune* 19 February 2006.

Edgar GJ (2011) Does the global network of marine protected areas provide an adequate safety net for marine biodiversity? *Aquatic Conservation: Marine and Freshwater Ecosystems* **21**, 313–316.

Edgar GJ, Barrett NS (2010) Biotic affinities of rocky reef fishes, invertebrates and macroalgae in different zones of the Port Davey marine protected area, south-western Tasmania. *Aquatic Conservation: Marine and Freshwater Ecosystems* **20**, 282–296.

Edgar GJ, Barrett NS (2012) An assessment of population responses of common inshore fishes and invertebrates following declaration of five Australian marine protected areas. *Environmental Conservation* **39**, 271–281.

Edgar GJ, Samson CR (2004) Catastrophic decline in mollusc diversity in eastern Tasmania and its concurrence with shellfish fisheries. *Conservation Biology* **18**, 1579–1588.

Edgar GJ, Stuart-Smith RD (2009) Ecological effects of marine protected areas on rocky reef communities – a continental-scale analysis. *Marine Ecology Progress Series* **388**, 51–62.

Edgar GJ, Russ GR, Babcock RC (2008) Marine protected areas. In *Marine Ecology* (Eds SD Connell and BM Gillanders) pp. 533–555. Oxford University Press, Melbourne.

Edgar GJ, Banks S, Bernsted-Smith R, Calvopiña M *et al.* (2008a) Conservation of threatened species in the Galapagos Marine Reserve through identification and protection of marine key biodiversity areas. *Aquatic Conservation: Marine and Freshwater Ecosystems* **18**, 955–968.

Edgar GJ, Langhammer PF, Allen G, Brooks TM *et al.* (2008b) Key biodiversity areas as globally significant target sites for the conservation of marine biological diversity. *Aquatic Conservation: Marine and Freshwater Ecosystems* **18**, 969–983.

Edgar G, Barrett N, Crane K, Bancroft K (2009a) 'Ecosystem monitoring of subtidal reefs in different management zones of the Jurien Bay Marine Park 1999–2007'. Internal Report. Tasmanian Aquaculture and Fisheries Institute, Hobart.

Edgar GJ, Barrett NS, Stuart-Smith RD (2009b) Exploited reefs protected from fishing transform over decades into conservation features otherwise absent from seascapes. *Ecological Applications* **19**, 1967–1974.

Edgar GJ, Last PR, Barrett NS, Gowlett-Holmes K *et al.* (2010) Conservation of natural wilderness values in the Port Davey marine and estuarine protected area, south-western Tasmania. *Aquatic Conservation: Marine and Freshwater Ecosystems* **20**, 297–311.

Fairweather PG, McNeill S (1992) Ecological and other scientific imperatives for marine and estuarine conservation. In *Protection of Marine and Estuarine Areas: a Challenge for Australians.* (Eds AM Ivanovici, D Tarte and M Olsen). pp. 39–48. Occasional Paper No. 4, Australian Committee for IUCN, Sydney.

Fernandes L, Day J, Lewis A, Siegers S *et al.* (2005) Establishing representative no-take areas in the Great Barrier Reef: large-scale implementation of theory on marine protected areas. *Conservation Biology* **19**, 1733–1744.

Freeman DJ, MacDiarmid AB (2009) Healthier lobsters in a marine reserve: effects of fishing on disease incidence in the spiny lobster, *Jasus edwardsii*. *Marine and Freshwater Research* **60**, 140–145.

Gaylord B, Gaines S, Siegel DA, Carr M (2005) Marine reserves exploit population structure and life history in potentially improving fisheries yields. *Ecological Applications* **15**, 2180–2191.

Gladstone W (2007) Requirements for marine protected areas to conserve the biodiversity of rocky reef fishes. *Aquatic Conservation: Marine and Freshwater Ecosystems* **17**, 71–87.

Goldsworthy SD, Page B, Shaughnessy PD, Linnane A (2010) 'Mitigating seal interactions in the SRLF and gillnet sector SESSF in South Australia'. FRDC Project 2007/041 Report. SARDI Publication No. F2009/000613-1. SARDI (Aquatic Sciences), Adelaide.

Goñi R, Hilborn R, Diaz D, Mallol S *et al.* (2010) Net contribution of spillover from a marine reserve to fishery catches. *Marine Ecology Progress Series* **400**, 233–243.

Government of South Australia (2011) Government submission to the Marine Parks Select Committee. <www.parliament.sa.gov.au/committees/Pages/Committees.aspx?CTld=3lcld=263> (Accessed January 2012).

Halpern BS, Selkoe KA, Micheli F, Kappel CV (2007) Evaluating and ranking the vulnerability of global marine ecosystems to anthropogenic threats. *Conservation Biology* **21**, 1301–1315.

Halpern BS, Lester SE, Mcleod KL (2010) Placing marine protected areas onto the ecosystem-based management seascape. *Proceedings of the National Academy of Sciences of the United States of America* **107**, 18312–18317.

Hobday AJ (2011) Sliding baselines and shuffling species: implications of climate change for marine conservation. *Marine Ecology (Berlin)* **32**, 392–403.

IUCN (2001) *IUCN Red List Categories and Criteria, Version 3.1.* Species Survival Commission, IUCN, Gland, Switzerland.

Jenkins GP, Fowler AJ, Black KP (2003) Metapopulation structure of a temperate fish in relation to spatial variation in hydrodynamics: implications for selection of marine protected areas. In *Aquatic Protected Areas: What Works Best and How Do We Know?* (Eds JP Beumer, A Grant and DC Smith) pp. 181–190. Australian Society for Fish Biology, Brisbane.

Jones PJS (2007) Point-of-View: arguments for conventional fisheries management and against no-take marine protected areas: only half the story? *Reviews in Fish Biology and Fisheries* **17**, 31–43.

Kearney R, Buxton CD, Farebrother G (2012) Australia's no-take marine protected areas: appropriate conservation or inappropriate management of fishing? *Marine Policy* **36**, 1064–1071.

Keough MJ, Ross DJ, Knott NA (2007) 'Ecological performance measures for Victorian Marine Protected Areas: review of the existing biological sampling program'. Technical Paper Series No. 51. Parks Victoria, Melbourne.

Kinlan BR, Gaines SD (2003) Propagule dispersal in marine and terrestrial environments: a community perspective. *Ecology* **84**, 2007–2020.

Kirkman H (2013) Choosing boundaries to marine protected areas and zoning the MPAs for restricted use and management. *Ocean and Coastal Management* **81**, 38–48.

Kleczkowski M, Babcock RC, Clapin G (2008) Density and size of reef fishes in and around a temperate marine reserve. *Marine and Freshwater Research* **59**, 165–176.

Kuparinen A, Merila J (2007) Detecting and managing fisheries-induced evolution. *Trends in Ecology & Evolution* **22**, 652–659.

Lauck T, Clark CW, Mangel M, Munro GR (1998) Implementing the precautionary principle in fisheries management through marine reserves. *Ecological Applications* **8**, S72–S78.

Leopold A (1964) *A Passing in Cincinnati.* US Government Printing Office, Washington DC.

Lipcius RN, Crowder LB, Morgan LE (2005) Metapopulation structure and marine reserves. In *Marine Conservation Biology: the Science of maintaining the Sea's Biodiversity.* (Eds EA Norse and LB Crowder) pp. 328–345. Island Press, Washington, DC.

McGarvey R (2004) Estimating the emigration rate of fish stocks from marine sanctuaries using tag-recovery data. *Fishery Bulletin* **102**, 464–472.

McKinley AC, Johnston EL (2010) Impacts of contaminant sources on marine fish abundance and species richness: a review and meta-analysis of evidence from the field. *Marine Ecology Progress Series* **420**, 175–191.

McKinley AC, Miskiewicz A, Taylor MD, Johnston EL (2011a) Strong links between metal contamination, habitat modification and estuarine larval fish distributions. *Environmental Pollution* **159**, 1499–1509.

McKinley AC, Ryan L, Coleman MA, Knott N *et al.* (2011b) Putting marine sanctuaries into context: a comparison of estuary assemblages over multiple levels of protection and modification. *Aquatic Conservation: Marine and Freshwater Ecosystems* **21**, 636–648.

McLeay LJ, Sorokin SJ, Rogers PJ, Ward TM (2003) 'Benthic protection zone of the Great Australian Bight Marine Park: Literature review'. Report to National Parks and Wildlife, South Australian and Commonwealth Department of the Environment and Heritage. SARDI Aquatic Sciences, Adelaide.

McLeod E, Salm R, Green A, Almany J (2008) Designing marine protected area networks to address the impacts of climate change. *Frontiers in Ecology and the Environment* **7**, 362–370.

McNeill SE (1994) The selection and design of marine protected areas: Australia as a case study. *Biodiversity and Conservation* **3**, 586–605.

Mondoux S, Pitcher T, Pauly D (2008) Ranking maritime countries by the sustainability of their fisheries. In *A Comparative Assessment of Biodiversity, Fisheries and Aquaculture in 53 Countries* (Eds J Alder and D Pauly) pp. 13–27. Fisheries Centre, University of British Columbia, Vancouver.

Mora C, Aburto-Oropeza O, Bocos A, Ayotte AA *et al.* (2011) Global human imprint on the linkage between biodiversity and ecosystem functioning in reef fishes. *PLoS Biology* **9**(4), e1000606.

Morgan L, Shepherd SA (2006) Population and spatial structure of two common temperate reef herbivores: abalone and sea-urchins. In *Marine Metapopulations* (Eds JP Kritzer and PF Sale) pp. 205–246. Elsevier, Amsterdam.

Murphy H, Jenkins GP (2010) Observational measures used in marine monitoring: a review. *Marine and Freshwater Research* **61**, 236–252.

Nardi K, Jones GP, Moran MJ, Cheng YW (2004) Contrasting effects of marine protected areas on the abundance of two exploited reef fishes at the sub-tropical Houtman Abrolhos Islands, Western Australia. *Environmental Conservation* **31**, 160–168.

Nevill J (2010) *Overfishing Under Regulation: the Application of the Precautionary Principle and the ecosystem Approach to Australian Fisheries Management.* VDM Verlag, Saarbrucken. <www.onlyoneplanet.com.au> (Accessed May 2012).

Nevill J (2012) Fisheries management: opportunities for reform. In *New Zealand and Australia in Focus: Economics, the Environment and Issues in Health Care* (Eds JR Barker and R Walters) pp.84–106. Nova Publishers, New York.

Nevill J, Ward T (2009) The National Representative System of Marine Protected Areas: comment on recent progress. *Ecological Management & Restoration* **10**, 228–231.

Nursey-Bray M (2011) Social contexts and customary fisheries: marine protected areas and indigenous use, Australia. *Environmental Management* **47**, 671–683.

Parks Victoria (2006) *Wilsons Promontory Marine National Park and Wilsons Promontory Marine Park Management Plan.* Parks Victoria, Melbourne. <www.parkweb.vic.gov.au> (Accessed February 2012).

Pederson HG, Barrett NS, Frusher SD, Buxton CD (2008) Effect of predator–prey and competitive interactions on size at emergence in the black-lip abalone *Haliotis rubra* in a Tasmanian MPA. *Marine Ecology Progress Series* **366**, 91–98.

Pérez-Ruzafa A, Martin E, Marcos C, Zamarro JM *et al.* (2008) Modelling spatial and temporal scales for spill-over and biomass exportation from MPAs and their potential for fisheries enhancement. *Journal for Nature Conservation* **16**, 234–255.

Peterson MN, Peterson MJ, Peterson TR (2004) Conservation and the myth of consensus. *Conservation Biology* **19**, 762–767.

Pritchard K, Edmunds M, Stewart K, Davis S (2011) 'Victorian subtidal monitoring program: the rich reef biota at Bunurong Marine National Park'. Parks Victoria Technical Series No. 69. Parks Victoria, Melbourne.

RLS (Reef Life Surveys) (2011) Reef Life Survey Website. <http://www.reeflifesurvey.com/> (Accessed February 2012).

Roberts CM, Hawkins JP, Gell FR (2005) The role of marine reserves in achieving sustainable fisheries. *Philosophical Transactions of the Royal Society of London. Series B, Biological Sciences* **360**, 123–132.

Ryan KA (2008) Small, no-take marine protected areas and wave exposure affect temperate, sub-tidal reef communities at Marmion Marine Park, Western Australia. PhD thesis. University of Western Australia, Perth.

Shepherd SA, Breen PA (1992) Mortality in abalone: its estimation, variability and causes. In *Abalone of the World: Biology, Fisheries and Culture.* (Eds SA Shepherd, MJ Tegner and SA Guzmán del Próo) pp. 276–304. Blackwell, Oxford.

Shepherd SA, Brown LD (1993) What is an abalone stock: implications for the role of refugia in conservation. *Canadian Journal of Fisheries and Aquatic Sciences* **50**, 2001–2009.

Shepherd SA, Rodda KR (2001) Sustainability demands vigilance: evidence for serial decline of the South Australian greenlip abalone fishery and a review of management. *Journal of Shellfish Research* **20**, 829–841.

Shepherd SA, Rodda KR, Vargas KM (2001) A chronicle of collapse in two abalone stocks with proposals for precautionary management. *Journal of Shellfish Research* **20**, 843–856.

Shepherd SA, Svane I, Kirkegaard I, Harbison P (2008) The Gulf now and its management. In *Natural History of Gulf St Vincent* (Eds SA Shepherd, S Bryars, I Kirkegaard, P Harbison *et al.*). Royal Society of South Australia, Adelaide.

Shepherd SA, Brook JB, Xiao Y (2010) Environmental and fishing effects on the abundance, size and sex ratio of the blue-throated wrasse, *Notolabrus tetricus*, on South Australian coastal reefs. *Fisheries Management and Ecology* **17**, 209–220.

Smith SDA (2005) Rapid assessment of invertebrate biodiversity on rocky shores: where there's a whelk there's a way. *Biodiversity and Conservation* **14**, 3565–3576.

Smith DG, Fulton B, Johnson P, Jenkins G *et al.* (2011) 'Developing integrated performance measures for spatial management of marine systems'. Final Report FRDC Project 2004/005. CSIRO, Hobart.

Stobart B, Warwick R, González Cl, Mallol S *et al.* (2009) Long-term and spillover effects of a marine protected area on an exploited fish community. *Marine Ecology Progress Series* **384**, 47–60.

Vandeperre F, Higgins RM, Sánchez-Meca J, Maynou F *et al.* (2011) Effects of no-take area size and age of marine protected areas on fisheries yields: a meta-analytical approach. *Fish and Fisheries* **11**, 1–15.

Vanderklift MA, Ward TJ, Phillips JC (1998) Use of assemblages derived from different taxonomic levels to select areas for conserving marine biodiversity. *Biological Conservation* **86**, 307–315.

Ward TJ, Vanderclift MA, Nicholls AG, Kenchington RA (1999) Selecting marine reserves using habitats and species assemblages as surrogates for biological diversity. *Ecological Applications* **9**, 691–698.

Ward TJ, Heinemann D, Evans N (2001) *The Role of Marine Reserves as Fisheries Management Tools: A Review of Concepts, Evidence and International Experience.* Bureau of Rural Sciences, Canberra.

Watson DL, Harvey ES, Kendrick GA, Nardi K *et al.* (2007) Protection from fishing alters the species composition of fish assemblages in a temperate-tropical transition zone. *Marine Biology* **152**, 1197–1206.

Watts ME, Ball IR, Stewart RS, Klein CJ *et al.* (2009) Marxan with zones: software for optimal conservation based land- and sea-use zoning. *Environmental Modelling & Software* **24**, 1513–1521.

Wells FE, Keesing JK (1990) Population characteristics of the abalone *Haliotis roei* on intertidal platforms in the Perth metropolitan area. *Journal of the Malacological Society of Australia* **11**, 65–71.

Wescott G (2006) The long and winding road: the development of a comprehensive, adequate and representative system of highly protected marine protected areas in Victoria, Australia. *Ocean and Coastal Management* **49**, 905–922.

White JW, Botsford LW, Baskett ML, Barnett LAK *et al.* (2011) Linking models with monitoring data for assessing performance of no-take marine reserves. *Frontiers in Ecology and the Environment* **9**, 390–399.

Index

www.ingramcontent.com/pod-product-compliance
Lightning Source LLC
LaVergne TN
LVHW070829220926
848486LV00029B/213
9781486300099